Differential Equations Using MATLAB

An Introductory Guide for Students, Engineers, and Beginners

by
Peter I. Kattan

BOOKBOUND PRESS

Peter I. Kattan, PhD

Correspondence about this book may be sent to the author at one of the following two email addresses:

info@petrabooks.com

pkattan@petrabooks.com

Differential Equations with MATLAB: A Guide for Beginners, written by Peter I. Kattan.

ISBN: 979-8-2954-4151-6

Preface

Differential equations play a central role in mathematics, engineering, the sciences, and many real-world applications. Yet for countless students and self-learners, they remain one of the more challenging subjects to master—especially when transitioning from theory to practical computation. I wrote this book to bridge that gap. By combining clear explanations of differential equation concepts with the powerful computational tools offered by MATLAB, this text aims to provide a smooth, accessible, and intuitive learning experience.

My goal is to help readers of all backgrounds—engineering students, science majors, professionals, and complete beginners—understand not just the "how," but also the "why" behind solving differential equations. MATLAB offers a supportive environment for experimentation, visualization, and symbolic manipulation, making it an ideal companion for anyone seeking to learn differential equations efficiently and confidently.

Throughout this book, you will find step-by-step examples, annotated MATLAB commands, symbolic solutions, and dozens of applications drawn from mechanics, physics, biology, and engineering. Whether you are tackling your first ordinary differential equation or refining your computational skills for advanced study, this guide is structured to serve both as a learning tool and as a long-term reference.

This book is built upon years of teaching, writing, and helping students navigate subjects that initially appear overwhelming. My aim has always been to make learning clear, practical, and enjoyable. I hope this book gives you the confidence not only to solve differential equations, but to apply them meaningfully in your academic, professional, or research journey.

Thank you for choosing this book as part of your learning experience. I wish you success, insight, and satisfaction as you explore differential equations through the power of MATLAB. There are 42 exercises at the ends of chapters for students to practice. The solutions of the all the exercise are shown at the end of the book.

— Peter I. Kattan, PhD November 2025

Differential Equations Using MATLAB

An Introductory Guide for Students, Engineers, and beginners

Table of Contents

Differential Equations Using MATLAB

An Introductory Guide for Students, Engineers, and beginners

1. Introduction

In this introductory chapter a short MATLAB tutorial is provided. This tutorial describes the basic MATLAB commands needed. More details about these commands will follow in subsequent chapters. Note that there are numerous free MATLAB tutorials on the internet – check references [1-28]. Also, you may want to consult many of the excellent books on the subject – check references [29-47]. This chapter may be skipped on a first reading of the book.

In this tutorial it is assumed that you have started MATLAB on your computer system successfully and that you are now ready to type the commands at the MATLAB prompt (which is denoted by double arrows ">>"). For installing MATLAB on your computer system, check the web links provided at the end of the book.

Entering scalars and simple operations is easy as is shown in the examples below:

```
>> 2*3+5

ans =

    11
```

The order of the operations will be discussed in subsequent chapters.

```
>> cos(60*pi/180)

ans =

    0.5000
```

The argument for the `cos` command and the value of `pi` will be discussed in subsequent chapters.

We assign the value of 4 to the variable `x` as follows:

```
>> x = 4

x =

    4

>> 3/sqrt(2+x)

ans =

    1.2247
```

To suppress the output in MATLAB use a semicolon to end the command line as in the following examples. If the semicolon is not used then the output will be shown by MATLAB:

```
>> y = 30;

>> z = 8;

>> x = 2; y-z;

>> w = 4*y + 3*z


w =


    144
```

MATLAB is case-sensitive, i.e. variables with lowercase letters are different than variables with uppercase letters. Consider the following examples using the variables x and X:

```
>> x = 1


x =
```

```
     1

>> X = 2

X =

     2

>> x

x =

     1

>> X

X =

     2
```

Use the `help` command to obtain help on any particular MATLAB command. The following example demonstrates the use of `help` to obtain help on the `det` command which calculated the determinant of a matrix:

```
>> help det

 DET    Determinant.

    DET(X) is the determinant of the
square matrix X.


    Use COND instead of DET to test for
matrix singularity.


    See also cond.


    Overloaded functions or methods
(ones with the same name in other
directories)

       help sym/det.m


    Reference page in Help browser

       doc det
```

The following examples show how to enter matrices and perform some simple matrix operations:

```
>> x = [1 2 3 ; 4 5 6 ; 7 8 9]

x =

     1     2     3
     4     5     6
     7     8     9

>> y = [1 ; 0 ; -4]

y =

     1
     0
    -4

>> w = x*y

w =

   -11
   -20
```

```
-29
```

Let us now see how to use MATLAB to solve a system of simultaneous algebraic equations. Let us solve the following system of simultaneous algebraic equations:

$$\begin{bmatrix} 3 & 5 & -1 \\ 0 & 4 & 2 \\ -2 & 1 & 5 \end{bmatrix} \begin{Bmatrix} x_1 \\ x_2 \\ x_3 \end{Bmatrix} = \begin{Bmatrix} 2 \\ 1 \\ -4 \end{Bmatrix} \tag{1.1}$$

We will use Gaussian elimination to solve the above system of equations. This is performed in MATLAB by using the backslash operator "\" as follows:

```
>> A= [3 5 -1 ; 0 4 2 ; -2 1 5]

A =

     3     5    -1

     0     4     2

    -2     1     5


>> b = [2 ; 1 ; -4]
```

```
b =

     2
     1
    -4

>> x = A\b

x =

   -1.7692
    1.1154
   -1.7308
```

It is clear that the solution is x_1 = -1.7692, x_2 = 1.1154, and x_3 = -1.7308. Alternatively, one can use the inverse matrix of A to obtain the same solution directly as follows:

```
>> x = inv(A)*b

x =

   -1.7692
```

```
    1.1154
   -1.7308
```

It should be noted that using the inverse method usually takes longer than using Gaussian elimination especially for large systems of equations.

Consider now the following 4 x 4 matrix D:

```
>> D = [1 2 3 4 ; 2 4 6 8 ; 3 6 9 12 ; -
5 -3 -1 0]

D =

     1     2     3     4
     2     4     6     8
     3     6     9    12
    -5    -3    -1     0
```

We can extract the sub-matrix in rows 2 to 4 and columns 1 to 3 as follows:

```
>> E = D(2:4, 1:3)
```

```
E =

     2     4     6
     3     6     9
    -5    -3    -1
```

We can extract the third column of D as follows:

```
>> F = D(1:4, 3)

F =

     3
     6
     9
    -1
```

We can extract the second row of D as follows:

```
>> G = D(2, 1:4)

G =
```

```
     2     4     6     8
```

We can also extract the element in row 3 and column 2 as follows:

```
>> H = D(3,2)

H =

     6
```

We will now show how to produce a two-dimensional plot using MATLAB. In order to plot a graph of the function $y = f(x)$, we use the MATLAB command `plot(x,y)` after we have adequately defined both vectors `x` and `y`. The following is a simple example to plot the function $y = x^2 + 3$ for a certain range:

```
>> x = [1 2 3 4 5 6 7 8 9 10 11]

x =

  Columns 1 through 10
```

```
     1     2     3     4     5     6     7
8     9    10

  Column 11

    11

>> y = x.^2 + 3

y =

  Columns 1 through 10

     4     7    12    19    28    39    52
67    84   103

  Column 11

   124

>> plot(x,y)
```

Figure 1.1 shows the plot obtained by MATLAB. It is usually shown in a separate window. In this figure no titles are given to the `x`- and `y`-axes. These titles may be easily added to the figure using the `xlabel` and `ylabel` commands. More details about these commands will be presented in the chapter on graphs.

In the calculation of the values of the vector y, notice the dot "." before the exponentiation symbol "^". This dot is used to denote that the operation following it be performed element by element. More details about this issue will be discussed in subsequent chapters.

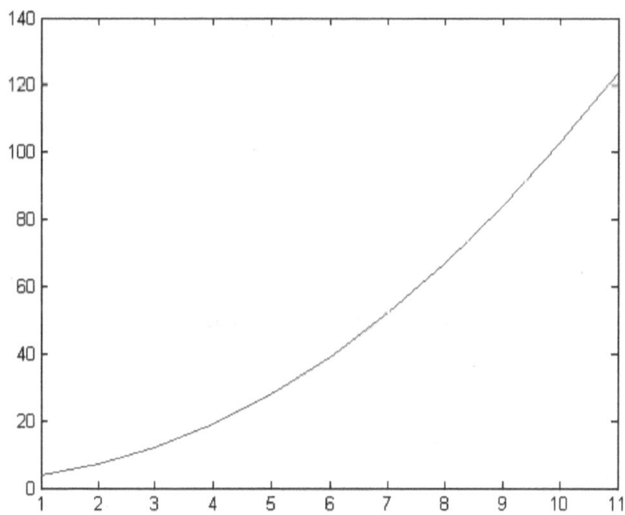

Figure 1.1: Using the MATLAB "plot" command

Finally, we will show how to make magic squares with MATLAB. A magic square is a square grid of numbers where the total of any row, column, or diagonal is the same. Magic squares are produced by MATLAB using the command `magic`. Here is a simple example to produce a 5 x 5 magic square:

```
>> magic(5)

ans =

    17    24     1     8    15
    23     5     7    14    16
     4     6    13    20    22
    10    12    19    21     3
    11    18    25     2     9
```

It is clear that the total of each row, column or diagonal in the above matrix is 65.

Exercises

Solve all the exercises using MATLAB. All the needed MATLAB commands for these exercises were presented in this chapter.

1. Perform the operation 3*4+6. The order of the operations will be discussed in subsequent chapters.

2. Perform the operation cos(5). The value of 5 is in radians.

3. Perform the operation $3\sqrt{6+x}$ for $x=4$.

4. Assign the value of 5.2 to the variable `y`.

5. Assign the values of 3 and 4 to the variables `x` and `y`, respectively, then calculate the value of `z` where $z=2x-7y$.

6. Obtain help on the `inv` command.

7. Generate the following matrix: `A`

$$A=\begin{bmatrix} 1 & 0 & 2 & -3 \\ 0 & 5 & 2 & 2 \\ 1 & 2 & 3 & 4 \\ -2 & 0 & 1 & 3 \end{bmatrix}$$

8. Generate the following vector `b`

$$b = \begin{Bmatrix} 1 \\ 2 \\ 3 \\ 4 \end{Bmatrix}$$

9. Evaluate the vector c where $\{c\} = [A]\{b\}$ where A is the matrix given in Exercise 7 above and b is the vector given in Exercise 8 above.

10. Solve the following system of simultaneous algebraic equation using Gaussian elimination.

$$\begin{bmatrix} 5 & 2 \\ 1 & 3 \end{bmatrix} \begin{Bmatrix} x_1 \\ x_2 \end{Bmatrix} = \begin{Bmatrix} 3 \\ -1 \end{Bmatrix}$$

11. Solve the system of simultaneous algebraic equations of Exercise 10 above using matrix inversion.

12. Generate the following matrix X:

$$X = \begin{bmatrix} 1 & 0 & 6 \\ 1 & 2 & 3 \\ 4 & 5 & -2 \end{bmatrix}$$

13. Extract the sub-matrix in rows 2 to 3 and columns 1 to 2 of the matrix X in Exercise 12 above.

14. Extract the second column of the matrix X in Exercise 12 above.

15. Extract the first row of the matrix X in Exercise 12 above.

16. Extract the element in row 1 and column 3 of the matrix X in Exercise 12 above.

17. Generate the row vector x with integer values ranging from 1 to 9.

18. Plot the graph of the function $y = x^3 - 2$ for the range of the values of x in Exercise 17 above.

19. Generate a 4 x 4 magic square. What is the total of each row, column, and diagonal in this matrix.

2. Differential Equations –I-

The analytical closed-from solutions of ordinary differential equations[1] are presented in this chapter. Ordinary Differential Equations (ODEs) were briefly mentioned with one example at the end of Chapter 11. More details about their solution using the Symbolic Math Toolbox are shown here.

Ordinary differential equations are classified according to the order of the derivative that appears in the equations. First-Order Ordinary Differential Equations have the highest derivative appearing in the equation of the first order. Second-Order Ordinary Differential Equations have the highest derivative appearing in the equation of the second order, and so on. In addition to solving single ordinary differential equations, this chapter also shows how to solve systems of ordinary differential equations using MATLAB.

First-Order Ordinary Differential Equations

In this section we deal with first order differential equations which are of the general form:

[1] Other types of differential equations are called Partial Differential Equations (PDEs) but these are more complicated and are beyond the scope of this book. PDEs involve partial derivates instead of ordinary derivatives.

$$\frac{dy}{dx} = f(x, y)$$

The above equation is called a first-order differential ordinary equation because the derivative $\frac{dy}{dx}$ appears raised to the first power only. The function f is a function of the two variables x and y. In particular, x is the independent variable and y is the dependent variable.

The solution of the above differential equation is obtained by finding an expression for y in terms of x. We will use the MATLAB Symbolic Math Toolbox in this chaper to solve such differential equations. Using this toolbox will enable us to obtain analytical solutions to these equations without resorting to numerical methods. In particular, we will make extensive use of the MATLAB command `dsolve` in the solution of ordinary differential equations. In order to use this command, make sure that you have installed both MATLAB and the MATLAB Symbolic Math Toolbox successfully on your computing system.

Let us first solve the following simple first-order differential equation

$$\frac{dy}{dx} = a \quad , \quad \text{where } a \text{ is a constant.}$$

The above equation is solved using the following use of the `dsolve` command (which is available in the MATLAB Symbolic Math Toolbox):

```
>> dsolve('Dy = a')

ans =

a*t+C1
```

Here are some comments about the above solution that was generated by MATLAB. Note how the differential equation is written between apostrophes in the argument[2] of the `dsolve` command. Not also the use of the letter D to denote the derivative of the variable. Note also that the solution obtained is written in terms of the independent variable t. In order to use x for the independent variable, we need to write x explicitly in the argument of the command as follows:

```
>> dsolve('Dy = a', 'x')

ans =
```

---

2 The argument of a command or a function is the expression appearing inside the parentheses.

```
a*x+C1
```

Note that the solution is generated successfully by MATLAB as follows:

$$y = ax + C_1$$

where C_1 is the constant of integration. In order to find the value of the constant C_1, we need to know the initial condition. However, the intial condition is not given in the example, so we leave the final result in terms of C_1 as shown above.

Next, we solve the following differential equation:

$$\frac{dy}{dx} = x$$

The following are the needed MATLAB command `dsolve` to get the solution[3]:

```
>> dsolve('Dy = x', 'x')
```

[3] Note that `Dy` is used for the first derivative, while `D2y` is used for the second derivative.

```
ans =

1/2*x^2+C1
```

Thus, the solution is correctly obtained as:

$$y = \frac{1}{2}x^2 + C_1$$

Note that the result shown above leaves the constant of integration C_1 undetermined because we are not given an intial condition in this example. Note also that the constant C_1 of this example is not the same as the constant C_1 of the previous example. It is just that MATLAB uses the same notation for the constant of integration in each example.

Next, let us solve the following differential equation:

$$\frac{dy}{dx} = 5x + 4$$

Here is the MTLAB command:

```
>> dsolve('Dy = 5*x+4', 'x')

ans =

5/2*x^2+4*x+C1
```

The solution is correctly obtained as:

$$y = \frac{5}{2}x^2 + 4x + C_1$$

Let us next solve the more complicated differential equation:

$$\frac{dy}{dx} = y$$

The following are the MATLAB commands to solve this equation:

```
>> dsolve('Dy = y', 'x')

ans =
```

```
C1*exp(x)
```

We correctly obtain the exponential function as the solution as follows:

$$y = C_1 e^x$$

Let us find next the solution to the following first-order differential equation[4] that comes with an initial condition:

$$y' + 3y = e^{-x} \quad , \quad y(0) = 0.45$$

The following are the MATLAB commands to solve the above equation:

```
>> dsolve('Dy +3*y = exp(-x)', 'y(0) =
0.45', 'x')

ans =
```

---

[4] The derivative is denoted by dy/dx, or equivalently, by y'.

```
(1/2*exp(2*x)-1/20)*exp(-3*x)
```

Note in the above output the use of the initial condition y(0) = 0.45 in the argument[5] of the `dsolve` command. Thus, the solution is obtained as follows:

$$y = \left(\frac{1}{2}e^{2x} - \frac{1}{20}\right)e^{-3x}$$

Note that the result shown above does not include the constant of integration C_1. The reason is that this example comes with an intial condition, and MATLAB used the intial condition to evaluate the constant of integration, and finally to display the final result that accounts for the intial condition. We show below how the above expression satisfies the initial condition $y(0) = 0.45$:

$$y(0) = \left(\frac{1}{2}e^{0} - \frac{1}{20}\right)e^{0} = \frac{1}{2} - \frac{1}{20} = \frac{9}{20} = 0.45$$

Next, let us find the solution to the following first-order differential equation:

---

[5] The argument of a command is the expression shown between the parentheses.

$$y' + 4xy = x \quad , \quad y(0) = 0$$

The following are the MATLAB commands to solve the above equation:

```
>> dsolve('Dy +4*x*y = x', 'y(0) = 0',
'x')

ans =

1/4-1/4*exp(-2*x^2)
```

Note how the intial condition $y(0) = 0$ is used in the `dsolve` command. The solution is obtained as follows:

$$y = \frac{1}{4} - \frac{1}{4} e^{-2x^2}$$

Note that the above result does not include a constant of integration simply because an intial condition is used in the example. It is clear that the above expression satisfies the initial condition $y(0) = 0$. This is shown below:

$$y(0) = \frac{1}{4} - \frac{1}{4}e^0 = \frac{1}{4} - \frac{1}{4} = 0$$

Let us plot the above function on a graph to show the behavior of the solution in this example. For this purpose, use is made of the MATLAB command `ezplot`[6] that is specially designed to plot functions such as $y = f(x)$. Here are the MATLAB commands needed to plot the above function. The resulting plot is shown in Figure 2.1.

```
>> ezplot(ans,[-2 2])
>> hold on
>> xlabel('x')
>> ylabel('y')
```

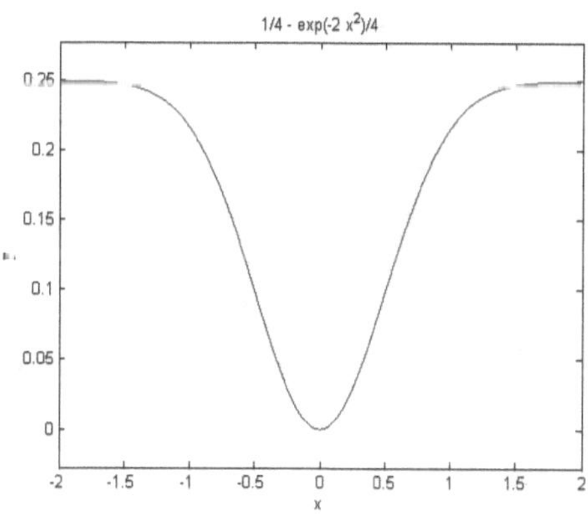

Figure 2.1: Plot of the Resulting Function in the Previous Example

[6] This is the first use of the MATLAB command `ezplot`. This command is not discussed in other chapters.

Next, let us consider the same differential equation as above, but change the initial condition to be in terms of an unspecified constant b, as follows:

$$y' + 4xy = x \quad , \quad y(0) = b$$

The MATLAB commands to solve the above differential equation are listed below:

```
>>
dsolve('Dy+4*x*y=x','y(0)=b','x')

ans =

(exp(-2*x^2)*(4*b - 1))/4 + 1/4
```

It is seen from the above MATLAB output that the solution to the differential equation is the following function:

$$y = \frac{1}{4} + \frac{1}{4}(4b - 1)e^{-2x^2}$$

We show next how the above expression satisfies the initial condition $y(0)=b$. Here are the details:

$$y(0)=\frac{1}{4}+\frac{1}{4}(4b-1)e^{0}=\frac{1}{4}+\frac{1}{4}(4b-1)=b$$

Our next example is to solve the following differential equation:

$$x^{2}\frac{dy}{dx}=y\ln y+2\frac{dy}{dx}$$

The needed MATLAB commands to solve the above equation are shown below. Note the use of the MATLAB command `simplify` on the third command line:

```
>>
dsolve('x^2*Dy=y*log(y)+2*Dy','x')

ans =

1
```

```
 exp(exp(C11 - (2^(1/2)*log(x +
2^(1/2)))/4 + (2^(1/2)*log(x -
2^(1/2)))/4))


>> simplify(ans)


ans =


1
 exp((exp(C11)*(x -
2^(1/2))^(2^(1/2)/4))/(x +
2^(1/2))^(2^(1/2)/4))
```

It is seen from the above solution that there is a constant of integration C11. This is clear since there is no initial condition given in this example. The solution is written as follows:

$$y = e^{\frac{e^{c11}\left(x-\sqrt{2}\right)^{\sqrt{2}/4}}{\left(x+\sqrt{2}\right)^{\sqrt{2}/4}}}$$

Second-Order Ordinary Differential Equations[7]

---

[7] There are also higher-order ordinary differential equations of third-order, fourth-order, etc. – check Chapter 3 for solving these types of differential equations.

In this section we deal with second order differential equations which are of the general form:

$$\frac{d^2y}{dx^2}=f\left(x,y,\frac{dy}{dx}\right)$$

The above equation is called a second-order differential equation because the second derivative $\frac{d^2y}{dx^2}$ appears raised to the first power only. The function f is a function of the two variables x and y, as well as the first derivative $\frac{dy}{dx}$. In particular, x is the independent variable and y is the dependent variable.

Let us first solve the following second-order ordinary differential equation:

$$\frac{d^2y}{dx^2}+4\frac{dy}{dx}=2e^{3x} \quad , \qquad y(0)=1 \ , \ y'(0)=0$$

Note that the above second-order ordinary differential equation comes with two initial conditions; one for y and one for $y^{/} \equiv \frac{dy}{dx}$. The following are the MATLAB commands needed to solve the above equation:

```
>>
dsolve('D2y+4*Dy=2*exp(3*x)','y(0)
=1','Dy(0)=0','x')

ans =

exp(-4*x)/14 + exp(-
4*x)*((5*exp(4*x))/6 +
(2*exp(7*x))/21)

>> simplify(ans)

ans =

 (2*exp(3*x))/21 + exp(-4*x)/14 +
5/6
```

It is noted that the MATLAB command `simplify` is used to simplify the resulting function. Thus, the solution to the above equation is written as follows:

$$y = \frac{5}{6} + \frac{1}{14}e^{-4x} + \frac{2}{21}e^{3x}$$

Let us show that the above expression satisfies the initial condition $y(0) = 1$. The details follow:

$$y(0) = \frac{5}{6} + \frac{1}{14}e^{0} + \frac{2}{21}e^{0} = \frac{5}{6} + \frac{1}{14} + \frac{2}{21} = \frac{42}{42} = 1$$

Let us now solve the following second-order ordinary differential equation:

$$5\frac{d^2y}{dx^2} + 3\frac{dy}{dx} = 2\sin x \qquad , \qquad y(0) = -1 \ , \ y'(0) = 1$$

The MATLAB commands used to solve the above equation are:

```
>>
dsolve('5*D2y+3*Dy=2*sin(x)','y(0)
=-1','Dy(0)=1','x')


ans =
```

```
4/3 - (3*cos(x))/17 -
(5*sin(x))/17 - (110*exp(-
(3*x)/5))/51
```

```
>> simplify(ans)
```

```
ans =
```

```
4/3 - (3*cos(x))/17 -
(5*sin(x))/17 - (110*exp(-
(3*x)/5))/51
```

It is noted from the MATLAB output above that the use of the MATLAB command `simplify` did not result in a simplified expression. Thus, the solution to the above differential equation is listed as follows:

$$y = \frac{4}{3} - \frac{3}{17}\cos x - \frac{5}{17}\sin x - \frac{110}{51}e^{-3x/5}$$

Let us check that the above expression satisfies the initial condition $y(0) = -1$. The details follow:

$$
\begin{aligned}
y &= \frac{4}{3} - \frac{3}{17}\cos 0 - \frac{5}{17}\sin 0 - \frac{110}{51}e^{0} \\
&= \frac{4}{3} - \frac{3}{17} - \frac{110}{51} = -\frac{51}{51} = -1
\end{aligned}
$$

Let us now plot the above solution function using the MATLAB command `ezplot`. The needed MATLAB commands are listed below. The resulting plot is shown in Figure 2.2.

```
>> ezplot(ans)
>> hold on
>> xlabel('x')
>> ylabel('y')
```

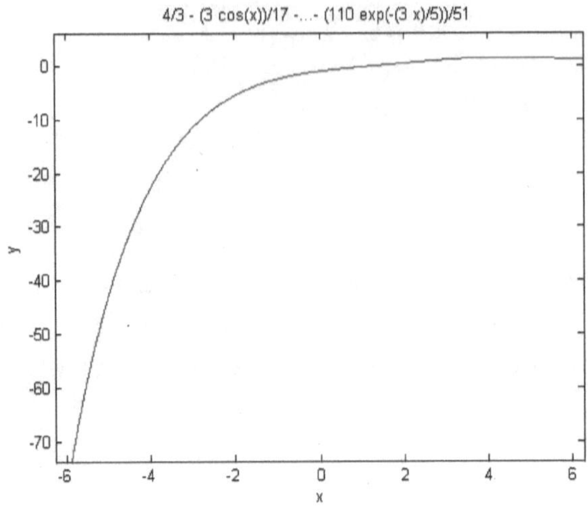

Figure 2.2: Plot of the Reulting Function of the Previous Example

Systems of Ordinary Differential Equations

In this section we deal with several ordinary differential equations that must be solved simultaneously. Such equations are called systems of ordinary differential equations. We will limit our presentation to linear first-order systems of ordinary differential equations. For this purpose, one example is illustrated.

Let us solve the following linear first-order system of three ordinary differential equations:

$$x'(t) = 2x(t) + 3y(t) - z(t)$$

$$y'(t) = x(t) + 2z(t)$$

$$z'(t) = 4x(t) - 4y(t) + 7z(t)$$

It is noted from the above system of equations that the dependent variables are x, y, and z, while the independent variable is t. It is noted also that all three equations are linear equations. All the derivatives appear on the left-hand-sides of the equations (but this is not a necessary requirement). The above system of equations can be solved using the MATLAB command `dsolve` as follows:

```
>> [x y z] = dsolve('Dx=2*x+3*y-
z','Dy=x+2*z','Dz=4*x-4*y+7*z')

x =

-(exp(-(t*(3*(4 - (27^(1/2)*368^(1/2))/27)^(2/3) - 18*(4 -
(27^(1/2)*368^(1/2))/27)^(1/3) + 4))/(6*(4 -
(27^(1/2)*368^(1/2))/27)^(1/3)))*(264*C19*cos((t*(4*3^(1/2)
- 3*3^(1/2)*(4 - (27^(1/2)*368^(1/2))/27)^(2/3)))/(6*(4 -
(27^(1/2)*368^(1/2))/27)^(1/3))) -
264*C21*sin((t*(4*3^(1/2) - 3*3^(1/2)*(4 -
(27^(1/2)*368^(1/2))/27)^(2/3)))/(6*(4 -
(27^(1/2)*368^(1/2))/27)^(1/3))) + 72*C19*cos((t*(4*3^(1/2)
- 3*3^(1/2)*(4 - (27^(1/2)*368^(1/2))/27)^(2/3)))/(6*(4 -
(27^(1/2)*368^(1/2))/27)^(1/3)))*(4 -
(27^(1/2)*368^(1/2))/27)^(1/3) + 504*C19*cos((t*(4*3^(1/2)
- 3*3^(1/2)*(4 - (27^(1/2)*368^(1/2))/27)^(2/3)))/(6*(4 -
(27^(1/2)*368^(1/2))/27)^(1/3)))*(4 -
(27^(1/2)*368^(1/2))/27)^(2/3) + 27*C19*cos((t*(4*3^(1/2) -
3*3^(1/2)*(4 - (27^(1/2)*368^(1/2))/27)^(2/3)))/(6*(4 -
(27^(1/2)*368^(1/2))/27)^(1/3)))*(4 -
```

```
(27^(1/2)*368^(1/2))/27)^(4/3) - 72*C21*sin((t*(4*3^(1/2) -
3*3^(1/2)*(4 - (27^(1/2)*368^(1/2))/27)^(2/3)))/(6*(4 -
(27^(1/2)*368^(1/2))/27)^(1/3)))*(4 -
(27^(1/2)*368^(1/2))/27)^(1/3) - 504*C21*sin((t*(4*3^(1/2)
- 3*3^(1/2)*(4 - (27^(1/2)*368^(1/2))/27)^(2/3)))/(6*(4 -
(27^(1/2)*368^(1/2))/27)^(1/3)))*(4 -
(27^(1/2)*368^(1/2))/27)^(2/3) - 27*C21*sin((t*(4*3^(1/2) -
3*3^(1/2)*(4 - (27^(1/2)*368^(1/2))/27)^(2/3)))/(6*(4 -
(27^(1/2)*368^(1/2))/27)^(1/3)))*(4 -
(27^(1/2)*368^(1/2))/27)^(4/3) - 528*C20*exp((t*(9*(4 -
(27^(1/2)*368^(1/2))/27)^(1/3) + 3*(4 -
(27^(1/2)*368^(1/2))/27)^(2/3) + 4))/(3*(4 -
(27^(1/2)*368^(1/2))/27)^(1/3)))*exp((t*(3*(4 -
(27^(1/2)*368^(1/2))/27)^(2/3) - 18*(4 -
(27^(1/2)*368^(1/2))/27)^(1/3) + 4))/(6*(4 -
(27^(1/2)*368^(1/2))/27)^(1/3))) -
264*3^(1/2)*C21*cos((t*(4*3^(1/2) - 3*3^(1/2)*(4 -
(27^(1/2)*368^(1/2))/27)^(2/3)))/(6*(4 -
(27^(1/2)*368^(1/2))/27)^(1/3))) -
264*3^(1/2)*C19*sin((t*(4*3^(1/2) - 3*3^(1/2)*(4 -
(27^(1/2)*368^(1/2))/27)^(2/3)))/(6*(4 -
(27^(1/2)*368^(1/2))/27)^(1/3))) -
2*27^(1/2)*368^(1/2)*C19*cos((t*(4*3^(1/2) - 3*3^(1/2)*(4 -
(27^(1/2)*368^(1/2))/27)^(2/3)))/(6*(4 -
(27^(1/2)*368^(1/2))/27)^(1/3))) +
2*27^(1/2)*368^(1/2)*C21*sin((t*(4*3^(1/2) - 3*3^(1/2)*(4 -
(27^(1/2)*368^(1/2))/27)^(2/3)))/(6*(4 -
(27^(1/2)*368^(1/2))/27)^(1/3))) - 144*C20*exp((t*(9*(4 -
(27^(1/2)*368^(1/2))/27)^(1/3) + 3*(4 -
(27^(1/2)*368^(1/2))/27)^(2/3) + 4))/(3*(4 -
(27^(1/2)*368^(1/2))/27)^(1/3)))*exp((t*(3*(4 -
(27^(1/2)*368^(1/2))/27)^(2/3) - 18*(4 -
(27^(1/2)*368^(1/2))/27)^(1/3) + 4))/(6*(4 -
(27^(1/2)*368^(1/2))/27)^(1/3)))*(4 -
(27^(1/2)*368^(1/2))/27)^(1/3) + 504*C20*exp((t*(9*(4 -
(27^(1/2)*368^(1/2))/27)^(1/3) + 3*(4 -
(27^(1/2)*368^(1/2))/27)^(2/3) + 4))/(3*(4 -
(27^(1/2)*368^(1/2))/27)^(1/3)))*exp((t*(3*(4 -
(27^(1/2)*368^(1/2))/27)^(2/3) - 18*(4 -
(27^(1/2)*368^(1/2))/27)^(1/3) + 4))/(6*(4 -
(27^(1/2)*368^(1/2))/27)^(1/3)))*(4 -
(27^(1/2)*368^(1/2))/27)^(2/3) - 54*C20*exp((t*(9*(4 -
(27^(1/2)*368^(1/2))/27)^(1/3) + 3*(4 -
(27^(1/2)*368^(1/2))/27)^(2/3) + 4))/(3*(4 -
(27^(1/2)*368^(1/2))/27)^(1/3)))*exp((t*(3*(4 -
(27^(1/2)*368^(1/2))/27)^(2/3) - 18*(4 -
(27^(1/2)*368^(1/2))/27)^(1/3) + 4))/(6*(4 -
(27^(1/2)*368^(1/2))/27)^(1/3)))*(4 -
(27^(1/2)*368^(1/2))/27)^(4/3) +
72*3^(1/2)*C21*cos((t*(4*3^(1/2) - 3*3^(1/2)*(4 -
(27^(1/2)*368^(1/2))/27)^(2/3)))/(6*(4 -
```

```
(27^(1/2)*368^(1/2))/27)^(1/3)))*(4 -
(27^(1/2)*368^(1/2))/27)^(1/3) +
27*3^(1/2)*C21*cos((t*(4*3^(1/2) - 3*3^(1/2)*(4 -
(27^(1/2)*368^(1/2))/27)^(2/3)))/(6*(4 -
(27^(1/2)*368^(1/2))/27)^(1/3)))*(4 -
(27^(1/2)*368^(1/2))/27)^(4/3) +
72*3^(1/2)*C19*sin((t*(4*3^(1/2) - 3*3^(1/2)*(4 -
(27^(1/2)*368^(1/2))/27)^(2/3)))/(6*(4 -
(27^(1/2)*368^(1/2))/27)^(1/3)))*(4 -
(27^(1/2)*368^(1/2))/27)^(1/3) +
27*3^(1/2)*C19*sin((t*(4*3^(1/2) - 3*3^(1/2)*(4 -
(27^(1/2)*368^(1/2))/27)^(2/3)))/(6*(4 -
(27^(1/2)*368^(1/2))/27)^(1/3)))*(4 -
(27^(1/2)*368^(1/2))/27)^(4/3) +
4*27^(1/2)*368^(1/2)*C20*exp((t*(9*(4 -
(27^(1/2)*368^(1/2))/27)^(1/3) + 3*(4 -
(27^(1/2)*368^(1/2))/27)^(2/3) + 4))/(3*(4 -
(27^(1/2)*368^(1/2))/27)^(1/3)))*exp((t*(3*(4 -
(27^(1/2)*368^(1/2))/27)^(2/3) - 18*(4 -
(27^(1/2)*368^(1/2))/27)^(1/3) + 4))/(6*(4 -
(27^(1/2)*368^(1/2))/27)^(1/3))) +
2*3^(1/2)*27^(1/2)*368^(1/2)*C21*cos((t*(4*3^(1/2) -
3*3^(1/2)*(4 - (27^(1/2)*368^(1/2))/27)^(2/3)))/(6*(4 -
(27^(1/2)*368^(1/2))/27)^(1/3))) +
2*3^(1/2)*27^(1/2)*368^(1/2)*C19*sin((t*(4*3^(1/2) -
3*3^(1/2)*(4 - (27^(1/2)*368^(1/2))/27)^(2/3)))/(6*(4 -
(27^(1/2)*368^(1/2))/27)^(1/3)))))/(864*(4 -
(27^(1/2)*368^(1/2))/27)^(2/3))


y =


(exp(-(t*(3*(4 - (27^(1/2)*368^(1/2))/27)^(2/3) - 18*(4 -
(27^(1/2)*368^(1/2))/27)^(1/3) + 4))/(6*(4 -
(27^(1/2)*368^(1/2))/27)^(1/3)))*(168*C19*cos((t*(4*3^(1/2)
- 3*3^(1/2)*(4 - (27^(1/2)*368^(1/2))/27)^(2/3)))/(6*(4 -
(27^(1/2)*368^(1/2))/27)^(1/3))) -
168*C21*sin((t*(4*3^(1/2) - 3*3^(1/2)*(4 -
(27^(1/2)*368^(1/2))/27)^(2/3)))/(6*(4 -
(27^(1/2)*368^(1/2))/27)^(1/3))) + 72*C19*cos((t*(4*3^(1/2)
- 3*3^(1/2)*(4 - (27^(1/2)*368^(1/2))/27)^(2/3)))/(6*(4 -
(27^(1/2)*368^(1/2))/27)^(1/3)))*(4 -
(27^(1/2)*368^(1/2))/27)^(1/3) + 360*C19*cos((t*(4*3^(1/2)
- 3*3^(1/2)*(4 - (27^(1/2)*368^(1/2))/27)^(2/3)))/(6*(4 -
(27^(1/2)*368^(1/2))/27)^(1/3)))*(4 -
(27^(1/2)*368^(1/2))/27)^(2/3) - 27*C19*cos((t*(4*3^(1/2) -
3*3^(1/2)*(4 - (27^(1/2)*368^(1/2))/27)^(2/3)))/(6*(4 -
```

```
(27^(1/2)*368^(1/2))/27)^(1/3)))*(4 -
(27^(1/2)*368^(1/2))/27)^(4/3) - 72*C21*sin((t*(4*3^(1/2) -
3*3^(1/2)*(4 - (27^(1/2)*368^(1/2))/27)^(2/3)))/(6*(4 -
(27^(1/2)*368^(1/2))/27)^(1/3)))*(4 -
(27^(1/2)*368^(1/2))/27)^(1/3) - 360*C21*sin((t*(4*3^(1/2)
- 3*3^(1/2)*(4 - (27^(1/2)*368^(1/2))/27)^(2/3)))/(6*(4 -
(27^(1/2)*368^(1/2))/27)^(1/3)))*(4 -
(27^(1/2)*368^(1/2))/27)^(2/3) + 27*C21*sin((t*(4*3^(1/2) -
3*3^(1/2)*(4 - (27^(1/2)*368^(1/2))/27)^(2/3)))/(6*(4 -
(27^(1/2)*368^(1/2))/27)^(1/3)))*(4 -
(27^(1/2)*368^(1/2))/27)^(4/3) - 336*C20*exp((t*(9*(4 -
(27^(1/2)*368^(1/2))/27)^(1/3) + 3*(4 -
(27^(1/2)*368^(1/2))/27)^(2/3) + 4))/(3*(4 -
(27^(1/2)*368^(1/2))/27)^(1/3)))*exp((t*(3*(4 -
(27^(1/2)*368^(1/2))/27)^(2/3) - 18*(4 -
(27^(1/2)*368^(1/2))/27)^(1/3) + 4))/(6*(4 -
(27^(1/2)*368^(1/2))/27)^(1/3))) -
168*3^(1/2)*C21*cos((t*(4*3^(1/2) - 3*3^(1/2)*(4 -
(27^(1/2)*368^(1/2))/27)^(2/3)))/(6*(4 -
(27^(1/2)*368^(1/2))/27)^(1/3))) -
168*3^(1/2)*C19*sin((t*(4*3^(1/2) - 3*3^(1/2)*(4 -
(27^(1/2)*368^(1/2))/27)^(2/3)))/(6*(4 -
(27^(1/2)*368^(1/2))/27)^(1/3))) -
2*27^(1/2)*368^(1/2)*C19*cos((t*(4*3^(1/2) - 3*3^(1/2)*(4 -
(27^(1/2)*368^(1/2))/27)^(2/3)))/(6*(4 -
(27^(1/2)*368^(1/2))/27)^(1/3))) +
2*27^(1/2)*368^(1/2)*C21*sin((t*(4*3^(1/2) - 3*3^(1/2)*(4 -
(27^(1/2)*368^(1/2))/27)^(2/3)))/(6*(4 -
(27^(1/2)*368^(1/2))/27)^(1/3))) - 144*C20*exp((t*(9*(4 -
(27^(1/2)*368^(1/2))/27)^(1/3) + 3*(4 -
(27^(1/2)*368^(1/2))/27)^(2/3) + 4))/(3*(4 -
(27^(1/2)*368^(1/2))/27)^(1/3)))*exp((t*(3*(4 -
(27^(1/2)*368^(1/2))/27)^(2/3) - 18*(4 -
(27^(1/2)*368^(1/2))/27)^(1/3) + 4))/(6*(4 -
(27^(1/2)*368^(1/2))/27)^(1/3)))*(4 -
(27^(1/2)*368^(1/2))/27)^(1/3) + 360*C20*exp((t*(9*(4 -
(27^(1/2)*368^(1/2))/27)^(1/3) + 3*(4 -
(27^(1/2)*368^(1/2))/27)^(2/3) + 4))/(3*(4 -
(27^(1/2)*368^(1/2))/27)^(1/3)))*exp((t*(3*(4 -
(27^(1/2)*368^(1/2))/27)^(2/3) - 18*(4 -
(27^(1/2)*368^(1/2))/27)^(1/3) + 4))/(6*(4 -
(27^(1/2)*368^(1/2))/27)^(1/3)))*(4 -
(27^(1/2)*368^(1/2))/27)^(2/3) + 54*C20*exp((t*(9*(4 -
(27^(1/2)*368^(1/2))/27)^(1/3) + 3*(4 -
(27^(1/2)*368^(1/2))/27)^(2/3) + 4))/(3*(4 -
(27^(1/2)*368^(1/2))/27)^(1/3)))*exp((t*(3*(4 -
(27^(1/2)*368^(1/2))/27)^(2/3) - 18*(4 -
(27^(1/2)*368^(1/2))/27)^(1/3) + 4))/(6*(4 -
(27^(1/2)*368^(1/2))/27)^(1/3)))*(4 -
(27^(1/2)*368^(1/2))/27)^(4/3) +
72*3^(1/2)*C21*cos((t*(4*3^(1/2) - 3*3^(1/2)*(4 -
```

```
(27^(1/2)*368^(1/2))/27)^(2/3)))/(6*(4 -
(27^(1/2)*368^(1/2))/27)^(1/3)))*(4 -
(27^(1/2)*368^(1/2))/27)^(1/3) -
27*3^(1/2)*C21*cos((t*(4*3^(1/2) - 3*3^(1/2)*(4 -
(27^(1/2)*368^(1/2))/27)^(2/3)))/(6*(4 -
(27^(1/2)*368^(1/2))/27)^(1/3)))*(4 -
(27^(1/2)*368^(1/2))/27)^(4/3) +
72*3^(1/2)*C19*sin((t*(4*3^(1/2) - 3*3^(1/2)*(4 -
(27^(1/2)*368^(1/2))/27)^(2/3)))/(6*(4 -
(27^(1/2)*368^(1/2))/27)^(1/3)))*(4 -
(27^(1/2)*368^(1/2))/27)^(1/3) -
27*3^(1/2)*C19*sin((t*(4*3^(1/2) - 3*3^(1/2)*(4 -
(27^(1/2)*368^(1/2))/27)^(2/3)))/(6*(4 -
(27^(1/2)*368^(1/2))/27)^(1/3)))*(4 -
(27^(1/2)*368^(1/2))/27)^(4/3) +
4*27^(1/2)*368^(1/2)*C20*exp((t*(9*(4 -
(27^(1/2)*368^(1/2))/27)^(1/3) + 3*(4 -
(27^(1/2)*368^(1/2))/27)^(2/3) + 4))/(3*(4 -
(27^(1/2)*368^(1/2))/27)^(1/3)))*exp((t*(3*(4 -
(27^(1/2)*368^(1/2))/27)^(2/3) - 18*(4 -
(27^(1/2)*368^(1/2))/27)^(1/3) + 4))/(6*(4 -
(27^(1/2)*368^(1/2))/27)^(1/3))) +
2*3^(1/2)*27^(1/2)*368^(1/2)*C21*cos((t*(4*3^(1/2) -
3*3^(1/2)*(4 - (27^(1/2)*368^(1/2))/27)^(2/3)))/(6*(4 -
(27^(1/2)*368^(1/2))/27)^(1/3))) +
2*3^(1/2)*27^(1/2)*368^(1/2)*C19*sin((t*(4*3^(1/2) -
3*3^(1/2)*(4 - (27^(1/2)*368^(1/2))/27)^(2/3)))/(6*(4 -
(27^(1/2)*368^(1/2))/27)^(1/3)))))/(864*(4 -
(27^(1/2)*368^(1/2))/27)^(2/3))


z =


exp(-(t*(3*(4 - (27^(1/2)*368^(1/2))/27)^(2/3) - 18*(4 -
(27^(1/2)*368^(1/2))/27)^(1/3) + 4))/(6*(4 -
(27^(1/2)*368^(1/2))/27)^(1/3)))*(C19*cos((t*(4*3^(1/2) -
3*3^(1/2)*(4 - (27^(1/2)*368^(1/2))/27)^(2/3)))/(6*(4 -
(27^(1/2)*368^(1/2))/27)^(1/3))) - C21*sin((t*(4*3^(1/2) -
3*3^(1/2)*(4 - (27^(1/2)*368^(1/2))/27)^(2/3)))/(6*(4 -
(27^(1/2)*368^(1/2))/27)^(1/3))) + C20*exp((t*(9*(4 -
(27^(1/2)*368^(1/2))/27)^(1/3) + 3*(4 -
(27^(1/2)*368^(1/2))/27)^(2/3) + 4))/(3*(4 -
(27^(1/2)*368^(1/2))/27)^(1/3)))*exp((t*(3*(4 -
(27^(1/2)*368^(1/2))/27)^(2/3) - 18*(4 -
(27^(1/2)*368^(1/2))/27)^(1/3) + 4))/(6*(4 -
(27^(1/2)*368^(1/2))/27)^(1/3))))
```

It is seen from the above MATLAB output that complicated and very long expressions are obtained for the three functions $x(t)$, $y(t)$, and $z(t)$. No attempt is made here to simplify the above expressions, write them out explicitly, or even plot them on a graph. The interested reader may wish to do so with much care.

Exercises

Solve all the exercises using MATLAB. All the needed MATLAB commands for these exercises were presented in this chapter. Note that all of these exercises (1-12) require the use of the MATLAB Symbolic Math Toolbox.

1. Solve the first-order ordinary differential equation:

$$\frac{dy}{dx} = b$$

2. Solve the first-order ordinary differential equation:

$$\frac{dy}{dx} = 3x$$

3. Solve the first-order linear ordinary differential equation with an initial condition:

$$\frac{dy}{dx} = 2x - 3 \qquad , \qquad y(0) = 1$$

4. Plot the solution function obtained in Exercise # 3 above.

5. Solve the first-order linear ordinary differential equation with an initial condition involving an unspecified constant b, given as follows:

$$\frac{dy}{dx} = x + 2y \qquad , \qquad y(0) = b$$

6. Solve the following ordinary differential equation:

$$y' - 2y = \sin(3x)$$

7. Solve the following ordinary differential equation:

$$y' + 3xy = 2x - y$$

8. Solve the following ordinary differential equation:

$$x^2 \frac{dy}{dx} = 2y \ln y$$

9. Solve the following second-order ordinary differential equation with two initial conditions:

$$2\frac{d^2y}{dx^2}+3\frac{dy}{dx}=e^x \quad , \quad y(0)=1 \quad , \quad y'(0)=1$$

10. Solve the following second-order ordinary differential equation with two initial conditions:

$$\frac{d^2y}{dx^2}-2\frac{dy}{dx}=\sin(2x) \quad , \quad y(0)=0 \quad , \quad y'(0)=1$$

11. Plot the solution function obtained in Exercise # 10 above.

12. Solve the following system of two linear first-order ordinary differential equations with two initial conditions:

$$x'(t)=3x(t)-y(t)$$

$$y'(t)=x(t)+y(t)$$

The initial conditions are $x(0)=1$ and $y(0)=1$.

3. Differential Equations –II-

In this chapter, we continue solving more ordinary differential equations. In particular, we solve differential equations of the third-order and fourth-order. In addition, we solve some nonlinear ordinary differential equations also. All the differential equations in this chapter are solved analytically using the MATLAB command `dsolve` which requires the MATLAB Symbolic Math Toolbox.

Higher-Order Ordinary Differential Equations

In this section, let us solve three higher-order ordinary differential equations. First, let us solve the third-order linear ordinary differential equation:

$$\frac{d^3y}{dx^3} = x + 1$$

The MATLAB commands needed to solve the above differential equations are shown below along with the output:

```
>> dsolve('D3y = x+1','x')

ans =
```

```
x^4/24 + x^3/6 + (C8*x^2)/2 + C9*x + C10
```

Based on the above MATLAB output, the solution function is written explicitly as follows:

$$y = \frac{1}{24}x^4 + \frac{1}{6}x^3 + \frac{1}{2}C_8x^2 + C_9x + C_{10}$$

where C_8, C_9, and C_{10} are constants of integration. Note that there are three undetermined constants because we are dealing with a third-order ordinary differential equations. In order to determine the three constants, we need three initial conditions (but this is not required in this example).

Next, let us solve the fourth-order linear ordinary differential equation:

$$\frac{d^4y}{dx^4} = y$$

The MATLAB commands needed to solve the above equation are shown below along with the resulting output:

```
>> dsolve('D4y = y','x')
```

```
ans =

C12*cos(x) + C15*exp(x) + C13*sin(x) +
C14*exp(-x)
```

Based on the above MATLAB output, the solution function is written explicitly as follows:

$$y = C_{12}\cos x + C_{15}e^{x} + C_{13}\sin x + C_{14}e^{-x}$$

Note that we have four undetermined constants of integration because we are dealing with a fourth-order ordinary differential equation. In order to find the values of the four constants, we need four initial conditions, but these are not given in this example.

Next, let us solve the fourth-order linear ordinary differential equation:

$$\frac{d^4 y}{dx^4} = -y$$

The MATLAB commands needed to solve the above equation are shown below along with the resulting output:

```
>> dsolve('D4y = -y','x')

ans =

C17*exp((2^(1/2)*x)/2)*cos((2^(1/2)*x)/2) +
C19*exp(-(2^(1/2)*x)/2)*cos((2^(1/2)*x)/2) +
C18*exp((2^(1/2)*x)/2)*sin((2^(1/2)*x)/2) +
C20*exp(-(2^(1/2)*x)/2)*sin((2^(1/2)*x)/2)

>> simplify(ans)

ans =

exp(-(2^(1/2)*x)/2)*(C19*cos((2^(1/2)*x)/2)
+ C20*sin((2^(1/2)*x)/2) +
C17*exp(2^(1/2)*x)*cos((2^(1/2)*x)/2) +
C18*exp(2^(1/2)*x)*sin((2^(1/2)*x)/2))
```

Based on the above MATLAB output, the solution function is written explicitly as follows (note the use of the `simplify` command):

$$y = e^{\frac{-\sqrt{2}x}{2}} \left[C_{19} \cos\left(\frac{\sqrt{2}x}{2}\right) + C_{20} \sin\left(\frac{\sqrt{2}x}{2}\right) + C_{17} e^{\sqrt{2}x} \cos\left(\frac{\sqrt{2}x}{2}\right) + C_{18} e^{\sqrt{2}x} \sin\left(\frac{\sqrt{2}x}{2}\right) \right]$$

Note that we have four undetermined constant of integration because we are dealing with a fourth-order ordinary differential equation. In order to find the values of the four constants, we need four initial conditions, but these are not given in this example.

Nonlinear Ordinary Differential Equations

In this section we show the solution of several nonlinear ordinary differential equations. Let us start and solve the following nonlinear ordinary differential equation that comes with an initial condition:

$$\frac{dx}{dt} = k\,x(1000 - x) \quad , \quad x(0) = 1$$

where k is a constant. The MATLAB commands needed to solve the above equation are shown below along with the resulting output:

```
>> dsolve('Dx = k*x*(1000-
x)','x(0)=1','t')


ans =
```

```
1000/(exp(log(999) - 1000*k*t) + 1)
```

Based on the above MATLAB output, the solution function is written explicitly as follows:

$$x(t) = \frac{1}{1000}\left(e^{(\ln 9999 - 1000\,k\,t)} + 1\right)$$

Let us check that the above solution function satisfies the initial condition $x(0) = 1$. The details follow:

$$x(0) = \frac{1}{1000}\left(e^{(\ln 9999 - 0)} + 1\right) = \frac{1}{1000}\left(e^{\ln 999} + 1\right) = \frac{1}{1000}(999 + 1) = 1$$

Next, let us solve the following nonlinear ordinary differential equation[8]:

$$\frac{dx}{dt} = k(200 - x)(50 - x)$$

---

[8] A generalization of this equation is given in Exercise # 4 at the end of this chapter.

where k is a constant. The MATLAB commands needed to solve the above equation are shown below along with the resulting output:

```
>> dsolve('Dx = k*(200-x)*(50-x)','t')

ans =

                                        50

                                       200

 50 - 150/(exp(C27 + 150*k*t) - 1)
```

Based on the above MATLAB output, the solution function is written explicitly as follows (note that the first two solutions of 50 and 200 are trivial; we are interested in the third nontrivial solution):

$$x(t) = 50 - \frac{150}{e^{(C_{27} + 150kt)} - 1}$$

Next, let us solve the following nonlinear ordinary differential equation:

$$u'' - k(1-u)u' = 0$$

where k is a constant. The MATLAB commands needed to solve the above equation are shown below along with the resulting output:

```
>> dsolve('D2u = -k*(1-u)*Du = 0','t')

ans =

C52
 (k + tan((C55 + t/2)*(- k^2 +
2*C52*k)^(1/2))*(- k^2 + 2*C52*k)^(1/2))/k
```

Based on the above MATLAB output, the solution function is written explicitly as follows:

$$u(t) = \frac{k + \sqrt{-k^2 + 2C_{52}k}\,\tan\left[(C_{55} + t/2)\sqrt{-k^2 + 2C_{52}k}\right]}{k}$$

Note that there are two undetermined constants of integration above. This is because the differential equation is of the second-order and thus needs two intial conditions. But no intial conditions are given in this example. Thus the two constants remain undetermined.

Next, let us solve the following nonlinear ordinary differential equation:

$$\frac{dr}{dt} = \frac{2k}{r} + 2h$$

where k and h are constants. The MATLAB commands needed to solve the above equation are shown below along with the resulting output:

```
>> dsolve('Dr = 2*k/r + 2*h','t')


ans =


-k/h

 -(k + k*lambertw(0, -exp(-((C67 + 2*t)*h^2
+ k)/k)/k))/h
```

Based on the above MATLAB output, the solution function is written explicitly as follows (neglecting the trivial solution $-k/h$ that is shown above):

$$r(t) = -\frac{k + kW\left(0, -\frac{e^{\frac{-(C_{67}+2t)h^2+k}{k}}}{k}\right)}{h}$$

where W is the *Lambert W* function[9]. Note that there is only one undetermined constant of integration. This is because the differential equation is of the first order but nonlinear. Only one initial condition is needed to determined the constant, but this is not given in this example.

Finally let us solve the following nonlinear ordinary differential equation:

$$\frac{dy}{dx} = -\frac{y}{s^2 - y^2}$$

where s is a constant. The MATLAB commands needed to solve the above equation are shown below along with the resulting output:

```
>> dsolve('Dy = -y/(s^2-y^2)','x')
```

```
ans =
```

[9] The *Labmert W* function is one of the Special Functions of mathematics. Consult a book on "Special Functions" for details about this function.

```
0

 exp(- wrightOmega(log(-1/s^2) - (2*(C75 +
x))/s^2)/2 - (C75 + x)/s^2)
```

Based on the above MATLAB output, the solution function is written explicitly as follows (disregarding the zero trivial solution shown above):

$$y = e^{-\omega\left[\left(\ln\left(\frac{-1}{s^2}\right)-\frac{2(C_{75}+x)}{s^2}\right]-\frac{(C_{75}+x)}{s^2}}$$

where ω is the *Wright omega* function[10]. Note that there is only one undetermined constant of integration. This is because the differential equation is of the first order but nonlinear. Only one initial condition is needed to determined the constant, but this is not given in this example.

Exercises

Solve all the exercises using MATLAB. All the needed MATLAB commands for these exercises were presented in this chapter. Note that all these exercises (1 - 6) require the use of the MATLAB Symbolic Math Toolbox (in particular, the MATLAB command `dsolve` is needed).

---

[10] The *Wright omega* function (or just the *Wright* function) is one of the Special Functions of mathematics. It is usually defined in terms of the *Lambert W* function. Consult a book on "Special Functions" for details.

1. Solve the following third-order linear first-order ordinary differential equation[11] using MATLAB:

$$y''' - y'' - y' + y = 2e^{-x} + 3$$

2. Solve the following fourth-order linear ordinary differential equation using MATLAB:

$$\frac{d^4y}{dx^4} + 2\frac{d^2y}{dx^2} + y = 0$$

3. Solve the following nonlinear ordinary differential equation[12] using MATLAB:

$$\frac{dP}{dt} = P(a - b\ln P)$$

where a and b are constants.

4. Solve the following nonlinear ordinary differential equation using MATLAB:

---

[11] Note that y''' is the third derivative, also denoted by d^3y/dx^3.

[12] This equation is called a *modified logistic* equation.

$$\frac{dx}{dt} = k(a - x)(b - x)(c - x)$$

where a, b, and c are constants, and $a \neq b \neq c$.

5. Solve the following nonlinear ordinary differential equation using MATLAB:

$$x\left(\frac{dy}{dx}\right)^2 + 2\frac{dy}{dx} = x$$

6. Solve the following nonlinear ordinary differential equation using MATLAB:

$$m\frac{dv}{dt} = mg - kv^2$$

where k and m are constants.

4. Differential Equations –III-

In this chapter several applications of differential equations are presented in various fields. These application exercises are solved here using MATLAB. The example applications include the mechanics of free-falling objects, the vibration of a mass attached to a spring, the oscillatory motion of a pendulum, electric circuits, population growth/dynamics, the spread of infectious/contagious diseases, and the tension in a hanging wire. Furhter applications that appear in the set of Exercies at the end of the chapter include the cooling of a body, the process of radioactive decay, the infusion of a drug in a patient, rocket launch and rocket motion, and compound interest. In all the examples that follow, time is denoted by the letter t.

The first example application is the mechanics of a free-falling body. Consider a body of a constant mass that is falling under the effect of gravity, i.e. a free-falling body. Let g denote the constant acceleration of gravity, t is the time elapsed, s is the height of the body above the surface of earth, $v = ds/dt$ is the velocity of the body, and $a = d^2s/dt^2$ is its acceleration. The governing differential equation in this case is an ordinary differential equation of the second order that can be written in the following form (note that the mass of the body does not play a part in the governing differential equation and we assume also that there is no air resistance):

$$\frac{d^2s}{dt^2} = -g$$

with the initial conditions:

$$s(0) = s_o \qquad \text{initial height}$$

$$\frac{ds}{dt}(0) = v_o \qquad \text{initial velocity}$$

It is seen that the governing differential equation is a linear second-order ordinary differential equation. The solution of the above differential equation with initial conditions is performed in MATLAB using the following command, followed by the output:

```
>>                                    dsolve('D2s=-
g','s(0)=s0','Ds(0)=v0','t')

ans =

s0 + t*v0 - (g*t^2)/2
```

The solution as shown above is obtained as the following function which satisfies the intial conditions:

$$s(t) = -\frac{1}{2}gt^2 + v_0t + s_0$$

The <u>second example</u> to be examined is the vibration of a mass attached to a spring. Let m be the mass of the object and let k be the elastic stiffness of the spring. Let also x be the distance the mass moves from a initial setting, then the differential equation that governs the vibration of the mass attached to the spring is given by (check Figure 4.1):

$$m\frac{d^2x}{dt^2} = -kx$$

with the initial conditions:

$$x(0) = x_0$$

$$\frac{dx}{dt}(0) = 0$$

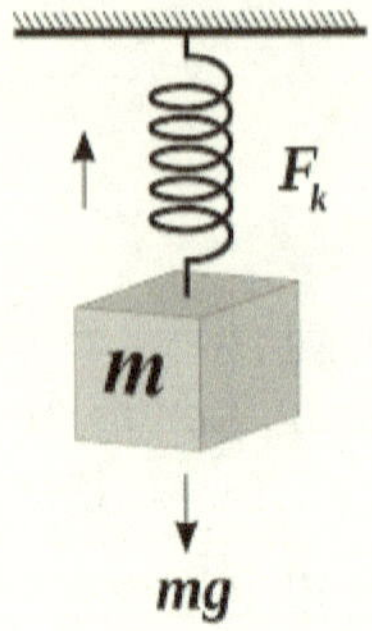

Figure 4.1: Mass Attached to a Spring

The governing differential equation in this case can be re-written in the following standard form:

$$\frac{d^2x}{dt^2} + \omega^2 x = 0$$

where $\omega^2 = \frac{k}{m}$. It is seen that the governing differential equation is a linear second-order ordinary differential equation. The solution of the above differential equation is performed in MATLAB using the following commands:

```
>> dsolve('D2x = -
omega^2*x','x(0)=x0','Dx(0)=0','t')


ans =
```

```
(x0*exp(-omega*t*i))/2 +
(x0*exp(omega*t*i))/2


>> simplify(ans)


ans =


x0*cos(omega*t)
```

It is seen from the above MATLAB output that the solution is written as the following function which satisfies the intial conditions (note the use of the MATLAB command `simplify`):

$$x(t) = x_0 \cos(\omega t)$$

The <u>third example</u> to be studied is the motion of a simple pendulum. Let L be the length of the pendulum and m be the mass of the object that is attached to it. Let also θ be the angle that the pendulum moves through from the vertical position, and let g be the constant acceleration of gravity. In this case, the differential equation that governs the motion of the pendulum is given by (check Figure 4.2):

$$\frac{d^2\theta}{dt^2} + \frac{g}{L}\sin\theta = 0$$

with the initial conditions

$$\theta(0) = 0$$

$$\frac{d\theta}{dt}(0) = v_0 \qquad \text{initial velocity}$$

It is clear that the above differential equation is nonlinear because of the presence of $\sin\theta$. For the special case of small oscillations, i.e, when the angle θ is small, we can use the approximation[13] $\sin\theta \approx \theta$, and thus the above differential equation can be linearized, i.e. re-written in the following linear form:

[13] This approximation results from taking the first term of the Taylor series expansion of the *sine* function.

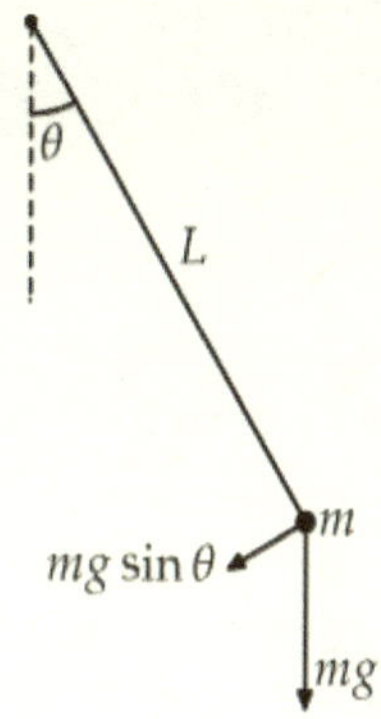

Figure 4.2: Vibration of a Simple Pendulum

$$\frac{d^2\theta}{dt^2} + \omega^2\theta = 0$$

where $\omega^2 = \frac{g}{L}$. It is seen that the differential equation governing the motion of a simple pendulum given in the third example has exactly the same form as the differential equation governing the vibration of a mass attached to a spring given in the second example. It is also seen that the governing differential equation is a linear second-order ordinary differential equation. The solution of this differential equation is performed in MATLAB using the following commands (we solve only the second linear equation, not the first nonlinear equation):

```
>>
dsolve('D2pheta+omega^2*pheta=0','pheta(0
)=0','Dpheta(0)=v0','t')
```

```
ans =

(v0*exp(-omega*t*i)*i)/(2*omega) -
(v0*exp(omega*t*i)*i)/(2*omega)

>> simplify(ans)

ans =

(v0*sin(omega*t))/omega
```

The solution is obtained as the following function which satisfies the initial conditions (note the use of the `simplify` command above):

$$\theta(t) = \frac{v_0}{\omega}\sin(\omega t)$$

In the fourth example, we consider a simple electric circuit. Let $I(t)$ denote the current in the circuit at time t (measured in Amperes), R denote the resistance (measured in Ohms), C denote the capacitance (measures in Farads), and L denote the inductance (measured in Henrys). Note that R, C, and L are assumed to be constants in this example. Let also $E(t)$ and $Q(t)$ denote the voltage[14] at time t (measured in Volts) and the total charge on the capacitor at time t (measured in Coulombs), respectively. The governing differential equation for the total charge $Q(t)$ at time t is given by the following second-order ordinary differential equation (check Figure 4.3):

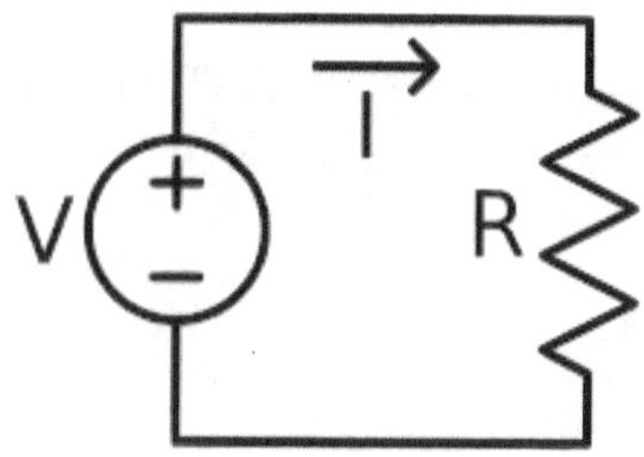

Figure 4.3: Simple Electric Circuit

$$L\frac{d^2Q}{dt^2}+R\frac{dQ}{dt}+\frac{1}{c}Q=E(t)$$

with initial conditions:

$$Q(t_0)=Q_0$$

---

14 Although the voltage is denoted by the letter Q in the text, it is denoted by the letter V in Figure 18.3.

$$\frac{dQ}{dt}(t_0) = I(t_0) = I_0$$

It is seen that the differential equation in this example is a linear second-order ordinary differential equation. Note that the current in the circuit is given by $I = \frac{dQ}{dt}$. The solution of this differential equation using MATLAB is performed by following the following commands:

```
>> dsolve('L*D2Q+R*DQ+Q/C = E', 'Q(0)=Q0', 'DQ(0) =
I0','t')


ans =


C*exp(1) - (exp(-(t*((-C*(- C*R^2 + 4*L))^(1/2) +
C*R))/(2*C*L))*(2*C*I0*L - Q0*(-C*(- C*R^2 +
4*L))^(1/2) + C*Q0*R + C*exp(1)*(-C*(- C*R^2 +
4*L))^(1/2) - C^2*R*exp(1)))/(2*(-C*(4*L -
C*R^2))^(1/2)) + (exp((t*((-C*(- C*R^2 + 4*L))^(1/2)
- C*R))/(2*C*L))*(Q0*(-C*(- C*R^2 + 4*L))^(1/2) +
2*C*I0*L + C*Q0*R - C*exp(1)*(-C*(- C*R^2 +
4*L))^(1/2) - C^2*R*exp(1)))/(2*(-C*(4*L -
C*R^2))^(1/2))


>> simplify(ans)


ans =


C*exp(1) - (exp(-(t*((-C*(- C*R^2 + 4*L))^(1/2) +
C*R))/(2*C*L))*(2*C*I0*L - Q0*(-C*(- C*R^2 +
4*L))^(1/2) + C*Q0*R + C*exp(1)*(-C*(- C*R^2 +
```

```
4*L))^(1/2) - C^2*R*exp(1)))/(2*(-C*(4*L -
C*R^2))^(1/2)) + (exp((t*((-C*(- C*R^2 + 4*L))^(1/2)
- C*R))/(2*C*L))*(Q0*(-C*(- C*R^2 + 4*L))^(1/2) +
2*C*I0*L + C*Q0*R - C*exp(1)*(-C*(- C*R^2 +
4*L))^(1/2) - C^2*R*exp(1)))/(2*(-C*(4*L -
C*R^2))^(1/2))
```

It is seen from the above MATLAB output that the solution is very long and complicated (consisting of seven lines of output). It is left to the reader to write the solution explicity based on the above output. However, let us try to obtain a somewhat simplified version of the solution for a very special case. Let us consider the case when the initial conditions are zero, i.e. $Q_0 = I_0 = 0$. Thus, let us solve the differential equation again but this time with the new initial conditions. The MATLAB commands are as follows:

```
>> dsolve('L*D2Q+R*DQ+Q/C = E', 'Q(0)=0', 'DQ(0) =
0','t')

ans =

C*exp(1) - (C*exp(1)*exp(-(t*((-C*(- C*R^2 +
4*L))^(1/2) + C*R))/(2*C*L))*((-C*(- C*R^2 +
4*L))^(1/2) - C*R))/(2*(-C*(4*L - C*R^2))^(1/2)) -
(C*exp(1)*exp((t*((-C*(- C*R^2 + 4*L))^(1/2) -
C*R))/(2*C*L))*((-C*(- C*R^2 + 4*L))^(1/2) +
C*R))/(2*(-C*(4*L - C*R^2))^(1/2))

>> simplify(ans)

ans =
```

```
C*exp(1) - (C*exp(1)*exp(-(t*((-C*(- C*R^2 +
4*L))^(1/2) + C*R))/(2*C*L))*((-C*(- C*R^2 +
4*L))^(1/2) - C*R))/(2*(-C*(4*L - C*R^2))^(1/2)) -
(C*exp(1)*exp((t*((-C*(- C*R^2 + 4*L))^(1/2) -
C*R))/(2*C*L))*((-C*(- C*R^2 + 4*L))^(1/2) +
C*R))/(2*(-C*(4*L - C*R^2))^(1/2))
```

It is seen from the above MATLAB output that the solution is still long and complicated, but it consists now of only five lines of output. The solution function can now be written expicity as follows (which satisfies the zero initial conditions):

$$Q(t) = Ce - Ce\frac{(\beta - CR)}{2\beta}e^{\frac{-(\beta+CR)t}{2CL}} - Ce\frac{(\beta + CR)}{2\beta}e^{\frac{+(\beta-CR)t}{2CL}}$$

where the constant β is given by the following expression:

$$\beta = \sqrt{-C(-CR^2 + 4L)} = \sqrt{C^2R^2 - 4CL}$$

we show below how the above function satisfies the intial condition $Q(t_0) = Q_0 = 0$, where $t_0 = 0$:

$$
\begin{aligned}
Q(0) &= Ce - Ce\frac{(\beta - CR)}{2\beta}e^{0} - Ce\frac{(\beta + CR)}{2\beta}e^{0} \\
&= Ce - Ce\frac{(\beta - CR)}{2\beta} - Ce\frac{(\beta + CR)}{2\beta} \\
&= Ce - Ce\frac{(\beta - CR + \beta + CR)}{2\beta} \\
&= Ce - Ce\frac{(2\beta)}{2\beta} \\
&= Ce - Ce \\
&= 0
\end{aligned}
$$

The fifth example involves the problem of population growth or population dynamics. Let P be the population of a certain region at time t, with an initial population P_o. Let k be a constant relating to population growth in the region, where $k > 0$ for an increasing population. The governing differential equation in this case is written as follows:

$$\frac{dP}{dt} = kP$$

with the initial condition:

$P(0) = P_o$ initial population

It is seen that the governing differential equation in this case is a linear first-order ordinary differential equation. The solution of the differential equation is performed in MATLAB using the following commands:

```
>> dsolve('DP=k*P','DP(0)=P0','t')

ans =

(P0*exp(k*t))/k
```

It is seen that the solution is is given as the following function (which satisfies the initial condition):

$$P(t) = \frac{P_0}{k} e^{kt}$$

The <u>sixth example</u> involves the spread of infectious or contagious diseases. Let $x(t)$ denote the number of people at time t that have contacted the disease, and let $y(t)$ denote the number of people not exposed yet. Then we have the following relationship:

$$x + y = n + 1$$

where n is the number of people in the total population, and one infected person is introduced to a fixed population of people. In this case, the governing differential equation can be written in the following form:

$$\frac{dx}{dt} = kx(n+1-x)$$

with the initial condition:

$$x(0) = 1$$

where k is a constant related to the rate of spread of the disease. It is clear that the governing differential equation is a nonlinear first-order ordinary differential equations. It is nonlinear because of the presence of the term x^2 (the right-hand-side becomes $kxn + kx - kx^2$ when expanded.

The solution of the above differential equation using MATLAB is given by the following:

```
>> dsolve('Dx=k*x*(n+1-x)','x(0)=1','t')

ans =
```

```
(n + 1)/(exp(-(k*t - log(n)/(n + 1))*(n +
1)) + 1)


>> simplify(ans)


ans =


(exp(k*t + k*n*t)*(n + 1))/(n + exp(k*t +
k*n*t))
```

The solution based on the above MATLAB output is given by the following function (which satisfies the initial condition):

$$x(t) = \frac{(n+1)e^{kt(n+1)}}{n+e^{kt(n+1)}}$$

We show below how the above function satisfies the initial condition $x(0)=1$:

$$\begin{aligned} x(0) &= \frac{(n+1)e^{0}}{n+e^{0}} \\ &= \frac{(n+1)(1)}{n+1} \\ &= 1 \end{aligned}$$

It should be mentioned that the nonlinear differential equation that governs the spread of infectious or contagious diseases is a special case of what is known as the *logistic equation*, which has the following general form:

$$\frac{dP}{dt} = P(a - bP)$$

with the intial condition:

$$P(0) = P_o$$

In the logistic equation[15] shown above, a and b are constants. The solution of the logistic equation is given by the following MATLAB commands:

```
>> dsolve('DP=P*(a-b*P)','P(0)=P0','t')

ans =

a/(b + exp(-a*(t - log((a -
P0*b)/P0)/a)))
```

[15] Check Exercise # 3 in Chapter 17 for a modified form of the logistic equation.

The solution based on the above MATLAB output is given by the following function (which satisfies the initial condition):

$$P(t) = \frac{a}{b + e^{-a\left(\frac{t - \ln(a - P_0 b)/P_0}{a}\right)}}$$

We show next how the above function satifies the initial condition $P(0) = P_o$:

$$P(0) = \frac{a}{b + e^{-a\left(\frac{0 - \ln(a - P_0 b)/P_0}{a}\right)}} = \frac{a}{b + e^{\ln(a - P_0 b)/P_0}}$$

$$= \frac{a}{b + \dfrac{a - P_0 b}{P_0}} = \frac{a}{\dfrac{P_0 b + a - P_0 b}{P_0}} = \frac{a}{\dfrac{a}{P_0}} = P_0$$

The <u>seventh example</u> is concerned with the tension in a hanging wire. Consider a hanging wire between two points, with tension T at its lowest point. The length of the wire s between the two points is given by the following integral equation:

$$s = \int_0^x \sqrt{1 + \left(\frac{dy}{dx}\right)^2}\, dx$$

where x and y are the coordinates of an arbitrary point on the wire.

It should be noted that the derivative $\frac{dy}{dx}$ represents the slope of the wire at the specified point (x, y). Finally the governing differential equation for the hanging wire can be written as follows (check Figure 4.4):

$$\frac{d^2y}{dx^2} = \frac{w}{T}\sqrt{1 + \left(\frac{dy}{dx}\right)^2}$$

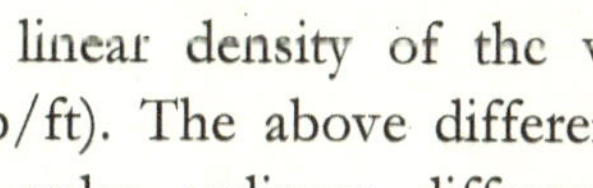

Figure 4.4: Tension in a Hanging Wire

where w is the linear density of the wire (measured in gram/meter or lb/ft). The above differential equation is a nonlinear second-order ordinary differential equation that governs the tension in a hanging wire. The solution of the above differential equation is performed using MATLAB as follows (where we have used an initial height y_0 and and an initial slope v_0):

```
>>
dsolve('D2y=w*sqrt(1+Dy^2)/T','y(0)=y0','Dy(0)=v0','
x')


ans =


   (exp((w*(x + (T*log(T*(v0 + (v0^2 +
1)^(1/2))))/w))/T) - 2*T*(v0^2 + 1)^(1/2) + 2*w*y0 +
T^2*exp(-(w*(x + (T*log(T*(v0 + (v0^2 +
1)^(1/2))))/w))/T))/(2*w)

   (exp((w*(x + (T*log(T*(v0 - (v0^2 +
1)^(1/2))))/w))/T) + 2*T*(v0^2 + 1)^(1/2) +
T^2*exp(-(w*(x + (T*log(T*(v0 - (v0^2 +
1)^(1/2))))/w))/T) + 2*w*y0)/(2*w)

 (exp(-(w*(x - (T*log(-T*(v0 + (v0^2 +
1)^(1/2))))/w))/T) + 2*T*(v0^2 + 1)^(1/2) + 2*w*y0 +
T^2*exp((w*(x - (T*log(-T*(v0 + (v0^2 +
1)^(1/2))))/w))/T))/(2*w)

 (exp(-(w*(x - (T*log(-T*(v0 - (v0^2 +
1)^(1/2))))/w))/T) - 2*T*(v0^2 + 1)^(1/2) +
T^2*exp((w*(x - (T*log(-T*(v0 - (v0^2 +
1)^(1/2))))/w))/T) + 2*w*y0)/(2*w)


>> simplify(ans)


ans =


 y0 + (T*exp(-(w*x)/T)*(exp((w*x)/T) - 1)*(v0 +
v0*exp((w*x)/T) + exp((w*x)/T)*(v0^2 + 1)^(1/2) -
(v0^2 + 1)^(1/2)))/(2*w)

 y0 + (T*exp(-(w*x)/T)*(exp((w*x)/T) - 1)*(v0 +
v0*exp((w*x)/T) - exp((w*x)/T)*(v0^2 + 1)^(1/2) +
(v0^2 + 1)^(1/2)))/(2*w)

 y0 + (T*exp(-(w*x)/T)*(exp((w*x)/T) - 1)*(v0 +
v0*exp((w*x)/T) - exp((w*x)/T)*(v0^2 + 1)^(1/2) +
(v0^2 + 1)^(1/2)))/(2*w)
```

```
y0 + (T*exp(-(w*x)/T)*(exp((w*x)/T) - 1)*(v0 +
v0*exp((w*x)/T) + exp((w*x)/T)*(v0^2 + 1)^(1/2) -
(v0^2 + 1)^(1/2)))/(2*w)
```

Although the `simplified` command is used above, the final solution is still complicated and long (consisting of twelve lines of output). No attempt will be made to simplify it further, or even to write it out fully. However, there is a way to obtain a simpler solution as illustrated below, but only for a special case.

If the slope $\frac{dy}{dx}$ of the wire is small, the above governing differential equation can be approximated using the Taylor series expansion of the square root function $\sqrt{1+x} \approx 1 + \frac{1}{2}x + ...$ as follows (check Chapter 11 to see how to evaluate a Taylor series using MATLAB):

$$\frac{d^2y}{dx^2} = \frac{w}{T}\left[1 + \frac{1}{2}\left(\frac{dy}{dx}\right)^2\right]$$

The resulting approximate differential equation is still a nonlinear second-order ordinary differential equation, but it is significantly a lot easier to solve than the original one. The solution of the approximate differential equation is performed as follows using MATLAB:

```
>>
dsolve('D2y=w*(1+Dy^2/2)/T','y(0)=y0','Dy(0)=v0','x'
)


ans =


(T*log((exp((w*y0)/T)*(tan((2^(1/2)*w*(x +
(2^(1/2)*T*atan((2^(1/2)*v0)/2))/w))/(2*T))^2 +
1)*(cos(2*atan((2^(1/2)*v0)/2)) + 1))/2))/w


>> simplify(ans)


ans =


(T*log((exp((w*y0)/T)*(cos(2*atan((2^(1/2)*v0)/2)) +
1)*(tan((2^(1/2)*(w*x +
2^(1/2)*T*atan((2^(1/2)*v0)/2)))/(2*T))^2 +
1))/2))/w
```

The final simplified solution based on the above MATLAB output is shown above and it is much shorter than the first output. The solution function is written as follows based on the above output:

$$y(x)=\frac{T}{w}\ln\left\{\begin{array}{l}\frac{1}{2}e^{wy_0/T}\left[\cos\left(2\tan^{-1}\left(\sqrt{2}v_0/2\right)\right)+1\right]* \\ \qquad *\left[\tan\left(\frac{\sqrt{2}\left(wx+\sqrt{2}T\tan^{-1}\left(\sqrt{2}v_0/2\right)\right)}{2T}\right)^2+1\right]\end{array}\right\}$$

In order to check if the above function satisfies the initial conditions, let us write the above function for the special case when $v_0 = 0$. Substituing this value in the above function results in the following much simplified expression:

$$y(x) = \frac{T}{w} \ln\left\{ e^{wy_0/T} \left[\tan\left(\frac{\sqrt{2}(wx)}{2T} \right)^2 + 1 \right] \right\}$$

Now, it can be clearly seen that the above simplified expression satisfies the intial condition $y(0) = y_0$, by substituting the value $x = 0$ in the above expression. This is shown in detail below:

$$\begin{aligned} y(0) &= \frac{T}{w} \ln\left\{ e^{wy_0/T} [0+1] \right\} \\ &= \frac{T}{w} \ln\left\{ e^{wy_0/T} \right\} \\ &= \frac{T}{w} \frac{w y_0}{T} \\ &= y_0 \end{aligned}$$

With these seven application examples, we complete this chapter on differential equations. More types of applications of ordinary differential equations can be found in the Exercises below.

Exercises

Solve all the exercises (1 - 5) using MATLAB. All the needed MATLAB commands for these exercises were presented in this chapter.

1. Consider the cooling of a body. Let T denote the temperature of the body at time t. Let T_o denote the initial temperature of the body. The governing differential equation for the cooling of the body can be written as follows:

$$\frac{dT}{dt} = kT$$

with the initial condition $T(0) = T_o$ is the constant temperature of the outside medium, and k is a constant related to the cooling of the body. Note that $k < 0$ since the temperature T is decreasing. The above differential equation is a linear first-order ordinary differential equation. Solve this differential equation using MATLAB.

2. Consider the problem of radioactive decay. Let A denote the amount of a radioactive material remaining at time t. The governing differential equation for radioactive decay of the material is written as follows:

$$\frac{dA}{dt} = kA$$

with the initial condition $A(0) = A_o$ is the initial amount of the material, and k is a constant related to the decay of the material. Note that $k < 0$ since the amount A is decreasing. It is noted that the governing differential equation is a linear first-order ordinary differential equation. Solve this differential equation using MATLAB.

3. Consider the problem of drug infusion in a patient. Let $x(t)$ denote the amount of the drug that remains in the patient at time t. Let r denote the constant rate of infusion in gm/s, and let k denote the rate of removal of the drug which is proportional to $x(t)$. In this case, the governing differential equation can be written in the following form:

$$\frac{dx}{dt} = r - kx$$

with the initial condition $x(0) = 0$. The above differential equation is a linear first-order ordinary differential equation. Solve this differential equation using MATLAB.

4. Consider the problem of rocket launch and motion. In this case, the governing differential equation is given in the following form (check Figure 4.5):

$$-mg = m\frac{dv}{dt} + V\frac{dm}{dt}$$

with the initial conditions $v(0) = 0$ and $m(0) = m_0$

.

In the above equation, m is the mass of the rocket at time t, v is the velocity of the rocket at time t, and V is the constant velocity of exhaust gases relative to the rocket. The symbol g denotes the constant gravitational acceleration, and m_0 denotes the initial mass of the rocket. The above governing differential equation is a linear first-order ordinary differential equation. Solve this differential equation using MATLAB.

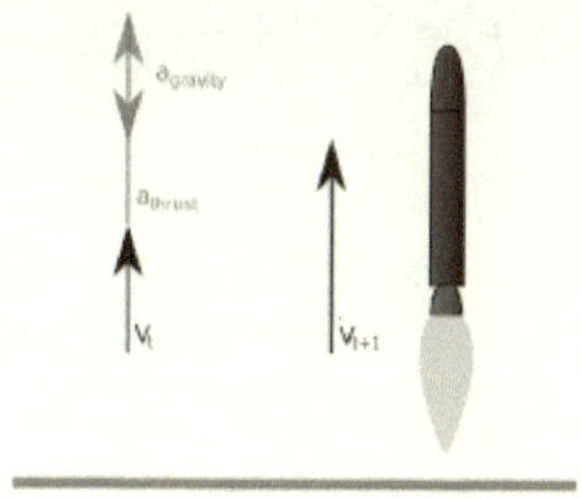

Figure 4.5: Rocket Lanuch and Motion

5. Consider the compound interest that is obtained based on an initial sum of money. Let S_0 be the initial amount of money, and let $S(t)$ be the amount of money at time t. Considering that interenst is compounded continuously, the amount $S(t)$ is governed by the following differential equation:

$$\frac{dS}{dt} = r\,S \qquad , \qquad S(0) = S_0$$

where r is the interest rate. Solve the above differential equation using MATLAB to obtain an explicit expression for the amount of money $S(t)$ at any time t. Compare the result that you obtain with the following expression for $S(t)$, obtaind assuming that the interest is compounded m times per year.

$$S(t) = S_0\left(1 + \frac{r}{m}\right)^{mt}$$

Solutions to Exercises

1. Introduction

1. Perform the operation 3*4+6. The order of the operations will be discussed in subsequent chapters.

```
>> 3*4+6

ans =

    18
```

2. Perform the operation $\cos(5)$. The value of 5 is in radians.

```
>> cos(5)

ans =

    0.2837
```

3. Perform the operation $3\sqrt{6+x}$ for $x = 4$.

```
>> x = 4

x =

     4

>> 3*sqrt(6+x)

ans =

    9.4868
```

4. Assign the value of 5.2 to the variable y.

```
>> y = 5.2

y =

    5.2000
```

5. Assign the values of 3 and 4 to the variables x and y, respectively, then calculate the value of z where $z = 2x - 7y$.

```
>> x = 3;

>> y = 4;

>> z = 2*x - 7*y


z =


   -22
```

6. Obtain help on the inv command.

```
>> help inv

 INV    Matrix inverse.

    INV(X) is the inverse of the square matrix X.

    A warning message is printed if X is badly scaled or

    nearly singular.


    See also slash, pinv, cond, condest, lsqnonneg, lscov.
```

```
    Overloaded functions or methods
(ones with the same name in other
directories)

       help sym/inv.m


    Reference page in Help browser

       doc inv
```

7. Generate the following matrix: A

$$A = \begin{bmatrix} 1 & 0 & 2 & -3 \\ 0 & 5 & 2 & 2 \\ 1 & 2 & 3 & 4 \\ -2 & 0 & 1 & 3 \end{bmatrix}$$

```
>> A = [1 0 2 -3 ; 0 5 2 2 ; 1 2 3
4 ; -2 0 1 3]


A =


     1     0     2    -3

     0     5     2     2

     1     2     3     4

    -2     0     1     3
```

8. Generate the following vector b

$$b = \begin{Bmatrix} 1 \\ 2 \\ 3 \\ 4 \end{Bmatrix}$$

```
>> b = [1 ; 2 ; 3 ; 4]

b =

     1
     2
     3
     4
```

9. Evaluate the vector c where $\{c\} = [A]\{b\}$ where A is the matrix given in Exercise 7 above and b is the vector given in Exercise 8 above.

```
>> c = A*b

c =
```

```
     3
    -1

>> x = A\b

x =

    0.8462
   -0.6154
```

11. Solve the system of simultaneous algebraic equations of Exercise 10 above using matrix inversion.

```
>> x = inv(A)*b

x =

    0.8462
   -0.6154
```

12. Generate the following matrix X:

$$X = \begin{bmatrix} 1 & 0 & 6 \\ 1 & 2 & 3 \\ 4 & 5 & -2 \end{bmatrix}$$

```
>> X = [1 0 6 ; 1 2 3 ; 4 5 -2]

X =

     1     0     6

     1     2     3

     4     5    -2
```

13. Extract the sub-matrix in rows 2 to 3 and columns 1 to 2 of the matrix X in Exercise 12 above.

```
>> X(2:3,1:2)

ans =

     1     2

     4     5
```

14. Extract the second column of the matrix X in Exercise 12 above.

```
>> X(1:3,2)

ans =

     0
     2
     5
```

15. Extract the first row of the matrix X in Exercise 12 above.

```
>> X(1,1:3)

ans =

     1     0     6
```

16. Extract the element in row 1 and column 3 of the matrix X in Exercise 12 above.

```
>> X(1,3)

ans =
```

```
     6
```

17. Generate the row vector x with integer values ranging from 1 to 9.

```
>> x = [1 2 3 4 5 6 7 8 9]

x =

     1     2     3     4     5     6
7     8     9
```

18. Plot the graph of the function $y = x^3 - 2$ for the range of the values of x in Exercise 17 above.

```
>> y = x.^3 - 2

y =

    -1     6    25    62   123   214
341   510   727

>> plot(x,y)
```

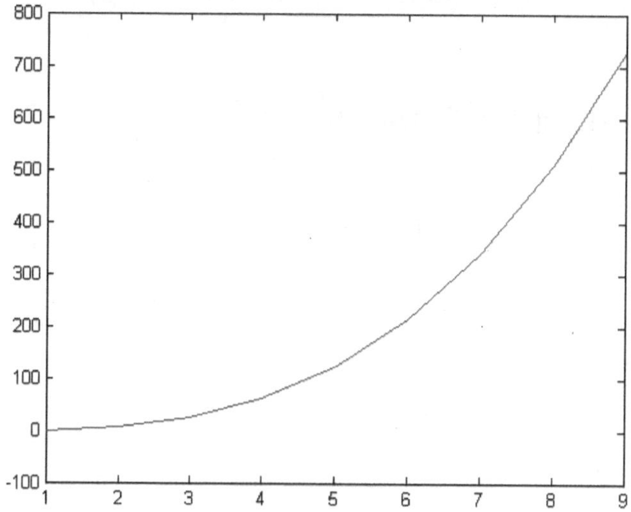

19. Generate a 4 x 4 magic square. What is the total of each row, column, and diagonal in this matrix.

```
>> magic(4)

ans =

    16     2     3    13
     5    11    10     8
     9     7     6    12
     4    14    15     1
```

The sum of each row, column, or diagonal is 34.

2. Differential Equations –I-

1.
```
>> dsolve('Dy = b','x')

ans =

C2 + b*x
```

The resulting function is $y = bx + C_2$. If you do not specify x as the independent variable in the argument of the command above, MATLAB will use t by default.

2.
```
>> dsolve('Dy = 3*x','x')

ans =

(3*x^2)/2 + C4
```

The resulting function is $y = \frac{3}{2}x^2 + C4$.

3.
```
>> dsolve('Dy=2*x-3','y(0)=1','x')

ans =

x*(x - 3) + 1
```

The resulting function is $y = x(x-3)+1$.

Let us show that the above expression satisfies the initial condition $y(0)=1$. The details follow:

$$y(0) = (0)(0-3)+1 = 0+1 = 1$$

4.
```
>> ezplot(ans)
>> hold on

>> xlabel('x')

>> ylabel('y')
```

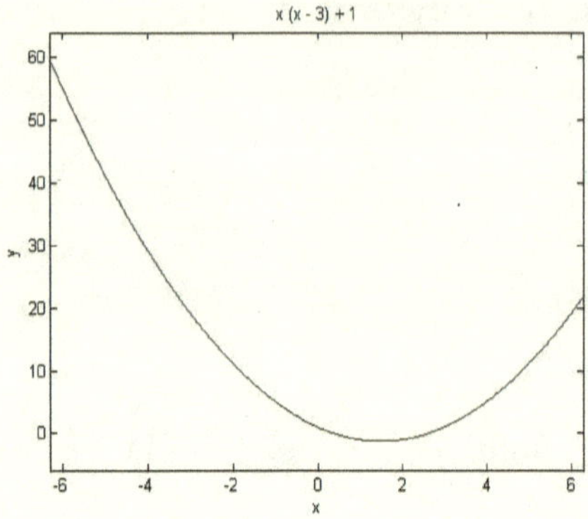


```
5. >> dsolve('Dy=x+2*y','y(0)=b','x')

ans =

(exp(2*x)*(4*b + 1))/4 - x/2 - 1/4
```

The resulting function is

$$y = \frac{1}{4}(4b+1)e^{2x} - \frac{1}{2}x - \frac{1}{4}.$$

Let us show that the above expression satisties the initial condition $y(0) = b$. The details follow:

$$y(0)=\frac{1}{4}(4b+1)e^{0}-\frac{1}{2}(0)-\frac{1}{4}=\frac{1}{4}(4b+1)-\frac{1}{4}=\frac{4b}{4}=b$$

6.
```
>> dsolve('Dy-2*y=sin(3*x)','x')

ans =

C32*exp(2*x) - (2*sin(3*x))/13 -
(3*cos(3*x))/13
```

The resulting function is $y=C_{32}e^{2x}-\frac{2}{13}\sin(3x)-\frac{3}{13}\cos(3x)$.

7.
```
>> dsolve('Dy+3*x*y=2*x-y','x')

ans =

C34*exp(-(x*(3*x + 2))/2) + exp(-(x*(3*x +
2))/2)*(((-3)^(3/2)*3^(1/2)*exp((3*x^2)/2 +
x)*2*i)/27 + (2^(1/2)*3^(1/2)*pi^(1/2)*exp(-
1/6)*erf((2^(1/2)*3^(1/2)*(3*x + 1)*i)/6)*i)/9)

>> simplify(ans)

ans =
```

```
C34*exp(-(x*(3*x + 2))/2) +
(6^(1/2)*pi^(1/2)*erf(6^(1/2)*(x/2 + 1/6)*i)*exp(-
1/6)*exp(-(x*(3*x + 2))/2)*i)/9 + 2/3
```

Note the use of the MATLAB command `simplify` above. The final resulting function is long and involves complicated terms. It will not be shown explicity here, except as presented in the MATLAB output above (Note that `erf` represents the *Error Function*[16], `i`[17] represents an imaginary number, and `pi`[18] is the ratio of the circumference of a circle to its diameter).

8.
```
>> dsolve('x^2*Dy = 2*y*log(y)','x')

ans =

                1

 exp(exp(C7 - 2/x))
```

The resulting function is $y = e^{e^{\left(C7-\frac{2}{x}\right)}}$, where C7 is the constant of integration..

9.
```
>>
dsolve('2*D2y+3*Dy=exp(x)','y(0)=1','Dy(0
)=1','x')
```

---

[16] The *Error Function* is one of the special functions in mathematics.
[17] The imaginary number i is equal to the square root of -1.
[18] The value of pi is equal to 3.14….. (never ending decimals) and is approximately equal to 22/7.

```
ans =

exp(x)/5 - (8*exp(-(3*x)/2))/15 + 4/3
```

The resuling function is $y = \frac{1}{5}e^{x} - \frac{8}{15}e^{-3x/2} + \frac{4}{3}$.

Let is show that the above expression satisfies the initial condition $y(0)=1$. The details follow:

$$y(0) = \frac{1}{5}e^{0} - \frac{8}{15}e^{0} + \frac{4}{3} = \frac{1}{5} - \frac{8}{15} + \frac{4}{3} = \frac{15}{15} = 1$$

10.
```
>> dsolve('D2y-
2*Dy=sin(2*x)','y(0)=0','Dy(0)=1','x')

ans =

(5*exp(2*x))/8 - exp(2*x)*((3*exp(-2*x))/4
(cos(2*x)*exp(-2*x))/8 + (sin(2*x)*exp(-2*x))/8)

>> simplify(ans)

ans =

(5*exp(2*x))/8 - sin(2*x)/8 + cos(x)^2/4 - 7/8
```

Note the use of the MATLAB command `simplify` above. The resulting function is $y = \frac{5}{8}e^{2x} - \frac{1}{8}\sin(2x) + \frac{1}{4}\cos^2 x - \frac{7}{8}$.

Let us show that the above expression satisfies the initial condition $y(0) = 0$. The details follow:

$$y = \frac{5}{8}e^0 - \frac{1}{8}\sin(0) + \frac{1}{4}\cos^2(0) - \frac{7}{8} = \frac{5}{8} + \frac{1}{4} - \frac{7}{8} = 0$$

11.

```
>> ezplot(ans)
>> hold on
>> xlabel('x')
>> ylabel('y')
```

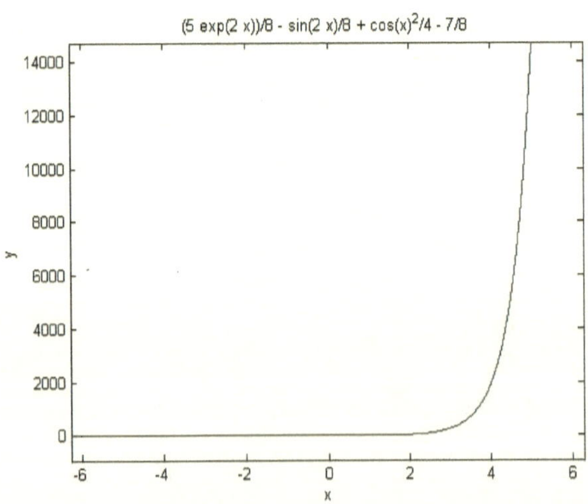

12.
```
>> [x y] = dsolve('Dx=3*x-
y','Dy=x+y','x(0)=1','y(0)=1','t')

x =

exp(2*t)


y =

exp(2*t)
```

It is seen from the MATLAB output above that the two solution functions are given by $x(t) = e^{2t}$ and $y(t) = e^{2t}$. Clearly, these two functions satisfy the two initial conditions, as well as the two simultaneous differential equations.

3. Differential Equations –II-

1.
```
>> dsolve('D3y - D2y - Dy + y = 2*exp(-x)
+3','x')
```

```
ans =


exp(-x)*(x/2 + (3*exp(x))/4) +
exp(x)*(exp(-2*x)*(x/2 + 1/2) + exp(-
x)*((3*x)/2 + 9/4)) + C78*exp(x) +
C80*exp(-x) - (x*exp(-x)*(3*exp(x) +
1))/2 + C79*x*exp(x)


>> simplify(ans)


ans =


exp(-x)/2 + (x*exp(-x))/2 + C78*exp(x) +
C80*exp(-x) + C79*x*exp(x) + 3
```

According to the MATLAB output above, the solution function is written explicitly as follows (note the use of the `simplify` command):

$$y = \frac{1}{2}e^{-x} + \frac{1}{2}xe^{-x} + C_{78}e^{x} + C_{80}e^{-x} + C_{79}xe^{x} + 3$$

Note that there are three undetermined constants. These constants can be determined using three initial conditions, because the differential equation is of the third-order. However, it is not required to find the constants as the initial conditions are not given.

2.
```
>> dsolve('D4y + 2*D2y + y = 0','x')

ans =

C2*cos(x) + C4*sin(x) + C5*x*sin(x) +
C3*x*cos(x)
```

Based on the above MATLAB output, the solution function is written explicitly as follows (with four undetermined constants):

$$y = C_2 \cos x + C_4 \sin x + C_5 x \sin x + C_3 x \cos x$$

Note that there are four undetermined constants. These constants can be determined using four initial conditions, because the differential equation is of the fourth-order. However, it is not required to find the constants as the initial conditions are not given.

3.
```
>> dsolve('DP = P*(a-b*log(P))','t')

ans =

                      exp(a/b)

 exp((a - exp(-b*(C86 + t)))/b)
```

According to the above MATLAB output, the solution function is written explicitly as follows (disregarding the first trivial solution $e^{a/b}$ shown above):

$$P(t) = e^{\frac{a - e^{-b(C_{86}+t)}}{b}}$$

4. >> dsolve('Dx = k*(a-x)*(b-x)*(c-x)','t')

```
Warning: Explicit solution could not be found;
implicit solution

returned.

> In dsolve at 201


ans =



c


b


a

 solve(a*(log(b - x) - log(c - x)) - b*(log(a - x) -
log(c - x)) + c*log(a - x) - c*log(b - x) == (a -
b)*(a - c)*(b - c)*(C90 + k*t), x)
```

Based on the above MATLAB output, an explicit solution function is not obtained. Instead an implicit solution is given in terms of an algebraic equation that needs to be solved. The algebraic equation is listed above and will not be pursued further here. It is left to the reader to write the algebraic equaton explicitly and try to solve it. However, it seems that resort to numerical methods is needed but this is beyond the scope of this beginner's book.

5.
```
>> dsolve('x*(Dy)^2 + 2*Dy = x','x')

ans =

 C95 + asinh((1/x^2)^(1/2)) - log(x) - (x^2 +
1)^(1/2)

 C93 - asinh((1/x^2)^(1/2)) - log(x) + (x^2 +
1)^(1/2)
```

According to the above MATLAB output, there are two solution functions obtained (this is because the ordinary differential equation is quadratic, i.e. one of the derivatives is raised to the power 2; thus two solutions exist). The two solution functions are written explicitly as follows:

$$y = C_{95} + \sinh^{-1}\left[\left(\frac{1}{x^2}\right)^{1/2}\right] - \ln x - \sqrt{x^2+1}$$

$$y = C_{93} - \sinh^{-1}\left[\left(\frac{1}{x^2}\right)^{1/2}\right] - \ln x + \sqrt{x^2+1}$$

Note that there is only one undetermined constant for each solution function. This is because the differential equation is of the first order but nonlinear. An initial condition is needed to find the value of each constant, but this is not given in this exercise.

```
6.  >> dsolve('m*Dv = m*g - k*v^2','t')

    ans =


    (g^(1/2)*m^(1/2))/k^(1/2)

                                                    -
    (g^(1/2)*m^(1/2))/k^(1/2)

     -(g^(1/2)*m^(1/2)*tan(g^(1/2)*k^(1/2)*m^(1/2)*(C98
    + t/m)*i)*i)/k^(1/2)


    >> simplify(ans)


    ans =


    (g^(1/2)*m^(1/2))/k^(1/2)

                                                    -
    (g^(1/2)*m^(1/2))/k^(1/2)

     (g^(1/2)*m^(1/2)*tanh((g^(1/2)*k^(1/2)*(t +
    C98*m))/m^(1/2)))/k^(1/2)
```

Based on the above MATLAB output, the solution function is written explicitly as follows (we disregard the trivial solution $\sqrt{mg/k}$ that is given above) – note the use of the `simplify` command above:

$$v(t)=\frac{mg}{k}\tanh\left[\sqrt{\frac{gk}{m}}\left(t+C_{98}m\right)\right]$$

where the above expression was written after further simplification. Note that there is only one undetermined constant of integration. In order to determine this constant, we need one initial condition, but this is not given in this exercise.

4. Differential Equations –III-

1.
```
>> dsolve('DT=k*T','T(0)=T0','t')

ans =

T0*exp(k*t)
```

The solution as shown above is given by the following function (which satisfies the initial condition):

$$T(t)=T_0e^{kt}$$

2.

```
>> dsolve('DA=k*A','A(0)=A0','t')

ans =

A0*exp(k*t)
```

The solution as shown above is given by the following function (which satisfies the initial condition):

$$A(t) = A_0 e^{kt}$$

3.

```
>> dsolve('Dx=r-k*x','x(0)=0','t')

ans =

(r - r*exp(-k*t))/k
```

The solution as shown above is given by the following function (which satisfies the initial condition):

$$x(t) = \frac{1}{k}\left(r - re^{-kt}\right)$$

4. >> dsolve('-m*g=m*Dv+V1*Dm','t')

```
ans =

C127*exp(-(v(t) + g*t)/V1)
```

The solution as shown above is given by the following function (which satisfies the initial condition):

$$m(t) = C_{127} e^{-(v(t)+gt)/V}$$

The above solution was obtained without including the initial conditions in MATLAB. When including the initial conditions manually, we obtain:

$$m(0) = C_{127} = m_0$$

Thus, the solution function becomes;

$$m(t) = m_0\, e^{-(v(t)+gt)/V}$$

5.
```
>> dsolve('DS = r*S', 'S(0)=S0','t')

ans =
```

```
S0*exp(r*t)
```

Thus, the solution function representing the amount of money $S(t)$ at any time t (based on continuous compounding) is given as follows using the MATLAB output above:

$$S(t) = S_0\, e^{rt}$$

Note that the above expression is valid when the interest is compounded continuously. However, as stated in the text of the Exercise in Chapter 18, when interest is compounded m times per year, then the correct expression to use for $S(t)$ is the following:

$$S(t) = S_0 \left(1 + \frac{r}{m}\right)^{mt}$$

To find the relationship between the above two expressions of $S(t)$, let us take the limit of the second expression as $m \to \infty$.

$$\lim_{m \to \infty} S_0 \left(1 + \frac{r}{m}\right)^{mt} = S_0\, e^{rt}$$

It is left to the reader to verify the above result (check Chapter 11 to see how to evaluate a limit using MATLAB). In conclusion, it is seen that the limit of $S(t)$ when compounded m times per year, as $m \to \infty$, is equal to the expression of $S(t)$ for continuous compounding. This mean that continuous compounding is compounding an infinite number of times per year.

References

MATLAB Tutorials

1. http://www.cyclismo.org/tutorial/matlab/

   This is an online tutorial in HTML format from Union College, New York. The tutorial emphasizes vectors, matrices, vector operations, loops, plots, executable files (scripts), subroutines (functions), if statements, and data files.

2. http://www.math.ufl.edu/help/matlab-tutorial/

   This is an online tutorial in HTML format from the University of Florida, Gainesville. The tutorial emphasizes matrices, variables, functions, decisions, loops, and scripts. This tutorial has a nice list of important commands of MATLAB with a short description of each command.

3. http://www.mines.utah.edu/gg_computer_seminar/matlab/matlab.html

   This is an online tutorial in HTML format available at the website of University of Utah, Salt Lake City (Original by Kermit Sigmon, University of Florida). This tutorial emphasizes matrices, decisions, loops, scalar functions, vector functions, matrix functions, strings, and graphics. You may also download the tutorial as a postscript file (39 pages that include a comprehensive reference section).

4. http://web.mit.edu/afs/.athena/astaff/project/logos/olh/Math/Matlab/Matlab.html

   This is an online tutorial in HTML format hosted at MIT. It emphasizes matrices, arithmetic and logical operators, control structures, selective indexing, polynomial operations, signal processing functions, graphics, scripts, and functions.

5. http://www.mit.edu/people/abbe/matlab/main.html

   This is an online 3-day tutorial at MIT in HTML format. It covers matrices, various MATLAB commands, help, plotting, polynomials, variables, scripts, functions, loops, debugging, differential equations, vectorization, three-dimensional graphics, and symbolic math.

6. http://texas.math.ttu.edu/~gilliam/ttu/m4330/m4330.html

   This is a tutorial hosted at Texas Tech University, Lubbock. They provide six PDF files for download in the form of six lessons for a mathematics course they teach there. These lessons are somewhat advanced and specialized and are not recommended for beginner students of MATLAB.

7. http://www.engin.umich.edu/group/ctm/basic/basic.html

   This is a basic tutorial from Carnegie Mellon and the University of Michigan. It covers the basics of vectors, functions, plotting, polynomials, matrices, printing, help, and M-files.

8. http://www.mathworks.com/academia/student_center/tutorials/launchpad.html

   This is an online tutorial from the MathWorks, the company that develops and sells MATLAB. This tutorial covers variables, calculations, plotting, scripts, and files.

9. http://www.math.utah.edu/lab/ms/matlab/matlab.html#starting

   This is an online short tutorial from the University of Utah, Salt Lake City. This is a very simple tutorial recommended for beginners. It covers matrices, vectors, systems of equations, loops, and graphing in one and two dimensions.

10. http://www.math.mtu.edu/~msgocken/intro/intro.html

    This is an online tutorial in HTML format hosted at Michigan Tech and written by Mark S. Gockenback. It covers calculations, graphs, programming, advanced matrix calculations, advanced graphics, solving nonlinear problems, and advanced data types like structures, cell arrays, and objects.

11. http://www.eece.maine.edu/mm/matweb.html

    MATLAB Educational Websites: This is an online HTML page containing links to several educational websites on various topics using MATLAB. The page contains also links to numerous MATLAB tutorials.

12. http://web.cecs.pdx.edu/~gerry/MATLAB/

    MATLAB Hypertext Reference: This is an online tutorial in HTML format written by Gerald Recktenwald and hosted at Portland State University, Portland. This tutorial covers variables, plotting, and programming.

13. http://www.facstaff.bucknell.edu/maneval/help211/helpmain.html

Helpful Information for Using MATLAB: These pages of helpful information are maintained by Jim Maneval and hosted at Bucknell University, Lewisburg.

14. http://www.indiana.edu/~statmath/math/matlab/gettingstarted/index.html

    Getting Started with MATLAB: This is an online tutorial in HTML format. The tutorial may also be downloaded and printed as a PDF file (14 pages). This tutorial covers syntax, matrices, graphics, and programming.

15. www.math.udel.edu/~driscoll/teaching/**matlab**_adv.pdf

    Crash Course in MATLAB: This tutorial is in the form of a PDF file download. It is written by Tobin A. Driscoll and hosted at the University of Delaware, The tutorial is 66 pages that can be downloaded and printed. This tutorial covers arrays, matrices, scripts, functions, errors, 2D and 3D graphics, color, handles and properties, vectorization, advanced data structures (strings, cell arrays, structures), linear algebra, optimization, data fitting, quadrature, and differential equations.

16. www.maths.dundee.ac.uk/~ftp/na-reports/**MatlabNotes**.pdf

An Introduction to MATLAB: This tutorial is in the form of a PDF file download. It is written by David F. Griffiths and is hosted at the University of Dundee. The tutorial is 37 pages that can be downloaded and printed. This tutorial covers numbers, formats, variables, output, vectors, 2D plotting, scripts, vector operations, matrices, matrix operations, systems of linear equations, strings, loops, decisions, 3D plotting, files, and graphical user interfaces.

17. www.geosci.uchicago.edu/~gidon/geosci236/organize/**matlab**Intro.pdf

An Introduction to MATLAB: This tutorial is in the form of a PDF file download. It is written by S. Butenko, P. Pardalos, and L. Pitsoulis. It is hosted at the University of Chicago. The tutorial is 28 pages that can be downloaded and printed. This tutorial covers matrices, matrix operations, workspace, formats, functions, vector functions, matrix functions, polynomial functions, programming, loops, decisions, M-files, and 2D and 3D graphics.

18. http://www.math.colostate.edu/~gerhard/classes/340/notes/index.html

MATLAB Notes: This tutorial is in the form HTML pages online. It is hosted at Colorado State University. This tutorial covers arithmetic, array operations, scripts, functions, plotting, polynomials, matrices, vectors, linear systems of algebraic equations, and symbolic computation. There are some links to more tutorials on the internet that are hosted at other places.

19. http://www.indiana.edu/~statmath/math/matlab/gettingstarted/index.html

    Getting Started with MATLAB: This tutorial is in the form of online HTML pages that are hosted at Indiana University. It may also be downloaded as a PDF file (14 pages) and printed. This tutorial covers syntax, matrices, graphics, and programming.

20. http://www-ccs.ucsd.edu/matlab/toolbox/symbolic/symbmath.html

    This is an online reference guide for the MATLAB Symbolic Math Toolbox. It is hosted at the University of California, San Diego. The tutorial covers about 85 MATLAB symbolic math commands that are listed alphabetically with examples.

21. http://www.phys.ufl.edu/docs/matlab/toolbox/symbolic/ch1.html

    Using the Symbolic Math Toolbox: This tutorial is in the form of HTML pages that is hosted at the

University of Florida, Gainesville. This tutorial seems to be comprehensive and covers calculus, simplifications, substitutions, variable-precision arithmetic, linear algebra, solving equations, special mathematical functions, using Maple functions, and the extended symbolic math toolbox.

22. http://en.wikipedia.org/wiki/Matlab

    This is a small overview in the form of an HTML page hosted at the website of Wikipedia, the free internet encyclopedia. The page has a brief history of MATLAB and some useful links.

23. http://www.facstaff.bucknell.edu/maneval/help211/exercises.html

    MATLAB Exercises: This web page provides some MATLAB exercises for practice. The site is hosted at Bucknell University. The exercises cover syntax, arrays, relational operators, logical operators, decisions, loops, and programming.

24. http://www.math.chalmers.se/~nilss/enm/INTRO_matlab_I.html

    MATLAB Exercises: This web page provides three sets of MATLAB exercises for practice. The site is hosted at Chalmers University of Technology and Gteborg University. The exercises cover arithmetic, formats, variables, vectors, matrices, decisions, scripts, loops, computer arithmetic, strings, and graphics.

25. http://college.hmco.com/mathematics/larson/elementary_linear/5e/students/matlabs.html

MATLAB Exercises: These are MATLAB exercises taken from the book "Elementary Linear Algebra", fifth edition, by Ron Larson. Each set of exercises is downloaded as a PDF file that can be printed. These exercises cover systems of linear equations, matrices, determinants, vector spaces, inner product spaces, linear transformations, eigenvalues and eigenvectors, and complex vector spaces.

26. http://www.math.umn.edu/~ROberts/math5385/matlabEx1.html

Supplementary MATLAB Exercises: These exercises are hosted at the University of Minnesota. They are written specifically for a mathematics course at the university. These exercises mostly deal with the graphics capabilities of MATLAB.

27. www.kom.aau.dk/~borre/**matlab**7/exercise.pdf

This is a PDF file download of a set of 8 pages of MATLAB exercises. The exercises are somewhat advanced and are not recommended for beginning students of MATLAB.

28. www.chaos.swarthmore.edu/courses/Phys50L_2006 /**Matlab/MatlabExercises**.pdf

This is a PDF file download of 27 pages of MATLAB exercises. These exercises are written for a physics course and are somewhat advanced. They are not recommended for beginners.

MATLAB Books:

1. Gilat, A., *MATLAB: An Introduction with Applications*, Second Edition, John Wiley & Sons, 2004.

2. Pratap, R., *Getting Started with MATLAB 7: An Introduction for Scientists and Engineers*, Oxford University Press, 2005.

3. Hanselmann, D. and Littlefield, B., *Mastering MATLAB 7*, Prentice Hall, 2004.

4. Palm, W., *Introduction to MATLAB 7 for Engineers*, McGraw-Hill, 2004.

5. Moore, H., *MATLAB for Engineers*, Prentice Hall, 2006.

6. Chapman, S., *MATLAB Programming for Engineers*, Thomson Engineering, 2004.

7. Davis, T. and Sigmon, K., *MATLAB Primer*, Seventh Edition, Chapman & Hall, 2004.

8. Higham, D. and Higham, N., *MATLAB Guide*, Second Edition, SIAM, 2005.

9. King, J., *MATLAB for Engineers*, Addison-Wesley, 1988.

10. Etter, D., *Introduction to MATLAB for Engineers and Scientists*, Prentice Hall, 1995.

11. Magrab, E. et al., *An Engineer's Guide to MATLAB*, Prentice Hall, 2000.

12. Etter, D., Kuncicky, D. and Hull, D., *Introduction to MATLAB 6*, Prentice Hall, 2001.

13. Recktenwald, G., *Introduction to Numerical Methods and MATLAB: Implementation and Applications*, Prentice Hall, 2001.

14. Biran, A. and Breiner, M., *MATLAB 5 for Engineers*, Addison-Wesley, 1999.

15. Part-Enander, E. and Sjoberg, A., *The MATLAB 5 Handbook*, Addison-Wesley, 1999.

16. Etter, D., *Engineering Problem Solving with MATLAB*, Prentice Hall, 1993.

17. Chen, K., Giblin, P, and Irving, A., *Mathematical Explorations with MATLAB*, Cambridge University Press, 1999.

18. Mathews, J. and Fink, K., *Numerical Methods Using MATLAB*, Third Edition, Prentice Hall, 1999.

19. Fausett, L., *Applied Numerical Analysis Using MATLAB*, Prentice Hall, 1999.

Differential Equations Books:

20. Zill, D. G., *A First Course in Differential Equations with Applications*, Third Edition, Prindle, Weber & Schmidt, 1986.

21. Boyce, W. E. and DiPrima, R. C., *Elementary Differential Equations and Boundary Value Problems*, Fifth Edition, John Wiley & Sons, 1992.

22. Tenenbaum, M. and Pollard, H., *Ordinary Differential Equations*, Dover Books on Mathematics, 1985.

23. Bronson, R. and Costa, G., *Schaum's Outline of Differential Equations*, Fourth Edition, McGraw-Hill, 2014.

24. Holzner, S., *Differential Equations for Dummies*, For Dummies Series, 2008.

25. Farlow, S. J., *Partial Differential Equations for Scientists and Engineers*, Dover Books on Mathematics, 1993.

26. Blanchard, P. and Devaney, R. L., *Differential Equations*, Fourth Edition, Cengage Learning, 2011.

27. Nagle, R. K. and Saff, E. B., "*Fundamentals of Differential Equations*, Eighth Edition, Pearson, 2011.

28. Polking, J. and Boggess, A., *Differential Equations*, Second Edition, Pearson, 2005.

29. Coddington, E. A., *An Introduction to Ordinary Differential Equations*, Dover Books on Mathematics, 1989.

30. Stroud, K. A. and Booth, D., *Differential Equations*, Industrial Press, 2004.

31. Strauss, W. A., *Partial Differential Equations: An Introduction*, Second Edition, Wiley, 2007.

32. Farlow, S. J., *An Introduction to Differential Equations and Their Applications*, Dover Books on Mathematics, 2006.

33. Ince, E. L., *Ordinary Differential Equations*, Dover Books on Mathematics, 1956.

34. Ross, S. L., *Introduction to Ordinary Differential Equations*, Fourth Edition, Wiley, 1989.

Differential Equations/MATLAB Books:

35. Hunt, B. R. and Lipsman, R. L, *Differential Equations with MATLAB*, Third Edition, Wiley, 2012.

36. Polking, J. C. and Arnold, D., *Ordinary Differential Equations Using MATLAB*, Third Edition, Pearson, 2003.

37. Lopez, C., *MATLAB Differential Equations*, Apress, 2014.

38. Coleman, M. W., *An Introduction to Partial Differential Equations with MATLAB*, Second Edition, Chapman & Hall – CRC, 2013.

39. Golubitsky, M. and Dellintz, M., *Linear Algebra and Differential Equations Using MATLAB*, Cengage Learning, 1999.

40. Li, J. and Chen, Y.-T., *Computational Partial Differential Equations Using MATLAB*, Chapman & Hall – CRC, 2008.

41. Stanoyevitch, A., *Introduction to Numerical Ordinary and Partial Differential Equations Using MATLAB*, Wiley-Interscience, 2004.

42. Cooper, J. M., *Introduction to Partial Differential Equations with MATLAB*, Birkhauser, 2000.

43. McKibben, M. and Webster, M. D., *Differential Equations with MATLAB: Exploration, Applications, and Theory*, Chapman and Hall – CRC, 2014.

44. Boyce, W. E., *Elementary Differential Equations 8th Edition with Differential Equations MATLAB 2nd Edition Set*, Eight Edition, Wiley, 2005.

45. Davis, P, *Differential Equations Modeling with MATLAB*, Prentice Hall, 1999.

46. Butt, R., *An Introduction to Differential Equations Using MATLAB*, Alpha Science International Limited, 2017.

Installation of MATLAB

In this book, it is assumed that you have already installed MATLAB on your computer system. To go through the eleven chapters of the book, you need to have MATLAB running on your computer. For help on installing MATLAB on your computer, check the following web links:

MATLAB Installation for Windows:

```
http://www.itc.virginia.edu/research/ma
     tlab/install/installwin.html
```

MATLAB Installation for Linux:

```
http://www.freebsd.org/doc/zh_TW/books/
     handbook/linuxemu-matlab.html
```

MATLAB Installation for Macintosh:

```
http://www.itc.virginia.edu/research/ma
     tlab/install/installmac.html
```

MATLAB Installation for Other Systems:

```
http://shum.huji.ac.il/cc/matlR13linuxh
     omeinst.html
```

Index

www.ingramcontent.com/pod-product-compliance
Lightning Source LLC
LaVergne TN
LVHW090412210926
848480LV00002B/539
* 9 7 9 8 2 9 5 4 4 1 5 1 6 *

microRNA: A New Era for Diagnosis and Therapies

Utpal Bhadra

microRNA: A New Era for Diagnosis and Therapies

Utpal Bhadra

Mission for Protocols, Technologies, and Applications

Academica Press
Washington~London

Library of Congress Cataloging-in-Publication Data

Names: Bhadra, Utpal (author)
Title: MicroRNA : a new era for diagnosis and therapies | Bhadra, Utpal.
Description: Washington : Academica Press, 2022. | Includes references.
Identifiers: LCCN 2022931309 | ISBN 9781680539134 | 9781680538595 (e-book)

Dedicated to
My parents and parents in law

Table Contents

This microRNA book is truly a living legend and immortal epic for microRNA early diagnosis and cutting-edge therapies production. The readers engulf and engross in reading this book even in the last content at a stretch once they are started. It is essential for all medical students, biotech researchers, interested readers for popular bioscience, medicine and vaccine developers. In broad sense this book allures for scientific and medical reading. The major beauty of the content of this book are different unusual and interesting topics that contribute to microRNAs therapy development and most popular but general microRNA subjects. Author imagines the readers will fully engross and digest the contents of the books at a sporadic speed. I request the reader to draw their attention to the content of the books. This book is specially designed for describing the Rare Asian Communicable Diseases (RACD) and most prominent African Contagious Diseases (ACD) upon the request of Asian and African well wishers and the role of microRNA for their immediate remedy. Author hopes innumerable readers of these two large continents find this book is the indispensable material as the strongest solution against those deadliest infections.

Although numerous specific reviews have been published in the last two decades covering these tiny microRNAs, However, no single volume on MicroRNAs focusing on early diagnosis and therapy development is available in the current market. This book focuses and establishes the elegant medical concept of microRNAs biology. The volume integrated all major concepts including efficient extraction process from human blood, their contribution in heart failure, kidney failure even encompassing the modern delivery in the human. Even detailed description of clinical critical diseases is included. It covers the ten separate unique chapters related to industrial biotech, pharma companies and medical biology of the microRNAomes field. The book is an essential ingredient of all valuable executive scientific plans and summaries for new investors and provides a new avenue for those responsible for business development in the biotech industries. I imagine that this volume will serve both as a prologue to the field for newcomers as well as provides a blazing update to those already active in the microRNA field for a long time.

Contents*

(*Contribution of miRNAs is not only inclusively Important but also humongous in the cancer field. It is impossible to accommodate all aspects in a narrow Chapter. Due to the benefit of miRNA readers and meeting their quests, we will not try to accommodate them here. In future, we will discuss their contribution to different cancers in detail.)

```
-5

24

30

13
```

10. Solve the following system of simultaneous algebraic equation using Gaussian elimination.

$$\begin{bmatrix} 5 & 2 \\ 1 & 3 \end{bmatrix} \begin{Bmatrix} x_1 \\ x_2 \end{Bmatrix} = \begin{Bmatrix} 3 \\ -1 \end{Bmatrix}$$

```
>> A = [ 5 2 ; 1 3]

A =

     5     2

     1     3

>> b = [3 ; -1]

b =
```

Chapter X

Acknowledgment

I would like to take an opportunity to offer immense thanks to all my scientific colleagues, friends, and well-wishers, who helped me extensively in sketching the book. I also indebted a talented group of early research postdocs, project assistants to support my work in the past several years by contributing many review articles. My special thanks go to my family members, my wife Manika Pal Bhadra and our both sons, Kaustav (Shambo) and Utsav (Arjun) for continuous encouragement on writing; they also offer me valuable suggestions, positive comments throughout the implementation of this challenging task. I would like to acknowledge the valuable help from Mr. Govindaraj Giridharan for all drawing and cover page designing and production and the entire team of Academica press publishers for their financial and administrative support for the project. Without their active help and participation, it is not possible to publish. We also apologize to colleagues whose relevant primary contributions are not cited in the book due to space constraints.

Preface

The discovery of three important operational devices and their products related to RNA emerged as a breakthrough for Central Dogma and illustrated a new mission for RNA. These gene silencing events include 1) unimaginable occurrence of cosuppression in animals and its by-product RNA interference that controls gene regulation, cellular differentiation, development, and existence of 98% non-coding small regulatory RNAs including micro-RNA. The ability of microRNA modulates the gene function and their abundance in the circulatory system is a powerful new tool that revolutionaries the early detection of different diseases and treatments. To date nearly 2500 active miRNA found in human cells. It has been predicted that 30–80% of human genes are modulated by at least one miRNA. In humans, the biogenesis process, function in gene regulation and their rapid degradation are the long-standing topics for intense microRNA research. Since they act as cellular differentiation, rheostats subtly modulate disease related gene expression and various developmental processes. Numerous mature miRNAs have been implicated in a wide range of physiologic and pathologic processes relevant to human diseases.

Tiny microRNAs (miRNAs) are endogenous ~22 nt long RNAs that can target mature mRNAs for cleavage or translational repression. It also destroys many protein coding genes. During biogenesis, most of the miRNAs are transcribed from template DNA including non-coding introns to primary miRNAs. Subsequently processed into precursor miRNAs, and finally mature miRNAs. miRNAs can be secreted into extracellular fluids and transported to target cells via exosomes vesicles. The microRNA guides for base-pairing with target mRNA sequences to destroy mRNAs that have protein coding ability. The strong complementarity between microRNA and their mRNA target determines cleavage of base pairing mRNA. It subsequently degrades or inhibits the translational process. In most cases, miRNAs interact with the 3′ untranslated region (3′ UTR) of target mRNAs to induce mRNA degradation and translational repression. However, interaction of miRNAs with other regions, including the 5′ UTR, coding sequence, and gene promoters, have also been reported. The interaction of microRNAs to their target mRNAs is dynamic and dependent on many factors, such as subcellular location of miRNAs, the abundance of miRNAs and target mRNAs, and the affinity of miRNA-mRNA interactions etc. Extracellular miRNAs function as chemical messengers to mediate cell-cell communication.

Individual chapters of the book are bonded strongly to each other and cover a wide range of areas regarding molecular characterization of microRNA. Especially its synthesis, mechanism to operate, function and instant degradation and two separate milestones for clinical use including early diagnostics and therapy development. We provide an update on canonical and non-canonical miRNA biogenesis pathways and various mechanisms underlying microRNA-mediated gene regulations. The major knowledge of their functional importance in human health are also accommodated in the book. Aside from the biological aspect, the clinical aspect and major role in microRNA therapies of different diseases was also discussed within this volume. We strongly believe that the book captures not only microRNA essences in modern biology but vividly discusses excitement and most challenging tasks for microRNA ingredients for drug development and early detection for human diseases.

Introduction

Utpal Bhadra, Ph.D., D. Sc. FRS (b+c)
Deputy Director ex, CCMB, India
Fellow of Royal Society UK (b+c)
Senior WELLCOME TRUST fellow (U.K.)

Fellow Investigator, Human Frontier Science Program
1841 Shirley Lane, Ann Arbor, Michigan-48105, USA

All physical events abide strictly by the third law of "Newton." Similarly, the continuity of genetic flow is our perfect thumb rule to flow our lives in Biology. Popularly, it is stated as "DNA makes RNA, and RNA makes protein" although it is not its core inner meaning. More precisely, the transfer of genetic information including the precise determination of sequence from nucleic acid to nucleic acid, or from nucleic acid to protein may be eligible, but transfer from protein to protein, or from protein to nucleic acid is absurd and impossible. To follow this law of biology we cannot shoot messenger elegantly to our live cells. However, some cells happen to be adamant, absurd, and bully in nature. They do not follow the disciplined rules. They react differently following their inquisitive nature. They can shoot their non-translatable RNA randomly and other cells accept those messages or silence their activity. Multiple post transcriptional silencing processes are in active in different tissues in addition to another level of control to overrule the existing dominating rules. Such mechanisms are fully governed by small noncoding pieces of RNA, small interfering RNA and more predominantly microRNA in the medical field. It is nowadays established that an extension of microRNA work has no limit in the sky and flying its unseen genetic rocket sporadically beyond our imagination. These small, noncoding RNA molecules were initially discovered in 1993 more than just two decades away for fine regulation of gene expression.

Practically MicroRNAome's is a dominating sub-section of RNomics that illustrates the biogenesis and mechanisms of tiny RNA regulators and their involvement during development, differentiation, cell proliferation, cell death, chromosomal segregation, and metabolism. Their sophisticated fine tuning regulatory functions add a new tier of our understanding of modern gene expression. In addition, the strategy of RNA interference (RNAi) mediated gene silencing and its associate mechanisms propelled an elucidation of microRNA governing. The fundamental development and sporadic progress in microRNAs research over the past two decades allowed us to engulf the application of gene silencing at a greater height. Therefore, it is now the most contribution of appropriate time to jot down the entire contribution of microRNA in early diagnosis, prognosis, and therapy formulation in the form of a book to save mankind from different uncontrollable diseases. The present book is a tiny integrated effort in this direction. Diverse role of microRNAs development to disease biology encompassing quantitation, prognosis, and diagnostics as well drug development and validation. I imagine that this volume will be an easy solution as a prologue to the field for newcomers as well as an update to those already active in the field.

Journey of microRNA started in the mid 70s, when World renowned scientist Sydney Brenner in Cambridge selected a peculiar field of biology and opened a new vision using *C. elegans*, a nematode worm as his experimental tool, when the *lin-4* mutant was originally identified. Later his post-doctoral fellow Howard Robert Horvitz from MIT discovered *lin-14* locus in the study of developmental timing defects of the worm. Further two Horovitz's post-doctoral fellows, Ambros and Ruvkun identified the nature and function of those genes. In 1993, Victor Ambros characterized *lin-4* as a short non-coding microRNA in *C. elegans.* Meanwhile another post doc Gary Ruvkun characterized *lin-14* as a possible first microRNA target gene. Together, these two seminal discoveries identified their mutual genetic interaction. A successful mutual collaboration and subsequent independent contributions by two successful groups of scientists paved a revolutionary and initiating way to a novel mechanism of post-transcriptional gene regulation via microRNA. The microRNAs world took off to a major limelight by the discovery of RNAi and *let-7*, a second microRNA in 2000. miRNAs are thought to regulate many more steps beyond routine translation. Despite considerable attention having been paid to microRNAs via different reviews over the past decade a lot more questions remain unanswered. The present volume raises those issues and gives a clear indication how microRNA can be used successfully encompassing various diseases diagnosis and therapy initiative drives. As Ambros noted earlier, the *lin-4* discovery is one of persistent curiosity, luck, timing, and the generosity of colleagues. Indeed, it is a leading break-through of modern juvenile molecular biology. In the coming years, microRNA may overcome the routine concept of the central dogma of Biology.

When a small, double-stranded RNA (dsRNA) discovers in nematode larvae that complements the sense strand of a larger mRNA and is bound to its 3' untranslated region, finally inhibiting translation. Since then, numerous mechanisms for translational regulation by small RNAs have been detected. The mRNA is either directed for cleavage by an siRNA-protein complex, or translation is prevented by miRNA. Either way, the mRNA is eventually destroyed within the cell. A microRNA-induced silencing complex (miRISC) relates to the mature miRNA, the complex targets mRNA and physically blocks translation. Many miRNAs form imperfectly complementary stem-loop structures on the interacting sense strand of mRNA, which require near-perfect matches.

The biggest hurdle to turning microRNA invention from bench to bedside are not exclusively in the advancement of technology, but also the necessary stringent regulatory approvals. In a nutshell, genome wide microRNA array technology has effectively accommodated all required assay formats on one testing goal and to execute better reagents for pathological diagnosis and therapeutic ingredients in the future. It is speculated that entire clues and strategies of microRNA functions in all forms of human diseases and their subsequent progress as theragnostic, so we will enter in a new era of personalized 'translational medicine'.

In general, only one miRNA is produced from one precursor. A microRNA-induced silencing complex (miRISC) links to the mature miRNA, the complex binds to mRNA and physically blocks translation. Many miRNAs generate partial complementary stem-loop structures for interacting sense strands of mRNA that require near-perfect matches. In curiosity, how evolutionary changes in gene-modulating networks affect phenotypic changes in an organism, an appropriate role of miRNA in cell differentiation is established. Despite till now there is little extension of miRNAs' involvement in cell differentiation, in current scenario they function to reinforce more powerful factors for controlling different developmental processes, specially, many transcription factors are strictly conserved between distant species, while miRNA is not found in some species. Evolutionary studies also indicate that humans alone might have over 1,000 species-specific (or primate-specific) miRNAs, each of which can bind to hundreds of different mRNA strands. However, in animals, miRNA-mediated control of gene expression is often relatively weak compared to repression by transcription factors. To what extent, then, do we depend on miRNAs to control gene expression that we need to have so many? Some theories are that over time, new miRNAs were acquired in sync with the development of new tissue types and organs.

Additional roles that miRNA and siRNA have in gene expression involve control of the inheritance of epigenetic modifications during cell division and in activation of translation, in certain circumstances, depending on the cell cycle stage.

Different evolutionary changes in gene-regulatory networks affect phenotypic changes in an organism. Phenotypic modulations have given scientists an imaginary sketch of miRNA contribution during cell differentiation. Although, there is little evidence of the rate of evolutionary changes linked with the miRNAs' involvement in cell differentiation. The current understanding is that they act to reinforce more powerful factors that control developmental processes, particularly because many transcription factors are highly conserved between distant species while miRNA is not found in some species, invariably in animals, miRNA-mediated fine tuning of gene expression is often relatively weak compared to repression by transcription factors. To what extent, then, do we depend on miRNAs to control gene expression that we need to have so many? Some theories are that, over time, new miRNAs were acquired in sync with the progression of new tissue types and organs.

Dysregulation of miRNAs upset the delicate gene networks in numerous disease states for example metabolic diseases, kidney and heart failure, neurodegenerative diseases; thus, miRNA therapy could normalize gene expression in the complex human cells. Their dysregulation may allow maximum progress of potential therapeutic targets for treating various diseases. Promising preclinical trials in the biomedical/clinical field demonstrate constructive critical cure strategy for treating immediate defects including multi-organ failure. However, several hurdles including tissue specificity, targeted delivery remain to be overcome. Indeed, the ratio of success stories is low, due to costly technologies and approaches used in miRNA discovery, identification of biomarkers and therapeutic markers. Until now, miRNAs have been extensively used, but their specific functions remain largely not known. Nevertheless, numerous unexplored missions about the precise mechanism of miRNA specific function and action during complex diseases and as a regulator could be modulated in chronic disease in a tissue-specific manner. More understanding and elucidating potentiality of miRNAs is required for miRNA profiling assays related to human disease, combined with sensitive, accurate and flexible detection techniques, which is elaborately described in the book. These dramatic steps in the book could bring upcoming essential results for further clarification of miRNA-mediated pathways in diseases. Therefore, the cutting-edge applications of miRNA acts as the future therapeutic ingredients is too bright and impulsive at this moment as described. Nevertheless, with rapid expansion of miRNA invention since more than a decade is certainly promising, hopeful and allure us in this direction in near future.

Finally, I am also thankful to all my teachers throughout my life for their excellent teaching, guidance, and mentorship, which caused me to become a scientist. Thanks to you, the reader, the librarians, and people who have little interest in science and who make the hours I have spent putting together this volume worthwhile, if you find value in the hours you spend with this book. I am indebted to the staff of Academica Press, and in particular the people who efficiently edit this entire book.

Chapter I

Introduction of Unique microRNAs: Placental and Chromosomal miRNAs

Abstract

MicroRNAs (miRNAs) are tiny 21-23nt long regulatory RNAs, required for post-transcriptional silencing of gene expression in multiple biological contexts. The discovery of small RNA-guided gene depletion followed by an expansion of the miRNA repertoire is dramatically related to animal complexities. The miRNAs become indispensable agents for animal development processes including placental development during pregnancy, chromatin structure, dynamics and open chromatin context in the human cells. Indeed, the miRNA ablation in biogenesis machinery that is related to the silencing process causes multiple strong defects during animal embryogenesis and pregnancy. It is of great importance that mRNA annotation framework should be revisited nowadays for discovering new miRNA species via a rapid variation of mature miRNAs (iso-miRNAs) and organellar miRNAs: cipo-miRNAs and mito-miRNAs. The key terminology and nomenclature denoting the uniqueness, the evolution of miRNAs is crucial potentially. In this chapter, we try to discuss a better and elaborated understanding of miRNA features in humans for harnessing these broad classes of regulators. The miRNA not only evolves a spatiotemporal specificity manner within the human, but also to promote chromatin architecture. The canonical mechanisms of miRNA biology also contribute to miRNA roles in the development and chromosomal physiology. Finally, the emergence of diversified miRNA families is encouraged through the lens of evolution in humans. The early discovery, classification for modern nomenclature is also discussed for the restructuring of classification of this unique component in the therapy development.

Introduction

An intricate gene regulatory mechanism is required for the segregation of different functions in many specific cell types that leads to an ultimate complexity in humans. Diversified miRNA distributions in all cell types promote versatile repressors of gene expression for originating different cell types with specialized physiology.[1] It finds no surprise that the evolution of vertebrate organisms has been always accompanied by a diversified complication of the miRNA methods and the miRNA repertoires.[2, 3] The versatilities of miRNA repressors have been exploited adding multiple mechanisms that are responsible for the production of specific developmental timing, multiple cell types.

Moreover, the increasing number of genomic microRNAs has replaced normal miRNAs for their upgradation and the associated proteins during evolution. A wide spectrum concept in phenotypic complexity of microRNA during the recent era in many organisms have been led by these methods of understanding. A comprehensive concept of adopting diverse mechanisms opens how humans have generated this broad class of regulators are also led to utmost emphasis. Finally, an updated introductory view including its brief definition, nomenclature, discovery, evolution and their role in embryogenesis, placental development and their nuclear and chromosomal contribution in the microRNA world teach us so far about the functionality of microRNAs in humans.

Soon after the discovery, numerous miRNAs of different animal and plant species were reported.[4-6] The short RNA molecules revolutionize our knowledge of post-transcriptional gene regulation. Following discovery, a rapid explosion of new miRNA isolation has been associated with a central registry requirement, repository database and their systematic annotation. The first 'microRNA Registry' (miRbase) was initiated in 2002 to prevent mis-annotations and overlapping miRNA names.[7, 8] The first nomenclature includes 218 miRNAs from five separate species. After thorough revision, the new names of miRNA are assigned in the current nomenclature. It has shown a rapid explosion of microRNA registry that belongs to 35000 mature miRNA sequences from 223 species in the world.[8]

Many small RNAs open up a pivotal role in the modification of physiological and pathological processes during embryonic development. Although miRNAs that functions undisputed instrumental agents for changes in gene expression, underlie versatile development mechanisms. Indeed, their repressing strategies restrict expression patterns in time and space. Their controlling role is noticed for shaping many controlled expression profiles in many cell types during development.[9]

It is also a general custom that miRNAs control several pathological processes and physiology.[10-12] During the pregnancy, placenta is required for gaseous exchange to the developing fetus and their nutrient uptake. It also regulates temperature, hormone production and protects several fetuses from internal infections. The cellular activities of all placenta forming cells including trophoblast proliferation, differentiation, invasion, mesenchymal cell differentiation, decidualization and angiogenesis play a pivotal role in a healthy pregnancy.[13,14] The placental miRNAs control maternal inherited genes, nutrition and physiological inflammation, hypoxia factors in pregnancy.[15-16] Thus, miRNAs regulated gene expressions have a great responsibility to protect physiological conditions during pregnancy for helping successful pregnancy processes.

Lastly, the eukaryotic DNA is tightly wrapped outside of histone octamers to form nucleosome complexes. Chromatins consist of series of nucleosome arrays with few dynamic characters, for example, DNA methylation, histone tail modifications, chromatin-remodeling complexes, and modified binding proteins etc.[17-19] It is not surprising that miRNA and chromatin regulation interact in a well-coordinated manner. In this view, miRNA-epigenetic machinery functions as an instrumental network for gene regulation. The four distinct comprehensive relationships between miRNA and chromatin architectural modular provide a better understanding of the realization of the complexity of transcriptional control.

What is microRNA?

Around 98% of the nucleic acid content of healthy human cells do not produce any functional protein. This concept has been unexplored till the last decades of the last century. In the last three decades, small regulatory RNAs have been coming to the limelight at an exponential rate of their isolation. The miRNAs are tiny non-coding RNAs with an average of 22nt in length. They are distributed in all cell types for potential fine-tuning the post-transcriptional gene regulatory devices and carry a lion share among all small regulatory RNAs. Most miRNAs originate initially from DNA sequences into primary miRNAs (pri-miRNAs), later processed into precursor miRNAs (pre-miRNAs) and ultimately convert into mature miRNAs. Although micro in size they have a powerhouse for macro gene regulation. In many instances, miRNAs interact normally with the 3′ UTR of target mRNAs using base-pairing either degrade mRNAs or push for translational inhibition.[20] However, the interaction of miRNAs with 5′ UTR coding sequences and gene promoters have also been noticed.[21] Many such interactions are important to activate miRNAs-based gene expression.[22] It is indicated that miRNAs are shuttled between many separate subcellular compartments to regulate the level of translation inhibition and degradation of transcription.[23]

The rapid changes in developmental processes exhibit numerous miRNAs-based gene control. They participate in all varieties of developmental processes in every cell type.[24] Numerous aberrant expressions of miRNAs instigate many human diseases.[25, 26] Their extracellular secretions mainly in circulatory blood are utilized as biomarkers for early diagnosis, classification, stage and prognosis of different diseases. The miRNAs also serve as signaling molecules governing cell-cell communications.[27, 28] Here, a brief overview expanding miRNA biogenesis in humans and their complexity for target gene regulation provides a new dimension for early diagnosis and pushing towards our personalized medicines. Moreover, intracellular localization and function of miRNAs explore many dynamics. Finally, the secretion and circulation of miRNAs are also reported for their extracellular roles in mediating intercellular communications.

The interplays between miRNA abundance/localization, miRNA-guided regulation, different cell types and cell state are still under a positive experimental scanner. The miRNA expression levels are highly variable among distributed miRNAs in many cell types. Nearly, half of the miRNAs are enriched in expression in a cell type, one quarter is expressed intensely, and the remaining quarter has low-level expression. These results promote the general roles that miRNA play within cells.

Generally, miRNAs can control gene networks dynamically and/or transiently with the help of feedback and feedforward loops [29] and stable gene regulation, e.g., upregulation throughout differentiation.[30] In this pathway, miRNAs steady state expression by buffering out transcriptional stochastic fluctuations. The high miRNA to protein ratios that are a common indicator of expression noise also downregulates translational rates resistant to prevent random fluctuations of mRNA concentration.[31]

The low expression profile of functional miRNAs enhances RNAi activity synergistically on a subset of target mRNAs thereby activating the cognate miRNAs concentration.[32, 33] In contrast, the activities of low miRNA levels can be compensated for high levels of target mRNAs that activate high loading of cognate miRNAs into AGO, thereby promoting RNAi activity.[34]

The cellular stress and serum starvation are also altered by the miRNA/mRNA dynamics[35] for affecting miRNA activity. It is found that an urgent transportation of steady-state vesicular or vesicle-free miRNA and simultaneous decrease of intracellular miRNA levels are required for serum starvation in multiple human cell lines.[36]

Nomenclature

The miRNAs are well spread throughout different cell types.[37-38] In humans there have been nearly 1900 miRNAs synthesizing so far [39] but in a separate study claimed that the number is only limited to 600.[40] They are responsible for targeting about 60% of the human and other mammalian genes.[41, 42] The miRNAs are evolutionarily conserved in different mammals for implementing many biological functions. Nearly 100 miRNA families are found in humans, in which one or multiple members are required for transcription repression. The key terminology of existing nomenclature facilitates multiple advantages for miRNA recognition including the novelty, evolutionary history or biogenesis of miRNAs, identification in the literature and dissemination in a central miRNA registry. However, in the present context, miRNA nomenclature, curation and stems of central miRNA registry are essential miRNA nomenclature and terminology revision.

The names of miRNAs are assigned experimentally before they appear in publication[43,44] by a standard nomenclature protocol. A prefix "miR" is followed by a dash and a number indicating an order of their discovery. For an example, miR-124 was identified earlier than miR-456. The Uppercase letter R denotes the mature form of miRNA, whereas the low case refers to pre-miRNA and the pri-miRNA.[45]

The prefix three letters of the nomenclature miRNAs are denoted for conserved genes of different species such as the official miRNAs prefix is "*mir-1* in *C. elegans, Mir-1* in mouse and *MIR-25* in the human. The same family miRNAs show nearly identical sequences in the same species with one or two nucleotides are replaced with an additional Uppercase letter. As found in the sequence analysis, human miR181a and 181b are closely related.

hsa-miR-181a: aacauucaACgcugucggugAgu
hsa-miR-181b: aacauucaUUgcugucggugGgu

The 100% identical mature mRNA sequences are derived from the same Pre-miRNAs. The pri-miRNAs and genes from different genomic locations are referred with an additional dash-number suffix. The pre-miRNAs 'hsa-mir-194-1' and 'hsa-mir-194-2' are identical mature miRNAs but are originated from genes at different genomic sites. Three additional prefixes represent the origin of species of mature miRNAs. The 'hsa' in front of miR in 'hsa-miR-124' is originated from humans (*Homo sapiens*) and 'oar-miR-124' is a sheep (*Ovis aries*) miRNA. Other common prefixes include "d" for *Drosophila* miRNA (a fruit fly) and "v" miRNA encode by a viral genome.

When the same pre-miRNA synthesized equal amounts of two mature microRNAs from the opposite arms of a pre-miRNA they are represented as -3p or -5p suffix denoting 3' and 5' arms that refer to as sense and antisense strands. The mature microRNAs synthesize basically from one arm of the hairpin. This amount is more than miRNAs produced from the other arm.[46] An example, an asterisk following the name of miRNA is used for low levels of mature miRNAs that are produced by the opposite arm of a hairpin. The miR-124 and miR-124* share same pre-miRNA hairpin, in which miR-124* synthesizes at a low amount.

An elaborate guideline for miRNA annotation, identification and isolation annotation was initially established in 2003.[47] The uses of additional criteria of miRNA biogenesis are also established in the original guidelines.[47] The close biochemical and functional similarities of miRNAs to other classes of small regulatory noncoding RNA are viewed that present miRNA nomenclature is not sufficient for miRNAs reliable identification. So additional parameters are also added for better emphasis of original guidelines.[47] A better guideline would provide the confirmation of miRNA biogenesis that may occur one of the three important pathways: (1) the lowest free energy stem-loop structure of the pre-miRNA may yield mature miRNA from one arm of the hairpin, (2) providing the secondary structure of precursor Pre-miRNA and mature miRNA, (3) allowing precursor miRNA accumulates less Dicer activity in backgrounds.

Mis-concept in nomenclature: The strong similarities in siRNAs structure, their biogenesis and function make the possible contamination in miRNA identification and nomenclature. High-throughput RNA library sequencing is a powerful tool for miRNA identification. Indeed, the biogenesis of miRNAs via a DCL1 cleavage is the same as for secondary or natural antisense siRNAs generation.[48] Despite the related features miRNAs and siRNAs also have three major differences in their characteristics: 1) the difference in average length, for example miRNAs are 21nt long, while siRNA is approximately 24nt in length. 2) endogenous siRNAs are produced by the action of RNA POLYMERASE IV/RNA POLYMERASE V (PolIV/PolV) enzymes, whereas miRNAs are generated through RNA POLYMERASE II-dependent synthesized pathway. 3) the miRNAs are produced by imperfect base-pairing hairpin structures, while siRNAs are yielded from double-stranded and perfectly base-pair to exogenous or endogenous RNA precursors.[49] These critical differences between siRNAs and miRNAs require a sophisticated accurate annotation to distinguish themselves. Further, siRNAs are evolved from transposon-derived and repeat-related sequences, however, in repeat related miRNAs putative precursors are required for miRNA identification and annotation.[48]

The abundance of four regulatory RNAs, transfer RNAs (tRNAs), ribosomal RNAs (rRNAs) and small fragments of mRNAs, long noncoding RNAs,[50-51] are the predominant contaminants for mRNAs in general. These small RNAs may impose a serious threat to the specificity and reliability of pure miRNA sequence analysis. During sequencing, miRNA sequencing data sets often evolve from those contaminating RNAs. They must be separated from miRNA sequences with the help of accurate miRNA identification and annotation. A single mutation is sufficient to separate the pure miRNAs from the miRNAs originated from tRNA or ribozyme sequences [52] for example, the miR-712 in mice is originated from pre-ribosomal RNA.[53] Therefore, processing for the elimination of the similar contaminated sequences, a cautious measure would employ before miRNA annotation avoiding unexpected origin of miRNAs.

miRNA nomenclature and related concerns: All independent miRNAs apart from sRNAs are equally critical to avoid redundancies or mis-annotations. There are distinct differences in naming animal and plant originated miRNAs. For instance, animal pre-miRNAs are denoted with italic characters and in lowercase (e.g., *mir-14*). Identical mature miRNA sequences within a species are referred with numbered suffixes after a 'dash' symbol.[47] In exception, identical miRNA members in miRbase notation have not to carry the exact same sequence (hsa-miR-19b-2-5p: AGUUUUGCAGGUUU GCAUUUCA and hsa-miR-19b-1-5p: AGUUUUGCA GGUUUGCAUCCAGC). However, special care must be adopted in miRNA genes (MIR) for encoding identical miRNAs from each species generating from completely different primary transcripts or an intraspecific duplication event. (**Figure 1**) Besides, mature miRNAs do not always carry the same corresponding sequences.

The relationship between canonical and noncanonical miRNA production methods and the current nomenclature of related microRNAs are found on several occasions. **(A)** In recent annotation, the marked mature miRNA sequences originate from the same hairpin structure (miRNA-like RNA) and miR1a-3p.2. sequence **(B)** Multiple Iso-miRNAs like miR-1a, miR-1b and miR-1c have distinct sequences in the miRbase annotation. However, Iso-miRNAs are used for distinct names (miR1a-3p.3 and miR1a-3p.4) for simplified genomic loci annotation that is derived from sequential processing. The same sequence of Iso-miRs process from different miRNA family members aggravate serious confusion. Therefore, multiple isoform prefixes are synthesized from the additional numerical suffixes like miR1a-3p.1: isomiR1a-3p.1.1 and isomiR1a-3p1. 2) by utilizing the same precursor. **(C)** The mir-1a-2 and miR-1a-3 have identical sequences, but they are encoded by distinct genomic loci. Similarly, MIR1a-3/mir-1a-3: Paralogous miRNA: mature miRNA and same primary sequences derive from the intragenomic duplication events. **(D)** miR-1 family members synthesize from different genomic locations (MIR1d/mir-1d). **(E)** Nat-miRNAs: nat-miRNAs are derived from the antisense strand of the precursor target in most-transcription for controlling the same product.

(F) miRtrons: cleaves from the intronic regions and processed by DCL1. Future miRNA naming should be followed by organellar miRNAs also. Common precursors of pre-miRNA produce a certain group of miRNA species. They are originated from sequential processing referred to as a number suffix after a dot (e.g., aly-miR774a-3p.2: and aly-miR774a-3p.1 or osa-miR159a.1 and osa-miR159a.2).[50] Such naming is inconclusive because of the formation of almost the same canonical iso-miR, miRNA-like RNAs and some other divergence mature sequences. (**Figure 1**) The iso-miRs having the same sequences of miR1-3p.2 and miR1-3p.3 are generated from different parts of the same precursor molecule that are however different from miR1-3p.1. In a new miRbase, all iso-miRs are considered as distinct sequences having an unrealistic plethora in miRNA repertoires. To reduce these complications, different 'iso-miR' are further renamed the prefix with an additional number suffix (isomiR1: isomiR1-3p.1.1, isomiR2: isomiR1-3p.1.2). In miRbase, a single record for iso-miRs is denoted differently nowadays.

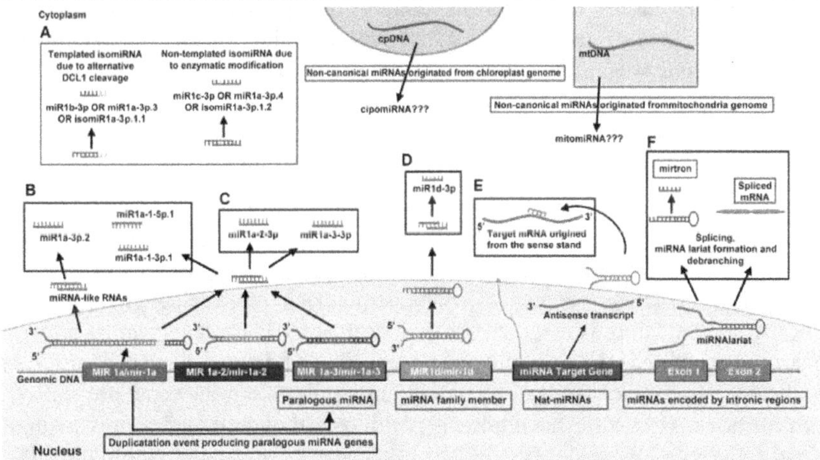

Figure-1. A simplified diagram of canonical and noncanonical miRNA biogenesis pathways and current nomenclature of associated microRNAs. **(A)** The miRNAs originating from the MIR1a/mir-1a genomic loci through canonical miRNA biogenesis pathway are presumed as the first identified miRNAs and named as miR1a-1-3p.1 and miR1a-1-5p.1. According to the annotation criteria, a distinct mature miRNA sequence excises from the same hairpin structure (miRNA-like RNA), resulting from sequential processing is named as miR1a-3p.2. **(B)** Iso-miRNAs of miR1a is named as miR1b and miR1c as the miRbase currently acknowledges these isoforms as distinct sequences. However, the naming scheme used for the products of the sequential processing can be applied to iso-miRs as well (miR1a-3p.3 and miR1a-3p.4). This approach will also simplify the genomic loci annotation. It must be noted that these isoforms may have exact same sequences as different miRNA family members, worsening the confusion further. Therefore, we additionally propose the usage of 'iso-miR' prefix with the additional numerical suffixes for the mature miRNA isoforms diced from the same precursor (isoforms of miR1a-3p.1: isomiR1a-3p.1.1 and isomiR1a-3p1.2). **(C)** Identical mature miRNA sequences encode by distinct genomic loci (mir-1a-2 and miR-1a-3). MIR1a-3/mir-1a-3: Paralogous miRNA: The same primary and mature miRNA sequences result from the intragenomic duplication events. **(D)** A miR1 family member encode by a different genomic location (MIR1d/mir-1d). **(E)** Nat-miRNAs: nat-miRNAs are transcribed from the miRNA target gene antisense strand and post-transcriptionally regulate the same genes' products. **(F)** miRtrons: The miRNAs that are diced from the intronic regions and processed by DCL1 after the splicing of an mRNA transcript. Although they have not been fully established yet, future miRNA naming schemes should be designed to embrace organellar miRNAs as well. (Loosely Adapted, Redraw and Modify using skeleton from Budak et al., 2016)

==

Similar miRNA names with unrelated sequences that are also targeted similar sequences further complicate the annotation.[54] Basically, during functioning specific miRNAs orchestrate gene modulation or knockdown by cleaving the targets or translation inhibition depending on the annotation of sequence similarities. Although new miRNAs either evolve or alter from earlier old miRNA names and sequences, more refinements are required for new annotation efforts bringing forth outdated nomenclature uses.[55] An independent new database impede dramatically the use of updated naming criteria. Therefore, a new miRbase tracker must use for controlling the modern online tools.[55]

Restructuring of miRNA nomenclature guidelines

The current scrutiny of nomenclature and miRNA regulation has never coped up with the continuous expansion of miRNAs features, variety and target selection. For such consequences, the restructuring of existing guidelines for miRNA nomenclature is essential for their features. Their nomenclature patterns in those days impede the ambiguity and confusion. The nomenclature of miRNAs synthesized from two stands of an unwinding duplex from the same pre-miRNA are processed by Dicer. This duplex is previously thought to carry only a single strand reported as a 'guide strand'. The guide strand is inserted within the RISC complex for maturation, while the opposite 'passenger strand' is believed to be degraded. But it is found that they are still accumulated but less than a level from guide strands, which is denoted by adding an asterisk in its back. (e.g., miR172*). Similarly, miRNAs originated from 3'and 5' strands of pre-miRNA hairpins as discussed earlier are called differently.

It is also found that non-overlapping miRNAs are derived from the same hairpin,[56] which are not known during the formation of the previous guideline of miRNA annotation. These are suspected to be believed as siRNAs. Alternatively, if they carry the miRNA features as alternative processed products of the same gene, they are referred to as miRNA-like RNA.[56] Thus, a restructuring of standard nomenclature rule is essential to avoid such confusions and to impose for nonoverlapping mature miRNA.

The mature miRNA sequence variants iso-miRNAs (iso-miRs) originated from the same genomic locus have been isolated nowadays. The template of these variants is either shifted their start to end positions by the action of an alternative DCL1 enzyme. Otherwise, non-templated miRNAs either originate via nucleotide addition removals or modifications by other enzymes.[57] The iso-miRs exert a new level of miRNA-based control during stress responses.[56] The confusion and ambiguity in iso-miRs are still to be resolved via restructuring.

Moreover, a new class of miRNAs is isolated. They are reported as 'Organellar miRNA' particularly mitomiRs, residing inside the mitochondrial matrix.[58] These miRNAs either initially generate from the nuclear genome and later transport to mitochondria or originate from the mitochondrial genome transcription. They are required for mitochondrial genome evolution.[59, 60] The new guideline for the annotation of miRNAs is needed for the recognition of many organellar miRNAs from their nuclear homologs. Like mitomiRs, cipomiRs are generated from the chloroplast genome, which must be annotated without any ambiguity.

Among two different miRNA annotation guidelines,[47, 54] the miRbase Web site (http://www.mirbase.org/help/nomenclature.shtml) is referred for an online clarification. For better annotation an initiative must act for restructuring the miRNAs classification for covering all-new criteria. Besides protecting against current uncertainty, a correct adoption and usage of a new central registry database–miRbase can be synthesized and follow efficiently.

Early Discovery

Macro functions of miRNAs have been revolutionized, particularly early diagnosis and their role in therapy development that dominate all aspects of recent medical fields. The miRNAs are running in the formation of our personalized medicine. However, three decades earlier nobody knew what miRNAs are? Until then, people-focused keenly on the classical central dogma, in which noncoding RNAs have no role to play in the molecular ground of human live cells. Only in 1993 the existence and importance of miRNA function began to be explored.[61, 62]

In 1993, the discovery of lin-4 appeared as the first miRNA in nematodes [61,63] that is required for controlling the temporal development pattern at the larval stages. After the discovery of this small noncoding developmental RNA, the growing knowledge has been expanded exponentially. The loss of function of lin-4 gene generates certain deformities in adult structure unable to lay eggs and loss of early developmental programs at the late larval stages. Particularly, it is essential for L1 to L2 stage transition in the larval development.[64, 65] It also reported that loss of function lin-14 mutation can convert the null lin-4 mutational phenotype[65,66] suggesting that lin-4 and lin-14 are antagonistic to each other. Lin-4 controls negatively the lin-14.[65] The lin-4 gene is 700 bp long without start and stop codons and does not code any protein.[61, 65] Nevertheless, lin-4 introduced a putative mutation in the open reading frame but is not changed functionally. Further, it was observed that lin-4 contains two shorter transcripts 61nt and 22nt in length.[61] In the meanwhile, it was also established that lin-14 is post-transcriptionally regulated via 3' UTR terminal sequences.[62] These two independent results derived a relevant same conclusion that lin-4 sequences are complementary to the 3' UTR of lin-14 and derived a unique result that lin-4 controls lin-14 transcription via its 3' UTR. This report claimed an unexpected cell transcriptional mechanism involving small non-protein-coding transcripts.[61, 62, 65]

After seven years later, the second miRNA was isolated. The let-7 is a 21-nt long-short RNA required for L4-to-adult transition during larval development.[67] Following the above concept, loss of function let-7 gene regerminated the larval cell fates at the time of adult development, similarly enhanced let-7 gene activity reciprocates the adult fates in the larvae.[67] Further, it was established that let-7a activity partially suppresses the lin-41 gene that was identified as to let-7 potential target.[67] It was also found that let-7 sequences are complementary to two closely spaced sites in lin-41 3′ UTR. Therefore, two different criteria either mutation of let-7 or elimination of the entire lin-41 3' URT, which eliminate the lin-41 silencing. It is suggested that both conditions are essential for the silencing mechanism.[67-69] Both lin-4 and let-7 noncoding tiny RNAs are conserved wildly from flies to humans.[70] The let-7 is responsible for late temporal transitions during animal phylogenic development. The let 7 exist also in species including vertebrate, ascidian, hemichordate, mollusk, annelid, and arthropod etc. Interestingly, different levels of let-7 products are available in many human organs for instance, brain, heart, kidney, liver, lung, trachea, colon, small intestine, spleen, stomach, and thymus.[70] Twelve members of the let-7 family are derived from the same complete 5' end sequences of the let-7 gene, referring to as seed sequences.[71] Indeed, the members act upon redundantly to control L2-to-L3 transition separately by inhibiting hbl-1.[72]

The isolation of let-7 after lin-4 has emerged a new branch of a small non-coding RNA network known as microRNAs. Just within three decades nearly one thousand nine hundred (1900) miRNAs have been evolving in humans. The repositories of these classified miRNAs are presented in the miRbase database.[73-75] Therefore, present tools and computer software allow us to identify the existence of newly emerge miRNAs and their functional network.[76, 77]

Evolution

miRNAs are the most important evolutionary components for fine-tuning gene regulation. This is well-spread in the plant and animal kingdom.[78,79] An explosion of miRNA biogenesis and the production levels are conserved widely in two parallel kingdoms plants and animals. They have been emerging by different independent modes of action.[80, 81]

These phylogenetic biomarkers have a low level of evolution.[82] The RNAi machinery, which works as a defense mechanism against infectious parasites and exogenous virus particles are the main driving forces for microRNA biogenesis.[83] Their origin and function are required for the evolution of many phylogenetics trees, development of morphological innovation, the genesis of complex organs[84] and ultimately complex life form.[79] Rapid bursts of morphological innovation and a high rate of microRNA accumulation are integrated into different species.[82, 84]

New microRNAs can be evolved by following many ways. They have synthesized from the random formation of hairpin loops produced by noncoding DNA template, duplication, or sequence modification of the existing miRNAs.[85] They also generate from inverted duplications of protein-coding sequences by the formation of foldback hairpin structure.[86]

The rate of nucleotide substitution during evolution implicates the evolution by neutral drift. However, older microRNAs have a low level of change for instances less than one substitution per hundred million years.[79] Various sectors of independent miRNA structure confront many evolutionary pressures undergoing highly conservative function and their processing.[87] In fact, old and mature miRNAs are rarely lost and well distributed in animals,[79] but new miRNAs that are non-functional or evolved for specific function are sensitive for rapid degradation.[85] The evolutionary emergence of many valuable phylogenetic markers can involve possible solutions to outstanding phylogenetic problems.[88] On the contrary, many microRNAs do not correlate with phylogeny.

The difference in functioning and processing of many eukaryotic miRNAs indicates that microRNAs are originated following independent processes in plants and animals during the evolution of canonical microRNA biogenesis.[89] The nuclear proteins *Drosha* and *Pasha* are the main enzymes required for canonical microRNA biogenesis. Across all species above of 6000 different miRNAs have been isolated so far.[90] Many evolutionary pressures may be in action for the synthesis of new miRNAs specific for well-diverted function within the species.

Evolution and Emergence of Canonical miRNAs

In canonical miRNAs, a stringent analysis is required to eliminate all other contaminated small regulatory RNAs that adulterate annotated miRNAs. In this analysis, the pure annotated miRNA carries three major features, a potential hairpin corresponding to both strands of the miRNA duplex, a consistent 5' terminus and a diagnostic 2-nt 3' over-hangs.[91] Using these features in a recent authentic annotation is identified for canonical miRNA genes that possess 475 miRNAs in mouse [92] and 519 in human.[93]

Numerous studies have adopted many additional gene annotations without imposing stringent criteria. Those lead to 1,193 miRNAs in mice and 1,881 in human[94] according to miRbase v21 lists. These huge amounts of miRNAs raise the question of whether the miRNAs are isolated by some additional annotations missing any stringent authentic criteria. However, a large majority of them are false positives.[92] It implies that if an essential criteria miRNA function is included then stringent criteria seem to be satisfactory.

The human genome possesses 296 out of 500 plus miRNAs in which 177 seed families are conserved strongly in placental mammals and other animals. The list of conserved miRNAs number has been closing to complete. A notable exception is that all human miRNA genes were highly conserved fish at least 12 years earlier before the beginning of their isolation. Indeed, the exception is the *MIR1306* gene,[95, 96] with a suppressive control element that overlaps the *DGCR8* coding sequence.[97] Therefore, a weak accumulation of conserved miRNA, miR-1306 presents as a negative-feedback element for pri-miR-1306 hairpin for controlling DGCR8 expression.[97] A list of very lowly or narrowly expressed conserved miRNAs are reported. According to a recent tally, only lower bound canonical miRNA genes are isolated from the human genome that exceeds 296 family miRNA genes in humans like common ancestor of placental mammals. However, these poorly conserved miRNAs that are not isolated and/or annotated because either of their very low level of expression or fail to identify as a tiny fraction of the miRNA molecules in many tissues.[98, 99] One of the main reasons for poor conservation of these miRNAs is that they are relatively young or isolate from duplication and divergence of an ancestral gene.[93, 100] Indeed, poorly conserved miRNAs have rapidly lost their expression in the course of metazoan evolution, because they are not yet acquired targeting interaction that enhances fitness or loss in absence of interaction promoting their selective retention.

microRNAs in Development

The progressive changes during development are contributed by modulation of gene expression via microRNAs (miRNAs). The collective depletion of all miRNA functions shows in embryonic lethality in all animals, but the loss of important individuals causes severe deformities during development. It suggests fine tuning of miRNA-based gene regulation leads to robustness in development defects. The unique hierarchical and canalizing series of gene controlling networks that are not present in general reviews in the chapter. Interestingly, only a fraction of embryonic miRNAs that are in the peak of the hierarchy is involved in broad developmental defects, whereas many specific miRNAs play a role at the periphery of development, mainly altering the terminal features of specialized development cells.

Two classes of miRNA in development

The founder miRNA members are divided into two broad categories. The members, *lin-4* and *let-7* are required for setting the developmental timing in *C. elegans*,[101, 102] and their inhibition impair larval development dramatically. On the other hand, loss of expression of *bantam* and *mir-14* in *Drosophila* causes abrupt lethality and/or reduction of body size.[103-105] They are involved in most cases in cellular division, differentiation or to die. Their defective expression leads to a bizarre mutant phenotype also. In the second group, *lsy-6* miRNA in nematodes regulates a single sensory neuron expression,[106] which is necessary for neuron development without adding any animal processes.[107] Similarly, mouse *mir-96* also involve in CO_2 sensing and hearing respectively.[108, 109]

A comprehensive network of many regulatory genes can explain the development of multiple biochemical and structural functions during animal embryogenesis.[110, 111] In these hierarchical regulatory networks (GRNs), certain factors are reached to the highest peak of the hierarchy and are responsible for initial morphological patterning events at the top, while many more genes inducing multiple functions of cell differentiation locate at the lower section or periphery of the hierarchy. (**Figure 2**) GRNs are also canalized and are required for initializing different regulatory phases in a region of the embryo. They also block many active subsequent processes within the domain. Despite fine-tuning regulatory factor-a miRNA or others regulate development processes markedly and keeping an intact hierarchical position of the networks, the mutated core miRNAs severed broad consequences in embryo development relative to the loss of function of peripheral miRNAs. However, in the periphery certain important miRNAs are essential for organism viability and deformities in the developmental process and ultimately leads to lethality. Indeed, peripheral miRNAs or other regulators act upon in the later stage of development for instance, *mir-451* involve in red blood cell development postnatally.[112, 113]

In this context, embryonic miRNAs have two major implications. 1) many peripheral miRNAs are involved in cellular differentiation, though some useless candidates in the categories show subtle or no defects.[114-116] Nonetheless, cell-specific miRNAs including *lsy-6* in *C. elegans* are promoting stressful conditions or inhibit its target and define a cell fate. These mRNAs are known as specific cell-fate markers.[107, 117] Second group of miRNAs act up on hierarchical development by a redundant and synergistic effect of other miRNAs. Thus, embryonic miRNAs are the major guiding force for showing the functional contribution of other miRNAs during embryogenesis.

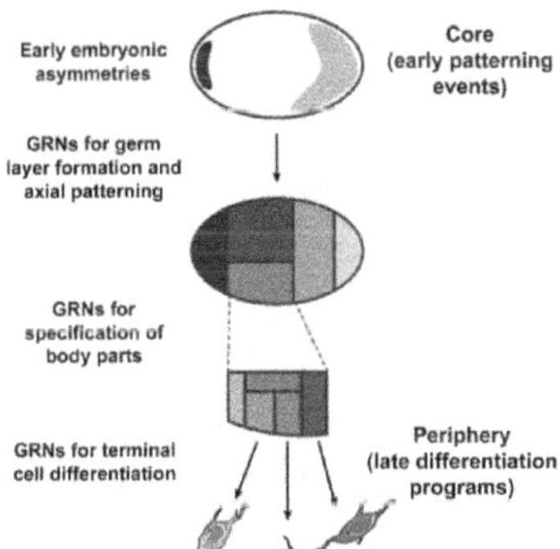

Figure-2. Hierarchical organization of the gene regulatory networks that control embryonic development. At the onset of embryogenesis, gene regulatory events are triggered by asymmetries in the oocyte and sperm that result in regionalization of maternal and eventually early zygotic regulatory factors. These represent the input for new gene regulatory networks (GRNs) that progressively narrow down the potential identities of cells until the point of terminal differentiation. 'Core' events are those required to set up the basis for embryogenesis and place the appropriate progenitors in the right place at the right time. Further specification of those progenitors and differentiation into different specialized cell types is driven by GRNs at the 'periphery.' (Loosely Adapted, Redraw and Modify using skeleton from Alberti and Cochella 2017)

miRNAs in the periphery of development

Genes at the hierarchical peak are required for earlier development processes. The fine-tuned miRNAs are classified depending upon a positive correlation between the functions and their expression pattern. They are expressed broadly and abundantly. In contrast, peripheral miRNAs function in the later stage of development in a cell type-restricted manner. Many miRNAs are shown as spatiotemporal patterns of expression specifically.[116, 118-119]

To understand what amounts of embryonic miRNAs are peripheral. A collective view throughout embryogenesis exhibits that all miRNAs must be expressed in all cell types. However, such expectation is never followed and full filled, a small fraction of miRNAs is expressed abundantly limited to developmental stage, whereas the majority appears to be shown at a very low level of expression. (**Figure 3**) In later categories, few miRNAs have a considerable function. They are expressed widely at a low level in each developmental cell.[120] Conversely, other low expressed miRNAs are the peripheral miRNAs. They are found at high concentrations, albeit within restricted cell types of embryos.

The low expressing miRNAs found a delicate pattern of expression. For example, *lsy-6* extends its expression from embryogenesis to adulthood at a single neuron.[106] The *mir-791* is expressed in three pairs of neurons during entire embryogenesis. Three out of four miRNAs in nematode only express in few cells participating in cell-type-specific developmental roles.[121] In contrast, around 75% of all embryonic miRNAs belong to *mir-35* and *mir-51* families are expressed in GRN and are responsible for embryonic survivalism.[122] Many lower abundant miRNAs are involved in the regulation of highly restricted cell populations, (**Figure 3B**) but it is not marked enough that these miRNAs are expressed widely in the later stage of development.

Numerous miRNAs occupy intermediate stages of developmental hierarchy. (**Figure 2**) They control tissues or organs in the border of embryos without involving in early patterning. The *mir-1*, *mir-9* and *mir-124* are required for broader muscle cells and neuronal differentiation in vertebrate embryos.[123, 124] However, the exact position of miRNAs within the hierarchy are not conclusive. Moreover, it is reported that the hierarchical locational require not to be exclusive: few core miRNAs involve in regulating peripheral genes. Using the same strategy, several transcriptional regulators and signaling factors are required for early patterning and later reinvolved in cellular differentiation. Below, we also add the roles of different peripheral RNA, which are used to activate the core of GRNs as a key component for cell and development regulatory processes.

miRNAs in core development

Many embryonic miRNAs function for replacing maternal components with zygotic genome ingredients in embryogenesis. They also regulate different cell divisions, intercellular signaling, differentiation and apoptosis etc. requiring for indispensable coordinated growth and patterning. These conserved miRNAs are controlled by the core developmental process during embryogenesis.

Maternal RNA clearance

Many miRNA expressions are involved in the clearing of maternal contributed mRNAs during early embryogenesis. Multiple copies of *mir-430* in Zebrafish express in a huge amount at the onset of zygotic component inclusion.[125] The loss of maternal and zygotic *dicer-1* as an alternative for replacing null mutation of *mir-430* accommodate hundreds of maternal transcripts at the end of *mir-40* binding sites. Further, reinsertion of *mir-430* miRNA reduces these maternal transcripts via deadenylating and eventually destabilizes the *mir-430* specific targets.[126] Functionally, similar miRNA *mir-427* in *Xenopus* activates the decannulation process of many maternal mRNAs during maternal to zygotic transition in the frog embryos.[127] However, the pinpoint role of *mir-427* in maternal mRNA clearance at the time of embryogenesis remains inconclusive. As earlier the maternal/zygotic *Dicer*-deficient zebrafish embryos are still able to activate the zygotic program despite the long prolongation in maternal mRNA clearance.[126] Further, *mir-430* functions independently in other morphogenesis processes.

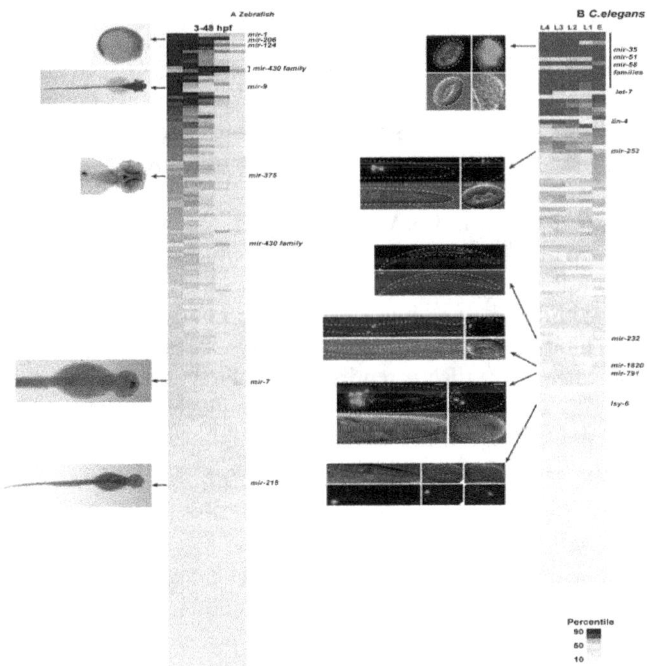

Figure-3. The abundance and expression of miRNAs. **(A)** *C. elegans* miRNA sequencing data ranked by the maximal expression at any developmental stage. miRNAs are sequenced from whole embryos **(E)** or larvae (L1 to L4). Data are replotted. The miRNAs noted in the main text are highlighted and the expression patterns of some of these (based on transcriptional reporters) is shown to the right to highlight the correlation between expression level and pattern. The high abundance of the *mir-35* cluster, for example, correlates with its broad expression. Several seemingly low abundance miRNAs, such as lsy-6 and mir-791, are expressed in a highly restricted manner. The lsy-6 expression pattern is reproduced with permission. The mir-791 expression pattern is reproduced. **(B)** miRNA sequencing data from zebrafish embryos ranked as for *C. elegans*. Data is further replotted. Expression patterns of selected miRNAs (as detected by in situ hybridization) are reproduced. hpf, hours post-fertilization. (Loosely Adapted, Redraw and Modify using skeleton from Alberti and Cochella 2017).

Numerous miRNAs play important constructive roles during cell proliferation and rapid cell death events of vertebrate embryogenesis.

Other lineage-specific miRNAs

Numerous miRNAs expressed in neurons and glia are involved in lineage-restricted and cell type-restricted manner expression in heart cells.[124] They also function during cartilage and bone formation.[128] In mice *mir-375* is required for murine pancreatic islets and pancreatic cell lines formation but not in other tested tissues.[129] In zebrafish, the same miRNA is expressed in the pituitary gland and pancreatic islets.[130] The morpholino triggered loss of *mir-375* function in fish forming deformities in pancreatic islet structure via activating unknown targets. Nonetheless, insulin is produced by expressing its known markers.[131] Also, loss of insulin production is observed in pancreatic cells by the inhibition of *mir-375* miRNA. Many findings also show that these miRNAs are responsible for the development and function of the pancreas.

The cell-specific miRNAs are generated from the fertile breeding ground in mice hematopoietic lineage. The *mir-150* is critical for cell transition from the pro-B to the pre-B stage at the time of mature B-cell production through affecting transcription factor Myb.[132, 133] Interestingly, *mir-150*$^{-/-}$ mice are morphologically healthy and exhibit lineage expansion.[132] The *mir-223* is required for granulocytes differentiation also. The miRNA represses a vital transcription factor for myeloid development such as Mef2c for controlling granulocyte production.[134] Moreover, the normal viability and fertility never hamper in *mir-223* mutant mice and the miRNA also activate the granulocyte progenitors and circulating neutrophils numbers.[134]

Pregnancy regulation

In human tissue, natural loss or artificial deletion of *Dicer* gene develop hypertrophy and disorganization of uterus and elimination of oviduct in female sterility.[135] The deletion of DICER in the human placenta explains the subsequent loss of maximum miRNAs activated in the pro mitogenic signaling pathways [136] via cytotrophoblast proliferation.

Pregnancy-related miRNAs are divided broadly into two categories, tissue-resident and circulating miRNAs based on their physiological localization and transportation. The formation of the placenta and uterus is important for pregnancy-related organs. They found in many tissue-resident miRNAs in placenta-specific, associated phenomena and uterine miRNAs. (**Table 1**) The different complications including intrauterine growth restriction/fetal growth restriction (IUGR/FGR) activating and preterm birth related to miRNAs expression and pregnancy loss preeclampsia are also noticed.

Placental Development

The regulatory roles of few microRNAs in many pathophysiological conditions appear to be a new branch for the microRNA network field. The role of microRNAs in pregnancy and related complications including preeclampsia (PE), fetal growth restriction (FGR) and preterm labor is a budding field for proliferation. Over the last decade, miRNA regulatory mechanisms in pregnancy-related diseases have been exposed exponentially. The existent and regulation of microRNAs associated with placenta-specific, placenta-related, placenta-derived circulation are described vividly that open up a secret weapon against pregnancy-related disorders for developing early diagnostic biomarkers and therapeutic targets.

The existing terminology provides serious confusion regarding the location of miRNA synthesis. The placental miRNAs are found typically in the placenta, whereas the placental associated miRNAs are generated from the placenta and other neighboring tissues ubiquitously. On the other hand, the placenta releases circulatory miRNAs from the placenta and later is transported to the circulatory blood. Their circulatory profile is the essential factor for regulating mechanisms in normal and complicated pregnancy. (**Figure 4**)[137] against pregnancy-related disorders for developing early diagnostic biomarkers and therapeutic targets.

Table-1. Multiple pregnancy-related miRNAs.

Pregnancy related function	Related microRNA
Placenta-specific microRNA	
Enriched in placenta; trophoblast cell proliferation; target Pleiomorphic adenoma gene 1 (PLAG1)	miR-23a, miR-136, miR-141
Abundant in developing embryo and placenta	C14MC cluster
Expression restricted in reproductive system and placenta; trophoblast cell invasion, migration; implicated in complete hydatidiform moles (CHM), fetal growth restriction (FGR)	C19MC cluster (miR-516, 517, 518, 519 etc.)
Predominantly expressed in placenta	miR-371, miR-372, and miR-373
Inhibits embryonic and extraembryonic cell lines proliferation	miR-675
Found in mammal placenta	miR-1302 family

Placenta-associated microRNA	
Regulates NF-κB pathway and IL-6	Let-7
Regulates angiogenesis and is increased in preeclampsia; differentially expressed in preterm birth placenta	miR-15a, miR-15b
Downregulated in response to smoke exposure, angiogenesis associated, and decreased in the small for gestation (SGA) placentas	miR-16, miR-21, miR-146a
Regulate syncytio-trophoblast differentiation	miR-17~92 cluster
Differentially expressed in gestation age; downregulated in early pregnancy loss placenta	miR-17, miR-27, miR-92
Increased in preeclampsia and regulate hydrogen sulphide)	miR-17, miR-20a, miR-20b
Inhibits invasion and promotes apoptosis of trophoblast cells	miR-18a
Downregulated in early pregnancy loss placenta	miR-19b
Regulate renin-angiote Attenuates to help placenta adapt to hypoxia	miR-27a, miR-199b, miR-429
Regulates apoptosis, invasion, and angiogenesis of trophoblast cells	miR-29b
Attenuates mesenchymal stem cells (MSCs) mediated immune response in preeclampsia	miR-30a
It is decreased in placenta accreta patients; and suppression of miR-34a increased trophoblast invasion	miR-34a
Differentially expressed in primary trophoblasts exposed to hypoxia and downregulate eNOS (miR-335)	miR-93, miR-205, miR-224, miR-335, miR-424, miR-451, miR-491
Regulates apoptosis of trophoblast cells	miR-101
Proangiogenic factor and decreased in preeclampsia	miR-126
Overexpressed in recurrent spontaneous abortion (RSA) and downregulates human leukocyte antigen (HLA)-G	miR-133a
Affects proliferation and migration of placenta trophoblast cells	miR-137
Inhibits IL-12, IL-6, and TNF-α and mediates immune tolerance; regulates HLA-G	miR-148/152 family
Regulates trophoblast function, including proliferation, migration, invasion, and differentiation; downregulates angiogenic factors and implicated in preeclampsia; targets	miR-155

angiotensin II type 1 receptor; modulates eNOS expression.	
Regulates TGF-β pathway and IL-6; attenuates the immunosuppressive properties in placenta; inhibits proliferation and immunosuppressive properties of MSCs; is differentially expressed in preterm birth placenta.	miR-181
Altered expression in preeclampsia; affects trophoblast cell invasion	miR-195
Increased expression in response to the hypoxic placenta; regulates trophoblast cell migration and invasion; modulates mitochondrial respiration in placenta; targets critical steroidogenetic enzyme; modulates inflammation-related pathway, potassium channel modulatory factor, and thrombospondin. increased in FGR; differentially expressed in preterm birth placenta	miR-210
Elevated in preeclampsia and correlates with blood pressure	miR-325
Promotes trophoblast cell proliferation and invasion	miR-376c
Promotes trophoblast cell survival, migration, and invasion and contributes to preeclampsia	miR-378a-5p
Regulates trophoblast differentiation	miR-424
Inhibits MSCs proliferation and angiogenesis	miR-494
Regulates placental trophoblast cell proliferation	miR-675
Placenta-derived circulating microRNA	
Identified in the circulation of pregnancy women	C14MC cluster
Identified in exosome and released from human primary trophoblast; upregulated in preeclamptic plasma; increased at early gestational age in circulation	C19MC cluster
Uniquely correlated in the maternal and fetal circulation cluster	miR-127, miR-134, miR-498
Detected in maternal plasma during pregnancy and decreased in postdelivery plasma	miR-135b, miR-141, miR-149, and miR-299-5p
Uterine microRNA	
Induction on the process of embryo invasion during implantation; gradually increased in uteri to inhibit uterine stromal cell proliferation	Let-7a, let-7b
Overexpressed in spontaneous term labor	miR-34b, miR-34c, and miR-223

Differentially expressed during pruriceptive and receptive phase; regulator of progesterone receptor	miR-96, miR-219-5p, and miR-375
Regulate critical gene for implantation	mmu-miR-101a and mmu-miR-
Decreased in laboring myometrium and in an inflammatory preterm labor mouse model	199a* miR-199a/miR-214
Implicated in uterus of preterm labor; involved in progesterone/progesterone receptor pathway	miR-200
Regulates endometrial stromal cells (ESCs) differentiation	miR-222

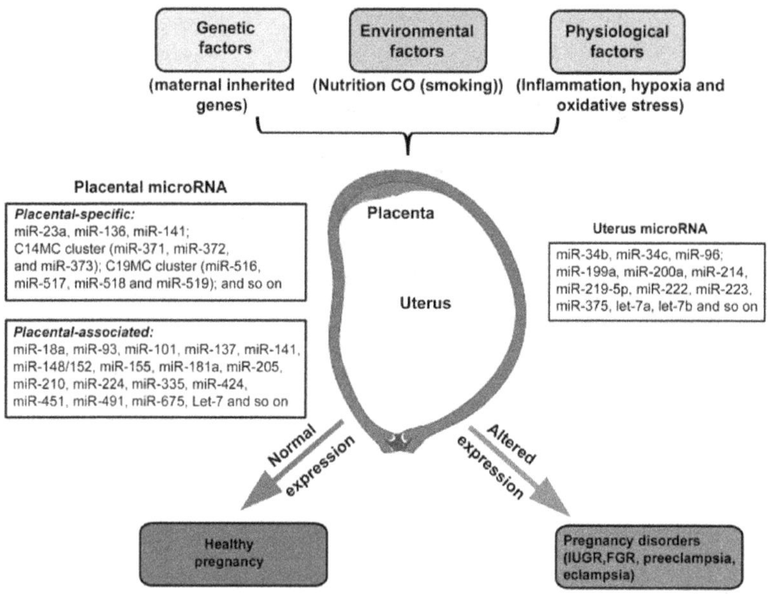

Figure-4. The pregnancy event is controlled by three major factors including genetic, environmental, and physiological features. The placental and uterine miRNAs liked to the change of these features at the time of pregnancy. The change in expression of these miRNAs develops different pregnancy disorders. (Loosely Adapted, Redraw and Modify using skeleton from Meng Cai *et al.*, 2017)

Placenta specific miRNA clusters

Three miRNAs miR-23a, miR-136 and miR-141 are abundant in nearly 100 copies in the mammalian placenta.[138,139] Most placenta-specific miRNAs are closely aggregated to the neighboring chromosomes in a cluster. C19MC a notable cluster on chromosome 19 is limited to its presence in the reproductive system and placenta.[139,140] Also, C14MC, a cluster on chromosome 14, is accommodated close to 52 genes expanding 40 kb region in placental and embryonic tissue.[141] The last miR-371-3 cluster belongs to the same chromosome 19.[141] Their three important members are originated from the placenta.[142] In contrast, the *Sfmbt2* placenta-specific miRNA cluster is located from a single 10th intron. A total of 36 members of the *Sfmbt2* cluster are produced by embryonic stem cells and placenta.[143] Moreover, placenta specific miR-1302 family clusters carry as much as 58 orthologs. They are produced by a MER53 transposon from various chromosomes within the mammalian placenta.[144] Placenta-specific miRNAs have possible regulatory roles in normal and complicated pregnancies via a limited tight pathway of basic bio-characters. A high throughput profiling array indicates that C19MC members are expressed strongly in the primary human trophoblasts.[145] The placenta-derived mesenchymal Stromal cells (PDMSC) represent their hardcore role in the stem/progenitor cells.[146] Further the temporal expression of C19MC members miR-141 and C19MC family are expressed differently in many development steps based on regulatory demands of pregnancy.[147,148]

Recently, the individual members of C19MC cluster miR-141 and miR-519d-3p control intercellular communication and trophoblast cell proliferation, invasion, migration.[149-151] Further upregulation of placenta-exclusive miR-675 inhibits the embryonic and extraembryonic cell proliferation.[152] Notably the downregulation of four C19MC members, miR-517a, miR-517b, miR-518b, and miR-519a is found complete hydatidiform moles (CHM).[153] Likewise, seven members miR-515-5p, miR-516b, miR-518b, miR-519d, miR-520h, miR-526b and miR-1323 are repressed significantly in the placenta of FGR patients, meanwhile four members miR-518b, miR-519d, miR-520h, and miR-1323 are restricted to FGR.[154] In contrast, placenta-specific miR-141 is involved in regulating insulin-like growth factor 1 (IGF-1 of pleomorphic adenoma gene 1 (PLAG1) in FGR.[155]

Placenta-Associated MicroRNAs

The placenta-associated miRNAs express in various gestational ages of trophoblast cells, placental tissues, and maternal plasmas. Total 45 miRNAs are isolated in the trophoblast cells of first-and third-trimester placentas. Among them 58% microRNAs are placenta-associated miRNAs.[156] Recently a total of 191 microRNAs expressed differently in between first and third-trimester placentas.[157] In fact, angiogenic and anti-apoptotic miRNAs are expressed strongly in the first-trimester placentas only oncogenic. Conversely, a predominate of expression of cell differentiation and tumor suppressor miRNAs is observed in the third-trimester placentas. Also, only considerable amounts of placental-associated miRNAs are expressed at different levels in the maternal plasma during the first-, second- and third-trimester gestation.[158]

miRNAs alter Inflammation and Hypoxia in Pregnancy: The miRNAs express widely based on the needs of inflammation and hypoxia during gestation.[159, 160] The let-7 miRNA works the downstream of NF-κB signaling pathway and negatively controls IL-6 expression, whereas an increased expression is observed in the placenta and amnion involving in placental inflammation control.[161] Identically, miR-181a blocks the increase of the TGF-β signaling pathway and activates IL-6 expression. Thus, activation of miR-181a in the placenta reduces the stem cell immunosuppression in the mesenchyme and is responsible for the abnormal pregnancy.[162] Conversely, the miR-148/152 family controls innate immune responses negatively regulating immune tolerance for a healthy pregnancy.[163, 164]

The placenta-associated miRNAs are required for the hypoxic challenge that is shown in early-stage and pathological conditions of pregnancy.[165,166] The emerging of hypoxia treatment does not control the global miRNA biosynthetic pathway in primary trophoblast but certain individual miRNA is shown different expression profiles during hypoxia.[167] Seven important miRNAs miR-93, miR-205, miR-224, miR-335, miR-424, miR-451, and miR-491 are required for different levels of primary trophoblasts expression linked to hypoxia. Besides miR-205 is involved in placental development factor MED1.[168] Moreover, three miRNAs miR-16, miR-21 and miR-146a in cigarette smoke-exposed placenta are reduced strongly.[169] In a separate study, it is found that placenta-associated miRNAs are required for renin-angiotensin system adaptation to hypoxia and mitochondrial electron transport. The miR-210 is highly sensitive to HIF that is elevated in the hypoxic placenta for replacing mitochondrial electron transport chain function and energy metabolism.[170] On the other hand, three miRNAs miR-27a, miR-199b and miR-429 are suppressed in antenatal maternal hypoxia (AMH) for activating renin-angiotensin system factors for promoting hypoxia-like adaptation.[171]

miRNAs in Pregnancy-related Cellular Activities

The potential regulators are found in pregnancy-related trophoblasts and endothelium cells. The miR-17~92 cluster members and its two paralogs, miR-106b~25 and miR-106a~363 are reduced for facilitating syncytial-trophoblast differentiation.[172] Besides, miR-675 is upregulated the trophoblast proliferation.[173, 174] In contrast, activation of certain miRNAs including miR-376c and miR-378a-5p are not able to instigate multiple cellular mechanisms such as trophoblast proliferation, migration and invasion processes.[175, 176] Basically few members miR-424,[177] miR-101,[178] miR-18a,[179] miR-335,[180] miR-137,[181] and miR-155 [182] have been involved in soluble vascular endothelial growth factor receptor 1 (sVEGFR-1 or sFlt-1) that targets in trophoblast dysfunction and endothelial nitric oxide synthase (eNOS). During early pregnancy placental angiogenesis is one of the important physiological phenomena.[183] In gestational age, modulation of three miRNAs miR-17, miR-92 and miR-27[184] is noticed in orcin placenta. A proangiogenic miRNA miR-126 is reduced in VEGF expression of preeclampsia patients,[185] while negative regulator miR-15b miRNA is activated strongly in preeclamptic placentas.[186]

MicroRNAs in Pregnancy Complications and pathogenesis

Nowadays, two miRNAs miR-155 and miR-210 function in downstream targets in preeclampsia. Among them, miR-155 repress angiogenic factors for contributing to preeclamptic patient endothelial cells.[187] Later, the miRNA also targets angiotensin II type 1 receptor a critical preeclamptic pathogenic factor.[188] Also, miR-155 is related to different trophoblast functions for example proliferation, invasion and differentiation etc.[189, 190] It also modulates eNOS expression in trophoblast cells.[191] Also, miR-210 is known as an iron metabolism regulator for responding to defective placentation.[192] The miR-210 is activated in preeclampsia that acts as an inverse regulator for trophoblast cell migration and invasion.[193,194] On the other hand, miR210 functions for the alteration of mitochondrial respiration in preeclampsia [195] and alter in inflammation pathway, potassium channel and thrombospondin regarding contribution to preeclampsia.[196-198]

Apart from these miR-325 is related to the blood pressure of preeclamptic patients.[199] They also conduct stem cell function and deregulation of trophoblast in preeclamptic pathophysiology. Three miRNAs miR-29b, miR-195 and MiR-378a-5p are involved in trophoblast cell proliferation, invasion, migration, angiogenesis and preeclamptic placental deregulation.[200-202] A dramatic change of many miRNAs is found in mesenchymal stem cells (MSCs) in preeclamptic placentas. The MIR-494 expression is reduced for MSCs proliferation and angiogenesis potential,[203] while upregulation of miR-30a is required for MSCs-conducted immune response.[204] On the other hand, four angiogenesis-specific miRNAs miR-17, miR-20a, miR-20b and miR-21 the negative regulators of hydrogen sulfide generation are activated in preeclamptic placentas strongly.[205, 206]

Also, seven miRNAs are involved in differential expression of small-for-gestational-age (SGA), FGR, or IUGR between preeclampsia and control placentas. (**Figure 5**) [207] However, nearly 2 folds activation of many placentas related miRNAs is found in the FGR and control patients.[208] Besides two miRNAs miR-16 and miR-21 are reduced in the SGA placentas significantly indicating an additional expression in SGA pathogenesis.[209]

Many placenta-related miRNAs are involved in maternal-related pregnancy disorders, preterm birth and abortion. As many as twenty differentially expressed miRNAs, including a few important preeclamptic or SGA-associated miRNAs are found also.[210] Two miRNAs miR-17 and 19b are reduced markedly in early pregnancy loss, placental expression.[211] Conversely, MiR-133a is overexpressed in the recurrent spontaneous abortion (RSA) patient villi for downregulating HLA-G.[212] Moreover, a significant decrease of miR-34a expression is reduced in trophoblast cells but increases trophoblast invasive potential.[213]

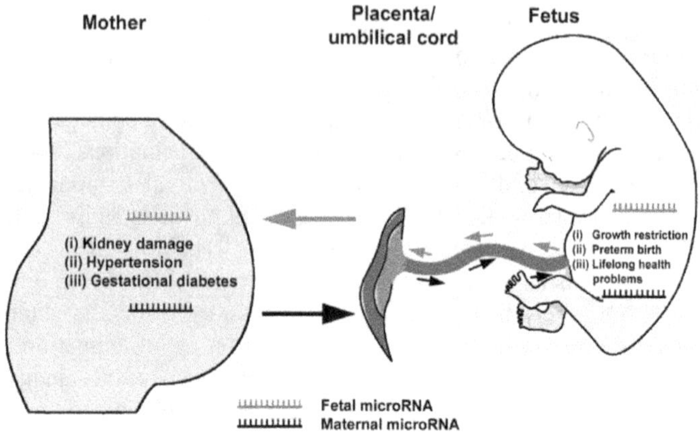

Figure-5. A mutual link between fetus and mother miRNAs during pregnancy related predominant pregnancy-dependent maternal and fetal disorders. The abnormal exchange of fetal and maternal miRNAs at the time pregnancy allows long term problems in the fetus and mother. (Loosely Adapted, Redraw and Modify using skeleton from Meng Cai *et al.*, 2017)

Circulating miRNA as Biomarker in the placenta

The expression of many circulatory miRNAs reflects the forecast for placental diagnosis potential.[214, 215] The placental-specific C19MC members were first identified in the maternal blood and for release from the trophoblasts via exosomes.[216] Upregulation of circulating C19MC miRNAs is identified as a consequence of preeclampsia onset.[217] Seven microRNA expressions are activated in the circulatory plasma of women having early gestational age and later develop pregnancy disorder, [218] whereas reductions of certain C19MC miRNAs are also found in the placenta of pregnancy-related complication carrying patients.[219] The circulation related miRNA clusters miR-127 (8 individual microRNAs), miR-134 (41 individual microRNAs) and miR-498 (46 individual microRNAs) are involved in a dramatic loss of C19MC and C14MC clusters. These differences in miRNA clusters are found in maternal and fetal circulation but not observed in men or nonpregnant women circulation. Interestingly, certain miRNA profiles reflect a marked difference in primary isolated trophoblast cells and trophoblast cell lines for miRNA transportation and exchange to neighboring cell types.[156]

Placenta-related circulating miRNAs are utilized as biomarkers for pregnancy monitoring. The placental miRNAs miR-135b, miR-141, miR-149 and miR-299-5p are increased in the serum of nonpregnant women relative to pregnancy stages of women.[220] In fact, a group of hypoxia-related and placenta-sensitive miRNAs increase nearly two folds in the plasma of women with FGR.[168] Importantly, the expressions of important miRNAs are reduced in the FGR women placenta relative to normal pregnant women. It reveals an inverse correlation of their occurrence in circulation and placenta. Interestingly, four C19MC members have the same expression between FGR and normal pregnancy,[154] suggesting that miRNA pools are appropriate biomarkers rather than individual miRNA. In view, the profile of maternal circulating miRNAs utilizes as a potential biomarker both in pregnancy-related disorders and fetal congenital heart defects (CHD).[221]

Uterine miRNA in Pregnancy Process

Two uterine miRNAs mmu-miR-101a and mmu-miR-199 are isolated earlier to regulate post-transcriptionally an important implantation gene.[222] A cancer-related miRNA let-7a is isolated for the temporal and spatial expression in the uterus during the embryo invasion process.[223] On the other hand, let-7b is related to preimplantation stages in epithelial cells and in the uterus for reducing uterine stromal cell proliferation.[224] The 49 miRNAs are shown a marked fluctuation in between the noninduced endometrial stromal cells (ESCs) and induced ESCs.[225.] Among miR-222 is required for ESCs differentiation via targeting of cyclin-dependent kinase.

The miRNAs are also responsible for progesterone/estrogen signaling pathways. A miRNA family miR-200 a direct target for STAT5b and a transcriptional repressor of the progesterone-metabolizing enzyme that is involved in the uterus of human preterm labor and is utilized as progesterone/progesterone receptor- (PR-) mediated regulators.[226, 227] Another miRNA cluster miR-199a/miR-214 is inhibited markedly in the laboring myometrium of pregnant humans.[228] Additionally, at least three miRNAs, miR-96, miR-219-5p and miR-375 are involved directly for the progesterone receptor.[229] Conversely, progesterone or estrogen stimulation alters uterine miRNA expression.

MicroRNA in the nucleus

Numerous microRNAs reside within the nucleus, while putative miRNAs are limited in the cytoplasm after their biogenesis and transport.[230-232] The nuclear miRNAs are involved in different phenomena in a variety of cells. The microRNA in nuclear localization is considerable phenomenon their profiling in the nucleus varies across many tissue types.

Transcriptional regulation

Three Ago proteins dependent miRNA degradation model is found distinctly, in which miRNA promoters interact with siRNA for transcriptional regulation within the nucleus. (**Figure 6**)[233, 234] However, the exact member of the Ago protein family is still not known. May be either Ago-I or Ago-2 is involved.[235] Further depending on the nature of regulatory target sites either the TATA box motif and CpG island region or DNA methylation are produced appropriate signals whether either repression or activation should occur in such cases.[236] Moreover, miRNAs target sites have a long way from the transcription initiation site. Such as the targeting site of an effective small RNA is resided to 1611-bp upstream of the transcription start site (TSS).[237] Certain miRNAs-based histone modifiers have been recruited to the promoter region. However, the role of miRNAs governs the regulatory processes via the complex epigenetic modifications [238] and conducting through DNA methylation is still not known.[235, 239, 240]

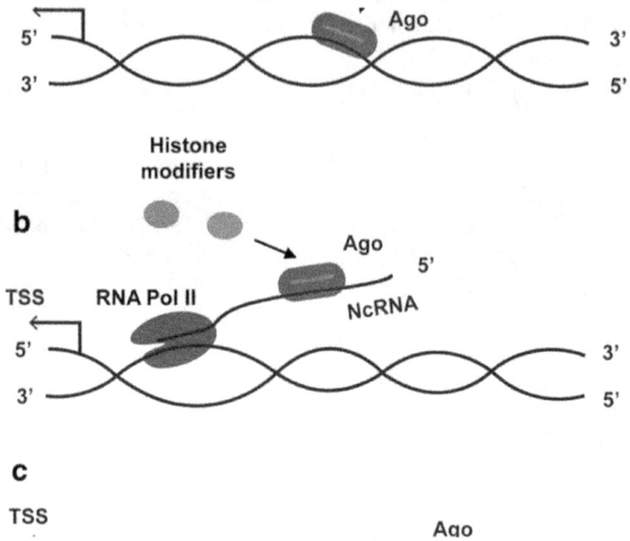

Figure-6. MicroRNA-promoter interaction mechanisms. There are three models for microRNA-promoter interaction, all of which underscore the importance of Ago protein, though there are some conflicting views on whether Ago1 or Ago2 is involved. The resultant interaction effect can be either transcriptional activation or transcriptional suppression. In most cases, the RNA-DNA and the RNA-RNA hybrid model rely on histone modifiers and the recruitment of RNA polymerase II, while whether microRNA can mediate DNA methylation remains dubious. Moreover, in the RNA-DNA-DNA triplex model, the interaction is not related to epigenetic modification. **a)** In the RNA-DNA model, microRNA-Ago complex directly targets to one of the DNA strands which always contains the TATA box motif or transcription factor binding sites when this region is in open configuration for transcription. Then some transcription factors and/or epigenetic modifiers are recruited to the promoter region leading to RNA polymerase II recruitment and/or epigenetic modification. **b)** The RNA-RNA model is related to the non-coding transcripts derived from the promoter region. Either sense or antisense transcript can be a target for microRNA-Ago complex, serving as a scaffold for microRNA-Ago complex and recruiting the histone modifiers and/or transcription factors. **c)** In the RNA-DNA triplex model, microRNAs form triple-helical structures with DNA, altering the topography of the chromatin, which is rendered either more accessible or inaccessible to distinct transcription factors. **TF**-transcription factor; **ncRNA**-non-coding RNA, **RNA pol II**-RNA polymerase II; **TSS**-transcription start site; **Ago**-Argonaut 1 or Argonaut 2; Histone modifiers, such as histone methyltransferase enhancer of zeste homolog 2 (EZH2) and euchromatic histone lysine methyltransferase 2 (EHMT2). (Loosely adapted, Redraw and Modify using skeleton from Hongyu Liu *et al.*, 2018)

In this model, the noncoding sense and antisense transcripts are attracted by the microRNA-Ago complex directly. They form a molecular scaffold for accumulating H3K4me3 and H3K27me3 with an epigenetic factor that facilitates epigenetic modifications.[241-243] This complex can also accommodate more RNA polymerase II at the promoter region.[241, 244, 245] However, bidirectional gene regulation serves as a model for miRNA-promoter interaction.[245] Decreased antisense RNA activates loss of H3K27me3 and Ago-1 at the p21 promoter as a result, downregulating the promoter-related RNA and upregulating the sense RNA are found.

In the RNA-RNA model, the gene located upstream to the promoter is activated by miRNAs through targeting its promoter, miRNA. These phenomena are related to the noncoding RNA that overlap the promoter and gene looping for considering two linearly distant promoter regions having proximal spatially.[242, 246]

In the RNA-DNA hybrid model, the TATA-box motif or the regions of transcription factors are the main promoter target for the miRNAs-Ago protein complex. This preinitiation complex (PIC) is formed either by histone epigenetic alteration or the binding of transcription factors.[247-249]

In the RNA-DNA triplex model, purine or pyrimidine rich miRNAs alter their triple-helical structures with purine-rich duplex DNA in the major groove of the DNA duplex via Hoogsteen or reverse Hoogsteen interaction. Such interactions change the DNA topography and facilitate transcription factors binding for transcriptional activation or suppression.[250, 251] In the meantime a transient interaction between purine riches hsa-miR-5p and a double-stranded DNA is detected. However, the presence of triplex formation of miRNAs *in vivo* is not identified so far.[251, 252]

Nuclear microRNA as enhancer trigger

A separate group of miRNAs, miR-24-1, miR-24-2, miR-26a-1, miR-339 and miR-3179 acts as cis-regulatory elements for upregulation of neighboring gene transcription. The miR-26a-1 regulates protein-coding genes *ITGA9*, *CTDSPL*, *VILL* and *PLCD1* in the surrounding. The miR-26a-1 is required for activation of *ITGA9* and *VILL* genes in HEK293T. A deletion of the seed sequence of miRNA that examines positive base pairing between miRNA and target gene enhancer is required for its action. Similarly, miR-24-1 is also upregulated the neighboring genes *FBP1* and *FANC* expression. The increment of three factors, RNA polymerase II, p300/CBP, and enhancer RNAs are required for miRNAs regulatory function. But such influence in neighboring gene regulation is absent if Ago-2 is inhibited.[253] So, miRNAs require active Ago-2 for transporting mature miRNAs from the cytoplasm to the nucleus.

Nuclear microRNA in immunity

Cytokine

Nuclear microRNAs can control immune cells through the expression of cellular cytokines. The let-7i microRNA conjugates with TATA Box for influencing the interleukin-2 (IL-2) in CD4+T-lymphocytes [242] transcripts as an early symptom for activating CD4+T cells.[254] The miR-205 promotes the enhancement of two parallel promoters of IL-24 and IL-32 genes of the cytokine family.[255] The tumor suppressor IL-24 exhibits melanoma differentiation-associated-7 in non-hematopoietic tissues as an inducer of cell death,[256] whereas a proinflammatory cytokine IL-32 is required for cell differentiation and induction for the maturation of NK and NKT cells, immature DCs and infection control etc.[257] Thus, a single miRNA is involved in both immune responses and signal transduction pathway by targeting cytokine genes.

Asthma

Transmembrane glycoprotein CD44 is important for three immune processes, lymphocyte activation, cell proliferation and assembly of inflammatory cells. It is required for airway inflammation, particularly asthma.[258, 259] A miRNA miR-31 can bind *CD44* promoter in a higher efficiency for upregulation. This overexpressed CD44 further upregulates the asthma-associated molecules including IL-6, IL-8 and intercellular adhesion molecule 1 (ICAM), for the progression of asthma.

Inflammation

The miR-589 upregulates two neighboring inflammatory genes, cyclo-oxygenase-2 (*COX-2*) and phospholipase A2 (*PLA2G4A*). The miR-589 binds to the *COX-2* promoter and forms a gene looping structure that is important for its interaction with the *COX-2* and *PLA2G4A* enzyme. The upregulation has a tremendous impact on inflammatory factors. The *PLA2G4A* enzyme degrades the membrane phospholipids to release arachidonic acid, whereas *COX-2* elevation leads to conversion of arachidonic acid to an inflammatory factor prostaglandin for playing a vital role in vasodilation.[242]

Granulopoiesis

In many stages, hematopoietic differentiation [260, 261] a specialized group of microRNAs in nuclear myeloid cells miR-467a* miR-223, miR-690, miR-706 and miR-709 have been involved for granulopoiesis modulators.[262, 263] In the nucleus, miR-223 binds to NF1A promoter for accumulating *Polycomb* repressive complex that increases the DNA methylation and repressive histone markers.[263] It reveals that miR-223 is an essential factor for hematopoietic proliferation and differentiation.

MicroRNA in Chromatin structure

Chromatin architecture and miRNA transcription

Mounting of evidence has been reported that Chromatin architecture appears to be controlling the miRNA transcripts in the same manner as they control the protein-coding genes.[264] The 12 genome-wide chromatin trackers including 10 tracks of histone modification marks, one track of DNA methylation, one track of chromatin components assembly and its accessibility are identified in six human cell lines. On average, general histone modifications are estimated across the 4,000 bp window covering the center of each pre-miRNA sequence. It suggests that enrichment of signals of multiple chromatin structure is noticed in and around human pre-miRNA genomic sequence than from intergenic regions. **(Figure 7)**

Next, enrichments and distribution of many chromatins structural features are found in highly expressed miRNAs relative to lowly expressed miRNAs and silenced miRNAs in human embryonic stem cells and multiple cell lines. Notably, the tight bond between chromatin regulation and miRNA biogenesis is found. In such cases, the histone activators are present in higher numbers in chromatin histone tails that is highly expressed miRNAs. Inconsistent with earlier data DNA methylations around pre-miRNA regions are occupied distinctly during miRNA expression levels. **(Figure 8)**

For correlation of chromatin features and miRNA expression, many modifications in chromatin structure around pre-miRNAs genomic sequences are used as positive input. For estimating miRNA expression, the SVM classifier model is assessed by cross-validation. In the model, the result produces by measuring the area under the receiver operator characteristic curve, AUC in different cell lines for example $AUC_{H1\text{-}hESC}$=0.64, $AUC_{GM12878}$=0.65, AUC_{HepG2}=0.69, AUC_{K562}=0.65, $AUC_{HeLa\text{-}S3}$=0.74, AUC_{A549}=0.71, respectively are found. On the other hand, a high expression of silenced miRNAs shows a higher level of classification accuracy with the AUC values ranging from 0.83 to 0.91 as found in $AUC_{H1\text{-}hESC}$=0.87, $AUC_{GM12878}$=0.83, AUC_{HepG2}=0.91, AUC_{K562}=0.85, $AUC_{HeLa\text{-}S3}$=0.89, AUC_{A549}=0.86, respectively. However, this is not the causal effect between chromatin architecture and miRNA expression. A prominent finding is evolved in such cases where chromatin architectural changes are involved in miRNA transcription.

The miRNA expressions vary dramatically in many human cell lines. Therefore, only the existence of expressed miRNAs in each cell line is considered to examine the relationship between miRNA expression and chromatin. Multiple chromatin structural features including DNA methylation, chromatin accessibility and histone modification of the gene promoter regions encompassing 4,000 bp long window relative to the transcription start site are considered to evaluate their relationship.

Moreover, DNase I hypersensitive sites (DHS) data are utilized for examining the relationship in different human cell lines. Mainly DHS peaks that appear to be a quantitative marked for an open architecture of chromatin, are preferentially accumulated in the promoters of miRNA targets than non-targets. **(Table 2)**

The DNase I hypersensitive signals are estimated in their promoters of miRNAs in each group. For instance, in six human cell lines the DNase I hypersensitive signals R_{K562}=0.63, $R_{H1\text{-}hESC}$=0.68, $R_{GM12878}$=0.69, R_{A549}=0.7, R_{HepG2}=0.73, $R_{HeLa\text{-}S3}$=0.74 are shown as a marker correlation with the miRNA target rate. These results reveal that promoter regions that are well exposed to the open chromatin structure are highly sensitive to miRNA targets.

miRNA control and chromatin features

DNA methylation

Next histone methylome data from different human cell lines investigates to identify the relationship between miRNA and promoter methylation. In this study, the DNA methylation of miRNA promoters is decreased considerably than the promoter of miRNA non-target indicating their functional efficacy. (**Table 2**) The ratio of CpG content ($CpG_{o/e}$) is a reliable proxy for the DNA methylation status in the human genome. The genes divided broadly based on DNA methylation into a hypermethylated group that carries a low amount of $CpG_{o/e}$ and the hypo-methylated group has a higher number than expected $CpG_{o/e}$. Despite the DNA methylation in hyper-methylated promoters have no distinct differences between miRNA targets and non-targets.

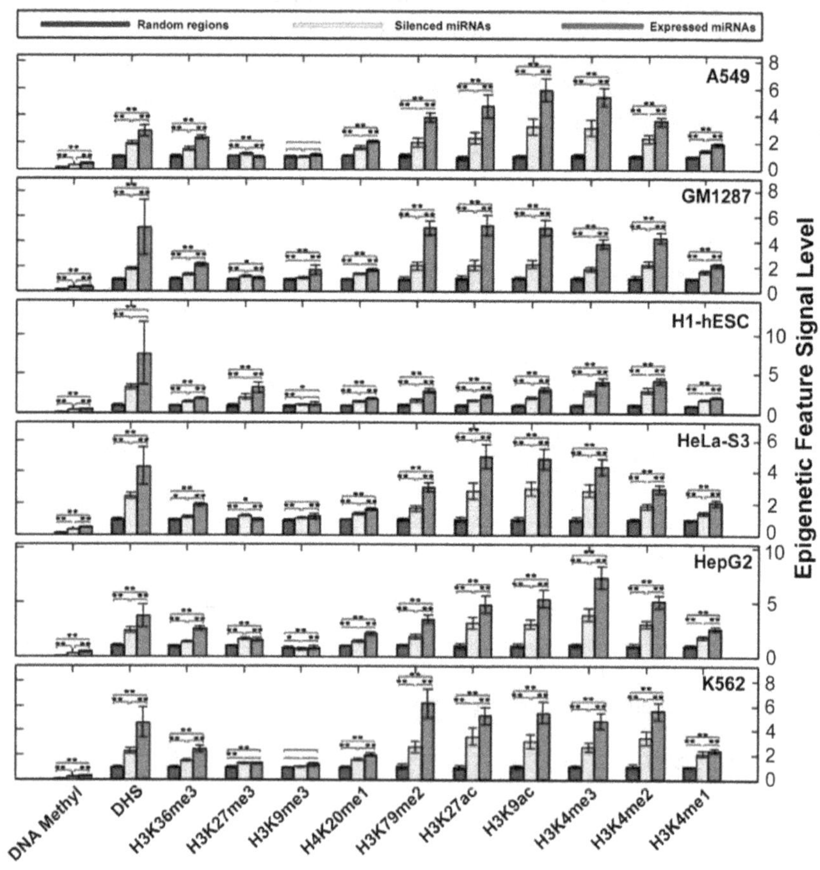

Figure-7. Enrichments of chromatin features around pre-miRNA sequences. We compare the observed prevalence of chromatin features in pre-miRNA sequences against random sequences control. Statistical significances are assessed by *Wilcox rank-sum* test, and error bars are estimated by using 95% confidence intervals. '*' represents *P-value*< 0.05 and '**' represents *P-value*< 0.01. (Loosely Adapted, Redraw and Modify using skeleton from Bang-Bao Tao *et al.*, 2017)

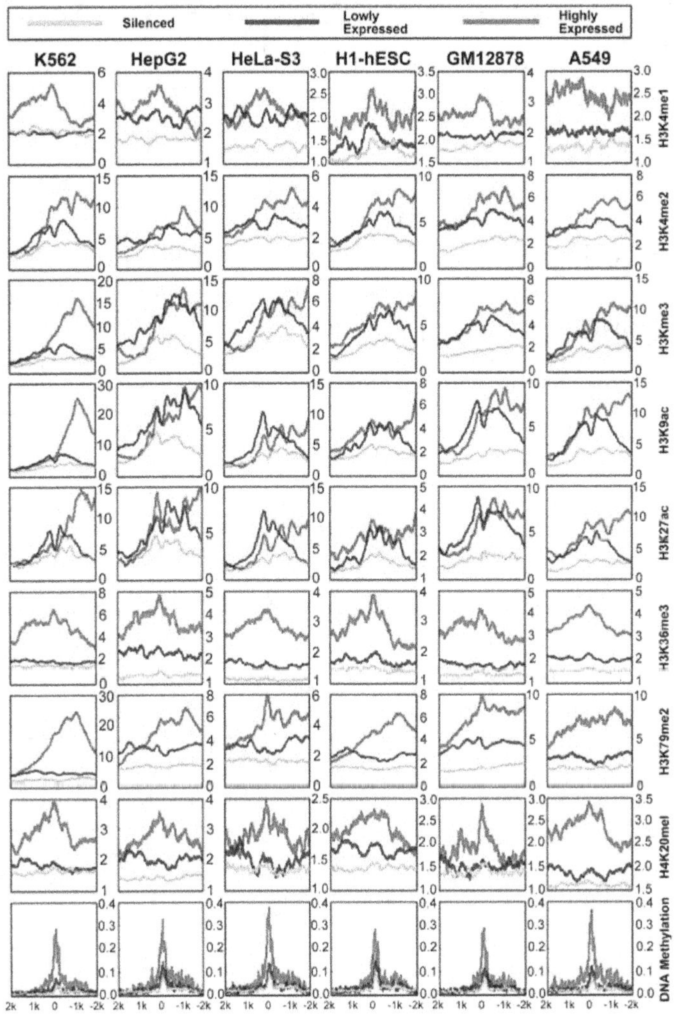

Figure-8. Profiles of chromatin features along pre-miRNA genomic sequences in human embryonic stem cells. The y axis represents the average values of chromatin feature coverage, and the x axis represents the distance relative to the center of pre-miRNA. (Loosely Adapted, Redraw and Modify using skeleton from Bang-Bao Tao *et al.*, 2017)

Table-2. Average promoter signals of the chromatins between miRNA targets and non-targets

Chromatin Signals	H1-hESC	Hela-S3	HepG2	A549	GM12878	K562
DNA methylation	0.22/0.36 (2.7E-143)	0.35/0.40 (8.9E-19)	0.22/0.27 (3.7E-40)	0.26/0.34 (1.1E-71)	0.21/0.28 (3.8E-71)	0.19/0.2
						(3.0E-06)
DHS	0.35/0.25 (6.0E-112)	0.17/0.14 (1.2E-31)	0.23/0.19 (4.7E-42)	0.13/0.10 (2.0E-43)	0.14/0.12 (6.1E-21)	0.17/0.14 (2.9E-32)
H3K4me1	3.26/2.62 (4.3E-86)	3.10/2.67 (1.0E-20)	3.62/3.41 (1.8E-09)	2.28/1.92 (3.2E-49)	2.74/2.58 (8.8E-12)	03.66/3.35 (5.2E-07)
H3K4me2	15.15/10.31 (2.2E-153)	8.57/6.61 (5.6E-41)	16.05/12.38 (2.7E-50)	10.97/7.75 (2.0E-107)	10.36/8.33 (1.4E-47)	17.78/14.25 (2.9E-26)
H3K4me3	14.46/10.06 (1.6E-108)	16.41/12.05 (1.2E-51)	26.61/19.20 (9.9E-63)	21.50/14.58 (1.3E-96)	9.72/7.65 (2.4E-49)	13.73/10.98 (7.5E-25)
H3K9ac	7.05/5.11 (1.9E-78)	16.51/12.03 (1.2E-40)	16.77/12.07 (1.1E-39)	19.66/13.52 (6.3E-79)	12.74/9.90 (5.4E-34)	14.23/10.99 (9.3E-20)
H3K27ac	3.79/2.90 (3.8E-45)	15.10/10.84 (8.3E-35)	14.32/10.25 (2.5E-28)	11.47/7.73 (1.4E-62)	10.65/8.22 (2.0E-32)	13.70/10.47 (1.2E-18)
H3K79me2	4.65/2.91 (3.2E-91)	5.30/3.81 (3.2E-43)	6.01/4.19 (1.4E-41)	6.10/3.96 (2.0E-74)	9.39/6.85 (7.3E-39)	10.85/7.76 (3.4E-25)
H4K20me1	1.91/1.49 (1.1E-133)	1.37/1.23 (1.0E-11)	1.93/1.40 (8.1E-91)	1.86/1.50 (4.0E-69)	1.58/1.40 (2.7E-33)	2.06/1.74 (6.5E-38)
H3K9me3	0.78/0.86 (4.2E-09)	0.86/0.89 (0.09)	0.46/0.48 (0.96)	0.79/0.83 (0.83)	1.12/1.11 (0.66)	1.06/1.09 (0.3)
H3K27me3	5.94/3.68 (1.0E-19)	0.88/1.00 (1.4E-10)	3.08/3.54 (4.2E-26)	1.45/1.44 (086)	1.54/1.29 (8.0E-07)	1.41/1.45 (0.075)
H3K36me3	1.03/1.04 (0.78)	0.94/0.92 (0.49)	1.06/1.03 (6.7E-05)	1.40/1.29 (1.4E-16)	1.01/0.99 (0.34)	1.52/1.46 (0.0063)

The values in the parenthesis are the *P-values* calculated by the *Wilcoxon signed-rank* test.

Histone modifications

For better understanding the relationship between histone modifications and miRNA regulation is estimated via the profiles of 10 different histone tail modification markers including H3K9me3, H3K27me3 and H3K36me3 between miRNA targets and non-targets. The histone tail modifiers are increased markedly across different cellular contexts. (**Table 2**) The histone modifiers H3K36me3 are responsible for transcriptional activation, while H3K9me3 and H3K27me3 are prominent markers for transcriptional repression.[265] A cumulative effect of activated histone cluster and repressing histone modifiers are required for miRNA transcription. At least two H3K4me3 and H3K27ac histone tail modifiers are found in an active cluster (Cluster A), while two important histone methylations H3K9me3 and H3K27me3 are the integral components for a repressed cluster (Cluster R). A balancing act between cluster A and Cluster R expression might function as preferential targets for miRNA transcription. Thus, an internal signal for chromatic architectural modification is the prime importance for miRNA regulation.

Conclusion

A brief snapshot of the microRNA features and their important field of contributions are presented in rapid development and possible roles in chromatin morphology related to their function. It is also clear that miRNAs are essential for the correct development and function of numerous cell types. In most cases, however the functional relationships of miRNAs and their targets do not explain. These consequences do not open up many new miRNAs and their regulatory devices and how the effect on specific cell types impacts the organism. Given these miRNAs, introductions are not well enough to understand their positive implication of miRNAs in physiology and diseases. Later, a deeper mechanism concerning the roles of miRNAs within humans is highly desirable and discussed eventually.

Over the last decade, the functions of miRNAs have been isolated in development. They have high expression profiles for targeting multiple target mRNA at the same time. For instance, the 72 members of the mir-430 miRNA family can interact with many targets simultaneously in Zebrafish.[126] The miRNAs in embryogenesis have been classified into two categories many miRNAs work at top of their hierarchy are focused on active core processes, while less predominant miRNAs function in the periphery, Moreover, the loss of the limited number of miRNAs distributed many broad effects.[115, 266]

The evolution of many peripheral miRNAs holds the clue for their high specific expression. Most newly evolved miRNAs at their initial appearance are harmful to the regulatory process. However, their depleted level of expression can be overcome in many cell types.[267] These miRNAs are required for the evolution of new cell types during development.

Pregnancy is a particularly regulated by a complex regulated process. Dislodging of balance in physiological factors including inflammation generates disorders or complications and relevant oxygen concentration. The chronic hypoxia state in the pathological placenta damages endothelium instigating myocardial and microvascular dysfunctions. It induces preeclampsia event and IUGR.[268, 269] MiRNA as a "fine-tuner" is the best candidate for these precise regulations. Indeed, nearly 60% of mRNA targets control their physiological and pathological processes by miRNAs. The miRNAs are involved actively in immune tolerance, inflammation, hypoxia, fetus development, preeclampsia, IUGR during pregnancy regulation and TGF and progesterone/estrogen signaling pathways. Recent functions of primitive pregnancy-associated miRNAs have participated mainly in trophoblast and mesenchymal stem cell formation. The circulatory miRNAs in placental tissue and blood from pregnant women act as critical biomarkers of early pregnancy disorders. Further serum miRNAs measure in the first trimester of pregnancy. Although miRNA evolves as a budding state in pregnancy and its associated disorders. Nevertheless, it is easy to say that the implication of pregnancy-related miRNAs in the diagnosis and pregnancy-related disorder treatments is of utmost promise.

The miRNA triggers enhancers [270] within the nucleus. The disturbances of nuclear miRNAs functions have multiple implications of diseases, disorders and normal physiology in immunity and tumor.[271-273] Nowadays miRNA therapy and related functions are becoming the most common features of predominant cytoplasmic miRNAs.[274] However, massive exploitation of nuclear miRNA functions is also found that deregulates the multiple numbers of clinical settings.[275]

References

References: chapter wise are cited before the Index at the end of the Book. (1-275)

Chapter II

Biogenesis principle, Control, Circulation and Function of miRNA

Abstract

MicroRNAs (miRNAs) are unique types of tiny regulatory non-coding RNAs that function as major fundamental roles in regulating gene expression and translation inhibition. Multiple steps in biogenesis of microRNA include initial transcription from noncoding DNA sequences into precursor primary miRNAs and subsequently processed into precursor miRNAs, transported from nucleus to cytoplasm, and ultimately transformed into mature miRNA duplexes. Predominantly, miRNAs interact with the 3′ untranslated region (3′ UTR) of target mRNAs via base pairing for induction mRNA degradation and translational inhibition. In few cases, the interaction of miRNAs with the 5′ UTR, coding sequence and gene promoters, have also been found. These abnormal interactions instead of suppressing miRNAs can induce translation or regulate transcription. The base-paired interaction of miRNAs along with target genes is rapid, dynamic, and related to many unique factors including their subcellular location, their abundance distribution pattern, selection of target mRNAs and the affinity of miRNA-mRNA base pairing. The miRNAs secrete inside extracellular fluids and transport to target cells via vesicles including exosomes, or by binding to Argonaute proteins. Extracellular miRNAs behave as chemical messengers relating to cell-cell communication. An updated description of canonical and non-canonical miRNA biogenesis pathways, long-standing transport and circulation, different mechanisms underlying miRNA-mediated gene regulations are described. The updated knowledge of the dynamics of miRNA action, the secretion, transfer, and uptake of extracellular miRNAs are discussed.

These small RNAs play a significant role in the modulation of an array of physiological and pathological processes ranging from embryonic development to neoplastic progression. Moreover, the interconnection between miRNAs and their enhancer involved in the highly sensitive cell type mediated expression patterns and the evolution of miRNA–target relationship. The function opens a new light for understanding the changes of sensitive miRNA expression in different human diseases. The versatile miRNA signatures of many inherited, metabolic, infectious, and neoplastic diseases shed a new dimension to pathogenesis and open new sources for reliable biomarkers. The possible changes in miRNA function and novel delivery methods must modify and activate as novel approaches for therapeutic candidates in many diseases.

Introduction

The innovation of lin-4 at the early age of miRNA research speculates that it was a worm-specific unique curiosity but the subsequent additional discovery of conserved let-7 miRNA[1, 2] provide a substantial greater notice to researcher imaginary interpretations. They imagine that there may be more similar to small regulatory RNA that will exist in the future in the animal kingdom. Since then, numerous miRNAs have been isolated in humans at a greater pace and distribution manner that exerts a huge impact on shaping transcriptomes and proteomes along with extensive eukaryotic organisms.[3-5] Conventional wisdom holds the notion that miRNAs have a master controller for specific gene modulations typically using base pairing to the 30nt translated regions (30 UTRs) of messenger RNAs (mRNAs). They also involve the reduction of the targeted message, either by transcript degradation, translational inhibition, or both.[6] Interestingly, translational inhibition alone is not the single rule for miRNA variable features, but it is a frequent exception.[7, 8] In fact, miRNAs can also be associated with multiple regions of a protein-coding exon.[9] Based on total or partial complementary base pairing a single miRNA can bind on average a hundred mRNAs. Estimating that over 60-70% of human protein-coding genes carry predicted 30 UTRs miRNA-binding sites.[10] Close to 1900 miRNA genes have been isolated in the human genome alone that target potentially nearly 13% of all human protein-coding genes. Thus, miRNAs are the most abundant fine tuner for gene regulation in humans. They are involved in a broad range of physiological and developmental machineries with a greater emphasis on human diseases.[11-13] A reasonable proportion of existing miRNAs are limited to control tissue-specific and developmental stage-specific gene expression altered during different diseases. Especially, the ultimate miRNA biogenesis rate can be guided in principle, at any step of multi-stepped synthesis from transcription to mature miRNA turn over, post-transcriptional regulation and activity.[14, 15] In general, miRNA biogenesis terms two subsequent and sequential processing reactions: one in the nucleus and other in the cytoplasm. They occur the processing from pri-miRNA transcript to mature miRNA in a well-directed narrow streamline. In the nucleus, the pri-miRNA turns into mature pre-miRNA via a multiprotein complex called the Microprocessor. In the cytoplasm, the pre-miRNA further cleaves into a 22nt long miRNA mature duplex using Dicer, another RNase III enzyme, in collaboration with the dsRBD.

The main critical miRNA contribution in the human body is to control the genes [16] by fine-tuning for altering mRNA expression and also by inhibiting transcription and translation via canonical and non-canonical mechanisms.[17] The canonical process defines that the miRISC complex along with miRNA guided strand is conducting its action by simple complementary binding to the selected target mRNA using its 3′-untranslated region (3′-UTR).[18] This process appears to accommodate the seed sequence of the miRNA, particularly the first 2-7 nucleotides from the 5′ end. It is responsible for mRNA de-adenylation, translation inhibition and ultimately degradation. [19-21] In fact, nearly 60% of the interactions between mRNA and the miRISC complex in human cells are non-canonical.[22] In this separate method their chains are always partially complementary to each other.[23] As a result, a single miRNA could potentially target hundreds of mRNAs. Therefore, one mRNA can bind with multiple miRNA sequences converting into a wide spectrum of biological processes.[18]

The miRNA is also critical for intercellular signaling. Despite most miRNAs are inside the cell how remaining miRNAs transport outside the cell those are usually present in extracellular circulatory fluids?[24-28] These so-called circulatory miRNAs are involved in tissue damage, apoptosis, and necrosis [17] or by their active passage in streamline in micro-vesicles exosomes or conjugated to a distinct protein.[29, 30] It is speculated that the existence of exogenous miRNA in the human blood mainly originated from,[31, 32] bacteria, food and fungi primarily from the gut for their pathological consideration. Approximately 10% of the circulating miRNAs are exosome and the remaining 90% form complexes with multiple proteins including Argonaute 2 (Ago2), nucleophosmin 1 (NPM 1,) and high-density lipoprotein (HDL).[30, 33, 34] This type of complex formation is necessary to protect against RNase degradation that is available in body fluids.[35] The complexes ultimately reach the extracellular space guided by the prominent protein-specific transporters.[26] Recently, exponential progress of origin, functions and uses of miRNA in research and clinical practice is noticed. The amounts of many circulating miRNAs in cardiac muscle contractility, inflammation, and angiogenesis are involved directly by the miRNA intensity. Therefore, they contribute a constructive role in guiding the physiological cardiac adaptation to exercise.[36-38]

Biogenesis of microRNA

Outlines of Biogenesis

The microRNA precursor genes are widely distributed in the genome. They are classified into two broad categories: 1) The microRNAs synthesize from precursor overlapping introns of protein-coding transcripts and 2) they are synthesized from usual exons, that are simplistically associated with miRNA maturation.[39] During a complex cluster formation, miRNA genes are co-expressed polycistronically from a single repeat unit.[40, 41] Nearly 50% of miRNAs are originated from nonprotein coding transcripts.[42] The remaining miRNAs are mostly synthesized from the introns of coding genes. They are generally co-expressed with their host genes and are processed separately.[42] A significant overhaul of the current understanding of miRNA genesis is observed in this rapidly evolving field. (**Figure 1**) Various mature miRNAs can synthesize from diverse locations within the genome.

Canonical pathway

The miRNAs are processed frequently by a canonical biogenesis pathway. As described earlier pri-miRNAs are synthesized from their DNA sequences and then completed multiple steps for converting into pre-miRNAs via microprocessor complex. The multiple steps of canonical biogenesis pathway are transcription of primary miRNA transcripts, nuclear processing by Drosha, nuclei-cytoplasmic export, cytoplasmic processing by Dicer and formation of RNA-induced silencing complex (RISC) with Argonaute (Ago) proteins. (**Figure 2**)[43-46] The biogenesis of canonical miRNAs begins with the generation of long primary miRNA transcripts (pri-miRNAs) mainly synthesizing by RNA polymerase II. Animal miRNAs are encoded as individual miRNA genes (mono-cistronic), as miRNA clusters (polycistronic), or in introns of protein-coding genes (intronic).[44, 47] Pri-miRNAs are processed to hairpin-structured RNAs termed as precursor miRNAs (pre-miRNAs) by the Drosha complex in the nucleus by forming 2 nt 3′ overhang on pre-miRNA. The Drosha complex consists of the RNase III enzyme Drosha and RNA binding protein DiGeorge Syndrome Critical Region 8 (DGCR8) and a ribonuclease III enzyme Drosha.[74] DGCR8 identifies an N6-methyladenylated GGAC and other motifs along with the pri-miRNA. Following Drosha processing the pre-miRNAs are exported to the cytoplasm by exportin-5 (XPO5) and finally converted into mature RNA by the RNase III endonuclease Dicer cleavage.[74, 77] This conversion removes the terminal loop formation to a mature miRNA duplex.[78] Once they enter cytoplasm, Dicer in the cytoplasm liberating a 21–24nt mature miRNA duplex with help from dsRNA-binding proteins, protein kinase RNA activator and transactivation response RNA. Several Dicer-associated proteins, iIntc.lJu. Mdionl. gScTi. R20B2P0, 2P1A, xCFTO, RanPEdEAR RDEAVRIE1W are involved in the process The direction of two miRNA strands designates the mature miRNA name. The 5p strand determines from the 5′ end of the pre-miRNA hairpin, while the 3p strand refers from the 3′ end. Two strands of the mature miRNA duplex are prerequisites for binding with the Argonaute (AGO) family proteins (AGO1-4 in humans) in an ATP-dependent manner.[79] The proportion of AGO-loaded 5p or 3p strand in a miRNA depends mainly on the cell type or cellular environment encompassing from near equal proportions to predominantly one or the other.[80] The selection of the 5p or 3p strand recognizes by the thermodynamic stability at the 5′ ends of the miRNA duplex or a 5′ U at nucleotide position 1.[81] During normal processing the strand with lower 5′ stability or 5′ uracil is loaded into AGO selectively and is referred as a guided strand. The unloaded strand is deemed as the passenger strand that will be unwound from the guide strand be the help of multiple mechanisms based on the degree of complementarity. The passenger strands with no mismatches are diced by AGO2 and degrade by a cellular machinery that generates a strong strand bias. On the other hand, miRNA duplexes with central mismatches or non-AGO2 loaded miRNA are unwound frequently and degrades.[82]

Multiple experiments claim a better understanding of how Drosha/DGCR8 convert from pri-miRNAs to pre-miRNAs. (**Figure 2**) In this analysis of pri-miRNA variants, several sequence features of pri-miRNAs are important for efficient pri-miRNA processing. It includes a UG motif at the base of the pri-miRNA hairpin, a UGU(GUG) motif. This motif with the help of (GW182) proteins shortens the poly(A) tails of mRNAs. It also represses translation and destabilizes mRNAs. [45, 48] On the other hand, target recognition and selection of miRNAs typically depends on complementary base pairing between seed sequence of the different factors for example thermodynamic asymmetry of the duplex and base-pairing stability at the 5' end is responsible for 2 miRNA strand separation. For processing, miRNA forms a unique instrumental complex using micro-ribonuclear proteins called RNA-induced silencing complex (RISC). (**Figure 3E**)[56] The guide strand associated with multiple RNA binding proteins and trinucleotide repeat-containing gene 6A (TNRC6A) along with catalytic Argonaute (AGO) proteins form this protein complex (miRNP) among two stands in miRNA duplex. The most unstable base pairing at the 5' end usually functions as the guide strand, whereas the stable base pairing at the 5' end is designated as the passenger or miR* strand that is normally degraded or often even conjugate with AGO proteins in functional miRNAs.[57, 58] The guide strand is involved in targeting mRNA via base pair complementarity required for translational repression. (**Figure 3F**) For conducting miRNAs associated base pairing with target miRNAs the Ago2 proteins along with cytoplasmic bodies called GW/P-bodies known as processing bodies to form a complex for degradation or translational repression. (**Figure 3G**)[59]

Non-canonical pathway

On the other hand, it is found that miRNA biogenesis is microprocessor independent. Multiple Drosha- or Dicer-independent biogenesis routes are required in different classes of non-canonical pre–miRNA-like hairpins called mirtrons.[60-62] They are derived from spliced and disbranched short hairpin introns, some small nucleolar RNAs (snoRNAs) and endogenous short hairpin RNAs (shRNAs), tailed mirtrons,[63] tRNA fragments,[64, 65] snoRNA fragments,[66-69] 5′-capped miRNA precursor-derived 3p miRNAs,[70] Dicer-independent erythrocyte-specific miR-451,[71,72] and transcription start site miRNAs (TSS-miRNAs),[73] etc.

Multiple non-canonical miRNA biogenesis pathways have been identified recently. (**Figure 2**) The protein combinations in this pathway are different entirely from that in the canonical pathway including mainly Drosha, Dicer, exportin 5, and AGO2, etc. The non-canonical biogenesis is independent of the Drosha/DGCR8 combination and Dicer alone. The pre-miRNAs like mirtrons are derived by the Drosha/DGCR8-independent manner but resemble Dicer substrates. The mirtrons are isolated from the introns of mRNA during splicing.[83, 84] Moreover, the nascent RNAs of the 7-methylguanosine (m^7G)-capped pre-miRNA are directly transported from nucleus to cytoplasm via exportin 1 without Drosha cleavage. Indeed, a strong 3p strand biasness is found for the presence of m^7G cap. It prevents 5p strand loading into Argonaute.[85] In a separate instance, sometimes Dicer-independent miRNAs are involved in Drosha from endogenous short hairpin RNA (shRNA) transcripts.[86] The pre-miRNAs require AGO-2 related completion during their maturation inside the cytoplasm. Because they do not provide sufficient length for Dicer-substrate executions.[86] Ultimately. Noncanonical process, in turn, involves loading of the entire pre-miRNA into AGO-2 and AGO-2-dependent slicing of the 3p strand. Thus only 3′-5′ trimming miRNAs (nucleotides 2–7) and mature transcripts.[49, 50] The short recognizable sequences of the seed strand of a miRNA helps them to target hundreds of target mRNAs. However, target inhibition is regulated by multiple RNA-binding proteins (RBPs) and other RNA molecules including circular RNAs, RNA sponges and competing endogenous RNAs (ceRNAs).[51-55]

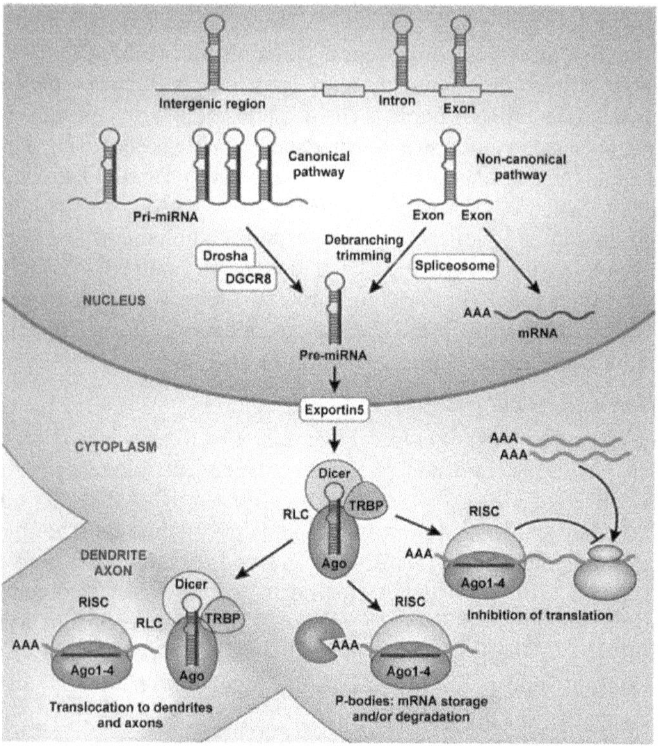

Figure-1. Detailed scheme of miRNA biosynthesis. The canonical miRNA pathway generates pri-miRNA transcripts from miRNA genes residing in exonic, intronic, or intergenic regions, followed by Drosha/DGCR8 processing of the pri-miRNA transcripts into pre-miRNAs. On the other hand, Intronic pre-miRNA hairpins derived from the noncanonical mirtron pathway are generated by splicing, debranching, and trimming of short introns without Drosha processing. Pre-miRNAs either produced by the canonical or noncanonical pathways are released and are exported from the nucleus to cytoplasm using Exportin 5. Subsequently, Dicer cleavage within the RISC loading complex (RLC) are occurred by unwinding of the miRNA/miRNA* duplex using Argonaute, and TRBP-dependent loading into the RNA-induced silencing complex (RISC). Different miRNAs in RISC target ideal mRNA via inhibition of translation and/or mRNA degradation within P-bodies in the cytosol. The exportation of RLC and RISC inside the dendritic and axonal compartments occurs via a still unknown and unconventional mechanism. (Loosely Adapted, Redraw and Modify using skeleton from O'Carroll[1] and Schaefer 2013)

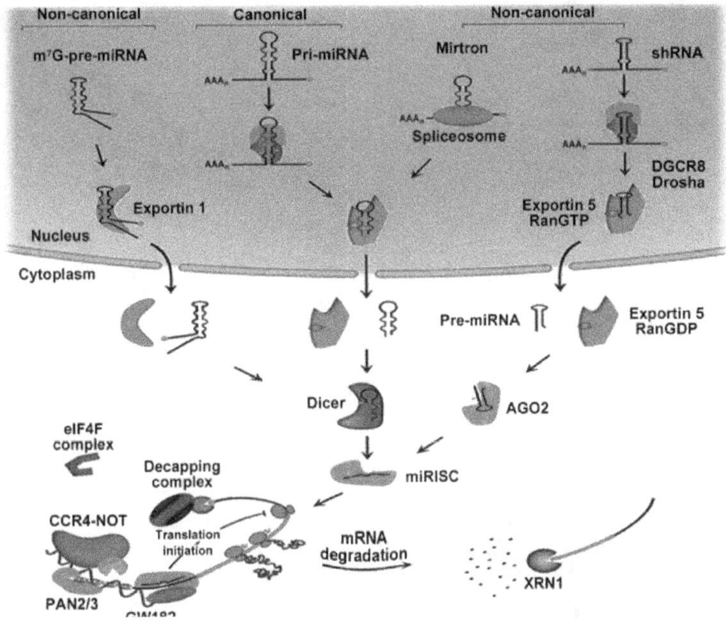

Figure-2. The biogenesis of MicroRNA and their mechanism of function. The pri-miRNA transcript is the initial product of Canonical miRNA biogenesis. The Drosha and DiGeorge Syndrome Critical Region 8 (DGCR8) containing microprocessor complex slices the pri-miRNA for transforming the precursor-miRNA (pre-miRNA). The pre-miRNA is further transported from nucleus to the cytoplasm using Exportin 5/RanGTP-mediated manner and processed ultimately the mature miRNA duplex production. Finally, either the 5p or 3p strands of the mature miRNA duplex is conjugated to the Argonaute (AGO) family proteins for producing a miRNA-induced silencing complex (miRISC). In a separate process, in the non-canonical pathways, small hairpin RNA (shRNA) is initially diced by the microprocessor complex and migrated from nucleus to the cytoplasm by the help of Exportin 5/RanGTP. They are further processed using AGO2-mediated, but Dicer-independent cleavage. Later Mirtrons and 7-methylguanine capped (m7G)-pre-miRNA based on Dicer for completion of their cytoplasmic maturation, but they differ in their nucleocytoplasmic shuttling. Mirtrons are transported via Exportin 5/RanGTP while m7G-pre-miRNA are exported via Exportin1. Ultimately all pathways proceed to a functional miRISC complex. In most cases, miRISC conjugates to target mRNAs for translational silencing induction, most possibly via eIF4F complex interference. Next, GW182 family proteins conjugate to Argonaute for recruiting the poly(A)-deadenylases PAN2/3 and CCR4-NOT. PAN2/3 deadenylate initially while the CCR4-NOT complex completes the process, that removes m7G cap on target mRNA via a specified decapping complex. Decapped mRNA may then undergo 5′−3′ degradation using exoribonuclease XRN1. (Loosely Adapted, Redraw and Modify using skeleton from Jacob O'Brien *et al.*, 2018)

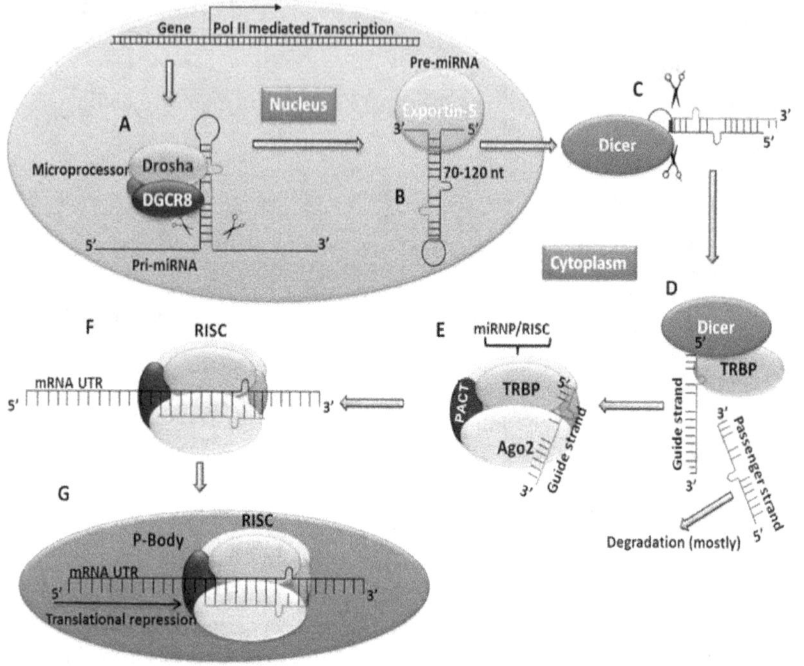

Figure-3. Canonical microRNA (miRNA) biogenesis pathway: miRNAs are initially synthesized as long, variable-length hairpin RNA substrates called primary miRNAs (pri-miRNAs) directly off the DNA template in the nucleus by RNA polymerase II (**A**). In the nucleus, these pri-miRNAs are further processed by the Microprocessor, a large complex that includes the RNase type III enzyme called Drosha and the RNA binding protein DGCR8, into *70- to 120-nucleotide-long premature (pre) hairpin precursor forms termed as pre-miRNAs (A). The pre-miRNAs are then migrated from nucleus to the cytoplasm by exportin 5 (**B**) in the cytoplasm pre-miRNA loop region is chopped by the RNAse type III enzyme called Dicer into *18- to 23-nucleotide-long mature miRNAs (**C**). A cellular protein called transactivation response RNA binding protein (TRBP) (**D**) promotes Dicer-miRNA complex entry into the RNA-induced silencing complex (RISC) that carries Argonaute 2 (Ago2), protein kinase RNA activator (PACT), trinucleotide repeat-containing gene 6A (TNRC6A), and other RNA binding proteins (**E**). After conjugating to the mature miRNAs into the RNA-induced silencing complex (RISC), the ''passenger'' strand is degenerated, and the ''guide'' strand is guided to the 30 UTR of the target messenger RNA (mRNA) to either destruction (in case of perfect base complementarity) or bring about translational inhibition (in case of multiple sequence mismatches between the target mRNA and the miRNA) (**F**). 'Numerous evidences', suggest that transcripts linked to miRNA-mediated Ago2 are channeled into P bodies structures that results in translational repression (**G**). (Loosely Adapted, Redraw and Modify using skeleton from Bhaskaran and Mohan 2013)

Biogenesis Regulation

Control of miRNA Maturation

Multiple and/or single miRNA biogenesis is regulated stringently and in certain cases is executed by the feedback regulation. Certain miRNAs control few transcription factors that in turn control miRNA expression by forming a double-negative feedback loop.[14,15, 88, 89] This coordinated and sequential expression restricts cell-type-specific miRNAs misexpression and to promote the underlying gene expression programs during development robustly. Multiple steps of the biogenesis pathway are specifically involved during differentiation and tumor progression processes. Several pri-miRNAs are expressed in the early development phases but are not processed into mature miRNAs efficiently.[90] Similarly, reduction of multiple transition processes is involved in widespread miRNA downregulation in numerous human tumors[13, 90, 91] These examples provide a strong sign for post-transcriptional mechanisms of miRNA biogenesis. Independent regulation is also required for individual miRNA processing. Several miRNAs form clusters inside the genome and synthesize as long polycistronic primary transcripts. Interestingly one specific miRNA of a cluster can process independently from the other miRNAs in the cluster.[92] This separate signature of each miRNA is processed post-transcriptionally by numerous factors. They regulate many points during the biogenesis pathway by different specificity and functions.

Limited processing at the "Cropping" level

Two microRNA features, activity, and specificity of Microprocessor-based pri-miRNA cleavage are important for core regulations. Occasionally the specific pri-miRNAs editing from Adenosine-to-inosine (A-to-I) by ADAR (adenosine deaminase acting on RNA) enzymes inhibits Drosha mediated effective processing. In some cases, A-to-I editing guides RNA molecules for degrading Tudor-SN (Tudor staphylococcal nuclease homolog).[93] The miRNA seed region editing allows producing altered mature miRNAs that identify a wide range of target sets relative to their original miRNA counterparts.[94] These findings reveal that RNA editing is an important and essential mechanism for converting biogenesis and miRNA activity generation. The Microprocessor complex suppresses a specific set of miRNAs level due to significant reduction of the RNA helicases p68 or p72.[95] indicating a potential role for p68 and p72 in controlling Drosha cleavage of a subset of miRNAs. p68 and p72 interact with multiple protein members.[15] During DNA damage the tumor suppressor protein p53 is found in both Drosha and p68 complexes and induces Drosha processing in a particular miRNA set including miR-16-1 and miR-143.[96] Conversely, bone morphogenetic protein 4 (BMP4) or transforming growth factor-beta (TGF−β) induces SMAD binding to p68 in human vascular smooth muscle cells that enhances Drosha processing of pri-miR-21 and pri-miR-199a.[97] Moreover, Era based E2 interacts with Drosha in a p68-/p72-dependent manner for induction separation of pri-miRNAs from the Microprocessor complex. In some cases, it alters a conformational change of the Microprocessor along with subsequent reduction before pri-miRNAs interactions. Thus, it blocks Drosha based processing.[98] These findings suggest that signaling cascades control the biogenesis of individual miRNAs strictly by interacting with p68 and/or p72 along with accessory p53 and SMAD protein factors. It either alters the specificity and activity of Drosha/DGCR8 processing particularly via a conformational change of the Microprocessor or pri-miRNAs or by shifting the depleted RNA-binding proteins. (**Figure 4A**) It will recognize different factors that allow the specificity of the Microprocessor complex for identifying a specific set of pri-miRNAs and factors that lead specificity for choosing the Microprocessor complex on another set of pri-miRNAs among different Microprocessor complexes. This recurring theme is important for many signaling pathways that control Microprocessor complexes by changing the activity of the p68-/p72-interacting protein.

Regulation during Cropping

The miRNA loop miR-18a is involved in polycistronic pri-miR-17-92 transcript processing that carries five additional miRNAs. An mRNA splicing regulator hnRNP A1 induces independent maturation of the only miR18a from the miR-17-92 cluster. Only Pri-miR-18a carries two similar consensus regions of hnRNP A1-binding site, UAGGGA/U inside its terminal loop and stem.[92, 99] A close and specific interaction of hnRNP A1 with the pri-miR-18a hairpin terminal loop and stem stimulates a structural rearrangement of the hairpin for yielding a favorable Drosha/DGCR8 binding and cleavage site. (**Figure 4A**) It indicates that hnRNP A1 functions as a chaperone for the recognition and cropping of specific pri-miRNAs using Drosha/DGCR8. Nearly14% of human miRNAs have conserved strongly in loop sequences demonstrating that processing regulation using hnRNPs and other nuclear RNA-binding proteins are well extended beyond miR-18a.[99] Besides, the KH-type splicing regulator protein (KSRP, also known as KHSRP) identifies G-rich regions including a GGG triplet that present in the terminal loops of a section of pri-miRNAs to DGCR8 based processing.[100] The positioning and/or recruitment of the miRNA precursor processing complexes via KSRP binding leads to protein-protein interactions. The KSRP is accommodated with the terminal loop of the target miRNA precursors during nucleo-cytoplasmic transit and stimulates the processing of the target pre-miRNAs via Dicer in the cytoplasm. This finding reveals that a unique recognition of the terminal loop using RNA-binding proteins is an important factor for allowing reduction of miRNA processing relative to individual miRNAs. Special sequences of the terminal loop regions of miRNA precursors along with the binding of RNA-binding proteins potentially either compete for binding or induce the assembly of processing factors. Many signaling pathways alter the activity of a few RNA-binding proteins. The RNA-binding motifs are associated with conserved protein-protein interaction domains and consensus phosphorylation motifs.[15] Alteration of RNA-binding activity along with protein partners and subcellular localization of RNA-binding proteins in response to extracellular signals control individual miRNAs biogenesis.

Global-Regulation

The accumulation of global miRNA along with other Nuclear RNA Processing events increase cell density. The accumulation induces Drosha processing activity without altering Drosha or DGCR8 protein levels.[101] This indicates that miRNA biogenesis is related to cell proliferation and processing activity that controls the same stoichiometric amount of the complex using changing the single Microprocessor complex turnover rate. Arsenate resistance protein 2 (ARS2) that is downregulated during cell quiescence and stimulates during proliferation, is an essential factor for the nuclear RNA cap-binding complex (CBC). It is associated with and induces the activity and fidelity of the Microprocessor complex.[102. 103] Moreover, the nuclear CBC along with ARS2 leads to conjugate prior-miRNA processing to other RNA processing events. The CBC/ARS2 adds the Microprocessor to 50-capped nascent pri-miRNA transcripts. It also either reduces the vulnerability of pri-miRNAs or induces the cleavage activity and specificity of Drosha/DGCR8. The recognition of Drosha/DGCR8 allows rapid cytoplasmic transport of pri-miRNA transcripts and cleavage in the nucleus. Therefore, CBC and ARS2 may lead to a reduction of nuclear transport of pri-miRNA transcripts thereby increasing Microprocessor recognition and processing of target hairpins. Moreover, nearly 40% of miRNAs are synthesized from the introns of genes,[104] and RNA Pol II transcripts from pre-mRNAs and subsequent product pri-miRNAs. This indicates that the physical connection between the Microprocessor and the pre-mRNA splicing machinery including SE like hnRNPA1 and KSRP leads to pre-mRNA splicing. Besides, the CBC not only conjugates from export factors to nascent pre-mRNAs but also leads to splicing factors[105] directly. Therefore, it plays an integral and pivotal role in coordinated pre-mRNA and pri-miRNA processing. Once processing is completed, pre-miRNAs are exported to the cytoplasm by rate-limiting Exportin-5 enzymes in the next step[106, 107] by which pre-miRNA export is regulated to a reasonable extent.

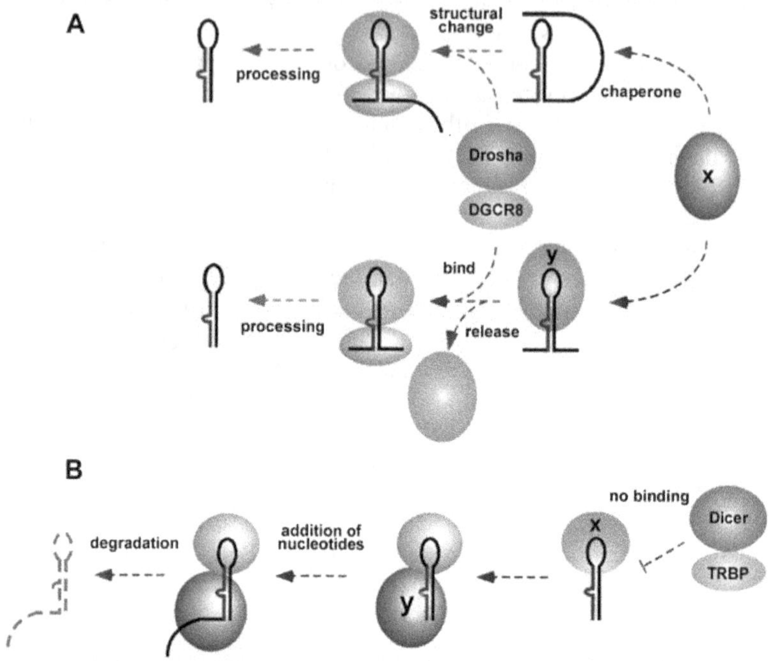

Figure-4. Posttranscriptional regulations of miRNA Biogenesis Mediated by RNA-Binding Proteins The conjugation of RNA-binding proteins to miRNA precursors, specially to the stem-loop structures within them, either prevents, inhibits, or activates additional processing by competing with or linking to miRNA processing complexes. In (**A**), ideal binding of RNA-binding proteins (X in the figure) to miRNA precursors activates either a specific conformational change (chaperone activity, e.g., hnRNP A1) or a dislocation of inhibitory RNA-binding proteins (Y in the figure). (**B**) Multiple RNA-binding proteins (X in the figure) also are linking many auxiliary factors (Y in the figure) on special miRNA precursors, thereby controlling processing, stability and subcellular location of miRNAs. (Loosely Adapted, Redraw and Modify using skeleton from Siomi,and Siomi 2010).

Autoregulation of miRNA Cropping Machinery

Drosha and DGCR8 are also restricted strongly in an intricate feedback circuit. It leads to the homeostatic controls of miRNA biogenesis.[108] Invariably DGCR8 makes Drosha stable using a direct interaction. In turn, Drosha cleaves two hairpin structures in the 50 UTR and DGCR8 mRNA coding sequences. It folds the pri-miRNA structure leading to miRNA degradation. These findings indicate a double-negative feedback loop relying upon a solid bonding of the core Microprocessor components. Two enzymes perform two separate works for example DICER-LIKE1 (DCL1) pre-mRNA carries the pre-miR838 hairpin, while DCL-mediated processing of DCL1 pre-mRNA liberates pre-miR838 for producing nonproductive fragments of DCL1 pre-mRNA. Drosha stabilization is performed by changing it associated partner DGCR8 or using transforming pri-miRNA-like hairpins in the DGCR8 mRNA via associating an RNA-binding protein. This event modifies global miRNA biogenesis during the cell proliferative or in extracellular stimuli. In certain issues Microprocessor with the help of DGCR8 targets directly and destabilizes coding.[109] This interaction imagines that a series of mRNAs along with pri-miRNA-like hairpins forms a network that balances global miRNA biogenesis using attenuators. The attenuators are also involved in competing pri-miRNAs with Microprocessor binding.

Regulation of Dicer activity

The low abundance of pre-miRNA levels relative to pri- and mature miRNA allows the efficient function of Dicer in Pre-miRNA processing demonstrating that Dicer dicing is a speedy substrate turnover rather than regulation. However, recently it shows an opposite interpretation. It claims that the Dicer expression and activity functions a pivotal and important role at the controlling point in global and individual miRNA biogenesis. The level of Dicer activity determines and reduces pre-miRNAs nuclear retention of pre-miRNAs and thrusting a quick export to the cytoplasm.[110] Moreover, an unidentified specific inhibitor that is involved in pre-miRNA binding with Dicer is reduced significantly.[111] In fact, Dicer level controls the miRNA synthesis in an autoregulatory negative feedback loop. For example, let-7 targeting sites inside the Dicer coding sequence show that let-7 miRNA interacts with Dicer directly for controlling the total rate of miRNA biogenesis.[112]

The Importance of the miRNA Loop in Biogenesis

Nine members of the human *let-7* miRNA family evolve from 12 precursor transcripts. The developmental distributed mature *let-7* exhibits post-transcriptional control for miRNA biogenesis.[113] The amount of an RNA-binding protein LIN-28 influences pluripotency and relates inversely with mature *let-7* expression during development.[114] Though LIN-28 prevents mature *let-7* accumulation by reducing both Drosha and Dicer activities,[115-118] LIN-28 can increase the pre-let-7miRNA 3′ terminal uridylation of the embryonic stem (ES) cells for blocking Dicer cleavage and *let-7* maturation.[115] LIN-28 adds multiple features including a noncanonical poly (A) polymerase (PAP), terminal uridylyl transferase 4 (TUT4), also referred to as ZCCHC11 in mammals to pre-*let-7* identifying a tetra-nucleotide sequence motif (GGAG) in the terminal loop. (**Figure 4B**)[119, 120] The TUT4 also recruits a uridine (U) tail in a template-independent manner to the 3′ end of pre-miRNA. Dicer does not cleave the long 3′ extended hairpin RNAs. On the other hand, Uridylyl groups are added from 3′-to-5′ exonucleases. It suggests that 3′ terminal uridylation promotes degradation of the U tail of pre-let-7. An additional miRNA including the GGAG motif in the terminal loop also responsible for LIN-28-dependent uridylation.[120] This indicates that the capacity of LIN-28 and TUT4 increases uridylation for blocking pre-miRNA processing is a broad spectrum. LIN 28,_and other specific factors may also help TUT4 (and other TUTases) for identifying the distinct pre-miRNAs. These issues demonstrate that the terminal loop serves as an instrumental structure or functions launching pad for miRNA processing "activators" and (for example hnRNP A1 and KSRP) as well as "repressors'"(for example, LIN-28) function in a coordinated fashion for transforming proliferating, apoptotic or differentiating cues during altering the miRNA expression. (**Figure 4B**) The LIN-28 influences let 7 pre-miRNA uridylation using PUP-2 summarizing that LIN-28 and let-7 use an ancient regulatory switch.[121]

Signal-Mediated Regulation of Dicer Activity

The TRBP can directly control the amount of Dicer protein. Deleted mutations of TRBP reduces TRBP protein expression in rapid and hereditary carcinomas. They are conjugated both with miRNAs processing defects and Dicer protein destabilization.[122] Two kinase enzymes in the mitogen-activated protein kinase (MAPK) and extracellular signal-regulated kinase_(ERK) pathway stimulate cell growth, phosphorylates TRBP, and induces miRNA production with the help of the Dicer-TRBP complex stabilization.[123]

The ERK signaling pathway is an interconnecting step of phosphorylation from a MAPK kinase to MAPK ERK. The presence of TRBP causes phosphorylated ERK1/2 and its phosphorylation involves MKK1 and ERK2. Phosphorylation of TRBP is restricted by the MKK inhibitor U0126, and concomitantly reduces miRNA growth. The reduction of miRNA growth stimulates U0126 treated cancer cell lines. These findings demonstrate a tight miRNA-controlled program able to respond to mitogenic signals positively and reduction of these signals negatively. This event claims that the amount of Dicer-TRBP complex requires for pre-miRNA processing preferentially. This complex is responsible for or reduction of growth-suppressing miRNAs in specific cancer cells. However, the determined level of the DicerTRBP complex for selective target pre-miRNA is not defined elaborately, but it uses implicates the important therapeutic significance.

miRNA Accumulation in miRNA Turnover

Maintenance of the ideal steady-state of miRNA levels is critical for normal development. Both overexpression and/or under expression of miRNAs are mainly related to different human diseases. The steady-state levels of miRNAs are the net result for opposite actions of miRNA biogenesis (transcription and processing) and degradation (Inhibition). The miRNAs exist inside the mi-RISC complexes thus both the 5′ and 3′ ends are possibly bound to Argonaute [124] that causes resistance to exonucleases. In some instances, cells somehow dislodge miRNAs from Argonaute apparently for exposing the miRNAs to XRN-2 for degradation. The miRNA degradation using XRN-2 prevents miRNA addition to target RNA. It infers that target may change the stability of mature miRNAs individually. These reports explain that miRNA blocks detrimental overexpression of miRNAs that requires for degradation of functional miRNA homeostasis. A coordinated complex between miRNA and their target allows miRISC reprogramming or turnover when the presence of potential targets is less.

The stability of miR-122 mature miRNAs is controlled by single important components in mammalian liver cells the cytoplasmic poly(A) polymerase GLD2 (also known as TUTase2) particularly mono adenylates. [125] Besides, non-templated adenine recruitment is found in several types of animal miRNAs.[126-127] This predicts that despite the presence of same sized miRNA, miRNAs need additional sequence elements for regulating their behavior via specific recognition by RNA-binding protein cofactors and ultimately contribute to the efficiency of miRNA turnover.[128, 129]

The Argonaut protein governs the presence of numerous mature miRNAs. An enhanced amount of Argonaut proteins in the mammalian cells increase mature miRNAs level proportionately. [130] Therefore, a direct binding between Argonaut proteins and miRNAs must exist. It indicates that the presence of Argonaut proteins conducts miRNA stabilization. This coordinated bonding opens a feedback mechanism for titrating mature miRNA with Argonaut proteins to form functional, stoichiometric mi-RISCs. Therefore, the total amount of Argonaut proteins inside the cell requires for global miRNA turnover. One important protein is the type I collage nprolyl-4-hydroxylase (C-P4H[I]). Human AGO2 interacts with these enzymes and is hydroxylated by the help of C-P4H(I) for increasing AGO2 stability.[131] C-P4H(I) stimulates many miRNAs that are overexpressed by different factors including hypoxia and TGF−β, factors, etc. The endosomal pathway regulates the formation and turnover of mi-RISC complexes. These pathway factors require endosome formation for controlling miRNA loading into RISC.[132, 133] Therefore, turnover inside the endosome is an important criterion for mature miRNA net accumulation.

The regulation of final mature miRNA requires coordination of separate machinery and their ultimate net effects. The determination of multiple factors that control both global and individual miRNA biogenesis. Their dramatic turnover imposes a crucial challenge in near future. MiRNAs are an integral part of large multimolecular protein complexes, mi-RISC, and their aligned multiple RNA-binding proteins. It suggests as a future challenge how RNA-binding proteins are responsible for the outcome of miRNA biogenesis. Moreover, how these agents require specific expression, function, and modulation of individual miRNAs. Thus, the development of novel biochemical approaches is critical essentially for identifying complexes including miRNA, their precursors, and associated proteins *in vivo*.

Recently, the development of high-throughput sequencing along with Chip analyzes A-mRNA interactions.[134] It is useful for the identification of different miRNA biogenesis regulators. Further studies unravel multiple mechanisms and the interactions for coordinating the limited expression of many miRNAs at posttranscriptional levels. Such understanding will not only increase our knowledge of how they modulate in pathological conditions but also provide exciting opportunities for manipulating multiple aspects of the miRNA machinery for therapeutic innovation.

Circulation

It is a custom that miRNAs release to the extracellular fluids that can be used for important and efficient biomarkers of all essential human diseases. The extracellular miRNAs can be released based on selective target cells and function as autocrine, paracrine, and/or endocrine regulators for regulating versatile cellular activities like hormone-like activities.[135]

miRNA

Many miRNAs are distributed inside different body fluids.[136-138] They are participated in cell-to-cell communication and transport from one cell and utilize by another cell similar to endocrine hormones.[139-142] The circulatory miRNAs highly stable, prevents RNase degradation in a greater pH and activities in via de range of temperatures relative to conventional RNA molecules.[143, 144] They are also protected against cytoplasmic ribonucleases by employing a concrete packaging inside the lipid vesicles for example, micro-vesicles, and exosomes. In general, miRNAs form complexes with many RNA binding proteins.[139, 145, 146] The expression of miRNAs isolated from serum and plasma used as protein markers for almost all diseases.

The serum miRNAs consist of especially the tumor-associated miRNAs, miR-155, miR-210, which are exported from diffuse large B-cell lymphoma (DLBCL). The DLBCL patients along with high miR-21 in serum levels increase relapse-free survival but not overall survival. A defined analysis of serum miRNAs from lung cancer, colorectal cancer, and diabetes patients display unique miRNA profiles.[143] The oral squamous carcinoma patients have an overexpressed miR-31in their plasma but reduce markedly in their saliva upon the removal of oral tumor.[147] The plasma miR-132 of rheumatoid arthritis and osteoarthritis patients are increased significantly.[148] In summary, the expression of specific circulating miRNAs are correlated highly with their expression in designated tissues particularly in a tumor developing tissues.[149] These properties demonstrate that circulatory miRNAs are the reliable source for potential biomarkers of almost all human diseases. Nonetheless, occasionally single miRNA and a small group of miRNA families cannot use an effective biomarker in certain malignancies, Therefore, additional features should be executed for discovery and validation for single mRNA panels for the use as effective markers in complex malignances.[149] Several key steps including RNA extraction, normalization, quantitation, and analysis of serum miRNA profiles are required in cell-free biological fluids for the reliable source for markers in different diseases.

In Biological Fluids

The extracellular/circulating miRNAs are extracted from different biological fluids for example plasma and serum,[144,150] cerebrospinal fluid,[151] saliva,[152] breast milk,[153] urine, tears, colostrum, peritoneal fluid, bronchial lavage, seminal fluid,[154] and ovarian follicular fluid.[155] The extracellular miRNAs are stable that are started degradation at room temperature for up to 4 days and even in abnormal dreadful conditions including boiling, multiple freeze-thaw cycles, and a wide range of pH.[144, 156]

Two broad categories of extracellular miRNAs isolate from biological fluids. One class exists inside exosomes, micro-vesicles, and apoptotic bodies.[152,157] On the other hand, second classes are conjugated with AGO2 and other allied proteins.[157, 158] However, two opposite results are found for their existence. According to one study, extracellular miRNAs are abundant in exosomes/micro-vesicles relative to their association with AGO-2.[150, 158] Other studies reported that extracellular miRNAs are distributed predominantly in human exosomes from the blood samples.[152] In fact, miRNA categories are restricted to only a few plasma samples. Apart from them exosome or vesicle-free miRNAs also found. Some individual miRNAs are derived from different cell types and other factors. Moreover, extracellular miRNAs are associated with high-density lipoprotein (HDL)[159, 160,] and nucleophosmin 1 (NPM1).[150, 161, 162] The vesicular miRNAs along with allied proteins are involved in protecting miRNAs from a wide range of RNases and stimulate their stability against extracellular milieu.[152]

Secretion and Uptake of microRNAs

The release of extracellular miRNAs is a tight control process. Some miRNAs are considered as a product of cellular injury or death.[162] The exosome miRNAs are secreted via a ceramide-dependent pathway. They also stimulate controlled growth in the target cells.[163] Recently, it is found that athero-protective laminar shear stress export vesicle-free miR-126-3p and other miRNAs conjugated with AGO-2 from endothelial cells via activating vesicle-associated membrane protein 3 (VAMP3) and synaptosomal-associated protein 23 (SNAP23).[164] In fact, miRNAs from the endothelium regulate smooth muscle cell activity.[164] Moreover, miRNAs inside large dense-core vesicles (LDCVs) export using the exocytosis in the neuroendocrine cells via vesicle fusion. This process is conducted by the SNARE complex and is accelerated by Ca^{2+} stimulus.[165] Transportation of exosome miRNAs has also been controlled using different signaling molecules including interleukin-4 (IL4)[166] and Docosahexaenoic acid (DHA).[167] The DHA carrying anticancer and anti-angiogenic activities enhance the secretion of miRNA-containing exosomes for stimulating inhibitory effects on tumor angiogenesis.[167]

Biologically extracellular miRNAs in the recipient cells control their activity, thereby activating intracellular signaling molecules. A tight junction protein, zonula occluden 1 (ZO-1) targets directly for exosome miRNA export. The miR-105 transports from metastatic breast cancer cells to endothelial cells. It destroys the barrier function of endothelium and metastasis promotion.[168] Moreover, an enriched exosome miRNA miR-21-3p from umbilical cord blood increase proliferation, migration of fibroblasts, and instigates the angiogenic activities of endothelial cells during accelerated wound healing.[169] Two metastatic miRNAs, miR-342–3p and miR-1246, from the human oral cancer cell line are involved in inducing metastasis of a poor metastatic tumor cell line.[170] Moreover, extracellular miRNAs are associated with Toll-like receptors[171] and increase downstream signaling events and ultimately allow the induction of tumor growth and metastasis[172] and neurodegeneration.[173] These events claim that miRNAs work as chemical messengers for controlling cell-cell communications.

The mechanisms related to extracellular miRNA uptake are still not known. The vesicle-associated extracellular miRNAs transport into cells using endocytosis, phagocytosis, or direct fusion methods via the plasma membranes. Besides, vesicle-free secreted miRNAs are exported using specific receptors on the cell surface.[174] Indeed, miRNAs penetrate inside the recipient cells using endocytosis and micropinocytosis[175, 176] in a clathrin-dependent manner but not through caveolae or lipid rafts in PC12 cells.[176] However, sometimes endocytosis of exosome miRNAs occurs via caveolae- and lipid raft-dependent pathways in the A549-P cells.[177] Moreover, HDL along with vesicle-free miRNAs are transported using HDL receptor and scavenger receptor BI (SR-BI) receptor in the plasma membrane of the target cells.[159,160] The miRNAs export in between co-cultured cells using direct cell-cell contact and gap junctions.[178] In fact, it is demonstrated that extracellular miRNAs are communicated with recipient cells by multiple mechanisms using many potential factors that are yet to be identified.

Remarks

It is demonstrated that miRNAs are efficient gene regulators for transcription along with their stability and translation inhibition. Similarly, the use of miRNA opposite actions that influenced translational activation requires repeat validation. Also, a series of *in vivo* studies identify different chemical tags to miRNA that are involved in miRNA: MRE interaction or AGO: miRNA interactions particularly with 5′ and 3′ miRNA nucleotide modifications. The anchoring of AGO proteins along with first and last miRNA nucleotides may provide miRNA functioning in unpredictable fashions.

The dynamic miRNA nature is correlated with the complexity of miRNA-dependent gene regulation. Numerous factors for instance cell type/state and the availability of various mi-RISC components are also present. The activity of miRNAs including subcellular location, miRNA/mRNA abundance, miRNA: MRE affinity, miRNAs in the nucleus claims an important role in transcription regulation and alternative splicing. The location of different compartments along with Cytosolic mi-RISC, the levels of miRNAs, target mRNAs, and the affinity of the miRNA-mRNA interaction is essential for gene regulation success. Moreover, the movement of single miRNA and/or mRNA along with spatial and temporal resolution allows us to understand the complex dynamics unimaginably. Thus, large-scale investigations of miRNA global interactomes facilitate a field forward approach that instigates the powerful mathematical models for complex regulatory networks.

It is well accepted that extracellular/circulating miRNAs play instrumental roles in intercellular communication. miRNAs explicit host cell activities upon their secretion and transportation to recipient cells. In a few instances, a series of miRNAs behave hormone-like activities and are associated with Toll-like cell surface receptors. Apart from Toll-like receptor association, many miRceptors remain to be identified[171] *in vivo*. In the majority, the secreted mechanisms and their release to the target cells are not discussed elaborately and that's remained to be investigated.

Clinical function

In development and organogenesis

Animals without any expressed miRNA fail to survive or reproduce abnormally.[179-181] Impairment of the entire miRNA pool in different animals finds three drastic symptoms including their embryonic lethality, abnormal deformities in nearly all organs, ultimate unnatural death, and lack of stem cells.[182-186] Elimination of embryonic stem cell differentiation in the developing of organs is changed by miRNAs.[187] Studies demonstrate that miRNAs can either facilitate or reduces the self-renewal of stem cells based on their cell types or cultural environment.

Dislodging of differentiation ability of embryonic stem cells severely results in significant defects in cellular morphology.[188] Members of the miR-290 family cluster lead to control of the embryonic stem cell (ESC) cycle. They are designated as the ESCC family of miRNAs (ESC cell cycle–promoting miRNAs).[187,189] ESCC miRNAs interact directly with two keys cell cycle inhibitors p21 and LATS2 promoting the G1-S phase transition phase.[189] Apart from them many crucial pluripotency factors Oct3/4, Nanog, and Sox2 are associated with the promoter of the miR-290 cluster. Their stable expression involves self-renewal and thereby in the maintenance of the pluripotent phase.[190, 191] The elimination of ESC miRNA expression using depleted Dicer or DGCR 8 in vertebrates alters normal cell cycle profile along with extended G1 stage.[192, 193] Multiple ESCC miRNA expressions reduce gradually during ESC transition of self-renewal sections. MiRNA let-7 suppresses pluripotency and performs an opposite effect of the miR-290 cluster. The overexpression of let-7 occurs to help in differentiated states and stability for the antagonizing effect of the miR-290 cluster.[187, 194] Species sets of miRNAs expression are also essential in proliferation regulation during the differentiation of hematopoietic stem cells. Interestingly, miR-125b performs dual functions. On one hand, it controls stem cell hematopoiesis. On the other sector, its specific expression alters inflammation and innate immunity with a specific increase of the differentiation and activation of macrophages.[195, 196] This pro-inflammatory effect of miR-125b is involved using the nuclear factor (NF)–kB control predominantly. Besides, the aberrant miR-125b expression is involved in human cancers including leukemia that results in acute myeloid and lymphoid leukemias in primate models.[197, 198]

miRNAs play an instrumental and efficient role in harmonizing complex coordinate organ development. Besides, a stable and uniform temporal and spatial expression of marked groups of tissue-specific miRNAs is essential in orchestrating tissue development and differentiation. Moreover, miR-273 is involved in left-right asymmetry establishment during neuronal development.[199]

The miR-1 (miR-1-2) expression plays a pivotal role in cardio genesis regulation.[200] The wide distribution of conserved miR-1 expression in the adult heart is important in controlling three events including cardiac morphogenesis, electrical conduction, and the cell cycle. The miR-1 participates in the fine-tuning expression of an essential transcription factor Hand 2 for cardiac development.[201] On the other hand, miR-1 targets insulin-like growth factor1, calmodulin, and myocyte enhancer factor 2A. They are all essential for cumulative cardiac hypertrophy development.[202, 203] In mice, deletion of miR-208 markedly impaired the ability of the heart to respond to stress stimuli.[204] Elimination of two miRNAs miR-208 and muscle-specific miRNA miR-133a in primates enhances four crucial states proliferation, myocyte apoptosis, ventricular septal defects, and embryonic lethality by developing different cardiac dilatation and failure.[205] The miR-27 overexpression results in a reduction of Pax3 target proteins. This protein is essential for myocyte proliferation reduction and promotes myogenic differentiation in the skeletal muscle. Besides, miRNAs also involve specific functions in skin development in cardiac and skeletal muscle. Moreover, miR-203 stimulates mouse skin differentiation and stratification. It regulates the basal to suprabasal transition by suppressing p63. The p63 is a crucial member of the p53 family of transcription factors.[206, 207] Conversely, p63 reduces miR-34a/c expression for stabilizing cell cycle progression, which is antagonistic to the miR-203 effects.[208-209] Moreover, the miR-124 expression is essential in the proper development of the nervous system.[210, 211]

A systematic miR-127 expression is required for proper branching of the lung. Its ultimate increase functions for defective branching and lung bud malformation.[212] The enhancement of miR-17-92 cluster stimulates proliferation but reduces normal differentiation in lung epithelial progenitor cells.[213] On the other hand, miR-375 expression is important for insulin secretion. Its direct expression in pancreatic islets modulates exocytosis of insulin from pancreatic beta cells.[214] Also miR-143 controls adipocyte differentiation and miR-122 regulates lipid metabolism by changing the expression of multiple target genes in primates.[215, 216] The miR-192 reduces enzymes Naþ/Kþ-ATPase. It is a significant transporter of solutes and fluids in renal epithelial cells that executes to renal handling of fluid balance during increased sodium/water intake.[217] These examples together with high-throughput miRNA profiling, spatial and temporal expression patterns in specific organ systems have undoubtedly established the essential role and relevance of miRNAs in animal development and organogenesis.

In diseases

The programmed cell death is indispensable for development and disease. Its non-programmed and random existence has been related to the many neoplastic process development. Apoptosis and cell death-controlled miRNAs are known as apopto-miRs. They can be either pro- or anti-apoptotic. Following the same paths, several apoptosis related miRNAs are involved in the initiation and progression of different neoplastic processes. Nearly 50% of miRNAs distributed in human cells are either inhibited or overexpressed in numerous cancers.[218] The miRNAs have been playing a pivotal role in the tumor (known as oncomiRs) isolated in 2002 in initial evidence.[219] The tumor suppressor genes are located at chromosome 13q14, which are deleted mainly in chronic lymphocytic leukemia (CLL). The existence of CLL increase numbers of predominantly nondividing malignant B cells considerably by upregulating the anti-apoptotic B-cell lymphoma 2 (Bcl2) protein. The tumor suppressor loci on chromosome 13q14 are modulated frequently of the CLL. However, two potential miRNA genes miR-15a and miR-16-1 replace a tumor suppressor protein. Their upregulation reduces oppositely an antiapoptotic Bcl2 gene at the posttranscriptional level. Later, the tumor-suppressing role of many miRNAs has been isolated throughout the decade.

For instance, the miR-34 family promotes considerable suppressor activity by increasing a potent tumor suppressor/cell cycle regulator p53. The overexpression of miR-34 causes G1 blockage that stimulates p53 parallelly and complementarily.[220, 221] The miR-34 reduces the silent mating information regulator 1 (SIRT1) gene causing overexpression of p53, p21, and PUMA (p53-upregulated modulator of apoptosis). Their activities involve cell cycle regulation, apoptosis, and working as a tumor suppressor agent for changing the SIRT1-p53 pathway.[222] Moreover, miR-34 mediated growth arrest controls many cell cycle regulatory directly factors like cyclin E2 (CCNE2), cyclin-dependent kinase 4 (CDK4), E2F3, the hepatocyte growth factor receptor (c-Met), and finally increasing caspase-dependent cell death.[221, 223] In an additional study miR-34 reduces proliferation/growth of human pancreatic tumor-initiating cells in one hand. On the other hand, it increases p53-deficient human pancreatic cancer cells that restore the tumor-suppressing function of p53.[224]

A member of the BCL-2 family MCL-1 is involved in miR-29a, b, and c. posttranscriptional regulation.[218, 221] A transient expression of miR-29b promotes tumor cell apoptosis by inhibiting MCL-1 expression as a novel intervention in cancer therapy. Similarly, let-7a miRNA also promotes tumor suppressor activity by targeting 2 widely recognized oncogenes RAS and HMGA2 expression directly.[225, 226] Additionally, many tumor suppressor miRNAs include miR-7, miR-124, miR- 137, miR-146b, miR-15b, miR-128, and miR-326 are also suppressed.[227] All overall knockdown of miRNAs by depletion or elimination of Dicer1, RNASEN, and its cofactor DGCR8 enhances the oncogenic potential of transformed cell lines. It implicates for speed up of tumor formation in K-RAS–mediated lung cancer and Rb-governed retinoblastoma.[228- 230]

MiRNAs also control tumor development (oncogenes) based on regulated target protein functions. The oncomiR clusters miR-155 and miR-17-92 cluster enhance tumor development in B-cell lymphomas.[231-232] Ectopic miR-155 expression in transgenic mice causes pre–B-cell expansion, splenomegaly, and lymphopenia that accelerate the development of lymphoblastic leukemia and lymphoma.[233] Truly mir-155 participates in a pivotal function in lymphoma development possibly by altering upstream and downstream regulatory pathways.

The common retroviral integration site generated a noncoding RNA from the "bic" locus before miRNA discovery in mammalian cells.[234] Later, the mature miR-155 coding sequence produces a potential function of bic[234] after the discovery of miRNAs. It is also shown that transcript harbored many members of the mir-17-92 cluster as activators of cell proliferation. They are upregulated in numerous neoplasms, including lymphoma, multiple myeloma, medulloblastoma, and cancers of the lung, colon, breast, and prostate frequently.[235] The tumor activator miR- 21 activates glioblastoma, lymphomas, and cancers of the breast, ovary, colon, rectum, pancreas, lung, liver, gallbladder, prostate, stomach, thyroid, and cervix, etc.[234] In one end miR-21 induction is found in glioblastoma tumors and cell lines. On the opposite end, it functions cell death induction. These symptoms demonstrate that miR-21 plays an instrumental role in cell death and oncogene regulation.[236] Furthermore, in glioblastoma cells systematic depletion of miR-2 induced many cancerous pathways like caspase-3 activation, transforming growth factor–b, p53, and mitochondrial apoptotic pathways via influencing its many validated targets, heterogeneous nuclear ribonucleoprotein K, p53-related TAp63, and PDCD4. The miR-21 plays a synergic role with the above-mediated proteins.[234, 237-238]

In a few instances, a single miRNA can perform dual roles as an oncogene and tumor suppressor potentially. It participates in both antiproliferative genes and growth-enhancing genes respectively. The miR-26 acts as an oncogene in glioma and glioblastoma multiform by regulating PTEN intimately. It also functions in the opposite Akt pathway molecularly causing the reduction of RB1and MAP3K2/ MEKK2 expression and JNK-mediated apoptosis.[239, 240] On the other hand, miR-26 acts as a tumor suppressor. Its function is either reduced or its suppression is involved in overexpression in anaplasia or various metastasis in neoplasms like hepatocellular carcinoma, breast cancer, squamous cell carcinoma, thyroid cancers, rhabdomyosarcoma, Myc-induced lymphomas, etc.[241] Interestingly, miRNA expression in cancers sometimes regulate many epigenetic mechanisms like DNA methylation, histone tail methylation.

The miR-9-1, miR-9-3, miR-34b/c, and miR-193a are instrumental for methylation of CpG islands in genes in primary lung tumors for exhibiting reduction or a concurrent promotion in different target gene RAR-b2 and NKIRAS1 expression.[242] Moreover miR-91, miR- 124a-3, miR-148, miR-152, and miR-663 in human breast cancer use for hypermethylation. They function in aberrant DNA methylation and depletion of miR-127 in prostate and bladder cancers [243, 244] In a few cases, miR-127 restore chromatin-modifying drugs successfully and knock down its predicted proto-oncogene target BCL. Thus, it is used as a therapeutic agent of chromatin-modifying drugs for changing miRNA expression.

The profile of miRNA microarrays showing unique miRNA expression signatures like miRNomes in multiple cancers and their subclassifications are used as important markers for diagnostic and classical therapeutic candidates. The miRNAomes appear to be more reliable than mRNA profiles in some tumor subclassifications. In miRNAomes arrays of *200 miRNAs results in an inner classification of tumors in different type and source.[245] In fact, miRNAs are more stable than mRNAs in both body fluids and routinely collected formalin-fixed paraffin-embedded tissues.[245] Their extraction from plasma and serum provides a more convenient and noninvasive method for miRNA profiling. These developments in circulatory body fluids, other unique miRNA properties provide a distinct sign for exciting new avenues of classification, diagnosis, prognosis, and treatment of almost all cancers.

The miRNAs can trigger many manifestations and pathogenesis of infectious diseases in a great multitude. Such as altering the (1) pathogenicity, (2) the efficacy of host innates and adaptive immune response, and (3) the level and resolution of inflammatory responses. The miRNAs play a pivotal and instrumental role in immunity and inflammation. However, the changes in miRNA expression and their contribution to inflammation/immune response are regulated not only by fine-tuning the gene transcription. Nevertheless, they are instruments for interacting inflammatory/immune responses along with the structural and functional ingredients of the miRNA biogenesis.

In the viral pathogenesis, viral infections can modulate various host miRNAs for controlling antiviral mechanisms, viral latency, and lytic cycles. These miRNAs either attack the viral RNA directly or the host cell factors vital for their replication. In a few cases, many viruses encode their miRNAs (v-miRNAs) for controlling their propagation inside the host cells.[246] v-miRNAs attack and deplete the related host genes thereby generating a new cellular environment for promoting virus replication. It is advantageous and rapid to synthesize a miRNA complementary to a new gene than to synthesize a regulatory protein responsible for the same function. Interestingly, many v-miRNAs are not conserved similar to eukaryotic miRNAs.[238] Numerous unique v-miRNAs generate from the Bam H1 fragment H open reading frame 1 (BHRF1) and the Bam H1-A region rightward transcript (BART) in the viral genome when an Epstein-Barr virus (EBV) infected B cells latently.[246] In a few instances, v-miRNAs appear to interact with lytic genes, reduce viral proliferation, and persist latency. EBV miRNAs are extracted from many EBV-positive lymphomas. EBV is also involved in increasing the copies in latently infected cells by hijacking the functions of the host cellular miR-155 miRNA.[247, 248] Other five miRNAs miR-21, miR-23a, miR-24, miR-27a, and miR-34a are also overexpressed during latency, while miR-96 and miR-128a/b are reduced dramatically in EBV-positive classic Hodgkin lymphoma.[249, 250]

The human immunodeficiency virus 1 (HIV-1) represses different miRNA-mediated inhibition during host cell replication.[251] Two key enzymes for host miRNA processing Dicer and Drosha are involved essentially in the reduction of HIV-1 replication in two separate cells either in HIV-1–infected peripheral blood mononuclear cells and/or latently infected cells. In this study, virus replication kinetics is overexpressed significantly when Dicer and Drosha are depleted using small interfering RNAs (siRNAs). The polycistronic miRNA cluster miR-17/92 are instrumental for inhibiting HIV-1 viruses by targeting histone acetyltransferase Tat cofactor PCAF. Along similar lines transcriptional enhancer associated with a fundamental RNA binding domain of the HIV-1 Tat protein depletes the interferon response. It also inhibits miRNA-siRNA processing directly using Dicer interfering functions. Further, five crucial miRNAs, miR-28, miR-125b, miR-150, miR-223, and miR-382 downregulate HIV-1 replication complementing with the viral genome sites.[252, 253] These blocking for HIV miRNAs proliferation are noticed by increasing resting CD4þ T cells. Thus, they act in pro-viral latency development. Besides, these miRNAs are also expressed differentially between monocytes and macrophages. It demonstrates a positive argument that why macrophages and not monocytes are sensitive for HIV infection and replication.[254]

Also, four cellular miRNAs miR-323, miR-491, miR-654, and let-7c change influenza viral expression. Many viruses also modulated by different cellular miRNAs for instances 1) by miR-32 for primate foamy virus 1; (2) by two miRNA miR-24 and miR-93 for vesicular stomatitis virus; (3) by miR-125a-5p, miR-199a-3p, and miR-210 for hepatitis B virus and (4) by miR-196, miR-296, miR-351, miR-431, and miR-448 yy for hepatitis C virus (HCV). Despite HCV exhibits unique and unexpected interaction in which a liver-specific miRNA miR-122 involves directly in RNA sequence for enhancing virus replication.[255] This interaction shows complete opposite findings from the centralized general features that host miRNAs generally block viral replication[256] representing a major deviation. The miR-122 blocking function occurs via an experimental DNA-mediated drug SPC3649 (locked nucleic acid–modulation antagomir to miR-122) successfully. It prevents HCV replication.[257] A high dose of SPC3649 administration reduces HCV viremia in blood and liver for a long time. Besides, this medicine promotes HCV-induced liver pathology and exhibits no potential side effects. This drug is currently being in phase II clinical trials in humans eliminating any sign of projected side effects indicating a huge potential for functioning as the first miRNA-based therapy for HCV infection.

The significant modulation of many miRNAs expression that participates within the immune response is found in versatile bacterial diseases. Multiple host miRNA following infection alter in wide varieties of extracellular and intracellular (e.g., Salmonella enterica) pathogens. Abnormal expression of miRNAs controls the level of sepsis via bacterial infections. The miR-155 overexpression and miR-146a reduction also increase the amount of sepsis in murine models. Further miR-155 downregulates 2 negative regulators of inflammation SHIP1 and SOCS1. However, a similar depression of miR-146a involves an enhancement of its 2 proinflammatory signaling proteins IRAK1 and TRAF6, which are linked to the TLR/IL-1R processes.[258-260] The miR-155 overexpression causes a potential immune response that occurred during Salmonella typhimurium infection.[261]

Extraordinary tuberculosis guided host evasion mechanism is controlled by miR-99b. The miR-99b upregulation in infected dendritic cells and macrophages and its marked reduction act as bacterial growth inhibition.[262] The miR-99b marked inhibition upregulates many proinflammatory cytokines including interleukin (IL)–6, IL-12, and IL-1b and increases two different tumor necrosis factor-a (TNF-a) and TNFRSF-4 production. In a separate study, miR-21 overexpression after Bacillus Calmette-Guérin (BCG) vaccination by NF-kB increment.[263] On the other hand, miR-21 reduces IL-12 production via influencing IL-12p35, which in turn deregulates anti-mycobacterial T-cell responses and influences dendritic cell apoptosis via Bcl-2 targeting. A versatile miRNAs expression pattern opens a fundamental function in the lepromatous form of leprosy development relative to the tuberculoid form. Moreover, a prominent increase of miR-21 in Mycobacterium leprae–infected monocytes in the lepromatous form was observed.[264] In this analysis miR-21 inhibit directly the TLR2/1-mediated CYP27B1 and IL-1b expression. It also overexpresses IL-10 indirectly. Finally, miRNAs alter the reduced expression of 2 genes encoding vitamin D–dependent antimicrobial peptides, CAMP, and DEFB4A. It produces significant changes in bacteria escaping from the vitamin D–mediated antimicrobial processes. This function of M. leprae that overexpresses miR-21 is involved in progressing from the self-limiting tuberculoid form to the positive lepromatous form of leprosy. The miRNA profiles of numerous infectious diseases via fungal pathogens are observed nowadays. Therefore, the functional implementations of miRNA profiles are used for digging different pathogenic processes and are involving in miRNA-mediated antimicrobial therapeutics at an aggressive speed.

Abnormal miRNA expressions have been involved in multiple developmental defects in many organisms and organ systems as discussed earlier. They also use a diverse array of multiple noninfectious diseases for instance autoimmune diseases, metabolic disorders, and genetic diseases. Two miR-155 and miR-326 are increased in human multiple sclerosis (MS).[265-267] The miRNAs silencing in autoimmune encephalomyelitis implicate that different miRNA functional capability increase the Th17 response and modify T-cell developmental pathways by modulating the expression of a complicated array of targets.[267] The response of excess Th17 copies opens up a key role in multiple autoimmune diseases manifestation. The miRNA expression profiles in many disease tissues along with systemic lupus erythematosus (SLE) cause multiple fluctuations in their signatures. Particularly, miR-146a is downregulated in SLE patients. It is also related intimately to many clinical diseases and overexpresses the type I interferon (IFN) pathway.[268, 269] The downregulation of miR-146a causes abnormal accumulation of multiple target proteins including STAT1, IRF5, TRAF6, and IRAK1. They promote a long-lasting hyperactivation of the IFN pathway and disease symptoms.[270] The aberrant DNA hypomethylation related to SLE Patients increases 2 miRNAs miR-21 and miR-148 expression that is involved in T cells.[269, 270]

The miRNAs are important regulators of metabolism. Their therapeutic implications involve treating multiple metabolic disorders. The miRNAs play a versatile instrumental role in regulating glucose homeostasis and insulin signaling. Abnormal miRNA expression in multiple cardiometabolic disorders like obesity, fatty liver disease, insulin resistance, type 2 diabetes, and coronary artery disease executes their specific roles in the pathological manifestation. The liver expressing miRNA miR-122 was the first miRNA related metabolic control.[271] The miR-122 is important in cholesterol and lipid regulation. Eventually, transient silencing of miR-122 causes a *25% to 30% reduction in plasma cholesterol levels. It also reduces hepatic cholesterol, fatty acid biosynthesis, and an enhancement in fatty acid b-oxidation.[272] The miR-122 depletion causes hepatosteatosis reduction in high-fat diet-fed mice. A similar effect is found to silent miR-122 on circulating cholesterol levels in African green monkeys.[273] Invariably, the true mechanisms for a cholesterol-reducing response for miR-122 silencing remains obscure because liver sited genes related to cholesterol and lipid metabolism are not a direct target of miR-122. The other miRNA miR-33a regulates cholesterol homeostasis via a cholesterol-genic transcription factor SREBP2 to enhance intracellular cholesterol levels.[274] Other miRNAs are important for metabolic disorders like(1) miR-29 that promotes insulin secretion. The miR-9 and miR-375 are linked to the reduction of insulin secretion through their effect on monocarboxylate transporter (1) one cut homeobox 2, sirtuin 1, phosphoinositide-dependent kinase 1, and myotrophin, etc; (2) alteration of insulin signaling in adipose tissue via miR-103 and miR-107 by targeting caveolin 1 and miR-29 and (3) miR-34a in humans for hepatic metabolic diseases, including obesity, type 2 diabetes, nonalcoholic fatty liver disease. The function of miR-34a effects on nonalcoholic steatohepatitis, invariably via a core regulatory network of miR-34a, SIRT1 (a key sensor and regulator of metabolic states), farnesoid X receptor, and p53[274,] etc. Moreover, two miRNAs miR-375 and miR-223 are also distributed in circulating high-density lipoproteins revealing that disease-oriented miRNAs can be exported in lipoprotein particles thereby evolving their controlling effect in a paracrine fashion on distantly related target tissues.[275]

The involvement of miRNAs in genetic diseases plays their fundamental roles in development and organogenesis. The miRNA mutations or their sensitive target genes are related to many defective phenotypes. Only in a few instances, a distinct link between miRNAs and specific genetic disorders is noticed. A unique point mutation of miRNA is important to etiopathogenic contribution in Mendelian disease. It includes 2 different nucleotide substitutions in the seed sequence of miR-96 (13G>A and þ14C>A) that modulates the autosomal dominant form of deafness, DFNA50.[276.] Furthermore, a single base alteration (A>T) in the seed sequence of miR-96 causes a progressive hearing loss phenotype. It is also shown that loss of miR-96 target gene regulation as a result of its point mutations in the seed sequence promoting hearing loss phenotype in both human and mouse.[277] Similarly, mutations and resulting sequence variabilities at the 3' UTR of mRNAs modify miRNA recognition/binding sites. These mechanisms demonstrate an autosomal dominant form of hereditary spastic paraplegia (SPG31). Therefore 2 different mutations at the miR-140 binding sequences on the 3' UTR of the REEP1 gene fail to regulate its putative target gene expression.[278-279] At least two independent studies exert mutations in miRNA processing/ biogenesis. Ultimately, a component of the Drosha complex DGCR8 mutation participates in DiGeorge syndrome and loss of function of the FRM1 gene for coding RNA binding protein inside the fragile X syndrome.[280, 281]

microRNA Dynamic Function

The miRNA-guided gene regulation is dynamic and promotes fine-tuning the gene expression at a steady-state level. Nowadays an elaborated but comprehensive study on miRNA dynamics opens a light on versatile and robust miRNA-guided gene regulation. Two broad factors include ultimate compartmentalization and shuttling of mi-RISC within the cells for contributing to the dynamic of miRNA-mediated gene regulation. The distribution, availability, abundance of miRNAs and their selective target mRNAs are also critical factors for miRNA-based gene regulation. MiRNA represses most mRNA targets in a ubiquitous manner between cell types. Nevertheless, multiple events including alternative splicing and alternative polyadenylation affecting 3′ UTRs and cell type-specific RNA binding proteins for targeting mRNA secondary structures, modulate the MREs available pool.[282-284] These agents promote subsets of mRNAs sensitive or insensitive control of genes in a cell type/state-specific fashion.

Compartmentalization of miRNA

The mi-RISC complex and selective target mRNAs are distributed in different subcellular compartments for examples the trans-Golgi network (TGN), early/late endosomes,[285] multivesicular bodies (MVB),[286] rough endoplasmic reticula (rER),[287] processing (P)-bodies,[288] stress granules (SG),[289] lysosomes,[286] mitochondria,[290, 291] and the nucleus[292, 293], etc. (**Figure 5**) These compartments help miRNAs in post-transcriptional gene regulation based on miRNA activities. The miRNA is crowded strongly at the miRISC: mRNA complexes sites. Their distribution patterns are important for the enrichment of spatial mRNA, mi-RISC concentrations over time, and facilitate efficient gene control.[287]

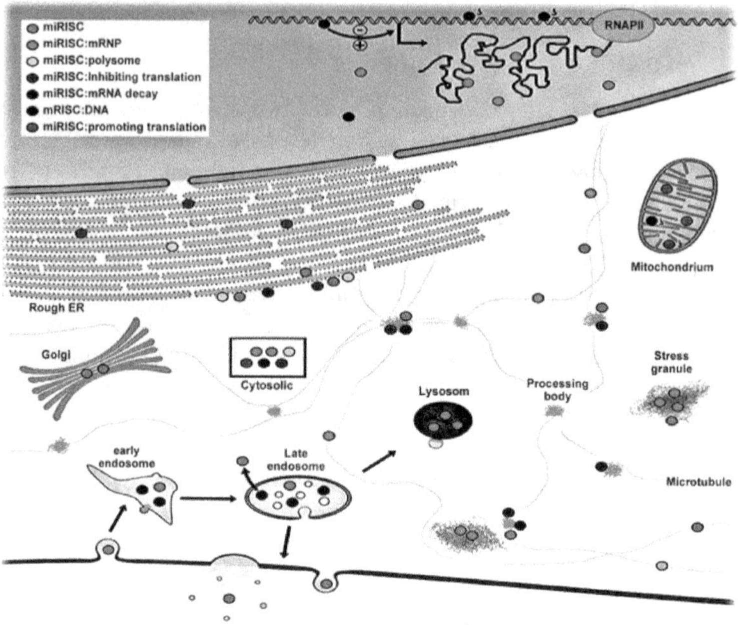

Figure-5. Hypothesis for miRNA localization and function. miRISC has been identified in multiple subcellular locations. In the nucleus, miRISC is accumulated at the active transcription sites, where it preferably interacts with DNA for promoting active or inactive chromatin states. It possibly interacts with nascent mRNA for encouraging more effective splicing or separate splicing profiles. miRISC links together with nuclear messenger ribonucleoprotein (mRNP) to induce its destruction or remain in the formation A miRISC: mRNP complex as it is liberated from the nucleus. In the cytoplasm, miRISC spreads entirely the cytosol or undergo shuttling, possibly using microtubules. Inside the cytosol, miRISC can link to polysomes, reduce translation initiation, facilitate mRNA decay, or encourage translational induction. miRISC inside the rough endoplasmic reticulum can conjugate translating mRNA for translation silencing. Further, unbound miRISC can also accumulate around the rER for linking with newly rER-bound mRNA. Rough ER miRISC: mRNP complexes that are translationally repressed can spread to the early/late endosomes to complete mRNA deadenylation and decay.

miRISC is then further recycled into the cytosol or speeded to the lysosome for clinical destruction. miRISC sometimes resides inside transient, membrane-free processing bodies which are responsible for targeting mRNA translational silencing and storage or decay. Under certain different cellular environments, miRISCRNPs possibly accumulate to stress granules for storage and/or degradation. miRISC sometimes reside in the mitochondria for translational induction or mRNA translational silencing and decay. miRISC inside the Golgi is possibly derived from vesicles secreted from the early endosome. Moreover, endocytosed miRISC possibly spread the Golgi or recycled into the cytosol. Lastly, vesicular, or vesicle-free miRISC possibly exocytosed from at least the late endosome into the extracellular milieu to mediate cell-cell communication. (Loosely Adapted, Redraw and Modify using skeleton from Jacob O'Brien *et al.*, 2018).

===

P-bodies is a possible designated site for miRNA suppressing function.[294] These cytoplasmic foci deplete ribosomes[295] and are full of enzymes required for mRNA degradation machinery for example multiple mRNA decapping proteins, the CCR4-NOT complex, XRN1, and GW182 family proteins.[294] Besides, P-bodies form as a consequence of RNAi and RNA degradation in an RNA-dependent manner.[296] P-bodies along with active mi-RISC and SYNE1 (nesprin-1) amalgamates to Argonaute proteins and other P-body components with the microtubules.[297] The nesprin-1 depletion and inhibition of RNAi markedly reduce miRNA expression of target mRNA and P-body formation. On the other hand, steady state levels of microtubules along with Taxol cause translation inhibition. However, P body accumulation does not function for mRNA decay of *HIF1A*.[298] In fact, MMI-RISC-dependent-body accumulation inverts the recovery after Taxol treatment that leads to dissociation of P-body close to control levels and re-association of *HIF1A* polysome. The relationship of microtubule stability with the increase in P-body count is not known. The repression of polysome occupancy of *HIF1A*[298] acts as a buildup of *HIF1A* mRNA on the static microtubule network encouraging an enhancement in *HIF1A*-bound mi-RISC. In turn, these increase mi-RISC on the static microtubule network and nucleate many P-body formations. Typically, the instability of the microtubule reduces microtubule-associated mi-RISC by blocking nucleation site enrichments. However, the P-bodies levels are essential for miRNA-mediated silencing is not determined elaborately, because mRNA degradation machinery leads to a potential cytoplasmic diffusion, at other subcellular locations, and able to confer RNAi mechanisms even in the elimination of P-bodies.[294] Polysomes in mRNA complexes along with multiple translating ribosomes exist freely inside the cytoplasm or conjugated to the cytoskeleton or membranous subcellular organelles including rER. The mi-RISC copurifies in polysomes. The miRISC: mRNA complexes conjugated with polysomes are associated with enriched translational inhibition and mRNA degradation.[299] It is found that miRNAs, AGO2, and target mRNAs are accommodated within rER-membrane-bound polysomes. They also present before miRNA-guided translational inhibition.[287] Moreover, rER-bound mRNA required for miRNA-mediated degradation allows within endosomes and MVBs that follow mi-RISC recycling into the cytoplasm.[285] In fact, this interaction executes multiple functions including miRNA-dependent translation inhibition, micrococcal nuclease degradation of cytoplasmic mRNA, or mRNA mutants that devoid of cognate MRE and leading the potential loss of polysome/miRISC copurification.[287, 300] Interestingly, the level of polysome presence of target mRNAs is involved in positive affinity between miRNA seed region and MRE, but not on miRNA abundance.[300] Moreover, the average ratio between polysome-bound and -unbound miRISC at any given time is reported as polysome occupancy that is below 40% level. About a high rate of miRNA bound AGO sampling of targets observed *in vitro.*[301] For instance miRISC is capable of the probe for MREs quickly but does not form stable interactions on its own indicating a key aspect of miRNA functioning. A sensitive and critical equilibrium is maintained between MRE bound and unbound states. This equilibrium is essential for the spatiotemporal dynamics of mi-RISC that allows miRNA to respond to change in a short time in subcellular environments and control multiple target mRNAs dynamically.

miRNA at the Cellular Level

An interplay between cell type, cell state, miRNA frequency/ distribution, and miRNA-guided gene regulation is still under intense experimental scanner. Recently, it is estimated that the top five expressed miRNAs exist in nearly ~50% of the overall miRNA pool on an average for any tested human cell type.[302] Moreover, miRNA can control gene networks dynamically and/or transiently via feedback and feedforward loops[303] and steady state gene expression. In some cases, miRNA overexpression occurs throughout differentiation.[304] In both occasions, increased effects on gene regulatory networks extend further[305] for example control of transcription factor expression alters in transcriptional profiles. Intricately, miRNA neither alone nor even dominantly acts as target-mediated regulators but play important roles in the post-transcriptional silencing of expression noise.[306, 307] In this way miRNA facilitates steady-state gene expression by buffering out stochastic transcriptional fluctuations. A common indication of expression noise regulation contents high mRNA to protein ratios is promoting reduced translational rates that prevent random fluctuations in mRNA concentration.[308] In some cases, a small fluctuation of highly abundant mRNA has a negligible effect on protein output for choosing regulated protein output rather than the coupling of mRNA/protein levels [309] Besides, the strongest protein level for any tested gene is involved in the rate of translation via mRNA levels.[310]

This interaction dilutes cellular miRNAs among many potential targets. That executes a small proportion of each target mRNA is conjugated to a cognate miRNA at any given time. [311] Besides, mRNAs with higher affinity MREs exhibit a greater sensitivity to miRNA-guided post-transcriptional inhibition.[297, 311] Nevertheless, an accurate balance between MRE load and cognate miRNA copies involves versatile effects. The gene overexpression causes enrichment of the MRE load of a miRNA for sequestering miRNA away from other targets.[311, 312] In this method modification of MRE load along with endogenous RNAs (ceRNAs) for example circular RNAs (circRNAs),[313] can sequester mi-RISC from the target mRNA that leads to derepress target mRNA.[314] During growing compilation, the MRE load can be altered by changing subcellular miRISC localizations. A quick and unique RNAi machinery involves due to the colocalization of target mRNAs and cognate miRNAs at locations of synaptic formation in neurons.[315]

In classical miRNA features, a single mRNA may carry many MREs.[316] They contribute to many MRE loads for a variety of miRNA. To a certain extent single mRNA along with multiple MREs functions as a ceRNA and multiple

MREs also works to enrich mi-RISC occupancy. A classical bonding of many proximal MREs on target mRNA overexpress RNAi activity and prolongs retention times of miRISC complexes on those mRNA. It can occur most likely via three AGO binding sites on the GW182 family proteins.[311, 317] In this process, the low expressing miRNA function enhances RNAi machinery synergistically on a subset of target mRNAs that effectively increase the cognate miRNAs concentration for specific mRNAs.[318, 319] Conversely, a lower amount of miRNA compensates by enhancing the loading of cognate miRNA into AGO when target mRNA quantity is relatively high. Thus, it improves RNAi activity.[320] Furthermore, miRNA synergistic effects are important for three important biological cellular processes denoting neurogenesis,[321] embryonic stem cell pluripotency, and cell differentiation.[322] The considerable syngenesis of miRNA regulation of target mRNA are involved in oncogenesis. For instant, a tumor suppressor cyclin-mediated kinase inhibitor 1A (CDKN1A is reduced in many cancers).[323] At 28 targeted miRNAs along with others miRNA overexpression in cancers where CDKN1A has been implicated.[324] It indicates that overexpression of miRNA in many cancers prevents CDKN1A expression. Thus, miRNA synergism can be used as therapeutic agents for different strategies.[325]

Moreover, special miRNA/mRNA dynamics modulate in global mi-RISC localization related to modifications in cellular environments including stress or serum starvation.[326] It also changes the miRNA activity. There is a rapid release of stable, vesicular, or vesicle-free miRNA during serum starvation of many human cell lines.[327] It also reduces the decrease of intracellular miRNA levels.[327] Concomitantly, increased cell stress produced by physical heat shock and molecularly translation inhibition along with treatment with hippuristanol or cycloheximide influences translocation of mi-RISC from the nucleus and cytoplasm to transient cytoplasmic SG.[289, 328] SG are involved to work as intermediate storage of messenger ribonucleoproteins (mRNPs). They are in action during translation or following viral infection.[329] Occasionally SG components can back also to the cytoplasm, exchange with P-bodies, or be degraded by lysosomes.[294, 329] SG bounded mi-RISC during cell stress carry many target mRNAs that lead to downregulate RNAi activity. Moreover, the exportation of miRNA during cell stress and distribution of miRNA:mRNA complexes to SG indicates an mRNA-protective role.

Conclusion

Recent inventions and structural and functional developments in miRNA biology have been revolutionizing intensely their structural landscape and have developed immense interest within the scientific community. These functions estimate their functional variety and significantly brush up and modify the classical concepts about gene silencing, gene control, and RNA dependent function. miRNA based RNA silencing and protein inhibition challenge the "central dogma of molecular biology," dramatically. Conventionally RNA performs a mere intermediary function to the flow of information between the genetic material (DNA) and the ultimate product (protein). The isolation of miRNA function also refutes the faulty "junk DNA hypothesis." It stands more than 80% of cellular DNA designs for no biological function and is "junk" that has accommodated inside the cells as evolutionary fossils in a prolonged time.[330] The implementations of the core miRNA studies are far-reaching. Many knockouts or overexpression studies were performed in the past to determine. The fluctuation of gene function sometimes depends on their intron's deletion. A significant number of miRNAs are evolved from the introns of a gene. Their frequent and random deletion particularly in the miRNA coding sequences may evolve a dramatic phenotypic consequence during random knockout overexpression studies that frequently occur during the loss of protein-coding genes. Therefore, loss of key microRNA regulators can occur during random deletion, which should be reexamined again.[331]

Besides fundamental gene control miRNAs evolve as ligands for signal transduction through the binding to cell surface receptors. In this issue, two cancer mediated miRNAs miR-21 and miR-29a are formed a complex with TLR7 and TLR8 in immune cells. The binding increases TLR-guided pro-metastatic inflammatory response for enhancing tumor growth and metastasis. Moreover, loss of function of one or multiple miRNAs modulates the inherent ability of any biological system. Also, miRNAs on multiple occasions are ''fine-tuners'' for gene expression, Therefore, a marked modulation in spatial and temporal miRNA expression involves fluctuating rapidly and functions in minimal expression changes levels unlike the change in protein-coding sequences. Thus, these delicate changes are highly sensitive and are sophisticated in a gene controlling system. Despite havoc progress that has been attributed to considering multiple bottlenecks still more future work is required for a better understanding of their biological role.

The miRNAs therapeutic implications are humongous. On many occasions, a single miRNA is involved in disease development that can target many anti-miRNAs. They are antisense oligonucleotides with specific structures.[332] In cancer therapy, a series of cholesterol-conjugated anti-miRNAs have been promoting better cellular uptake and serum protein binding refers to antagomirs. They are used for preventing oncomir interactions.[332, 333] On the other hand, multiple inhibitory miRNAs can bind locked nucleic acid (LNA) oligos or 20-O-methoxyethyl phosphor-thioate (MOE) modification efficiently.[333] During tumor suppression the delivery of synthetic RNA duplexes as a mimic for miRNA duplexes can then be identified by the RISC complex that process finally into mature miRNAs. Despite steady-state stability and profound efficiency of the miRNAs face a major challenge in *vivo* delivery. The rapid progress in this direction opens up an entirely new field of small RNA-based therapeutics. Thus, miRNA development as therapy together with their role in biomarker devices emerges them indispensable tools for resolving multiple aspects of a disease process that includes diagnosis, classification, and treatment.

References

References: chapter wise are cited before the Index at the end of the Book. (1-333)

Chapter III

Unique small Regulatory RNAs in mammalian therapy

(Among non-coding small regulatory RNAs, siRNA covers a lion share in the literature, whose vivid descriptions have not allowed us to fit in this book space. So, it is deliberately eliminated in the present context despite of our strong temptation to write)

Abstract

In recent decades, the extensive search of early and effective biomarkers has been critical for extending all adequate treatment plans against most dreadful and devastating diseases. The biomarkers also monitor core mechanisms for disease progression. Nowadays, gene mutations, Central Spinal Fluid (CSF) and blood protein markers along with their imaging technologies are dominating the field of diagnosis. Despite the presence of these techniques, confusing and conflicting data still exist that demands the need of new classes of early biomarkers, particularly at the budding stages of diseases. Small non-coding RNAs MicroRNA, small interfering RNA, antisense RNA and Piwi-interacting RNA are considered a "relatively" new class of cutting-edge molecules that are proven for application as modern markers for differentially controlled diseases.

Non-coding antisense transcripts generate from the strand opposite the sense strand. More than 70% of the human genome produces non-coding antisense transcripts. However, less than 2% of the genome codes structural proteins. Antisense transcripts control gene expression and genome integrity via interacting with sense transcription and a dramatic change in histone tails or DNA methylation. They claim their effective contribution in pathological and physiological events. Truly, antisense transcripts are required for multiple diseases like cancer, diabetes, cardiac and neurodegenerative disorders. They also involve in achieving the practical application for prognostic and early diagnostic markers and therapeutic components. However, their contribution is not clearly defined due to initiating many epigenetic controls. This inside knowledge possibly accounts for novel insights for antisense transcription control and ultimately early detection of disease and with therapeutic agent development. A limited description of antisense transcription/transcript control gene expression, genome integrity, disease pathogenesis and therapeutic implementation.

Piwi-interacting RNAs (pi-RNAs) are a unique class of small noncoding regulatory RNAs. They are tiny and only 26–30 nt long. They also interact specifically with Piwi proteins. These tiny RNAs were initially generated in germline cells and are critical regulators for germline maintenance. A considerable amount of evidence suggests that pi-RNAs have enormous biological significance. They allow to control gene expression in somatic cells also via transposon silencing, epigenetic programming, DNA rearrangements, mRNA turnover and translational control etc. Mounting evidence demonstrates that pi-RNAs dysregulation leads to epigenetic modification for treatment against diverse diseases. Pi-RNA biogenesis, mechanisms behind pi-RNA-guided gene control and alteration of pi-RNAs expression are related to different diseases, especially in cancer. The emergence of predominant small gene regulatory RNAs like miRNA and pi-RNA alter our conventional concepts of RNA's contribution in gene regulation in different organisms.

Nowadays, a novel small non-coding RNA invents from tRNA or pre-tRNA, cleavage known as tRNA-derived small RNA (ts-RNA) that contribute as a unique regulator of gene expression. The ts-RNAs are required in some specific physiological and pathological process controls. Despite the growing results on specific biological roles of miRNA and pi-RNA is expanding robustly, however the identical role of ts-RNAs is not elucidated vividly. Currently, the processes, progresses and biological action of ts RNAs in intracellular, extracellular, and intergenerational inheritance are reported. The ts-RNAs are the contributing factors for ideal application of ts-RNAs in human diseases and multiple research developments.

Introduction

Complex eukaryotic transcription of the protein-coding sequences is a well-coordinated and complicated process that initiates through the assembly of all routine transcription factors and RNA Polymerase II at the promoter using different activator proteins (activator). It leads via multiple elongation processes, and ultimately termination.[1-3] This process is tightly bonded with many epigenetic factors and multiple epigenetic processes for example DNA methylation, histone modifications, and/or ATP-dependent chromatin remodeling. These epigenetic processes and transcriptions are eventually regulated by many non-coding RNAs such as mainly siRNAs (small interfering RNA), miRNAs (microRNA), antisense non-coding transcripts, pi-RNAs (Piwi-interacting RNA), ts-RNAs (tRNA derived RNA). These small noncoding regulatory RNAs serve nearly all valuable roles in numerous biological processes and medical diagnosis and therapy designs and developments.

Non-coding regulatory antisense transcripts are derived from the strand opposite the sense strand and regulate sense transcription and ultimately gene expression. Nearly, 72% of the human genome produces antisense transcripts.[4-5] Antisense transcripts are essential for physiological and pathological processes. The oligonucleotide-based antisense therapy controls gene expression during clinical trials for treatment of numerous diseases such as cancer, hypertension, respiratory illness, neurological and muscular disorders, and HIV infection etc.[6-9] Despite antisense transcripts/transcriptions are effective largely in disease treatment and pathogenesis, however, it is still not clear enough how antisense transcriptions are emerged and controlled via chromatin architecture. These findings may generate excess/additional avenues for antisense transcription/ transcripts regulation leading towards therapeutic development and disease pathogenesis. Only a few studies expose our understanding towards the mechanisms of antisense transcription initiation and its modulation in chromatin architecture. Further, the essentiality of antisense transcription/transcripts for the gene regulation and genomic stability along with disease pathogenesis and therapeutic mechanisms implementation are also narrated.

Two protein subfamilies AGO and PIWI segregate Mammalian Argonaute proteins that are essential ingredients for generation of small non-coding RNA (snoRNA) function.[10] The AGO clade of Argonaute proteins is an important factor that interacts with microRNA (miRNA) and short interference RNA (siRNA) guided processes.[11] Three components of AGO protein, siRNA and miRNA are essential for interacting with targeted RNAs properly and directed the cleavage. The damage mRNA target formation. The strong AGO-miRNA bonding associates with 3'UTR region interacting mRNAs imperfectly that causes their de-adenylation, translational inhibition and ultimate degradation.[12, 13] Truly, PIWI clades of the Argonaute proteins are essential components in the formation, progression and meiosis of germline stem cells. Invariably, pi-RNAs combination with miRNAs and siRNAs is the great motivation for converting RNA to PIWI proteins. Moreover, a novel class of 24~31 nt long RNA is formed that is longer than characterized 21~23 nt of miRNAs or siRNAs.[14-17] The pi-RNA can be separated from miRNAs, by the presence of 2′-O-methylated at the 3′-termini.[18] The PIWI-interacting RNA (pi-RNA) is a special class of small noncoding RNAs along with many PIWI family proteins. They are particularly germline limited and are mediated on the germline cell after its emergence. Humongous level of germline-mediated pi-RNA is involved functionally in germline progression, maintaining germline genome integrity, inhibiting selfish DNA element etc.[19-21]

The pi-RNA often functions beyond germline cells as reported by one icebreaking research. PIWI-pi-RNA complex leads for interacting somatic cells containing genomic sites and interacts DNA regulation epigenetically. Also, PIWI interacts with Heterochromatin Protein 1A (HP1A) and is involved in the methylation of H3K9 at several genomic sites in somatic cells. Thus, pi-RNA is indispensable for many epigenetic genomic modulations in somatic cells.[22] An additional experiment shows that pi-RNAs localize in the mouse hippocampus [23] using neurons and dendrites. The silencing of One special pi-RNA, DQ541777 emerges strong deformities in morphogenesis of the spine.[23] Till to date multiple functions of pi-RNAs have been spreading broadly in different fields including metabolism, cardiology, and clinical oncology. The biological functions of disease-related pi-RNAs are the absolute aim for evaluating underlying processes in numerous diseases and many insights for the future pi-RNA research arena. There are four existing isoforms of Piwi proteins in human such as PIWIL1 (Hiwi), PIWIL2 (Hili), PIWIL3 and PIWIL4 (Hiwi2); three in rodents, i.e., PIWIL1 (Miwi), PIWIL2 (Mili), and PIWIL4 (Miwi2); and three in Drosophila, for examples PIWI, Aub, and Ago3.[24,25] Aberrant and abnormal pi-RNA expressions have been identified in numerous diseases particularly in neoplasms and reproductive system diseases. pi-RNAs are valuable account for novel biomarkers during early detections and their function as therapeutic targets for precise and personal medicine.

Apart from above small regulatory RNA, tRNA-derived small RNAs (ts-RNA), a unique and novel type of small noncoding RNAs have recently drawn a lucrative and humongous attention. They are not basically random degradation byproduct, but well directed cleavage products required extensively for numerous pathological and physiological conditions.[26] Taken together, there are broadly two categories of tsRNAs derived from ancestral tRNAs. Initially, both 5′ and 3′ tRNA halves, also known as tRNA-derived stress-induced RNAs (ti-RNAs). They are 30–40 nt long. ti-RNAs are derived from trimming of tRNAs using closer ribonucleases Rny1 and angiogenin (ANG) or anticodon loop under pressure.[27, 28] Different natural under pressures like amino acid deficiency, hypoxia, UV radiation, oxidative damage, heat shock, phosphate starvation and viral infection.[27, 29-32] are involved in targeted tRNA breakage. Nonetheless, other than generating pressures, tRNA can rarely trim under normal conditions for generating a rare ti-RNAs.[33] This kind of tsRNAs is yielded from many tRNA-derived fragments (tRFs). The tRFs are 18–22nt in length and smaller than ti-RNAs. tRFs are derived from mature and tRNA precursors using nucleases Dicer or RNase Z. There are four categories of tRFs exist including 5-tRFs, 3-tRFs, 1-tRFs, and 2-tRFs based on their specific cleavage site and recognize sequence location on tRNAs.[34-38] (**Figure-1**) Two noted characters of ts-RNA length and mechanisms are identical with miRNAs. The important functions of ts-RNA that are identical and differ from clinical mechanisms between miRNAs and ts-RNAs, are reported.

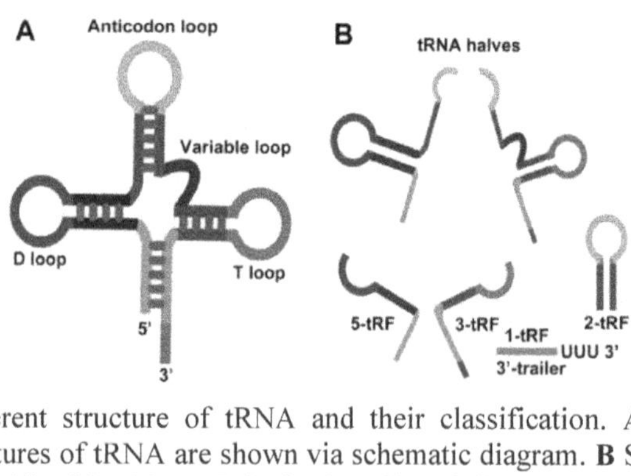

Figure-1. Different structure of tRNA and their classification. **A.** The different secondary structures of tRNA are shown via schematic diagram. **B** Separate types of tsRNAs: 1- tRF (diced by RNase Z in the 3′ trailer of ribosomal RNA precursor), 2-tRF (produced by unknown ribonuclease), 3-tRFs and 5-tRFs, 5′ ti-RNA and 3′ ti-RNA are trimmed from the anticodon loop. (Loosely Adapted, Redraw and Modify using skeleton from Jin and Guo 2019)

Antisense RNA

Human genome yields nearly 72% antisense transcripts, the predominant epigenetic factor for gene regulation.[39, 40] They are produced by antisense strands of transcripts and control the sense stand according to physiological and pathological needs. So, they are referred as "Antisense RNA."[41, 42-46] The advantages of noncoding Antisense RNA for gene regulation have been utilized in clinical trials for multiple major diseases including cancer, cardiovascular diseases, hypertension, diabetes, respiratory illness, neurological and muscular disorders and HIV infection etc.[47-50] However, the function of antisense RNA is not sufficient for disease pathogenesis and basic chromatin architecture and dynamics. Therefore, additional drives will be euploid to understand the anti-sense RNA mediated gene control during therapy development and these are discussed in subsequent sections.

Antisense Transcription Initiation

More than 70% of the transcripts in human antisense transcripts.[39, 40] accumulate in the nucleus.[51] A few antisense transcripts are distributed in the cytoplasm and the mitochondria.[52] Three separate promoters, independent promoters, bidirectional promoters or cryptic promoters produce antisense transcripts.[53-56] Apart from sequence in antisense orientation, antisense RNA does not possess any biochemically modulated characteristics. Generally, non-coding antisense RNA sequence is constrained sometimes by overlapping sense transcripts. However, there is some evidence for partial pairing of sense and antisense overlapping transcripts having protein-coding activity.[57-59] In addition, antisense transcripts have specific domains with DNA, RNA or proteins interaction forming unique functional complexes to execute cellular activities.[60-62] The wide-spread distribution of antisense RNA in higher eukaryotic genomes is not technically visible like yeast GAL gene clusters because the antisense RNA always functions in the background of sense transcription. Despite the permissive antRNA anti-sense RNA mechanism is more prominent in the yeast GAL gene clusters but such yeast mechanism is not the matter of discussion in the present context.[63, 64]

Briefly it is reported that RNA polymerase II can be tracked in association with *GAL10* antisense transcription in the absence of sense transcription in yeast growth medium.[63] Although *GAL* locus deciphered the antisense transcription mechanism initiation, the role of cryptic or bidirectional promoters in the initiation of antisense transcripts is difficult to understand.

Chromatin regulation

Antisense transcripts are essential for chromatin architecture. An antisense transcript uses by itself and/or by function of its transcription in *cis* that regulate the DNA strand of controlled gene locally. It also acts in *trans* for controlling other DNA strands of the target genes. The *cis* action is delivered by the action of antisense transcription, while the *trans* effect is guided by the antisense transcript routinely.[65] The location of antisense transcription interacts with other *trans* loci using three-dimensional chromatin organization. Moreover, an antisense transcript is accumulated at their synthesis sites using accumulated RNA polymerase, R-loops, or triple helices, to enhance its function in *cis*. Thus, both antisense transcription/transcripts act in *cis* than in *trans* in a frequent manner.[60]

Antisense transcription affects DNA methylation at the promoter CpG islands dramatically for regulating sense transcription.[66, 67] For example, the hemoglobin α1 gene (*HBA1*) in α-thalassemia patients is downregulated by antisense transcription. During antisense synthesis a Luc7-like putative RNA-binding protein LUC7L transcribes aberrantly the antisense copies of the *HBA1* locus. It also increases CpG island methylation at the promoter for inhibiting the *HBA1* gene expression.[66] Antisense transcription is also essential in gene imprinting.[68, 69] For instance, AIRN antisense transcription transcribes insulin-like growth factor 2 receptor (IGF2R) non-coding RNA in antisense stands, but AIRN transcript does not inhibits *IGF2R* using DNA methylation and transcriptional interference.[69] On the other hand, antisense transcription also increases sense transcription by preventing the active promoter via R-loop formation and *de novo* methylation.[70, 71] Sometimes, sense transcription is also controlled antisense transcription using histone modifications. X chromosome silencing occurs using a dramatic histone modification by antisense expression. Antisense transcription also controls antisense transcript via histone modification in *trans*. In HOTAIR (HOX transcript antisense intergenic RNA) controls histone modification using PRC2 (polycomb repressive complex 2) silencing that are required in histone H3 K27 methylation for regulating sense transcription.[71, 72] Furthermore, antisense transcription directly controls chromatin modifications. Antisense transcription from the internal cryptic promoters alters the sense genes associated with chromatin and therefore, sense transcription.[53, 73, 74] Furthermore, a triple RNA-DNA helix structure formation at the promoter in *cis* and in *trans* has been implemented in sense transcription silencing.[75-77] In fact, antisense transcription/transcripts all together control gene expression by three distinct inhibitory methods promoter methylation,[66, 69, 78] histone modifications,[61, 79-82] or silencing sense transcription.[41, 76, 83, 84] Antisense transcription also involves mRNA splicing,[85-87] mRNA stability [88] or translational stability by accumulating additional factors.[89] Thus, antisense transcription/transcripts control gene expression by those activities. Antisense transcription/transcripts are also participating in the expression/generation of non-coding RNAs in general. Naturally, antisense transcription synthesizes siRNAs from double-stranded sense-antisense hybrids.[90, 91] Furthermore, lnc TAM34a, antisense transcript able to modify miR34a expression related to cancer silencing. It demonstrates that antisense transcription/transcripts are essential for expression/generation of coding and non-coding miRNAs.

Optimized antisense RNA chemical modification

The worst part is that antisense RNAs are sensitive and fragile for degradation against all nuclease activity. Therefore, multiple stringent methods of chemical modification are required for their protection against nuclease activities. An attempt is made initially to exchange phosphodiester bonds to phosphorothioate (PS) linkages.[92] (**Figure 2A**) This chemical modification protects against nuclease degradation largely. The PS linkages perform the dual roles. It not only improves their serum protein binding but increases the half-life and efficient tissue deliveries.[93, 94] It also develops a noble class antisense RNA inside the cells that affects less efficiently.[95] (**Figure 2.**) Exclusively less efficiency against potential target sequences and less effectiveness leads toward an opportunity for using high oligonucleotides concentration. It also reduces the target selection among the major off-target effects predominantly. The off target leads to select the binding sequences of other genes or conjugate with the proteins directly.[96] These routine obstructions of antisense RNA are eliminated completely by their chemical modifications that facilitate potency and selectivity of ideal complementary sequence binding.

Three important selective chemical modifications are, 2′-*O*-methyl (2′-O-Me) [97], 2′-fluoro (2′-F) [98-101], and 2′-*O*-methoxyethyl (2′-MOE) [102, 103] RNA (**Figure 2B**) and locked nucleic acid (LNA) that generates a methylene bridge between the 2′ and 4′ position of the ribose.[104, 105] The LNA in the ribose is optimal for associating with complementary sequences efficiently.[106] An identical high binding capacity is observed via bridge between different nucleic acid (BNA) compounds.[107-111] The introduction of 2′-O-Me, 2′-MOE, 2′-F, or LNA promotes DNA or RNAs synthesis. This compatibility facilitates special series of clinical applications via fine tuning of antisense RNA.

Enhance of Aso effectiveness

The RNA strand of DNA-RNA hybrids is diced by novel RNase H enzymes that are distributed predominantly eukaryotic nucleus and cytoplasm.[112] This catalytic cleavage can be utilized and increases potency using synthetic oligonucleotides.[113, 114] The "Gapmer" oligonucleotides carries 2–5 chemically modified nucleotides (e.g., LNA or 2′-MOE) on each terminus encapsulating a central 8–10 base "gap" of DNA.[101, 115] The chemically modified oligonucleotides enhance affinity for target sequences and nuclease resistance, while the DNA gap allows a DNA-RNA hybrid formation indicating a potential RNase H substrate. Many ASOs act as gapmers during clinical development. The use of chemically modified oligonucleotides mimic DNA structure and are possible to recruit RNase H for binding with complementary RNA strongly. The 2′-Fluoroarabinonucleic acid (2′F-ANA, **Figure 2B**) is an ideal notable example of this oligonucleotide class.[116, 117]

Translation blockage

ASOs eliminating DNA stretches (or DNA-like nucleotides) can base pairing with RNA and prevent normal gene expression. They are referred as "steric blockers" because they obstruct the ribosome rather than RNA cleaving. The steric blockers derived from oligomers that are distinct from DNA or RNA. (**Figure 2C**) Peptide nucleic acid (PNAs) is a unique series of oligonucleotides that mimic their bases and are associated with amide bonds. The PNA synthesis is identical with multiple features with peptide synthesis.[118] Invariably the amide backbone is unaltered, they are associated with greater affinity and rate.[119, 120] Phosphor-odiamidate morpholino oligomers (commonly called PMOs or "morpholinos") are other unique stable DNA analogs.[121] PMOs do not associate with their opposite targets. It has strong affinity feature with PNA binding. It is also an effective agent inside the cells that functions as silencing agents strongly.[122, 123]

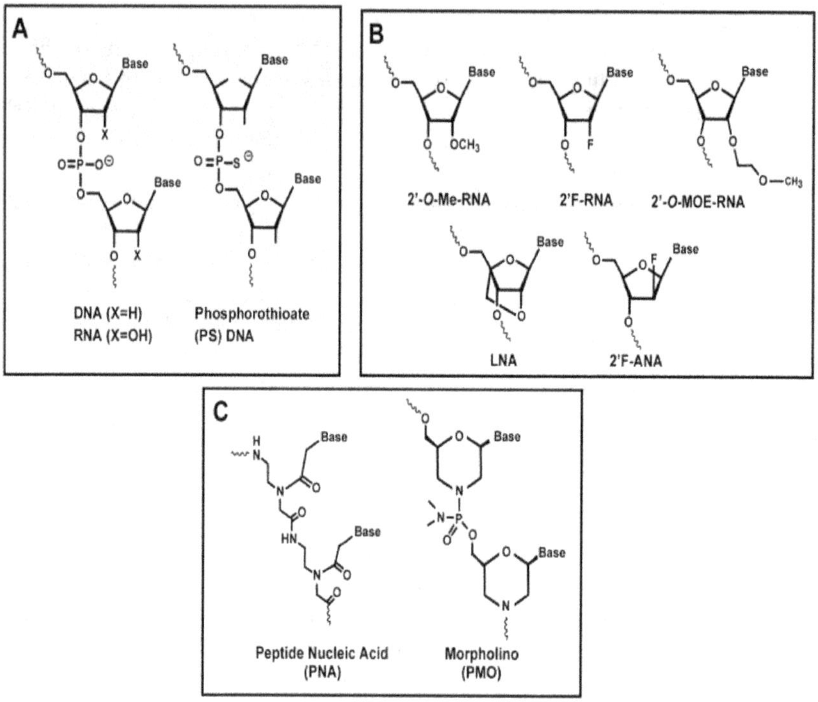

Figure-2. Some usual structure of chemically modified nucleotides. **(A)** A non-bridging phosphate oxygen is replaced with sulfur that gives the phosphorothioate (PS) linkage for dramatically increasing nuclease stability. **(B)** Multiple sugar modifications are used for compatibility with unmodified DNA/RNA synthesis that dramatically increase binding affinity and nuclease stability. **(C)** A neutral backbone structure carries PNA and PMO modifications that is strongly different from the sugar-phosphate backbone of natural oligonucleotides. (Loosely Adapted, Redraw and Modify using skeleton from Watts and Corey 2012)

Splicing modulator

mRNA is derived from intervening sequences or introns containing precursor pre-mRNA. These introns must be eliminated from the mature mRNA employing a complex splicing mechanism. The polycistronic gene employs nearly 95% of alternative splicing methods, in which the pre-mRNAs are cleaved intron differently for generating many types of mature mRNA.[124] Oligomers target many pre-mRNA as an initial step for splicing machinery and promote many desired isoforms production.[125-129] The oligonucleotides employ different methods that do not use 2′-modified oligonucleotides or LNA carrying RNase H or unaltered analogues like PNA or PMO oligomers.

The splice-switching oligonucleotides target an extraordinary variety of genes and their pre-existing conditions.[125, 126] A large 2.6 Mb long dystrophin gene carrying 97 exons encodes a protein essential for healthy muscle cells is a most complex example for splicing. Abnormal dystrophin gene expression in mutation instigates muscular dystrophy. The most severe form of muscular dystrophy, Duchenne muscular dystrophy (DMD) mutations act as a frameshift mutation functioning with no dystrophin protein. However, antisense oligonucleotides are required to eliminate the mutant exon and restore the ideal reading frame. They also generate a truncated partially functional dystrophin protein required for leading the milder symptoms.[129-131] Most potential modifications in many splicing mechanisms occurs of multiple other therapeutic targets including β-globin (for β-thalassemia),[127, 132, 133] SMN2 (for spinal muscular atrophy),[128] HER2 or Bcl-x (for cancer),[134, 135] and others.[125, 126]

MicroRNA targeting

MicroRNAs (miRNAs) are tiny endogenous RNAs that control normal physiological processes and recover treatment for many diseases.[136-139] A growing number of miRNAs is required for different physiological processes that control diseases. They interact with many selective target genes for controlling sensitive disease genes and utilize them for many therapeutic leading agents. In many cases, ASOs complementary to miRNAs blocks their function.[140-142] The liver specific miRNA miR-122 is responsible for treating cancer and hepatitis C.[143] Antisense oligonucleotides complementary to miR-122 modify liver metabolism[144, 145] and block hepatitis C virus replication.[146,147] Different oligonucleotides PNA,[148] LNA,[146-149] 2′-*O*-Me,[150] 2′-MOE [145] and morpholino[151] and their allied chemistries work as inhibitors of miRNA function.

Modification and duplex RNA

The duplex RNA is a strong stable molecule. The chemically altered siRNAs are not required for gene silencing in cultured cells. (**see Table 1**) Stable and normal siRNAs are not active strongly *in vivo*, whereas the targeted and selection modulation of siRNA chemistry enhances their interactive properties markedly.[152] The modification of siRNA chemistry and structure can promote the nuclease stability and prolong the duration of action.[153-155] The normal RNA is also cleared quickly.[156] The chemical modification of ASO, their complexity with interacting agents and local transport and injection to a disease target is also enhanced *in vivo* results.

Many chemical modifications enhance the efficacy of multiple sequences.[155-158] The double-stranded RNA inside the cells can influence the innate immunity.[159-161] This increased immunity may be responsible for therapeutic advantages.[162-164] On the other hand, its delivery also enhances the dose-dependent toxicity.[165] The modulation in ASO chemistry represses the immunogenicity.[158, 165, 166] However, its use is important in progress particularly towards clinical applications.[167, 168] The architecture of an ASO duplex also changed. The lengthy RNA duplexes are the potential target for Dicer, an enzyme of the RISC complex. It has a greater potential relative to 21-mer counterparts.[169, 170]

Detecting ASO in gene silencing

The initial manipulation of gene expression related to a biological problem or disease state provides a unique phenotype or physiologic stresses. Later changes of one or more target genes are found. If such symptom never exists, commercially available duplex antisense RNAs inhibit the target gene. If commercial RNAs are not present and tremendously expensive investigators tried to produce ASOs experimentally.

The algorithms of ASO RNA design[171-173] are not perfect. Therefore, the examination of many duplex RNAs sequences is required for their activity and selectivity. Several groups have suggested the computational[174-176] or experimental[177-180] designs for ideal ASOs identification. Different features of mRNA secondary structure can be estimated roughly, but their algorithms like mFOLD are not used for their tertiary structure detections. They do not involve RNA binding proteins suggesting a narrow utility in practice.[181] Finally, for a serious inquiry and identifying a potential ASO many oligonucleotides can be tested based on their timing and budget permit.

Table 1. ASOs and siRNAs of different methods to overcome similar obstacles

Main methods for reaching goals		**Probable goals**
ASOs	**siRNA**	
Chemical Modification	Duplex Structure, Protein Association Chemical Modification	Nuclease Stability
Chemical Modification	Seed sequences preorganization Guided by Argonaute	Binding affinity, target structure disruption
Chemical Modification	Chemical Modification	Overcoming immune stimulation
Careful sequence selection	Careful sequence selection Chemical Modification	
Chemical Modification	Conjugation, Formulation	Cellular Uptake
Conjugation, Formulation		

Uptakes of ASOs in culture cells

The uptake of ASOs in cell culture transfects nucleic acids via a series of cationic lipids. An effective mixture of cationic lipid with negatively charged nucleic acids is an effective complex that can easily penetrate cell membranes and deliver active oligonucleotide into the cell cytoplasm.

Numerous varieties of cationic lipids are commercially available depending on the specific cell line used. In a specific cell line, the selection of lipid may vary based on a transfected ASO. All cell lines are efficient equally for using cationic lipids. It may be always useful to test different cationic lipids experimentally if the relevant literature is not available. To select the perfect one the interested cell line is transfected successfully. For transportation oligonucleotides in the live cells can be electroporated occasionally.[184-186] This method is purely effective in the cell lines for lipid transfection. In such cases, a special series of equipment and expertise are required.

In recent days, often active ASOs can enter some cell lines freely without any additional lipids.[187, 188] The transport protocol is simplified, efficient and lacks any off-target effects in absence of lipids. The protocol is utilized within different compatible cell lines using lipid-guided transfection. However, in such cases large amounts of higher concentrations of ASO are required compared to the amounts used in lipid-guided transportations.

Delivery of ASOs in Vivo

As gene silencing machinery extends its hand from cultured cells to animal models and ultimately for recent clinical application, the potential challenge of perfect delivery methods employs nowadays. Numerous blockage and barriers emerge during the delivering oligonucleotides in whole organisms.[189, 190] Three main obstacles, clearance by the kidney, rapid degradation by serum nucleases, and/or abnormal biodistribution totally block oligonucleotide from ever reaching its routine targets. For proper distribution, oligonucleotide must penetrate through the blood wall barriers and guided export to the interstitial space and extracellular matrix. Ultimately if the oligonucleotides can transport and enter through the appropriate cell membrane, an active uptake by the endosome should occur. In fact, ASOs are required saline solutions for efficient delivery and depend on their dramatic chemical modifications. Their phosphor-rothioate backbone conjugates to serum proteins allowing for efficient excretion via kidney.[191] Moreover, different aromatic nucleobases are also associated with many sera hydrophobic molecules on the cell surfaces. The different cell types accommodate surface receptors *in vivo* for efficient uptake of oligonucleotides. Interestingly these receptors are often absent in cultured cells. It demonstrates that for this absence lipid is applicable for efficient ASOs delivery in culture cells relative to *in vivo.*[192]

Invariably, delivery of duplex RNAs is more challenging relative to single-stranded oligonucleotides. In an siRNA, natural insider aromatic nucleobases liberate only heavily hydrated phosphates on the outside. This hydrated surface associates with cell surfaces at a trace level and is excreted in the urine quickly. This drawback of delivery vehicles is improved for their further development.[190, 193-196] The emerging technologies allow to form a complex between RNA with cationic and neutral lipids for delivering.[193, 197, 198] Nevertheless an encouraging result is emerged in the application of peptide transduction domains [199] and cationic polymers.[200] It also uses PEGylated lipids in the formation of prolonged particle half-life for circulation.[201] Bonding of cholesterol with one strand of siRNA provide essential and improver and effective knockdown version in mice liver,[202] but many folds of increased quantities are required (50 mg/kg) relative to present lipid-based formulations (as low as 0.01 mg/kg).[203]

Lipid-based formulations are appropriate for targeting the liver. The lipid nanoparticles uptake inside the liver cells efficiently. Moreover, a systemic conjugation with different ligands to the siRNA itself are required for targeting other organs following systemic administration. Scientists have conjugated various ligands to the siRNA as part of a formulation. Further, the application of aptamers, [204] antibodies/fragments,[205] vitamins [206] and other targeting ligands [200, 207] are essential for development of different promising strategies.

Clinical progress

The clinical progress of antisense oligonucleotide medicines has been described at a slower pace because of examining the ASOs potency. The ASO duplex RNAs discover a unique series and models for many pharmaceutical therapies. The highly charged molecules may synthesize economically for target tissues distribution, export in cells. They act as toxic materials to an acceptable limit. In fact, Oligonucleotides are used as routine small molecule drugs (<500–700 molecular weight) and have been required to understand their properties and ideal optimization. The oligonucleotide therapies can be easily compared to antibody therapeutics. This type of ASOs is now a predominant source for cutting aged medicines, but they also required many years to generate and maturate. Only one ASO has been approved by the FDA and nearly 22 oligonucleotide drugs are in Phase II or III clinical trials (**Table 2**). However, many ASOs are in the process of clinical development and markets.[168, 208-211]

Table 2. Various Oligonucleotide drugs in advanced clinical trials (Phase II or higher). Information taken spefically from different company websites

Target gene and disease	Drugs	Company	Phase	Notes
mRNA1 guided Antisense Oligonucleotide Drugs				
ApoB for hypercholesterolemia	Mipomersen	ISIS/Genzyme	III	PS-MOE gapmer, Intravenous delivery

Bcl2 for melanoma. Leukemia Lymphoma	Gene sense (oblimersen)	Genta	III	PS-DNA, Intravenous delivery of Naked DNA
Clustering for prostate NSCLC, Breast cancer	OGX-011	ISIS/Teva/Oncogenex	III	PS-MOE gamper
TGF-b2 for breast Cancer	AP-12009 (Trabedersen)	Antisense-Pharma	III	PS-DNA
Insulin receptor substrate 1 For corneal neovascularization	GS-101 (Agnirsen)	Gene Signal	III	PS-DNA topical delivery (eyedrops)
Ribonucleoside reductase For cancer	LOR-2040	Lorus	II	PS-DNA
Multiple targets for allergic For asthma	ASM8	Pharmaxis	II	Two PS-DNA, ASOs delivered by inhalation
AKT-1 for Cancer	Archexin	Rexahn	II	PS-DNA
Survivin for Cancer	LY2181308	ISIS/Lilly	II	PS-MOE gamper
elF4E for Cancer	ISIS/EIF4E	ISIS/Lilly	II	PS-MOE gamper
Hsp27 for Cancer	OGX-427	ISIS/OncoGenex	II	PS-MOE gamper
I-CAM 1 for ulcerative Colitis	Alicaforsen	ISIS/Atlantic	II	available for inflammatory Bowel disease
XIAP for Cancer	AEG35156	Aegera	II	PS-2'-O-Me-gamper
Splice switching Oligonucleotide drug candidates				
Dystrophin (exon 51 skipping) For DMD	PRO051	Prosenza/GSK	III	20 mer PS fully 2'-O-Me ASO
Dystrophin (exon 51 skipping) For DMD	AVI-4658	AVI-BioPharma	II	30 mer Morpholino
Dystrophin (exon 44 skipping) For DMD	PRO044	Prosenza	I/II	PS fully 2'-O-Me ASO
anti-miR Oligonucleotide drug candidate				
miR-122 for Hepatitis C	Miravirsen	Santaris	II	LNA modified anti miR Oligos deliver intravenously

siRNA drug candidates				
RSV Nucleocapsid	ANL-RSV01	Alnylam/ Cubist/ Kyowa Kirin	II	Unmodified siRNA, intra-nasal or inhalated
RTP801 wet for AMD and diabetic macular edema	PF-655	Quark/Pfizer/Silence	II	Chemically modified, intraocular
P53 for acute renal failure	QPI-1002	Quark/Novartis/Silence	II	Chemically modified, Intravenous
SYK kinase for Asthma	Excellair	ZaBeCor	II	inhaled

Regulation of anisense RNA

Regulation of DNA repair

Different molecular events are associated with DNA damage and repair mechanisms. These include chromatin structure, mRNA splicing and stability, translation, and generation/expression of non-coding RNAs. Many genotoxic factors are associated with antisense transcription/transcripts of the cells. The DNA lesions and damage are rectified by the transcription-conjugated DNA repair.[212-215] DNA damage response activates checkpoints for cell cycle arrest and DNA repair. In the absence of DNA repair, cell mutations do not enhance the apoptosis.[216] Therefore, cell cycle blockage and programmed cell death are essential for handling devastating genotoxic stress. These essential methods of DNA repair are controlled using multiple transcription factors and antisense transcripts/transcription.

The antisense transcripts regulate downstream gene expression for DNA repair by modifying the DNA damage mechanism. The production of antisense non-coding RNA generates genotoxic stress particularly from the upstream region of the *CCND1* (cyclin D1, a cell cycle regulator).[217] Thus, a hypo-acetylated chromatin state along with binding of RNA binding protein TLS (translocated in liposarcoma that inhibits CBP/p300 histone acetyltransferase activity) is involved in the same damage-guided antisense transcripts for *CCND1* sense expression suppression.[218] The reduction of *CCND1* is linked to cell cycle arrest and checkpoint activation for DNA repair. Another cell cycle regulator, a long intergenic non-coding RNA-p21 (or lincRNA-p21) can originate from the opposite strand of p21 (*CDKN1A*), transcription.[219] The lincRNA-p21 reduces gene transcription globally. It differs from locally active *CCND1*antisense RNA and are related to apoptosis and DNA repair.[219] Another antisense non-coding RNA is ANRIL (a 3.8 kb transcript in the opposite orientation of *INK4B-ARF-INK4A* gene cluster) at the *INK* locus referring as CDKN2B-AS1. It also promotes DNA damage using transcription factor E2F1 (E2F transcription factor 1) in an ATM (ataxia-telangiectasia mutated)-mediated manner.[220] The activation ANRIL transcript downregulates the *INK4A* expression (also referred as *CDKN2A* and p16). *INK4B* or *CDKN2B* and p15 and *ARF* alternate p14 reading frame.[221] These modified transcriptions of *INK4A*, *INK4B* and *ARF* involves returning to their normal DNA repair using the impediment of cell cycle checkpoints and stimulation of cell cycle progression.[221] Moreover, ANRIL also participated in homologous recombination-oriented DNA repair pathways. [221] A different antisense RNA, DLX6-AS1 (DLX6 antisense RNA 1 or Evf2 lncRNA) is required for DNA repair control using the catalytic BRG1 (Brahma-related gene 1). That is the major subunit of the SWI/SNF chromatin remodeling complex via its ATM-mediated phosphorylation. Further the absence of BRG1is linked to impaired homologous recombination.[222-224] The DNA damage activator PANDA (p21-associated non-coding RNA), antisense RNA is present 4.5 kb upstream of the cell cycle regulator *CDKN1A* (p21) transcriptional start site. It is also promoted to p53-dependent DNA damage responses.[225] PANDA involves in strong interaction with the transcription factor NF-YA and reduces the pro-apoptotic transcripts.[225] Therefore, induction PANDA during DNA damage blocks apoptosis using NF-Y recruitment. Moreover, PANDA is essential in DNA damage to stabilize p53. Similarly, numerous examples of antisense transcripts related to DNA repair are also found.[226-228] Antisense transcription/transcripts open a new era in genomic integrity regulation. Thus, aberrant control of antisense transcription/transcripts would modify genomic integrity, leading to cellular pathologic.

Regulation of cancers

Antisense transcription/transcripts are essential components for prognostic and diagnostic markers and for therapeutic cancer agents. Most promising component, HOTAIR, is upregulated considerably in different cancers like breast, colorectal, hepatocellular and pancreatic cancers.[229-234] A 2.2 kb long lncRNA HOTAIR derived from the *HOXC* locus is the antisense product of the *HOXC11* and *HOXC12* genes.[230] This antisense RNA binds to PRC2 for chromatin silencing by accumulating histone H3 K 27 methylation at the histone tail.[229] HOTAIR increases PRC2 accumulation at the *HOXD* locus and reduces *HOXD* gene transcription by changing the chromatin architect. A significant upregulation of HOTAIR stimulates genome dependent retargeting of PRC2 to hundred genes that alters histone H3 K27 methylation for cancer progression and malignancy.[229 -234] A classical depletion of HOTAIR represses cell migration, proliferation, increased programmed cell death and preventing cell cycle in many cancer types.[229-234] Further, antisense H19 RNA is H19 transcript from *H19/IGF2* on chromosome 11 that upregulates its related cellular migration and invasion in many cancers like stomach, breast, liver, lung and pancreas cancers.[235] Conversely, H19 antisense RNA depletion is a unique effective therapeutic strategy for pancreatic cancer treatment. Other antisense RNA, MAPT-AS1 (MAPT antisense RNA 1) is an essential therapeutic agent for ER (estrogen receptor)-negative breast cancers.[236] It is also expressed extensively in breast cancer cells. Overexpression of that antisense RNA is related to metastasis in breast cancer and other cancers, whereas its inhibition represses the proliferation and migration of cancer cells demonstrating that MAPT-AS1 should imply as an important therapeutic component for ER-negative breast cancer treatment.[236]

Other antisense RNAs are also essential targets for cancer therapy. Four notable examples are WRAP53 (WD repeat carrying antisense to TP53), HOXA-AS2 (*HOXA* cluster antisense RNA 2), HOXA11-AS (*HOXA11* Antisense RNA), PANDA, and ANRIL. Among them, WRAP53 controls tumor suppressor p53 and is upregulated in many tumor cell-lines.[237] HOXA-AS2 overexpresses normally in breast cancer and its significant depletion represses breast cancer development.[238] It reveals essential prognostic marker and therapeutic breast cancer target. HOXA11-AS is normally overexpressed in human gastric cancer cells.[239] PANDA is overexpressed significantly in gastric and breast cancers but shows a opposite down regulation in non-small cell lung cancers.[240-242] Similarly, ANRIL is upregulated considerably in non-small cell lung cancer and cervical cancer and its inhibition is found in cell proliferation.[243, 244] Moreover, other potential antisense transcripts are related to many cancers.[245-250] All those antisense RNA works as versatile biomarkers and/or therapeutic components for cancer therapy.

Regulation of neurological defects

Antisense transcripts are also required essentially in neurological disorders. The mode of action and the transcripts leads to diagnosis, evaluating disease progression and selecting therapy components in many neurological disorders. The BACE1-AS (*β*-site amyloid precursor protein-cleaving enzyme-antisense) is linked to Alzheimer's disease. A 2 kb long BACE1-AS transcript derived from antisense strand of *BACE1* (*β* secretase 1)[129] requires for increasing the BACE1 mRNA stability intimately using a RNA duplex formation. It is also involved in increasing the BACE1 protein that ultimately influences the β-amyloid production.[251] In contrast, the systemic depletion of this antisense transcript reduces the BACE1 level and finally inhibits amyloid formation and their accumulation in the brain. The BACE1-AS is normally overexpressed and increases amyloid formation in Alzheimer's disease by promoting BACE1 stability.[252] Therefore, BACE1-AS is recognized as a unique biomarker and instrumental therapeutic component for Alzheimer's disease treatment. A 1.2 kb long lncRNA UCHL1-AS (ubiquitin carboxy-terminal hydrolase L1-antisense) directs UCHL1 mRNA to heavy polysomes for increased translation and ultimately upregulates UCHL1 protein level.[193] Moreover, UCHL1 upregulation is linked to less amount of amyloid β production and prolongs Alzheimer's disease progression.[253] Besides, two UCHL1 and UCHL1-AS are reduced markedly in Parkinson's disease.[254] The PINK1-AS (*PINK1* antisense RNA) transcripts synthesize PTEN (phosphatase and tensin homologue deleted on chromosome 10)-mediated putative kinase 1. Its mutation is required for early onset of Parkinson's disease. The PINK1-AS is required for steady state level of a PINK1 splice variant expression, svPINK1 using double strand RNA-guided mechanism.[255] The PINK1-AS transcript depletion is responsible for less expression of svPINK1 in neuronal cells.[255] Thus, the opposite fluctuation of PINK1-AS transcript levels has a direct effect on Parkinson's disease. A natural antisense transcript, HTTAS (Huntingtin antisense) is linked to Huntington's disease (HD). It is located at the Huntingtin CAG trinucleotide repeat locus. An important splicing method produces HTTAS-v1 (exons 1 and 3) and HTTAS-v2 (exons 2 and 3) transcripts. The HTTAS-v1 upregulation inhibits endogenous HTT transcript production (Huntingtin). Conversely, functional silencing of HTTAS-v1 increases the HTT transcript level using different cell systems. The silencing of HTTAS-v1 is located in human HD frontal cortices indicating the requirement of HTTAS-v1 in the HTT expression control and the propagation of HD.[256, 257] An antisense lncRNA, TUG1 (Taurine upregulated gene 1) is overexpressed in HD patients.[257] TUG1 binds to EZH2 (enhancer of Zeste homolog 2) an essential component of PRC2. Therefore, epigenetically silent the target genes expression.[258,259] The silencing of TUG1 encourages programmed cell death.[260] Thus, the aberrant TUG1 expression is linked to HD and other neurological disorders.[257] In addition, there are other effective antisense transcripts related to different neurological disorders.[261-264]

Control in Diabetes, Cardiovascular and other Diseases

The antisense transcripts are also instrumental for numerous other diseases predominantly in diabetes and cardiovascular disorders. Many diversified transcriptomes produce many antisense transcripts that are the components for diabetes mellitus, a delicate metabolic disorder related to high blood glucose content. The aberrant expression of these antisense transcript is related intimately to type 1 and type 2 diabetes.[265-269] The antisense transcript ANRIL is in a hot spot region binds to type 2 diabetes, diabetic nephropathy, diabetic retinopathy, diabetic cardiomyopathy and coronary artery disease (CAD) etc. Besides, ANRIL expression is also found in retina and retinal endothelial cells at a high level due to hyperglycemia. The increased expression of ANRIL control function and transcription of VEGF (Vascular Endothelial Growth Factor) employing an interaction with epigenetic factors, p300 and PRC2, in diabetic patients.[265] The ANRIL functional knockout in mice causes production of VEGF and extracellular matrix proteins at a trace level.[265] Thus, ANRIL regulates the heart, kidneys and eyes in diabetes by expressing VEGF and the extracellular matrix proteins.[265] These characteristics define a novel therapeutic strategy for regulating diabetes and associated complications using an RNA-guided approach. Other antisense lncRNA, MEG3 (Maternally Expressed Gene 3) in microvascular dysfunction plays a pivotal feature in diabetes complications.[266] MEG3 expression found at a trace level in the retinas of streptozotocin-mediated diabetic mice and in endothelial cells influence dominant glucose stress. The MEG3 silencing exacerbates considerably in retinal vascular abnormalities. It is also responsible for endothelial cell proliferation, migration, and tube formation.[266] In fact, MEG3 overexpression designs as a new therapeutic agent in diabetes-guided microvascular complications treatment. Similarly, several selective antisense transcripts are involved in diabetes mis-regulation and its allied complications.[267-269]

In addition, antisense RNAs are required for diabetes based cardiac diseases. Antisense transcription/transcripts are also essential for only cardiovascular diseases directly. Antisense RNAs open up a new era for early and late diagnostic markers and therapeutic components for cardiovascular diseases.[262-264] The H19 antisense transcript is related to CAD cardiovascular. The routine H19 silencing influences cardiomyocyte hypertrophy[270] suggesting that H19's contribution in cardiac health maintenance. Furthermore, H19 functions as a precursor for miR-675 for reduction of cardiac hypertrophy.[271] A downstream target of H19-miR-675, CaMKIId (Calcium/Calmodulin-Mediated protein Kinase II Delta) is a serine/threonine protein kinase linked to cardiac electrical conduction. It shows that the aberrant expression of CaMKIId using H19-miR-675 are related to cardiac electrical conduction defects. Further, overexpression of H19 in atherosclerosis enhances H19 amount in VSMC (Vascular Smooth Muscle Cells) and HUVSC (Human Umbilical Vein Endothelial Cells) causing cellular proliferation and apoptosis inhibition.[272] Likewise, antisense RNA, MALAT1 (Metastasis-Associated Lung Adenocarcinoma Transcript 1) is related to cardiovascular diseases, hypertension and diabetic cardiomyopathy.[273, 274] Moreover, MALAT1 is required for promoting angiogenesis using VEGFR2 (Vascular Endothelial Growth Factor Receptor 2).[275] Similarly, ANRIL is also bound to cardiac diseases like myocardial infarction (MI) and CAD. The promoter of ANRIL is expressed for methylation of a high risk of CAD.[276, 277] The single nucleotide polymorphism (SNP) ANRIL variants (e.g., rs10965215 and rs10738605) are linked MI risk only in the Chinese Han population.[278] The ANRIL variant rs3217992 is associated with the adverse cardiovascular events,[279] while ANRIL SNP variant rs10757278 is connected to selective adverse cardiovascular development in hemodialysis patients.[280] The ANRIL transcript is expressed in a trace level in the blood of CAD patients demonstrating a direct relationship between ANRIL expression and the emergence of disease symptom. The HOTAIR is connected to acute MI and its plasma concentration for detecting and monitoring acute MI.[281] HOTAIR is overexpressed inside the cardiac tissues and plasma of patients of acute MI and CAD.[281, 282] In summary, antisense transcripts are important for cardiac health and functions. Many other antisense transcripts are used as diagnostic markers and therapeutic agents for cardiovascular diseases.[262-264, 283-285]

Antisense transcripts regulate muscular diseases. MALAT1 is related to muscular dystrophy.[286,287] MALAT1 in *SUV39H1* (Su(var)3-9 Homolog 1) binds to muscle differentiation regulator MyoD in proliferating myoblasts. Its presence in histone H3 K9 trimethylation expression in MyoD target in muscle differentiation.[286] In contrast, MALAT1 silencing enhances myogenic differentiation in cultured cells.[286] Similarly, MALAT1 depleted mice overexpress muscle regeneration. MALAT1 is also involved in depletion of novel myostatin targets[287] a negative regulator of muscle growth.[288] It suggests that MALAT1 is related to myogenic differentiation and muscle regeneration control.[286] Abnormal expression MALAT1 is connected to muscular disorders.[286, 287] The antisense lncRNA, SIRT1 AS (Sirtuin 1 Antisense RNA) controls myogenesis. The SIRT1 silencing emerges for myoblasts differentiation in C2C12 and human skeletal muscle cells. In contrast, SIRT1 AS upregulates NAD+-dependent histone/protein deacetylase level using RNA duplexes formation and promoting SIRT1 translation connected to miR34a (that can bind with SIRT1) for reducing muscle formation.[289, 290] Thus, SIRT1 AS aberrant regulation modifies myogenesis promoting muscular diseases. Similarly, other antisense RNAs are important for muscle diseases [291,292] and in immune diseases.[293-295] H19 and HOTAIR are overexpressed in rheumatoid arthritis.[293, 294] MALAT1 is also expressed highly in rheumatoid arthritis fibroblast-like synoviocytes.[295]

Since gene expression is core ingredients for cellular processes, the antisense transcription/transcripts appear to be clinical agents for cellular gene expression and health using protein-coding gene regulation. The abnormalities of antisense transcription or antisense transcripts generation are related to many common diseases. Further, antisense transcription/transcripts are also connected intimately with aging, metabolic disorders, stress, thalassemia, and spinocerebellar ataxia diseases.[262-264]

Piwi-interacting RNA

Among 98% non-protein coding human genome short chain non-coding RNAs for example small interfering RNAs (siRNAs), microRNAs (miRNAs) and piwi-interacting RNAs (pi-RNAs) claim the major share.[296] It has been reported that 26–31 nt long piwi RNAs [297, 298] are required for epigenetics modulation and control chromosome related gene expression during cellular differentiation.[299-301] The pi-RNA functions actively in live human cells [302] that regulate an association of piwi proteins with the polycomb group proteins (PcGs) via PcG response elements. The pi-RNAs are divided broadly into 3 categories at their origin for example transposon-derived pi-RNA, lncRNA-derived pi-RNAs and mRNA-derived pi-RNAs. Besides, there are two sub-clusters for pi-RNAs, one is assembled in premeiotic germ cells (pre-pachytene pi-RNAs) and another is in meiosis and haploid spermatid stage (pachytene pi-RNAs). Though the molecular characters of two subtypes of pi-RNA are the same but pre-pachytene pi-RNAs form clusters totally in repetitive sequence elements.[303] Usually, pi-RNAs bind to PIWI proteins for active silencing of Argonaute family members to their target genes.[304]

The pi-RNA-induced complexes function also in transcription and post-transcriptional silencing of transposons for protecting the genomic integrity of the germline cells. The PIWI–pi-RNA pathway acts both inside and outside the germline for regulating cellular genes.[305-307] In addition to the usual role in gonad development and male and female fertility, pi-RNAs are involved in many cancerous cells and other diseases.[308] The disease creating abilities of pi-RNAs can be utilized for early diagnostic tools as prominent biomarkers. The transposon-dependent pi-RNAs transcribe from both sense and antisense of DNA strands. The mRNA-generated pi-RNAs are formed from mRNAs 3′ untranslated sequence and the lncRNAs-dependent pi-RNAs derived from the whole lncRNA transcript.[309] They are mediated mostly on RNase III Dicer for their maturation identical to miRNA and siRNA.

Pi-RNA Biogenesis

For intensive synthesis of the clinical roles of pi-RNAs the biogenesis in mouse germline tissue is briefly reported initially. The pi-RNA synthesis is mostly followed by two pathways: the primary synthesis methods and the secondary ping-pong cycle. The precursor pi-RNAs are initially transcribed that lead long single-stranded transcripts from pi-RNA clusters unidirectionally carrying several sequences complementary to transposons (transposon-dependent pi-RNA). Later, it is reported that few pi-RNAs function due to separate origin besides the transposon-mediated pi-RNA. It is categorically classified into three separate pi-RNA pathways: transposon-mediated, protein coding region-dependent and intergenic region-mediated pi-RNA. The ratio of all three separate pi-RNAs generation varies during separate phases of spermatogenesis.[310] The long single-stranded transcripts sequentially slice the mature pi-RNA pieces having different lengths, the inner process remains still elusive. The PIWI protein is firmly binding with small RNA sequences preferentially that contains uridine residue located in the 5′ ends due to the loop formation within the MID domain of PIWI proteins.[311, 312] The 3′ end of the small RNA sequences are then cleaved by exonuclease after PIWI protein binding.[313] The length of various PIWI proteins forms the length of the mature pi-RNAs.

The ancestral pi-RNA precursor is featured with uridine residue at 5′ ends tendency, and multiple length pi-RNAs associate with specific PIWI protein species. Ultimately, the pi-RNA synthesis of 2′-O-methylation at the pi-RNA 3′ ends using Hen1 methyltransferase for developing PIWI-pi-RNA maturation.[314] The synthesis methods of secondary pi-RNAs are referred to as ping-pong cycles. The basic pi-RNA identifies mRNAs that are transcribed inversely from transposon via base-pairing. Mili protein slices the complementary mRNA at 5′ ends of primary pi-RNA in between 10 and 11 nt position and leads to an RNA product placing a dominating biasness for adenine at the 10th position. The small RNA sequence then associates with miwi2 and goes via the same 3′ end trimming. Small RNA altered processes in primary pi-RNA production that generates mature secondary pi-RNA. The pi-RNA binds to miwi2, and follows the same procedures for directing mRNA recognition, 5′ end cleaving, association with mili protein, 3′ end slicing, and alteration. A lion share of pi-RNAs is generated by ping-pong cycle repeats. (**Figure-3**)[314, 315]

There are two broad functions of pi-RNA. The miwi-2/pi-RNA complex undergoes the nuclear regions during transcription. It recognizes the chromosomal loci for following transcription, employs enzymes and changes the methylation level of the corresponding region that causes in the silencing of either transposon genes or protein coding genes during transcription. (**Figure-4**) A larger mRNA synthesized using transposons in the genome is cleaved during pi-RNA process generation for confirming genome stabilization. The pi-RNA identifies many mRNAs beyond their synthesis from transposon using imperfect base pairing. It is a procedure like microRNA (miRNA) synthesis from its target mRNA allowing the subsequent mRNA de-adenylation and degradation. In fact, pi-RNA targets multiple genes simultaneously. (**Figure-3**)[316-318] The usual pi-RNA functions for treatment of several diseases noted above during pi-RNA functioning. In many diseases, the pi-RNAs modulated expression alters several gene expressions using either transcriptional inhibition or epigenetic modification. The selective protein encoded by those genes are participating in numerous diseases mediated signaling cascades that control several cellular events like proliferation, apoptosis, migration, invasion etc.

Uni-strand cluster

The transcription of Uni-strand clusters is identical to the canonical mRNA transcription. Uni-strand clusters contain the transcription-linked to promoter with histone 3 lysine 4 demethylation (H3K4me2) mark. Moreover, pi-RNA precursor molecules are characterized with 50 methyl-guanosine containing capped sequences and 30 terminators.[319, 320] Uni-strand pi-RNA cluster transcribes and evolutionary conserved in several metazoans including sponges, hydras, silkworms, fruit flies, zebrafish, mice, and humans.

Dual-strand cluster

Dual-strand clusters do not contain a prominent of RNA polymerase II (RNA Pol II) mark at the promoters. Dual-strand clusters are transcribed from pi-RNA precursors including multiple proteins like RNA Pol II, Rhino-Cutoff-Deadlock complex (RDC complex), Moonshiner, TATA-box binding protein-related factor (TRF2) and three prime repair exonucleases (TREX). Initially, RNA Pol II is accumulated onto pi-RNA clusters. Later, the transcription is promoted by the RDC complex containing three subunits: Rhino (Rhi), Deadlock (Del), and Cutoff (Cuff). Rhi, the paralog of heterochromatin protein HP1 link to histone 3 lysine 9 trimethylation (H3K9m3) mark. Though the H3K9me3 mark is inhibited transcription in usual heterochromatin, nevertheless it is necessary for the pi-RNA precursor biogenesis process. In the meantime, Rhi reduces splicing inaccurately between dual-strand cluster transcripts and inhibits the canonical cleavage and polyadenylation sequence motifs.[321] Moreover, cuff blocks premature termination and prevents precursors from degradation.[319, 322] Moonshiner, the paralog of the germline-specific transcription initiation factor IIA subunit 1 also promotes transcription initiation and inhibits the H3K9me3 mark combining with the RDC complex and TRF2.[322] TREX to protects R-loop formation, the hybrids between nascent RNA and DNA.[323-325] The nuclear DEAD box RNA helicase 56-kDa U2AF-linked protein (UAP56) inhibits the splicing of dual-strand cluster precursors. (**Figure 5A**)[321, 326] The dual-strand pi-RNA cluster transcriptions are found mainly in *Drosophila,* but their existence in other organisms needs to be explored.[320]

Mature pi-RNAs Generation

Once the transcription is completed, pi-RNA precursor molecules are exported out of the nucleus. (**Figure 5B**) Initially, pi-RNA secondary structures are involved through the RNA helicase Armitage (Armi). The pi-RNA precursors are synthesized into pre-pi-RNAs along with a 5'mono-phosphate after decapitalization using the cleavage of the mitochondria-mediated endonuclease Zucchini (Zuc). Thus, pre-pi-RNAs are conjugated into PIWI proteins and are cutting at the 3'ends using a 3' to 5' exonuclease, Nibbler (Nbr). Subsequently, the newly generated 3' terminal ends are methylated at the 2' oxygen using small RNA 2'-O-methyltransferase Hen1.[327-331] Indeed, 2'-O-methylation are induced pi-RNA stability.[330, 332-334] This process is referred primary biogenesis of pi-RNAs, and pi-RNAs synthesize as primary pi-RNAs in this way. Primed by primary pi-RNAs, the generation of pi-RNAs is amplified with the involvement of Ago3 and Aub proteins. Aub associates with anti-sense-strand pi-RNAs and then trims sense pi-RNA precursors and generating sense pi-RNAs combined with Ago3. Conversely, Ago3 interconnects with sense-strand pi-RNAs and slices to antisense pi-RNA precursors and providing antisense pi-RNAs carrying with Aub.[335, 336] The repeated slicing round synthesizes several pi-RNAs in a short period. In this way, a large level of pi-RNAs is synthesized that is known as secondary biogenesis of pi-RNAs and the ping-pong cycle. These pi-RNAs are further conjugated with PIWI proteins and export back to the nucleus for target genes inhibition. The exact procedure pi-RNA producing process comes from Drosophila germline cells. A similar procedure is noticed in mice for involving multiple Piwi proteins. In fact, the human mediated procedure is identical to Drosophila.[337, 338]

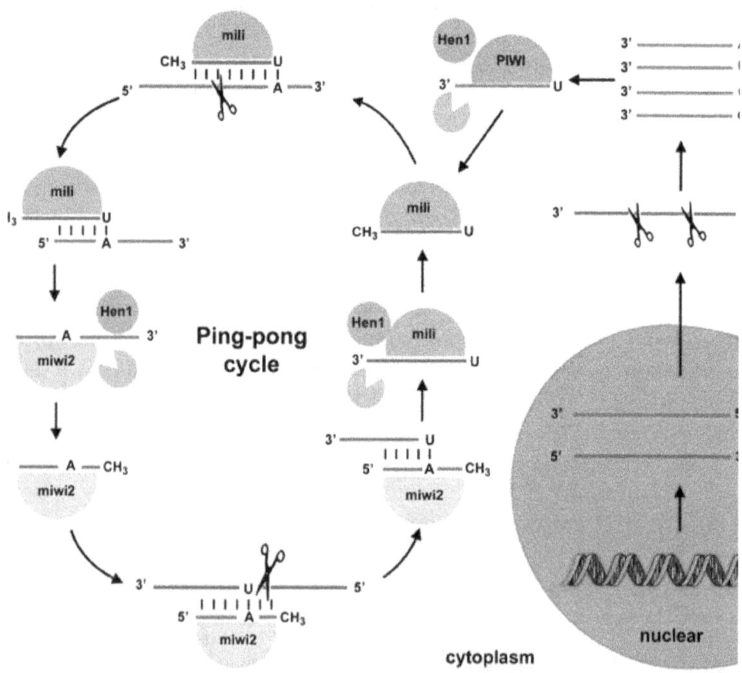

Figure-3. Potential biogenic pathways of pi-RNA in mice. Pi-RNA precursor is derived from pi-RNA cluster, protein coding region or intergenic region. It is later cleaved into small pieces with variable in length. After PIWI links to separate RNA pieces, they enter 3′ end trimming process. The 3′ end is later modulated by the RNA methyltransferase Hen1 for demarking the primary pi-RNA maturation. The secondary pi-RNA biogenesis process is referred as ping-pong cycle: Initially primary pi-RNA is linked to mili that is followed by the conjugation between primary pi-RNA and mRNA via base-paring. Mili cleaves the paired mRNA between 10 and 11 nt position at 5′ end. These RNA pieces associate with miwi2 and convert through the same 3′ end trimming using an alternative procedure. Only miwi2-linked pi-RNA identify target RNA at a similar manner that is undergone further 5′ end trimming, mili association, 3′ end trimming and modification. Following cycle is then repeated multiple times. (Loosely Adapted, Redraw and Modify using skeleton from Sun and Han 2019)

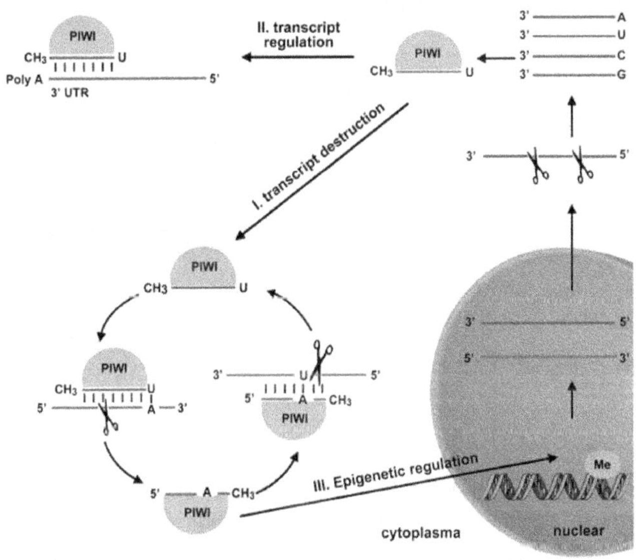

Figure-4. The functional pathways of pi-RNA. (**1**) The PIWI-antisense pi-RNA complex trims sense transposon, pi-RNA precursor and protein-coding transcripts, the important process of which leads to the suppression of transposon and gene expression. (**2**) PIWI-pi-RNA complex could suppress translation of target mRNA through imperfect binding to the 3’UTR region, a process like how miRNA functions. (**3**) PIWI-pi-RNA complex enter possibly to the nucleus and induces DNA methylation level in the promoter region of target gene enriching “Me,” methylation. (Loosely Adapted, Redraw and Modify using skeleton from Sun and Han 2019)

Functions of pi-RNAs

Pi-RNAs protect against transposon modification in fly germline cells that are highly conserved ranging from hydras to humans. Transposons referred to as jumping genes or mobile elements are close to endogenous viruses. They are susceptible to gene stability as "copy and paste." The pi-RNAs DNA in the host genome for protecting themselves leads a series of consequences. Exon insertions inhibit the coding sequence and intron insertion alter splicing patterns that produce unique and deleterious fusion proteins considerably.[339] If transposons transcriptions are incorporated into promoter or enhancer regions, gene transcriptions are transformed. In the meantime, insertions into 50 UTRs or 30 UTRs leads to post-transcriptional gene degradation. Transposon insertions also allow DNA nicks and double-strand breaks. Moreover, mistakes in these lesions repair protect recombination between transposon repeats, triggering chromosomal duplication, deletion, translocation, and inversion.[340]

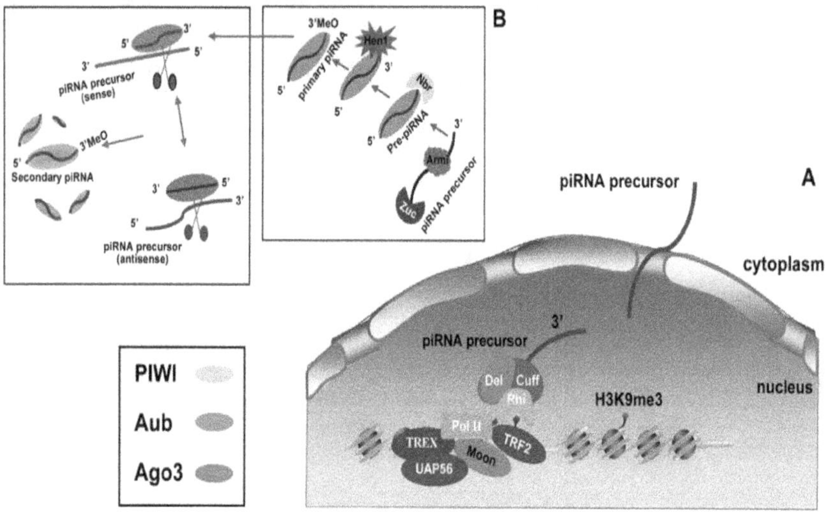

Figure-5. The method of pi-RNA Production **(A)** Biogenesis of pi-RNAs precursors from dual-strand clusters. pi-RNA precursors are produced with the involvement of RNA polymerase II, the Rhino (Rhi)-Cutoff (Cuff)- Deadlock (Del) complex, Moonshiner (Moon), TATA-box binding protein (TBP)-related factor 2 (TRF2), three prime repair exonuclease (TREX), and 56-kDa U2AF-associated protein (UAP56). Rhi binds to the H3K9me3 mark and inhibits transcript splicing, as well as represses the use of canonical cleavage and polyadenylation sequence motifs. Cuff prevents premature termination and protects RNA from degradation. Moonshiner promotes initiation of transcription and represses the H3K9me3 mark by interacting with the RDC complex and TRF2. TREX prevents the formation of R-loops. UAP56 suppresses the splicing of dual-strand cluster. **(B)** Biogenesis of mature pi-RNAs. The red dotted box represents the primary biogenesis of pi-RNAs. First, secondary structures of pi-RNA precursors are resolved by the RNA helicase Armitage (Armi). After des-piralization, precursors are processed into pre-pi-RNAs with a 5° monophosphate by the endonuclease Zucchini (Zuc). Then, pre-pi-RNAs are loaded onto the PIWI and trimmed by the 3° to 5° exonuclease Nibbler (Nbr). Concomitantly, the newly formed 3° terminal ends are methylated at the 2° oxygen by the small RNA 2° -O-methyltransferase Hen1. This procedure is called primary biogenesis. The dark blue dotted box represents the secondary biogenesis of pi-RNAs (the ping-pong cycle). Aub binds to antisense- strand pi-RNAs and then cleaves sense pi-RNA precursors, giving rise to sense pi-RNAs loaded into Ago3. In contrast, Ago3 binds to pi-RNAs from sense clusters. The Ago3/ pi-RNA complex cleaves antisense pi-RNA precursors and produces antisense pi-RNAs that are loaded into Aub. The round of cleavage and trimming repeats and produces a few pi-RNAs in a short time. (Loosely Adapted, Redraw and Modify using skeleton from Wu *et al.*, 2020)

Pi-RNAs and PIWI proteins directly change chromatin architecture and histone proteins in the nucleus at the transcriptional level using the DNA methyltransferase (DNMT) control. In promoter DNMT methylation, CpG islands reduce the initiation of transcription. PIWI proteins guide DNMT for binding the transposable elements or interacting genes. For these consequences, transposable elements or interacted genes are inhibited. However, the exact mechanisms of pi-RNAs and PIWI proteins control DNMT protein expression. pi-RNAs and PIWI proteins also target histone methylation machinery and control the methylation of histone lysine residues (H3K and H4K). The histone methyltransferases are required for interacting transcripts for transcription inhibition.[341, 342] However, the relationships between histone modification and pi-RNAs are still not known. pi-RNAs and PIWI proteins induce mRNA degradation at the post-transcriptional level using mRNA decay machinery. The major ways of mRNA decay and improves the translation is to reduce the poly(A) tail of mRNAs de-adenylation.[338] As pi-RNAs are responsible in gene control, however it is our budding interest for defining their roles in human diseases. Many studies demonstrate that the pi-RNAs control can both facilitate or inhibit the presence and development of numerous diseases, especially cancers.[343]

Pi-RNAs in numerous Cancers

A limited studies demonstrates the link between pi-RNAs and lung cancer.[344, 345] The pi-RNA profiles are found in eight non-small-cell lung cancer (NSCLC) tissue samples and three human bronchial epithelial cell samples. [346] piR-L-163 in NSCLC cells, the top inhibited one, reduces cell migration and invasion using maintaining the function of phosphorylated Ezrin-Radixin-Moesin (p-ERM) that joins transmembrane proteins including ERM-binding phosphoprotein 50 (EBP50) and filamentous actin (F-actin).[346] In another study, piR- 55490 is reduced lung cancer cell proliferation by related to the 30 UTR of mammalian target of rapamycin (mTOR) mRNA.[347] The enhanced piR-651 induces cell proliferation, migration and invasion in opposite reduces cell death in NSCLC A549 and HCC827 cell lines, but the exact mechanism explores in additional study.[348] Moreover, piR-57125 is decreased metastatic lung adenocarcinoma cells considerably related to primary cancer cells.[349] The important remaining pi-RNAs like piR-34871, piR-35127, piR-46545, piR-52200, and piR-57125, are also reduced in lung cancer cells. They induce the proliferation of lung cancer cell lines comprehensively.[350] Therefore, aberrant pi-

RNA induction is required in the tumorigenesis, invasion, and metastasis of lung cancer. It produces effective and functional targets and novel biomarkers applicable as precise medicine. piR-823 reduces gastric cancer tissues markedly. Cell proliferation is silenced by piR-823 mimics. piR-823 reduces tumor growth in a xenograft nude mouse model. [351] In addition, gastric cancer patients show reduced piR-651 and piR-823 expression in peripheral blood. piR-32105, piR-58099, and piR- 59056 are increased in gastric cancer tissues.[352] Although at the budding stage of gastric cancer are treated by surgery or endoscopic treatment, however the prognosis of advanced gastric cancer is not sufficient. Immunotherapy identical to anti-programmed cell death 1/programmed cell death ligand 1 (PD-1/PD-L1) therapy is most effective therapy in cancer. In the cases, the advanced gastric cancer patients are not very delicate to immunotherapy alone or associate with chemotherapy in first-line therapy defense. Additional studies are needed to find whether pi-RNAs are linked to immune escape as piR-823 that are identified in peripheral blood. It functions as most promising biomarkers for indicating immunotherapy outcomes and for detecting gastric cancer.

The liver cancer is classified in many steps: cirrhotic nodules (CNs), low-grade dysplastic nodules (LGDNs), high-grade dysplastic nodules (HGDNs), early hepatocellular carcinoma (eHCC), and progressed HCC (pHCC).[353] Each stage is governed by multiple pi-RNAs. The piR_LLi_24894 is mediated only in CNs; piR-LLi-30552 and hsa-piR-020498 are regulated mainly in HGDNs, eHCCs, and pHCCs; and hsa-piR-013306 is accommodated only in HCC. Reduction in pi-RNAs is involved in early HCC-directed genes including tumor necrosis factor (TNF) receptor, phosphatidylinositol 3-kinase (PI3K)/AKT (serine/threonine kinase), WNT/b-catenin, growth arrest and DNA damage-inducible protein 45 (GADD45), adenosine 50 -monophosphate (AMP)-activated protein kinase (AMPK), high mobility group box 1 (HMGB1), and phosphatase and tensin homolog (PTEN). These pi-RNAs control cell cycle regulation, telomerase activity, protein ubiquitination, DNA methylation and apoptosis.[353] Further, piR-823 activate a-Smooth Muscle Actin (a-SMA) production and Collagen type I alpha 1 (COL1a), and components of extracellular matrix that ultimately causes in cirrhosis.[354] piR-Hep1 reduces cell death by inducing the signal transducer and transcription activator 3 (Stat3)/Bcl-xL signaling procedures.[355] Virtually, mainly HCCs induces cirrhosis indicating hepatitis B virus infection and alcohol abuse are the more relevant causes. It examines effective relationships between pi-RNAs and the two kinds of cirrhosis for preventing cirrhosis to transform HCC. piR-1245 upregulates prominently in Colorectal cancer (CRC) tissues and control all crucial genes that regulate many cellular processes including cell death and survival, DNA replication and repair, and cell-cell communication.[356] These genes require induction of transcription factor 3 (ATF3), BTG anti-proliferation factor 3 (BTG1), dual specificity phosphatase 1 (DUSP1), Fas cell surface death receptor (FAS), nuclear factor kB (NF-kB) inhibitor a (NFkBIa), uridine phosphorylase 1 (UPP1), sestrin 2 (SESN2), and tumor protein p53 activating nuclear protein 1 (TP53INP1).[356] In consistent piR-1245 silencing activates cell apoptosis. Conversely, piR-1245 upregulation induces cell proliferation.[356] piR-54265 overexpression induces CRC cell proliferation and invasiveness via the STAT3 procedures that are improving of metastasis-causing molecules including Matrix Metallopeptidase (MMP)2 and MMP9. In addition, piR-54265 hampers chemotherapy. The larger serum piR-54265 amount hinders response considerably to chemotherapy.[357] piR-823 is induced of Heat Shock transcription Factor 1 (HSF1) expression by induction its phosphorylation at Ser326 and activates colorectal carcinogenesis.[358] Many pi-RNAs piR-5937, piR-001311, piR-004153, piR-017723, piR-017724, piR-020365, piR-28876, piR-32105, piR-58099, and piR-59056 are also reduced in CRC patients.[359-361] Patients carrying different primary CRCs sites are responsible for clinical manifestations, therapeutic responses, prognosis and other characteristics. Although multiple pi-RNAs are required for CRC development, however pi-RNAs involve versatile expression patterns in left colon cancer, right colon cancer or rectal cancer. They are related to prognosis and are valuable for further study.

Involved in shorter Disease-Free Survival (DFS) and Overall Survival (OS) piR-823 is reduced markedly in different tumor tissues and a trace amount is related to a longer DFS.[362, 363] Additionally, piR-32051, piR-39894 and piR-43607 are overexpressed considerably in embryonic renal cancer and all ccRCC cell lines that are linked to a shorter cancer-specific survival and a higher probability of tumor metastasis.[364] These immune checkpoint silencers are for advanced kidney cancer treatment potentially. However, there is no studies still to be documented regarding pi-RNAs and kidney immunotherapy.[365] In fact, the relationship between pi-RNAs and immunotherapy are still to be examined.

Till date, very few studies have been reporting the roles of pi-RNAs in bladder cancer. As many as 197 pi-RNAs are found in different expression patterns in bladder cancer cells relative to normal cells. Among them piR-60152 reduces at a trace level.[366] It results in piR-60152 silencing in bladder cancer EJ cell apoptosis by interacting with TNFSF4. It is the frequently related to the TNF superfamily member, OX40. However, it is not obvious that piR-60152 overexpressed cell cycle progression, cell migration and invasion.[366]

Breast cancers are classified broadly into different molecular subtypes according to versatile cellular markers expressions including estrogen receptor (ER), progesterone receptor (PR), human epidermal growth factor receptor-2 (HER2), Ki-67 and basal cytokeratin (CK). The broad six subtypes include luminal A, luminal B, luminal-HER2, HER2-enriched, basal-like and TNP (Triple-Negative Phenotype)-nonbasal.[367] piR-36712 silences breast cancer cell proliferation by Seleno-Protein W1 (SEPW1) expression reduction via a retro-processed pseudogene of SEPW1, SEPW1P, SEPW1. It also silences p53 and p21 expression by enhancing mRNA degradation. The SEPW1 silencing results in G1 cell cycle arrest. A weak piR- 36712 amount is linked to a shorter Developed-Free Survival (PFS) and a larger risk of axillary lymph node metastasis.[368] Moreover, piR-36712 is responsible for inducing the effect of chemotherapeutic agents.[369] piR-021285 enhances invasion using methylating the 50 UTR first exon of the pro-invasive Rho GTPase activating protein 11A (ARHGAP11A) gene.[370] In another experiment, overexpressed piR36026 enhances cancer cell proliferation through silencing Serpin Family A member 1 (SERPINA1) and Lecithin Retinol Acyltransferase (LRAT) gene expression that are involved in tumor-silencing function.[371] Moreover, piR-932 is upregulated markedly in breast cancer cells that allows tiny level of a type of tumor-suppressing protein latex in expression.[372] However, overexpression of three important pi-RNAs piR-20365, piR-20485, and piR-4987 is parallelly related to metastasis.[373] Moreover many pi-RNAs , piR-23662, piR-26526, piR-26527, piR-26528, piR-30293, piR-34377, piR-34736, piR-35407, piR-36318, and piR-36249 are silenced dramatically, in contrast piR-651, piR-1282, piR-21131, piR-23672, piR-31106, piR-32745, and piR-36743 are overexpressed in breast cancer cells. (**Figure 6**)[374, 375] However, the special methods behind those modification are still not known. Though many pi-RNAs are silenced in breast cancers, however the experiment is not histologically classified. Still many questions remain still to be unanswered. Any modification in pi-RNA expressions in many luminal subtypes causes abnormal pi-RNA expressions. that are related to the phenotypic impact of genomic alterations. The pi-RNAs are responsible for treating many subtypes of breast cancer.

In Endometrioid Ovarian Cancer (ENOCa), the overexpression of piR-52207 silences four interacting genes, nudix (Nucleoside Diphosphate linked moiety X)-type motif 4 (NUDT4), Methionine Synthetase (MTR), Eukaryotic translation Initiation Factor 2 Subunit 3 (EIF2S3), and M-Phase Phosphoprotein 8 (MPHOSPH8) that are involved in 3-phosphoinositide biosynthesis, D-myo-inositol-5-phosphate biosynthesis, folate transformation, vascular endothelial growth factor (VEGF) and EIF2 signaling pathways, respectively.[376] In Serous Ovarian Cancer (SOCa), piR-52207 inhibits Actin-Related protein 10 (ACTR10) expression and Pleckstrin Homology domain-containing family A member 5 (PLEKHA5) that are related to the PI3K/Akt signaling methods and dynactin's pointed-end complex biogenesis, respectively.[376] In addition, piR-33733 is directed to modulate the Lipoic Acid Synthase (LISA) gene that is required for lipoic acid synthesis, an established known antioxidant for enhancing apoptosis.[376] It is reported that breast cancer mutations are sensitive to genes BRCA1 and BRCA2 regarding for almost 40% of ovarian cancer cases in women.[377] It is worthwhile of examining that pi-RNAs are linked to BRCA mutation. If they are linked together, they allow them to function as axillary factors for genomic evaluation and anti-BRCA drug resistance markers.

The primary malignant brain tumor Neuroglioma are broadly divided into four separate grades (I–IV) depending on their malignant behavior.[378] piR-598 functions as a tumor suppressor because its interacts genes that are linked to BAX like cell death genes for controlling p53-guided cell death.[379] Further, piR-DQ593109 controls blood-tumor barrier (BTB) permeability using the MEG3/miR-330-5p/RUNX3 axis.[380] Moreover, piR-39980 induces tumor growth and promotes drug resistance. Despite piR-8041 silences glioma growth.[381] Neuroblastoma is a major complicated and therapeutic potential for children with malignant cancer. Seven important pi-RNAs including piR-32287, piR-32512, piR-36095, piR-38581, piR-52205, piR-52206, and piR-57816 are overexpressed excessively in neuroblastoma cells and their interacting genes like BRCA1, Eukaryotic translation initiation Factor 2 Alpha Kinase 2 (EIF2AK2), Histone Deacetylase 2 (HDAC2), and Peroxisome Proliferator induced Receptor Alpha (PPARA), leading to promote tumorigenesis.[382] Nowadays, gliomas are treated routinely in surgery because it does not have any effective drugs or drugs do not penetrate and incorporate via blood-brain barrier. In expectation. if sufficient pi-RNAs are produced in cerebrospinal fluid or encapsulate with nanoparticle bio-interfacing using platelet membrane cloaking including piR-DQ593,[383] this pi-RNA allow to induce the permeability of the blood-brain barrier. Therefore, the medicine for neurological cancer will be more effective.

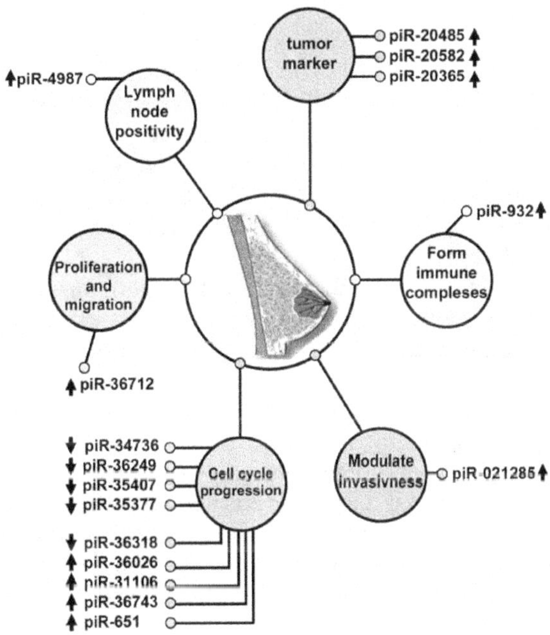

Figure-6. Multiple Problems related to PIWI proteins and pi-RNAs in breast cancer induction. (Loosely Adapted, Redraw and Modify using skeleton from Maleki Dana *et al.*, 2020)

Multiple Myeloma (MM) is detected usually as a hematological neoplasm. It is the most progressive and incurable cancer disease.[384] MM cells are permeated into end organs and develop tumors using Endothelial Cells (ECs). It is a critical type of cells in tumor micro-environment formation. Overexpression piR-823 enhances EC proliferation using silencing the caspase-3/Bax signaling procedures. It promotes angiogenesis by activating the secretion of VEGF and interleukin (IL)-6.[383] Moreover, piR-823 activates Intercellular Adhesion Molecule 1 (ICAM-1) and C-X-C motif chemokine Receptor 4 (CXCR4) secretion to enhance EC invasion.[383] In a separate study, pi-RNA-823 activate DNA methylation via DNMTs (DNMT3A and DNMT3B) and silences a methylation-mediated tumor suppressor, p16INK4A.[385] In the meantime, pi-RNA-4987 and pi-RNA-Hep1 are upregulated in MM cells.[386] As pi-RNAs causes MM progression via DNA methylation and histone-lysine N-methyltransferase EZH2 suppressors are required for MM treatment application. The pi-RNAs promote with EZH2 in MM pathogenesis is valuable addition for future study.

Head and Neck Squamous Cell Carcinoma (HNSCC) grows in oral cavity, oropharynx, hypopharynx, larynx, and nasopharynx. Virus infection and tobacco and alcohol consumption are the major ingredients for HNSCC.[387] In HPV16/18-related HNSCC tissues, pi-RNA FR140858 silencing overexpressed Mini-chromosomal maintenance complex component 7 (Mcm7) that regulates initiation, elongation of eukaryotic DNA replication.[388] In smoking-dependent HNSCC tissues, four pi-RNAs overexpression including NONHSAT015828, NONHSAT081250, NONHSAT113708, and NONHSAT123636 are related to advanced tumor stage and metastasis. Further NONHSAT067200 silencing is linked to longer patient survival.[389] In the same way, three pi-RNAs piR-34946, piR-38034, piR-43219, and piR-58510 are overexpressed in the alcohol-related HNSCC tissues and piR-70732 is silenced.[390] Numerous efforts are applied for deeper relationship between pi-RNAs and HNSCC are identified, and more therapeutic procedures are established.

Osteosarcoma is a predominant important primary solid tumor of children bone and in adolescence and possibly develops in the second decade of life.[391] Current therapeutic methods include surgical resection of all cancer tissues and systemic chemotherapy. Many tests demonstrate that pi-RNAs are related to osteosarcoma. piR-39980 overexpresses the apoptosis of fibrosarcoma cells but represses cell proliferation by silencing ribonucleoside-diphosphate reductase subunit M2 expression. It is a unique enzyme essential for DNA synthesis and repair.[392] Usually almost all sarcomas are not susceptible to chemotherapy, therefore it is a far-reaching experiment to find out how pi-RNAs reduce the systemic chemotherapy to sarcomas transformation efficiency.

Overall, cancer cells are related to different hallmarks including sustaining proliferative signaling, evading growth suppressors, promoting angiogenesis, enhancing invasion and metastasis, and others.[393] In every state, the aberrant expression of pi-RNAs carry utmost importance and the acquire many hallmark capacities. (**Figure** 7)

piR-L-163 silences the cell migration and invasion by acting on p-ERM and F-actin; piR-823 induces angiogenesis by activating the secretion of VEGF and IL-6.[383] pi-RNAs also are firmly executed tumor cell malignant phenotype, clinical stage, as well as clinical response. They are the most presumptive factor for prognosis. Most pi-RNAs are restricted to individual cancer related action, while multiple pi-RNAs are common for many cancer types including piR-823. piR-823 silences the progress of gastric cancer and kidney cancer and establishes the development of liver cancer, MM, and CRCs. Pi-RNAs in specific cancers acts as individual biomarkers, however those common candidates are complementary to other tests. The specific role of pi-RNAs leading to tumorigenesis is still in dark specially whether pi-RNAs influence their interacting targets transcriptionally or post-transcriptionally. In addition, the interconnection between cancer susceptibility to chemoradiotherapy transformation and pi-RNAs is also unique and valuable for further discussion. Nowadays, immunotherapy is effective in NSCLC, malignant melanoma, and kidney cancer, however it has no detectable effect in gastric, colorectal, and breast cancer. The changes in pi-RNA expression profiles between immunotherapy-sensitive and insensitive cancers possibly are important. Further experiments are required to determine the underlying coherent molecular mechanisms. Invariably a deeper understanding is needed for functioning as pi-RNAs or therapeutic agents in cancer.

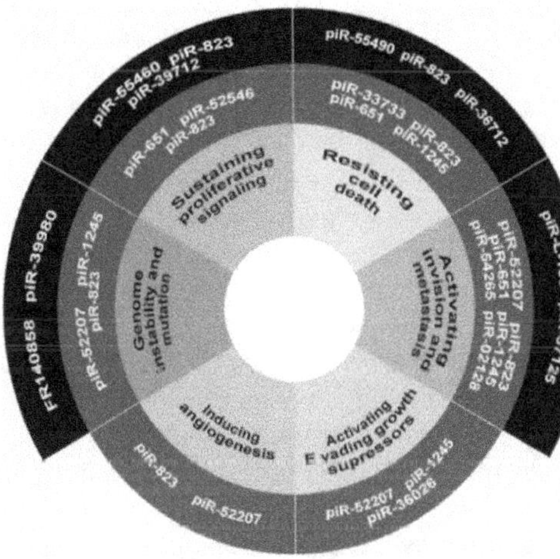

Figure-7. Several hall marks for pi-RNAs Related with Cancer. pi-RNAs in inner circle activate normal cells to get these cancer hallmarks, sign, capabilities. Different others in outer circle block normal cells from producing these cancer hallmarks. (Loosely Adapted, Redraw and Modify using skeleton from Wu *et al.*, 2020)

ts-RNA

Many small noncoding regulatory RNAs have emerged as vital players in various biological processes. Recently, tRNA-derived small RNAs (ts-RNA), a new type of small noncoding RNAs, have drawn great attention. The ts-RNA are the specific cleavage products, playing vital roles for gene regulation in many pathological and physiological environments.[394] Broadly ts-RNA are classified in two major categories. The tRNA derived stress induced RNAs (ti-RNAs) are made up of 5' or 3' tRNA halves with 30-40 nt long. They are constructed by the specific cleavage from near or in an anticodon loop. The cleavage is instigated by the ribonuclease enzyme under many pressures[395,396] including amino acid deficiency, hypoxia, UV radiation, oxidative damage, heat shock, phosphate starvation and viral infection.[395, 397-400] In rare occasions, cleavage is executed in normal condition of ti-RNAs.[401] Second types of ts-RNA are made up from tRNA-derived fragments (tRFs). They are 18-22 nt long identical to miRNAs. The tRFs are derived from precursor or mature tRNA by an active DICER or RNase Z nuclease action. Four separate types for tRFs, 5-tRFs, 3-tRFs, 1-tRFs, and 2-tRFs are usually constructed.[402-405] (**Figure-1**) The ts-RNAs carry multiple common features with miRNA, but still a lot more differences are noticed to their functional aspects.

Functional mechanisms

The common translational inhibition or mRNA post transcriptional degradation are the functional mechanism for miRNAs that is instigated by the base pairing of complementary sequences between 3'-UTR of target mRNA and sense strand of miRNAs.[406] Unlike post-transcriptional silencing of miRNAs in certain cases, ts-RNAs bind to a physiological ligand Toll-Like Receptors (TLRs) for triggering similar biological function.[407, 408] To date, the common degradation policies of ts-RNAs have not been defined, but it seems plausible that many ts-RNA follow the same generalized path of miRNAs or pi-RNAs for their degradation of target mRNAs. Few reports claimed that 3-tRFs carry the miRNA-like structure and function. They base paired with target mRNA directly resulting in mRNA post transcriptional degradation or translational inhibition in mature B lymphocytes or non-small cell lung cancer cells.[409, 410] A recent report established that tRFs may combine with Argonautes (Ago) 1, 3 and 4 but not Ago2 in human cells. However, lack of tRFs binding to main effector protein Ago-2 for miRNA inhibition like function did not hamper to follow similar silencing pathway.[411]

To harmonize multiple cellular processes and modifying cell phenotype 1-tRFs and 3-tRFs combined with many Ago family proteins for inducing small RNA-like silencing to a greater extent.[412] In addition, typical 2-tRFs generated from tRNA(Asp), tRNA(Tyr), tRNA(Gly) and tRNA(Glu) precursors, bound to YBX-1 for suppressing the stability for many oncogenic mRNAs in breast cancer (BRC) via transposing their 3′ UTRs.[413] A few occasions, ti-RNAs could repress translation of a terminal oligo-guanine motif via displacing elF4G/elF4A from uncapped to capped RNAs.[414] The ti-RNAs further conjugated with structurally similar G-quadruplex and form a competitive eIF4G/eIF4A complex for translation repression and mRNA silencing.[415] Sometimes tRNA Val-GAC-produces 5-tRF that construct ribosomal subunits close to mRNA channels and silent the protein synthesis integrally.[416] These few examples are not sufficient for ts-RNA function evaluation; a major explanation for digging their actual mechanism will yet to come.

A group of 18 or 22 nt long tRFs inhibit reverse transcription, retrotransposon mobility (18nt) and promote the long terminal repeat (LTR) retrotransposon silencing or post-transcription silencing similar to miRNAs by the help of effective epigenetic factor transposable element.[417] Furthermore, 5-tRFsgenerated via Dicer-like 1 (DCL1) cleavage are combined with Ago-1 for targeting transposon elements in regulation of genome stability. The ts-RNAs also involved in regulating retrovirus regulation via reverse transcription. The selective tRF-3019 derived from tRNA-proline A notice a perfect sequence complementarity to the binding sites of human T-cell leukemia virus type 1 (HTLV-1), then influenced in HTLV-1 reverse transcription.[418] The respiratory syncytial virus (RSV) infection can induce ANG for cleaving ti-RNAs in host cells. Basically, these host ti-RNAs can enhance RSV replication infection efficiency. [419-421] Therefore, both tRFs and ti-RNAs can regulate RNA reverse transcription via their primer function. In future, the valuable mechanisms that yet to come, add the valuable functional mechanism of ts-RNAs for exploring their greater biological functions.

Biological functions

Intracellular ts-RNA

The stress stimulation, including UV irradiation and heat/cold shock, hypoxia, oxidative stress are the sole ingredients for promoting tRNA halves fragment expression.[422] A ribonuclease of RNase A superfamily member is required for transforming mature tRNAs into tRNA derived ti-RNA fragments.[395, 397] The cleavage by stress induced tRNAs is observed in oxidative stressed humans [422] that are functionally involved in protein synthesis inhibition. Only 5′-tRNA fragments overexpression, but not 3′-tRNA fragments, are involved translation global inhibition in human U2OS cells [397] and require for cytoplasmic foci and stress granules (SGs) for induction of translation initiations.[423] Earlier it was reported that ti-RNAs are associated with YB-1 for controlling stressed cells related translation.[414]

During hypoxic stress and in breast cancer a variety of tRFs are generated for inhibition of many oncogenic mRNA stability in BRC cells by replacing YBX1 gene containing 3′ UTRs. Suppressing of tRFs by-pass different sensitive pathways of metastatic cells.[413] (**Table-3**) Two ts-RNAs ts-3676 and ts-4521 are down-regulated in lung cancer and chronic lymphocytic leukemia (CLL) by interacting with PIWI proteins. It established that ts-RNAs are dysregulated in lung cancer and CLL.[424] Multiple copies of same ts-RNA are found in estrogen receptor (ER)-positive BRC, prostate cancer cell lines and positive androgen receptor (AR) by activating sex hormones and their receptors. A miRNA-like tRNA-derived fragment represses in B cell lymphoma and controls DNA damage response via repressing mRNA transcripts in a sequence-specific manner.[409] Only 5′- ts-RNAs are involved during cell proliferation. Moreover four 5′-ti-RNAs are downregulated in renal cell carcinoma tissues.[425] Although an essential tumor suppressor role of many ts-RNA is executed during cancer progression and tumorigenesis, the major pinpoint mechanism and their overall functions in cancer need to evaluate further.

Multiple neurological disorders exhibit numerous defects in ts-RNA and isolate their processing enzymes. The defective enzyme ANG decreases RNase activity and provides a predisposing gene for amyotrophic lateral sclerosis (ALS) due to lack of motor neurons onsets.[426] The ANG have an important for RNase activity during processing of many ts-RNAs[427-429] that facilitate motor neuron survival via inhibiting apoptosis.[430]

Many ts-RNAs are also involved in neurodegeneration diseases.[431-433] The RNA kinase enzyme CLP1 mutation reduces the kinase activity for accumulating un-spliced tRNA precursors. The 5-tRFs accumulate in the CLP1 kinase-free models that activate cell sensitivity to oxidative stress and further promote cell apoptosis in a p53-dependent form.[431] Although it is not clear sufficiently the interaction between CLP1 activities, tRNA splicing and maturation of tRNA precursor-derived tRFs functions. Beside an idea is also floating finding deregulated tRNA metabolism and neurodegeneration development. It is suspected that ts-RNAs are potential immune signaling molecules.[434] Viral infection induces huge production of ts-RNA molecules for inhibiting target mRNA and facilitating RSV replication.[420] The pseudo-uridylation of specific ts-RNAs influences translational regulation in stem cells as a key developmental process.[435] The ts-RNA also is associated with RPS15 and RPS28 ribosomal protein to enhance the translation and fine-tuned ribosome biogenesis.[436]

Table-3 Existence and establish of multiple tsRNAs

tRNA name	ID	Type	Sequences
Chr10 tRNA2 SerTGA	1001	tRF1	GAAGCGGGTGCTCTTATTTT
Rsph17025_R0021-Gln	3019	tRF-3	CGAATCCAGCCGCCCCAGCCA
Chr17.trna29-CysGCA	3003b	tRF-3	TCAAATCCGGGTGCCCCCTCCA
Chr16.trna19-GlyGCC	3027b	tRF-3	TCGATTCCCGGCCAATGCACCA
Chr2.trna234-ThrAGT	1028	tRF-1	GTCTGCAGTTGGCTGTGCTTT
Chr21.trna2-GlyGCC	5003b	tRF-5	GCATGGGTGGTTCAGTGGTAGA

Extracellular tsRNA

It is indicated that ts-RNAs are useful for stability in the circulation. The 5′-ti-RNAs in the blood and ti-RNA levels are stable with age and calorie restriction.[437] The ti-RNAs regulate stability in blood by binding to a nucleoprotein complex that changes 5′-ti-RNAs amounts associated with breast cancer.[438] Besides, high percent (~ 70%) of ts-RNAs are not aggregated in exosomes in the serum but they remained in the supernatant due to protection against rapid degradation through usual protein binding and nucleotide modifications.[439] Many further studies focus on serum related ts-RNA for their numerous diagnostic values and biological functions.

It is also speculated that ts-RNAs and miRNAs provide mobile signals for alterations of paternal environment and promote intergenerational crosstalk. The 5′ end of tRNAs produces 29-34 nt long unique ts-RNAs claiming a lion share of a small RNA pool in the sperm.[440-442] Sperm ts-RNAs expression profiles and RNA modifications are required for metabolic disorder and change many metabolic pathways for early embryo gene regulation. Basically ts-RNAs in sperm are paternal epigenetic factors that control intergenerational inheritance of diet induced metabolic disorders. Besides, the ts-RNAs contributions in sperm and their relationship with acquired traits transfer is just the first step of contribution. It seems that the relationship is more complex and for evaluation numerous experiments should be conducted for better understanding.

The study of ts-RNA is still in its infancy and budding world of small RNAs. Numerous functional aspects of ts-RNA will be explored soon including their biogenesis and clinical application. Despite the universality and specificity of ts-RNA are still not explained vividly. The ts-RNA may control translational processes resulting in some abnormal function. Besides, better evaluation of ts-RNA function yet to come that explains valuable views on human physiology and pathophysiology.

Conclusion

Antisense transcripts are world-wide transcriptions using eukaryotic genomes and are derived from independent, bidirectional, or artificial cryptic promoters. Antisense transcription/transcripts control gene expression and genome integrity using transcriptional interference, histone modification and/or DNA methylation. Antisense transcripts lead to many macromolecules along within the three-dimensional aspects of the cell for providing transcriptional, post-transcriptional, and epigenetic mechanisms. Thus, antisense transcription/ transcripts are required in multiple biological processes and are abnormally controlled in different diseases through cancer, neurological diseases, diabetes and cardiovascular disorders. An understanding of the regulatory processes of antisense transcription, antisense transcript features. The mode of functions is indispensable factors for the diagnosis, monitoring and targeted therapies of various diseases. Indeed, several antisense oligonucleotides are in clinical trials for the direction of many diseases.[443] Two antisense oligonucleotide-guided therapies are essential for the treatment of Duchenne muscular dystrophy and spinal muscular atrophy.[443] Thus, quick antisense transcription-guided therapy is related to a large promise for many diseases treatment of the near future.

Despite antisense transcription/transcripts are provided as essential and critical contributors in the gene regulation and genomic integrity during their participation in multiple diseases. Truly, the role of antisense RNAs is still not unclear how antisense transcription is emerged and epigenetically controlled through histone covalent modifications. Further, a thorough knowledge of antisense transcription may light-up novel insights of antisense transcription/transcript controls that encourage the understanding the etiologies of different diseases, therefore facilitating the emergence of early diagnostic markers and important therapeutic agents in the treatment plan.

In the last few decades, pi-RNA studies have been holding humongous development. pi-RNAs are widely conserved from protozoans to humans that are largely distributed in both germline cells and somatic cells, [444-446] the abrupt pi-RNAs regulation controls almost all types of diseases. (**Figure- 3, Table 1**) They are also required for the most effective class of biological agents for early diagnostic and prognostic markers. However, a large number of pi-RNAs mis-regulation emerges and relates to different abnormal pi-RNAs regulation in many clinical events without examining insight into molecular processes. Therefore, many relevant questions are still unresolved and require to be explained. Most blazing questions are: If a protein is placed upstream of a pi-RNA does it intervene with the pi-RNA itself or the PIWI protein? Second, what is/are the downstream directed protein(s) that causes abrupt pi-RNAs control in every pathological condition? Truly, are pi-RNAs possibly control their well-directed proteins in each case using induction of mRNA decay or via epigenetically by changing methylation status on the gene promoter or from pre-mRNA decay? Despite a mechanism explained earlier, [447] still many more experiments on cellular level are required to explain the questions properly. The ideal answer will emerge for identification of proper drug targets for each affected disease exclusively. Interestingly, pi-RNAs also exist in many body fluids including blood, urine, and saliva.[448] Importantly, most pi-RNAs are present in the exosome, whereas some unique pi-RNAs are not conjugated with exosomes within saliva.[448] The presence in exosomes possibly raises two following thoughts: 1) pi-RNA in the body fluid can be utilized as early diagnosis and prognostic markers for few diseases. 2) pi-RNAs may be secret from one site but affect the distant target site using exosomes.

Moreover, PIWI/pi-RNA have unique therapeutic value. Invariably many PIWI expressions are limited to stem cells and germline cells. In humans, HIWI, the ortholog of PIWI, exists in hematopoietic stem cells but devoid in the stem cell-guided differentiated cells.[449] The PIWI plays an ultimate crucial role in the stem cell self-renewal process found in different organisms.[450] that is expected to be the possible link between PIWI and cancer stem cells. Indeed, a new idea emerges, PIWI eventually targets in cancer therapy.[451-454] Recently a positive correlation exists in between PIWI proteins and cancer stem cell markers in colorectal cancer.[451]

Mechanistically most challenging task of rare pi-RNAs is that many pi-RNAs form as a pi-RNA cluster in close genomic locations.[455] In one ultimate case, only 32nt long chromosomal locations emerge in three different pi-RNAs using highly overlapped sequences. In fact, some tightly bonded pi-RNAs are essential for investigating combined functions of many separately expressed pi-RNAs simultaneously. In short, evidence of pi-RNA field during the past decades as well as the emergence of precision medicine age it must be in near futuristic approach for the ideal and realistic application of pi-RNAs in the early diagnostic prognostic as well as therapeutic health care.

The discussion of tsRNA is still in a budding stage. It is expected that versatile ts-RNA aspects will be elucidated dominantly soon, in particular biogenesis process and the clinical application. Moreover, the versatility and classical identity of ts-RNA are still not clear enough. The ts-RNA are descended products from tRNA. It is hard to believe to alter ts-RNAs particularly without mature tRNAs modification. Possibly ts-RNA modulates translational process along with certain unique and abnormal functions. Better understanding the ts-RNA function due to inside understanding no doubt produces valuable insights on human physiology and pathophysiology.

References

References: chapter wise are cited before the Index at the end of the Book (1-455).

Chapter IV

Easy, efficient, and selective miRNA isolation protocols and tools for quantifications

Abstract

The existence of free nucleic acids from the blood was first identified around 60 years ago. In those initial days, it was considered that the presence of an increased concentration of many nucleases inside the plasma prevents the utilization of RNA as potential biomarkers from the blood samples. This notion is discouraged by the discovery of miRNAs from the blood as unique biomarkers for numerous diseases. In recent years, extraordinary signs of progress regarding clinical isolation, enrichment and quantification of miRNAs have been noticed to identify the exact functions of miRNA in the extracellular fluids due to better accessibility, high specificity, and sensitivity. It focuses the essential utilization in challenging clinical and fundamental research. This chapter aims for providing the best and easy individual miRNA isolation, quantification and massive miRNAs profiling that helps to demonstrate the knowledge of these miRNA molecules in different human diseases and their applicability in the therapeutic fields. Despite devoid of optimized protocols regarding the miRNAs isolation and quantification in current clinical practice, they still constitute a dependable tool in several RNA pharma industries. The selective miRNAs isolation from extracellular fluids and their multi-dimensional quantification leads important criteria for a used of biomarker for early diagnosis. These novel miRNAs and their profound distribution possess usual multiple advantages that convert them optimized candidates for biomarkers in a wide range of affiliations. As noted, the ideal and accessible condition are easy extraction procedure for potential biomarkers that implicates miRNAs can ideally isolated via liquid biopsies from blood, urine, and other bodily fluids. The current techniques emerge that many miRNAs will allow more specific therapeutic interventions as a usual approach for producing personalized medicine.

Most importantly, in this chapter the numerous published protocols from many commercials industries did not discuss in toto. Only their applicable features, major criteria and other pros and cons of the protocol are described for the betterment of different techniques used for mRNA extraction and quantification.

Introduction

Gene regulation of the human body is the sole and major challenge of many miRNAs.[1] They mediate the degradation of mRNAs and inhibit protein-producing translation via canonical and non-canonical mechanisms.[2] The predominant canonical mechanisms design that miRNA guided stands inside the mi-RISC complex base pair with target mRNA via its 3′-untranslated region (3′-UTR).[3] This conjugation appears following the seed sequence of the miRNA. At least the first 2-7 nucleotides from the 5′ end can undergo mRNA de-adenylation and finally degrade or inhibit translation.[4-6] In human cells, approximately 60% of interaction has occurred between the mi-RISC complex and mRNA via a non-canonical manner[7] suggesting that their chain sequences are not always exactly complementary.[8] This mechanism allows us to believe that a single miRNA can be able to target potentially numerous mRNAs, whereas one transcript has multiple miRNAs binding sequences. These possibilities believe that a vast number of biological interactions can occur at the same time for regulating the multiple transcripts.[3]

Intercellular signaling is a vital miRNA function also. Even though most of the miRNAs reside inside the cell, their big portion transport outside in the extracellular body fluids [9-13] as circulatory miRNAs. Blood, urine, saliva, seminal fluid, breast milk are the routine residing place for variety of miRNAs [10, 14] and other fluids through tissue damage, apoptosis, and necrosis.[2] Besides, they also migrate to micro-vesicles, exosomes or conjugated to a protein.[15,16] In these considers exogenous miRNA can be used as biomarkers for potential early diagnosis of several human diseases.

Around 10% out of the entire miRNA population in the human body are circulatory miRNAs that are secreted inside the exosomes whereas the remaining 90% form complexes with Ago-2 proteins, nucleophosmin 1 (NPM 1) and high-density lipoprotein (HDL).[16-18] These packaging's are useful for protecting miRNA from digestive juices and concentrated RNases reside inside the extracellular body fluids.[19]

The nano-sized exosomes are originated from micro-vesicles fused with the membrane. They carry many miRNAs secreted by the numerous cells *in vitro* and *in vivo*. The miRNAs transport inside many fluids of the human body.[10,14] In fact, many aggregations of exosomes with neighboring cells follow the idea of "message in a bottle" type intercellular communication.[20-22] In recent years, extraordinary progress regarding easy and efficient miRNA extraction, its many quantification strategies and massive miRNA profiling have been implemented their potential use as biomarkers in basic and fundamental clinical practice for diseased patients and healthy individuals.

Biomarker is a term defining multiple types of potential indicators of health or diseases. These precise and reliable indicators were first identified in pulse, the look and even the taste of urine as a medical sign. Recently the biomarkers are detected in many extracellular fluids of the human body. However, the best-known protein biomarkers faced multiple difficulties including time-consuming and expensive operative natures. These low amounts of clinically significant proteins are complex in their structure and struggle in search of appropriate detection methods. In the present context a massive hunt has been conducted for easy and efficient detection and quantification of multiple miRNAs related to many human diseases.

Moreover, to synthesize personalized medicine the design for miRNA exaction is identified as a new and more accurate biomarker. Thus, miRNA extraction and quantification procedures isolate many ideal biomarkers using minimally invasive procedures that require easy accessibility and sensitivity. Their presence allows to detect many prior clinical symptoms that should alter according to the disease progression in response to treatment. Finally, it should run towards translatable research to the clinical market.[23] No doubt their assays deliver less time and lower costs relative to the production of new antibodies as protein biomarkers. The multi-marker for perfect diagnosis, mediated accurate treatment and evaluating the sensitive responsiveness to treatment plans are provided another major advantage of miRNAs detection. The detected protein markers are expensive and time-consuming. The multi-marker miRNAs isolate by a non-invasive method for diagnosis and predict disease progression. Indeed, a urinary miRNA signature is detected early symptoms of renal fibrosis.[24] However, the research of miRNAs as biomarkers are still in its budding stages. One of the drawbacks is the lack of identical reproducibility. To resolve this issue, easy and efficient protocols are developed and described. In these protocols multiple processes like sample collection, transport, storage and data analyzing for many versatile technologies are often used.

Designs for detection

To sketch the design of multiple diagnostic processes that execute the optimize treatment plans several molecular testing's have been generated in many efficient and cost-effective miRNA identification methods. They are used for most common diseases related to pathogenesis worldwide. Nevertheless, their quest effects have rather utmost challenging because isolation of miRNA and their quantification phases have been relatively new and suffer with multiple limitations. Many parameters including age, gender, health/disease have a tremendous modulating effect in the blood fluid concentration.[25]

The popular easy and cheapest miRNA quantifications are used in quantitative reverse transcriptase PCR (RT-qPCR).[25-27] Stem-loop reverse transcription (RT)-based TaqMan or SYBR Green microRNA assays are the main PCR technique used for high sensitivity and specificity rates of diagnostics.[28] They have two-stepped methods, the first step miRNA primers at the 3′ end is required to bind to proceeding stem-loop reverse transcription,[29] while in the second step the miRNA target is measured by real-time PCR analysis.[29] On the other hand, poly (A) tailing-based SYBR miRNA assays and direct RT-based methods are applied as additional option.[28] Usual drawbacks of this technique have high risk of contamination during amplification steps or sensing errors during sample preparation.[25]

To bypass two following problems RTPCR is replaced by Northern blot hybridization that is widely used and most popular miRNAs quantification. This method follows several cumbersome steps before following the accurate scale of miRNAs measurement. The method involves an initial separation of total RNA by a polyacrylamide gel-based RNA denaturation step. Later it follows RNA transfer on a nylon membrane, a process of UV cross-linking and finally enters hybridization with radioactive probes.[30-33] However, this complicated cumbersome multi-stepped technique is very strenuous, laborious, expensive and requires a large quantity of total RNA species. Moreover, this procedure eliminates the low level of rare miRNA.[33] For betterment, this method requires the use of low quantity of RNA and shorten the processing time.[34-36]

At least five other methods are also used popularly for the same purpose measurement of individual miRNAs extraction protocols or generation of many microRNA profiles. Among them in situ hybridization or ISH uses radioactive, fluorescent or digoxygenin probes to complement with desired RNA sequences that are highly expensive.[37] The expensive steps and lengthy methods with a predisposition towards errors are included in the ISH.[36] The remaining methods next-generation sequencing (NGS), miRNA microarray or bead-array analysis, multiplex profiling and use of integrated microfluid platform are accurate techniques for multiple miRNAs analysis that are able to detect small difference of many miRNAs. However, their high performing cost make them unpopular, prevent against wider and general accessibility.[33]

Selective microRNA isolation protocols

The tiny miRNAs from circulatory blood are the promising candidates for disease biomarkers. These biomarker signatures are highly abundant in circulation and stable in many biological fluids. The efficiency of ideal miRNA isolation that is isolated from a low level of plasma using traditional RNA extracted reagents is a huge challenging task. To avoid such problems numerous commercial kits available in the market for readymade use. In this study, a comparative account of selected popular six extraction kits suitable for utilizing miRNA extraction from either fresh or frozen plasma samples are identified by different manufacturers. Application of these kits across samples and operators for future miRNA profiling and effective potential biomarker studies are described. The selected kits have potential variation in miRNA extraction specificity, however, are included for more broadly applicability, cheaper and more widely used in procedural settings. Kits A (miRNeasy Serum/Plasma) and B (miRNeasy Serum/Plasma Advanced) both extract and enrich small RNA populations <200nt from the total RNA. Kit C and Kit D (Quick-*cf*RNA Serum and Plasma) are used popularly for both RNA and mRNA in a cell-free system, The Kit E (Isolate II) accommodates small RNA fractionation based on size from the total RNA populations. Both Kit E (PureLink RNA) and Kit F (Monarch Total RNA) are isolated for total RNA populations without any enrichment of small RNA. Additionally, these protocols also employ the small RNA capturing from the total RNA extract.

Evaluation for extraction kits

The selective commercial kits are also sensitive to subjective operational factors. Based on easy use and small time input the selective kits are also considered three additional parameters including proven time-efficiency, diversified technical skill and applicable for high-throughput experiments. Use of these parameters ultimately generated high quality miRNA from minimal volume of clinical sample. Across these operational factors, miRNeasy Serum/Plasma, miRNeasy Serum/Plasma Advanced kit, and the Quick-*cf*RNA Serum and Plasma kit are the ideal three top performing popular kits.

To evaluate thermodynamic stability and GC containing range from the isolated microRNAs ten miRNAs are isolated by six different kits separately. The extracted miRNAs have GC content ranging from 31 to 66% and thermodynamic stability varies 4.6 to 1.8 kcal/mol respectively. These miRNAs are the rare samples for high GC content of their entire isolation procedures which are required for optimum qPCR measurement. However, the GC content of these miRNAs do not cover the total pool of GC content of isolated miRNAs required for better performance of the extraction kits.

Two Kits miRNeasy Serum/Plasma kit and miRNeasy Serum/Plasma Advanced kit show a little variation when fresh and frozen blood samples are used for extraction from a minimal volume of sample. The Quick-*cf*RNA Serum and plasma kit is the only third-best kit, because it functions a worse extraction from frozen plasma samples. Also, the protocol requires the use of extra vacuum and manifold instruments. It reduces its rating to use and produces multiple obstacles during laboratory use. It has extended incubation time and sample digestion (between 1-2 hours) but the time is reduced when multiple samples are processed at the same time. These two kits, miRNeasy kits and the Quick-*cf*RNA are only found in serum and plasma. In the commercial market these kits have fair competition with the kits, the PureLink and Monarch Total RNA kits those who accommodate many sample types for extraction. The kits require the least amount of plasma sample (100 μl), with higher Cq values. These kits suffer with multiple protocol steps including column clogging and higher cellular debris from the samples that cause a lower efficiency for extraction as a major challenge.[38] Two miRNeasy kits have isolated miRNA with high GC content during the extraction process. The PureLink RNA and Monarch Total RNA kits have a faster and simple isolation methods, conversely showing poorer performance, less reliability and greater variability in detecting miRNA from fresh and frozen plasma. Each kit shows a wide range of elution volumes and RNA concentration during repeat isolation. However, their reverse transcription reactions require identical levels of cDNA quality.

The stability of secondary structures and the GC content are two important factors governing the performance of each extraction kit. Earlier traditional guanidinium thiocyanate/phenol-chloroform reagents eliminate of low GC/high G miRNA populations during extraction. It is important that this issue should be critically examined in each commercial kit. Before use they should promote an extraction bias of low GC/high G miRNA during their selection of proper reagents. The Quick-*cf*RNA kit shows higher Cq values across all miRNAs. Conversely it has low GC% and high G in the frozen samples indicating that multiple freeze/thaw processes are involved in its performance. Truly, an impact of low GC/high G in miRNA detection is found in all other kits including PureLink RNA, Isolate II and Monarch Total RNA kits. In fact, it is found that addition of carrier RNA or glycogen increases eventual yields and improves amplification performance [39-42] that may be a better viable option for increased miRNA recovery. Both qPCR extraction performance and ideal operational factor are found in only two top performing kits (miRNeasy Serum/Plasma and Serum/Plasma Advanced kits). The advantage of Advanced kit has a non-toxic, phenol-free containing option. It does not require any special protective devices but differentiates clear phase separation that shortens the required extraction timing. Two kits are suitable for miRNA extraction and qPCR quantification assays. These superior q PCR performing kits also employ column-based extraction methods ideal for complicated biological samples.[40, 43, 44] In fact, the Advanced kit is the most reliable and ideal kit in the market because it does not need any long incubation and digestion. Conversely, the Qubit microRNA assay kit use for concentrated and pure miRNA from a minimum volume of either fresh or frozen plasma. In this case Qubit fluorometer used in microRNA assay is the reliable in miRNA measurements.

The impact of assay system, their experimental techniques, the quality of the data generation are the major aspects for optimization of nucleic acid extraction step.[45, 46] The efficient miRNA extraction steps from small volumes of plasma and serum samples face an utmost challenge because they mix with high concentration of multiple RNA, albumin and globulin contaminating proteins. Recent column-based extraction kits are advantageous over traditional guanidinium thiocyanate/phenol-chloroform RNA extraction procedures. The procedures isolate selective miRNA populations and lower the amount of adulterate and impure smaller sample volumes.[47-49] However, both commercial extraction kits and traditional TRIZOL extraction methods follow with on-column clean-up. [50] They do not able to recover small RNAs efficiently from both plasma and cerebrospinal fluid. Nonetheless, commercial kits have advantageous for recovery from small sample of plasma and serum from the human blood. These kits are ideal for miRNA profiling and isolation of miRNA biomarkers from the clinical samples.[51, 52]

In addition, plasma is the most reliable source for these three methods for small RNA quantification, in which small amounts of fresh and frozen plasma are used for testing. Further, six kits (Kit A-F) are used in a large scale for commercial quantitative PCR purposes. Kits A and B are useful for best detection of the miRNA qPCR reference genes both in fresh and frozen samples ($p<0.001$). Their performances are followed by Kit C, whereas two assay protocols Qubit and microRNA assay showed a least variation during miRNA quantitation (% CV 5.47, SEM±0.07), followed by the Nanodrop (% CV 7.01, SEM±0.92) and Agilent Bioanalyzer (% CV 59.21, SEM±1.31). Indeed, Kit A is the ideal for efficient miRNA isolation from lower volumes of fresh and frozen plasma, but Kit B is the top performing kit for miRNA detection and operational factors.

miRNA enrichment and quantification

For miRNA extraction from plasma or serum samples, human blood is collected and treated following protocols guidelines. Both fresh blood and frozen blood kept at -80°C are used. An ideal comparison between six popular different commercial kits using for miRNA extraction from fresh or frozen plasma is conducted (Kits A-F). Additionally, three methods of RNA enrichment and sample quantification are performed. (**Figure 1**) After enrichment of small RNAs less than 150 nt long, the samples are further estimated using 3 different appropriate methods.

Three miRNA extraction methods are ready to measure the RNA from frozen blood plasma. The poorly expressed small RNA samples with least sample variation are measured efficiently by Qubit microRNA assay (average % CV 5.47, SEM±0.0). (**Figure 2**) In fact, more than one repeat nucleic acid extraction at a time has failed to identify and purify of any samples by two sophisticated RNA quantified instruments NanoDrop and the Bioanalyzer. The Bioanalyzer and the Qubit assay are allowed to measure in similar levels of samples, whereas the NanoDrop is measured at a higher amount. Both Bioanalyzer and Nanodrop find a greater variation between readings compared to Qubit (average % CV 59.21, SEM±1.31 and % CV 7.01, SEM±0.92 respectively)

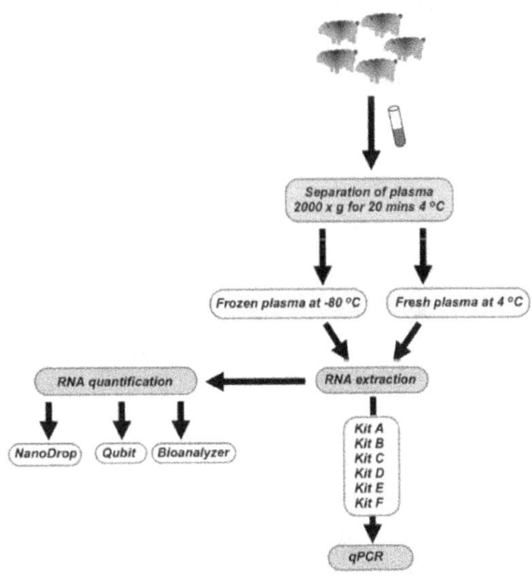

Figure-1. Display of experimental design. Whole blood was collected from 5 healthy animals and plasma stored at either−80°C or used fresh. RNA was isolated from both the fresh and frozen plasma using the 6 isolation kits (Kit A-F) and compared using qPCR. RNA isolated from frozen plasma samples was also estimated using 3 different methods relative to the most ideal method for estimating quantity and quality. (Loosely Adapted, Redraw and Modify using skeleton from Wright *et al.*, 2020)

The Qubit for a microRNA assay shows at least less variation in measurement of small, enriched RNA samples isolated from less amount of plasma. Among three instruments the Nanodrop spectrophotometer shows the least variation relative to quantification performed by Qubit and Bioanalyzer at the same sample. Basically, a high measurement values and low level of apparent of sensitivity identify in Nanodrop assay are due to contaminants with phenol and silica associated with extraction material. However, this dye-based method can overcome these contaminations in the sample. In contrast, the Bioanalyzer exhibited a high variation of the same sample measurements relative to the Qubit and Nanodrop. The Nanodrop shows a great reliability of the same sample. On the other hand, only two measurements per sample are completed by Bioanalyzer, while at least 3 repeats are performed in other two methods that are contributing the higher variation between readings. However, these advantages of bioanalyzer although apply exceptionally, many major disadvantages of bioanalysis are associated with for its high cost and other expensive reagents. Conversely, the reagents of Qubit machine are cheaper and highly accessible.

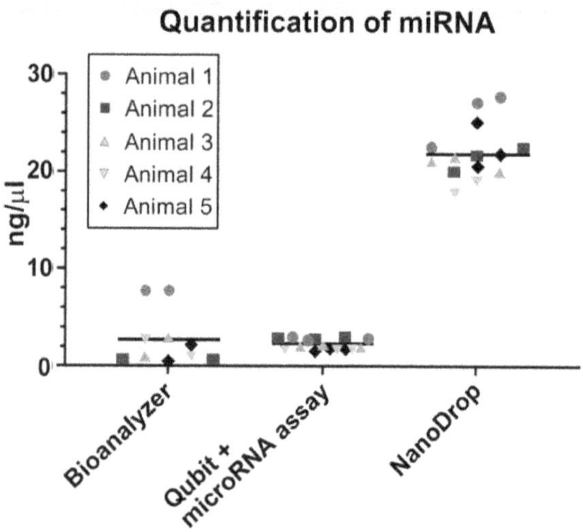

Figure-2. Relativity of different miRNA quantification methods. RNA isolates from 5 frozen ovine plasma samples were further measured using 3 separate methodologies. Multiple estimations were calculated for each sample, with the mean for each method represented by a horizontal line. (Loosely Adapted, Redraw and Modify using skeleton from Wright *et al.,* 2020)

Interestingly all spectrophotometric methods produced similar ideal results for miRNA quantification methodologies from plasma samples. These spectrophotometric methods are not sufficient for small quantities of plasma miRNA.[53] In other cases, the Qubit fluorometric assays are better suited from miRNA studies especially when it is analyzed from the multiple samples. In contrast, the microfluidic methodology of the Bioanalyzer demonstrated a consistent measurement and lower variation.[54] Particularly, Agilent Bioanalyzer only analyses a detailed quality measurement of the RNA integrity.[55] Interestingly nanodrop claims a top notched operational device for quantification and the variability of small sample sizes. However, such advantages are not sensitized when Qubit fluorometers are provided less room for operator error than the Bioanalyzer. The bioanalyzer requires multiple technical steps including chip loading and priming. It is also applicable for multiple users and reduces time for measurement analysis.

Detection of miRNAs

The miRNA extractions from multiple samples are analyzed in a similar manner by the Kits A and B that outperform Kits D, E, and F ($p<0.001$). The Kit A is the best overall both in fresh and frozen plasma. The Kits B, C and D function at an inferior level of extraction in the frozen samples ($p<0.001$). The Kit C isolates best extraction quality than both Kits A and B for fresh plasma samples, whereas its extraction from the frozen samples remains worse than others. ($p<0.001$). The Kits E and F do not show any significant difference in RNA extraction from the fresh and frozen samples however they have a higher Cq (quantification cycle) values among multiple miRNAs relative to other top performing kits.

Comparison of operational methods

The use and application of each kit have a wide range of operational factors for miRNA isolation. (**Table 1**) These factors contain multiple criteria including RNA yield, ease of use, time input, sample volume input and cost per sample etc. Indeed, the average yield of each kit is extracted based on one of the key factors. Kits A and B appear to isolate the highest amount of RNA. Kit C is used for extraction of good quality of RNA from fresh plasma. In this kit an equal amount for RNA is isolated from fresh and/or frozen plasma samples. This kit isolates 18-20 ng RNA from 200µl of plasma. Kit F accommodates larger elution volumes (50µl) that produces RNA in a lesser concentration than kit A that has lower elution volume (12µl). On other hand, Kit A (12µl) and Kit D are used for lowest minimum sample input of 100µl of plasma relative to other remaining kits to 200µl, while many other kits (A-C) generated a higher maximum input in a volume up to 600µl.

Considering the use and time required per extraction Kits A and B are employed the most efficient top performing methods with minimum time input (approximately 40-45 minutes for entire extraction). Conversely, Kits C and D require extra equipment or low usability of kit components for completion of extraction procedure. Kits A and B are easy to use that shows best time input and sample input and have a high performance relative to Kits D, E and F. Considering the cost, Kit A is the most expensive, while Kit B is moderate and reasonable in cost, and it has the best performing kit across the operational categories. However, cost wise Kit D and Kit F are cheapest and the most affordable choices, but they produce the lowest level of miRNA from plasma.

Table-1. Comparison of commercial miRNA extraction kits based on clinical factors

Variable kits	Sample needed	Time required	Easy to use	Cost/Sample
Kit A	+ + (200 µl)	+ (45 mins.)	+ + +	+ + + ($20.88)
Kit B	+ + (200 µl)	+ (40 mins.)	+ + +	+ + ($15.52)
Kit C	+ + (200 µl)	+ + + (2.5 hrs.)	+	+ + ($ 14.32)
Kit D	+ (100 µl)	+ (40 mins.)	+.	+ + ($ 13.40)
Kit E	+ + (200 µl)	+ (25 mins.)	+ +	+ + ($ 19.20)
Kit F	+ + (200 µl)	+ (25 mins.)	+ +	+ + ($ 7.08)
Best in Category	D	A/B/D/E/F	A/B	F
Best overall Kits KITA+B. + = Low, + + = Medium + + + = High				

Two stepped miRNA quantifications

The miRNA quantification is ideally isolated using a two-step process. First, the stem-loop RT primer designed is hybridized to the miRNA molecule and then further reverse transcribed in a pulsed RT reaction. Finally, the RT product was amplified using two primer a miRNA-specific forward primer and the universal reverse primer. (**Figure 3A**) The amplification product is displayed on an agarose gel using ethidium bromide staining. The amplification is also done in real-time, using the SYBR Green I assay. (**Figure 3B**) Moreover, a Universal Probe Library (UPL; Roche Diagnostics) is developed for RT-PCR method using miRNA detection and measurement that increases specificity at the time of estimating of low abundant miRNA expression. (**Figure 3C**)

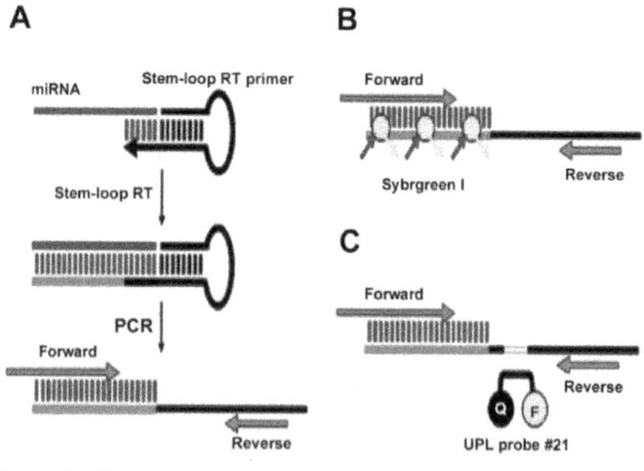

Figure-3. Schematic diagram of stem-loop RT-PCR miRNA assays. **A.** Stem-loop RT followed by end-point PCR. Stem-loop RT primers bind to the 3' portion of miRNA

molecules, initiating reverse transcription of the miRNA. Then, the RT product is amplified using a miRNA specific forward primer and the universal reverse primer. **B.** SYBR Green I assay. **C.** Universal Probe Library (UPL) probe assay. Highlighted in yellow is the UPL probe #21 binding site. (Loosely Adapted, Redraw and Modify using skeleton from Varkonyi-Gasic *et al.*, 2007)

Stem-looP pulse reverse transcription

Sensitivity

To determine the higher sensitivity of the stem-loop RT-PCR, a stepwise dilution series of total RNA isolated from *human blood* is used. Further an amplification is conducted in a semi-quantitative manner using 20–35 cycles The amplified volume shows that as little as 20 pg total RNA can be isolated by this two stepped miRNAs isolation procedure only after 35 cycles of PCR. No amplification or a weak band of primer-dimers is found at the same number of cycles in minus-RT control and from the water control. For amplification of the reverse complement (RC) strand of miRNA different primers in stem-loop RT-PCR reactions are used. However, some non-specific amplification in control reactions often produce at 40 or more in PCR higher cycles. Therefore, only 35 cycles is the only limit for all further end-point amplification reactions. The expression of most miRNAs from human blood samples are performed at standard 35 cycles. In some cases, miRNAs detected by RNA gel blot analysis that show a higher level of expression relative to the presence of low level in other miRNAs. In those higher expression cases only 28 cycles of amplification are allowed to conduct in stem-loop RT-PCR. To test the stem-loop RT-PCR measurement for extracting miRNA sequences, very small amounts of human samples are required. Interestingly, if RNA was isolated and purified at a concentration ranging from 300-400 ng/ml only as low as 0.1μl RNA is required for detection of miRNAs in stem-loop RT-PCR.

Specificity

General PCR amplification methods have a devoid of specificity for mature miRNAs. During processing of mature miRNAs, the large capped and polyadenylated transcripts (primary or pri-miRNA) are used for precursor molecules that first produce to short-lived hairpin intermediates (pre-miRNA) and are ready to export. Ultimately, they turn into mature single stranded miRNAs. In certain occasions, total RNA enriches with large amounts of pri-miRNAs. The Northern blot RNA gel analyses exhibit a reasonable expression of pri-miRNAs from the human blood. To test the existence of stem-loop RT-PCR assays, it is possible total RNA is differentiated into mature miRNAs and their pri-miRNAs. These reactions in circulatory RNAs are conducted both from the high molecular weight (HMW) and low molecular weight (LMW) RNA. The slight elevated level of amplification of total RNA relative to LMW RNA is determined the RNA losses during purification. In such cases, only 60–75% RNA is recovered. In normal cases, the amplification of HMW RNA is far more than LMW RNA.

The stem-loop pulsed RT-PCR extractions and measurement is the rapid, sensitive, class of device that are convenient for screening and profiling of many miRNAs in a speedy and inexpensive manner. A low tissue and time-consuming method relative to standard gel-based methods are conducted experimentally in this procedure. This assay also performs miRNA profiling from as little as 20 pg circulatory RNA. Further, this approach will be applicable in detection and miRNA measurements across kingdoms and for other small regulatory RNA sequences including artificial miRNAs and short interfering RNAs (siRNAs).

Advantages of Universal Stem-loop (USLP) primers

MicroRNA revolutionaries as a fine tuner for gene expression in several cellular and pathological processes including organ development, cell death, cell proliferation, hematopoiesis, and patterning of the nervous tissues.[56] For miRNA quantification (miR) Stem-loop RT-PCR is the most frequently used accepted technique. However, it is suffered by time-, reagent-consuming and inconvenient for scanning events. To avoid those difficulties in 2005 an innovative method named Universal Stem-looP primer' (USLP) was established that cut down reagent costs and time. Here, a specific sequence at the 39 end of the traditional stem-loop primer (TSLP) is replaced with 8 random nucleotides. This exchange claims a great advantage in miR profiling and to semi-quantify expression of miRs. The USLP method has high precision and delicate because of its low ICV (ICV,2.5%) and sensitive in 103 copies/ml miR concentration. In comparison with the TSLP, USLP saved 75% of the primer costs and 60% of the test time. The USLP method has a simple, rapid, precise, sensitive and cost-effective methods that is appropriate for quick screening miRNA profiles.

Method

To avoid non-specific amplification, the USLP 8 random nucleotides have added at the 39-end tail. The random nucleotides combine with all complementary mature miR sequences instead of 8 specific nucleotides. However, a specific sequence has more binding strength than 8 random sequences to avoid mismatch in base pairing formation at the 39 tailed ends. The use of 8 nt sequence are unique sequences in USLP, that are testing by the sequences presence in human Nucleotide BLAST Database.

USLP Primer sequence

To understand the exact identity, a specific USLP sequence is utilized by taking miR150 sequence as a notable example. Initially, the stem-loop primer is designed underlying the complementary miR-150 nucleotides. These are miR-150 stem-loop primer: 59GAAAGAAGGCGAGGAGCAGATCGAGGAAGAAGA CGGAAGAATGTGCGTCTCGCCTTCTTTCCACTGGTA-39.

The universal stem loop (USLP) primer: 59-GAAAGAAGGC GAGGAGCAGA

TCGAGGAAGAAGACGGAAGAATGTGCGTCTCGCCTTCTTTC NNNNN NNN-39 This sequence is modified using Primer Premier 5.

To amplify the following primers are used for each miRNA, miR-150

Forward qPCR primer for miR150: 59-GCTCTCCCAAC CCTTGTACC-39. The 2 nt was added to 59 nt that are underlined is to get the Tm to 58.5℃, being close to the reverse primer's Tm (57.9℃).

Reverse qPCR Primer: 59-CGAGGAAGAAGACGGAAGAAT-39. After reverse transcription of all miRs to cDNA, a universal PCR primer is derived from sequences within USLP.

During Real time PCR analysis if primers concentrations of USLP reached 1 mmol/L in the reverse transcription the PCR baseline reduces to a minimum of 1022, and the Cq values reaches into two cycles earlier. In fact, optimizing the concentration of USLP raises the earlier fluorescence value, that enhances curve slope and improves amplification efficiency. Therefore, rising the reaction concentration in 1 mmol/L is required for higher detection efficiency for PCR.

Comparisons USLP and TSLP primers

1. *Amplification efficiency*: The amplification efficiency of the USLP approach (88.6%) is found relative to the TSLP assay (107.2%), after seven gradient dilutions, but correlation coefficients (R2) in both cases are always greater than 0.98. In fact, the Cq values of the TSLP method are always lower than USLP. It is in between 2.2–3.4 range. These findings believe that USLP synthesizes lower cDNA concentration due to 8 random nt primer.
2. *Times and costs*: Only one kind of stem-loop primer is used in improved USLP method for all mature miRs in the reverse transcription reaction, whereas the TSLP method requires each set of specific stem-loop primers for each miRNA under the same condition. This primer reduces labor consumption and cost. In comparison, around 60% testing time and 75% primer cost is saved in the USLP method.

Advantages

Most popular and easy methods for miRNA quantification are largely based on microarrays and real time PCR, which are time-consuming and expensive.[57] Thus, for microRNA detection rapid and simple methods are required. An improved method including miR faster and more cost-effective is established by using stem loop-qPCR analysis. Among usual Stem-loop RT primers, the innovative USLP primers have higher sensitivity, specificity, and precision than conventional linear RT primers. Finally, the spatial constraint of the stem-loop structure possibly prevents the RT primer from binding double-stranded genomic DNA molecules and increases the thermal stability of the RNA-DNA heteroduplex.[58] Therefore, Universal stem-loop primers is designed the amplification in miRNAs screening and quantitative analysis.

First of all, TSLP method requires specific stem-loop primers (70, 80 bp) for each miRs that cannot save time and expenses. Using a USLP method reverse transcription of nearly hundred miRs can be conducted in one tube. In a separate instance, microarray is high-throughput profiling study that are expensive for screening all miRs in specific samples. The numerous miRs are detected advantageously via USLP method that is cheaper than microarray. Thus, this method makes it easy, simple cost-effective screening for numerous miRs that function as miR biomarkers of certain diseases. This method also used for specimen efficiency and reduces the number of samples

However, the sensitivity of the USLP method is lower than the normal TSLP. This difficulty blocks for accurate quantitative detection of low expressing miRs for the majority of miRs. This method is responsible extracting lower concentration of cDNA synthesized by USLP with 8 random nt primers. In spite of this drawback the USLP method is more appropriate for numerous miRs screening rather than measuring low-level expression of independent miRs.

End point Real time amplification

SYBR Green assay and UPL assay

The real time PCR SYBR Green I assay using 35 cycles is derived from an aliquot of cDNA previously used for sensitive of end-point PCRs and via Universal Primer Library (UPL, TaqMan primers) assay. The exact protocol for Real time PCR quantification using both assay systems do not discuss as noted earlier, but different parameters and features are described. In these assays, only 20 pg of total RNA is sufficient for respective miRNA amplification without minus RT control and the water control amplifications. A careful note is that a larger number of amplification cycles must be avoided because the contaminant nonspecific amplifications for 40-80 bp sequences cannot be distinguished from the desired amplification product. The nonspecific products are occasionally indistinguishable from the specific PCR products by melting-curve analysis.

Dual labeled hydrolysis probes like TaqMan (Applied Biosystems) and recent discovery UPL (Roche Diagnostics) probes are used routinely to enhance the specificity of real-time quantitative PCR (qPCR) assays for detecting human miRNAs.[59, 60] However, each particular miRNA sequence is required for amplification of unique TaqMan probe. It is highly expensive and is really impractical when large numbers of miRNAs are amplified. To sort out this problem a universal and efficient single hybridization probe is developed for universal hybridization. In fact, a short hydrolysis probe of 8 nucleotides together with locked nucleic acid (LNA) nucleotides at the end are used for enhancing binding specificity and melting temperature. Sometimes stem-loop oligonucleotides are redesigned to insert a UPL probe (Roche Diagnostics) in reverse complementary sequence in the stem region between and the universal reverse oligonucleotide sequences and the miRNA-specific sequence. Therefore, special Pulsed stem-loop RT reactions are employed for UPL qPCR assay on an RNA dilution series.

The amplified curves are obtained using SYBR Green I assay and the UPL probe assay. For a 40 cycles SYBR Green I assay amplification cloning and sequencing of amplicons is required to identify the presence of concatenated primer sequences generating from the minus-RT and water controls. In the UPL methods, amplification curves always correlates with RNA concentration even in 45 cycles of UPL probe assay amplification. However, in this case negative controls originated from minus-RT and water do not produce any detectable signal.

Other Quantitative assays

RNA Sequencing

An ideal approach for global miRNA profiling is employed by Next generation sequencing (NGS) that is the most suitable and sensitive method for unknown miRNA sequences relative to microarray. However, wider adoption of NGS (small RNA-Seq) is suffered by multiple biases for exact detection of miRNAs. The occurrence of many biases including sequencing bias during library preparation, adapter ligation involves in secondary structure formation that can be prevented by use of degenerated bases.[61-64] Thus, miRNA values are not calculated at an absolute level.

The dilute level of miRNAs in plasma raises a potential drawback for library preparation[65-68] particularly from small volumes. The use of Low RNA produces high frequency of adapter dimer and non-miRNA reads. The many non-target reads isolated from exogenous RNAs have a high proportion of contamination.[69-71] To identify the influence of diet, exogenous miRNAs are examined just before and after the diet intake. The gradual reduction of sequencing cost influences high throughput analysis. RNA-Seq is the essential technique for circulating miRNAs measuring. Indeed, the development of robust sequencing protocol addressing biases that lower RNA input for sequencing are useful for miRNA profiling as a potential tool to isolate biomarkers. So, these criteria should meet up during sequencing.

An ideal approach for global miRNA profiling is employed by Next Generation Sequencing (NGS) that is the most suitable and sensitive method for unknown miRNA sequences relative to microarray. However, wider adoption of NGS (small RNA-Seq) is suffered by multiple biases for identifying the miRNAs. The occurrence of several biases includes Sequencing bias during library preparation, adapter ligation involved in secondary structure formation that can be prevented by use of degenerated bases.[61-64] Thus, miRNA values are not calculated at an absolute level.

The small RNA-Seq library preparations are continually improving day by day[72, 73] considering various hurdles including low miRNA concentration and reduction of bases etc. Three commercially available small RNA libraries are popular for counteracting one or both drawbacks during the preparation. Clean Tag TM Small RNA Library Prep Kit (TriLink) prevents adapter dimer formation from low inputs of RNA.[73] NEXT flex® Small RNA Sequencing Kit v3 (Bio

Scientific) inhibits sequencing bias and adapter dimer formation from low RNA amount.[72] QIA seq miRNA Library Kit (QIAGEN) has ideal chemistry to protect against bias such as adapter dimer formation and contaminating non-miRNAs from low amounts of RNA. QIA seq is the only kit to insert unique molecular indices (UMIs) to each cDNA for rectifying PCR bias. Truly, the miRNA profile generated from plasma samples is greatly influenced by two parameters, selection of library preparation kit and ideal RNA extraction method at a lesser extent.

Design

The purified RNA is ideal for library preparation. Three library preparation kits, Clean-Tag (CT), NEXT-flex (NF) and QIA-seq (Q) are prepared from RNA extracted following the miRNeasy extraction kit. The exact amount of RNA isolated from human plasma differs in the preparation of these libraries. At least 5μl is needed for QIA-seq library. The sample input for NEXT-flex varies from 5 to 10.5 μl of plasma derived RNA, whereas only 2μl of RNA is maximally required for Clean-Tag library formation. In each case RNA is purified by free magnetic bead clean up and PAGE size selection.

The UMIs are a readymade solution for preventing PCR bias in only QIA-seq library preparation kit. Less than 5 million raw reads mapping are calculated to miRNAs in all QIA-seq libraries with and without UMI correction. The tally of average number of reads with and without UMI correction apply for a strong correlation for both mQ and MQ libraries. The identical count of reads mapping with or without UMI correction inhibits PCR bias introduced during library preparation. Indeed, the negligible PCR[61, 63, 74] shows amplicons with an identical length of all miRNAs that are amplified using the same flanking primers. Although amplification cycle numbers are the same for QIA seq libraries prepared from RNA isolated from two different extraction kits. In this library, the reads are 1.48 and 4.29 times higher respectively before UMI correction.

Further consideration

Induction of adapter ligation[61,62] largely develops three library preparation kits, while other small RNA kits prepare without ligation steps. Two kits, the CATS Small RNA-seq Kit (Diagenode) and SMARTer smRNA-Seq Kit (Takara Bio) are developed, in which 3′ polyadenylation and 5′ template switching to enable ligation-free library preparation are applied. It is shown that adapter ligation in the synthetic RNA miRNA pools has more bias that ligation-free library preparation.[75,76] In fact, bias has shown a significant impact all ideal reaction condition.[64] NEXT-flex protocol shows reduction in bias and utilizes degenerative adapters that may enable an additional reduction in bias in the NEXT-flex protocol beyond provided by degenerate adapters. The fixed sequence adapters of the QIA-seq protocol constructs the libraries with minimum bias, which is determined by diversity and its correlation with RT-PcR.

The small size difference in different miRNA-containing libraries produce adapter dimer. A gel purification step is required for most protocols. It is laborious, time consuming and increases more bias.[77] In addition, for consistent data analysis in all crucial protocols require small differences showing number of mismatches and length variations of the miRNA. The CT and Q library protocols show a gel free size selection during RNA concentration generated from plasma. The NF also needs a gel purification step that contributes lower diversity due to loss of lowly expressed miRNAs.

Three popular library preparation kits are reliable for detecting miRNAs. The selection of each library kit has a tremendous effect on miRNA profiles. However, RNA extraction and library preparation methods introduce dramatic variation relative to biological differentiation between individuals.

MicroRNA Bead-array analysis

Numerous miRNA features are consisting of their small size, devoid of polyadenylated tails, strong tendency for complementarity to their mRNA targets with imperfect sequence homology. The sequence homology among family members makes them venerable to quantification. Several methods have been generated for miRNA profiling for example quantitative PCR,[78] sequencing,[79-81] northern blotting [82, 83] and microarray analyses based on either direct hybridization[84-88] or hybridization associated with enzymatic extension.[89] While these methods still have some technical limitations with a narrow success in a variety of profiling studies.

To overcome these limitations a highly sensitive, specific, delicate, and reproducible microRNA (miRNA) estimation procedure is developed using the Bead-Array TM technology. This method is associated with an enzyme-assisted specific step, in which members of miRNA families and a universal PCR for amplifying all targets before array hybridization can be distinguished in a solid-phase primer extension method. The method has a 3.5–4 log (105–109 molecules) dynamic range and is equipped with1.2-to 1.3-fold-differences between the samples. It is highly comparable with RT–PCR (R2 = 0.85–0.90) and direct sequencing (R=0.87–0.89). In this method, for high-throughput miRNA expression procedure a 96-sample array matrix can be utilized for the large quantity samples.

Design

In this method, a capture sequence immobilized on the universal array platform is used for complementarity of a given miRNA. The input RNA containing miRNAs is well polyadenylated equally for hybridizing to their target cDNA sequences simultaneously. In this case, assay oligos are used for allele specific primer extension with the 3' end that are complementary to bases present in the cognate sequence in the cDNA target. In depth, this enzymatic step increases to avoid complementarity between homologous miRNA and mRNA sequences for providing a second level of specificity. These enzymatic steps are widely used in single nucleotide polymorphism and are involved for small sized miRNA measurement for optimal assay probes.[90, 91]

Later a universal PCR is performed for highly sensitive amplification of all extended products. The extended oligo population the PCR primers are used for unbiased amplification of all uniform sized target sequences. The labelled amplicon hybridizes with a universal array carrying complementary sequences.[92, 93] The signal intensity of each platform of the array corresponds to the relative to miRNA abundances targeted in the original sample.

Specificity

For specific assay, the perfect match and central mismatch probes for miRNAs are generated. Later, the intensity ratio of the perfect match versus the mismatch probes across all samples are calculated. Moreover, high correlation (R2 > 0.97) between low molecular weight (LMW) RNA and profiles generated with total RNA are evaluated. The microarray intensity data are very specific, the overall miRNA profile in each case is not contamination with mRNAs and ribosomal RNAs (rRNAs).

Advantages over microarray

A sensitive and specific miRNA estimated profile has a 3.5–4 log dynamic range that determines an average 1.2-to 1.3-fold differences with 95% confidence. Three features of the method are: (i) an enzymatic 3'-end discrimination method with specific DNA sequence-mediated hybridization; (ii) a solid-phase cDNA platform containing multiplexing capacity for miRNA profiling [92, 94] and (iii) a universal PCR amplification the method for similar amplicon size with universal PCR primers.[92] The method is generated feeding only 100–200 ng total RNA. It has an identical expression profile generated from total RNA and low molecular weight RNA. In fact, before sample labelling small RNA purification is not essential that is required for high-throughput miRNA analysis in a clinical setting.

In the method, detection of mature miRNAs is more effective than pre-miRNAs. First the cDNA synthesis short mature miRNA templates are functioned efficiently relative to pre-miRNAs. The mature RNA extension is advantageous because longer sequences devoid of any complete extension to the same degree under the ideal experimental condition. More importantly, the pre-miRNA stem-loop structure interfere with cDNA synthesis and assay oligo annealing that helps to increase the relative detection of mature miRNAs.

Here the direct sequence of small RNA and RT PCR is compared. This comprehensive digital miRNA expression profiling (DGE) has following features 1) It quantifies all RNAs completely in an unbiased manner. 2) The sequencing avoids all sorts of cross-hybridization that may grow by array versus array and array versus RT-PCR comparisons. The objective assessment of all kinds of miRNA platforms employs a comparative ratio between microarray versus DGE microarray. This comparison helps in generation both array and sequences in certain samples.

The inclusion of continuously discovered new miRNA sequences allows to generate a flexible array platform that help to generate highly desirable miRNA platforms. This platform contains universal array readout; therefore, inclusion of new miRNA assay probes is possible to add in the existing assay oligo pools without manufacturing new array platform. This feature adds array upgradation of existing microarray platforms. Thus, use of a 96-sample array matrix platform in this method provides a useful advantage for high-throughput estimation procedures of miRNAs from large numbers of samples and helps to expose fundamental disease mechanisms.

Multiplex miRNA profiling

The alteration of miRNA expression levels commonly exists in several major diseases including cancers, cardiovascular diseases, rheumatic diseases, and neurological disorders.[95-97] In each disease, an aberrant expression of a group of miRNAs is noticed rather than a single miRNA.[96, 98-100] Therefore, multiplex analysis of miRNA expression is highly useful not only to test miRNA regulated biological methods, but also to identify potential early diagnostic markers for many clinical settings.

Multiple features of miRNA like small size, higher sequence homology among the family members and multiple intrinsic properties raises the major setbacks for miRNA expression profiling. Many methods have been employed to test the miRNA expression patterns like widely used Northern blotting, quantitative RT-PCR (qRT-PCR), rolling circle amplification as well as rare and delicate methods loop-mediated isothermal amplification (LAMP), and isothermal exponential amplification reaction (EXPAR).[101-104] However, these methods are not able to analyze multiple miRNAs simultaneously in a single assay. Among them in certain instances qRT-PCR has been successfully used only for single plex assays.

A rare combination between high-resolution capillary electrophoresis single strand conformation polymorphism (CE-SSCP) analysis and a modified isothermal EXPAR is used for multiple miRNA profiling. (**Figure 4**) In this combination, CE-SSCP is used for identifying multiple amplified probes using a single fluorescent tag. Nonconventional CE requires the probes to have different lengths,[105,106] CE-SSCP utilizes Pluronic polymer as the separation matrix that increases the resolution dramatically to assay up to 20 targets simultaneously in a single analysis.[105-109]

Design

In this method an exponential amplification miRNA signal is developed in three steps, 1) In the first step, the repeated strand extensions and nicking steps develop multiple trigger molecules from a single miRNA molecule. In this case, the DNA polymerase strand displacement activity is able to generate initial extension of the hybridized stem-loop probe. The additional extension develops after nicking and releasing of triggers. The stem-loop probes have a nicking enzyme recognition sequence between the instigated sequence at the 5′ end and an anti-miRNA sequence at the 3′ end. In stem-loop structure a base stacking interaction develops the RNA-DNA heteroduplex that increases reverse transcription efficiency and thermal stability in relatively short RNAs. 2) In the second step, the spatially constrained stem-loop structure facilitates the specificity of mature miRNA primer only to 3′ end for inhibiting the primary miRNA (pri-miRNA) and precursor miRNA (pre-miRNA) binding.[110, 111] 3) Finally, the stem-loop structure reduces the non-specific interactions with genomic DNA in a negligible level. The stem-loop primer detaches transcribed miRNA sequences from the long genomic DNA fragments. As a result, it does not require any prior DNase treatment. In second step, the repeated nicking and extension of the amplifiers amplifies the signal barcodes. The reaction in this step is only instigated by the amplifier when hybridized with first step triggers. The signal barcodes carry a distinct region with variable sequence and labeling primer sequences. Later separate assay products can be utilized from each miRNA species in the final CE-SSCP analysis. In the final step, additional triggers are designed to generate via further extension of signal barcodes of the step 2. Lastly, the stem-loop probes can be hybridized the last 15 bases of the signal barcodes using initiating additional extension and nicking steps.[112]

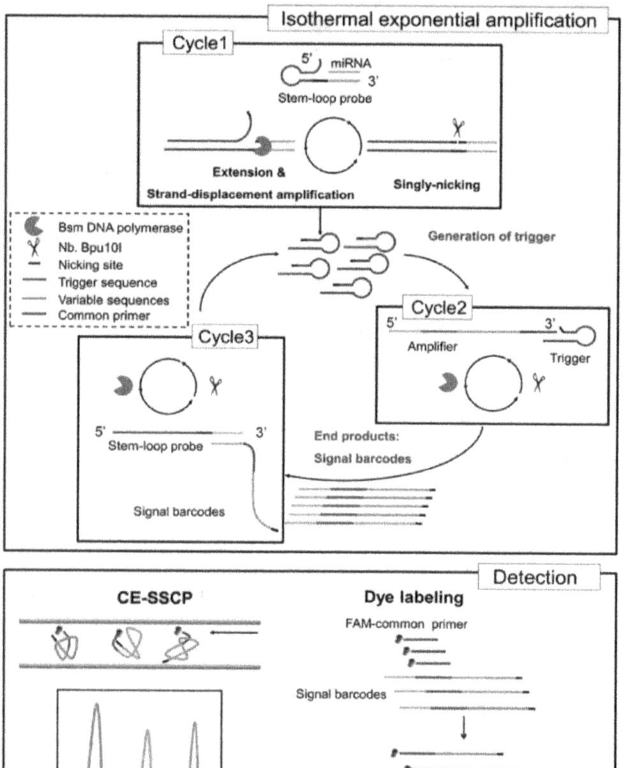

Figure-4. Schematic illustration of isothermal exponential amplification (EXPAR) and CE-SSCP analysis. Exponential amplification of miRNA signals under isothermal conditions through Cycles 1, 2, and 3.

Cycle 1: Generation of triggers from miRNAs by using stem-loop probe. When miRNAs hybridize with stem-loop structured probe, triggers are generated through repetitive extension, single-strand nicking, and strand- displacement by Bsm DNA polymerase and Nb. Bpu10I. **Cycle 2:** Amplification of signal barcodes from triggers generated in Cycle 1. When the trigger hybridizes with the amplifier, signal barcodes are generated in the same way as triggers are generated (i.e., repetitive extension, single-strand nicking, and strand displacement). **Cycle 3:** Further generation of triggers for exponential amplification of signal barcodes. Triggers can be generated from both signal barcode and stem-loop probes. Signal barcodes are detected by dye labeling and CE-SSCP.

For CE-SSCP analysis, the signal barcode must be labeled with a fluorescent dye. Products can be analyzed simultaneously via CE-SSCP using fluorescently labeled common primer in the subsequent linear amplification step. (Loosely Adapted, Redraw and Modify using skeleton from Jeongkyeong Na *et al.*, 2017)

===

During exponential amplification the second step is labeled with fluorescence and analyzed by CE-SSCP simultaneously. (**Figure 4**) The variable sequences of signal barcodes carry several single-strand DNA (ssDNA) folding stability. In CE-SSCP analysis, according to their folding stability similar sized ssDNA molecules at multiple rates is transported with the help of free energy difference (ΔG) between denatured and folded states. The free energy differences (ΔG) of secondary folding suggest a good correlation with CE-SSCP mobility.[112] The MFOLD software is used popularly to interpret secondary structure of 30-40 bp ranging sequences in absence of triple helical structure predicted tools.

Optimal Assay Condition

To generate the highest level of signal amplification of miRNA assay the concentrations of enzymes, probes and reaction period are examined. An ideal mixture of Nb. Bpu10I nicking enzyme and the Bsm DNA polymerase is required for optimal efficiency achievement.[104,113] Among various combinations of these two enzymes, 0.5 U of Bsm DNA polymerase accompany with 7 U of Nb. Bpu10I nicking enzyme optimally for yielding the ideal for maximum signal production. Interestingly, 7 U of the nicking enzyme that possibly shows non-specific endonuclease activity reduces the signal intensity markedly.

Subsequently, the stem-loop probes and amplifiers concentration along with reactive time are tested for optimized level by a series of experiments. For fixing only 30 fmol of miRNA several possible ratios between stem-loop probes and amplifiers (1:1, 1:2, 1:3, 1:4, 2:1, 3:1, and 4:1) are examined. To optimize maximum signal intensity a combination of concentration between stem-loop probe (60 fmol) and amplifier (30 fmol) are used for a 3-h reaction period. Utilizing the optimized conditions, a minimum detection limit nearly 2 fmol is identified that is comparable to isothermal EXPAR methods.[103, 114] In addition, the variety of the multiplex assays are determined by analyzing single-species miRNA samples. The variable miRNA peaks in the samples are also assayed, because the intensity of the peaks alters accordingly with the input miRNA amount. The quantity of miRNA assay along their biologically relevant dynamic range is also determined. A statistical correlation is noticed between the measured and expected miRNA fold change values that only show 5% average standard deviation reproducible within triplicate experiments.

For high specific multiplex quantification, a combination between a modified isothermal EXPAR method and CE-SSCP are used together with stem-loop probes. The power multiplexing is induced to an eight plex assay for utilizing high-resolution CE-SSCP as the detection method. Further the combined multiplex method carries universal amplifiers that amplify different sets of miRNA targets. The isothermal EXPAR involves amplification without including any target miRNA sequence. The fold changes of six miRNAs in both synthesized miRNA mixtures and in total RNA extract can be validated by this combined multiplex. Later high-resolution CE-SSCP introductions express multiple miRNAs without length-dependent bias. Thus, multiplex is involved in precise quantification of multiple miRNAs related to several biological methods.

Integral microfluid devices

The microRNAs (miRNAs) carrying extracellular vesicles (EV) have tremendous and un-comparable potential as early biomarkers detection of disease. The rapid and simple PCR-free integrated microfluidics platform for absolute measurement (<10% uncertainty) of both plasma free-floating miRNAs and EV-miRNAs having higher sensitivity. This microfluid device has tremendous reduction of assay processing time just 30 mins relative to 13 hrs as well as use of a minimum 20 μl sample in opposition of 1 ml sample that are used in conventional RT-PCR techniques. In this platform, a surface acoustic wave EV lysis microfluidic chip is integrated with a sensing microfluidic chip that incorporates an electrokinetic membrane sensor based on non-equilibrium ionic currents. This platform does not require multiple steps including any EV extraction, RNA purification, reverse transcription, or end point amplification. Therefore, it uses a vital and advantageous screening tool for early disease diagnosis, staging prognosis, and indicating therapeutic potentials. Similarly, conventional EV-miRNA extraction and quantification method suffer with many inefficiencies including multiple steps of purification, EV isolation, using antibody capture,[115] ultracentrifugation,[116] or polymeric precipitation (e.g., the commercial Exo-Quick® kit)[117] etc. Although microfluidic diagnostic platforms have gained significant advances in recent years, it is not absolute for ideal miRNA extraction and measurement devices. It suffers with rigorous sample treatment steps before downstream target detection using PCR or other fluorescence-based sensing techniques.[118-120] These multiple steps produce low yield and employ large detection biases and poor sensitivity. Particularly, when small sample volume is used a chemical-free EV lysing and label-free biomarker detection platform is required for current microfluidic systems.

The integrated chemical-free EV lysis and membrane sensor technologies together with PCR-free microfluidic diagnostic platform is formed multi-steps ahead for absolute quantification of EV-miRNAs from untreated plasma samples. This technology revolutionizes by accurate quantification of human plasma specific untreated target free-floating and EV-miRNA. The assay time and volume are reduced dramatically by hours, and in μl level as noted earlier. In this method, multiple quantification errors/biases are eliminated including various events of sample transfer, EV extraction, miRNA purification, ligation, and amplification etc.

Design

The architect of an integrated microfluidic platform is constructed to lyse EVs, efficiently and concentrate the extracted miRNA for detection and measurement the amount. The platform has two chips, one is surface acoustic wave (SAW) lysing chip for lyses the EVs mechanically and other is a second concentration/sensing chip along with two sets of ion-exchange membranes (IEMs) for extraction and measurement of the target miRNA respectively. The surface acoustic waves carry mechanical Rayleigh waves travelling along the surface of a piezoelectric crystal.[121] The platform construction occurs by exciting, interdigitated electrodes on the surface of the chip at radio frequencies. The SAW produces an acoustic wave by propagating away from the electrodes and refracts into liquid on the surface.

For exaction and measurement of miRNA, the external concentration polarization phenomena of IEMs and inherent negative charge of nucleic acids are utilized. IEMs are nano-porous membranes along with a high anion-exchange or cation-exchange surface for transporting of counter ions via the membrane under an applied electric field. This balance of ion exchange suppresses the ionic strength on one side and enhances the ionic strength on the other side of the membrane.[122]

For identification of target guided miRNAs, the depleted side of an anion-exchange membrane (AEM) is utilized. In anion exchange membrane the hybridization of miRNA along with oligo probes of the AEM inhibits the ion depletion action.[122,123] As a result, the voltage at the beginning for electro-convection involves over-limiting current in the CVC that is shifted by several volts.[122, 123] This dramatic voltage shift depleted ion current via hybridized miRNAs is larger markedly relative to voltage signals from electrochemical sensors that offers sensitive and sophisticated measurements.[123]

On a second polycarbonate chip the IEM concentrator and sensor construct a polycarbonate microfluidic channel that generates flow directly from the lysis chip via a microtube. There is an uninterrupted sample transfer in IEM sensor that helps to prevent any sample loss steps during analysis and promotes purification.

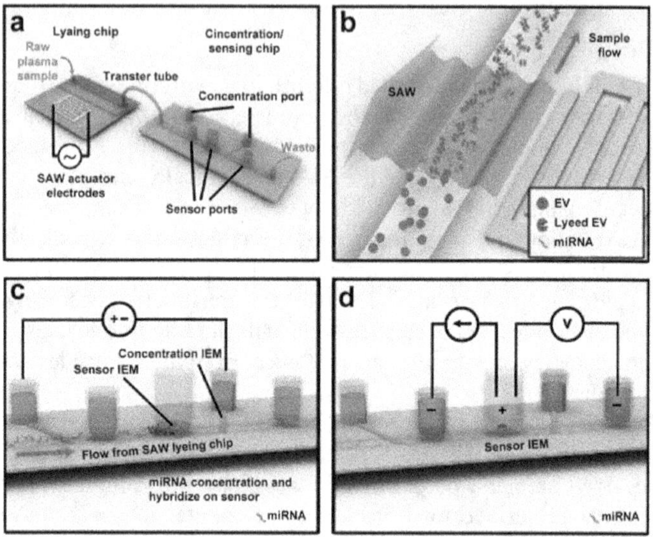

Figure-5. Design and operation of integrated microfluidic platform for EV-miRNA analysis. **a** The device consists of two chips, a lysing chip and concentration/ sensing chip, connected by a simple transfer tube such that sample is analyzed under continuous flow. **b** the principle of lysing using SAWs, where the SAWs refract into the microchannel and generate an acoustic pressure that shears the EVs open. SAWs are induced by 20 interdigitated electrodes on the surface of a 128° YX lithium niobate (LiNbO3) piezoelectric substrate with a polycarbonate microchannel attached to the surface using double-sided Kapton® tape. **c** The principle of IEM concentration, where a voltage is applied between two reservoirs separated by a CEM, creating an accumulation of negative space charge, including negatively charged miRNA, upstream of the CEM. **d** the principle of IEM sensing, where an AEM is functionalized with oligo probe complements of a target miRNA. After miRNA hybridization, the current–voltage relationship across the AEM sensor is shifted in the over-limiting region, and the voltage shift (ΔV) can be directly correlated with miRNA concentration. (Loosely Adapted, Redraw and Modify using skeleton from Ramshani *et al.*, 2019)

Sensor Calibration

A specific timing strategy is the main operational factor for integrated device. Its optimization ensures the exposure of the injected sample that is important for analytic step (lysing—concentration—sensing). Around ~20 μL of sample is injected into the microchannel on the lysing chip. The sample stays in the microchannel at a flow rate that is exposed to SAWs for 1 min. The sample present in the SAW lysing chip flows to the sensing chip directly via a narrow passage. In the SAW lysing chip flows to the sensing chip directly via a narrow passage. In the meanwhile, a 200V voltage is applied for 20 mins in between IEM reservoir concentration and a reference reservoir that induce the inhibition and subsequent miRNA concentration beneath the IEM sensor. The 20 mins long concentration step is involved in miRNA hybridization with IEM sensor with the help of complementary oligo probes. Later, the microfluidic channel is thoroughly washed with a PBS buffer to eliminate nonspecific miRNAs from the membrane surface. Finally, the existing voltage is terminated and a current–voltage sweep is participating in the sensing IEM reservoir and reference reservoir. The surface of a lithium niobate (LiNbO3) piezoelectric substrate forms a polycarbonate microfluidic channel inducing pressure in both cell and exosome lysis. As a result, the lysis of EV containing mRNA is taking place via the SAWs refract into the channel.

Performance

Three runs are conducted for each human plasma sample. Each run analyzes both free floating and total miRNA. The plasma is tested directly without adding any other materials. Voltage shifts are applied. They run before and after the EV lysis. It is found that voltage shift enhances the lysis due the continuous release of EV-miRNA. To measure the increased voltage, shift due to sole miRNA released from EVs is controlled with a negative control. For accuracy of miRNA as a potential biomarker, it is essential to compare EV-miRNAs with free-floating miRNAs. To calculate the voltage shift in free-floating miR, keeping the following control experiments set aside.

For accuracy of miRNA as a potential biomarker, it is essential to compare EV-miRNAs with free-floating miRNAs. To calculate the voltage shift in free-floating miR the following control experiment set aside. To get the free RNAs, the plasma is digested with RNase. The free RNA is injected into the integrated device. In this case, calculated voltage shift is due to the presence of free-floating samples. But many negatively charged proteins are associated with EV lysis by SAW. However, the voltage shift does not change due to presence of contamination that are release from the EV lysis. To eliminate the contamination Proteinase is routinely added to SAW lysed sample before introducing the sample to the IEM sensor. A similar voltage shift is calculated with a nonprotease-treated sample. This comparison of voltage shift validates the specificity of the IEM sensing system against protein interference demonstrating the high selectivity of the integrated device.

For calculating of plasma miRNA concentration, three separate measurements are compared 1) first measurement is performed without any sample passed through the device with no pBS reagent I in the micro-channel. It determines a baseline CVC for oligo probes on the IEM sensor. 2) the device sets for 20 µl of sample is injected into the integrated platform when the SAW device is off. The sample is further passing via the lysing chip and directly mediating into the concentration/sensing chip. However, all other parts of the procedure remaining identical. In these cases, the resulting CVC and the voltage shift are thus due to presence of only free- floating target miRNA in the sample during this measurement additional 20µL of sample is injected with the active SAW device and Evs lysis. The subsequent measurement of CVC and voltage shift denotes both EV-miRNA and free-floating, target miRNA concentration. Thus, comparison between last two values displays an accurate value for target miRNA in EVs. Surprisingly, the integrated device can calculate both free-floating and target EV-miRNAs values using only 40μL of human plasma. Notably, our platform uses very little sample (20 μL versus 1 mL) for a less time (30 min versus 13 h) than RT-qPCR. Therefore, the superior sensitive platform in integrated microfluidic device calculates the values for easy and efficient for diagnostic and prognostic applications in a quick time.

A single integrated device requires accurate quantification of EV-miRNAs from plasma. It reduces one-tenth of the time and sample volume relative to conventional RT-qPCR methods. Even in PCR-free quantification the sensitivity is highly accurate. It leads to optimal quantification of free-floating and EV-miRNAs. The integrated system does not necessary to count pretreatment or analyte transfer for preventing sample loss. In fact, it has fast (30min), reliable, and repeatable features.

Limitation for Circulatory miRNAs

Nowadays, circulating miRNAs are isolated from many venous plasma or serum. The miRNAs profiles from plasma originated either in veins or arteries are identical without any marked differences. In cancer early detection the use of venous miRNAs is occasionally still challenging. Experimentally ten arterial and fourteen venous are identified from blood samples that express multiple miRNAs significantly.[124] In humans, identical findings between venous and arterial samples predict the conclusion that the blood samples should be carefully selected for specific miRNA biomarkers.[125, 126]

Different delicate and specific approaches, Northern blot analysis[127,128], in situ hybridization,[129] real-time PCR,[130, 131] miRNA microarray[132, 133] and next-generation sequencing (NGS) [134] are conducted for miRNAs extraction and measurement in a sample useful for both clinical and personalized medicine. Different parameters validate disease age that can prevent circulatory miRNAs detection as important and consistent markers.[135] To determine the consistent and accurate level of miRNA expression in a specific tissue, body fluid or cell type, pre-identifying of miRNAs and normalization of different experiments, data evaluation and optimization must be employed.[136] It occurs in the minor difference in miRNAs expression pattern between healthy normal people and patients.[137] Only selective normalization strategies in RT-qPCR data analyses [138] using exogenous spike-in of a synthetic miRNA are used at the initial stage of RNA extraction.[139] However, the use of a single reference gene is not well sufficient for accurate miRNA estimation. In fact, a combined normalize strategy including both exogenous and endogenous[140] samples are tested in these cases.

On few occasions poorly diagnosed and reproducing miRNA are isolated for potential biomarker. For such uses a standard evaluation prior to their use is needed as new biomarkers for *in vitro* diagnostics.[141] The use of miRNAs in *in vitro* diagnostics faced a new challenge. Different possible biomarker extraction and estimation processes are needed for accurate, simple and inexpensive methods from a broad range of patients. Although it appears that miRNAs are most promising molecules in biomarkers and therapy. Their major challenges are attributed by off target effects as major unanticipated responses. Further single miRNA can participate differently in response to many disease stages.[142] To eliminate the miRNA limitation, sample size, validating processes are important for epidemiological and functional studies.[143]

The circular miRNAs that have high sensitivity and specificity are used for diagnostic and prognostic biomarkers in clinical application. They are used to avoid false positive or negative diagnosis. Many circulatory biomarkers that are appropriate for specific diseases, are expressed differently during different stages of diseases. Three miRNAs miR-21-5p, miR-155-5p and miR-210-3p associated with many cancer types express differently in between malignant cancer and benign polyps[144] suggesting that expression of these types of miRNAs in cancers are different. In these circumstances a larger sample size can be utilized to distinguish between the healthy or diseased status. Several undesirable criteria like age, gender, ethnicity, lifestyle, pre-treatment, early symptoms, pedigree of diseases and more are the major hurdle if a small sample size is used for diagnosis. In fact, the small sample size can be influenced by measurement principle, method used, instrument, technicians.[137]

Conclusion

The progress for efficient extraction and quantification of multiple miRNAs for pathological cases are emphasized. Still many critical conditions in different extraction steps have remained in insufficiently characterize and still pass through a young age of isolation. However, great efficiencies have still been laid in isolation of disease-specific molecules and integrating their participation of diagnosis, staging prognosis, and treatment. A rapid, easy, and early diagnosis is required for all deadly diseases to reduce the death rate.

Several extracellular circulating miRNAs show a high stability and specificity in human blood. The measures for high and enrich distribution of blood miRNAs depend on encapsulation in lipid vesicles or generating multiple complexes with versatile proteins for their protection against rapid degradation and protection against denaturation. In fact, extraction procedure requires isolation of miRNAs from different specific sources including exosomes, micro vesicles, or high-density lipoprotein particles and from versatile biological fluids including blood, urine, saliva, peritoneal fluid, amniotic fluid, bronchial lavage, CSF, and tears etc.[145-147] Further microRNAs are liberated from many organs including brain, heart, endothelial cells, ovary, uterus, and mammary glands in the blood streams of healthy human and patients. The focus of the current research deals with the miRNAs isolated from both extra and intracellular environments as useful biomarkers for early diagnosis. However, miRNAs isolated from blood, plasma, CSF samples show a wide range of variability in their signatures. Further the size of the cohorts of patients with multiple co-morbidities influence the amounts of isolated molecules. In fact when the miRNA levels from Alzheimer patient and CSF are compared individual miRNA shows opposite expression profiles.[148-150] Therefore, a clear discrepancy is found based on employed preservation, processing of the sampling methods.

Recently microarray, nCounter Nanostring-like hybridization-based approaches and reverse transcription quantitative PCR arrays (RT-qPCR), next generation sequencing (NGS) like approaches for miRNAs quantification technology have been employed. Among them, an inexpensive screening is used via microarray based on RNA-DNA hybrid capture method while expensive but sophisticated NGS can be used for isolating new miRNAs and different isoforms.[151-153] Despite of RT-qPCR appears to be the most popular and useful technique for their sophisticated sensitivity, specificity, well advanced accessibilities, and reproducibility relative to expensive microarray and NGS results. Many impressive challenges have been executed for rapid and efficient ways for isolation and measurement of a wide variety of miRNAs for different pathological needs. The use of miRNAs as early biomarkers can fulfil this desideratum despite present limitations as being a burning research field. In the current technique miRNAs isolate by a usual approach for personalized medicine development and therefore leading for targeted therapeutic interventions.

References

References: chapter wise are cited before the Index at the end of the Book. (1-153)

Chapter V

Deadly public health hazards and miRNAs for medical solution

Abstract

Major advances in detecting potential molecular targets for generating effective therapeutics against all metabolic diseases have become increasingly challenging and daunting. Main reason is that independent alcohol- responsive genes only contribute a tiny effect to the overall mechanism suggesting a multi-genic factor for implementation and disease phenotype, as it behaves like complex traits. Current remedy trends are hence shifting toward the modification of gene networks rather than emphasizing individual genes. The invention of microRNAs and their inner mechanisms of action on regulation of transcription and protein translation have utilized these tiny non-coding RNA molecules that they function as central coordinators of multiple cross-communicating cellular processes. Distinct microRNAs ensure fine-tuned and/or redundant control over many cellular functions via cell exploitation. Using the same token, we utilize the same method of microRNAs for creating novel, interactive strategies to challenge and prevent complex disorders. A direction on recent innovation of microRNA signatures in the human brain of alcoholic patients. It supports our speculation that absolute alteration in gene expression using microRNAs is required for long-term neuroadaptations generated during development of alcoholism. The possible insights of epigenetic modulation by a subset of microRNAs are discussed. Taken together, microRNA activity controls several cellular mechanisms that already develop alcoholism and encourages potential target development for novel alcoholic therapeutic interventions.

Finally, different metabolic disorders are characterized by the invariable defects for proper utilization and/or storage of energy. The burdens of another metabolic disease, particularly obesity, is developed through a complex interplay between genetics and epigenetics, predisposition, environment, and nutrition. Therefore, evaluation of basic molecular strategies will introduce novel insights for prevention and treatment during onset of metabolic diseases. A growing concern regarding micro-RNAs (miRNAs) dysregulation in metabolic diseases will recently develop a budding solution for obesity treatment. Here recent data for potential involvement of miRNAs in early diagnosis and therapy development of this metabolic diseases, particularly obesity and type 2 diabetes are discussed in two separate chapters of the book.

Obesity carries increased amount of adipose tissue. It is a metabolic chronic defect close to pandemic proportion. Lifestyle modulations are the first line therapy for obesity and a large restriction of dietary intakes causes efficient improvement of weight loss. Besides diet restriction and physical exercises, bariatric surgery is the effective therapeutic strategy for morbid obese patients. Improve weight-loss depends on the high inter-independent variability, that is governed by multiple epigenetic factors. microRNAs open up a new avenue in metabolic defects and their control in genic and epigenetic regulation is always related to obesity. A differential circulatory microRNAs expression before and after different dietary intakes and bariatric surgery isolate multiple biomarkers of weight loss. Significant and separate changes in microRNA expression are found in visceral and subcutaneous adipose tissue. Interestingly, reasonable differences in microRNA expression between responders and non-responders have emerged for dietary intakes or after surgery. A significant range of microRNA expression changes interact with obesity pathways. It promotes pathophysiological mechanisms underlying obesity and obesity-related metabolic diseases. Despite further understanding is required to determine firm conclusions, an increasing evidence of microRNAs involvement in weight loss and in obesity are flourishing rapidly.

Introduction

In this chapter, the presence of the tiny 23-25 nt long RNA-microRNA (miRNA) population in the animal world is so vast that it is next to impossible to handle. So, we concentrate our narrative mainly on their existence in humans, particularly in major disease traits. Among the small regulatory RNAs, the unique and most abundant MicroRNAs regulate 30-80% of human genes. Human disease cell studies establish alternative control mechanisms and regulate vast majorities of normal physiological functions by breaking the conventional and old concepts of central dogma. In human disease states, from cancer to regenerative to infectious diseases microRNA alteration different pathologic states. A brief narrative describing the role of microRNA in public human health and disease and its fascinating tales open to unfold in front of pathologists, academicians and medical learner and researcher regulators continues to unfold. Among deadliest public health hazards in current era, alcoholism and obesity are growing rapidly that cause our everyday burden in daily life. The conventional treatment plan of these diseases is adequate to cure. Recent microRNA therapy and early detections came into limelight. Here we drew our prime focus addressing the possible cure via microRNA therapy.

Excessive use of harmful use of alcohol is one of the risky health hazards. Alcohol consumption leads to more than 60 major health hazards and injuries resulting in nearly 2.5 million deaths each year.[1] Alcohol dependency causes chronic brain disease with abnormal biological, behavioral, and socioeconomic components.[2] Nevertheless, the molecular mechanisms underlying epigenetic and gene regulation related to excessive alcohol consumption are in the budding scientific stage.

The longstanding consumption of alcohol allows changes in gene and protein expression that allow human neurons for homeostatic alterations in the prominent signaling pathways. Consumption causes a functional change in the receptor tyrosine kinases that are activated by the growth factors such as EGF, GDNF, BDNF and insulin etc., serine–threonine kinases and protein–protein interactions among scaffolding proteins and associated binding partners.[3] The changes in signaling pathway are often targeted. An alteration in a wide range of transcriptional factors and essential genes are also noticed. In addition, epigenetic reprogramming related the alcoholism modulate different methylation of DNA, acetylation, methylation, and phosphorylation of histone tails.[4]

The microRNAs (miRNAs) and their functions are revolutionized gene.[5,6] These tiny (~17–24 nucleotides long) non-coding RNAs base paired at the 5' end of mature mRNAs recognition elements to execute degradation of mRNAs and or translation inhibition.[5,6] A greater abundance in the brain alters multiple distinct biological processes including neuronal differentiation,[7] neurodegeneration[8,9] brain development,[10] synapse formation and plasticity.[11] Use of nicotine[12] cocaine,[13] opioids[14] and alcohol[15-18] causes a havoc change in miRNAs population in the brain. Here, we also described the major biological consequences in miRNA populations and their upregulation in the alcoholic brain of the human.

Apart from the alcoholism obesity is another type of metabolic syndrome spreading rapidly worldwide. It provides multiple complex interactions of numerous factors including genetics and environment. Such disturbance firmly establishes that diet is a prime environmental factor involving many metabolic changes. During livelihood consuming a high-calorie diet leads to excessive fat accumulation in adipose tissue and subsequently enhances circulating free fatty acid levels in the blood. This allows macrophage induction and rapid synthesis of pro-inflammatory cytokines via Toll-like receptors, which leads to harmful inflammation in adipose tissue.[19, 20] A high-fat diet (H.F.D) promotes insulin resistance and obesity.[21, 22] A positive change in lifestyle that involves enhanced energy expenditure and reduction of energy consumption are ideal recommendations for reducing obesity and induction of insulin sensitivity as a result for dietary obesity (DIO) and obesity-related diet intakes. In general, a strong recommendation leads to restrict in calories intake and fat diet reduction of body weight improves insulin sensitivity.

The miR-103/107 are responsible for insulin regulation via caveolin-1. The miR-107 requires in inflammation and obesity in a Toll-like receptor 4/myeloid differentiation primary response 88 manner (TLR4/MYD88) in the enhanced macrophages resided in the adipose tissue; it also is a lipid-modulated miRNA.[23, 24] Moreover mir-107 relates in adaptive and innate immune activation/polarization responsible for obesity-mediated inflammation.[23] The mir-30 and mir-125a instigates molecular mechanisms related to insulin resistance (IR) and inflammation in adipose tissue. The miR-130 is increased by the inflammation environment and overexpressed adipocyte dysfunction. The peroxisome proliferator-activated receptor γ (PPARγ) directing strategy is inhibited at the time of pro-inflammatory response and adipocyte dysfunction.[25-27]

Currently, the answers of the two obvious questions are not clearly understood 1) how modulation of miRNA profiles affect adipose tissue both in functional and molecular state and 2) to what extend they vary in response to weight-control strategies. This information is the core controller in the dietary anti-obesity progress for obese patients. This present chapter encompasses the detailed investigation including miRNA changes in adipose tissue during weight loss stages, expansion/reduction, detect their contribution in biomarkers and therapeutic fields for obesity. It also dictated the interactive genes and the marked dysregulation of each miRNA and subsequently, excess accumulation of their functional contribution in obesity onset and consequent progress.

Alcoholism

Alcohol-Induced miRNAs in the Human Brain

Recent report uncovered and established that nearly 35 human miRNAs are upregulated in the alcoholic brain samples. **(Table 1)** Interestingly, chronic alcohol consumption can alter a set of sensitive upregulated mRNAs which appears to be down regulated in the normal brain. This result supports miRNA-dependent inhibition of gene regulation in the human alcoholic brain. These multiple upregulated miRNAs apparently control a single putative target mRNA. Such a consequence either ensures inhibition of mRNA targets at any means or fine-tuned the regulatory inhibition of specific mRNAs.

The human derived neuroblastoma cells are highly sensitive to change the miRNA expression related to alcoholism.[28] Among them, upregulated miRNAs are predominant and sensitive to the physiological environment of the alcohol-exposed human brain. In fact, only 20 distinct upregulated miRNAs remain altered significantly in the alcoholic brain. Six out of these 20 miRNAs including miR-369-3p, miR-34c-5p, miR-203, miR-146a, miR-194, and let-7 family members are related to the alcoholic brain. This statistical underscore of up-regulatory miRNAs related to alcohol exposure predicts a consistent involvement of multiple miRNAs in neuroblast cells potentially conceiving of alcohol dependence.

Key miRNA changes related to alcohol intake down regulates the transcript of Cell death-related genes.[18, 29] The inhibition apoptotic pathway of cell death machinery may reflect unique molecular adaptations that counter targets alcohol neurotoxicity in cells.[30] In contrast, other groups found miRNA-regulated effects induce neuronal cell cultures treated in short exposure ethanol in humans (3–72 h), their results do not oppose the major changes in neuroadaptive phenomenon relative to chronic alcoholism. These findings suggest that miRNA-mediated neuroadaptations can only develop via miRNA-mediated activation of cell death pathways at the early stages of alcohol abuse.

It is observed that minor changes in miRNA and related mRNA expression are found in human alcoholism's that can increase expression in localized specific cellular sectors, e.g., the neuronal synapse. This tight compartmentalized, enhanced differential gene regulation may be diluted in unfractionated tissues. In contrast, distinct differential changes in gene expression are neither compartment-specific nor cell type-specific for example, immune response signaling in glial cells, alteration of neuronal genes in specific neuronal subtypes. These symptoms become diluted when RNA is extracted from total tissue including cell subpopulations.

==

Table -1. Modulation of Human microRNAs in alcohol consumed tissues

Alcoholic Human miRNAs	Function of the family members*
miR-1	Neurotransmitter receptor (nAChR) availability, expression of HDAC4/epigenetic gene activation
miR-7	Repression of α-synuclein/neurodegeneration, neurite outgrowth miR-15b Lymphocyte differentiation
miR-18a	Angiogenesis
miR-34c-5p	Epigenetically regulated
miR-92a	Lymphocyte differentiation, angiogenesis, synaptic signaling biomarker for TBI
miR-101	Repression of EZH2/epigenetic gene silencing
miR-135b	
miR-140	Endocytotic recycling of neurotransmitter receptors, upregulated by nicotine and LPS
miR-144	
miR-146a	LPS-induced TLR/NFκB/IL6 signaling
miR-152	TLR/IL6/IL12/TNFα/INFβ signaling, repression of DNMT1/ epigenetic gene silencing, epigenetically regulated
let-7f	Neurotransmitter receptor availability, LPS-induced TLR/NFκB /IL1/TNF signaling
let-7g	
miR-153	Repression of Jagged/Notch signaling, repression of α-synuclein /neurodegeneration
miR-194	Upregulated by LPS exposure
miR-196a	Virus-induced inflammation, epigenetically regulated
miR-196b	-
miR-20	AK/STAT/IL6/INFγ signaling, repression of Bmi-1/epigenetic gene silencing, epigenetically regulated miR-301a -
miR-339-5p	-

miR-369-3p	TNFα signaling miR-374b -
miR-376c	-
miR-380 -	-
miR-423-5p	-
miR-454-3p	-
miR-515-3p	-
miR-519b-3p	-
miR-553	-
miR-580	-
miR-586	-
miR-652	-
miR-665	-
miR-802	-

***Validated functions discussed in this review. See text for details of references.**

Alcohol-Responsive miRNAs and Neuroinflammation

Inflammation is intimately related to alcoholism-induced neuropathology.[31-33] Alcohol responds to Toll-like receptor (TLR) 4, an important factor for immune activation in the Central nervous system. TLR are important for nuclear factor kappa B (NFκB) in ethanol-induced activated cells. They also influence by activating transcription of pro-inflammatory chemokines, cytokines, oxidases, and proteases. TLRs have well recognized roles in pathogen detection and initiation of an innate immune during infection also.[34]

The role of innate immune-related genes in alcohol dependency is supported by genetic association studies in humans.[35-38] Based on various mice experiments it is hypothesized that the neurobiology of alcohol addiction is due to modified brain signaling operated by immune signaling. Implications of multiple experimental results finds that depending on NFκB TLR signaling can modulate miRNA expression. Generally, enhancement of TLR activation mainly upregulates the miRNAs.[39] Here the result of intense cross talk between a group of up-regulatory miRNAs related to excess alcohol consumption and neuroinflammatory processes is occurred and defined by a positive hypothetical model. (**Figure 1)**

Hypothetical model in microglia

The blood brain barrier is disintegrated by high content ethanol allows LPS for reaching easily to the brain and instigate TLR signaling cascades for activation of nuclear factor kappa B (NFκB), many pro-inflammatory genes (e.g., cytokines) and sensitive miRNAs. The activated pro-inflammatory factors govern a positive feedback loop in the same sets of cells and secrete a systemic neuroinflammatory response. Later, the altered cytokine receptors of the same activated cell are produced by a pro-inflammatory factor. The cytokine receptors transduced signal back to the nucleus via the JAK/STAT pathway for induction of additional pro-inflammatory factors (e.g., interferons) and miRNAs.

Meanwhile, upregulation of the miR-146, miR-152, and let-7 family members try to avoid over-amplification of the signals and excessive inflammation by suppressing TLR4/CXCR4 signaling with the help multiple cascade transducers inhibition including IL1 receptor-associated kinases (IRAKs), TNF receptor-associated factor 6 (TRAF6) and TLR4/CXCR4 themselves among others. As a compensatory reaction, miR-203 unregulated immune induced state of the same cells by inducing alternative JAK/STAT pathway. These up-regulations produce a benign inflammatory response in the brain. Subsequently, immune-response cells cause upregulation of miRNA-mediated epigenetic controls, such as DNA methylation enrichment and modulation histone methylation and/or acetylation in the histone tails procuring chromatin structure changes and gene modulation. In true sense, alcohol sensitive miRNAs target many epigenetic factors for controlling and/or fine-tuning of ongoing cellular epigenome modification.

The selective miR-15, miR-34, miR-92, miR-140, miR-146, miR-152, miR-194, miR-196, miR-203, miR-369, let-7miRNAs of the 35 miRNAs is upregulated by the alcohol consumption by modifying major immunity and regulating TLR signaling. Their modes of action on immunity are unique and different from each other. Among them miR-146 is used for preferred target TRAF2, the TNF receptor-associated factor 2.[40] The miR-146 regulatory circuit micro-modulates cytokine and TLR signaling for preventing excess inflammation.[41]

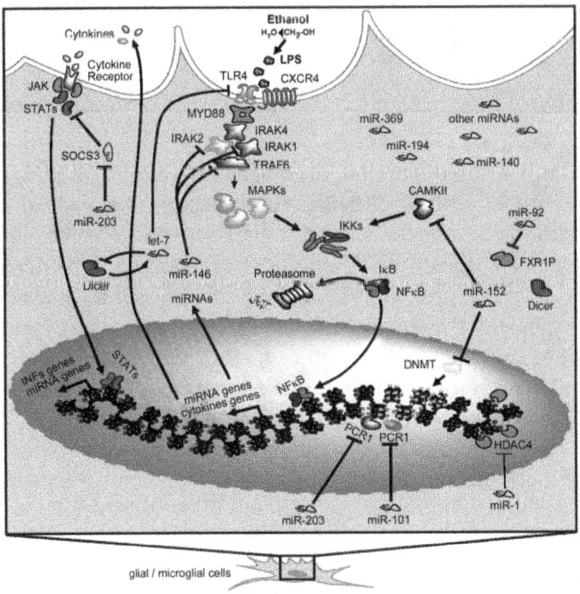

Figure-1. Predicted model for neuroimmune-mediated miRNA actions in the human alcoholic brain (e.g., in microglia). Bacterial lipopolysaccharides (LPS) from routine bacterial populations, for example, are contaminated in the bloodstream via gut pores facilitated by alcohol consumption. A compromised blood brain barrier due to huge ethanol concentration in the blood allowing LPS to reach the brain and instigate TLR signaling cascades that induce nuclear factor kappa B (NFκB) and activate the transcription of a variety of pro-inflammatory genes like cytokines and miRNA genes. Recently generating pro-inflammatory factors are transported to unleash a prominent neuroinflammatory response for maintaining a positive feedback loop in the same stimulated cell. This second round of pro-inflammatory factors is generated depending on the expression of alternative cytokine receptors in the induced cell. These signals return to the nucleus via JAK/STAT pathway for activating additional pro-inflammatory interferon factors production and miRNA genes. To avoid over-amplification of these signals and excessive inflammation due to hyper-responsiveness to LPS insult, specific miRNAs (e.g., members of the miR-146, miR-152, and let-7 families) are consequently upregulated to suppress TLR4/CXCR4 signaling through inhibition of multiple cascade transducers such as IL1 receptor-associated kinases (IRAKs), TNF receptor-associated factor 6 (TRAF6), and TLR4/CXCR4 themselves, among others. On the other hand, other miRNAs, miR-203 induces a separate JAK/STAT pathway for a reaction that overexpression for maintaining the immune-induced state of a special cell based on enhancing a benign inflammatory response. Subsequently, immune-induced cells develop miRNA-guided epigenetics strategies for example DNA methylation, histone methylation and/or acetylation that execute chromatin architectural changes in gene regulation for long-term homeostatic modulations and prevent cellular obstacles in specific environments. Finally, miRNAs control many epigenetic factors for enhancing and/or fine-tuning the ongoing changes of the several cellular epigenomes. (Loosely Adapted, Redraw and Modify using skeleton from Nunez and Mayfield, 2012)

===

Alcohol-guided miRNAs, neurotransmitter signaling and synaptic plasticity

Alcohol and other narcotic drug abuse develop long-term modification in the brain neuronal reward circuits. These changes promote drug tolerance, neuronal reward dysfunction, enhanced narcotic drug intake and eventually compulsive use.[42] Neuronal reward circuit is centered on dopaminergic neurons in the ventral tegmental area (VTA) of the midbrain and their extensions to the limbic system specially the nucleus accumbent (NAc), dorsal striatum, amygdala, hippocampus and regions of PFC.[43] Indeed, other neurotransmitters are involved in brain reward responses since animals possibly exhibit positive hedonic responses in the absence of dopamine.[44] Ethanol and other opioids, cannabinoids and nicotine are responsible to generate brain reward partially via non-dopaminergic mechanisms, for example opioid receptors expressed on NAc neurons that bypass dopamine inputs from the VTA. [44] Molecularly, delicate but complex gene expression methods govern long-lasting changes in dendrite and synapse structure and function. In these processes, the local mRNA translation regulators in neuronal dendrites are required for the tight spatial regulation of plasticity at the level of individual dendrites or spines.[45, 46] Recently, it is noticed that miRNAs provide extension of these local process regulation and the development of addictive animal behaviors. (**Figure 2**)

Multiple miRNA families let-7, miR-1, miR-7, miR-92, miR-135, miR-146, miR-339, miR-376, and miR-380 enrich or downregulate miR-34, miR-101, miR-144, miR-153, miR-301 and miR-652 in the human alcoholic brains.[47-49] Particularly, miR-92 regulates FXR1P. It is an RNA-binding protein present in dendrites miRNA biogenesis.[47-50] It indicates that dendrites and within dendritic spines miRNAs are possibly regulate local protein synthesis. Therefore, prime effect of miRNA overexpression in human alcoholic brain is predicted to be inhibition of targeted-gene expression and/or protein translation reduction[18] and prevent *de novo* protein synthesis in dendritic spines that attenuates long-term memory generation due to prevention of dendritic growth and remodeling.[46, 49] Thus, alcohol overexpress synapse-localized miRNAs that provide a major factor in the progress of new addiction-promoting, long-term memories. In fact, early or enhanced overexpression of alcohol-responsive miRNAs in the brain assumes to increase addiction development and could provide an essential therapeutic protocol.

Numerous modified synaptic miRNA families are participating in neurotransmitter signaling regulations that are functionally related to brain reward circuits induction and associative, synapse-specific memories generations.[44] As a notable example, a miRNA family, let-7 family that expressed differentially in PFC of human alcoholics is able to transform a variety of neuro-transmitter receptor activity. Few members of this family control different receptors including the μ opioid receptor, the β2-adrenergic receptor and the dopamine D3 receptor.[13, 14, 51] Three important let-7 family members, let-7a, let-7c, and let-7g are the conjugators of translocation and accumulate μ opioid receptor mRNA into P-bodies that involve translation inhibition without altering the mRNA levels. The μ opioid receptor is largely developed complex network along brain reward circuits by mediating the reinforced activities of morphine and many non-opioid drugs like alcohol, cannabinoids, and nicotine. The non-opioid drugs function at their own receptors, especially GABAA and NMDA receptors for alcohol, CB1 and CB2 receptors for cannabinoids and nicotinic acetylcholine receptor (nAChR) for nicotine. They enhance endogenous opioid peptides that, in turn, induce the μ receptors. These receptors function as a convergent molecular gate for initiating addictive behaviors.[52] Interestingly, another microRNA, miR-15, is overexpressed in human alcoholics that have predicted conjugated sites within the 3′ UTR of the β2AR mRNA.[53] Recently it is found that β-adrenergic receptors restore cocaine-related memories governed by norepinephrine acting at central β-adrenergic receptors.[54]

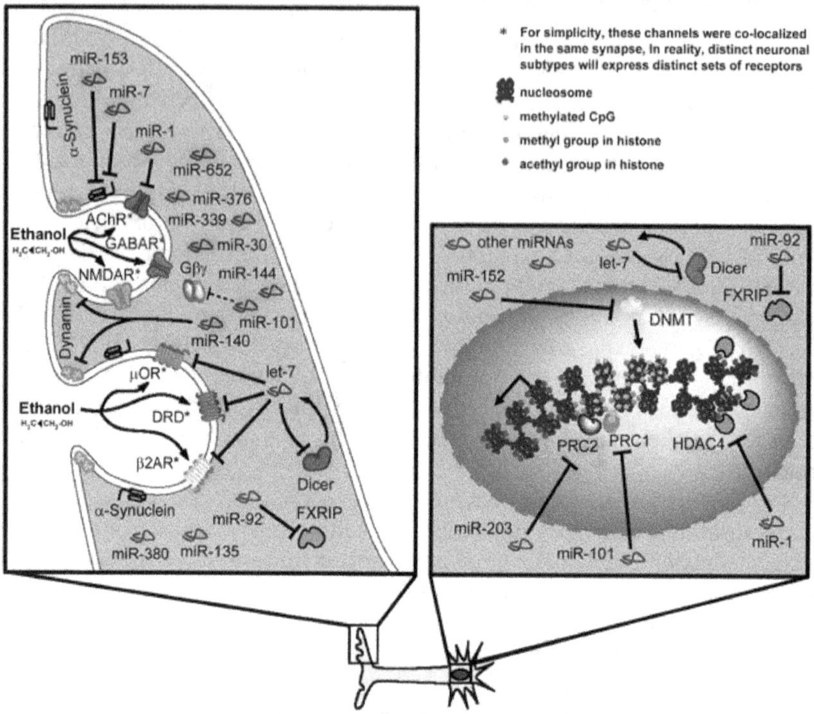

Figure-2. Predicted model for synaptic-mediated actions of miRNAs in the human alcoholic brain (e.g., in neurons). Ethanol can be directly or indirectly involved in changes of many neurotransmitter receptors at the neuronal synapse and induce brain reward circuits based on different forms of neuronal plasticity. It in turn, converts the drug-stimulation signals into long-term behavior alterations. (e.g., alcohol sensitivity). Receptors alter specific synapse based on the neuronal subtype and unique subset of receptors expression. Particularly multiple receptors like GABAR, NMDAR, AChR, μOR, DRD, and β2AR, are co-localized on the same synapse in the picture, however this is not real case. The miRNAs export and synthesis excessively in the synapses are locally generated by local miRNA-ribonucleoprotein (miRNP) complexes including Dicer and FXR1P factors and leads dominant inhibition on mRNA targets at the specific synapses. Multiple miRNA family members including let-7, miR-1, miR-101, miR-140, and several others inhibit neurotransmitter receptors by interacting respective mRNAs or by interfering with synaptic endocytosis via targeting of dynamin and α-synuclein mRNAs. Moreover, synaptic miRNAs involved in negative feedback loops via directing dicer and FXR1 mRNA and other miRNA biogenesis-mediated factors) that auto-regulate their excess levels. These feedback loops lead to an ideal homeostatic control of multiple synaptic functions. Later cells provide miRNA-related epigenetics factors like DNA methylation and histone methylation and/or acetylation for ensuring chromatin modification and global architectural changes in gene expression for long-term homeostatic modulation and cellular adaptations under unique environmental conditions. In fact, miRNAs direct epigenetic factors for controlling and/or fine-tune the ongoing remodeling of the cellular epigenome. (Loosely Adapted, Redraw and Modify using skeleton from Nunez and Mayfield, 2012)

==

Moreover, another let-7 family member, let-7d reduces strongly by cocaine in the mesolimbic brain and this miRNA is upregulated in hybrid NG108-15 cells during dopamine D3 receptor reduction.[13] Indeed, drugs of abuse activate consistently structural and functional alterations in the mesolimbic dopaminergic method that is required in high-risk, drug-seeking behaviors and relapse.[55] Conversely, another miRNA induction in the PFC of human alcoholics miR-101 supports changes of GABAergic transmission based on the sensitivity of alcohol consumption.[56] Basically, guanine nucleotide binding protein beta 1 subunit (Gnβ1) is the prime transcript that changes expression in all groups during high and low alcohol-drinking. Moreover, a quantitative trait locus for alcohol consumption, Gnβ1expression is inversely linked to the level of alcohol intake. Based on differentially expressed Gnβ1 transcripts, protein levels and 3′ UTR sequence analysis, the miR-101 controls the amount of Gβ1 protein. Gβ1 protein can induce a cascade of intracellular signaling, ultimately involving internalization of GABAA and other receptors.[56] Although studies examine the core interaction between miR-101 and the Gnβ1 transcripts.

A miRNA contribution in controlling of endocytic recycling of neurotransmitter receptors is noted. The miR-140* is overexpressed using continuous nicotine treatment and conjugates directly to dynamin (Dnm1) mRNA for inhibition.[57] Dynamin-1 is a cleaving of clathrin-coated vesicles, and it is indispensable for endocytic synaptic vesicle recycling only during a hard or consistent stimulation when exocytosis of neurotransmitter-filled vesicles is extreme. This rapidly retrieval of vesicles is essential for maintaining the synaptic vesicle pool.[58, 59] The dynamin-1 initiates a tough physical interaction together with the big conductance voltage- and calcium-induced potassium channel (BKCa) as a master player in alcohol actions.[60] Interestingly, multiple miRNAs, miR-101, miR-135b, miR-140, miR-203, miR-34c-5p, miR-376, miR-454, miR-519b-3p, miR-580, and miR-586 induced the expression in human alcoholics [18] are supposed to interact dynamin family members. This interaction underscores the contribution of these proteins and directing particularly miRNAs for interacting alcohol actions. Acetylcholine (ACh) is possibly another neurotransmitter signaling processes controlled by miRNAs in human alcoholic brains. miR-1, a conserved muscle-specific microRNA control both pre- and post-synaptic function at neuromuscular junctions.[61]

Taken together, the versatile essential effects of alcohol-sensitive miRNAs on neurotransmitter receptor-guided signaling processes. It suggests that a variety of neurotransmitter-regulated processes alter subsequently in response to alcohol and other drugs of abuse. Such activities of many essential miRNA families are responsible for this selective complex network of interactions. Strong miRNA upregulation is sensitive to human alcoholism.[18] They interact with different fundamental neurotransmitter receptors assuming a neuroadaptive reaction for counteracting drug-guided receptor induction related to multi type addiction. Conversely, miRNA-generated dynamin-1 reduction hampers neurotransmitter receptor endocytosis. That is responsible for alcohol generated addictive effects as well as other drugs of abuse.

Alcohol-sensitive miRNAs and Epigenetics

An explosion for unravelling numerous epigenetic mechanisms including chromatin architecture and histone tail modifications in response to different environmental cues and intake food material are the most challenging molecular dissections of human health. The drug addiction as well as repeated drugs abuse alter many epigenetic landmarks such as histone acetylation, phosphorylation, methylation, and DNA methylation.[62-64] The global DNA methylation related to excess alcohol consumption is a consequence for reduction of DNA methyltransferase 3b (DNMT3b). that a possible derangement of many epigenetic markers.[65, 66] Moreover, repeat alcohol consumption hypermethylated many nerve-growth factors (NGF) and pro-opiomelanocortin gene (POMC) in human blood and prodynorphin (PDYN) of the brain.[66-69] The miRNAs can control versatile epigenetic factors and vice versa[70] by the influence of alcohol in the human brain.

The different functions of few important miRNAs should be included due to their unique function and diversified mechanism of action. The miR-101 influences histone methylation and transcriptional inhibition of EZH2, a conserved catalytic subunit of Polycomb repressor complex 2 (PRC2).[71, 72] Conversely, miR-101 overexpression markedly altered cell proliferation and configured histone code of the cancer tissues procuring more benign cellular phenotypes.[71] In contrast, miR-203 targets important component of the polycomb repressor complex 1 (PRC1) Bmi-1 that stabilizes epigenetic gene silencing by working with PRC2,[70, 73] and miR-152 targets DNA methyltransferase DNMT1.[74] Many of related results of microRNA function predicted that[75, 76] once epigenetic control wipes away, cells are vulnerable to normal control and converted to become malignant and develop tumors. Therefore, it speculated that non-complicated alcoholic individuals protect against environmental stress and develop a defense shield via miRNA to overcome early development of malignancy.

Up-regulatory miRNAs in the alcoholics. human brain, anxiety related stress is developed by excessive alcohol intake enhanced HDAC activity and are decreased histone acetylation.[77] In shortcut, change of HDAC activity in alcoholism suggests that HDAC inhibitors may be strong therapeutic agents to prevent alcohol intake symptoms. However, later various experiments results predicted that alcohol consumption is not so easy to cure. Further, upregulation of miR-1, miR-34, miR-152, miR-196, and miR-203 in alcoholics are conducted many epigenetic changes by disrupting normal,[70] environmental factors, which happen to be in disease development. Alcohol causes a dramatic change in the central nervous system. The miR-34b/c and miR-203 both contain CpG islands change in methylation of CpG islands in various cancers.[78, 79] Many miRNAs involved in epigenetic remodeling can easily destroy the epigenetic mechanisms via miRNA feedback regulatory loops extending from havoc to fine tuning activity.

The etiology of alcohol abuse and alcohol dependence instigated brain-wide pathophysiological alterations via mechanistic crosstalk, we need to develop organ and system-wide models (even if initially incomplete and speculative to some extent) with the potential to capture the dynamism in such complexity. Once we discover main dynamic relationships conducive to stable states of the disease, design strategies may be created to detour or reverse disease progression, akin of chemical reactions transiting through multiple equilibrium states. From the analysis and hypothetical models presented here, the notion emerges that miRNA are efficiently being upregulated in response to alcohol insult, consequently inducing organ and system-wide homeostatic changes that produce long-term adaptation under the specific cellular environment. Remarkably, miRNAs that localize to and display activity at synapses in neurons are also capable of eliciting distinct and specific activities in other cell types, such as innate immunity-related functions in microglial cells, among others.

The wide range of miRNAs regulatory power generates strong and far-reaching knockdown and/or overexpression strategies in development by specific mRNAs targeting that eventually prevent or even reverse alcohol addiction. However, many so far untouching tissues require to address the understanding of miRNA existence and functions. This unexplored tissue organization related to miRNAs function might initiate to combat the alcohol abuses. Two important works of microRNA family members that conducted redundant specificity and biological function of targets and enjoying fine-tuning of post-transcriptional regulation or exert both types of control of mRNA targeting.

Finally, an extensive survey of miRNA population would benefit dramatically from epigenetic to development via high throughput technologies. The continuous prediction of bioinformatic platforms generate "miRNA code." That thrills our intense interest in the discovery of small molecules that can modulate select miRNAs, their function in different diseases like alcoholism in particular tissues, as invaluable clinical and therapeutic potentials.

Obesity

Obesity and overweight are two related common defects in our society. Obesity spreads a real public health problem because of its costs and promotes health defects. In fact, excessive body weight emerges several metabolic diseases in our likelihood such as heart defects, T2D, dyslipidemia, osteoarthritis, and certain types of cancer.[80]

Obesity is the deadliest metabolic syndrome progressing its existence worldwide. Obesity is a result of complicated interaction of multiple genetic and environmental factors. This abnormality is linked to excess incidence of many diseases like diabetes, cardiovascular disease, cancer and ultimately infects as a serious global epidemic of main health concerns. (**Figure 3**)[81, 82] The diet is the closest environmental factor for leading metabolic disorders. The abrupt increase of fat accumulation in adipose tissue by consuming high-calorie rich food also consistently induces circulating free fatty acids levels in the blood. This induction leads to macrophage enrichment and the production of pro-inflammatory cytokines via Toll-like receptors causing inflammation in adipose tissue.[83, 84] The mice have been consumed a high-fat diet (H.F.D) leading induction of insulin resistance and obesity in a manner like human disease development.[85, 86] Excessive energy expenditure with reduced energy consumption is the ideal recommendation in lifestyle for reducing obesity and developing insulin sensitivity during dietary obesity (DIO) treatment and preventing obesity-related disease. Generally, tight restriction in calories intake and fat diet percentage reduces body weight and induces insulin sensitivity that is also maintained by the loss of percentage of fat diet that is transformed from a HFD to a low-fat diet (LFD).[87]

MicroRNAs (miRNAs) open a potential new avenue in controlling adipose tissue pathways that regulate a variety of processes adipogenesis like insulin resistance and inflammation.[88, 89] The existence of numerous miRNAs in different metabolic organs can be utilized for most potential therapeutic agents in obesity and its consequences. Initially, let-7 microRNA functions as an effective controller of glucose metabolism and peripheral insulin resistance.[90] It leads a marked change in obesity and metabolic disorders.[91] Earlier it manifests its effects by interacting with high-mobility group AT-hook 2 (HMGA2), a transcription factor controlling growth and proliferation in other contexts. Thus, the let-7-guided silencing affects several components of the insulin-PI3K-mTOR pathway using IGF1R, INSR, and IRS2. Moreover, the mTOR inhibitor, rapamycin, abrogates Lin28a-guided insulin sensitivity and induces glucose uptake.[92]

Obesity and overweight emerge by a serious imbalance between energy intake and energy output. They are responsible for numerous factors interactions like caloric and food intake, physical inactivity, genetic predisposition, and individual metabolism. Obesity can be reduced by low-calorie low-fat diets, induced physical activity and multiple methods contributing lifestyle changes. However, weight loss contributes havoc lifestyle changes and restricts by high rates of recidivism and subsequent slowing metabolism. In this scenario, it is crucial to onset of genetic and epigenetic traits during obesity. Subsequently, multiple loci on the human genome are related to obesity and metabolic syndrome.[93-95] Interestingly, different omics approaches are required for a unique correlation between the miRNA expressions in multiple tissues including adipose tissue, liver, pancreas and obesity or metabolic diseases.[94, 96, 97] In a study 1736 loci are detected to obesity and 221 correspond to micro-RNAs.[94] In addition, many miRNA expressions are related directly with diet and lifestyle.[98, 99] Despite a long list of miRNAs related to a diet is essential, several miRNAs like miR-17/20/93 family, miR-21/590-5p family, miR-200b/c family, miR-221/222 family, let-7/miR-98 family and miR-203 are important in this context.[98]

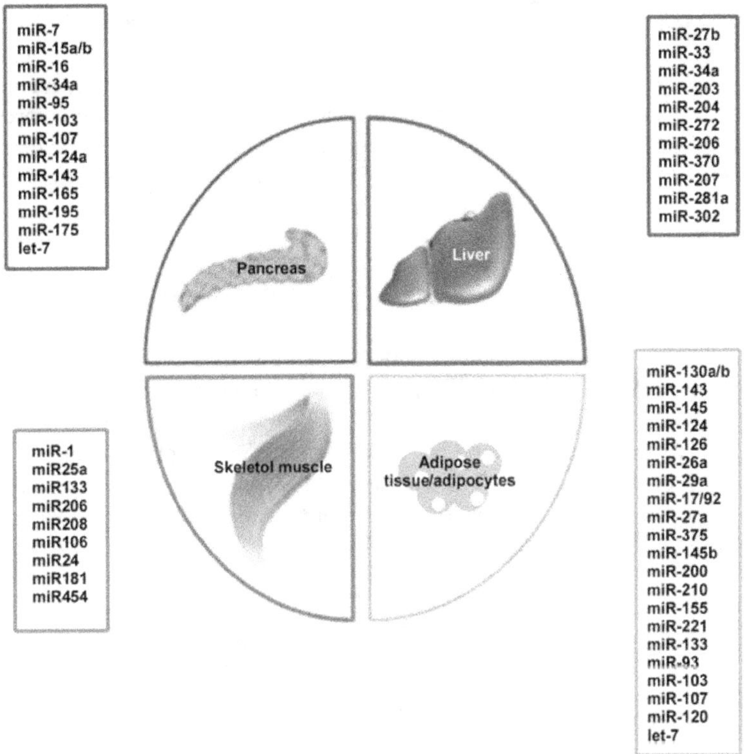

Figure-3. Possible miRNAs related to obesity and metabolic diseases in different tissues. (Loosely Adapted, Redraw and Modify using skeleton from Iacomino and Siani, 2017)

The energy homeostasis regulation is turned sharply by endocrine and neural mechanisms for cooperating the balance maintenance between caloric intake and energy expenses.[99, 100] In this relation, the central nervous system (CNS) has a watchful eye for continuous change monitoring in different metabolic parameters including blood glucose or free fatty acids levels or different hormone contents using insulin, leptin, ghrelin and illustrates adaptive responses including food intake regulation or autonomic nervous system alteration.[99, 100] Specially, leptin, an adipose-derived hormone is essential for normal body weight and insulin sensitivity maintenance by different nuclei functioning of the hypothalamus as arcuate or ventromedial nuclei.[101-103] Moreover, obesity is related to inflammation in adipose tissue intimately. The inflammatory cytokine productions interact with insulin signaling and ultimately related to T2D and various other obesity-related diseases.[104, 105] In relation, the endocrine function of the adipose tissue is required for maintaining normal weight and energy homeostasis control. Simultaneously, it is obvious that miRNAs may be a new contributor for controlling various functions of the adipose tissue in obesity.[106]

Many miRNAs are expressed differentially in white adipose tissue in obese patients relative to non-obese Individuals.[107-110] Adipose tissue particularly in the visceral area is essential for metabolic disorders than the subcutaneous tissue.[111] In this relation it is noticed that several miRNAs are expressed differently in subcutaneous and visceral white adipose tissue.[112] In humans, a correlation is emerging between the miRNA expression in adipose tissue and different metabolic parameters including BMI, adipogenesis, glycemia, leptinemia.[107, 113] In relation to this observation, the miR-17-5p and miR-132 expression are altered consistently between obese and non-obese omental fat.[114] In fact, these two miRNAs in omental fat and blood from obese patients are related invariably with glycosylated hemoglobin, leptin, body mass index and fasting blood glucose.[114] An activation of miR-21 expression is existed in the white adipose tissue (WAT) of obese patient related to slim controls and it is related to BMI positively.[115]

Currently, an estimation of miRNA expression in subcutaneous adipose tissue from 19 individuals with acute obesity (10 women and 9 men) before and after a 15-week weight loss scenario.[116] This intervention causes miR-29a-3p and miR-29a-5p activation and reduction of miR-20b-5p.[116] In addition, miR-221 is overexpressed in obese patients. This miRNA alters fat metabolism using leptin and tumor necrosis factor-α (TNF-α).[117] Curiously, miR-221 overexpression in the adipose tissue and liver from leptin deficient ob/ob and diet-induced obese (DIO) mice is observed.[117, 118] In relation a large number of miRNA expressions is modified in mouse models of obesity.[119] For example, overexpression of miR-342-3p, miR-142-3p, miR-142-5p, miR-21, miR-146a, miR-146b, miR-379 and the miR-122, miR-133b, miR-1, miR-30a, miR-192 and miR-203 reduction are noticed during obesity development.[120] The miR-21 is overexpressed in obese patients.[115] As a novel strategy use of locked nucleic acid (LNA) in miR-21 treatment result a dramatic weight loss and decreased adipocyte size and inhibition of targets such as TGFβ-receptor 2 (TGFBR2), phosphatase and tensing homolog (PTEN) etc.[121] Particularly, several the weight loss alters circulating miRNA amounts.[122, 123] Indeed, that numerous miRNAs in plasma found in obese relative to slim controls due to acute weight loss of obese women.[122]

Endogenous miRNAs within adipocytes have been largely characterized. However, currently large number of exogenous miRNAs are observed in various biological fluids including plasma, serum, urine and saliva.[124, 125] Multiple fluid-based miRNAs are present in exosomes, a small extracellular vesicle that are synthesized in different adipose cells. Adipose tissue liberates exosome miRNAs as signaling molecules.[126] (**Figure 3**) During obesity more exosomes are released rather than control patients.[127] By this mechanism, adipose tissue can alter genes expression related to other metabolic organs. (**Figure 3**) For example, exosome miRNAs discharge from adipose tissue that modulate their hepatic expression of fibroblast growth factor 21 (FGF-21) related to broader glucose homeostasis.[128]

Collectively miRNAs function as biomarkers for metabolic diseases for targeting pharmaceutical strategies.

Taken together miRNA roles are highlighted in obesity and related diseases fields. Specially, the havoc changes in miRNA expression activates the alteration of gene pattern for regulating a range of biological methods including adipogenesis, adipose tissue inflammation, lipid metabolism and insulin resistance etc. An emerging aspect relating to miRNA functions of miRNAs in the adipose tissue and the pancreas are found for interdependent relation between lipid and glucose metabolism.

The Adipokines and miRNAs

An important endocrine organ adipose tissue is released in 600 adipokines for acting as potential biomarkers in different clinical fields.[129] The essential adipokines are leptin, adiponectin, resistin, apelin.[130] These adipokines release and exhibit many effects on versatile organs. Interestingly, deterioration of adipose tissue function modifies the normal physiology function of these target organs.[130]

These adipokines change miRNA expression related to adipogenesis. [119] Leptin overexpresses miR-378 in human adipocytes.[131] Virtually miRNAs are secretory agents that effects adipokines including leptin, adiponectin. (**Figure 4**) For instance, miR-218 interacts with miRNA 3′UTR for the adiponectin receptor (AdipoR) for reducing the adiponectin effect on glucose uptake.[132] In a core level, multiple miRNAs are interacting leptin during their feeding behavior.[133-137] In contrast, miR-200a upregulation in the hypothalamus obese patients inhibits leptin receptor function in hypothalamus.[135] In a recent report, Dicer-derived miRNAs involved possibly in the hypothalamic sensitivity of leptin.[138]

The role of miRNAs in Adipogenesis

For better understanding, adipogenesis control is indispensable for novel therapies development during the onset of obesity. Indeed, mild obesity like hypertrophic obesity is linked to an increase of adipose cell size, while dramatic, acute and severe obesity or childhood obesity also symptomize for induction of adipocyte number specially in hyperplasia obesity. The adipogenesis functions as a complicated process requiring multiple hormones or transcription factors like peroxisome proliferator-activated receptor γ (PPARγ), CCAAT/enhancer-binding protein (C/EBP) family and Sterol regulatory element-binding protein 1 (SREBP1).[139] In recent days, it is noticed that miRNAs contribute an ideal role in adipocyte differentiation during obesity development. (**Figure 4**) The miRNAs function on adipogenesis appears to be highly complicated due to their multiple interacting targets. Nevertheless, miRNAs control adipogenesis by modifying the key factors expression of adipocyte differentiation.[140, 141]

A role for miRNAs in adipogenesis is that miR-14 and miR-278 control lipid metabolism.[142, 143] Among the several miRNAs, special focus has been drawn to miR-143. The miR-143 is a probable regulator of adipocyte differentiation during antisense oligonucleotides transfected into human preadipocytes.[144] In relation to this study, miR-143 in adipose tissue of obese patients are closely related to the expression of adipocyte differentiation markers like PPARγ.[145] Another miR-130a cluster suppresses adipogenesis via PPARγ silencing.[146] In relation to adipogenesis, the miRNA cluster miR-17-92 including miR-17, miR-18a, miR-19a, miR-19b-1, miR-20a, miR-92a, is overexpressed in adipocyte clonal expansion and induces adipocyte differentiation through an inverse control of the key cell cycle regulator and tumor suppressor gene Rb2/p130.[147] As noted earlier, many miRNAs affect adipogenesis targeting C/EBP. For example, miR-375 activates 3T3-L1 adipocyte cell line differentiation by overexpressing of C/EBP and PPARγ2.[148] On the other hand, a positive link between miR-519d and the PPARα protein amount present in fatty acid homeostasis are reported.[108] The miR-519d inhibits translation of the PPARα protein and accumulated lipid during preadipocyte differentiation.[108]

Inflammation of the Adipose Tissue and miRNAs

Chronic inflammation is an essential obesity character. Obesity-related inflammation is required for onsetting the metabolic syndrome.[149,150] Large number of miRNAs function as regulators for the inflammatory marker expression.[109,151,152] (**Figure 3**) For example, miR-132 induction in primary adipose-derived human stem cells improves the production of interleukin-8 and monocyte chemoattractant protein-1 (MCP1).[153] Moreover, miR-126 and miR-193b control the level and the export of MCP1 respectively.[154] On the other hand, miRNAs alter the amount and/or the synthesis of TNF-α. Particularly, miR-221 silencing is related to an excess level of TNF-α of adipose tissue-derived mesenchymal stem cells from obese woman.[155] The miR-145 induces the TNF-α expression in adipocytes by activating the nuclear transcription factor kappa-B (NF-κB) process.[156] Particularly, a proinflammatory connection is established between NF-κB and miR-155 for participating in the amplification of inflammatory stages in adipocytes.[157]

The miRNAs silencing is also found during the inflammation of the adipose tissue. The miRNA profiling is identical in the TNF-α-treated adipocyte 3T3-L1 cell line and in the adipocytes from the leptin deficient obese mice.[158] TNF-α or leptin activate miR-221 silencing and an overexpression of miR-335 in human preadipocyte.[155, 159] TNF-α also involves in overexpression of miR-130, miR-146a, miR-146b, miR-150, miR-221, miR-222 and miR-103 and miR-143 silencing in the murine adipocyte model.[158, 160-162] Collectively, the miRNA expression in adipose tissue possibly modifies using inflammation. Also, miRNA expression can be required in the inflammation pathway activated by adipose tissue during the emergence of obesity. (**Figure 3**) These inflammatory methods are related to glucose homeostasis defects leading to T2D. Specifically, the pro-inflammatory factors can destroy insulin secretion and the insulin sensitivity in peripheral organs. Currently, miRNAs are responsible for new agents in such pathophysiological aspects.

microRNAs and weight loss

Lifestyle Change studies

Interestingly, separate outcomes of dietary intake possibly linked to epigenetic factors.[163] The contribution of microRNAs for weight loss after dietary and lifestyle interventions are examined elaborately in numerous studies (**Table 2**), exploring microRNA guided gene regulation at a white blood cells, serum, and plasma.

miRNAs in white blood cells

In 2013, microRNA in peripheral blood mononuclear cells (PBMC) from 10 obese women undergoing an 8-week energy-restricted diet have shown. In the Non responder group (< 5% body mass loss, *n* = 5), miR-935 and miR-4772 are overexpressed, while miR-223, miR-224 and miR-376b are reduced strongly. Notably, miR-935, miR-4772 and miR-376b also link to weight loss, being therefore acts as potential biomarkers for weight loss and diet responses.[164] Subsequently, nine selected inflammation-mediated microRNAs of 40 patients white blood cells after an 8-week energy-restricted period along with Mediterranean diet. Two microRNAs miRNA-155-3p reduced dramatically, while let-7b is overexpressed consistently. Moreover, five selective microRNAs let-7b, miR-125b, miR-130a, miR-132-3p and miR-422b are required exclusively for Health Eating Index (HEI) improvement, a measuring device for diet quality and miR-155-3p is also responsible for weight loss.[165] On the other hand, few microRNA alter in MS enrolled in Metabolic Syndrome Reduction in Navarra during nutritional trials, after intaking separate energy-restricted dietary supplements. MicroRNA profiling including six selective microRNAs miR-1237, miR-1976, miR-642, miR-636, miR-612, miR-193B are both hypomethylated and overexpressed. Notably, miR-612 and miR-1976 are also hypomethylated and over-expressed strongly. Both are modified after the dietary intervention, suggesting a role of these microRNA/gene axes in MS and obesity.[166] Moreover, miR-548q and miR-1185-1 overexpression in obese patients irrespective to sex and gender. In a fundamental functional assay, miR-548q and miR-1185-1 silent glycogen synthase kinase-3 B (GSK3β) expression, interacting its mRNA.[167] GSK3β is involved in pro-inflammatory responses by inducing of pro-inflammatory cytokines expression such as IL-1ß, IL-6 or TNF-α.[168] Accordingly, a negative correlation between miR-1185-1 level and IL-6 amount in serum is exhibited.[167]

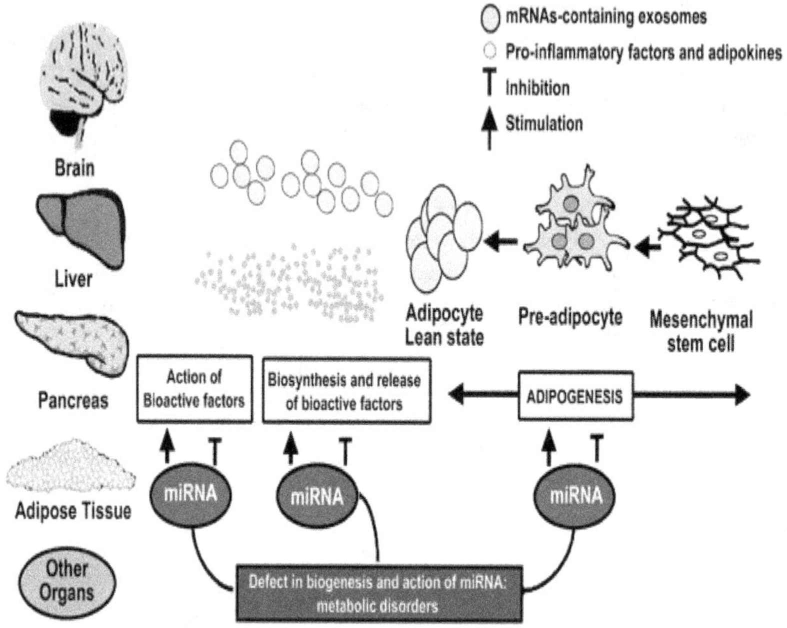

Figure-4. The role of microRNA (miRNA) function in the adipose tissue progress. miRNAs control adipogenesis, the liberated action of the many signaling evens including adipokines, inflammation and exosomes derived from adipose tissues. (Loosely Adapted, Redraw and Modify using skeleton from Landrier *et al.,* 2019)

Table -2. Alteration of miRNA expression after lifestyle intervention strategies

Different miRNAs	Roles/ targets	Sources	Intervention	Population
miR-935, miR-4772 (↑) miR-376b (↓)	Biomarkers of weight loss	PBMC	8-week hypocaloric diet	10 obese women
miR-155-3p (↓), let-7b (↑)	Biomarkers of improvement. Mediterranean diet in quality of diet and weight loss (155-3p)	WBC	8-week hypocaloric	40 MS
miR-612, miR-1976 (↑)	Biomarkers of high response to die/ TP53 and CD40	WBC	8-week hypocaloric diet	96 MS

miR-548q, miR-1185-1 (↑)	Biomarkers of high response. to diet/ GSK3β	WBC	12-week hypocaloric diet	96 MS
miR-222-3p (↑) miR-122-5p, miR-193a-5p (↓)	Biomarkers of weight loss. and MS/ terns/Lipid metabolism	Serum	12-week hypocaloric diet/overweig ht	85 obese
(↓) in LGI vs HGI diet: miR-139-3p, miR-411, miR-432, miR-99b, miR-340, miR-423-5p, miR-361, let-7c (↓) in LF diet: miR-139-3p	Association with specific dietary patterns/Lipid metabolism	Plasma	6-month LGI (n = 36), HGI (n = 36), LF (n = 31)	103 obese diet
LF diet: miR-130a-3p, miR 142-5p, miR-144-5p, miR-15a-5p, miR-221-3p, miR-29c-3p (↓), miR-22-3p (↑)	Biomarkers of response to LF diet/ PPAR-γ, PPAR- α, SIRT1 (miR- 22-3p)	Plasma	high-protein diet (n = 38) LF diet (n=40)	78 obese 25 CTRL
miR-126, miR-375, miR- 376, miR-499, miR-642 (↑) miR-208, miR-433 (↓)	Biomarkers of weight loss/. /glucose metabolism, inflammation angiogenesis, cell death	Plasma	4-week very-low-calorie diet	80 obese women 80 CTRL women
Women: miR-148b-3p, miR 26a-5p, miR-520h, miR-548d-3p Both sexes: miR-30e-5p, miR-502-5p, miR-590-5p, miR-644a, let-7b-5p, miR-143-3p, miR-504-5p	Sex differences in microRNA expression after KD/ inflammation, immunity, glucose metabolism	Plasma	6-week KD	36 obese

miR-133a-3p, miR-133b (↑)	Biomarkers of reduced Skeletal-muscle regeneration	Serum	28-day hypocaloric diet	16 elderly men
miR-140, miR-221, miR- 223 (↑)	Biomarkers of response to intervention	Plasma	16-week diet (HPHC HPMC/CON) and physical activity	111 obese
miR-126 (↑)	Biomarker of weight loss/endothelial endothelial function	Serum	6-week hypocaloric diet and aerobic exercise	37 obese male adolescents 10 CTRL
miR-29a-3p, miR-29a-5p (↑) miR-20b-5p, miR-454-3p (↓)	Glucose uptake, lipid metabolism, energy homeostasis	SAT	15-week hypocaloric diet and physical activity	38 obese
LGI diet: miR-551b, miR-221, miR-378, let7a (↓)	Association with metabolic and body composition parameters	SAT	6-month LGI (*n* = 3), HGI (*n* = 3), LF (*n* = 2) diet.	8 obese
HGI diet: miR-1276, miR-132, miR- 29a (↓) LF diet: miR-661, miR-1179, miR-132, miR-221, miR-29a, miR-378 (↓)				

MS, metabolic syndrome; PBMC, peripheral blood mononuclear cells; WBC, white blood cells; GSK3β, glycogen synthase kinase-3 Β; LGI, low glycemic index; HGI, high glycemic index; LF, low fat; CTRL, control subjects; PPAR-γ, peroxisome proliferator-activated receptor γ; PPAR-α, peroxisome proliferator-activated receptor α; SIRT1, sirtuin-1; KD, ketogenic diet; HPH, high dairy protein and carbohydrate; HPMC, high dairy protein and moderate carbohydrate; CON, low dairy protein and high carbohydrate; SAT, subcutaneous adipose tissue.

Circulating RNA

The changes of blood circulating microRNA are noticed after dietary intakes, comparing different dietary routine plans, with distinct proportions of macronutrients. Circulatory miR-222-3p overexpressed consistently after routine weight loss, while miR-122-5p and miR-193a-5p are decreased the levels. Indeed, miR-122-5p and miR-193a-5p amount are induced in the presence of MS and proportionately related to insulin resistance markers and body composition, such as HOMA- IR (homeostasis model assessment of insulin resistance), waist circumference and visceral fat mass.[169] Particularly, miR-122-5p overexpression is linked to NAFLD potentially and its silencing after weight loss diminishes liver defects.[170] As two miRNAs miR-122-5p and miR-193a-5p are linked to insulin resistance their inhibition after diet intake possibly a clear reflection for glucose metabolism improvement. MiR-222-3p is initially related to T2D and gestational diabetes (GDM), although its efficiency in glucose metabolism has yet to be demonstrated.[171-173] Taken together, all selective microRNA expression possibly the major hallmark metabolically unhealthy obesity.

However, the expression of few circulating miRNAs miR-520c-3p, miR-15a, miR-590-5p, miR-126, miR-636, miR-625 have no link to the potential weight loss. [174] Eight microRNAs miR-139-3p, miR-411, miR-432, miR-99b, miR-340, miR-423-5p, miR-361 and let-7c are differentially expressed during significant dysregulated a wider cohort of 103 patients (Low Glycemic Index, (LGI) n = 36; High Glycemic Index (HGI), n = 36; LF, n = 31). All these microRNAs are silenced in the LGI relative to HGI group, with miR-361showing the highest inhibition. Apart from this, miR-139-3p is silenced both in HGI and in LF group relative to the baseline. It is related positively with both LDL-cholesterol and total cholesterol. MiR-340 is inhibited in the HGI group and linked to triglycerides changes. MiR-432 and miR-423 are silenced only in the LF group and linked to miR-139-3p. Finally, let-7c relates to all lipid parameters.[175]

The expression of 86 obesity-related circulating microRNAs has been examined in three different sectors in obese patients including after taking a moderately high-protein diet (n = 38) randomly or a low-fat diet (n = 40) and in normal weight controls (n = 25). Seven microRNAs, miR-130a-3p, miR-142-5p, miR-144-5p, miR-15a-5p, miR-22-3p, miR-221-3p and miR-29c-3p are altered consistently in response to low-fat diet and expressed differentially between responders and non-responders. Particularly, a positive correlation between miR-22-3p expression and the entity of weight loss has been noticed.[176] Interestingly, miR-22 also interacts with many regulators of fatty acid metabolism for example peroxisome proliferator-activated receptor α (PPAR-α) and sirtuin-1 (SIRT1) etc.[177, 178]

A global microRNA profiling identified 21 microRNAs that eventually mis-controlled in obese women relative to lean women. All of them directed to the belonged in glucose metabolism pathway, inflammation, angiogenesis, and cell death. Interestingly, seven microRNAs including miR-126, miR-208, miR-375, miR-376, miR-433, miR-499, miR-642 are altered markedly in obese women after weight loss converting same as lean women. It reveals that an acute weight loss program reverses inclusively the marked alteration of blood microRNAs in obese patients.[179]

The stringent dietary plans increased the body mass index (BMI), plasma insulin and triglyceride levels in both sexes, while the microRNA modulation profile is different between men and women. In females, at the end of the Kinetic

Dietary (KD) program, many miRNAs including miR-148b-3p, miR-26a-5p, miR-30e-5p, miR-502-5p, miR-520 h, miR-548d-3p, miR-590-5p and miR-644a are modified consistently, while in males miR-30e-5p, miR-502-5p, miR-590-5p and miR-644a are altered. In fact, three microRNAs, let-7b-5p, miR-143-3p, miR-504-5p express equally between baseline and 6 weeks of KD in both sexes. These miRNAs interact with genes involved in different pathways such as cytokine signaling, inflammation and immunity, nutrient metabolism, oxidative phosphorylation, PPARs regulation and insulin signaling etc.[180] The selective skeletal-muscle related serum microRNAs like c-myomiR: miR-1, miR-133a-3p, miR-133b, miR-206 affect the protein synthesis rates. During Energy Restricted (ER) period, only overexpression of miR-133a-3p and miR-133b cause a reverse relation between c-myomiR score and entire body protein synthesis rate. It indicates that ER may have an inverse impact on skeletal-muscle regeneration, the top predicted target prediction. However, this study has multiple restrictions such as the small number of samples, the lack of muscle biopsy and body composition examination during fat-free mass.[181]

An increased miR-935 level is found in Low Responders compared to High Responders of body mass reduction at baseline and miR-140, miR-221 and miR-223 overexpress after16-week weight-loss intervention suggesting a putative role of these microRNAs as biomarkers of variability in individual response to weight-loss interventions. A positive Regression analysis between changes in serum miR-126 levels and BMI, serum levels of endothelium-derived nitric oxide (NO) and endothelin-1 (ET-1), before and after dietary intake of Experimental Group compared to Control Group is performed. These data demonstrate that miR-126 is possible in the development of microvascular endothelial function calculated by RHI index.[182]

Bariatric Surgery studies

The predicted relationship between microRNA regulation and bariatric surgery have been found in a wide spectrum of experimental results. (**Table 3**) It also focuses high level that surgical guided weight loss is important for rescuing microRNA signature.[183]

miRNAs in white blood cells

A consistent alteration of inflammatory toll-like receptor (TLR)/nuclear factor kB (NFkB) in relation with microRNAs regulation are observed in circulating monocytes of obese patients and lean controls. 133 differentially expressed circulatory monocytes microRNAs from obese are selected relative to lean control by a microRNA profiling. Nine microRNA families are dysregulated that are linked to TLR/ NFkB pathway. Moreover, only the miR-181 family is inhibited in obese patients, but its levels are recovered into normal range after bariatric surgery.[184]

Circulating RNA

During bariatric surgery, many circulating microRNAs are estimated before and after surgical procedure conducted. In 2013, a marked inhibition in miR-16-1, miR-122, miR-140-5p, (EX) and the other 11 to a control health education intervention (CON) is noticed. Interestingly, 94 plasma microRNAs, selected for their relationship with metabolism is examined. Specifically, miR-15a overexpresses, while miR-34a and miR-122 reduce both in the CON and in the EX-group (). It indicates that the modulation possibly a symptom of a routine weight loss response conducted by RYGB surgery. Specially miR-7 and miR-106 induce, while miR-221 inhibits only in the CON group and miR-135b, miR-144 and miR-206 reduce, while miR-149 upregulate in the EX-group (). Interestingly, 10 above microRNAs relate consistently with indices of HOMA-IR, β-cell function, body composition, plasma lipids and liver function. In addition, miR-15a, miR-7, miR-106b and miR-135b link to insulin sensitivity miR-15a, glucose efficient miR-7 and miR-106b and acute insulin response to glucose related miR-135b. It reveals that their role for predicted biomarkers of cardiometabolic modulations.[185]

===

Table -3. Distinct changes in miRNA expression after bariatric surgery

Distinct miRNAs	Regulation/ Target	intervention	Sources	Populations
miR-181 (↑)	TLR-NFkB pathway	RYGB	PBMC	9 obese, 6 CTRL
miR-16-1, miR-122, miR-140-5p, miR-193a-5p (↓) miR-221 and miR-199a-3p (↑)		RYGB	Plasma	6 obese
miR-15a (↑) miR-34a, miR-122 (↓)	Biomarkers of weight loss/glucose metabolism	RYGB	Plasma	22 obese
miR-7-5p, let-7f-5p, miR-15b-5p, miR-320c, miR-205-5p, miR-335-5p (↑), let-7i-5p (↓)	Inflammation, adipocyte proliferation, ß-cell function, thyroid and pituitary function	RYGB	Plasma	29 T2D
let-7a-5p, miR-16-5p women	Insulin signaling	RYGB	Plasma and serum adipocyte-derived exosomes	6 obese
miR-424-5p	Biomarker of weight loss	LSG (*n*=2) RYGB (*n* = 14)	Serum exosomes	16 obese 18 CTRL

SAT: miR-23a-5p, miR- 27a-5p miR-200c-3p, miR-223-3p, miR- 1246, miR-24-2-5p, miR-128, miR-421, miR-3178, miR- 1224-5p, miR-221, miR-22, miR-762 (↓) VAT: miR-223-3p (↓)	Inflammation, glucose uptake	RYGB	SAT VAT	15 obese women
VAT: miR-122 (↑)	PPAR-γ	LSG	SAT VAT	20 obese 8 CTRL
miR-146b-3p, miR- 146b-5p, miR-223-3p, miR-223-5p, miR-941 (↑)	BMPR2, FOXP1, IGF1R	Bariatric surgery	SAT	20 obese 7 CTRL
miR-155, miR-221, miR-130b (↓)	Inflammation	RYGB	SAT	16 obese
miR-19a/b, miR-146a/b, miR-155, miR-193b, miR-221, miR-222, miR-223, miR-376c, miR-411 (↓)	Glucose uptake, lipid metabolism, energy homeostasis	RYGB	SAT	9 obese women
miR-519d, miR-299-5p, miR-212, miR-671-3p (↓) miR-370, miR-487a (↑)	PPAR-α (miR-519d)	LAGB	SAT	3 obese 2 CTRL

CTRL, control samples; RYGP, Roux-en-Y gastric by-pass; PBMC, peripheral blood mononuclear cells; T2D, type 2 diabetes; SAT, subcutaneous adipose tissue; VAT, visceral adipose tissue; LSG, laparoscopic sleeve gastrectomy; BMPR2, bone morphogenic protein receptor 2; FOXP, fork-head box protein P1; IGF1R, insulin-like growth factor receptor 1; LAGB, laparoscopic adjustable gastric banding.

Six microRNAs, miR-7-5p, let-7f-5p, miR-15b-5p, miR-320c, miR-205-5p, miR-335-5 overexpress consistently, while let-7i-5p inhibits after surgery. Specially, a marked inhibition in blood glucose and glycated hemoglobin (HbA1c) amounts is noticed after surgery in majority of patients (more than 60%).[186] Regulated micro-RNAs are found residing in natural killer cells that are instrumental for obesity-induced inflammation.[187] However, the same microRNAs have been involved in different biological methods including adipocyte proliferation, ß-cell function, thyroid and pituitary function.[188, 189]

The plasma and serum adipocyte-derived exosome microRNAs are examined in six obese women before (T0) and 1-year (T1) after RYGB surgery. Nearly, 29 differentially regulated microRNAs between T0 and T1 are related markedly to HOMA-IR improvement. Moreover, two above microRNAs, let-7a-5p, miR-16-5p involve not only link to Insulin Receptor Signaling, but also induces branched chain amino acids (BCAAs) levels that are strongly connected to insulin dysregulation. Thus, the circulating adipocyte-derived exosome microRNAs modulation is possibly interconnected with post-surgery inductions in glucose homeostasis and insulin resistance.[190]

Circulatory microRNAs in serum exosomes are explored in obese patients (n = 16) before and 6 months after bariatric surgery and from healthy normal individuals (n = 18). Nine exosomes and 32 circulating microRNAs alter their expression in higher and lower levels in obese patients relative to normal controls. Among the nine overexpressed microRNAs, the amount of miR-424-5p before surgery strongly related to weight loss.[191] Due to their maximum stability, exosome microRNAs are ideal for biomarkers in multiple pathological processes. Since multiple exosomes microRNAs silent after bariatric surgery.

Challenge of obesity treatment

Despite rapid advances in therapeutic strategies, obesity controlling, and management remains highly challenging. Lifestyle alteration is the primary target for weight loss and fat mass reduction. Truly, popular strategies like dietary energy restriction and increased energy expenditure through physical activities are the well-known key pillars of obesity treatment.[192] Particularly, energy restriction is the short-termed measure for the critical weight-loss phase, while physical exercise is essential for long-term investments. At least 5% reduction of the initial body weight is proven clinically significant and can be achieved efficiently by reducing the 500 kCal of daily caloric intake.[193] Indeed, exercise without dietary restrictions is reduced only 3–5% weight loss in obese patients.[194] Variable dietary supplements such as Mediterranean diet, low carbohydrate diet, low-fat diet and ketogenic diet (KD) differ in their macronutrient proportions. They are extensively use for great efficacy in reducing body weight and in encouraging to prevent metabolic defects in obese patients.[195-197] In a separate strategy, bariatric surgery is the most efficient method for weight loss and comorbidities promotion in morbidly obese patients, when restricted lifestyle maintenance fails. Particularly, most popular available surgical methods like, laparoscopic sleeve gastrectomy (LSG) and laparoscopic adjustable gastric banding (LAGB) strictly restrict food intake, while Roux-en-Y gastric by-pass (RYGB) is mostly malabsorptive.[198] However, bariatric surgery frequently develops numerous health hazards like dumping, nausea, vomiting, diarrhea, infections, stenosis, bleeding, increased risk of alcohol abuse after surgery and ultimately perioperative death.[199, 200]

Despite wide range of therapeutic options, reaction to weight loss methods is major hallmarked by high inter-individual differences, partly by epigenetic factors.[201-203] Based on such criteria, efficient development of individual methods depending on nutrigenetic, and nutrigenomic information is of utmost hallmark in customized obesity management development.[204] Thus, numerous variable approaches are employed in response to diet or bariatric surgery. Recently microRNAs, a class of small non-coding RNAs that control obese related gene expression, have got strong momentum not only as regulators of obese biological processes but also as prognostic biomarkers exclusively required for obesity management.

Conclusion

Under persistent crosstalk brain-wide pathophysiology characters are altered based on the complicated etiology of alcohol abuse disorders and its dependence. It is argued that an organ and system-specific models should be developed for better understanding and treatment of alcohol abuse with the effectiveness to capture the dynamism of such complications. Once major dynamic relationships are governed by steady stable disease states the solution strategies conduct different detour or inverse disease development. It is a sole method of chemical reactions transiting via many equilibrium states. In this scenario, an idea establishes that miRNAs are overexpressed effectively due to excess alcohol intake and later activate organ and system mediated homeostatic changes as a prolonged adaptation under the special cellular environment. Curiously, localized miRNA activity within neuron synapses also demonstrates special and prominent activities in other cell types. For example, innate immunity-mediated functions in microglial cells, among others. This hibernated impact plays a potential role in molecular efficiency, signaling crosstalk and cellular economy during adaptation and evolution of cellular physiology, the learnt concepts for complex biological research.

Future progress of unique knock-down and/or overexpression strategies mediating miRNA activity due to the wide range and far-reaching regulatory power of miRNA. It is a deliberate but natural effort for protecting against reverse alcohol addiction. Nevertheless, several issues are not addressed for better explaining how miRNAs involve their actions and how they govern to fight against alcohol abuse. It is amazing that tiny sequence variations of miRNA family members affect their target genes and execute biological function. Also, whether miRNA family members emerge persistently random control over selective targets for fine-tuning alteration of post-transcriptional control. On the other hand, it is also equally thrilling for the discovery of small molecules that could direct and modulate numerous selective miRNAs in specific tissues for invaluable clinical and therapeutic fields.

The miRNAs establish themselves as crucial players for endocrine regulations. The miRNAs function with conjunction of versatile aspects of metabolic defects: organs development, release of hormones and hormonal sensitivity. The miRNA instigates as diagnostic early biomarkers of prognosis of almost all endocrine diseases. In this view, miRNAs packed with cell-generated exosomes are the most reliable cross-talk between multiple organs. Exosomes especially contribute to the communication between adipose tissue and the liver. In the presence situation the entire understanding of the novel mechanism of cell–cell communication is still not known. A clear dysregulation of miRNAs are the potential sites of energy storage, white adipose tissue, after a short time of high caloric diet consumption. The induction of microRNA defects sometimes switches upon normal diet. Moreover, the solid use of miRNAs as early biomarkers for adiposity changes and function as a therapeutic agent in white adipose tissue are found during diet-induced obesity proliferation.

Obesity and obesity-related defects are most rapidly spreading public health hazards world-wide during the last 20 years. Recently, microRNA focused many epigenetic modulators in obesity for understanding the pathophysiological progress. Several microRNAs lead to control obesity differently. Their expression alters many weight-loss approaches consistently like diet intake, physical exercise, and bariatric surgery. MicroRNAs modulation in many compartments like blood cells, serum, plasma, adipose tissue, and different available intervention methods regarding obesity treatment are found. Specially, dietary intakes along with marked changes in macronutrient composition are compared that highlight many altered microRNA expression profiles. Thus, weight loss produced by multiple dietary intakes possibly reflects a havoc change in microRNA signature. In bariatric surgery, huge intervention factors lead to the malabsorptive procedure RYGB, whereas results of other employed surgical strategies like LSG and LAGB are tightly restricted. Therefore, a healthy comparison between the effect of many surgical procedures on circulating and tissue microRNAs expression should be addressed. Basically, the microRNA profile estimation in adipose tissue before and after obesity treatment plans provide crucial findings regarding the metabolic changes related to weight loss. In fact, the fundamental principle of microRNAs control in adipose tissue metabolism is essential and utmost interest for leading consistent improvement during obesity treatment. Importantly, diverse microRNA expression patterns between VAT and SAT have been noticed in several studies, these indicate that weight loss possibly alters the metabolic tissue profile.

Considering the consistent variability in individual response to weight-loss interventions, circulating microRNAs might be valuable biomarkers of efficacy, possibly helping in the differentiation between responders and non-responders. Remarkably, a wide variety of dysregulated microRNA target pathways, some of them crucially involved in glucose and lipid metabolism, energy homeostasis, inflammation, immunity, endothelial function, are also detected for inducing the pathophysiological processes related to obesity and obesity-mediated metabolic disorders. Nevertheless, these discrepancies and confusing results contribute to many factors including heterogeneity of variable populations, separate samples collected from plasma, serum, adipose tissue, type of adopted diet composition, physical exercise, surgical procedure, and diversity variability in microRNA profile analysis. Later, a hard and straightforward effort is employed for microRNA estimation via solid quantification and RNA identification. Indeed, RNA measurement and its performance in strong lone molecule counting is possibly a critical advance in many analytical methodologies.

References

References: chapter wise are cited before the Index at the end of the Book. (1-204)

Chapter VI

miRNAs for critical care medicine diagnosis and therapy

Abstract

The noncoding tiny but functionally effective miRNAs emerge as a critical regulator of many cellular processes. The miRNAs degrade gene expression via base pairing to the 3′-untranslated termini of functional mRNAs. In clinical diagnosis, miRNAs function as a novel, important promising biomarkers as well as new frontier tools and targets for novel therapies. miRNAs recently enter the network of perioperative medicine, which may open new perspectives and dimensions in anesthesia, critical care, renal failure, and painkilling medicines. An elaborate promising instruction on miRNA-guided therapeutic opportunities and recent paradigms of miRNA-based effects in perioperative medicine is highlighted here.

An amazing finding that breaking the paradigm of a linear copy number correlating with the amounts of mature transcripts was invented nearly 25 years ago (Pal-Bhadra *et al.,* 1997). The inverse correlation means more copy number produced less product is the foundation for germinating small regulatory RNA. Around 5 years later, an active scientific group isolated small noncoding RNA let-7 in *Homo sapiens* assuming that functionally more similar molecules might be isolated in the future. Of such speculation within a year, many new microRNAs (miRNAs) were identified. To date, nearly1900 human miRNA sequences have been identified. The miRNA function in the regulatory aspect for virtually all relevant biologic processes. On relevancy, miRNAs control approximately 60% of genes constituted in the human genome. The microRNAs have really macro-effects that are involved in the improvement of pathological processes including critical care treatment plans.

The potent valid biomarkers miRNAs are an urgent class of paradigm that enables early and reliable detection of many life-threatening disorders including critical care defects. There is still an urgent need for valid biomarkers that enable early and precise detection of life-threatening disorders such as critical care diagnosis. A new set of biomarkers using single or multiple miRNAs have been required for sepsis, pneumonia, acute lung injury (ALI), renal failure and the failure of multi-organs. The implementation of biomarkers into the clinical routines is a challenging drive against many hurdles of critical disease diagnosis.

Introduction

The miRNA repertoire is expressed by each cell type that contributes to the shaping of multiple tissues specific features and functions. The expression of miRNAs in certain tissues or cell types combat a widely distributed spectrum of human diseases by dysfunction of the immune system.[1-4] The assignment of miRNAs to certain critical disorders has influenced only their correlative analyses. For instance, miR-146a inhibits inflammatory processing genes as it is by downregulating nuclear factor-κB (NF-κB) signals and ultimately repress the inflammatory defects.[5]

The gene inhibition is not the sole miRNAs criteria to the parental target cells, also influencing the signaling molecules facilitating intercellular communication. The miRNAs can be conjugated into exosomes or micro vesicles. Later, they are released from cells to surrounding tissues via circulation. These secreted moving miRNAs can control many mechanisms in the new cellular surrounding.[6-8] In fact, the membrane or lipid-bound extracellular miRNAs are further circulated stably in many body fluids such as blood, urine, liquor, breast milk saliva, and bronchial secretions, etc.[9]

The sole features, high stability, and miRNAs accessibility have gained momentum extraordinarily in the clinical biomarker field for early diagnosis, staging, and prognosis of specific diseases and to promote their treatment plans.

The convincing clinical biomarkers, their broad regulative capabilities are the potent characters of miRNAs that influence the high potential as therapeutic targets.[10] In the last decade, chemical formulation of protective mRNAs known as miRNA mimetics that downregulate specific pathology-driving genes. They are developed and delivered to patients. Besides these basic approaches are still in infancy and need excessive improvement that provides a great momentum toward the huge leap for drug discovery.

miRNA in Human critical care diseases

The valid markers executing an early and tight focused detection of life-threatening disorders are the urgent and indispensable need for critical care medicine against the frequently associated failure of organs, sepsis, and acute lung injury (ALI). In this regard, miRNAs are the first runner and gain positive momentum in the critical care diagnosis as remarkable new biomarkers. The expression of individual miRNAs or expression of miRNA families is utilized. Their implementations into the clinical routines particularly against critical care disorders are an ongoing process that prevents many hurdles from growing during different diseases. Multitude miRNA expression is the underlying force for numerous molecular diagnostics. A brief survey for the current status quo of sepsis, ALI, and acute organ dysfunction is discussed in the chapter.

Sepsis

The miRNAs can control both innate and adaptive immunity. Their multiple functions are found in many immune cells during proliferation and differentiation.[11] In fact, the magnitude for certain miRNAs function could deliver not only as efficient biomarkers for the detection of early sepsis as such but also differentiate the hyperinflammatory and the immunosuppressive phases of sepsis. The differential expressions of many miRNAs have been recognized in between sepsis versus normal controls and in non-survivors versus survivors in the circulatory blood. The miRNAs profiling from three blood samples plasma/serum or whole blood or in purified blood cells have shown that two miRNAs miR-21 and miR-155 are upregulated in sepsis, while miR-150 have reduced dramatically.[12-14] In a separate study three distinct miRNAs miR-34a, miR-143, and miR-223 are increased significantly in serum or plasma samples,[15, 16] while miR-15a and miR-146a are downregulated markedly.[15, 17] In few instances a specific approach is adopted to analyze adaptive immunity cells. In these studies, two prominent miRNAs miR-15a/16 and miR-223 show a clear elevation whereas two different miRNAs miR-3 and miR-146a downregulate their expression levels.[18-20] In a separate study five important miRNAs, miR-133a, miR-150, miR-233, and miR-547-5p expression are correlated relative to the severity of sepsis defects.[21, 22] that also marked early diagnosis in specific sepsis and the Systemic Inflammatory Response Syndrome.[23-25] All the differential expression of miRNAs expression is summarized in this comprehensive study.[26]

The precursor cell types that originate into free circulatory mRNAs are also aggregated of sepsis with another morbidity. Set of these free miRNAs may bias to sepsis-specific expression profiles strongly. For these consequences, a small sample size miRNAs pools are utilized. However, these small sample sizes do not utilize the role of the prominent biomarker in reliable tools for diagnosis and therapy development for sepsis in the future.

For better understanding the miRNAs molecular mechanisms in sepsis two diversified experiences are essential (i) To acquire an elaborative knowledge of the pathomechanisms in sepsis and (ii) to develop miRNA-based therapy plans and approaches. An elaborate insight for miRNA signaling networks in the immune system particularly in the sepsis pathophysiology emerges as new miRNAs hubs within the inflammatory regulatory circuits. The miRNAs network target multiple transcription factors such as NF-κB or hypoxia-inducible factor 1-alpha, toll-like cell surface receptors or mitogen-activated protein kinase pathway, etc. They can target cytokines and/or their receptors directly. In this scenario, both hyper inflammation and immunosuppression in sepsis affected the function, segregation, and development of both adaptive and innate immune cells. Many miRNAs are involved in the cell environment and allow them to function as clinical markers for modifying the regulators of the NF-κB-pathway. (**Figure 1, Table 1**) Further many miRNAs regulate multiple important pathways for example miR-155 modulates the transforming growth factor (TGF)-beta-activated kinase 1/MAP3K7-binding protein 2.[27] The miR-31 alters the NF-κB inducing kinase in sepsis[28] miR-15a/16 and miR-146a control the interleukin-1 receptor-associated kinase 1[29, 30] and miR-15a/16, miR-223, target the IκB kinase alpha.[31] Basically, the transcriptional factor from the NF-κB family can produce multiple inflammatory responses for instances proinflammatory cytokine induction, leukocyte insertion or cell survival, and the feedback control of inflammation.[32, 33] These miRNAs appear to be essential players for controlling NF-κB magnitude in the inflammatory networks and duration of inflammation during sepsis.

The miRNAs that are utilized for immune cell function are essential in sepsis. (**Table 1**) Thus, sepsis controlling miRNAs is considered as "guilty by association." However, the implication of sepsis-related miRNAs should be further evaluated for their inflammatory responses during the different phases.

Pneumonia

Pneumonia is the most dreadful frequently occurring critical disease in the world. Pneumonia causes ~1.3 million mortalities in children each year.[34, 35] community-acquired pneumonia via Human adenovirus (HAdV) infection occurs in children and young adults.[36, 37] The bacteria-based pneumonia is the most vulnerable and frequently delivered pneumonia in humans. A brief discussion of two types of pneumonia and their microRNA based regenerative therapy are described separately.

Bacterial Pneumonia

Pneumonia leads to long recovery and a frequent rate of mortality.[38] The bacteria *Streptococcus pneumoniae* (*Sp*) is the causative agent pneumonia generation following chronic flu infection.[39] The pathobiology of pneumonia leads to lung damage.[39-41] Its microbial infection correlates with bacterial virulence and host immune responses. The clearance of microbial pathogens is not the only remedy for bacterial pneumonia, the repair of airway epithelium damage is also equally important for chronic obstructive pulmonary disease (COPD), idiopathic pulmonary fibrosis, and emphysema.[42, 43] A positive drive for the reactivation process protects against lung injury in adults.[44, 45] The microRNAs (miRNAs) appear to involve multiple facets of tissue production and development.[46, 47] The miRNA cluster miR-302–367 known as the miR-302 family is overexpressed at the early stages of lung progenitor cell proliferation during embryogenesis.[48] A strong regenerative process influences proliferation and local stem differentiation or progenitor cell populations including the alveolar epithelial cells (AEC) and bronchiolar Club cells.[49-51] The expression of miR-302 upregulates AEC stimuli and correlates with AEC regeneration. The transient expression of miR-302 genes is important for local stem cell stimulation in the lung parenchyma to regenerate epithelium and enhance recovery following *Sp* infection. The existence of miR-302 mimics protection against lung functions enhances host recovery and survival by helping the production of alveolar epithelial progenitor cells. They also regenerate AEC and repair damaged alveolar epithelium for recovering bacterial pneumonia lung alveoli are made of alveolar epithelial type I cells (AECI) for gas exchange with alveolar epithelial type II cells (AECII). These cells are important to produce antimicrobial peptides, surfactant proteins, and lipids production that reduces alveolar surface tension. The expression of miR-302b and miR-302c is regenerated adult lung epithelium with subsequent lung infection related to bacterial pneumonia.

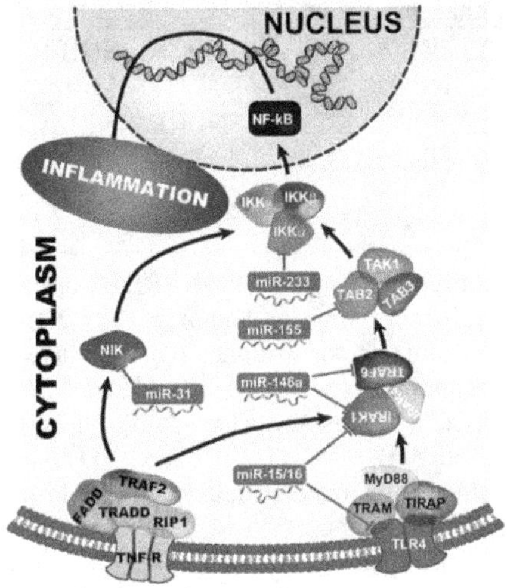

Figure-1. Inflammation activated sepsis miRNAs leads central ingredients of the nuclear factor-κB (NF-κB) signaling pathway for strong regulation. FADD represents fas-associated protein with death domain; IKK, IκB kinase alpha; IRAK, interleukin-1 receptor-related kinase; NIK, NF-κB activating kinase; RIP1, receptor-interacting serine/threonine protein kinase 1; TAB, TGF-beta-induced kinase 1/MAP3K7-related protein 2; TAK, tat-linking kinase; TGF, transforming growth factor; TIRAP, toll-interleukin 1 receptor domain containing adaptor protein; TLR4, toll-like receptor-4; TNF-R, tumor necrosis factor receptor; TRADD, TNF receptor type 1-related to death domain; TRAF, TNF receptor-linking factor 2; TRAM, toll-like receptor adaptor molecules. (Loosely Adapted, Redraw and Modify using skeleton from Kreth *et al.*, 2018)

Table- 1. Mechanisms of Action of miRNAs Identified as Sepsis Biomarkers

miRNAs	Regulated Signaling Pathways	Targets	Regulation in Sepsis Patients
miR-15a/16	TLRs, IL-1, NF-kB	TLR4, IRAK-1, IKKå	PBMCs downregulation
miR-31	HIF-1å, NF-kB, SLAM	FIH, NIK, SAP	T cells upregulation
miR-21	Apoptosis, STAT	PDCD4, SORBS2, IL-12	Blood Upregulation
miR-143	STAT, TLR Prostaglandins	IL13P, TLP2, COX-2	Serum upregulation
miR-146a	IL-1. TNF, NF-kB Ca Signaling	IRAK, TRAF6, PRKC8	Serum, T cell downregulation

miR-150	HF-1□, angiogenesis, signal inhibition	HIF-1å, VEGFA, ARRB2	Blood downregulation
miR-155	STAT, P13K, AKT, NF-kB	SOCS1, SHIP1, TAB2	Blood Upregulation
miR-223	IL-1, P13, AKT, STAT, NF-kB, DNA-binding	NLRP3, NFI-A, IGF-IR STAT3, FOXO1, IKK□	Serum PBMCs Upregulation

Abbreviations: AKT, Protein Kinase B; ARRB2, arrestin beta 2; COX-2, Cyclooxygenase-2; FIH, Factor inhibiting Hypoxia; FOXO-1, Fork head box Protein 01; HIF1a, Hypoxia-inducible factor 1-alpha; IGF-1R, insulin -like growth factor receptor-1; IKK-a, I-KappaB kinase alpha; IL, Interleukin; IRAK, Interleukin 1 receptor associated kinase; NF-kB, Nuclear factor -kB; NF1a, nuclear factor 1a; NIK, NF-kB inducible kinase; NLRP3, NACHT, LRR and PYD domain -containing Protein 3; PBMC, Peripheral blood mononuclear cells; PCDC4, Programmed cell death 4; PI3K, phosphatidylinositide 3 kinase; PRKC8. Protein kinase epsilon; PTGS2, Prostaglandin-endoperoxide synthase 2; SAP, SLAM-associated protein; SHIP1, Phosphatidylinositide-3, 4, 5-triphosphatase 1; SLAM, Signaling lymphocyte activation molecule; SOCS1, Suppressor of cytokine signaling 1; SORBS2, Sorbin and SH3domain containing protein 2; STAT3, signaling transducer and activator of transcription -3; TAB2, TGF-b activated kinase 1/ MAP3K7 binding protein; TLR, toll-like receptor; TNF tumor necrosis factor ; TRAF, TNF receptor associated factor; VEGFA, Vascular endothelial growth Factor A.

The presence of miR-302b/c mimics lower down the total protein and lactate dehydrogenase (LDH) activities within bronchoalveolar lavage fluid (BALF (**Figure 2**) that is important for less damage to lung tissue integrity. It also improves the regeneration of AEC, repair of the damaged alveolar epithelium, and influences pulmonary function, lower mortality rate, and faster recovery of surviving animals from pneumonia. A miRNA mimics approach can be used for microbial infection treatment by inducing the proliferation of lung progenitor cells. It also regenerates AEC to repair lung injury following bacterial pneumonia. The existence of miRNAs opens a new way in emerging therapies for curing microbial infection using regenerative medicine. Despite the regenerative approach, it does not eliminate microbial growth per se, it reduces suffering and reduces recovery time by improving tissue repair for serios pneumonic infection. Also, this approach is important for long-term benefit for severe chronic bacterial pneumonia.

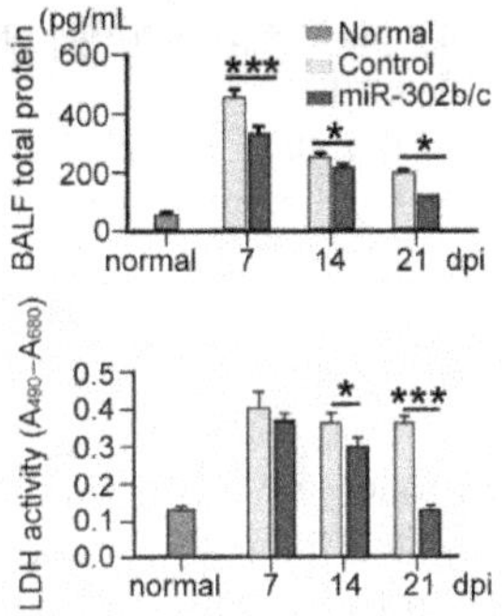

Figure-2. Response of miR-302b/c mimics treatment on SpT4-infected mice. Mice are treated with SpT4 on day0 and further treated with miR-302b/c or Control mimics at 5 and 6 dpi and observed daily for survival, Total protein levels (Upper) and LDH activity level (Bottom) in BALF at indicated dpi. Normal: uninfected/untreated control. Data are calculated using the Gehan–Breslow–Wilcoxon test of cumulative data (n = 14 for miR-302b/c; n = 16 for Ctrl mimics). Data displayed are means ± SEM (C–G, $n = 4$ per group; I, $n = 10$ per group). *$P < 0.05$; **P < 0.01; ***P < 0.001. (Loosely Adapted, Redraw and Modify using skeleton from Yan Wang *et al.*, 2019)

Pneumonia via Human Adenovirus (HAdV) infection

Human adenovirus (HAdV) infection is a dreadful causative agent for one of the frequently occurring community-acquired pneumonia in young adults and children.[52, 53] Adenoviruses in involved upper or lower respiratory and gastrointestinal tracts and the conjunctiva infections.[54-56] It has more than 50% Fatality rates for severe HAdV-associated pneumonia.[57] The conventional diagnosis of HAdV infected pneumonia is based on multiplex PCR and has faced many limitations that hinder it early and effectiveness of adenovirus-mediated lower respiratory tract infection. Therefore, to screen potential biomarkers for HAdV pneumonia, stable exosome serum miRNA profiles of HAdV-infected are tested. A pair of selected exosome miRNA are considered for clinical diagnosis of HAdV-infected children pneumonia. The major clinical symptoms of *HAdV* pneumonia are fever and cough, while few patients are symptomized with wheezing and tachypnea. Among the miRNAs, 18 miRNAs are overexpressed significantly while 10 are reduced markedly correlated with *HAdV* pneumonia. (**Figure 3B**) These are served as potential biomarkers for HAdV-infected pneumonia patients. Among nine differentially expressed miRNAs associated with *HAdV pneumonia,* the miR-450a-5p /miR-103a-3p and miR-103b-5p/miR-98-5p express following the opposite trends. miR-103b-5p and miR-450a-5p are overexpressed and miR-98-5p and miR-103a-3p are reduced dramatically in HAdV samples functioning as a potential biomarker for HAdV pneumonia. (**Figure 3**)

The relationship between serum miRNAs and HAdV pneumonia is hard to get. HAdV infection treatment is controversial because their therapeutic trials have not been involved frequently. Furthermore, severe HAdV pneumonia is found as long-term respiratory tracks infection including hyperlucent lung, bronchiectasis, and bronchiolitis obliterans.[58, 59]

Recently, the differential expression of exosome miRNAs has focused strong attention on finding an early diagnosis for HAdV pneumonia.[60] However limited information regarding its clinical knowledge and adenovirus-infected pneumonia management are evaluated about exosome miRNAs. These limitations incline for the section of serum miRNAs. The above two pairs of serum miRNAs miR-103b-5p/miR-98-5p and miR-450a-5p/miR-103a-3p are differentially expressed in the serum of HAdV-infected children that may function as the most promising non-invasive biomarkers for HAdV pneumonia. Among them serum miR-103a-3p participates in the differentiation of Th2/Th17 cells in severe equine asthma[61], expecting that miR-103a-3p is involved in the immune response of HAdV pneumonia patients.

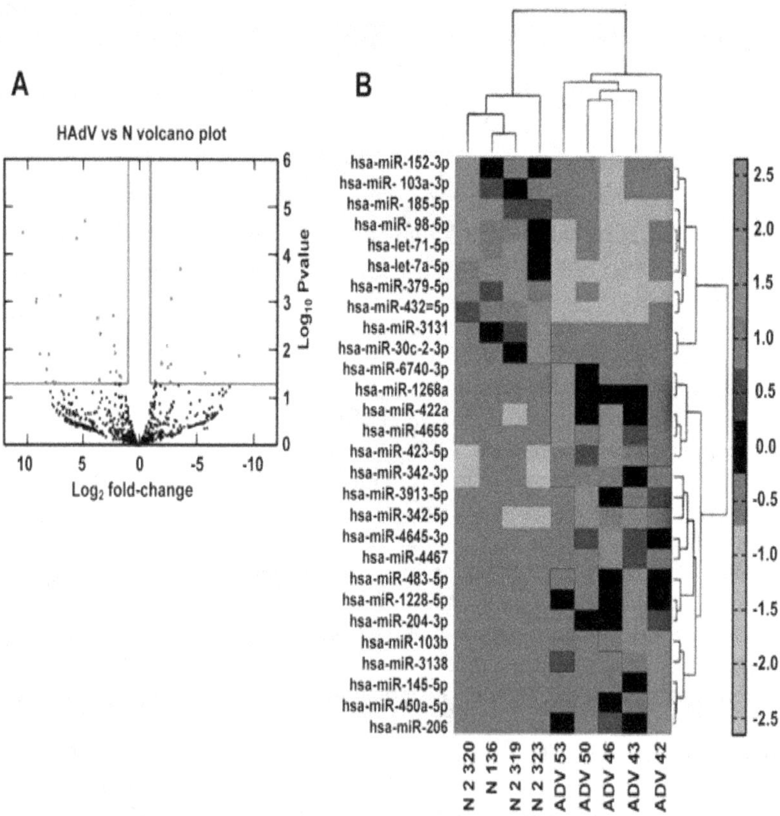

Figure-3. Differentially expression profile miRs derived from the HAdV group relative to the control group. **(A)** Expression profile of exosomal miRs isolated from the HAdV and control groups is measured by microarray analysis. **(B)** Heat map of miRs with differential modulation of the HAdV expression or control groups. miR, microRNA; HAdV, human adenovirus; N, healthy control. (Loosely Adapted, Redraw and Modify using skeleton from Feng Huang *et al.*, 2019)

==

Additional biomarkers

Pneumonia causes several infections in the lower respiratory tract. A total of 9,026 targets of the miRNAs are isolated for non-severe pneumonia. The overexpression of miRNAs in non-severe pneumonia can follow anatomical structure, morphogenesis, and metabolism regulation-associated functions of pneumonia. Apart from the above biomarkers, 11 differentially expressed miRNAs are important for developing innovative and improved early diagnosis and therapy candidates for treating pneumonia patients. Recognition of 11 overlapping's differentially expressed miRNA is noticed in severe pneumonia. Among them 6 miRNAs including hsa-miR-34a, hsa-miR-95, hsa-miR-193b, hsa-miR-200b, hsa-miR-455 and hsa-miR-483 are overexpressed and 5 miRNAs including hsa-miR-664b, hsa-miR-3161, hsa-miR-3617, hsa-miR-4485 and hsa-let-7f-1 are reduced their expression. **(Figure 4)** The distinct levels of their expression are involved in the three stages of pneumonia progression. **(Figure 5)**

==

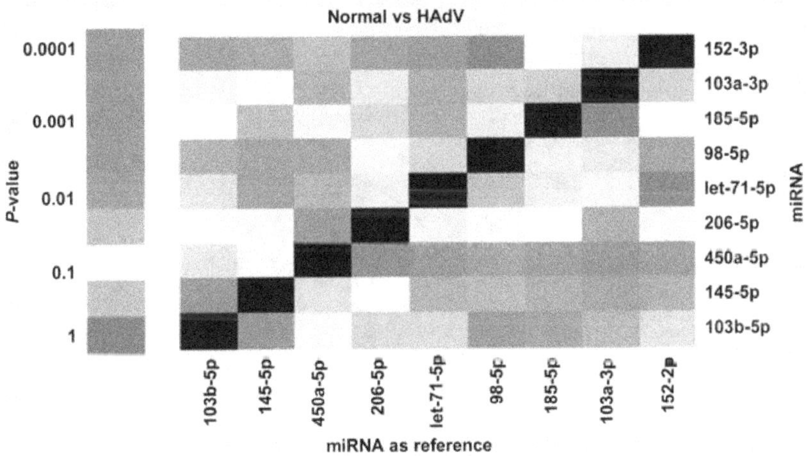

Figure-4. Correlation graph of nine miRNA expressions in the HAdV group relative to the control group using a pairwise t-test. Compares of five healthy children and five HAdV-infected children for examining the expression levels of these nine candidate miRNAs using qRT-PCR analysis. The P-values of the pairwise t-tests between each pair of groups are displayed in a color scale, where black displayed that the P-value cannot measure. HAdv, human adenovirus; miRNA, microRNA. (Loosely Adapted, Redraw and Modify using skeleton from Feng Huang *et al.*, 2019)

==

The human miRNA hsa-let-7f-1 enhances several pathways for pneumonia development. It also involves the 'adherents' junction' and 'Wnt signaling pathway'. The let-7b and let-7c have been isolated as critical miRNAs in the lung restoration at the late stage of influenza pneumonia.[62] Also, the hsa-let-7f-1 miRNA is reduced significantly in pneumonia. This distinct hsa-let-7f-1 reduction helps in the invasion and metastasis of pneumonia-related pathogens into lung epithelial cells or suppresses lung damage repair functions. The let-7f-1are involved in these signaling pathways for controlling pneumonia progression.

A novel miRNA, miR-200b-5p connects regulation of DNA repair mechanism and pulmonary injury.[63] The expression of the *KALRN* target gene reduces the miR-200b for insisting on pneumonia development. On the other hand, miR-455 downregulates NSCLC via mRNA encoding Zinc finger E-box-binding homeobox 1.[64] Its overexpression reduces pneumonia development.

Acute Lung injury

Two simultaneous activations for instance, increase of immune cells and activation of cytokine and protease release inside the alveolar space orchestrate the acute lung injury initiation. Like sepsis conditions many immune-modulatory miRNAs expression act as biomarkers for Acute lung injury (ALI) diagnosis. Moreover, ALI is involved in the alteration of epithelial and endothelial cells. Surprisingly so far, the testing of miRNA's involvement in acute lung injury diagnostics is rare. It is not as huge as sepsis. A study of only 45 blood samples for circulatory miRNAs exhibits differential expression that induces cardio-pulmonary bypass symptoms. The expression levels of miRNAs might be ideal for their use in ALI biomarkers.

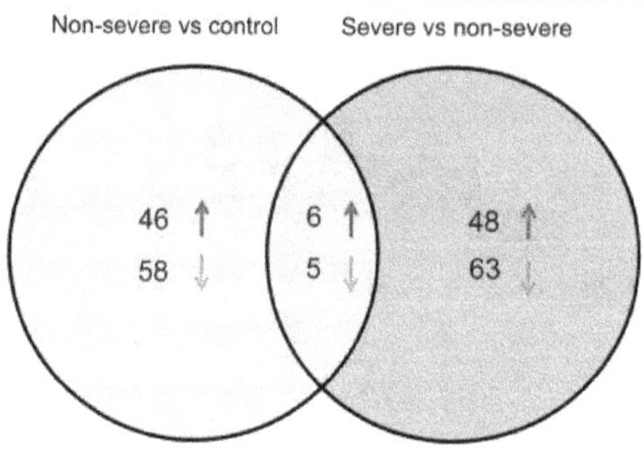

Figure-5. Venn diagram compares DE-miRs in-between non-severe pneumonia and control groups, and the severe and non-severe pneumonia groups. Upward arrows in red denotes overexpression and downward arrows in blue display silencing. DE-miRs, differentially altered microRNA expressions. (Loosely Adapted, Redraw and Modify using skeleton from Huang *et al.*, 2017)

Mainly ALI has instigated lipopolysaccharide (LPS), acid, or bacteria, or by ventilator trauma using systematic intratracheal injection. In these symptoms, a large population of miRNAs is involved in their differential expression and regulation. However, their exact roles have never been implicated in miRNAs in the development and resolution of ALI during wet-lab experiments.

Nevertheless, a high number of studies of these miRNA treatment approaches are still the hope for ALI patients because the expression of ALI related miRNAs is easy to handle. Particularly routine use of miR-mimics or anti-miRs is possible via the airways. No doubt few miRNAs expression have a strong capability to influence ALI symptoms for treatment.[65] Only 3 inflammation-related miRNAs related to sepsis indicate as promising therapeutic targets during LPS-mediated injury. The anti-miR-155 injection reduces inflammatory cell number and the proinflammatory cytokine amount in bronchoalveolar lavage markedly.[66] It reveals that miR-155 application places in the driver seat in lung inflammation.[67] Whereas miR-146a and miR-125b are involved in ameliorating lung injury. Invariably insertion of miR-146a-mimic inhibits both LPS- and acid-inductive action of proinflammatory cytokines and inducible nitric oxide synthase.[68, 69] On the other hand, overexpression of miR-125b suppresses the lung permeability and expression of proinflammatory mediators and appears to be improved ALI survival.[70] Thus the effect of both miRNAs appears to be potential therapeutic targets for ALI. These scanty roles of miRNA in ALI have been increasing day by day. Additional deep-rooted studies of miRNAs related to acute lung injury are essential for holding promises that miRNA has a phenomenal role in biomarker generation and execute personal medicine for therapeutic toll in near future.

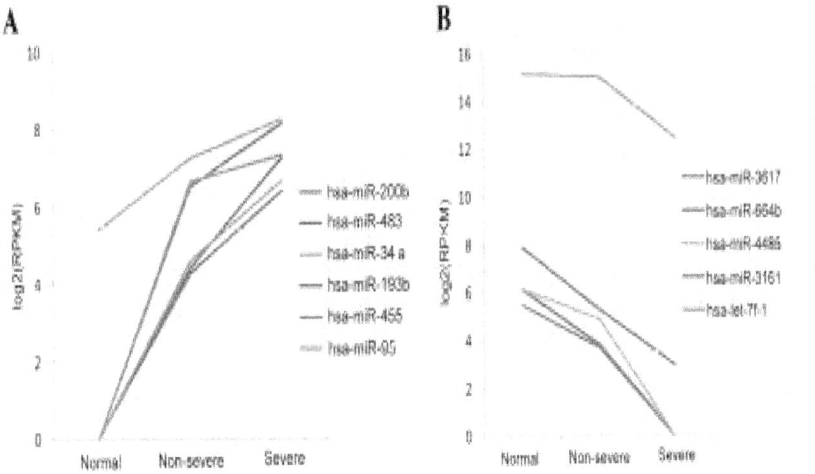

Figure-6. The level of Expression of 11 essential DE-miRs extracted from plasma of the peripheral blood of the severe pneumonia, non-severe pneumonia, and control groups. Essential DE-miRs that are **(A)** overexpressed and **(B)** repressed. DE-miRs, differentially altered microRNAs. RPKM, per kilobase per million reads. (Loosely Adapted, Redraw and Modify using skeleton from Huang *et al.*, 2017)

Acute multiorgan failure

Major clinical problems and complications in intensive care medicine related to high rates of mortality are associated with acute failure of three important organs, liver, kidney, and heart. Renal failure places at the center stage and needs to be discussed elaborately. Based on this demand acute renal failure is designed in a separate subsection. Right now, miRNAs are increasingly gaining positive momentum due to their role in potential biomarkers for identifying an early and exact diagnosis, prognostic estimation, and innovative therapy approaches.

Many miRNAs show liver-specific expression and enrichment. Among them most dominating expressions found in miRNA miR-122 inside the liver cells that are specifically expressed or enriched in the liver. Abnormal release of liver-specific miR-122 introduces both acute and chronic liver damage together with hepatocyte cell death. A significant increase of miR-122 and miR-192 in plasma causes acetaminophen-related acute liver injury inductions.[71] The amount of miR-122 outperforms many jobs in the liver using the international normalized ratio (INR), alanine aminotransferase (ALT), and acetaminophen plasma concentrations related to liver injury.[72] In fact, severe upregulation of miR-122 in liver steatosis is correlated with the severity of the liver disease.[73] Three typical inflammation-related miRNAs miR-125b, miR-146a, and miR-155 in chronic hepatitis C are upregulated in patients' plasma.[74] Moreover, these studies believe that together with miRNAs, miRNA mimics, and anta-miRs that exhibit liver-specific expression are the most promising therapeutic candidates for liver cirrhosis in the future days.

The intestine is the most vulnerable organ that exhibits multiple communications between the immune system, gut epithelium, and commensal microbiota. A blockage and degradation of homeostasis can contribute to inflammatory disorders in the perioperative context. The appropriate biomarkers indicating acute gut injury beginning are very rare. Although miRNAs are the most potent and sophisticated candidates for different biomarkers and one of the promising molecules is noticed in this field their role in inflammatory bowel disease (IBD). The pivotal role of miR-223 in IBD shed potential light on dextran sodium sulfate-induced colitis. Efficient delivery of miR-223 mimics inhibits the NLRP3 inflammasome, thereby downregulating interleukin-1β-mediated dextran sodium sulfate-induced colitis.[75] The expression of miR-223 in the gut appears to be a potent new biomarker for gut inflammation, reporting that miR-223 increases in IBD patients. This finding possibly transfers into the acute perioperative setting for the use of new biomarkers.

A rapid explosion in the discovery of miRNAs for different cardiovascular diseases as biomarkers has been coming into the limelight in recent years. The new biomarker entities miRNAs have been releasing certain additional information to establish them as protein-based markers including cardiac troponins and natriuretic peptide. The distinct feature of miRNAs related to miRNAs in myocardial infarction (MI) and heart failure is exploded exponentially

In acute MI, extensive-expression of the myocardial inside the peripheral circulation opens up the opportunity to enhance diagnostic power and/or activate diagnosis by unique determination of specific miRNAs within the peripheral blood of MI patients.[76] The cardiac-related miR-208b is identified and expressed profusely within 3 hours after MI suggesting a useful early biomarker of MI.[77, 78] Also, an alternate signature of 6 cardiac-specific miRNAs have established their reliable sign as MI diagnostic markers, in which an AUC extend troponin C and creatine kinase-MB considerably.[79] Three cohorts group of miRNA families miR-126, miR-197, and miR-233 are isolated as an authentic marker for MI patients carrying coronary artery disease.[80, 81]

Three miRNAs mR122*, miR-558 and miR-520-d-5p that are expressed during nonischemic heart failure are established as a potent and stable biomarker set.[82] In a separate study the expression of miR-182 in heart failure patients is identified as a predicted early sign that exhibits higher prognosis power relative to NT-proBNP and high-sensitive C-reactive protein.[83] The above miRNAs expression and/or modulation happen to be a clear sign on the verge of implementing diagnosis and prognosis of MI and heart failure that execute an obvious valuable addition in the future tool of intensive care medicine.

Acute pain

Pain places at the central stage in perioperative care. In a broader classification among numerous pains categories into two major classes Acute pain and chronic pain. The chronic pain is again subclassified into Inflammatory pains and neuropathy. Chronic pain shows a long and costly symptom in clinical treatment.[84, 85] However their diagnosis and treatment plan are very scanty, rare, and restricted because the pathophysiology of chronic pain is mischievous and difficult to understand. The downregulation of seven miRNAs was first identified which are related to the induction of inflammatory pain in the masseter muscle [86] nearly 15 years back. These unexpected findings appear to be an important sign to acknowledge that noncoding miRNA plays an indispensable role and involvement in chronic pain pathophysiology.

An essential structure associated with the pathophysiology of neuropathic pain processing[87, 88] is the dorsal root ganglion (DRG). The spinal nerve ligation seems to alter the proteome[89] and the transcriptome[90] of this structure. The miRNA expression plays a key and primary impact in inflammatory pain. An overall downregulation in a group of miRNA profiling and Dicer-depleted knockdown involves the reduction of several nociceptor-related proteins that are critical for the maintenance of hyperalgesia and their development.[91, 92] Afterword, a group of proteins related to pain recognition including CACNA2D1, SCN11A, and P2RX4, were affected by a set of miRNA expression.[87, 93-95]

The expression of miRNAs is important for the direct instigation of pain sensation. The expression of circulatory miRNAs in the human blood and the cerebrospinal fluid is responsible got neurodegeneration aggravation.[96] The onset of pain synthesized via instigation of rapid inward currents in DRG neurons is controlled by a group of extracellular miRNAs. This symptom is regulated by

Toll-like receptor-7 that recognize with the single-strand RNA motif GUUGUGU in the mature sequences of hsa-let-7b and hsa-miR-599.[97] Moreover another miRNA let-7b is accumulated richly in DRG neurons and it is liberated by the neuronal activity[97] and relates intimately with the complex regional pain syndrome.[98] Primarily the central nervous system is a deep core for pain processing that is controlled by versatile miRNA expressions. Marked upregulation of miR-155 and miR-223 causes the significant effect of inflammatory pain on the prefrontal cortex, which is converted into acute and chronic pain. Similarly, inhibition of miR-200b/miR-449 initiates neuropathic pain through the unleashed expression of DNA methyltransferase 3a.

Briefly, miRNAs are a major key factor related to pain development and maintenance.[85] The expression of hsa-miR-124 in CD4 T cells potentially supports the therapeutic decision for pain growth and aggravation that may be used for a potential treatment plan for chronic lower back pain.[99]

Acute Renal failure

Acute renal failure or acute kidney injury (AKI) is a devastating clinical problem that leads to several morbidity and mortality. The pathophysiology of AKI and the machinery related to post-AKI repair and fibrosis are not defined clearly. Many miRNA expressions promote inflammation, apoptosis, and fibrosis defects in human life. In opposite, few miRNAs exert an antigenic effect for instance anti-apoptotic, anti-fibrotic, anti-inflammatory, and pro-angiogenic effects. These features make them not only appear miRNAs as a novel early biomarker but also hold the potential key for use for potential therapeutic targets. Global suppression of miRNAs in the renal cortex via deletion of Dicer sequences leads to, low tissue damage, reduced tubular apoptosis, improved renal function, and enhanced survival rate, etc.[100] These miRNAs are normally involved against protection for ARF following bilateral renal ischeleads miareperfusion (IRI).[100]

The miRNA, miR-10a in renal tubules release from kidney renal IRI and streptozocin (STZ)-mediated induction of diabetic nephropathy during renal failure.[101-102] An aberrant elevation of miR-10a expect to target IL-12/IL-23p40 and the pro-apoptotic protein BIM[103] for protecting against kidney injury. The miR-17 is another member that is upregulated by pro-inflammatory cytokines during renal failure.[104, 105] Moreover, miR-21 exerts two opposite roles in the kidney. On one hand, it inhibits apoptosis and kidney inflammation, on the other hand, it induces the injury response and promotes fibrosis. The miR-21 suppression reduces the programmed cell death protein 4 (PDCD4), inhibiting member X (H2A.X) and heme oxygenase-1 (HO-1).

The miR-26a expression inhibits the IL-6 expression for inducing the regulatory T cells (Tregs) expansion dramatically. An artificial miR-26a overexpression attenuates renal IRI and increases renal recovery.[106] Conversely, systematic reduction of miR-26a and miR-27a are influenced in cardiac arrest and later in acute renal failure.[107] The miR-29a overexpression inducts a significant effect against fibrosis via tubular cells specific collagen reduction. Also, its similar downregulation in serum causes the existence of ARF symptoms and correlates with AKI severity.[107] The miR-30c is indispensable for normal homeostasis in the kidney and targets several genes of kidney structure and function. A miR-30c is induced in the tissue, blood, and urine of contrast nephropathy and IRI.[108] The family member miR-30d is liberated from the urine of kidney injury tissues. Its reduction involves the apoptotic proteins caspase 3 and p53 and may protect during IRI.[102] miR-101-3p is overexpressed normally in the kidney and down-regulates in the serum levels for predicting AKI occurrence.[107] The simultaneous miR-122 inhibition in kidneys causes cisplatin-induced AKI.[109] It instigates anti-apoptotic effects by reducing fork head box O3 (Foxo3).

The miR-127a is activated by HIF-1α and involves in protecting the cytoskeleton via combating with actin depolymerization, maintaining cell-matrix. It also participates in cell-cell adhesion maintenance by challenging focal adhesion complexes disassembly and tight junction disorganization and promote intracellular trafficking involving targeting kinesin family member 3B.[110] The miR-127a down-regulates AKI before cardiac surgery.[107] Two opposite expressions of miR-146a in the blood and kidney during AKI show differential effects. The miR-146a suppression in the blood levels exhibits the severity of AKI,[107] whereas profound inductions activate NF-kB and instigate anti-inflammatory effect by reducing TNF receptor-linked factor 6 (TRAF-6) and interleukin-1 receptor-related kinase 1 (IRAK-1).[111] The miR-192 is accumulated in the kidneys and the small intestine. During the stress response, TGF-β enhances miR-192 expression for promoting fibrosis by repressing SIP1. The miR-192 also reduces E3 ubiquitin ligase and murine double-minute 2 (MDM2). It causes de-repression of p53 and G2/M arrest.[112] The miR-194 expression is also increased profusely in kidneys and small intestine. It also increases during AKI in tissue, blood, urine, and the stress response.[101,112,113] The miR-199a expression influences the anti-inflammatory effect by reducing inhibitors of NF-kB kinases b (IKKb).[114] It also allows antiproliferative effect by inhibiting the proto-oncogene MET.[115] It also involves an anti-apoptosis effect by reducing extracellular signal-regulated kinase 2 (ERK-2) and HIF-1α[115, 116] for preventing kidney injury. The miR-126 and miR-296 expressions are distinct in micro-vesicles from endothelial progenitor cells and are instigated by Reno protective effects using their capabilities for reducing apoptosis and leukocyte infiltration. They also help in angiogenesis and tubular cell proliferation.[117] Hematopoietic miR-126 overexpression induces stromal cell-producing factor 1/chemokine receptor type 4 (CXCR4)-dependent vasculogenic progenitor cell mobilization. The expressions also help in 27a predict AKI and also reduce miR-26a and miR-27a expressions in AKI before cardiac surgery in the blood.[107]

The miR-200 family members are overexpressed in renal tubules, lungs, the small intestine, and various exocrine glands. Two anti-fibrotic family members miR-200b and miR-200c reduce TGFβR1 and zinc finger E-box-binding homeobox (ZEB1/ZEB2). They also work as transcriptional repressors of E-cadherin and combat the epithelial-to-mesenchymal transition (EMT) increasing by TGF-β.[118] The overexpression of miR-210 in HIF1-α and release extensively in renal endothelial cells. It controls angiogenesis by reducing ephrin-A3 and inducing vascular endothelial growth factor (VEGF) and their receptor 2 (VEGFR2). It is also involved in mitochondria ROS production. The miR-210 overexpression indicates post-AKI mortality of critical patients.[119] On the other hand, reduction in miR-210 expression correlates with AKI severity.[107] The presence of miR-214 enhances TGF-β and instigates fibrosis. The presence of miR-210 microRNA shows two opposite effects. It reduces PTEN and induces the AKT pathway and inhibits apoptosis of monocytes and macrophages. Another miRNA miR-214 induces in multiple models of AKI and renal fibrosis[120-122] and monocytes of chronic kidney disease patients. The expression of miR-210 is antagonized to miR-214 in renal fibrosis.[120]

The existence of mof iR-494 enhances early Abyith releasing urine levels of AKI patients. It helps apoptosis and inflammation by inhibiting the activation of transcription factor 3 (ATF3) and inducing IL-6, monocyte chemoattractant protein-1 (MCP-1), p-selectin.[123] It also influences adiponectin receptor 2 (ADIPOR2), BCL-2 facilitator, and insulin-like growth factor 1 receptor (IGF1R) finding via Pathway analysis. It also induces inflammation and causes more damage. The miR- 494 also influences pro-apoptotic proteins in the AKT pathway and increases protective effects. The miRNA governed mechanism is involved in controlling the balance between anti- and pro-apoptotic effects. Finally, a significant induction of miR-687 by HIF-1 and increases apoptosis by inhibiting PTEN. The miR-687 expression also blocks PTEN expression and attenuates cell cycle induction and reduces apoptosis for protecting against kidney injury.[124]

Many miRNAs have been involved in acute renal failure. Many miRNAs contribute to pathogenesis by regulating apoptosis, inflammation, and reducing acute injury responses dramatically. The remaining miRNAs in acute renal failure control fibrosis and angiogenesis for involvement in renal recovery or the progression to fibrosis. However, many miRNAs having biological and pathological functions in acute renal failure are still not fully known. Some miRNAs expression patterns across different samples, species, disease models, and time points are highly inconsistent. These multiple discrepancies warrant the reliability of investigations.

The existence of various miRNAs finds in serum, urine, saliva, and cerebral spinal fluid of different extracellular human body fluids. The miRNAs in exosomes are long-lasting and prolong their function. They involve two opposite actions: increase and reduction in response to injury and/or release into body fluids from resident tissues. These features of few miRNAs are the potential candidate for novel biomarkers for the early detection or prognosis of acute renal failure. In this case, the miRNA family profile is more advantageous in complex pathophysiology and the dynamic nature of AKI rather than a single miRNA.

Certain specific miRNAs may be therapeutic agents for AKI. A miRNA inhibitor can repress successfully during the replication of hepatitis C virus.[125] Artificially, systemic or local administration by site-directed injection of a few miRNAs mimics or antagonists *in vivo* can provide an essential strategy for preventing or ameliorating AKI or barricading chronic kidney disease progression. Indeed, miRNAs appear to a rising star in translational medicine. They also involve guiding a treatment plan for human acute renal failure and early individualized detection.

miRNA during anesthesia

Three popularly used anesthetic drugs including propofol, sevoflurane, and ketamine are responsible for induction different neurotoxic effects including neurodegeneration, neural apoptosis, and degradation for neural stem-cell self-renewal.[126-128] The miRNAs are responsible for the major instrument for regulating neurotoxic or protective effects.[129] The change in miRNA expression profiles is occurred by the simultaneous injection of propofol and sevoflurane. The opposite effect of two miRNAs containing simultaneous downregulation of miR-21 and up-regulation of miR-665 occurs by the injection of propofol that causes impairment of neuronal differentiation and induction of apoptosis.[130, 131] Similarly, injection of isoflurane downregulates miR-214 and let-7d that are involved in enhancing apoptosis through induction of Bax.[132, 133] Certain miRNA expressions are related to hippocampal neurodegeneration and memory impairment.[134, 135]

Taken together many miRNAs expression is altered by the injection of drugs required for induction and maintenance of anesthesia. The miRNA expression is associated with deteriorated neuronal integrity and neurocognitive processes including memory and learning.

A group of reliable biomarkers is required in postoperative care that are responsible for prediction or at least timely detection of different complications. The expression of many miRNA markers related to acute organ injury is considered for predictive signs in this post-operative setting. Truly few rare studies address that miRNAs expression is related to specific postoperative complications. The expression of a set of 3 miRNAs miR-208a, miR-208b, and miR-499 appears to be early biomarkers for postoperative myocardial damage.[136] Moreover, miRNA-499 is involved as a potential marker for postoperative MI in 30 patients suffered from coronary artery bypass grafting.[137] Many studies show that miRNAs suitably evolved as biomarkers in postoperative care including postoperative delirium prediction (ClinicalTrials.gov). This characterization of miRNAs eventually evolves that miRNAs are involved in important and valuable biomarkers in the postoperative setting in the coming days.

Conclusion

Despite the sporadic discovery and simultaneous isolation of miRNAs establish a reliable and stable biomarker in current times. The use of miRNAs has not been popular in its way into clinical practice so far. The essential features of miRNA including short size, strong homologous sequences among same family members, low concentration in body fluids provide major technical challenges for its universal and popular use as biomarkers. Indeed, miRNA does not carry the positive characters facilitating their global use, for instance, easy-to-handle, fast, reliable, and inexpensive determination. Moreover, miRNA profiling is suffered by genetic heterogeneity and multiple exogenous influences including medication, nutrition, or exposure to a few environmental conditions.[138, 139] Despite these difficulties multiple detection platforms are discovered and in current use for miRNA diagnostics and research (**Table 2, Figure** 7) available to date.

In perioperative medicine, multiple standardized protocols including sample extraction, miRNA identification and quantification, data analysis are involved clearly the measurement of valid biomarkers for future clinical use. **(Figure** 7) A clinical miRNA implication produces important complementary data on our roads toward a personalized medicine.

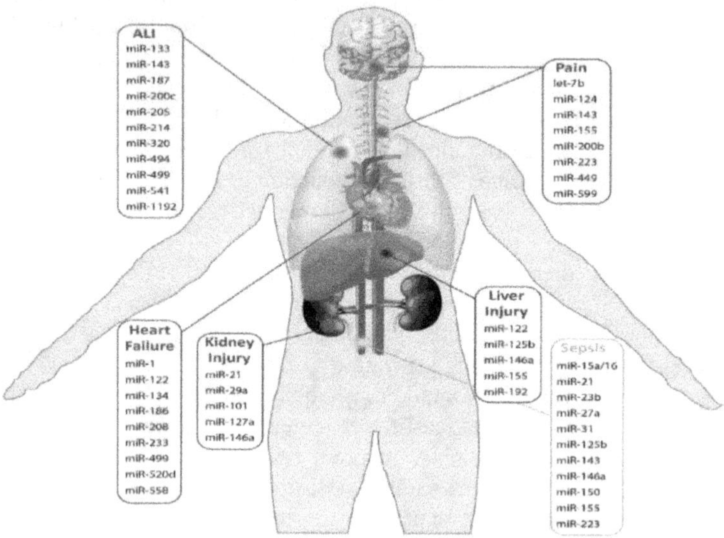

Figure-7. miRNAs responsible for organ failure, sepsis, and pain syndromes pathogenesis that are also involved as possible specific biomarkers. (Loosely Adapted, Redraw and Modify using skeleton from Kreth *et al.*, 2018)

The downregulation or overexpression of miRNAs opens a new frontier in the therapy of modern medicine. The use of miRNA mimics or miR-inhibitors has been involved as substitute therapeutic agents in perioperative medicine. Artificially reconstituted miRNAs that mimic natural miRNAs eliminate the disease symptoms or downregulating signaling pathways related to disease pathology. The synthetic miRNA mimics are double-stranded oligonucleotides encapsulated in liposomes, lipoprotein-based carriers, or nanoparticles for efficient delivery.[140, 141] Raising the multiimmune-related adverse effects the elusive use of mimics is blocked. Interestingly miRNA mimics MRG- 201 is recently utilized for phase 1 clinical study. This synthetic miRNA mimic is constituting to mimic miR-29b for downregulating collagen expression related to fibrous scar formation. It is injected in the intradermal space of the clinical healthy volunteers.

Table -2. Multiple detected methods for miRNA diagnosis and Research

Multiple features	Detected methods
Cost-efficient High sensitivity, accuracy, reproducibility Widely available Automatable Fast method	qRT-PCR
Multiple microRNA analyses in Parallel Discovery of microRNA signatures Not suitable for high throughput	Microarray
Highly Precise Discovery of unknown sequences Time consuming data processing Cost-intensive	NGS
Highly sensitive Very fast Suitable for point of care analysis Still in the stage of Development	SPR

Abbreviations: NGS, Next generation Sequencing; qRT-PCR, quantitative reverse transcription Polymerase chain reaction; SPR, surface plasmon resonance.

Two other therapeutic agents are anti-miRs or miRNA sponges. They block natural miRNA complementarity. It also silent natural miRNAs that are overexpressed signaling pathways during the occurrence of multi-organ failure. Anti-miRs are single-stranded oligonucleotides that are entered in multiple tissues and organs via injecting in the bloodstream with the help of the saline solution. They also enhance target affinity, stability, and tissue uptake.[140,141] The anti-miRNAs bind the target miRNAs complementarily for downregulating their activity. On the other hand, they specifically conjugate with endogenous miRNAs thus decreasing their availability. The miRNA Sponges carry multiple seed sequences for competitive inhibition of endogenous miRNAs through "hoovering."

The most ideal example of anti-miRNA therapy is miravirsen. An LNA-modified anti-miR-122 challenges effectively a hepatitis C virus infection.[142] Basically, Miravirsen targets the liver-specific miR-122 that is "hijacked" by the hepatitis C virus for base pairing 5' UTR sequences to the viral RNA, thereby overexpressing virus replication. Miravirsen does not show any adverse side effects during the clinical trial.

Till the present date, miRNA-based therapy is still in the budding stage. For broad-based reliable, stable, and safe clinical medicines above the commercially available conventional drugs, crucial and broad modification of many potential miRNA drug candidates is executed for critical care treatments. Better and easy entry, sophisticated development of delivery cargos for cell-specific uptake, the design of therapeutic candidates without toxic side effects, and eliminating off-target effects remain a major challenge. Executing the solutions of these hurdles in miRNA-based therapy candidates appear to have the most promising, potentials, and innovating perspective towards critical care medicines.

References

References: chapter wise are cited before the Index at the end of the Book. (1-142)

Chapter VII

Frequent infections in Asia, Africa and miRNA therapies and medicines

Abstract

To improve diagnostic experiments of different infectious diseases, numerous classes of molecules have been scrutinized as potential and effective biomarkers. microRNAs (miRNAs) now develop as the most promising candidates with excessive early diagnostic potential, due to their major contributions and early establishment of their stability within biofluids. Despite efficient use of miRNA signatures as potential diagnostic markers in various non-infectious diseases, their application in infectious disease has been induced slowly. The potential outlines of microRNA application for miRNA biomarkers promote diagnosis, management, and treatment of infectious diseases. The use of such small unique biomarkers could promote their application for personalized medicine in infectious disease treatment.

In addition, miRNAs can direct the cellular tropism of several viruses that are generated some strong resistance of individuals for infections. They are related to impaired vaccine resistance of older people. Pathogens are propagated in different ways to avoid the strike surveillance of miRNAs on immunity. miRNAs can be utilized as ideal infection biomarkers, and they are also applicable for restructuring new and attenuated vaccines, e.g., against the Dengue virus. This current knowledge of miRNA for host response against infection can be utilized for potential clinical applications. Most likely miRNAs also establish various challenges for overcoming the smooth propagation of bugs in their ideal hosts.

Introduction

Lethal Infectious diseases promote a large amount of morbidity. They appear to be responsible for about 15% of total global mortality.[1] Occurrence of easy global travel and effective global health epidemics, A crucial need for early diagnosis of infected diseases is in potential urgency for preventing the spread of diseases. In multiple infectious diseases early determination and followed by an extensive treatment plan allow a favorable prognosis. Two Important factors, clinical exposure or case history including epidemiological data are essential for diagnostic, however, the selection of an ideal course of treatment often depends upon the outcome of laboratory-mediated diagnostic testing. The diagnostic landscape in clinics that determine infectious diseases is diverse, versatile. It includes either selecting for the pathogen or looking for the host's response to the pathogen. Different strategies include *in vitro* culture and extraction, protein-dependent assays (for example ELISA and serology), microscopy (histological, pathological, and morphological assays), mass spectrometry[2] and molecular diagnostics are utilized for nucleic acid-based assays. These methods are expensive and require ample and prolonged time to perform and critical intensive sample preparation, expert users, and hampers technical restrictions.

In some instances, Diagnostic tests produce the wrong answer, either false-positive or false-negative results. Low sensitivity, low specificity, and wrong collection methods of the representative sample are the major causes for false-positive results. In some cases, infectious parasites reside in specific cell types or tissues and circulatory blood, however blood does not carry any detectable levels of pathogen. In contrast, antibodies can be detected easily, however, antibody responses can be expressed in weeks to months. Their specificity is sometimes reduced by cross-reactivity.[3] Moreover, certain viruses, in rabies[4] and cytomegalovirus[5] alter the immune response efficiently. Also, detections for some pathogens are not found in the early phases of infection. Looking for new diagnostic biomarkers to overcome these difficulties, multiple classes of various molecules have been examined. Amongst the most promising are microRNAs (miRNAs)-based early diagnosis and prognosis analysis.

Recently, several reports described alters in circulating miRNAs triggering by infectious diseases, initiating a probable new diagnostic tool against infections. MicroRNAs have been isolated as indispensable biomarkers of infections caused by a range of pathogens. In certain reports, modulation of miRNA expression profiles is found early in disease onset, before the pathogen can be recognized as a true infection especially before the onset of seroconversion.[6, 7] MicroRNAs have also been dictated for recognition in influenza infections[8] and rhinoviruses.[9] Therefore, an effective early miRNA diagnostic is noticed in other respiratory viruses including severe acute respiratory syndrome coronavirus 2 (SARS-CoV-2) outbreak. A recognized COVID-19 miRNA signature provides differential diagnosis to determine COVID-19 infections from separate lung-related viruses' infections including other coronaviruses. Additionally, miRNAs are prognostic, able to isolate the severe vs. mild infections and to recognize asymptomatic infections[10] that are involved in epidemic or pandemic spreading.

The efficient use of miRNAs as diagnostic biomarkers is promoted by their existence. propagation and stability in a wide variety of biological fluids, including peripheral blood circulation. They exhibit strong stability despite many freeze-thaw cycles or extreme pH corrosive environments.[11] After surviving in those corrosive fluids. They are routinely quantitated using next-generation sequencing (NGS) or qPCR.[12] In the short-term, self-limiting infections like the common cold or gastroenteritis miRNA diagnosis would be inappropriate, non-effective, or highly difficult to detect in clinical practice. On the Other hand, miRNAs have conspicuous implications, currently sub-par diagnostics, or long subclinical phases that would be advantageous greatly by the validated miRNA diagnostics. Therefore, based on disease classification and miRNA biomarkers would be ideally used.

Viruses synthesize several miRNAs inside their genome sequences that facilitate viral replication and control latency.[13-18] The parasite bugs including HIV, Ebola, Adenoviridae, Herpesviridae, and the BK polyomavirus utilize host cell machinery to produce their own miRNAs.[13-18] Viral miRNAs prevent cell apoptosis and inhibit many factors for reducing the inflammatory response and leukocyte recruitment.[19] Viral miRNAs may mimic host miRNAs that allow interferon response malfunctioning and create an unbalanced immune response for facilitating marked pathogenicity.[19] Manipulating viral miRNAs such as miRNA mimics or antagomiRs are potential agents for therapies. Many viral miRNAs are synthesized in exosomes of infected cells and modulate innate immune signals in recipient cells including dendritic cells and macrophages. However, there are few advantages in bacteria miRNAs. The mouse macrophage cell line (RAW264.7) infected with *Mycobacterium marina* generates a 23nt RNA molecule similar to stem-loop but does not exist in broth-grown bacteria.

A promising miRNA application promotes the immunomodulatory functions in antimicrobial pathways during infection and dysregulated inflammatory signals in sepsis. Truly miRNA therapies are effective treatment plans. A miRNA-dependent strategy that prevents the host from being resistant to their exploitation by pathogens. It opens a new venture for therapeutic developments. The antagomiR induces survival from 25 to 80% via facilitating an interferon-gamma response.[20] Moreover, bacterial infections are also treated using many miRNAs.

Overall, utilizing miRNA-dependent therapies to leverage immune responses that improve ideal adjuncts to standard antimicrobial therapies in multidrug-resistant gram-negative infections, or chronic viral infections. Despite that, multiple strong challenges for implicating miRNA-guided antimicrobial therapeutics are executed. They include conspicuous methods of drug design that protect miRNA mimics/antagomiRs from circulating RNase enzymes. Delivery systems are the main motto for effective miRNA delivery at the site of infection. The effective miRNAs exist in many cell types and perform many versatile functions generating many off-target effects, limited efficacy, and safety etc.[21-23]

Frequent infections in Asia

High humidity percentages and warm temperature of the entire tropical regions are the breeding ground of rare infecting bugs including bacteria and viruses. Moreover, enrichment in host biodiversity already harbors a large number of pathogens. It's an extraordinarily propagating ideal chance for developing numerous emerging pathogens. Using different livestock farming systems is the prime profession throughout Africa and Asia. The live stocks and human close interactions make humans vulnerable to different infections because the live stocks are prone to infection. On both continents, many families depend on subsistence farming that facilitates bugs to propagate easily via supply of many livestock hosts. Among rare diseases, a credible case determines that there are at least 10,000 diseases in the world, though there are likely more and only 500 diseases that have an appropriate treatment plan. Unfortunately, it is not possible to describe the diseases and treatment plans in this short space. Therefore, we only open a new light of those selective diseases that have a better and efficient microRNA-dependent diagnosis, prognosis, and therapy development.

Tuberculosis (TB)

Tuberculosis (TB) is the most devastated and fatal infectious disease. It ultimately leads to maximum frequency of mortality. Nearly 95% of these deaths occur in different Asian developing countries. The causative mediator, *Mycobacterium tuberculosis* (*Mtb*) has a widespread and confirmed ability to destroy the host's immune system at the intracellular level. Recently, it is observed that microRNAs are instrumental for the innate immune response against TB at a consistent level. Moreover, differential expressed miRNAs during *Mtb* infection results in disease development and may differentiate between active and latent TB infection (LTBI) for encouraging the potential use of miRNAs as effective early biomarkers. Similarly, miRNAs participate actively in autophagy modifications and apoptosis sensitivity against *Mtb* that causes an encouraging pathway for exploration of miRNAs as host-directed therapy (HDT) against TB. Nanoparticles guided miRNAs delivery has extensive potential for their use in TB. The in-ground reality, its exploitation of miRNAs as effective biomarkers and a proper agent in HDT is still in its budding stage. The contribution of miRNAs in TB as immune modifiers, miRNAs' capacity for distinguishing between active and latent TB and finally usage of possible agents as therapeutic targets against TB is described.

Many miRNA expressions in Mtb infection cell types

Currently numerous attempts have been attempted to identify the differentially expressed miRNAs in the *Mtd* infected host.[24-26] Today, *Mtb* infected mouse bone marrow-derived from macrophages (BMDMs) overexpress miR-155[27], while *Mtb* infection of peripheral blood mononuclear cells (PBMCs) derived from macrophages reduces the miR-155 expression.[28] Thus, separate cell types with the same infection may respond differently. Induction of let-7e, miR-29a and miR-886-5p in human *Mtb* infected monocyte-derived from macrophages (MDMs) is also found a distinct visual analysis of microRNA, differential mRNA expression, and various target prediction indicate caspases 3 and 7 as let-7e, and miR-29a target, respectively. Differential expression of hsa-mir-21 reduces vitamin D–dependent antimicrobial peptides, CAMP and DEFB4A expression in human monocytes derived from leprosy patients. Silencing of Toll-like receptor (TLR) 2/1 heterodimer (TLR2/1)-induced CYP27B1 and IL1B expression and indirect IL-10 overexpression also occurred during *Mtb* infection. The *Mtb*-induced miRNA differential expression profile in primary human macrophages infected with fetal *Mtb* H37Rv and a virulent *M. bovis* BCG is also found. It shows that differentially expressed miRNAs macrophages, including miR-155, miR-146a, miR-145, miR-222∗, miR-27a, and miR-27b. It also implies that miR-222∗, miR-27a, and miR-27b silent control inflammatory response and lipid metabolism drastically.[29, 30] Moreover miR-145 upregulates apoptosis,[31] however reduced activity of virulent *Mtb* strain to encourage apoptosis induction.[32-35] Silencing of miR-145 causes induction of its targets and apoptosis reduction.[34]

Massive global changes in miRNA expression profile demonstrate that nine miRNA genes miR-30a, miR-30e, miR-155, miR-1275, miR-3665, miR-3178, miR-4484, miR-4668-5p, and miR-4497 are altered their expression in THP-1 cells infected with *Mtb* H37Rv or *Mtb* H37Ra strains for executing numerous important functions. Among them miR-30e is induced by β-catenin [36], while miR-30a silent during epithelial to mesenchymal transition.[37] miR-1275 is also related to liver metastases of cancer [38], while miR-155 conjugates to a negative regulator related to TNF-α production.[39] An examination of PBMCs and pleural fluid mononuclear cells (PFMCs) noted that miRNAs expression is related to IL-6 levels, a cytokine involving a potential role in TB immunopathology. In addition, several miRNAs in Beijing/W TB strains infected human macrophages are noticed relative to the non-Beijing/W TB strains infected virulent cells for modulating the host response. Two other studies identified a group of miRNAs that are differentially expressed in PBMCs of Chinese patients with pulmonary TB using miRNA expression profiling. Moreover, multiple miRNA expressions are consistently changed in active TB patients. The miR-144∗ is overexpressed mainly in T cells. Functional analysis demonstrated that miR-144∗ reduces two important cytokines secretion INF-γ and TNF-α, but inhibits T cell proliferation. The miR-424 and miR-365 levels are induced consistently in active TB patients relative to healthy controls. Recently many differential expressions of miRNAs are found in response to mycobacterial infection in cell lines.[40, 41] However, these studies denote a solid foundation for reliable biomarkers development for TB diagnosis.

Important miRNAs in TB infection

M. tuberculosis is an ancient organism that has responded to macrophage during host cell survival.[42] Till today, a scanty knowledge has been developed that shows how the macrophage immune response alters during tuberculosis infection via host miRNAs. The miRNAs are the initial phagocyte immune response in pulmonary micro environments compared to *M. tuberculosis*.[42] To confirm survival and proliferation, pathogenic bacteria infect a broad spectrum of host cellular pathways and functions.[43] The modulation of miRNA expression by bacterial pathogens infection is an indispensable agent for the host infection response that may develop different novel strategies for controlling host cell pathways using bacteria. The macrophages are the direct target cells for *Mycobacterium* infection but are not altered by miRNAs at the time of infection. The dendritic cells resulting from the inherent and acquired immune responses can induce and polarize the topical T cell responses that are controlled by miRNAs.[44] The miRNAs contribute a pivotal and instructive role in controlling macrophages, dendritic cells, and Natural Killer Cell functions (NKCs).[45, 46] It is found that many modifications in macrophage and NKC expressions are due to latent and active TB, and in healthy individuals relative to TB infection.[47-49] Many miRNAs control T cell differentiation and their function.[50, 51] Even the induced pathway of the core macrophages could be regulated by numerous miRNAs. (**Figure 1**)

Further *M. tuberculosis* changes miR-26a, miR132 and other host miRNAs for affecting immune responses for ensuring survival. The miR-132 and miR-29a typically function as inverse regulators for macrophage function via interferon-gamma. Two induced miRNAs in alveolar macrophages restrict the immune response and deregulates the alveolar space in the pulmonary TB.[52] Oppositely, miR-361-5p is close to the level of bleomycin-induced fibrosis in the mouse lung. The miRNA is participated in exploring for lung injury and fibrosis.[53] The excess amount of miR-361 is produced in the serum of TB patients relative to healthy individuals. Thus, it reflects lung injury due to TB infection, although their related mechanism is still not explained. (**Table 1**)[54]

miRNA-29, miRNA-21 and miRNA-26a

The miR-29 expression instigates after *Mycobacterium* virulent infection.[54-56] Similar to listeria infection, the miR-29 is inhibited in interferon gamma-producing NKC cells at he onset of the *M. bovis* infection.[57] In fact, miRNA expression and regulation are cellular dependents.[58] The miR-29 silencing in mice is more resistant for *M. bovis* and *M. tuberculosis* infection,[57] it indicates that miR-29 activation during T cell infection promotes bacterial virulence. Moreover, miR-29 overexpression reduces the interferon-gamma expression.[57] Invariably, miR-29 reduces interferon-gamma, while excessive production of miR-29 transforms latent TB into an active TB.[57] Moreover, miR-29 is overexpressed in T cells of TB patients relative to LTBI and negative control.[59]

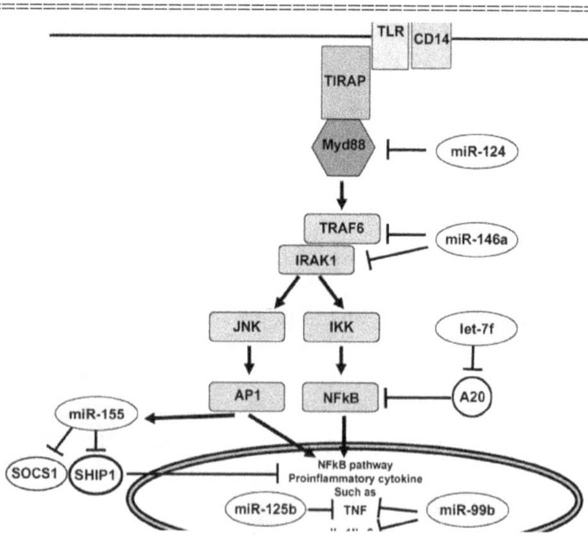

Figure-1. MiRNAs associated with immune response activation and macrophage defense at the time of *M. tuberculosis* infection. (Loosely Adapted, Redraw and Modify using skeleton from Behrouzi *et al.*, 2019)

Table-1. Different miRNAs and its regulatory target genes related to immunity against *M. tuberculosis*

Different miRNAs	**Contribution of miRNA**
Contribution of miRNA	Inhibition of interferon gamma (IFNγ)
MiR-21	Reduced expression of IL-1B IL-12P3 inhibition IL-10 increase inhibitory cytokine
MiR-99b	Target TNF-α mRNA transcri
MiR-125	Target TNF-α mRNA transcript Reduce inflammatory response
MiR-155	Target TNF-α mRNA transcript Inhibition of interferon gamma
MiR-144	Inhibition of interferon gamma and TNF-α
MiR-223	Inhibition of Interleukin 6
MiR-26a	Polarization induction of anti-inflammatory M2 macrophage phenotype.

Conversely, reduction of miR-29 expression is found in CD4 T cells of TB patients relative to LTBI (but not in the negative control group).[60] Moreover, the expression of 1223 miRNAs is investigated in the pooled serum from TB patients. Meanwhile, miR-29 overexpression occurs in sputum specimens.[54] In fact, *Mycobacterium* is a potent inducer of miR-21 expression that causes reduction of the macrophage induction and Th1-dependent immunity.[61] Although the ideal procedures for Bcl2 regulation via miR-21 are not elucidated. Silencing of miR-21 activates IL-12 production, enhancing anti-mycobacterial responses. The miR-21 employs as a useful strategy for mycobacterium for masking and escaping the host immune responses and initiating chronic infection.[62] By virtue, TB is a possible inducer of miR-26a. The overexpression of this miRNA results in a possible P300 silencing. P300 is in turn interferon-gamma-inducing transcription and later the macrophages respond to this crucial cytokine. Interestingly, Interferon-gamma inhibition in CD4, CD8 T cells via tuberculosis is possibly a unique survival strategy in the host cell.[52]

miRNA-125b and miR-155

A possible link between virulence *Mycobacterium* and TNF-α production, and a differential expression between miR-155 and miR-125b [63] are shown. The miR-125b directly involves TNF-α mRNA that causes related destabilization. The miR-125b is linked to the 3′-UTR of TNF-α transcript is and responsible for consistent silencing.[64] Oppositely, the permanent and sustained enhancer is KB2-Ras2 that is an NFkB signaling silencer in human macrophages, thus decreasing the inflammatory responses.[65] The miR-55 is possibly an enhancer of TNF-α production via targeting SHIP-1 that is an inverse controller of the P13K/AKT route. *Mycobacterium smegmatis* (*SmegLM*) is a miR-155 enhancer in macrophages that downregulates the SHIP1 expression and ultimately induces TNF mRNA stability and excess TNF production. Therefore, miRNAs are an indispensable component for TNF production at the time of Mycobacterial infection.[57] Interestingly, the cells instigation with lipomannan, the major ingredients of the bacterial cell wall, commences with virulent TB strain or *M. smegmatis* non-virulent strain, also responsible for opposite effects on the TNF-α synthesis. In this way, the lipomannan generated by TB is an effective silencer of TNF-α production, while the lipomannan produced via *M. smegmatis* is the effective enhancer of TNF-α expression. This executes the balance between the expression of miR-155 and miR-125b.[63, 66] On the other hand, transfection of murine macrophages using miR-155 causes a strong reduction in mycobacterium intracellular survival.[67] It is predicted that the miR-155 changes the antimicrobial activity via controlling two processes, like the macrophage apoptosis[68] and autophagy[69] for immunity. Moreover, miR-155 overexpression indicates TB infection in mouse macrophages through inducing the autophagy pathway[69] and inducing autophagy via silencing of the inverse regulator *Rheb* and other agents of the signaling pathway mTOR.[69, 70] *M. tuberculosis* involves an excessive amount of miR-155 and trace levels of miR-125b, while *M. smegmatis* is a unique enhancer of reduced levels of a miR-155 and a larger amount of miR-125b. The miR-155 activation in active or harmful TB infection is still not definitive. *M. tuberculosis* possibly changes the cell's environment in its favor in mouse macrophages. It also functions due to the miR-155 expression via the EAST-6 protein that is related to virulent bacteria.[67] The mutant strain of ESAT-6 TB has reduced instigation of miR-155 in macrophages than wild type.[67] miR-155 overexpression can induce the AKT pathway, responsible for *M. tuberculosis* survival in macrophages, and it is the silencer of the cytokine-induced proinflammatory IL-6.[67] It occurs by enhancing the TNF-α synthesis using SHIP1 pathway[63] and accepting some negative effects. However, miR-155 acts as the survival of mycobacteria within the host cell remain not clear enough. Considering these issues, it has been explored that mycobacterium has a unique strategy to deal with inverse effects of miR-155 that promotes mycobacterium to survival in the host, such as, the lipomannan from the cell wall of TBF-α is a potential repressor for TNF-α production and rather than the effect of miR-155 overexpression.[63] Conversely, TB is an inducible agent of miR-125b that interacts directly with the mRNA of TNF. Thus, miR-125b has also suppressed. TNF production and balances the effects of miR-155 overexpression.

miRNA-144 and miRNA-146a

Overexpression of miR-144 has been noticed in active TB patients.[71] The miR-144 is changed consistently in the PBMC of active TB patients.[72] In active TB patients many miRNAs are mainly overexpressed in the serum, while only selective seven miRNAs are silenced strongly. The miR-144 interacts directly with Janus/kinase (JAK) signal transducer genes, MAPK and TLR signaling methods, and Cyto-Cyto receptor interactions. The miR-144 also acts as an inhibitor of the TNF-α and interferon-gamma. These two components are essential in immunity production. On the other hand, inductive expression of the miR-144 in TB patients is found compared with the negative control group[73] , while no changes in expression in miR-144 are observed by the others. [74] Since miR-144 functions as an essential factor in T cells of TB patients, these different and vacillating results occur due to heterogeneity in PBMC samples.[75] Moreover, miRNA array decreases in the miR144 production in CD4 T cells in TB patients, relative to LTBI.[62] The miR-223 works as miR-146a for altering the IKK-α subunit of NFkB and controls inflammatory responses in phagocytic monocytes. In blood and lung, miR-223 is overexpressed considerably in TB patients.[76] In addition, miR-223 acts as a silencer of CCl3, CXCL2, IL- 6, and recently it is found that miR-223 deletion leads to hypersensitivity to TB infection.[77]

Mycobacterial infections in macrophages upregulate consistently in dose-dependent miR-146a expression.[78] This miRNA is responsible for two critical factors in the TLR/NFkB signaling processes, including IRAK1 and TRAF6; the induction of this miRNA at the time of infection production changes the TLR/ NF-kB methods and ultimately downregulates the cytokines TNF- α, IL-1b, IL-6, and chemokine MCP-1. Specifically, *M. tuberculosis* possibly used for mannose receptors for escaping for superoxide bactericidal effects.[79]

Other microRNAs

Eventually to survive and develop a disease *M. tuberculosis* is escaping the host defense and immune surveillance mechanisms. Specifically, tuberculosis is essential for survival within macrophage hostile environments. *M. tuberculosis* is proliferated in multiple processes for reducing macrophage antimicrobial effects for intracellular survival.[80] In one way it prevents phagosome maturation, but other measures overlook autophagy for avoiding the phagosome environment.[81-83] Autophagy has been introduced for killing pathogens. Autophagy is also an intracellular method responsible for self-digestion or self-eating. These cytoplasmic components are transmitted to the lysosome and are eventually disintegrated.[84] The autophagy linked methods are potentially regulating at post-transcriptional levels but their participation of miRNAs induction or reduction of autophagy during TB infection is mostly not known.[85] Certain reports explained how miR-33 induction in THP-1 and HEK-293 cells direct in autophagy, method reduction. They also reprogrammed host lipid metabolism for intracellular survival and TB stability.[85] Recently, miR-33 inhibits the autophagia via silencing of multiple autophagy factors, such as ATG5, ATG12, LC3B, and different transcription factors, including FOXO3 and TFEB.[76] The miR-125a-3p is overexpressed in TB infected macrophages relating to autophagy reduction by interacting UVRAG. Induction of miR-144-3p acts as an enhancer of the ATG4a gene, a gene required in autophagy silencing. Moreover, miR-23a-5p upregulation represses autophobic functions.[86] Further miR-3619-5p silencing by BCG, allows cathepsin S (CTSS) induction (Lysosomal Cysteine Protease), and the reduction of the CTSS expression that exhilarates autophagy. The miR-30a acts as an inverse regulator of autophagy that induces in TB infected macrophages. Although the miR-30a overexpression alone is not a possible cause of silencing of autophagy. Indeed, this miRNA functions as a complex system for regulating by multiple molecules, linked to autophagy. (**Figure 2**)[87]

miRNAs as a biomarker

Human miRNAs from the blood serum act as an important contributor in viral proliferation, restriction of antiviral responses, apoptosis suppressor and cell growth enhancement.[88] It is also a significant agent for immune response and inflammatory responses during bacterial infections.[57, 89] Broad Diagnosis of TB infection has a severe impact relative to many other bacterial infections.[42] Controlling the spread of TB infection is performed exclusively by early diagnosis. Till today multiple diagnostic tests cannot separate between active TB and LTBI, therefore miRNAs and their earliest existence concerning TB infection is the only reliable source for essential diagnostic biomarkers.[90] In recent days, many miRNAs act as a biomarker for TB diagnosis[72, 91, 92], using PBMC and serum[72] extracted miRNA from active TB patients.

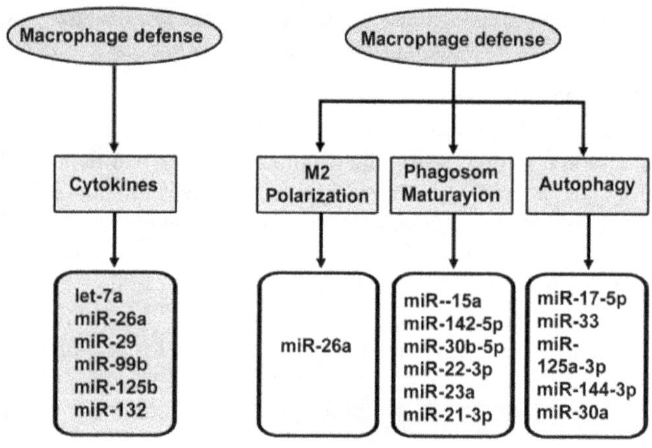

Figure-2. The contribution of immune system in *M. tuberculosis* infection: The response of innate immune system in *M. tuberculosis* infection involving alveolar macrophages and dendritic cells that function for first line defense, and then procures immunity also activated, as a second line of defense, in parallel. To eliminate intracellular bacterial infections by inducing macrophages, NKCs and granulocytes during infection, mycocidal activity initiates, the granuloma formation. After isolating and engulfment of the pathogen by phagocytic cells, for instances dendritic cells and macrophages, bacterial components that act as antigenic agents are liberated to lymphocytic cells. T lymphocyte identifies as antigenic agents via antigen-presenting cells, for example B cells, macrophages, dendritic cells, and then redirected to generate of cytokines (CD4+) or cytotoxic compounds (CD8+) after induction. (Loosely Adapted, Redraw and Modify using skeleton from Behrouzi *et al.*, 2019)

miRNAs in TB Immunity:

In Immuno-Modulatory and Inflammatory Cytokines

Many miRNAs illustrate to control the inflammatory and immune response signaling pathways (**Table 2**) against *Mtb* Infection.[93-98] The miR-155, miR-146a, miR-21, and miR-9 control innate immune cell induction.[99] In fact, miR-155 is an effective regulator of TLR signaling and is active in murine macrophages stimulation together with an interferon β (IFN-β) or TLR ligands.[100-101] The miR-125b reduces TNF biosynthesis in *Mtb*-infected human alveolar macrophages.[28] Moreover, miR-29 regulates innate and adaptive immune responses against *Mtb* via interacting interferon-γ and functions as a biomarker for pulmonary tuberculosis as it is linked to the clinical symptoms.[102]

Table 2 | miRNAs controls host immune response in tuberculosis.

miRNA	Targets	Function	Species/Tissue/ Cell examined
miR-145	TRAF6	Apoptosis	Mouse/Stem cells
let-7e	Caspase 3	Apoptosis	Human/ Monocyte derived macrophages
miR-29a	Caspase 7	Apoptosis	Human/ Monocyte derived macrophages
miR-155	FOXO3	Apoptosis	Human/ Monocytes
miR-20a-5p	JNK-2	Apoptosis	Human/ Macrophages
miR-21	Bcl-2	Apoptosis	Mouse/ RAW264.7 macrophages
miR-144* l	NF-γ and TNF-α	Cytokines	Human/ Whole blood
miR-146a	IRAK-1/TRAF-6 pathway	Cytokines	Mouse/ RAW264.7 macrophages
miR-146a	TNF-α	Cytokines	Human / Alveolar macrophages
let-7f	A20, TNF, IL-1β	Cytokines	Mouse/ RAW264.7 and BMDMs
miR-223	CXCL2, CCL3 and IL-6	Cytokines	Mouse/ miR-223– /– mouse
miR-125b	TNF	Cytokines	Human/ Monocyte derived macrophages
miR-27a	IRAK4	Cytokines	Human/ THP-1
miR-27a	IL-10 and TAB2	Cytokines	Mouse/ RAW264.7 and BMDMs
miR-27a ,miR-27b	IRF4	Cytokines	Computational study/ TargetScan database

miR-302c	IRF5	Cytokines	Computational study/ TargetScan database
miR-155, miR-132, and miR-455-5p	SOCS transcription factors	Cytokines	Computational study/ TargetScan database
miR-146a	NF-kB, MAPK	Nitric oxide suppression	Mouse/ RAW264.7 macrophages
miR-155	C/EBPb	Nitric oxide suppression	Mouse/ RAW264.7 macrophages
miR-155	SHIP1/protein kinase B (Akt) pathway	Increased bacterial Suppression Inhibition of antimicrobial peptides Autophagy	Human / Monocytes
hsa-miR-21	CYP27B1, IL1B	"	Human/ Monocytes
miR-155	Rheb	"	Mouse/ RAW264.7 and BMDMs
miR-142-3p primary human macrophages	N-Wasp	"	Mouse and human/ J774A.1 and
miR-33	ATG5, LAMP1	"	Human/ THP-1 and HEK293 cells
miR-125a-3p macrophages	UVRAG	"	Mouse/ RAW264.7 and J774A.1
miR-17-5p	ULK-1	"	Mouse/RAW264.7 macrophages
miR-144-3p	ATG4a	"	Mouse/RAW264.7 macrophages
miR-20a	ATG7andATG16L1	"	Mouse/RAW264.7 macrophages
miR-23a-5p	TLR2/MyD88/NF-κB	"	Mouse/RAW264.7 and BMDMs
miR-26a	KLF 4	"	Mouse/RAW264.7 macrophages
miR-17-5p	Mcl-1/STAT3	"	Mouse/RAW264.7 macrophages

Contribution of miR-146a and miR-155 in Host Immune Response control

The miR-146a and miR-155 are the most discussed miRNAs in TB promoting the host–pathogen interaction. The miR-146a expression is guided by the transcription factor NF-κB, assumed to be the 3′-UTRs base pair of the TNF receptor-related factor 6 and IL-1 receptor-mediated kinase 1 genes.[103] The miR-146a also inhibits mycobacteria-related inflammatory response and promotes bacterial proliferation in RAW264.7 macrophages via the IRAK-1/TRAF-6 process.[104] Moreover, its upregulation in *M. bovis* BCG macrophages. Conversely, its reduction is found in alveolar macrophages of pulmonary TB patients and controlling inversely TNF-α.[96] Further, miR-146a influences mycobacterial survival in RAW264.7 macrophages using nitric oxide (NO) suppression.[97]

The role of miR-155 also has been evaluated extensively and contributes an instrumental role in many physiological and pathological processes.[105-110] Its induction during murine macrophages infection with mycobacteria leads to an effective regulator of TLR signaling.[111] Different Pharmacological knockdown of the Janus kinase (JNK) process prevents the miR-155 induction response to either poly-riboinosinic acid or TNF-α. It indicates miR-155 influences via the JNK pathway.[112, 113] It also revealed that miRNA-155 deleted mice are very sensitive to *M. tuberculosis* infection and died abruptly with a greater frequency of CFU in the lungs. It influences the protective role of miRNA-155 against mycobacterial infection. In contrast, miR-155 induces the macrophages survival thereby generating a niche for *Mtb* replication. It facilitates the *Mtb* survival in specific T cell types for enabling a powerful adaptive immune response. The miR-155 influences in innate and adaptive immunity are related to the findings. It shows that miR-155 deleted mice execute better to control early *Mtb* infection relative to their ability to control late infection.[114] Further, miR- 155 reduces apoptosis in monocytes [115, 116] demonstrating an inverse correlation between serum miR-155 abundance and NK cell cytotoxicity. The amount of serum miR-155 is inversely correlated with the TB-repressing activity of NK cells. Similarly, miR-155 expression is upregulated strongly in macrophages after *M. marinum* infection. It causes reduction of NO synthesis and enhances mycobacterium load. Therefore, a single miRNA can interact with multiple target genes simultaneously and illustrate miRNA regulatory role.[117, 118]

Other miRNAs in Host Immune Response

Apart from the above discussed two miRNAs, different miRNAs contribute significantly to *Mtb* infection. miR-223 controls TB and other chronic inflammatory diseases effectively. The miR-223 expression utilizes as a target of Chemo-attractants like CXCL2, CCL3, and IL-6 in myeloid cells.[119] Similarly, endogenous miR-29 silencing induces resistance against *Mtb* infection.[57] Moreover, miR-27a acts as a restrainer of immune response in *Mtb* infection via directing IRAK4. Expression of IFN-γ, IL-β, IL-6, and TNF-α are reduced markedly following miR-27a mimics transfection.[120] Indeed, miRNAs expression and its transcriptional control multiple interacting genes effectively for regulating different immunological responses. These studies enrich us with the havoc contribution of miRNAs in host-pathogen interactions and open a new avenue for diagnostic tools improvement and act as TB therapy agents.

miRNAs for Apoptosis Regulation

MiRNAs function a tremendous role in apoptosis control in *Mtb* infected cells. The miR-20a-5p acts as inverse modifiers of mycobacterial-interacted apoptosis and miR-20a-5p downregulation led to an effective *Mtb* clearance.[121] MPT64 is an effective secreted protein by *Mtb* and its silent apoptosis of RAW264.7 cells possibly employing NF-kB-miRNA21-Bcl-2 process.[122] Further the miR-155 interacts with FOXO3 and controls apoptosis. Taken together the miRNAs contribution leads to autophagy and apoptosis regulation in TB. (**Figure 3**) Therefore, the manipulation of miRNAs in exogenous levels promotes multiplication and activates the *Mtb* clearance from the host cells.

miRNAs for Autophagy Regulators

Autophagy is a self-digestion or self-eating intracellular process, whereby cytoplasmic materials are migrated to and degenerated by lysosomes.[123] Apart from other essential cell functions, autophagy leads to an effective pathway in immune responses to combat viral and bacterial pathogens invasion.[124] The ability of *Mtb* survival and replication in host macrophages is the mainframe for *Mtb* pathogenesis and is related occasionally to its degree of virulence.[125] It is essential to investigate how *Mtb* hijacks host immune-modulatory signaling processes for survival and replication in macrophages for proper regulation and disease treatment. The *Mtb* employs multiple strategies for ignoring macrophage-related antimicrobial effects for surviving intracellularly.[126] One of the important methods is to prevent phagosome maturation and employ other counteractions to impede autophagy and involve the deadliest environment of phagolysosomes.[127-129] Indeed, phagosome maturation and mycobacterial killing are reinstated via exogenous autophagy activation exogenously for infected macrophages.[130]

The autophagy-related processes are controlled strongly at the post-transcriptional level in detail but the involvement in miRNAs induction or reduction of autophagy in *Mtb* infection is still obscured largely. However, many studies recently illustrate the miRNAs involvement in autophagy restriction during *Mtb* infection. The miR-33 activation in THP-1 and HEK293 cells reduces the integrated processes related to autophagy and reorganizes the host lipid metabolism for intracellular *Mtb* survival and persistence. The macrophage-related *M. bovis* BCG infection upregulates miR-144-3p that activates autophagy-related gene 4a (ATG4a) for preventing autophagy. On the other hand, miR-23a-5p upregulation prevents *Mtb*-activated autophagy drastically by altering TLR2/MyD88/NF-κb signaling in macrophages.[131] In the same manner, miRNA-20a interacts with ATG7 and ATG16L1 for autophagy prevention.[132]

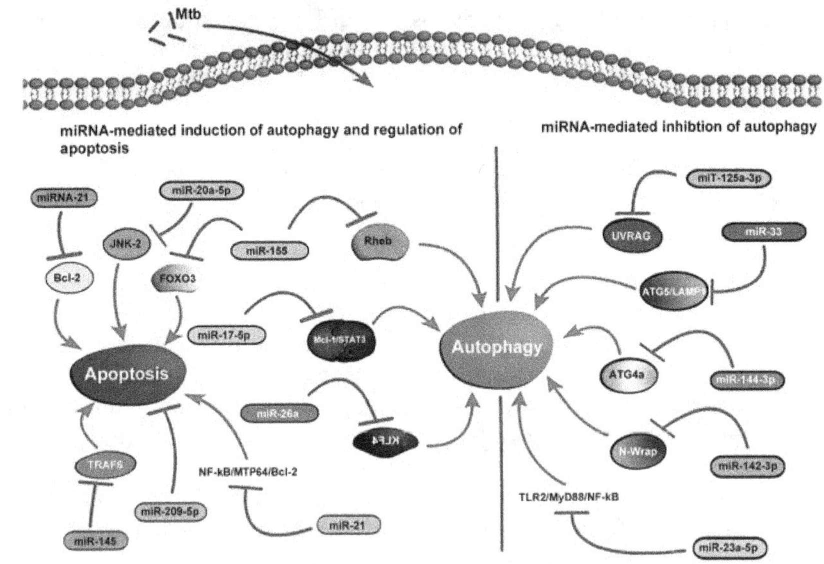

Figure-3. miRNA controls of autophagy and apoptosis. miR-155, miR-17-5p, and miR-26a interact Ras homolog induction in brain (Rheb) and Mcl-1/STAT3, KLF4. They play a positive contribution in control of autophagy during *Mycobacterium tuberculosis* (*Mtb*) infection. miR-155 also reduces apoptosis via directing FOXO3. miR-21 interacts with NF-kB-MTP64-Bcl-2 for silencing the apoptosis. miR-20a-5p, miR-21, miR-209-5p, and miR-145 also control apoptosis. All of these miRNAs work as effective agents in HDT. While miR-125a-3p, miR-33, miR-144-3p, miR-23a-5p, and miR-142-3p are suppressor of autophagy in *Mtb* infection. miR-125a-3p silents autophagy via interacting UV radiation resistance-associated gene (UVRAG), miR-33 guides ATG5/LAMP1, miR-144-3p directs autophagy-related gene 4a (ATG4a), miR-23a-5p reduces the TLR2/MyD88/NF-κB allowing autophagy reduction and miR-33 also silent by interacting some unknown factors. miR-142-3p guides Neural Wiskott-Aldrich syndrome protein (N-Wasp) facilitating phagocytosis reduction. (Loosely Adapted, Redraw and Modify using skeleton from Sabir *et al.*, 2018)

Conversely, the miRNA preventive roles in autophagy activation are observed. On the other hand, many miRNAs induce autophagy during *Mtb* infection. The miR-26a mimic attenuates in macrophage related *Mtb* survival using interaction with transcription factor KLF4 that prevents trafficking of *Mtb* to lysosomes. miR-155 enhances the autophagy-guided anti-mycobacterial response using interaction Ras homolog enriched in the brain (Rheb), an inverse controller of autophagy.[69] Similarly, a miR-17/PKCδ/STAT3 axis is controlled autophagy during *Mtb* infection. Neural Wiskott-Aldrich syndrome protein (N-Wasp) is an actin-conjugated protein responsible for phagocytosis during microbial challenges. *Mtb* instigated miR-142-3p expression interacts N-Wasp for causing decreased phagocytosis.[133]

miRNAs as Mtb therapeutic targets

Application of miRNAs are possibly directed therapeutically to reduce their maturation.[134] It is possible activation of reduced anti-mycobacterial miRNAs through synthetic oligos and reduced use. The effects of upregulation of pro-mycobacterial

miRNAs via antisense oligonucleotides or anti-miRNA complementary to mature miRNA are also observed.[135-137] Another alternative strategy is to change specific miRNA using gene therapy and viral vectors.[138]

miRNAs as Potential Therapeutic Targets

Recently many miRNAs are essential to treat TB. miR-155 is related inversely to the TB-reducing activity of NK cells.[116] The same miRNA also involves autophagy-guided mycobacterial blockage by directing Rheb.[69] The miR-33 and miR-33∗ silencing employing genetic or pharmacological strategies induce autophagy flux by suppressing potential autophagy effectors and AMPK-mediated induction of FOXO3 and TFEB transcription factors. The miR-23a-5p changes the host innate immune response by activating *Mtb* survival and preventing autophagy enhancement through TLR2/MyD88/NF-KB processes.[131] On the other hand, Cytokines are essential ingredients for inflammatory and immune responses. The miR-99b prevents pro-inflammatory cytokines secretion during *Mtb* infection. It implicates a new avenue for designing miRNA-guided therapies and TB control.[139] The miR-206 acts as an inflammatory regulator involving MMP9 expression directing TIMP3 in *Mtb* infection.[140] In the same way, miR-20b induces the inflammatory response during TB by interacting the NLRP3/caspase-1/IL-1β processes. The miR-32-5p upregulation increases the survival rate of *Mtb* dramatically via directing Follistatin-like protein 1 (FSTL1) through the TLR-4/miRNA-32-5p/FSTL1 pathway. However, most of the miRNAs are not entirely gene-specific for their limitations. As noticed, one miRNA may be able to target multiple mRNAs. It indicates that the administration of exogenous miRNA might exhibit off-target effects. In crude reality, miRNAs utilization in HDT is still in its budding stage but it opens the doors of an elegant avenue for the treatment of TB.

miRNA Manipulation in vivo

Apart from anti-miRNA technologies, considerable advances regarding miRNA delivery into lungs occur. It encourages results describing TGF-β1 altered expression during TB infection.[141] Normally lentiviral vectors, lipid conjugates, or small exosome-like vesicles are used *In vivo* miRNA delivery. Often Antigen-related exosome-like nanovesicles have been transfected using mimics or anti-miRNA to promote miRNA expression in target cells.[142] Similarly, primary B cells transfects also by a specific anti-miRNA occasionally deliver efficiently to antigen-induced T cells.[143, 26] It ultimately silent *in vivo* miR-195 expression via anti-miR-195. Particularly, *in vivo* miR-328 silencing using intra-tracheal inoculation of anti-328 leads 4-fold induction of non-typeable *Haemophilus influenzae* (NTHi) clearance from the lungs relative to normal lung.[144] Therefore, miRNAs in the lung can induce host immunity against microbial infections and execute an effective novel anti-mycobacterial approach for the treatment of TB. Furthermore, the most ideal route of miRNA delivery and pharmaco-vigilance problems, provides a solid and proper foundation for future miRNA-based HDTs of TB.

Nanoparticle-guided Delivery of miRNAs in TB

Technological know-how for the micro and nanoscale processes can change the immune response.[145] Multiple materials and strategies are employed to drug delivery and a broad cargo array to cells, covering nanostructured materials [146], synthetic nanoparticles[147], proactive biomimetic delivery systems[148], functional polymers [149] or cell-penetrating peptides [150] , among others. Multiple materials are used to promote immune diseases treatment with nanoparticles, covering silica, iron oxide, gold, liposomes, or polylactic-co-glycolic acid (PLGA).[151] Particularly in infectious diseases strategies, the encapsulate miRNAs and their delivery to the cells are the front runner in the immune response, i.e., macrophages or dendritic cells etc. It covers the profound advantage of the innate ability of these cells to phagocytes foreign bodies. The notion of miRNA delivery integrates into the normal concept of nanoparticle-guided immune responses induction by interacting phagocytes (i.e., macrophages). In contrast, an alternative method for interacting adaptive lymphocytes is found (i.e., T cells). The MiRNA deliveries are utilized for many applications related to inflammation[152] and its potential use in infectious diseases.[153] The miRNAs can be encompassed into both natural –i.e., chitosan- and synthetic – i.e., (PLGA)- particle-forming materials. In a nanoparticle- and liposome-load miRNAs have been discharged to macrophages.[154, 155] Recently, few limited reports are executed for implementation in TB. This application either discusses present general knowledge of nanoparticle uses in immunology or noting vividly in nanoparticle utility with miRNAs. These two views are combined briefly and apply nanoparticle strategies used elsewhere that can be implemented in miRNA delivery in TB.

Direct Delivery miRNA Deliveries for altering Host Immune Response

To eliminate or dramatically reduce mycobacterial functions interacting macrophages and dendritic cells, many strategies can be applied. One direct and obvious approach is miRNA delivery that increases host immune responses against TB infection rather efficiently promotes anti-inflammatory molecules like IL-10 activation. In different approaches, involvement of miR-155, miR-146a, miR-145, miR-99b, miR-19b-2*, miR-27a, or miR-27b are found. In addition, a second strategy delivers siRNAs that reduce pro-inflammatory cytokines expression for example TNF-α that are activated using multiple miRNAs. Those miRNAs let-7e, miR-29a, miR-886-5p, miR-147, or miR-3179 are upregulated during TB infection. Apart from integrated siRNAs, this strategy delivers small molecules (drugs) that possibly affect immune responses including dexamethasone or other small molecules for modulating the host's capacity to control miRNA expression.

The host molecular targets are silenced, upregulated, or inhibited by the Direct miRNA delivery during TB. The ideal strategy for direct delivery is the release of miR-223 from encompassed nanoparticles in macrophages for promoting phenotype transitions.[156] The efficacy of direct delivery depends on miRNA selection. The direct delivery of miR-155 using nanoparticles is an ideal option when innate immune cell induction is required as a positive regulator of TLR signaling [157] and its induction in autophagy and mycobacterial clearance.[69] In the same way, miR-155, nanoparticle-mediated miR-125b delivery in the *Mtb* infection possibly requires TNF biosynthesis inhibition. Generally multiple miRNAs, miR-155, miR-146a, miR-145, miR-222∗, miR-27a, or miR-27b [158-161] are used in direct delivery for their inhibition during TB infection. However very few reports are presenting detailed nanoparticle utilization for mi-R155, miR-223, miR-146, or miR- 125b required in TB infection.

Prospects

Tuberculosis (TB) is one of the important deadly communicable diseases in the world. The eradication of *Mtb* is next to impossible of its unique capacity for retaining activity within macrophages. Macrophages play an instrumental role in the host immune response during *Mtb,* a response to control strongly using multiple factors, including miRNAs. The budding miRNA contribution for controlling adaptive and innate immune responses in *Mtb* infection has allured improved attention in recent years. Differential miRNA expressions allow disease development that differentiates between active and LTBI. These findings emerge as potential miRNAs biomarkers for early detection of TB. However, their potential and restricted use as biomarkers are that almost all miRNAs are not gene dependent. Efficient miRNA involvement in autophagy alteration and apoptosis in *Mtb*, and the emergence of miRNA-mediated nanoparticles delivery provide an effective avenue for RNA exploitation in host-mediated therapy (HDT) development of TB.

A strong foundation for HDT development depends on miRNAs. Moreover, many useful applications of miRNA lead to better control of HDT and treatment of TB. Further a comprehensive survey elucidates the entire miRNA application as novel early TB biomarkers, and their adjunctive treatment for TB.

Inflammatory Bowl Diseases

Inflammatory bowel diseases (IBD) are chronic inflammatory gastrointestinal diseases, invariably consisting of ulcerative colitis and Crohn's disease. The complex disease has a limited therapeutic opportunity. The components of therapy are less effective and generate major side effects. These difficulties raise the development of novel strategies for generating therapeutic agents in IBD. IBD sensitivity and responses exist only in a certain percentage of IBD patients. It indicates that epigenetic modulation possibly influences disease pathogenesis. MicroRNAs (miRNAs) are small non-coding RNAs that regulate multiple molecular pathways involved in IBD pathobiology. In particular, miRNA silencer-directed IBD-induced miRNAs act as core therapeutic values for IBD patients. The recent advances in IBD pathogenesis-related diseases are reported that possible placed as pharmacological agents of miRNA guided IBD therapies.

miRNAs in IBD Colonic Tissue

Numerous identified miRNA expression profiles derived from IBD infected tissues [162] have been utilized to detect the differentially expressed miRNA levels in ileal Crohn's Disease (CD), ulcerative colitis (UC) and non-IBD patients. The miRNA profiles are identified in UC and/or CD often comparing between active vs. inactive disease, UC or CD samples vs. healthy controls, or UC vs. CD samples. Therefore, such variability in miRNA expression evolves from different groups of the same tissue type. In brief, considerable inhibition is noticed in many miRNAs including miR-124[163], miR-4284[164], miR-320[165], miR-141[166], miR-193a-3p[167], miR-200b[168], miR-10a [169], miR-19a [170], and miR-192. [171] Moreover, overexpression of these miRNAs, miR-21[172], miR-301a[173], miR-665[174], miR-155 [175], miR-133a[176], miR-210 [177], miR-31[178], miR-124[179], miR-223[180], Thus their potential contribution of the differentially expressed miRNAs are associated with IBD pathogenesis functionally.

The miR-21 is induced in both UC[172] and CD[181] patients and is responsible for multiple proinflammatory functions, including modification of the T-cell responses [182] and attenuation of tight junction integrity.[183] Oppositely, miR-21 silencing decreases the inflammatory responses but increases the survival rate in a dextran sodium sulfate (DSS)-mediated fatal colitis animal model.[184] Further miR-31 upregulates in both CD and UC patient colon tissue[185] and interacts directly *IL-25*. In fact, the miRNA-31/IL-25 process is altering Th1/Th17-guided inflammatory responses in colitis models.[178] The miR-141 is reduced in both UC [186] and CD [187] patient colon tissue. It interacts directly with *C-X-C motif chemokine (CXCL)5* and CXCL*12*, and its silencing in IBD leads to CXCL5 and CXCL12 for promoting leukocyte recruitment and induces an inflammatory response. Thus, three miRNAs miR-21, miR-31, and miR-141 have a dramatic impact on both UC and CD pathogenesis and function as essential interacting components for therapeutic intervention. These three miRNAs illustrate the functional implication of miRNAs related to IBD dysregulation in IBD and describe it as a causal agent for disease-related phenotypes particularly. The differentially expressed miRNA alters between whole colonic tissues and isolated cell types of the colon. Ironically, these differences impose a tight restriction when analyzing the potential contribution of therapy *in vitro* in the presence of non-cell-specific unforeseen side effects. However, an injection of the synthetic oligonucleotides directly into the colon restricts systemic side effects that possibly prevent therapeutic effectiveness and/or lead to harmful consequences.[188, 189]

microRNAs from IBD Blood

Despite miRNA signature isolation in colonic tissues is responsible for an invasive and expensive diagnostic method, therefore extraction circulatory miRNA panels are essentially derived from serum or plasma that distinguish UC from CD and contribute as a therapeutic agent for current IBD drugs. The circulating miRNAs isolated from peripheral blood function as essential diagnostic and/or therapeutic ingredients for IBD treatment. Indeed, the potential impact of exosome miRNAs on IBD [190-192] is important. The tissue-selective and vesicle-mediated transfer of miR-220b promote 2,4,6-trinitrobenzenesulfonic acid (TNBS)-guided colon fibrosis. Moreover, the consistent stability through Argonaute proteins binding and the encapsulation using exosomes promote excessive advantages for using serum miRNAs for IBD diagnostics.

The serum miRNA signatures involve proper distinction of UC from normal controls. Many miRNAs including miR-29b-3p, miR-122-5p, miR-192-5p, miR-194-5p, miR-375-3p, miR-150-5p, and miR-146a-3p are overexpressed in the serum. Interestingly, these miRNAs are linked to antitumor necrosis factor-a (TNF-a) therapy and sensitive to UC human samples along with 83.3% accuracy. Additionally, expression patterns of peripheral blood miRNA are required for early IBD diagnostics. In particular selection, miR-28-5p, miR-103-2, miR-151-5p, miR-340, and miR-532-3p are upregulated consistently in UC relative to CD patients, for the distinction of active IBD subclasses from each other and normal controls. Eight distinct miRNAs, miR-28, miR-103–2*, miR-149*, miR-151, miR-340*, miR-505*, miR-532, and miR-plus-E1153 possibly differentiate inactive UC from CD samples. The miRplus-E1153 also separates healthy control from UC active tissues.[194] Three more miRNAs, miR-146a, miR-21, and miR-31 are knockdown, while miR-19a, miR-101, miR142-5p, miR-223, miR-375, and miR-494 are enhanced when extracted from UC compared with control tissues.[185]

Many miRNAs profiling exposes certain miRNA dysregulation both in CD vs. non-IBD controls, and in selective CD-related complications. The miR-199a-5p, miR-362-3p, and miR-532-3p increases in CD patients' blood relative to controls, but miR-149 and miRplus-F1065 are reduced prominently.[193] Further, miRNA profiling from CD patient blood serum that is related to stricture formation silent miR-19a and miR-19b selectively from the serum of CD patients having strictures.[194] Collectively, changes in miRNAs contribute possibly to IBD pathobiology. However, it is important that miRNA profiling in greater cohorts of well-designated clinical samples usually detects robust miRNA diagnostic and/or prognostic signatures at an early stage. Therefore, differential miRNA expressions derived from serum from IBD patients are required for supporting the feasibility that produces greater sensitivity and selective specificity.

IBD miRNAs and the immune system

Activation of innate immunity

The innate immune system raises the initial defense against different pathogens and leads to early antigen recognition by pathogen-related molecular patterns. Pathogen-guided molecular patterns direct and induce extracellular receptors mediated toll-like receptors (TLRs) or intracytoplasmic nucleotide-binding oligomerization domain-containing protein (NOD)-like receptors. It induces reduction of signaling cascades via different pathways including NFκB, mitogen-activated protein kinase and interferon (IFN) regulatory factors.[195] miRNAs is essentially responsible for regulating the activity of these processes.

NOD-like receptors

Most relevant cause of IBD is NOD2, a section of NOD-like receptors family NOD2 has a dominating single genetic susceptibility locus in CD.[196] The miRNA-NOD2 conjugation demonstrates that different miRNAs like miR-192, miR-122, miR-29 and miR-146a are involved in IBD.[197-199] The conjugation of miR-192 and NOD2 is involved in IBD pathogenesis that acts as an SNP rs3135500 in the 3'UTR region of NOD2 for decreasing the miR-192 activity and NOD2 inhibition.[199] Polymorphisms in NOD2 also degrades the dendritic cells (DCs) activity for expressing miR-29. It exaggerates IL-23 activated inflammation.[197] In addition, miR-122 interacts with NOD2 expression upon LPS stimulation, albeit in various cell lines (HT-29 cells).[197] Lastly, miR-146a control NOD2 derived gut inflammation in IBD and instigate proinflammatory cytokines in MDP induced macrophages.[198]

Toll-like receptors

The miRNAs are required for interacting with a wide-varieties of molecules within the TLR signaling processes.[200] miR-146a/b and miR-155 are the most effective miRNAs for generating their respective knockout (KO) mice phenotypes.[201, 202] The miR-146a depleted mice develops autoimmune disorders, myeloid cell proliferation and tumorigenesis while knockout of miR-155 is required for degradation of DC function and do not develop adaptive immunity to pathogens.[201, 202] The miR-146 family and miR-155 activate the nuclear factor κ light chain enhancer of induced B cells (κκB) dependent pathways. These miRNAs function for negative feedback circuits for fine-tuning of inflammatory activity upon bacterial infection.[203-206]

On one hand miR-146 interacts with MyD88 adaptor proteins: tumor necrosis factor receptor-related factor 6 and IL-1 receptor-related kinase 1. On the other hand, miR-155 interacts with signaling proteins: inhibitor of cytokine signaling 1 and TAK1-binding protein.[207, 208] The miR-146 in cells develops tolerance to sub-inflammatory doses of LPS, however when interacts with proinflammatory doses of LPS, miR-155 induces a broad spectrum of the limit of inflammation.[209] The miR-146a expression procedure is dynamic and during the early stage of the inflammatory signals in macrophages. However, there is transient inversely effect of miRNA-guided silencing of inflammatory cytokines using AGO2 phosphorylation.[210] LPS activated TLR4 also induces the controlling PI-3K/Akt circuit containing let-7e and miR-155. It interacts with TLR4 and cytokine signaling 1.[211] Macrophages with Akt deletion inhibits let-7e but upregulates miR-155 leading to a hypersensitive phenotype to LPS.[211, 212] miRNAs have also been resulting different suggesting that miRNAs for innate immune response are always cell-specific. TLR4 activated NFκB promotes the miR-9 level in human monocytes and neutrophils, however their presence in murine macrophages, the NFκB feedback circuit is regulated by miR-210.[213, 214]

Chikungunya

Chikungunya virus (CHIKV) is a positive-sense, single-stranded deadly RNA virus of the *Togaviridae* family and *Alphavirus* genus.[215] The genome is 11.8 kb long carrying two open reading frames. The 5′ORF synthesized different non-structural proteins nsP1, nsP2, nsP3, and nsP4 and the 3′ORF synthesizes the structural proteins, two peptides (E3 and 6K), capsid (C), envelope (E1 and E2). Chikungunya virus was first detected in southern Tanzania [216, 217] and Kenya and later in Thailand Mauritius, and largely in India having 1.5 million infections. Till today three separate CHIKV genomes have been identified including West African, Asian, and the East Central and South Asian (ECSA) strains.[218] The word "Chikungunya" denotes "that which bends up" in Makonde language describing the stooped posture due to severe musculoskeletal pain. CHIKV causes multiple symptoms like CHIK fever (CHIKF), fever, headache, rash, and arthralgia etc.[219] Although infected patients generate chronic growing pain similar to arthritis, rheumatoid arthritis (RA).[219, 220] On certain occasions, CHIKV infection also develops encephalitic-like symptoms[221, 222] and bone mediated joint pain and arthritis-like conditions.

Viruses alter host miRNAs. They are related to the emergence of viral pathogenesis.[223] The routine feature of viruses also hijacks host gene expression during propagation and modifies cellular development for procuring infection and pathogenesis. Many reports have evaluated the imperative role of miRNAs during viral infections particularly[223, 224] in physiological functions. The abnormal expression of miRNAs is linked to pathogenesis including chikungunya viral infections[223-225] Many recent studies highlight numerous host mRNA with CHIKV infection.

The antiviral role of miRNAs

A significant number of miRNAs produce abnormal expression during early CHIKV infection, apoptosis and JAK-STAT signaling pathways.[226] Interestingly, the JAK/STAT pathway is a key signaling player in the interferon (IFN) response against chikungunya infection.[227] The *Ae. aegypti* mosquitos' infection induces resistance to Dengue virus (DENV) and Zika virus (ZIKV) infections using the JAK/STAT pathway.[228] In addition, CHIKV non-structural protein 2 (nsP2) has been related to JAK/STAT pathway.[229] Thus, emerging interaction among viral proteins, miRNAs and their participation in the JAK-STAT pathway leads to an important emerging report in CHIKV pathogens.

The miRNA profiling in CHIKV-infected skin fibroblasts produces numerous abnormalities in miRNAs during an early stage of CHIKV infection. [230] Both JAK/STAT and MAPK pathways are invariably related to CHIKV infection. [229, 231] The hsa-miR-15 and hsa-miR-16 expression are changed during CHIKV infection.[230] Interestingly, the modified hsa-miR-15 and hsa-miR-16 require for conducting instrumental roles in apoptosis induction for directing the anti-apoptotic protein BCL2.[232] Additionally, hsa-miR-15 knockdown is stringent in arthritic synovial tissue, whereas hsa-miR-16 is overexpressed in sera of RA patients.[233, 234] rno-miR-32–5 p is an inverse controller of phosphatase and tension homolog (PTEN).[235] In aggregate, the fundamental consequences of all miRNAs at the time of CHIKV infection instigates an idea to generate novel drug components.

A series of 152 differentially expressed miRNAs find in CHIKV infected human cell lines.[236] Among these, hsa-miRNAs, hsa-miR-744, hsa-miR-638, and hsa-miR-503 are overexpressed consistently. Further altered miRNAs of multiple miRNA clusters like hsa-miR-17-92, let-7e/99b, hsa-miR-191/425, hsa-miR-106b/25, hsa-miR-23a/24, and hsa-miR-15b/16 clusters are altered by the induction of CHIKV infection.[236] The miRNAs upregulate TGF-β genes (mothers against decapentaplegic homolog 6 (SMAD6), JUN, endocytosis processing genes (C-X-C motif chemokine receptor 4 (CXCR4), ski-related protein (SKIL) genes), heat shock 71kDa protein (HSPA8), adrenoceptor beta 1 (ADRB1)), but silenced two cell cycle-related genes, the cell division cycle 27 (CDC27) and CDC23). The cell's treatment along with a TGF-β inhibitor, SB-4315T42 induces CHIKV related cell death, demonstrating that TGF-β production is required for controlling CHIKV infection.[236] Importantly, TGF-β related processes are the most instrumental factor of CHIKV infection particularly in age-mediated complications.[237] Thus, miRNA-guided control of the TGF-β pathway is required for host response against CHIKV infection.

Pro-viral role of miRNAs

Occasionally viruses engage cellular miRNAs for interacting with different targets and silence their expression for inducing infection.[223, 224] The hsa-miR-146a is overexpressed in CHIKV-infected man synovial fibroblasts via targeting TRAF6 and IRAK1.[238] (**Table 3**) Moreover, hsa-miR-146a upregulation causes reduction of NF-KB phosphorylation during infection.[238] The hsa-miR-146a also upregulates DENV replication by interacting with the TRAF6-guided NF-KB pathway.[239] These results postulate that hsa-miR-146a-mediated interaction of the NF-κb pathway occurs during CHIKV pathogenesis.

A subset of 26 differentially expressed miRNAs (DEMs) are identified in human synovial fibroblasts during CHIKV infection. (**Table 3**)[240] Among the DEMS, hsa-miR-4717-3p, hsa-miR-4299, hsa-miR-1264, and hsa-miR-21-5p are overexpressed consistently. AKT3 is predicted as a target for hsa-miR-4717-3p. (**Table 3**) The AKT3 protein is an important modulator of the PI3K/AKT/mTOR signaling pathway.[241] The abnormal control of the PI3K/AKT/mTOR pathway increased [242, 243] during CHIKV infection considerably.[244] The hsa-miR-4717-3p conducts induction of massive inflammatory signals during CHIKV infection via directing AKT3.[240] The hsa-miR-4299 is induced during CHIKV infection in which the cytokine signaling 7 protein suppressor (SOCS7) acts as potential target. (**Table 3**)[240] SOCS7 is an inverse regulator of the STAT3 protein that either increases IL-6-related inflammation or IL-10-guided silencing of inflammation during CHIKV infection.[245] STAT3 is required for viral replication promotion and persistence.[246] The hsa-miR-4299 can silent SOCS7 causing an enhanced STAT3 expression that is required for reduction of immune signals during CHIKV infection. As noted, viruses utilize versatile strategies to execute host cellular processes for their servility and propagation.[247, 248] The hsa-miR-1264 upregulates during CHIKV infection and TRIM26 is the most predicted target. (**Table 3**)[240] CHIKV infection causes hsa-miR-1264-guided silencing of TRIM26 that leads to reduced antiviral response and oppositely induces viral replication and persistence.[240] Another E3 ubiquitin ligase, PELI1 is directed by hsa-miR-21-5p that leads to overexpression during infection. (**Table 3**)[240] PELI1can silence the NF-κb process via ubiquitination and degradation of an NF-κB inducing kinase (NIK).[249] The overexpression of miR-21 during CHIKV infection for reducing cytokine signaling by altering the NF-κB pathway as the NF-κB pathway exposes a crucial role in antiviral response.

Table 3. Altered miRNAs at the time of Chikungunya virus (CHIKV) infection

miRNA	Target mRNA	Changes of Expression
aae-miR-2b	3′ UTR of ubiquitin related modifier (URM), ubiquitin, and 3′ UTR of CHIKV	Up
hsa-miR-21-5p	B-cell lymphoma 2 (BCL2), chemokine ligand 1 (CCL1), FASLG, pellino E3 ubiquitin protein ligase 1 (PELI1), interleukin 12 A (IL12A), transforming growth factor beta (TGFβ1)	Up
hsa-miR-138-2-3p	Tripartite motif containing 26 (TRIM5), TGF-Beta activated kinase 1 (MAP3K7) binding protein 3 (TAB3), tumor necrosis factor receptor superfamily (TNFRSF19), mitogen-activated protein kinase 13 (MAPK13), Apoptotic protease activating factor 1 (APAF1), Forkhead Box O3 (FOXO3)	Up
hsa-miR-146	Tumor necrosis factor receptor (TNFR)-associated factor 6 (TRAF6), interleukin-1 receptor-associated kinase 1 (IRAK1/2)	Up

hsa-miR-216a-5p	Cluster of Differentiation 6 (CD6), Janus kinase 2 (JAK2)	Up
aal-miR-305-3p	ECM receptor interaction, endocytosis, and SNARE interactions in vesicular transport	Up
hsa-miR-382-3p	Beta-transducin repeat containing E3 ubiquitin protein ligase (BTRC), gap junction protein alpha 1 (GJA1), TRIM8, ubiquitin-conjugating enzyme E2 D2 (UBE2D2)	Up
hsa-miR-409-3p	DNA topoisomerase 2-beta (TOP2B)	Up
hsa-miR-491-3p	UBE2B	Up
hsa-miR-921	nicotinic acid uptake protein (NIAP)	Up
aal-miR-927	Soluble NSF attachment proteins receptor (SNARE) interactions in vesicular transport	Up
aae-miR-989	sh2/sh3 adaptor and vacuolar ATP synthase	Down
hsa-miR-1260a	BCL2 antagonist/killer 1 (BAK1), activating transcription factor 6 beta (ATF6B)	Up
hsa-miR-1260b	Receptor interacting serine/threonine kinase 1 (RIPK1), E3 ubiquitin-protein ligase NRDP1 (RNF41), suppressor of cytokine signaling 6 (SOCS6), NLR family CARD domain containing 5 (NLRC5), caspase 10 (CASP10)	Up
hsa-miR-1264	TRIM26, bone morphogenetic protein 2 (BMP2), baculoviral IAP repeat containing 6 (BIRC6), interleukin 6 signal transducer (IL6ST), listerin E3 ubiquitin protein ligase 1 (LTN1), itchy E3 ubiquitin protein ligase (ITCH), MAPK8, SOCS5, UBE2D3	Up
hsa-miR-3074	Integrin alpha-V (ITGAV), TNF receptor associated factor 3 (TRAF3)	Up
hsa-miR-4286	Interferon alpha and beta receptor subunit 1 (IFNAR2), interleukin 13 receptor, alpha 1 (IL13RA1), interferon regulatory factor 1 (IRF1), ubiquitin conjugating enzyme E2 Z (UBE2Z), IL6R	Up
hsa-miR-4299	SOCS7, signal transducer and activator of transcription 5 (STAT5B), TRIM28, mitogen-	Up

	activated protein kinase kinase kinase 7 (MAP3K7), TAB1, AKT serine/threonine kinase 1 (AKT1), MAP3K11, MAP kinase-activated protein kinase 3 (MAPKAPK3), CAMP responsive element binding protein 1 (CREB1)	
hsa-miR-4443	Interferon regulatory factor 3 (IRF3), mitogen-activated protein kinase kinase kinase 8 (MAP3K8), receptor interacting serine/threonine kinase 3 (RIPK3)	Up
hsa-miR-4695-3p	CASP8, chemokine C-X-C motif ligand 2 (CXCL2)	Up
hsa-miR-4717-3p	AKT3, UBE2M, sortilin (SORT1), ring finger protein 213 (RNF213), nerve growth factor receptor-associated protein 1 (NGFRAP1), MAPK10, IL11RA, C-C motif chemokine ligand 4 like 2 (CCL4L2), IFNAR1, IL7R, mitochondrial ubiquitin E3 ligase (MARCH5)	Up
hsa-miR-4762-5p	Nemo like kinase (NLK)	Up
hsa-miR-4775	TNFRSF10A, toll-like receptors (TLR1), CASP3, ATF2, TAB2, glycogen synthase kinase 3 beta (GSK3β)	Up
hsa-miR-4794-5p	UBE2S, STAT1, TRAF5, RIPK1, protein inhibitor of activated STAT 1 (PIAS1), AKT1	Up
hsa-miR-4878-3p	Diablo IAP-Binding Mitochondrial Protein (DIABLO)	Up
hsa-miR-5100	UBE2J1, axis inhibition protein 2 (AXIN2), IRAK4	Up
hsa-miR-5581-3p	MAPK6, IL4R, MAP3K1, CCL18, death inducer-obliterator 1 (DIDO1)	Up

A significant propagation of CHIKV is noticed during induction of hsa-miR-124, in opposite silencing of hsa-miR-124 that inhibits CHIKV infection. In certain cases, CHIKV infection show in encephalitic-like symptoms.[250, 251] hsa-miR-124 is predominantly found in neurons and acts as a key negative regulator of neuroinflammation.[252] A change hsa-miR-124 excessive expression in neurons is linked to encephalitic pathology during CHIKV infection. In a study, silencing of hsa-miR-124 is noticed in fibroblast-like synoviocytes (FLS) patients with RA, where hsa-miR-124a is involved in the inflammatory pathway in RA pathogenesis for interacting with the monocyte chemoattractant protein-1 (MCP-1) and cyclin-dependent kinase-2 (CDK-2).[253] Thus, hsa-miR-124 plays a critical role during inflammation along with CHIKV infection.

Aberrant Expression of Mosquito miRNAs

Viruses modify the transcriptional profile of the mosquito vector.[254] In Aag-2 cells, aae-miR-2944b-5p and aae-miR-2b bind the 3′ UTR sites of CHIKV.[255] When mosquitoes are treated. The presence of antagomiR-2944b-5p in mosquitoes is more susceptible during CHIKV infection indicating the role of antagomiR-2944b-5p in viral replication. Strong inhibition of aae-miR-2944b-5p during CHIKV infection enhances cellular MtMP. It demonstrates that aae-miR-2944b-5p targets VPS-13 for maintaining MtMP.[255] In humans, VPS-13 is required for post-Golgi apparatus sorting and trafficking. Therefore, the effect of hsa-miR-2944b-5p on VPS-13 expression in CHIKV infected human cell lines allows for detecting novel drug target candidates.

Eight miRNA expressions are changed during CHIKV infection.[256] The expressions of aae-miR-100, aae-miR-283, aae-miR-305-3p, and aae-miR-927 are overexpressed strongly and the expressions of aae-miR-1000, aae-miR-2b, aae-miR-2c-3p, and aae-miR-190-5p are inhibited. The aae-miR-100, aae-miR-283, and aae-miR-305-3p usually affect NK cell-mediated cytotoxicity target and protein processing in ER pathways. The aae-miR-305-3p interacts with target pathways indispensable for chikungunya viral entry like ECM receptor interaction, endocytosis, and SNARE interactions in vesicular transport. In another study, modulation of 13 miRNA expression during CHIKV infection exhibit that aae-miR-2b directs URM and ubiquitin, while aae-miR-100 targets CDC42 and sumo-ligase. Interestingly, during cells are interact with aae-antagomiR-2b, an enhanced CHIKV replication is noticed. The URM level is also excess in CHIKV infected cells. Moreover, CHIKV replication is reduced to half levels in URM silent cells. It suggests that aae-miR-2b-mediated control of URM explores an important role in chikungunya replication.

Usually, viruses develop an infection in the salivary gland during a blood meal in mosquito vectors.[254] For developing a deadly and fruitful infection, aae-miR-bantam, aae-miR-263a, aae-miR-125, and aae-miR-285 are overexpressed consistently in CHIKV-infected *Ae. aegypti* saliva.[257] In *A. albopictus* saliva, aal-miR-43b, aal-miR-43a, aal-miR-413a, aal-miR-5, and aal-miR-249 are overexpressed.[257] Moreover, inhibition of different selected miRNAs is often found in Aag-2 cells and BHK-21 cells of the salivary gland during changes in CHIKV replication. A series of miRNAs aae-miR-282-5p, aae-miR-34-3p, and aae-miR-11-5p can bind the different genotypes of CHIKV.[258] Moreover, aae-miR-11-5p is well preserved among various lineages of CHIKV and possibly interacted with the end of sub-genomic untranslated RNA region. It demonstrates that the CHIKV structural proteins control the yielding of a miRNA-viral RNA (vRNA) complex at the end of sub-genomic RNA untranslated regions. It thereby protects the binding of host different translational factors on viral RNA.

miRNAs in Bone Homeostasis

CHIKV infection is related to bony lesions and erosive arthritis as the first indication in CHIKV-infected IRF 3/7$^{-/-}$ mice.[259, 260] Bone is one of the most dynamic organs in the body that continuously undergoes remodeling. The most dynamic organ is bone, and its homeostasis is controlled strongly a delicate balance between osteoblast genesis and osteoclast genesis.[261] Osteoblast genesis is a bone formation process that differentiates mesenchymal stem cells (MSCs) into osteoblastic cell lineage and later forms the bone cells or osteoblasts (OBs) and finally into osteocytes, the mature OBs.[262] Conversely, osteoclast genesis is a bone resorption process that generates multinucleated osteoclasts (OCs) from the myeloid precursors fusions by the help of differentiating of hematopoietic stem cells (HSCs).[263] (**Figure 4 and 5**)

An array of transcription factors and complex signaling pathways thus, understanding how these pathways control Bone homeostasis that is altered during CHIKV infection in a massive extends importance. Recently, numerous miRNAs are required for altering several bone-mediated pathologies.[264] Moreover, the above-noted miRNAs expression has been changed during CHIKV infection. (**Table 3**) The aberrant expression of miRNAs in bone homeostasis during CHIKV infection are the most crucial factors for detecting novel biomarkers and for the development of miRNA-mediated therapeutics during infection.

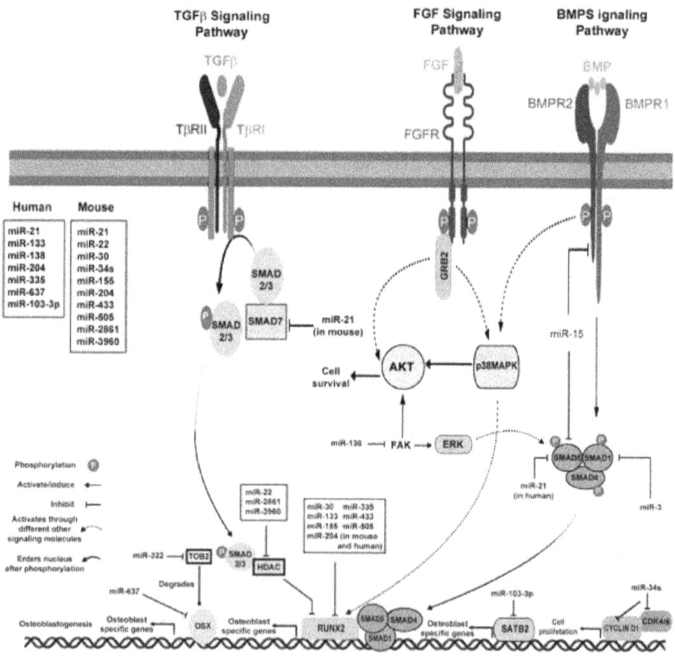

Figure-4. Diagram representing miRNAs that control different signaling pathways in osteoblastogenesis. (Loosely Adapted, Redraw and Modify using skeleton from Roy *et al.*, 2020)

During CHIKV infection, hsa-miR-21-5p is overexpressed, and PELI1, an E3 ubiquitin-protein ligase acts as potential target.[240] PELI1 inhibits the NF-κB signaling process during osteoclast genesis.[249] Moreover, miR-21 damages osteoclast genesis during CHIKV infection. The hsa-miR-138-2-3p is overexpressed during CHIKV infection by interacting with multiple genes including MAPK13.[265] MAPK13 encodes p38 MAPK that is important for bone homeostasis.[266] hsa-miR-138 is also silenced during osteoblast genesis by targeting FAK and reducing the FAK-mediated signaling pathway.[265] However, the exact role of miR-138 controlling any FAK-mediated MAPK-signaling pathway during CHIKV infection is yet not known. Furthermore, many joint inflammations are conjugated with arthritic-like signals during CHIKV infection.[267] An induction of hsa-miR-146a miRNA is noticed in synovial fibroblasts during CHIKV infection.[238] Furthermore, TRAF6 and IRAK1 induce the NF-κB signaling process for the production of pro-inflammatory cytokines as combating CHIKV infection.[239] However, the proper role of hsa-miR-146a on the NF-κB signaling pathway during CHIKV infection remains in dark.

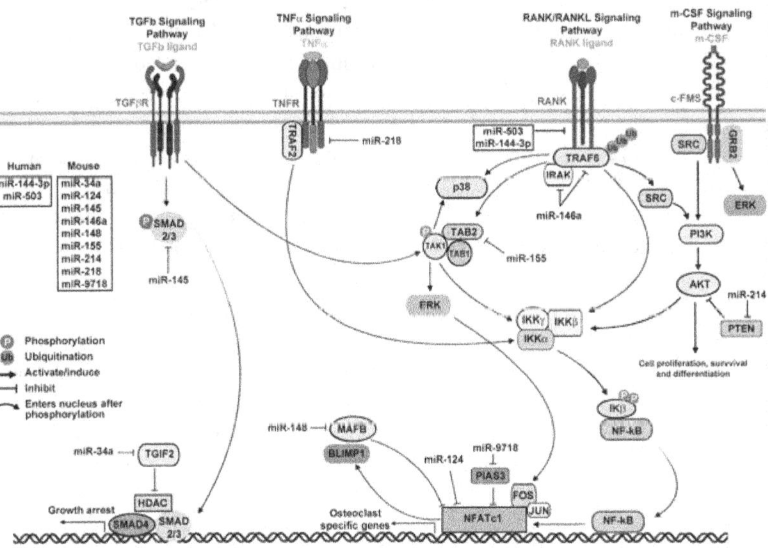

Figure-5. diagram representing miRNAs that guide different signaling pathways in osteoclastogenesis. (Loosely Adapted, Redraw and Modify using skeleton from Roy *et al.*, 2020)

Communicable diseases in Africa

Trachoma

Few pioneering results open fundamental contributions of miRNAs in many ocular infectious diseases especially trachoma in Africa and their most effective ideal therapeutic agents and diagnostic biomarkers. Trachoma is the defective and widely spreading communicable blinding eye disease worldwide predominant in Africa. (https://www.who. int/news-room/fact-sheets/detail/ trachoma). It will also enlighten dark knowledge gaps in the treatment of miRNAs in ocular infection.

Trachoma affects nearly ~150 million people living in African rural and poorest communities.[268] It damages conjunctival epithelium along with the obligate intracellular, Gram-negative bacterium, chlamydia trachomatis (Ct). There are five different stages of trachoma diseases that are most prevalent in Africa. (1) trachomatous follicular inflammation (TF), the existence of five or more follicles in the upper tarsal conjunctiva; (2) trachomatous intense inflammation (TI), a prominent and distinct inflammatory thickening of the tarsal conjunctiva that is distributed more than half of the normal deep tarsal vessels; (3) trachomatous scarring (TS) existence of scarring in the tarsal conjunctiva leading to upper eyelid margin turning-entropion; (4) trachomatous trichiasis (TT) minimally one eyelash rubbing on the eyeball or evidence of recent removal of in-turned eyelashes; and (5) blinding corneal opacity (CO).[269-271] Although chlamydia infection can be effectively treated by antibiotics, e.g., azithromycin, as routine health strategy for trachoma elimination. Multiple infections of the conjunctiva in childhood lead to the chronic inflammatory response that can maintain without active Ct infection, responsible for fibrosis development, scarring, and ultimately trichiasis and blindness.[271, 272]

Recent results show that multiple miRNAs are required in pathogenesis and trachoma development.[270, 271] At least eight miRNAs miR-155, -150, -142, -181a/b, -342, -132, -4728, and miR-184) are differentially expressed with follicular trachoma between TF (with or without detectable Ct) and normal controls during initial disease stages. (**Table 4**)[270] The miR-155 and miR-184 exhibit a direct relationship with the degree of clinical inflammation: miR-155 is overexpressed, while miR-184 is silenced strongly during induction of severe clinical inflammation.[270]. miR-150 and miR-142 are expressed to hematopoietic cells particularly.[273] miR-181b and miR-132 are silenced during inflammation further affecting toll-like receptors (TLR) or NF-κB induction.[270] miR-342 is required in the inflammatory response, cell proliferation, and cancer.[274, 275] miR-4728 is required for focal adhesion and wound healing.[276] The excessive expression of miR-184 found in corneal epithelium during corneal development and function.[277] An emergence of point mutations in the miR-184 seed sequence raise syndromes with chronic keratoconus.[278-280] The miR-184 expression is also reduced during a chronic corneal injury but recovered during wound healing.[277] The miR-184 is needed during ischemia-induced neovascularization by inversely controlling the Wnt pathway via directing a Wnt receptor, frizzled-7.[281] These findings refer to the host immune response to Ct infection and the wound healing development in early stages of trachoma.[270]

Massive miRNA profiling reports that 82 miRNAs are virtually expressed in different quantities in the TS stage tissue from the conjunctival swabs between healthy and diseased patients. The miR-147 and miR-1285 are overexpressed markedly.[271] The predicted interacting genes of these miRNAs in trachomatous scarring and normal conjunctival tissues in enrichment in pathways responsible for fibrosis and epithelial cell differentiation.[271] The miRNAs in inflammation and wound healing processes are required for trachoma pathogenesis. Furthermore, many studies identify exact molecular processes responsible for miRNAs influences in chronic inflammation, fibrosis, and scarring. Once the ideal procedure is exposed, a miRNA-guided therapeutic strategy is employed for regulating the inflammation and scarring that induces the healing process.

Table 4. Different miRNAs Involved in Trachoma

miRNA	Disease	Species/ Tissue/cell	Alteration	Function and/or pathway
miR-155–5p	Trachoma/ Follicular	human/ conjunctival swabs	Up	hematopoisis, immune cells development and function
miR-150–5p	Trachoma/ Follicular	human/ conjunctival swabs	Up	hematopoietic cells
miR-142–5p	Trachoma/ Follicular	human/ conjunctival swabs	Up	hematopoietic cells
miR-181a/b-5p	Trachoma/ Follicular	human/ conjunctival swabs	Up	hematoposis, inflammation
miR-342–3p	Trachoma/ Follicular	human/ conjunctival swabs	Up	cell proliferation, inflammation
miR-132–3p	Trachoma/ Follicular	human/ conjunctival swabs	Up	hematopoiesis, inflammation
miR-4728	Trachoma/ Follicular	human/ conjunctival swabs	Down	focal adhesion and wound healing, cancer Corneal development and function
miR-184	Trachoma/ Follicular	human/ conjunctival swabs	Down	wound healing, ischemia-induced neovascularization
miR-147b	Trachoma/S carring	human/ conjunctival swabs	Up.	fibrosis and epithelial cell differentiation
miR-1285	Trachoma/ Scarring	human/ conjunctival swabs	Up.	fibrosis and epithelial cell differentiation

River Blindness

River blindness is a unique tropical blinding disease found mainly in Africa responsible for filarial nematode *Onchocerca volvulus* (*O. volvulus*) infection and is gradually infected by blood-feeding arthropods. It is the second-most routine visual impairment and blindness due to ocular infection, after trachoma (*Parasites*. CDC. 21 May 2013).[282] In Africa, at least 120 million people are at risk sensitive for infection.[282-284] Its distribution and spreading are severe that nearly 17 million African pollutions have been infected with *O. volvulus*, predominantly and 1.2 million Africans have been suffering from vision impairment or *O. volvulus* mediated blindness .[285] The infection can affect each section of the eye, from conjunctiva and cornea anteriorly to the uvea and posterior segment along with retina and optic nerve. There is atrophy of the retinal pigmented epithelium and subretinal fibrosis in the posterior segment. River blindness is required for an autoimmune response. The larvae in the cornea and the anterior chamber can be identified by slit-lamp examination in the anterior segment.[286] The larvae are transported within the human body without inducing an immune response. It is the host's immune/inflammatory reactions to the dead or dying larvae of *O. volvulus* that lead major of the *Onchocerca*-mediated pathology along with onchocercal keratitis,[286, 287] Anti-helminthic drug, Mectizan TM (Ivermectin) eliminates onchocerciasis. Despite ivermectin treatment has reduced onchocerciasis [288], the most reliable single drug potentially induces the germination of ivermectin-resistant *O. volvulus.*[289]

Serum *O. volvulus* miRNAs act as diagnostic biomarkers. However, a set of 17 parasitic miRNAs including bma-miR-236–1, ov-miR-100d, ov-bantam-a are identified for the use as a diagnostic marker.[283] Furthermore, the plasma-derived parasitic miRNAs act as diagnostic biomarkers for river blindness is potential greatly. More susceptible systems have identified that the effective parasitic miRNAs function as non-invasive diagnostic biomarkers for river blindness.

Lymphatic filariasis

The lymphatic filarial parasites *Brugia malayi*, *Brugia timori* and *Wuchereria bancrofti* infects 120 million people in 80 countries worldwide predominantly in Africa.[290] The excessive infective third-stage larvae (L3s) of mosquitos are transmitted via bites and enter their vertebrate host. They propagate by molting L4s and ultimately mature to adulthood within 6–12 months.[291] Adult parasites easily settle in ideal preferred organ lymphatic vessels and produce infected microfilariae (**mf**). The mf can propagate in vertebrates and survive lifelong close to a year. Meanwhile they are migrating excessively in peripheral circulatory bloods and waiting for transmission by a mosquito during sucking of the blood meal.[292] Lymphatic filarial infections cause the raising of recurrent fevers, painful adenolymphangitis and elephantiasis.[293] Despite no fatality, the morbidity that leads to acute filarial infections greatly hampers the socio-economic progress of the affected communities.[294] Three popular conventional medicines Diethylcarbamazine (DEC), ivermectin and albendazole are treated ideally the acute lymphatic filarial infections. All three destroy microfilariae but only DEC show some efficacy against adult parasites.[295]

The drug resistance varieties against ivermectin[296] and mutant microfilaricides are the recent inclusion essential for new strategies development for fighting against this debilitate disease. The most completed filarial life cycle using two vectors and lack of capacity of genetically manipulated parasites makes biological procedural studies difficult. Recently, an RNA-mediated regulatory MicroRNA against filarial parasites may develop new avenues for treatment plan. Isolation of filarial-specific components of small RNA pathways or miRNAs possible advantageous development of novel anti-filarial therapies. 32 miRNA isolates initially in *B. malayi.*[297] *B. malayi* (90–95 Mb) likely encodes equal numbers of miRNAs roughly related to *C. elegans.*[298] An easy comprehensive miRNAs extraction in *B. malayi* forms for core understanding miRNA functions in light of the complicated *B. malayi* life cycle and use possibly for designing anti-miRNA compounds that are lethal to the parasites and provide useful drug candidates.

Detection of MicroRNA and Family Assignment

A total of 145 miRNAs are derived from 129 hairpin sequences of the *B. malayi* small RNA libraries. These include 96 dominating miRNAs and 49 are in rare existence. The *B. malayi* miRNAs separate into 99 different families each of them having a unique seed sequence.[299, 300] (**Figure 6**) Sixty-six miRNAs are divided into 20 families of paralogues depending on nucleotide identity in the 5′ seed sequences or by global identity of ≥70%.[301] The remaining 79 families carry a single miRNA. Five highly conserved miRNAs are detected by examining the small RNA libraries directly. However, highly conserved miR-1 is not isolated presumably because its DNA sequence is not found in the genome [297] but it is easily identified by directly searching the small RNA libraries.

The Worm base (www.wormbase.org) curates a brand-new version of the *B. malayi* genome. Therefore, multiple unidentified miRNAs will likely be isolated from this genome analysis of different developmental stages. In relative assay the miRNAs from two species *B. malayi* and in *B. pahangi.*[302] are isolated. Fifty-five of the 99 *B. malayi* miRNA families are conserved strongly in *B. pahangi* depending on the nucleotide identity in the 5′ seed region or by global identity of ≥70%.[301] This weak homology between the two closely related *Brugia* species is possibly employed in the different methodology to detect the *B. pahangi* miRNAs.[302] Therefore new *B. pahangi* miRNAs are isolated depending on *B. malayi* genome.[302]

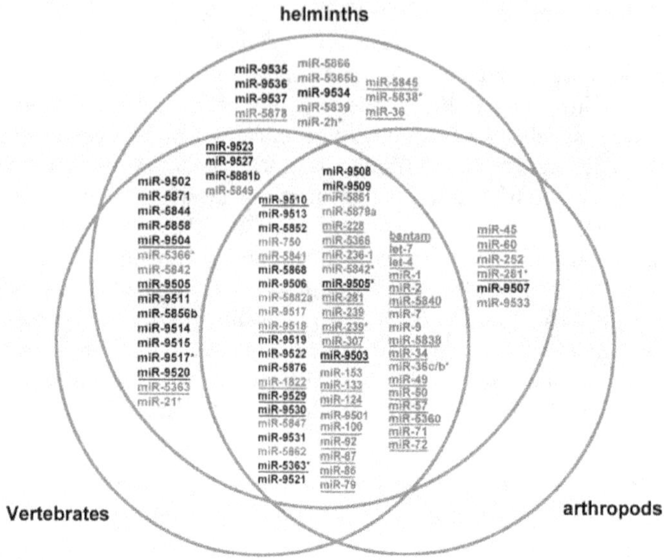

Figure-6. Conservation of multiple miRNAs families of *B. malayi*. (Loosely Adapted, Redraw and Modify using skeleton from Poole *et al.*, 2014)

Conserved MicroRNAs in Filarial Parasites

Out of 99 *B. malayi* miRNA families are conserved strictly in mosquitos and vertebrates, the prime hosts of filarial parasites. Among them 61 (2/3) of the *B. malayi* miRNA families are conserved in a wide range of organisms including arthropods, vertebrates, and other helminths. **(Figure 6)**

The bma-miR-5844, homologues are isolated in prime vertebrates infected by filarial parasites including humans, horses, cattle, pigs, and platypus. In contrast, homologues are detected in arthropods but not vertebrates for 6 *B. malayi* miRNA families. It suggests that three of the 6 microRNAs, miR-45, -252 and -281* highly have homologues in mosquitos, the invertebrate hosts of *Brugia*. However, Homologues are possibly not isolated in either arthropods or vertebrates for 12 *B. malayi* miRNA families that suggest their existence in helminth or even filarial-specific miRNA families.

Putative filarial-related miRNAs

Out of the 12 nine miRNA families in *B. malayi* finds only in helminths. 8 orthologues appear at least one of the specific filarial parasites but not in other worms. (**Figure 6**) In fact, Bma-miR-5866 is one of the most abundant filarial-related miRNAs. However, Orthologues of bma-miR-5866 are identified in *B. pahangi.*[302] Mature miR-5866 is also close to the miR-57 family. The miR-5866 seed sequence (TACCAT) is differed by single nucleotide sequence (TACCCT) at the 5′ end of *B. malayi* and *C. elegans* miR-57. Based on Clustal W alignments, bma-miR-5866 and bma-miR-57 are 67% identical. However, the filarial miR-5866 hairpin sequences only display 34–40% homology with the hairpin sequences of *B. malayi* and *C. elegans* miR-57. It is predicted that the miR-5866 and miR-57 families possibly evolve from the same common ancestor by simply gene duplication.

The location of putative filarial specific microRNAs, miR-9536 is unique because it resides within an intron of Bm1_03065, a *cut-1* cuticlin gene (genbank acc.#: XM_001892072). Cuticlin is a cross-linked and insoluble protein of the worm cuticle. The *C. elegans* cuticlin genes *cut-1*, *-3* and *-5* are tightly transcriptionally controlled and are required for body contour and Body shape formation.[303] miR-9536 transcriptions inside the Cuticlin intron is possibly coordinated with the cuticlin gene expression. It indicates that miR-9536 interacts with many components of cuticlin biosynthesis or is especially responsible for molting and cuticle synthesis. These putative filarial specific miRNAs possibly control some of the 20% of the proteome of *B. malayi.*[298]

MicroRNA clusters

A total of 19 miRNAs from eight clusters in *B. malayi*, are isolated ≤2 kbps in length. Truly, a smaller number of miRNA clusters have been isolated in *B. malayi* relative to *C. elegans.* However, new clusters may be isolated based on *Brugia* genome sequences. In *B. malayi*, the only family members notice clustered are miR-100a and-100d[297] as well as miR-2i and -2d. (**Figure** 7)

The miR-2 family containing 11 members is the largest in *B. malayi.* The 6 nt conserved seed sequence, ATCACA, is present in all family members except miR-2i and miR-2h. They contain ATCGCA as a seed sequence. Moreover, the similarities of the *B. malayi* miR-2 hairpins lead to a highly conserved nt blocks present in both the miR-2i (5/7) and miR-2h (6/7) hairpins. The miR-2 hairpin has consensus similarity with miR-2b, -2d, -2f, -2i and -2h based on their phylogenetic tree. They are derived from a common ancestor depending on an expansion event. Indeed bma-miR-2a and cel-miR-2 are the most closely related family members with 100% identity.[297]

In *B. malayi* miR-2 paralogues Bma-miR-2a, -2b and -2e are strongly expressed. But they exhibit a differential expression pattern. Interestingly, Bma-miR-2a carries only ~2% of the miRNA population in mf. It is nearly 7.5X higher than the amount of this miRNA both in male and female libraries. The bma-miR-2b, -2c, -2d and -2e are usually expressed in the adult libraries, however their existence is not at a consistent level in the mf library. The rest of the *B. malayi* miR-2 paralogues (miR-2f, -2g, -2h and -2i) are only present at a traced level (<100 ends) in the libraries.

Nearly 50% of the *B. malayi* miR-2 family members are oriented in cluster formations. The miR-2b, -36 clusters exist in probable repeat units with 94% similarities. (**Figure** 7) The miR-2d, -36d cluster on scaffold C is present in the repeat unit with only 72-68% similarities to the miR-2b, -36 clusters on scaffolds A and B respectively. A fourth miR-2 paralogue, miR-2i, is also observed in clusters on scaffold C with miR-2d and -36d. (**Figure** 7) Based on the existing *B. malayi* genome, miR-2 clusters A-C are classified into 2 groups together with a 23 kb scaffold and with a fourth miR-2 cluster consisting of miR-2h-2, miR-5839 and miR-2f.

The formation of the Cluster causes a synchronized expression. As an example, miR-2b and its cluster partners miR-36c and -36b are exclusively expressed in adult parasites especially in females compared to mf.[297] Moreover, miR-2d and -36d and miR-2c and -5841 expression patterns are also induced in adults relative to mf. However, it is to be identified whether these miRNAs clustered are processed from the same RNA transcripts. Although the miR-2, -36 clusters action in *B. malayi* must await target identification. The miR-2 family inhibits apoptosis during embryogenesis by targeting the 3′ UTRs of pro-apoptotic genes.[304] Moreover, the BH3-like pro-apoptotic gene, as a target of the miR-36 family in embryos are identified by the RISC immunoprecipitation studies.[306]

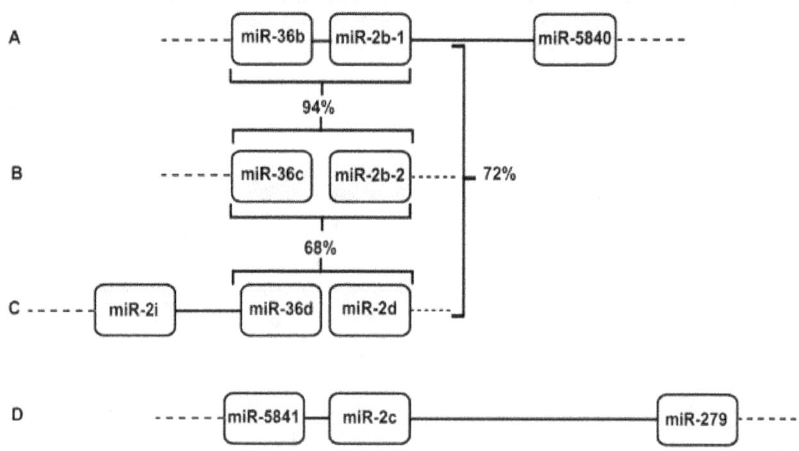

Figure-7. Several MicroRNA-2 clusters in *B. malayi*. (Loosely Adapted, Redraw and Modify using skeleton from Poole *et al.*, 2014)

Stage-specific expression of miRNAs in B. malayi

Deep sequencing is the most vital experimental technique for determining the relative expression of the same miRNA because it prevents any biases during the comparison of different sequences. [307, 308] Moreover, normalized readings are utilized for quantitative miRNA assays in multiple developmental stages. The relativity among miRNAs of different libraries, the miRNA raw read numbers in a library leads to normalization. Using this methodology, it is noticed strikingly, 31 miRNAs are at least 5X folds enriched in expression in adults than mf, whereas only 4 miRNAs miR-2a, -71, -92 and -5366* are enhanced expressed in mf.

The miR-71 normally expresses ~27% of the total miRNAs in mf and shows 4.6 and 7.2X folds of enrichment than the miR-71 amount found in either the male or female CIP libraries that is validated by qPCR analysis. miR-71 is one of the most enriched expressed miRNAs in *B. malayi* mf that the controls longevity of mf. These signals spread via the body and effectively more than a year waiting before ingulfing by a mosquito.[309, 310] The miR-71 targets multiple homologues proteins in *B. malayi* including *pdk-1* (genbank acc. EDP33519) and *Daf-16* (genbank acc. XP_001901487). Among 31 miRNAs expressed abundantly in adults than mf, 11 miRNAs are expressed extensively in females than males; 11 miRNAs have a larger expression in males while 9 miRNAs are equally expressed in both sexes.

Among the 4 miR-36 paralogues, miR-36a is isolated in *B. malayi* that contains 8% of the total miRNA population in the female CIP library but carries only ~0.4% of miRNAs in the male library. It is only at a trace level in mf. *B. malayi* miR-36c and -36b are also produced excessively in the female CIP library but at lesser levels relative to miR-36a. The bma-miR-36a has a striking similarity nearly 76% nt identity to miR-36 homologues from a wide variety of organisms. However, like other *B. malayi* miR-36 paralogues, miR-36a is not clustered with a miR-2 paralogue. (**Figure** 7)

In addition, Bma-miR-5838 is one of the important members of the 11 miRNAs that are more widely expressed in the male than the female CIP library. It contains ~0.7% of miRNAs in the male CIP library considering only .02% in the female library, The bma-miR-5838 shares 8 nt at its 5′ end that carries the seed sequence, GAGTA, with *Aedes aegypti* miR-12 and human miR-496.

Chemotherapy Development and Target Identification

Multiple developmental stages of filarial parasites exist in insect and vertebrate hosts that are interacted traditionally for chemotherapeutic processes. However, a predominant expression of some miRNAs in adult parasites promotes its usefulness for microfilaricides design.

Recently, it is found that miRNAs often interact functionally with various related mRNAs via a biochemical pathway [3 11-313] and are also often via feedback loops with transcription factors. They control functional transcription of closely linked genes.[314, 315] A link between miRNAs and transcription factors is found in *B. malayi* as well. Bma-miR-5855 is resided within an intron of Bm131015 as a putative sequence of transcription factors.

The chemotherapeutics production against filarial-specific miRNAs leads a unique way to destroy post-transcriptional regulation in parasites without hampering the host's miRNA control networks. Filarial–specific miRNAs are interacted possibly using modified antisense RNAs such as antagomirs.[316, 317] Recent developments in antagomir technology allow an intravenously injected 8 mer locked nucleic acid oligonucleotides in the mouse tissue. This injection silent whole miRNA families efficiently with a trace level of off-target effects.[318]

Among ~30 million small RNA sequencing only 145 miRNAs are cloned from *B. malayi* adults and mf. The miRNAs are classified into 99 families based on their unique seed sequence. The quantification identifies the quantity of four miRNAs from various stages of *B. malayi.* They showed a strong and close relationship between the sequencing read numbers and the libraries. The miRNA expression from three independent CIP libraries exhibits that nearly 33% of the *B. malayi* miRNAs are exclusively expressed differentially. Sixty-one of the *B. malayi* families are conserved broadly in arthropods, vertebrates, and helminths. About 50% of families have a homologous seed sequence match in *B. pahangi* and *C. elegans*. Interestingly, 20 *B. malayi* homologous miRNA families are well preserved in vertebrates but not in other helminths or arthropods. Among them only nine miRNA families are filarial-specific and not found in other organisms.

For determining the initial step of the miRNA regulatory network in filarial parasite *B. malayi* miRNAs are essential factors. The filarial parasites counteract the host's immune responses. It is possible that developing miRNAs perform crucial contributions in these processes particularly in anti-filaricidal. Future work isolates interacting miRNAs, specifically the filarial-specific miRNAs that open novel approaches for therapy development against this debilitating filaria disease.

Trypanosomiasis, human African sleeping sickness

African Trypanosomiases (AT) are devastating communicable diseases of livestock and African humans infected by Tsetse fly bites. Nearly 350,000 humans are infected by *Trypanosoma brucei gambiense* in West Africa.[319] However, the new infection rate of African trypanosomiasis (HAT) is relatively low approximately 10,000 cases per year [320] as a result of consistent surveillance efforts.[320] The routine symptoms of human African trypanosomiasis (HAT) initiate a hemolymphatic stage (stage I). [321] Later, ultimately progress stage II affects the central nervous system. Progress of infection is comparatively slower for *T. b. gambiense* than for infection by the East African form, *T. b. rhodesiense*. Although in certain cases the patients eventually become asymptomatic or even self-cure. [322, 323]

MicroRNAs (miRNAs) have an instrumental role in many systems, including inflammation, immunity and control of cell growth and differentiation. [324-326] A realistic drive is attempted whether miRNA changes affect peripheral circulatory diseases in both the cellular and the cell-free serum fractions.[327-330] Invariably, much fewer efforts are paid in the miRNAs contribution in infectious diseases related to intracellular pathogens especially viruses. The peripheral blood miRNA profiles of humans infected with *T. b. gambiense* are explored focusing on both biomarker and immune response processes. The consistent changes in miRNA profiles of blood lymphocytes in human African trypanosomiasis led to immune instigation or inflammation, However, they are not disease-specific, and cannot determine the different disease stages. Nevertheless, such an approach has a tiny promise in diagnostics but might generate disease pathology treatments.

State-specific miRNA Expression

Fourteen miRNAs are expressed differentially between all patients (groups HAT-1 and HAT-2) and the control group (group C). **(Table 5, Figure 8**) Among 14 miRNA members the expression 13 miRNAs are altered consistently between patients with stage-II disease (group HAT-2) and controls (group C), whereas ten miRNAs are modified between stage-I patients (group HAT-1) and controls (group C). However, not a single miRNA is used to differentiate between stage I (HAT-1) and stage II (HAT-2) patients.

Sensitive miRNA Targets

According to prediction and experimental analysis more than 3000 human genes are sensitive to possible targets. The consistent and marked differences of miR-195 are observed between all patients and controls, Next, a master gene susceptible to mRNA level alteration, possibly an ideal of the ten miRNAs. The miRNAs and their cognate mRNAs are expressed in the same cells suggesting that their expressions reduce the availability of target mRNAs.

Nearly 34 predicted targets, only one, RFXAP, is inhibited more than 2-fold at the steady-state mRNA level, but the cognate miRNA is also silent as well. Significant reduction of two miRNAs, miR-4291 and miR-454 leads to an activation of a metalloprotease inhibitor, TIMP2, at a considerable level. Consequently, overexpression of miRNA-193b reasonably inhibits its four potential target genes GPR146, EIF2S1, PLA2G4D and MAPK10.

Effect of miRNAs

Recently nine miRNAs have been identified whose levels are changed in the peripheral blood of HAT patients. In the "HAT-like" patients who are CATT-positive, but PCR-negative, miRNAs of them are detected. Interestingly, such miRNA profiles are even observed in trypanolytic-negative samples. When these people are infected with trypanosomes at a low, or no level. In this case trypanosome-mediated antigens expression is not detected by the conventional trypanolytic test.[331] Moreover, for people with trace or undetectable levels of parasite loads, miRNA profiles are not altered relative to uninfected controls. Therefore, the expression miRNA modification in HAT patients is highly non-specific and possibly related to immune induction or inflammation. Unfortunately, the steady-state level of the miRNAs is not altered inbetween stage I and stage II infection. Nevertheless, human *T. gambiense* infection alters the miRNA expression in peripheral blood leukocytes. Several consistently affected miRNAs are linked to cellular proliferation alteration, which might reflect the lymphocyte induction during active trypanosomes related diseases.

Table 5. Different miRNAs with modulated abundance in trypanosomiasis

miRNA	Log2FC	p value (1)	Log2 Fc.	p value (2)	Log2 FC all	p value all
miR-199a-3p	-6.9	2 E-4	-6.9	1 E-5	-6.8	2 E-6
miR-27b	-6.8	2 E-4	-6.7	6 E-5	-6.6	2 E-6
miR-126*	-6.7	3 E-4	-6.4.	2 E-4	-6.4	2 E-6
miR-98	--5.9	3 E-3	-6.3	6 E-4	-6.1	2 E-4
miR-409-3p	-5.5	7 E-3	-5.1	0.01	-5.2	1 E-3
miR-4291	-5.3	7 E-3	-6.1	6 E-4	-5.6	2 E-4
miR-146b-5p	-4.5	0.03	-4.2	0.02	n/a	n/a
miR-454	-4.1	0.04	-4.1	0.04	-4.1	0.02
miR-193b	4.4	0.04	4.1	0.05	4.2	0.02
miR-195			-5.1	0.01	-4.5	0.01
miR-144*			-4.7	0.01	-4.3	0.01
miR-22*			-5.1	3 E-3	n/a.	n/a
miR-374c			-4.6	0.02	n/a	n/a
miR-338-5p	`4.7	0.01			n/a	n/a

All miRNA having alteration in one stage of sleeping sickness are shown. Log 2FC denotes Log-2 of the arithmetic means of fold changes of trypanosomiasis patients compared to average value of healthy controls. The P value and student t test are utilized to compare paired and multiple values including a Benjamini-Hochberg correction for false discovery. A threshold of 0.05 is calculated for significance. Despite many miRNAs are changes their expression in one stage, significance rather analysis represents no significance difference between state 1 (1) and stage 2 (2).

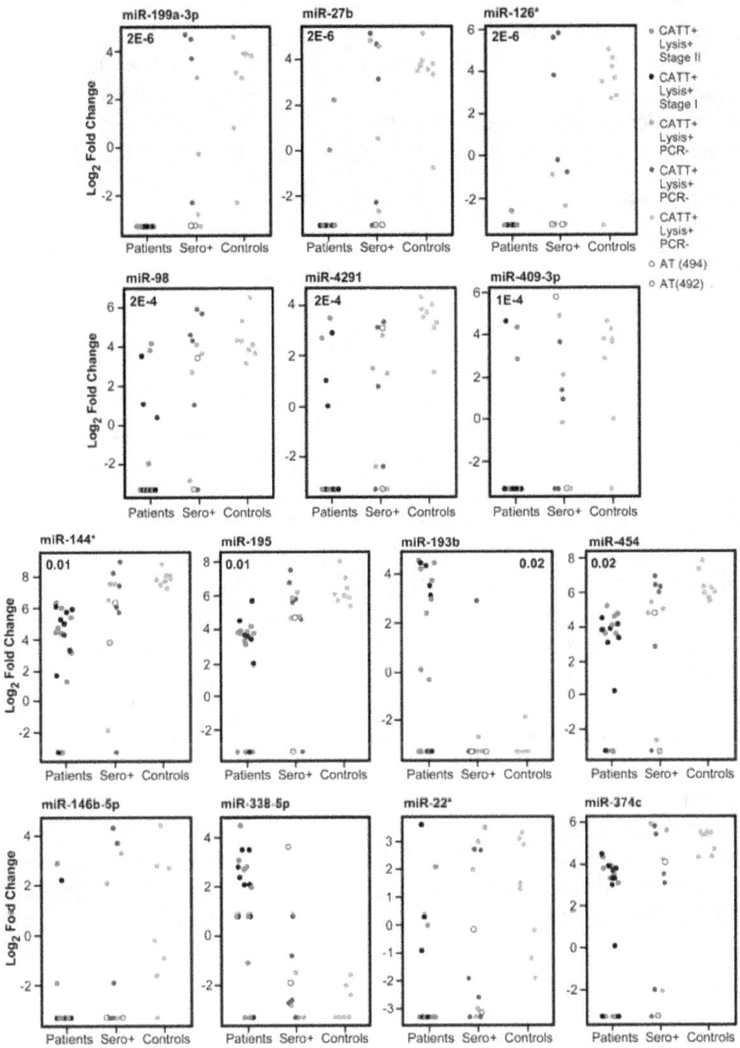

Figure-8. Multiple miRNAs with modulated abundance in trypanosomiasis. (Loosely Adapted, Redraw and Modify using skeleton from Leong *et al.*, 2013)

The alteration of miRNA and mRNA transcriptomes of peripheral blood cells resemble cell types and physiological modification. Unfortunately, no transcriptome miRNAs alter in response to known pathology. Few modifications of miRNAs, however, do not exhibit their effective links with cytokines or cell proliferation. The miR-199a-3p, miR-193b and miR-126 are inhibited strongly in cell proliferation. [332-335] Interestingly, miR-199a-3p and miR-126 silencing in

HAT samples involve an effective activation in leukocyte proliferation. The miR-193b, the only reproducible miRNA, sometimes enhances the expression that does not fit with the hypothesis.

Silencing of miR-144* in most HAT patients leads to a reasonable silencing of TNF-a and IFN-c production and T-cell proliferation.[336] Sometimes miR-27b is silent in HAT patient peripheral blood. However, its activation in liver cells, a negative regulator in lipid metabolism[337], leads to angiogenesis control.[338]

Crimean-Congo hemorrhagic fever

Crimean-Congo hemorrhagic fever (CCHF) is a tick-borne disease, which is produced by the incorporation of arbovirus belonging to the genus Nairovirus of the family Bunyaviridae in host cells. Crimean-Congo hemorrhagic fever (CCHF) is an acute viral hemorrhagic fever due to the insertion of a negative-sense single-stranded RNA (ssRNA) genome. The CCHFV genome produces three different sizes: large (L, 12kb), medium (M, 6.8kb), and small (S, 3 kb) RNA segments. These segments of CCHFV encode mainly the RNA-dependent RNA polymerase, viral glycoproteins, Gn and Gc, and the nucleocapsid protein. CCHF is increasingly spread as an endemic in part of Western and Central Asia, the Middle East, South-Eastern Europe, and Africa. CCHFV is usually inserted in the human host using bites of infected ticks or contact through blood, tissues, or body fluids of viremic animals or humans.[339] The ideal incubation period of CCHF (usually <7 days) that leads to sudden high fever, headache, muscle aches, dizziness, vomiting, and diarrhea.[340] A strong complex pathogenesis of CCHF is found between virus and host cells.

Currently, several miRNAs may play an effective role in both virus-host association and virus pathogenesis. host miRNAs influence and alter viral pathogenesis using host gene regulation and virus genome binding and controlling the translation and replication of viruses. Viral infections modify the host cell transcriptome for huge virus advancement. However, in this unique study the role of miRNAs is investigated in CCHF infections. Here we illustrate microRNA gene regulation profiles in CCHF patients that lead to an initial investigation using microarray analysis. Moreover, we initially also show the relative expression studies of miRNAs between fatal cases and nonfatal cases, and fatal cases and controls. In this study, no statistically significant differences between the patient and the normal healthy groups regarding age and sex are determined. The whole numbers of miRNAs in controls, cases, and fatal cases are 198, 192, and 205. 99 miRNAs are identified only in fatal CCHF patients. (**Figure 9A**) Moreover, relative calculation of viral load in fatal and nonfatal cases is found. The statistically significant difference between fatal and nonfatal cases is $P < 0.001$. The maximum viral load is 7.5×107 copies/mL in fatal samples, 22×106 copies/mL in nonfatal cases.

Case Vs Control

In the heat map fifteen miRNAs are statistically significant between the case and control individuals ($FC \geq 2$). (**Figure 9B**) Five miRNAs miR-1268a, miR-

3940-5p, miR-937-5p, miR-1249-3p, miR-296-5p are overexpressed, while 10 miRNAs miR-144-3p, miR-144-5p, miR-148a-3p, miR-101-3p, miR-126-3p, miR-96-5p, miR-18a-5p, miR-27a-3p, miR-590-5p, and miR-18b-5p are reduced consistently in the CCHF patients relative to controls.

A total of 75 miRNAs are differentially expressed between fatal and control individuals (FC ≥ 50). Relative to the normal individual, 10 miRNAs are overexpressed in fatal individuals, whereas 65 are reduced in their expression. (**Table 6**) Among overexpressed miRNAs only the highest Fatal case are in hsa-miR-4793-3p, hsa-miR-195-3p, hsa-miR-4440, hsa-miR-1909-5p, the highest Fatal cases in strongly reduced miRNAs are found in hsa-let-7f-5p, hsa-let-7a-5p, hsa-let-7g-5p, and hsa-miR-144-3p .

Moreover, miRNA expressions in CCHF patients are compared between fatal and nonfatal individuals. 66 miRNAs are expressed differentially in fatal cases relative to nonfatal cases. Fifteen miRNAs are overexpressed, whereas 51 miRNAs are silent strongly. (**Table 6**) Among the highest FC miRNAs hsa-miR-1909-5p, hsa-miR-4767, hsa-miR-548 am-5p, and hsa-miR- 6511a-3p are upregulated, while for downregulation miRNAs are in hsa-miR-7975, hsa-miR-142-5p, hsa-miR-532-3p, and hsa-let-7a-5p are found in fatal cases.

CCHF miRNAs and other functions

The incidence and geographical range are gradually spreading for CCHF fatal disease. Important CCHF cellular targets are mainly mononuclear phagocytes, endothelial cells, and hepatocytes.[341] The CCHF intracellular replication cycle initiates with the interaction between virions and cell surface receptors. miRNAs directly link to viral RNA replication and pathogenesis by conjugating with the RNA virus genome or by virus-guided modulation in the host transcriptome.[342] Multiple miRNAs utilize as positive or negative modifiers in the host antiviral response.

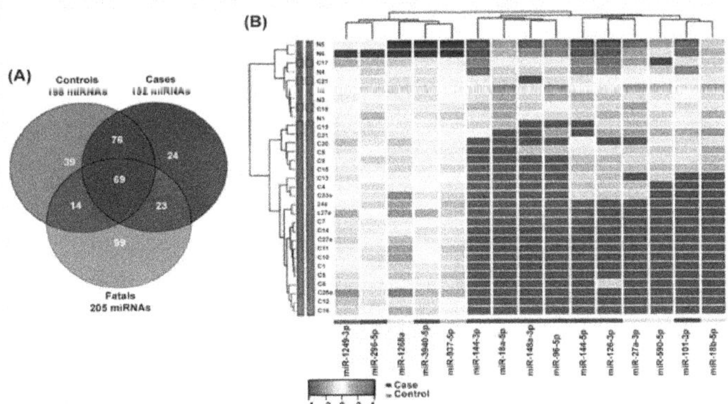

Figure-9. Multiple behaviours of differentially expressed miRNAs. A Venn diagram displaying number of miRNAs overlapped in controls, cases, and fatal. The total numbers of miRNAs produced in controls, cases and fatal cases are 198, 192, and 205, respectively. Twenty-four miRNAs are produced only in cases, whereas 99 miRNAs are identified only in fatal cases. B, Heat map demonstrating differentially expressed miRNAs in patients with CCHF vs controls. Red color for higher expression and blue color for lower expression in all samples. Five miRNAs are overexpressed (miR-1268a, miR-3940-5p, miR-937-5p, miR-1249-3p, and miR-296-5p), whereas 10 miRNAs are reduced (miR-144-3p, miR-144-5p, miR-148a-3p, miR-101-3p, miR-126-3p, miR-96-5p, miR-18a-5p, miR-27a-3p, miR-590-5p, and miR- 18b-5p) in the cases relative to the controls. CCHF-Crimean-Congo hemorrhagic fever; miRNAs-microRNAs. (Loosely Adapted, Redraw and Modify using skeleton from Arslan *et al,*. 2019)

Table- 6. Overexpressed and inhibited miRNAs in fatal relative to controls and nonfatal cases

	Fatal vs control			Fatal vs nonfatal case	
	P (corr)	FC (abs)		P (corr)	FC (abs)
Upregulated miRNAs			**Upregulated miRNAs**		
miR-4793-3p	<0.01	248.29	miR-1909-5p	<0.01	100.15
miR-195-3p	<0.01	183.12	miR-4767	<0.01	93
miR-4440	<0.01	182.82	miR-548 am-5p	<0.01	82.49
miR-1909-5p	<0.01	158.34	miR-6511a-3p	<0.01	72.47
miR-548 am-5p	<0.01	154.64	miR-4785	<0.01	72.44
miR-7850-5p	<0.01	137.56	miR-6775-3p	<0.01	71.27
miR-1908-3p	0.01	106.49	miR-7850-5p	<0.01	70.84
miR-6775-3p	<0.01	101.13	miR-6784-3p	<0.01	69.53
miR-595	0.01	70.77	miR-4793-3p	<0.01	68.87
miR-7108-3p	<0.01	55.11	miR-4440	<0.01	66.59
Downregulated miRNAs			miR-195-3p	<0.01	65.59
let-7f-5p	<0.01	574.3	miR-1910-5p	<0.01	65.59
let-7a-5p	0.01	566.09	miR-3935	<0.01	60.18
let-7g-5p	<0.01	488.04	miR-1470	<0.01	58.09
miR-144-3p	<0.01	288.14	miR-142-5p	<0.01	264.04
miR-142-5p	<0.01	372.01	**Downregulated miRNAs**		
miR-16-5p	0.02	362.1	miR-7975	<0.01	277.22
miR-144-5p	<0.01	288.14	miR-142-5p	<0.01	264.04
miR-20b-5p	<0.01	281.82	miR-532-3p	<0.01	234.4
let-7d-5p	0.01	273.09	let-7a-5p	<0.01	222.15
miR-15b-5p	0.03	269.2	miR-425-5p	<0.01	202.56
miR-7975	0.04	266.82	miR-550a-3-5p	<0.01	195.48
miR-363-3p	0.01	265.1	miR-19b-3p	<0.01	185.87
miR-101-3p	<0.01	254.97	let-7f-5p	<0.01	181.14
miR-130a-3p	<0.01	251.93	miR-223-3p	<0.01	180. 71
miR-103a-3p	0.01	232.79	miR-185-5p	<0.01	179.6
miR-223-3p	0.04	221.98	miR-150-5p	<0.01	173.18

miR-532-5p	<0.01	219.53
miR-107	0.01	215.09
miR-126-3p	<0.01	211.47
miR-106b-5p	0.01	195.03
miR-15a-5p	0.01	194.07
miR-148a-3p	<0.01	178.29
let-7c-5p	0.01	175.19
miR-21-5p	0.01	165.9
miR-19b-3p	0.04	157.56
miR-16-2-3p	<0.01	154.52
miR-550a-3-5p	<0.01	148.71
miR-26b-5p	0.02	140.8
miR-185-5p	0.03	138.1
miR-625-5p	<0.01	132.59
miR-7-5p	<0.01	131.51
miR-150-5p	0.02	130.44
miR-29c-3p	0.01	128.14
miR-19a-3p	0.01	118.67
miR-20a-5p	0.01	118.34
miR-17-5p	0.01	112.06
miR-24-3p	0.01	110.18
miR-23a-3p	0.02	105.39
let-7i-5p	0.02	97.51
miR-96-5p	<0.01	91.23
miR-23b-3p	<0.01	90.83
miR-93-5p	0.01	90.23
miR-183-5p	0.01	85.25
miR-130b-3p	0.01	83.88
miR-532-3p	0.02	77.34
miR-590-5p	<0.01	76.3
miR-502-3p	0.01	75.71
miR-26a-5p	0.03	74.3
miR-660-5p	0.01	72.98
let-7b-5p	0.01	69.81
miR-151a-5p	0.03	68.64
miR-7977	0.03	66.5
miR-18b-5p	<0.01	65.49
miR-6073	<0.01	64.55
miR-29a-3p	0.01	62.12
miR-500a-3p	<0.01	61.48
miR-378i	<0.01	61.36
miR-181a-5p	0.01	59.65
miR-30c-5p	0.03	58.51
miR-18a-5p	<0.01	55.76
miR-27a-3p	0.01	52.19
miR-210-3p	0.01	51.59
miR-151b	0.02	50.99

miR-3653-3p	<0.01	141.61
miR-16-5p	<0.01	136.15
miR-25-3p	<0.01	126.45
let-7c-5p	<0.01	123.29
let-7g-5p	<0.01	121.98
let-7d-5p	<0.01	119.49
miR-30b-5p	<0.01	109
miR-502-3p	<0.01	106.82
miR-22-3p	<0.01	103.31
miR-30c-5p	<0.01	103.06
miR-361-3p	<0.01	103.06
miR-140-3p	<0.01	101.76
miR-103a-3p	<0.01	100.7
miR-4732-5p	<0.01	97.45
miR-331-3	<0.01	95.77
miR-625-5p	<0.01	91.66
miR-342-5p	<0.01	91.25
miR-320e	<0.01	90.61
miR-423-5p	<0.01	83.45
miR-107	<0.01	82.73
miR-15b-5p	<0.01	77.61
miR-7977	<0.01	76.94
miR-106b-5p	<0.01	76.82
miR-500a-3p	<0.01	75.36
miR-451a	<0.01	75.3
miR-324-5p	<0.01	74.34
miR-130b-3p	<0.01	74.21
miR-24-3p	<0.01	72.13
miR-181a-5p	<0.01	63.78
miR-378i	<0.01	63.21
miR-151a-5p	<0.01	61.52
miR-192-5p	<0.01	57.1
miR-7641	<0.01	55.1
miR-652-3p	<0.01	54.87
miR-320d	<0.01	54.84
let-7b-5p	<0.01	51.96
miR-550a-3p	<0.01	51.43
miR-151b	<0.01	51.04
miR-4306	<0.01	50.5
miR-30d-5p	<0.01	50.13

miR-29b-3p	0.01	50.43
miR-505-5p	0.01	50.32

Abbreviations: abs, absolute; corr, correlation; FC, fold change.

Hsa-miR-148a-3p are statistically and differentially expressed in the CCHF patient group relative to the controls based on the microarray analysis. The miR-148a controls Iκβ-β expression, and the expression of nuclear factor-κB (NF-κB), interleukin 6 (IL-6), Tumor necrosis factor α (TNFα), IL-12, interferon-β (IFN-β), Matrix metallopeptidase (MMP), MMP14, and MMP16 etc. [343] A consistent silencing of miRNA-148a expression in CCHF patients is related to profound cytokine release. The CCHF pathogenesis occurred using a logarithmic and exponential upregulation in proinflammatory cytokine release and this situation follows a vascular collapse and endotheliosis in CCHF patients. Moreover, hsa-miR-126-3p expression reduces nearly 45-fold. The miR-126-3p is involved in type-I interferon production and alter the pathophysiology of the CCHF patients' immune response.[344] Three important miRNAs, miR-18a-5p, miR-27a-3p, and miR-590-5p are inhibited strongly in CCHF patients relative to the controls. Moreover, CCHF related MiR-18a-5p directs the transforming growth factor β (TGF-β) receptor, Notch2 pathway E2F transcription factor, and the interferon regulatory factor (IRF) in the signaling pathway.

Five miRNAs miR-1268a, miR-3940-5p, miR-937- 5p, miR-1249-3p, and miR-296-5p are overexpressed consistently relative to controls. Entire 65 miRNAs are inhibited constantly in fatal individuals relative to the control population. The miR-101-3p, miR-144-3p, miR-126- 3p, and miR-148a-3p reduce their expression in both patients and fatal cases. (**Table 6**) Some of these silent miRNAs are essential to their inflammatory response during viral infections. For example, miR-142- 5p, miR-15b-5p, miR-223, miR-532-5p, miR-107, miR-20b-5p, and miR-106b-5p. MiR-142-5p control TGF-β signaling pathway via targeting SMAD3.[345]

In comparison between fatal and nonfatal groups, 66 miRNAs are modulated strongly. Some of these miRNAs are related to the fatal condition of the CCHF patients. Among them miR-195, miR- 23b, miR-15, miR-140-3p, miR-16-5p, miR-150-5p, and miR-185-5p are vital. The miR-195, miR-23b, and miR-15 families are efficient for inflammatory response due to numerous gene targeting controlled using NF-κB signaling. [346] On the other hand, proinflammatory cytokine liberation is momentous for CCHF pathogenesis. It is observed that fatally infected CCHF patients carry a high amount of IFN-γ, TNFα, and IL-10. CCHF patients produce high serum levels of TNFα, soluble TNF-R, IL-6, and IL-10. [347] Further TNFα and IL-6 expression are related to disseminated intravascular coagulation and as a fatal outcome. Excessive serum TNFα levels of CCHF patients are linked to induced disease severity, while IL-6 levels are elevated in all CCHF patients. High serum TNFα levels are also associated with disease severity. CCHFV inserts in the host cells via the multivesicular body and is essential for the endosomal sorting complex responsible for transport regulators. Since immune response in the CCHF disease occurs after interaction between signaling proteins, these pathways are related to differently expressed miRNAs. They also have been confirmed to have a strong impact on CCHF pathogenesis.

Consequently, multiple miRNAs are associated with the inflammatory response of the host and proinflammatory cytokines production. it is determined that the miRNAs related to a CCHF fatal outcome are linked to excessive cytokine production. The miRNAs identification in CCHF patients will lead to therapeutic targets determination in diseases. Two important miRNAs miR-1908-5p and miR-939 are overexpressed strongly in fatal individuals relative to their controls. These miRNAs are associated with various genes that open an essential role in the inflammatory response. Indeed, miRNA identification in CCHF patients will be determined to use for potential therapeutic targets in CCHF diseases.

Rift Valley fever

Rift Valley fever virus (RVFV; *Phlebovirus*, *Bunyaviridae*) is an emerging and rapidly spread zoonotic mosquito-borne pathogen of high importance for human and animal health in Africa. Successful strategies of governing RVFV-vector interactions enhance mosquito vectors transmission and a by its mosquito vectors and a better understanding of human and veterinary disease mechanisms.

Rift Valley fever virus (RVFV) infection causes high fever, anorexia, and acute deaths in young ruminants and abortions in pregnant animals. [348-350] In humans, symptoms are not so strong but infection converts to retinitis, encephalitis, and hemorrhagic fevers that ultimately turn into nearly 1% fatality.[351-353] RVFV causes at least 30 different mosquito bites. Most of them lead to the spreading of RVFV into new areas via virus transmission.[354] The interacting relationship between the vector and RVFV remains obscure due to the wide variety of vectors and the paucity of genetic tools.

RVFV has a tripartite negative-sense or ambisense RNA genome. The large (L) segment has RNA-dependent RNA polymerase sequences. The medium (M) segment encodes the nonstructural protein NSm and envelopes glycoproteins Gn and Gc. This segment also synthesizes a 78-kDa protein and a 14-kDa protein generated from alternative start codons. [355] The small (S) segment synthesizes the nucleocapsid protein (N) and the nonstructural protein (NSs). These two proteins are synthesized in an ambisense manner where the N protein is derived from transcribed mRNA from the genomic RNA, while NSs protein is synthesized from an mRNA derived from the antigenomic S RNA.[356, 357]

The activation and contribution of small-RNA pathways are analyzed in controlling RVFV infection for therapy development initiatives in multiple insect systems. Various RVFV-specific small RNAs are generated in response to RVFV infection in multiple mosquito vector species, mosquito-derived cells. These small RNAs including microRNA's function antivirally for silencing key proteins of the RVFV viruses. The small-RNA pathways for RVFV infection in different insects are compared. Two initiative key markers of an antiviral RNAi response of RVFV virus occurred: (i) by the production of viral small RNAs and by (ii) increased viral replication following silencing of small-RNA pathway mediator proteins.

An infection with RVFV strain MP12 of *A. aegypti*-derived Aag 2 cells and subsequently sequencing identifies a strong enrichment of 21-nt to 22-nt fraction most likely corresponds to viral siRNAs or microRNAs (miRNAs), whereas the 27-nt to 30-nt fraction represents viral pi-RNAs. The ViRNAs are derived from all three genome segments, specially, nucleotide matching with the M segment sequence and followed by the S segment and the L segment. They are complementary to the viral genome and antigenome in equal ratios.

RVFV-related viRNAs including miRNAs (21 nt) and pi-RNA-like small RNAs (24 to 29 nt) are available in infected *A. aegypti* mosquitoes and *A. vexans* mosquitoes. The existence of RVFV-mediated viRNAs reads is equal to or marginally greater than that of pi-RNA-like small RNAs in both *Aedes* mosquito species. viRNA including miRNA distribution patterns in both *Aedes* species strongly enriches reads for the M segment and S segment and low read numbers for the L segment. However, pi-RNA signature is noticed in fragment sequence logos of all three mosquito species (**Figure 10A to C**) and 5′ end overlap of genome and antigenome 28-nt small RNAs. (**Figure 10D to F**)

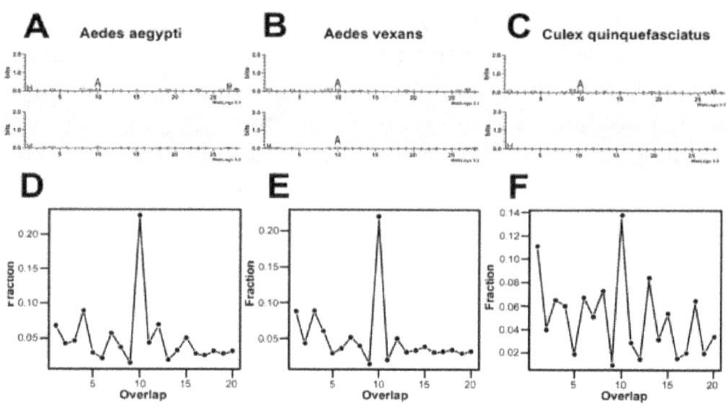

Figure-10. RVFV-related pi-RNA-like small RNAs (28 nt) are produced from infected mosquitoes. (A to C). *A. aegypti, A. vexans* and *C. quinquefasciatus* data are noted in panel A, B and C. (D to F) The panels demonstrate 5' end overlapping of the opposite viral small RNA (28-nt reads. The x axis represnts the overlapping nucleotide sequences between the antigenomic and genomic sequences; the y axis shows the mean values of overlapping reads. *A. aegypti, A. vexans* and *C. quinquefasciatus* data are presented in panel D-F. (Loosely Adapted, Redraw and Modify using skeleton from Dietrich et al., 2017)

Dengue and Severe Dengue

Dengue virus (DENV) is causative and etiological agent of dengue fever. Several Dengue basically causes a fatal disease that has no effective antiviral agent or vaccine except Dengvaxia, a low-profile vaccine against serotypes 1 and 2. Few miRNAs evade the host immune response against different viruses suppressing the IFN-α/β signaling, while others overexpress IFN-α/β induction and reduce the viral vigor. The experiences and knowledge of miRNA expressions and inhibitory effects in dengue infections help to develop of miRNA-guided therapeutics using miRNA antagonists or miRNA mimics. Truly, miRNAs prevent DENV replication by directly binding to the viral genome. Other miRNAs act as host factor modifications. In some cases, miRNAs alter immunopathogenesis. The efficiency of miRNA-based drug formulations and the success of delivery holds a great promise for the therapeutic applications.

miRNAs contribution to DENV infection

Modification of essential catalytic RNAi components allows activating DENV replication [358], and other virus infections in mammalian cells. The miRNA profiles in the blood of dengue patients produce a significant number of new miRNAs that change their expression during dengue infection. A profile for 348 differentially expressed miRNAs is found in dengue patients. Among them only 12 miRNAs are modified in a cohort manner during acute dengue infection. On the other hand, 17 miRNAs gradually change their expression from mild dengue to severe dengue-related complications. The levels of miR-24-1-5p, miR-512-5p and miR-4640-3p are altered in mild dengue together with multiple liver complications. While miR-383 is overexpressed markedly during mild dengue relative severe dengue and fluid accumulation. In few cases, 12 miRNAs are silent and 41 are overexpressed in DENV-1 patients' sera relative to healthy controls. Amongst them, hsa-miR-21-5p, hsa-miR-146a-5p, hsa-miR-590-5p, hsa-miR-188-5p, and hsa-miR-152-3p acts as promising non-invasive molecular markers for isolating DENV infection. miR-21-5p and miR146a-5p are associated with inflammation and cell proliferation. They are supposed to be used as strong sensitivity and specificity indicators of DENV infection. These findings indicate that miRNAs are the potential diagnostic markers for DENV infections. [359]

It is true that a wide range of host miRNA disruption is an important pathogenic strategy of viruses to neutralize their host response. Several viruses produce viral suppressors of RNAi proteins that possibly inhibit miRNA-governed silencing of the host RNAi machinery. In DENV, viral RNAi suppressors are effective for reducing host RNAi response and uncommon microRNA-like small RNAs are produced by DENV RNA viruses and other viruses. [360-362] A miRNA-like viral small RNA DENV-vsRNA-5 is involved in DENV replicating autoregulation directing the non-structural DENV protein 1 gene. It inhibits DENV replication.

Targeted miRNAs prevent DENV replication

The majority of miRNAs are involved in affecting DENV replication by interacting with the viral genome sequences. The miRNAs bind complementarily to the 5'UTR or 3'UTR of DENV genome and eventually prevent DENV replication. The miR-548 g-3p is the first miRNA that blocks DENV multiplication by targeted binding to the viral genome.[363] In the study, miR-548 g-

3p I conjugated with the Stem Loop A (SLA) promoter in the 5'UTR, a key element to DENV RNA synthesis and replication. It prevents DENV-1-4 replication. This suppressing effect occurs due to interactions between miR-548 g-3p and SLA that physically inhibit and disrupt the interaction between SLA promoter and NS5, a DENV protein carrying a C-terminal RNA-dependent RNA polymerase.

Three miRNAs miR-133a, miR-484 and miR-744 challenge and inhibit the DENV replication interacting with the 3'UTR of the DENV RNA. It is evident that the elements of the 3' stem-loop are responsible for genome circularization and viral viability.[364, 365] In fact, consistent upregulation of these three miRNAs miR-133a, miR-484 and miR-744 prevent DENV replication in Vero cells. In addition, miR-252 enhances its expression in a DENV-2 (strain New Guinea C) infection model. It also reduces DENV replication interacting with the DENV-2 E gene.[366] The miRNA-directed E protein has an essential role for cell attachment and viral entry.[367] Involvement of miR-252 mimics blocks E protein translation in humans that implicate it as an antiviral therapy without any minimal off-target effects.

miRNAs control host factors for DENV replication

Several miRNAs control DENV replication via a change in host factors or immune response. These indirect effects including an alteration of host cellular transcript lead to one or a few steps in the viral cycle. The differential expression of host receptors controls the entry of the virus, which includes changes in tropism and vital cofactors of replication complexes or translation. They also reduce or enhance viral replication and viral protein production. In addition, miRNAs are also activated or reduced differential cell responses during infection like immune response or defense mechanisms.[368]

An excessive expression of let-7c protects infected host cells from oxidative stress and inflammation response during DENV infection. [369] Let-7c targets directly the Basic Leucine Zipper Transcription Factor-1 (BACH1), a bolt inhibitor of the anti-inflammatory and antioxidant protein Heme Oxygenase 1 (HO-1). It also prevents both DENV-2 (strain New Guinea C) and DENV-4 (strain V3361–1956) infections. Due to its anti-inflammatory and antioxidant participation, let-7c protects the host from the viral-mediated excessive production of reactive oxygen and inflammatory response that is harmful to the host. [370-374] Further, a study demonstrates the antiviral effect of HO-1 against DENV.[375] The activation of HO-1 prevents DENV replication in host cells and prolongs DENV-2 infected disease and lethality. The HO-1-controlling antiviral effect further restores antiviral IFN response along with preventing DENV protease activity using biliverdin, a by-product of HO-1-catalyzed destruction of heme.

At the time of dengue infection, the predominant expression of miR-30e* inhibits DENV replication by activating the host antiviral immune response sensitivity.[376] When it is not activated, NF-κB is confined and floated inside the cytoplasm using IκBα, a consistent inverse regulator. The negative regulator once induces the NF-κB translocation into the nucleus during its degradation.[377] The miR-30e* exerts the antiviral effect via targeting the 3'UTR of IκBα instantly at the time of DENV infection that causes IκBα destruction. This provides sequential NF-κB pathway induction. It subsequently promotes the IFN-β expression and ISGs including OAS1, MxA and IFITM1 inhibition. [378] Four distinct miR-34 families including miR-34a, miR-34c, miR-449a and miR-449b conduct innate modulating roles via a DENV during infection.[378] An inhibition in DENV-2 (strain New Guinea C) replication is found during miR-34 family members induction. In this case the antiviral effect is involved due to Wnt signaling reduction. The Wnt signaling is responsible for silencing glycogen synthase kinase 3 beta (GSK3β) phosphorylation that controls the IFN signaling pathway directly. As a consequence, the miR-34 family activates the type I IFN and ISGs via Wnt signaling reduction.[378] Therefore, both miR-30e* and miR-34 family of miRNAs act as positive controllers for the antiviral immune response of cells during DENV infections. Two miRNAs overexpress the type I IFN production suggesting their use as therapeutic agents for protecting DENV infections *in vivo*.

Despite host immune response, the antiviral miRNA roles perturb other host factors. The miR-223 reduces DENV-2 (strain TR1751) replication. Its antiviral effect silences the expression of microtubule-destabilizing protein, stathmin1 (STMN1), a key microtubule controlling protein for regulating microtubule dynamics [379] The direct link of how STMN1 influences DENV-2 replication is still not clear. Nevertheless, based on an earlier study an intact microtubule network influences STMN1 sequestration. It is necessary for HCV for initiating an infection. In fact, STMN1 is playing an identical role in initiating restricted microtubule dynamics related to DENV infection.[380] Moreover, miR-133a silences DENV multiplication via targeting the polypyrimidine tract-binding protein (PTB). PTB is an essential host cellular protein that functions as RNA helicase via binding of conserved sequence 1 and long stem-loop structures of the 3'UTR of DENV.[381, 382]

A miRNA has a medium but prominent antiviral property against DENV. Here miRNA-3614-5p overexpresses in DENV-negative cells and its overexpression reduces DENV infectiveness. This miRNA-3614-5p silences DENV infectivity. The miR-3614-5p targets Adenosine deaminase acting on RNA 1 (ADAR1) and induces DENV infectivity at the onset of post-infection. Thus, it demonstrates that the contribution of human miRNAs is conducted by building the network of interactions between DENV and its human host cells.[383]

Conversely, miRNAs alter various host factors or immune responses for viral replication. For instance miR-146a is isolated for a proviral effect during dengue infection.[384] Indeed, this miRNA is synthesized profusely in the primary and transformed monocytes (THP-1 cells). These monocytes are the major and potent target cells for attaching DENV, thus implying a critical role of miR-146a for initiating an infection. Further, miR-146a acts as an enhancer for DENV-2 (strain New Guinea C) replication by reducing the host IFN-β production for interacting tumor necrosis factor receptor (TNFR)-associated factor 6 (TRAF6). Truly, a miR-146a repressor produces the inverse effect of the virus-guided immune response. Overexpression of miR-146a sign blocked DENV-2 (strain New Guinea C) markedly via stimulating autophagy and LNA-guided miR-146a silencing counteracts such effects.[385] The miR-146a antagonist might restore the host IFN activity against DENV. It is intimately related to dengue infection prominently excessive inflammation, the sign for fatal severe dengue. In another study, miR-21 is consistently upregulated on DENV-2 (strain 16,681) infection specifically facilitated for DENV-2 replication [386] similar to the miR-146a function.[384] An anti-miR-21 (AMO-21) leads a marked reduction in DENV-2 production. It suggests that miR-21 plays an important role in DENV-2 replication. However, the exact mechanism of how miR-21 activates DENV replication is still not known. Most probably miR-21 might interact directly with the NS1 protein sequence of the DENV-2 genome.[387]

miRNAs are mostly linked to DENV replication, either using direct interaction with the viral genome or via changes of host mediators instrumental for DENV multiplication. A few studies demonstrate that the DENV viral load had a proportional correlation with dengue severity and fatality.[388-390] Another study suggests that the dengue viral load is not directly related to DHF and dengue fatality. [367, 369, 370] Basically, dengue severity is predominantly linked to a systemic inflammatory response syndrome, commonly referred to as "cytokine storm." It occurs due to an imbalance between the pro-inflammatory and anti-inflammatory agents. However, whether the immune responses are skewed and related to either type 1 or type 2 is mainly paradoxical.[391] Pro-inflammatory cytokines TNF-α induced in endothelial cells and enhanced vascular permeability. Sera from dengue patients contain apoptosis-activated factors important for vascular damage.[392] Overexpression of T lymphocytes has a down regulatory effect on DENV infection. Based on diachronic T cell responses and clinical detection of severe dengue, the cytokine profiles during dengue infection open to play a potential role for the huge release of pro-inflammatory factors mainly monocytes, macrophages and other non-T-lymphocytes. Therefore, T cell induction is responsible for cytokine storm of a subset of patients.[391]

Certain studies show the contribution of miRNAs on cytokine dysregulation at the onset of dengue infection.[393] The miR-150 is overexpressed in DHF patients that regulate SOCS1 negatively. It is a distinct repressor of a cytokine signaling protein that expresses aberrantly for allowing cytokine dysregulation. They demonstrate an intensive reciprocal correlation between SOCS1 production and dengue severity. Considering a strong link between SOCS1 and miR-150, it is suggesting that restriction of overexpression of miR-150 during dengue infections restore SOCS1 expression and eventually revive the strong dysregulation in cytokine expression.[393] Indeed, SOCS1 is involved in immune regulation during VSV infection.[394] The SOCS1 inhibition using miR-155 induces type I IFN signaling, thereby reducing viral replication. This notion establishes the changes of particular miRNAs, non-optimal immune responses in dengue patients are invariably controlled.

To test a possible link between differentially produce miRNAs and strongly activated cytokines in DENV-2 (strain New Guinea C) infection,[395] the silencing of miR-106b, miR-20a and miR-30b during DENV-2 infection reduces target gene production. It ultimately activates pro-inflammatory cytokines. The miR-106b silencing allows downregulation of CCL5 . Therefore, it induces secretion of this cytokine. This study allows potential therapeutic application for facilitating the massive immune response of the host during DENV infection.[395]

Eventually the host miRNAs work as either proviral or antiviral via several mechanisms (**Table 7**). Based on the excessive existence of host miRNAs the proviral or antiviral effects can fluctuate. Overexpression of the proviral miR-146a promotes viral replication interacting with TRAF6 and inhibiting IFN-β production.[384] Dengue severity is related to the abrupt onset of the innate host response. This innate antiviral mechanism guided using IFN reveals a critical pathway of host defense during restricted viral replication. Apart from different miRNAs expressed at the onset of DENV infection are related to the severity of dengue via cytokine expressions. It demonstrates that cytokine storms promote extensive plasma leakage found in DHF and DSS. Many miRNAs control cytokine production. The miR-150 upregulation is linked to DHF together with SOCS1 silencing, a negative controller of several cytokines.[393] miR-let-7e also directs mRNAs of IL-6 and CCL3 while miR-106b, miR-451, and miR-4279 are interacting mRNAs of CCL5 and CXCL1. The differential expression of miRNAs leads to cytokine silencing, while different additional miRNAs interact with the epigenetic regulator of cytokine expression.[395] However, further molecular mechanisms can be conducted that activate cytokine storms during DENV infection as a crucial gap for demonstrating the limited outcomes of dengue diseases.

Control for dengue miRNAs delivery

Recently the role of miRNAs during dengue infection is limited by inadequate characterization of the miRNA-linked to molecular pathways in dengue infection, replication and host immunity.[396] The detection of miRNA targets within a living organism and in host defense processes with many steps of molecular and cellular elements is a major challenge for detecting miRNA interactions within host defense systems. Major challenges in miRNA-guided therapies are the off-target effects. A single miRNA can influence interaction with multiple mRNAs. Different pathophysiological signs found by changing their roles are linked to the levels of various target mRNAs.[396] Delivery of miRNAs is one of the critical issues that restrict miRNA mimics and silencers delivering to the infected and inflamed tissues. This limitation must be overcome as it is critical for disturbing the normal functioning of the surrounding tissues. Further, gradual development of modified miRNAs and suitable carriers are needed to combat these problems.[397] However, recently the delivery of most miRNAs has only been focused on the pathological tissues and organs of interest. Different side effects and toxicities are ignored. However, this must be addressed extensively for safety that is a crucial concern for new drug discovery.

Moreover, miRNA also induces a non-specific innate immune response that is linked to the structure, length, chemical modification, cellular localization and concentration based on base sequence. Immune activation enhanced by miRNAs produces a negative effect on the health. Thus, a few effective strategies are generated for miRNA silencing that induced immune responses using locked nucleotides (LNAs). The use of LNA-mediated miR-146a silencer against EV-A71 showed that it combats death in mice, restarting the type I IFN production during the onset of viral infection, the prophylactic introduction of LNA antagomiR-146a is highly efficient. Thus effective *in vivo* use of LNA for treatment against EV-A71 infection is found. In a few cases miR-146a in DENV is used for evaluating the therapeutic potential *in vivo.*

Table -7. Multiple miRNAs responsible for DENV infection

miRNA	Virus types of	Target	Effect
miR-548 g-3p	DENV-1 (strain Hawaii), DENV-2 (strain New Guinea C),DENV-3 (Philippine H87 Strain) and DENV-4 (GZ/ 9809/2012)	Direct: viral 5′ UTR	Antiviral
miR-133a	DENV-1, DENV-2 (strain New Guinea C) and DENV-4	Direct: viral 3′ UTR	Antiviral
miR-484	DENV-1, DENV-2 (strain New Guinea C) and DENV-4	Direct: viral 3′ UTR	Antiviral
miR-744	DENV-1, DENV-2 (strain New Guinea C) and DENV-4	Direct: viral 3′ UTR	Antiviral
miR-252	DENV-2 (strain New Guinea C)	Direct : viral gene E	Antiviral
Let-7c	DENV-2 (strain New Guinea C) and DENV4 (strain V3361–1956)	Indirect: BACH1	Antiviral
miR-30e*	DENV-1 (strain Hawaii), DENV-2 Indirect: (strain New Guinea C) and DENV-3 (strain H241)	IκBα	Antiviral
miR-34 family (miR-34a, miR-34c, miR-449a and miR-449b)	DENV-2 (strain New Guinea C)	Indirect: Wnt pathway	Antiviral
miR-223	DENV-2 (strain TR1751)	Indirect: STMN1 mRNA	Antiviral

miR-3614-5p	DENV-2 (strain 16,681)	Indirect: ADAR1 mRNA	Antiviral
miR-146a	DENV-2 (strain New Guinea C)	Indirect: TRAF6	Proviral
miR-21	DENV-2 (strain 16,681)	Direct: NS1 sequence	Proviral
mir-150	DENV-2 (strain New Guinea C)	Downregulated SOCS1 resulting [81] in lower IFN-γ production	Biomarker for severe disease

Abbreviations: DENV Dengue virus, UTR Untranslated region, BACH1 Basic leucine zipper transcription factor 1, STMN1 Stathmin1, ADAR1 Adenosine deaminase acting on RNA 1, NS1 Non-structural protein 1, SOCS1 Suppressor of cytokine signaling, IFN Interferon, TRAF6 Tumor necrosis factor receptor (TNFR)-associated factor 6

Conclusions

In several infectious diseases, conventional diagnostic technologies are inappropriate and inadequate. They mainly depend on the symptoms of pathogen-specific antibodies. They cannot be recognized until the pathogen is replicated sufficiently to a desirable titer. Different cutting-edge biomarkers are sorely required to underpin the next generation of diagnostic platforms. MicroRNAs are supposed to be the most potent biomarkers of infection. The miRNAs are frequently modulated in biofluids in the presence of infections with bacteria, parasites, viruses, and even prions. Their recognition as diagnostic or prognostic markers are not facing any new challenges, however, with robust ideal detection methods, their use alleviates the shortcomings of current experiments for improving patient outcomes. In addition, appropriate detection of predictive miRNAs, the abrupt progress in miRNA sensing technologies and profuse productions, commercial present for miRNA-guided diagnostics proclaim that miRNAs are the most inevitable source for powerful early tools in diagnostic, prognostic and even therapeutic strategies for infectious diseases.

The miRNA profiling and analysis is a promising tool for early biomarker detection and sub-classification to cancer and rare diseases therapy. Recently, circulating miRNAs harbors in serum/plasma emerge as revolutionary sources of biomarkers for diagnosis. In basic experiments answering the questions about tissue-specific functions and targets selection, dynamic time for active expression and evolution was explored. The existence of early invasive tests and monitoring for Rare Diseases reduce worldwide burdens. The identification of active microRNA and synthesis of antimiR small active components are extremely important designing the therapy ingredients which will govern treatment and prevention of Rare Diseases in the future era, Further, their functional roles and anti-miRNAs production up new era for opportunities of therapeutic agent generation. Although the use of miRNAs in Rare Diseases sets a new beginning for future studies. Finally, a tireless drive will apply quantification and normalization strategies of disease-related miRNA for far-reaching clinical practice.

MicroRNAs are an important interconnector between host response and pathogens. Their pleiotropic roles between host response and infectious disease are very interesting in successive generations. It recognizes key genes and their pathways that must be altered, induced, enhanced, repressed, or silenced to facilitate an effective immune response. Clinical applications of miRNAs almost touch the sky. The viral vaccines are promoted via miRNA target sequences incorporation during the preclinical stage. The use of miRNA biomarkers of infection holds a strong promise. The miRNAs have multi-gene targets and therefore any vaccines or treatments that harness miRNAs may reproduce off-target effects compromising safety. A large initiative for the use of miRNAs is responsible for diagnostic and therapeutic potential.

Based on the growing miRNA literatures, the usefulness for new miRNA therapeutics, diagnostics/prognostics, and vaccines comes closer tangibly. Fortunately, the Translating insights of miRNA studies induce patient care and their improvements that is the most critical next step in the future.

References

References: chapter wise are cited before the Index at the end of the Book. (1-397)

Chapter VIII

Early Detection in Cardiovascular diseases and miRNA therapy

Abstract

In a broad sense, cardiovascular diseases (CVD) are one of the inclusive leaders for morbidity and mortality in the world. Conventionally, reasonable progress has occurred at the diagnosis, treatment, and prognosis of CVD. However, more advanced diagnostic biomarkers are essential for the early detection of eliminating disease events. Recently, endogenous, stable, reliable single-stranded, tiny, non-coding circulating miRNAs (miRNAs) are utilized as early diagnostic biomarkers for CVD essentially. As an overview of heart failure, acute myocardial infarction, arrhythmias pulmonary hypertension, and separately the coronary artery diseases are presented for circulating miRNAs detection. The potential challenges and recommendations for their use during the diagnosis of CVD are included.

Coronary artery blockage is the most common member in the larger umbrella for hearing-related diseases known as cardiovascular diseases (CVD). Participation of circulatory microRNAs in CVD pathogenesis is reported either in a beneficial or detrimental section. Selective circulating microRNAs can also be utilized as potent biomarkers for the early detection of coronary artery disease. Particularly, microRNA-133a functions as an ideal biomarker due to its contribution in detecting early diagnosis, severity estimation, and prognosis. The combination of other circulating microRNAs also appropriate to estimate the power of coronary artery disease signature relative to single microRNAs.

As documented in the chapter, different noncoding RNAs use as classical instruments for physiological and pathophysiological functions. A wide variety of experiments suggest that they are responsible for therapeutic components. MicroRNAs exhibit fine-tuned post-transcriptionally control for gene expression based on their association with the 3′untranslated sequences of a potential target mRNA and ultimately preventing protein translation or virtually its degradation. In this chapter, the microRNAs a promising classical medicinal target for cardiovascular treatments including heart failure, acute myocardial infarction, fibrosis, and atherosclerosis are described. Further insights into available tools for delicate silencing or prompt upregulation of miRNA devices are discussed as a therapeutic challenge. Multiple protocols for improving miRNA therapeutics and dramatic reduction of toxicity to a trace level using induction of tissue specificity or cellular uptake are reported.

Introduction

Cardiovascular disease (CVD) is one of the most common lethal causes of death in the world, mainly in developed countries. For this consequence, there is an essential requirement to detect early prognostic and diagnostic biomarkers for protecting against CVD. With this perception, microRNAs (miRNAs) are used as ideal early markers of heart disease. The miRNAs are pivotal regulators of cardiovascular function and open up a new era for all aspects of diagnostic and treatment plans against cardiovascular biology. [1-4] An overview of the biology, therapeutic, and diagnostic potential and the limitations of miRNAs are found in coronary arteries transporting blood to the heart muscle. Two main arteries: the right and left coronary arteries and two branches, the left anterior descending artery and the circumflex artery mediate the blood supply (1) Normal coronary artery has three well-defined layers: the intima, media, and adventitia identical to other arteries. These three layers are separated by elastin layers. Internal elastic lamina places intima between media and external elastic lamina sandwiches between media from the adventitia (2). Moreover, Coronary artery disease (CAD) is one of the dominant causes of cardiovascular deaths (CVD) therefore a separate section is furnished for their description in chapter III. In 2020, this disease estimates for 11.1 million patient deaths globally. The wide clinical spectrum of CAD extends from stable angina pectoris (stable CAD) to acute coronary syndromes (ACS) that involve fragile angina (unstable CAD) and myocardial infarction. The myocardial infarction (MI) is classified into two categories ST-segment elevated MI (STEMI) and non-ST segment elevated MI (NSTEMI). The arterial occultation STEMI is characterized by a total occultation of epicardial coronary blood vessels and elevated ST in the electrocardiogram, while NSTEMI is known by a drastic coronary artery narrowing. However, they are responsible for the necrosis of myocardial cells. Consequently, the contribution of circulating miRNAs in the early detection, diagnosis, deuteriation of CAD, restenosis, and their important role in prognosis marker is also discussed.

Despite a lion share of the genome being transcribed, only a trace a ≈2% are protein-coding regions. The remaining non-coding parts are made up of small and long noncoding RNAs (lncRNAs). Among small noncoding RNAs, the predominance of circulatory microRNAs (miRNAs) plays a critical role in many different cellular functions including cell survival, differentiation, and proliferation. (**Figure 1**) A single miRNA can control a hundred mRNA expressions. As a result, they formulate different diversified biological roles in various tissues. The miRNAs lead to the most promising agents for therapy development. Any miRNA can be required for many stable diseases like cardiovascular or metabolic disorders.[5, 6] Even miRNAs fine-tune sensitive transcription and post-transcriptional gene control for modulating splicing or by sponging miRNAs. In recent days, two distinct approaches are evolved for modifying miRNA modulation. Its upregulation occurs via using miRNA mimics or viral vectors or silence selectively by using antisense oligonucleotides (ASOs) or by genetic silencing.

Multiple promises and challenges are applicable for miRNAs therapeutic targets in cardiovascular disease treatments. An overview can be focused on regarding the significant impact of miRNAs in controlling almost all cardiovascular disease symptoms. The toxicological obstruction of miRNA systemic interference and prevention against unwanted side effects via altered delivery systems of RNA therapeutics are reported.

miRNA in early diagnosis and prognosis

The common circulatory miRNAs from human blood are involved in heart, lung, liver, and kidney, diseases.[7, 8] The circulatory miRNAs inside different micro vesicles like exosomes, microparticles, and apoptotic bodies block RNase degradation and thus protect within circulation. **(Figure 1)**[7] Their stable features in the blood lead to use them as diagnosis and prognosis biomarkers of multiple diseases including CVD.

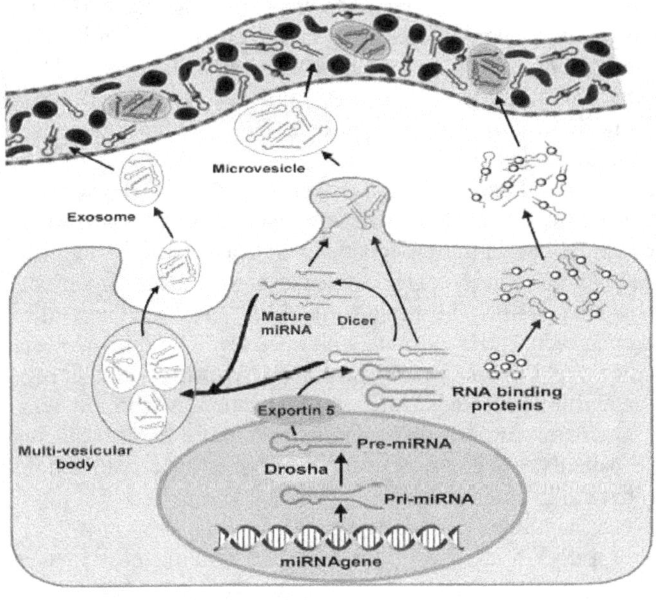

Figure-1. Formation of circulatory miRNAs. The miRNAs producing genes are transcribed systematically within the t nucleus as pri-miRNAs with extended 5'-caps and 3'-polyA tails. Drosha enzyme trims the 5' cap and 3' polyA tail and produces pre- miRNAs that are released from the nucleus using Exportin 5 to the cell cytoplasm. The cytoplasm Dicer cleaves the precursor pre-miRNA to newly converted mature miRNAs (**at the bottom).** Pre-miRNA and mature miRNA can **(A)** bind with exosomes and form multi-vesicular bodies and liberate directly from the cell, **(B)** accompanied into micro vesicles in a tight package **(C)** conjugate to RNA-binding proteins and are ultimately liberated. Pre-miRNA and mature miRNAs are accepted into the bloodstream by endocytosis for their distribution. They also bind to receptors or using dominant membrane fusion (**at the top**). (Loosely Adapted, Redraw and Modify using skeleton from Shan-shan Zhou *et al.*, 2018)

===

Cardiovascular diseases

Heart failure

Heart failure (HF) has an abrupt symptom of sudden clinical disease. Here, the heart fails to generate sufficient circulatory forces for compensating metabolic requirements.[9] Heart failure has initially explained of CVD and cardiac injury. It carries rare but critical etiologies like cardiomyopathies, valvular heart disease, prolonged arrhythmias, myocarditis, infections, and reaction against cardiotoxic drugs, etc.[10] The distribution and the amount of many circulating miRNAs are the unique biomarkers of HF. A pile of evidence found that miRNAs in circulatory blood-lead to the development and progression of HF.

Nearly critical 30 circulating miRNAs modulation between normal and disease states is responsible for HF and comorbid pathologies. (**Figure 2, Table 1**) A trace level of a bunch of circulating miRNAs, miR-18a, miR-27a, miR-30e, miR-26b, miR-199a, miR-106a, and miR-652, are associated with HF patients. Similarly marked decrease in circulating miRNAs let-7i, miR-18b, miR-18a, miR-223, miR-301a, miR-652, and miR-423 are the major diagnosis agent for acute HF admission and are linked to a general risk of 180-day mortality.[11] Both overexpressed miR-21 and miR-1 reduces in symptomatic HF patients. Moreover, the reduction of miR-1 levels is inversely correlated with N-terminal pro-brain natriuretic peptide concentration in patients in Class II/III.[12] The miR-210 levels are also parallel to improved HF. It leads to the downregulation of B-type natriuretic peptide (BNP). The plasma miR- 210 [13], miR-126, and miR-423 levels are reduced markedly in HF patients with trace levels of miR-423 related to >1-year mortality. [14-16] Two separate critical miRNAs miR-1254 and miR-1306 are linked to a high risk of death and hospitalization in chronic HF patients.[17] Moreover, circulatory miR-1306 is parallel to harmful clinical outcomes related to acute HF patients.[18] In contrast, overexpression of miR-208b and miR-499 are related strongly to the high risk of HF or death.

When combined circulatory miRNAs miR-30c, miR-221, miR-328, miR-146a, or miR-375 are used alone or as part of a diagnostic panel, the early diagnostic performance of BNP is improved markedly. Further, a mixture of two or more miRNAs linked to BNP is increased significantly for distinguishing HF measurement with preserved ejection fraction from HF and decreased ejection fraction relative to using BNP alone.[19] Early diagnosis is related to the alteration of miRNA expression. It is compared to stable HF patients, individuals with advanced HF with left ventricular (LV) assist devices. The miRNAs; miR-208b, miR-208a, and miR-499; and myomiRs miR-1 and miR-133b.[20] miR-208b and miR-499 are liberated from the coronary sinus after cardioplegia. Their reperfusion increases markedly relative to what existed before surgery. [21] It is found that LV inverses modulation after cardiac resynchronization therapy is related to circulating miRNA modulations. An excess level of circulating miR-

26b, miR-145, miR-92a, miR-30e, and miR-29a relative to non-responders are observed. [22] The basal levels of circulating miR-30d identical to miR-1306 is linked to LV remodeling in reply to cardiac resynchronization therapy in advanced chronic HF patients and a 1-year long all-cause mortality in acute HF patients.[23, 24] These findings isolated many miRNAs as potential candidates for use as important biomarkers for HF. They also show valuable clinical inclusion. They deliver early important signs in regulating the progress of therapeutic interventions. **(Figure 2)**

Acute myocardial infarction

Acute myocardial infarction (AMI) is a principal reason for human morbidity and mortality. It causes coronary artery occlusion along with myocardial tissue damage.[25] Myocardial modifications after AMI, heart chamber dilation, and ventricular wall thinning is the multiple causes for apoptosis, fibrosis inside cells and tissue. [26] Despite percutaneous coronary intervention and surgical revascularization are responsible for clinical outcomes of AMI, ischemia/reperfusion injury improvement. They influence revascularization protocols that can burden clinical complications promoting excess cardiac injury. [27, 28] Numerous circulatory miRNAs miR-1, miR-133, miR-21, miR-29b, miR-192, miR-194, miR-34a, miR-208, miR-499, miR-423, miR-126, miR-134, miR-328, and miR-486 are overexpressed in AMI patients linked to ischemia-related HF, and miR-106, miR-197, and miR-223 are downregulated. (**Figure 2, Table 2**)[21, 29-32] The apparent temporal modification in miRNAs along with myocardial growth, fibrosis and viability are related to post-infarct volume LV structural remodeling. [33-35] Therefore, circulating miRNAs can be utilized for diagnostic biomarkers for AMI. The amount of serum miR-1 improves myocardial infarct size. [30] The miR-1 expression in post-AMI patients, is linked to (a) the total modulation in infarct volume, (b) follow a trend along with LV ejection fraction, and (c) is finally related to AMI mortality.[36, 37]

==

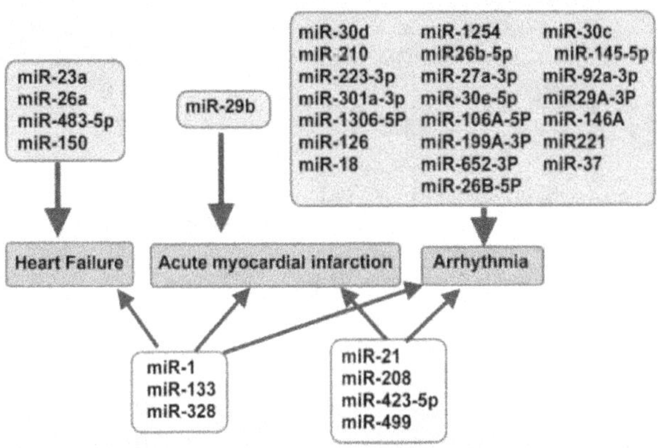

Figure-2. Selective miRNAs are linked to the diagnosis and prognosis of heart failure, acute myocardial infarction, and arrhythmia. miRNAs in violate colored boxes are related to a single pathology, while those in light, yellow-colored boxes with multiple pathologies. (Loosely Adapted, Redraw and Modify using skeleton from Shan-shan Zhou *et al.*, 2018)

==

==

Table-1 Early detected miRNAs in heart failure diagnosis

Sensitive miRNAs.	Potential purpose	Change in	Pathology
miR-1254	Diagnosis/Death/hospitalization	Up	CHF
miR-1306		Up	AHF
miR-30d	Diagnosis/response to CRT	Down	CHF
	Diagnosis/ Death	Down	AHF
miR-21	Diagnosis	Up	HF
miR-1		Down	
miR-210	Diagnosis	Up	HF, AHF
miR-126	Diagnosis/Death	Down	CHF
miR-423	Diagnosis/Death	Down	AHF
	Diagnosis	Up	HF
miR-1	Diagnosis	Up	HF, CHF
miR133a/b		Up	
miR208a/b		Up	
mir-499	Diagnosis/Predict: HF &Death	Up	MI
miR-18a	Diagnosis/Death	Down	AHF
miR18b		Down	
miR-223		Down	
miR301a	Predict: Death	Down	
miR-26b		Down	
miR-27a		Down	
miR-30e		Down	
miR-106a		Down	
miR-199a	Diagnosis	Down	
miR-652	Diagnosis/Death	Down	
miR-26b	Diagnosis/Predict: CRT	Up	HF
miR-145		Up	
miR-92a		Up	
miR-30e		Up	
miR-29a		Up	
miR-30c	Diagnosis	Down	HfpEF

miR-146a		Down	HfrEF
miR-221		Down	
miR-328		Down	
miR-37		Down	

HF, Heart Failure; CHF, Chronic HF; AHF, Acute HF; MI, Myocardial infarction; HfpEF Heart failure with preserved ejection fraction; HfrEF, Heart failure with reduced ejection fraction; CRT, Cardiac resynchronization therapy.

Table-2. Sensitive miRNAs in acute Myocardial infraction diagnosis

Effected miRNA	**Potential purpose**	**Change in regulation**	**Pathology**
miR-1	Diagnosis/Predict: LVEF	Up	AMI
	Diagnosis	Up	AMI
		Up	STMI
		Up	ACS
		Up	AMI
		Up	AMI
miR-21	Diagnosis/ Predict: LVEF	Up	AMI
	Diagnosis	Up	AMI
		Up	AMI
miR-29b	Diagnosis	Up	AMI
miR-133	Diagnosis/Death	Up	ACS
	Diagnosis	Up	STEMI
		Up	AMI
		Up	AMI
			AMI
miR-208	Diagnosis/ Predict: HF & Death	Up	AMI
	Diagnosis	Up	ACS
		Up	AMI
		Up	STEMI
miR-328	Diagnosis	Up	AMI
miR-499	Diagnosis/ Predict: AMI	Up	CABG
	Diagnosis	Up	STEMI
		Up	AMI
		Up	AMI
		Up	UA

			NSTEMI
miR-423	Diagnosis	Up	AMI
		Up	STEMI
		Up	AMI

AMI, Acute Myocardial Infarction; STEMI, ST-elevation Myocardial Infarction; ACS, Acute Coronary Syndrome; CABG, Coronary Artery Bypass Grafting; UA. Unstable Angina; NSTEMI. Non-ST elevation Myocardial Infarction; LVEF, Left Ventricular Ejection Fraction; HF Heart Failure.

The critical cardiac controller, improvement of skeletal muscle identical to miR-1, miR-133 are required in vascular smooth muscle cell biology.[38, 39] A bonding between miR-133 and AMI however is not found. AMI patients carry 4.4-fold increased miR-133 in plasma that returns to normal basal levels after 7 days relative to control. A further miR-133 expression is related to all-causal mortality markedly. [36, 40] Conversely, other studies show that no distinct difference in miR-133 plasma levels between AMI and healthy patients is found.[31] The amount of circulating miR-133 did not alter distinctly in AMI patients along with bradyarrhythmia or tachyarrhythmia and is not related to HF with the prediction of LV ejection fraction and BNP within 1-year post-AMI.[40, 41] An excess level of plasma miR-21 is found in Acute myocardial infarction patients relative to distinct healthy controls. The miR-21 is evolved to be a natural biomarker is predictive of LV changes after AMI. [29] Further miR-21 expression is related to multiple AMI traditional markers like creatine kinase-MB (CK-MB), creatine kinase (CK) and cardiac troponin I (cTnI), along with competitive diagnostic accuracy.[37, 42] The overexpression of circulating miR-208a is found in AMI patients that reach a peak 3 h after reperfusion along with fluctuating angina-patients.[43] The miR-208a upregulation is found along with the risk of mortality or HF within 6 months after the AMI.[36, 44] Despite miR-208b is not linked to the AMI clinical outcome after adjustment for cTnT separately, miR-208a expression is bonded with cTnI firmly and releases CK-MB from the infarcted area.[36, 43] The amount of cardiac myocyte-mediated miR-499 is overexpressed firmly along with cTnI in AMI patients for release from injured cardiomyocytes.[45]

The relative level of miR-1 or miR-208, miR-499 are the sensitive reliable AMI biomarkers of AMI, cTnI, and CK-MB.[46] The modified amount of circulating miR-499 is involved in unstable angina and non-ST elevation myocardial infarction (MI) in patients showing within 3 h of symptom onset. This supports serum miR-499 to open a role for strong novel biomarkers for diagnostic accuracy of acute coronary syndrome patients.[47] The reliability and specificity of miR-499 expression are increased at an excess level than cTnI demonstrating an excess risk factor for perioperative MI. These observations indicate that circulating miR-499 acts as an early biomarker.[48]

Significant overexpression of miR-423 is found at 1, 3, and 12 months after MI relative to basal levels.[49] The amount of miR-423 in AMI patients reduces

6, 12, and 24 h onset leading to percutaneous coronary intervention. Ultimately, the amount of miR-423 is competitive to the control group and reaches a basal level.[50] A correlation does not exist among miR-423 expression induces of LV function and changes echocardiographic parameters, the amount of cTn I or BNP; at any time-point during follow-up.[49-51] The miR-29b levels in plasma are related to the modification of infarct volume in post-AMI patients. Moreover, the miR-29b expression is related to the changes in LV end-diastolic volume over time.[37]

Despite the versatilities of many miRNA expressions are related to AMI. The peculiarities of AMI compared to comorbidities appears to be problematic. Multiple miRNAs along with AMI control non-cardiac protein expression. The muscle-specific miR-1 is effective for muscle damage. Therefore, it may be a routine muscle and tissue damage marker but not specific to AMI.[52-55] Among different miRNAs linked to AMI, miR-208a serve as a significant diagnostic biomarker. [56, 57] Initially, miR-208a exists in cardiomyocytes for controlling myosin heavy chain expression, Therefore, its heart-specific expression is influenced markedly in most AMI patients.[58] In fact, miR-208 is identified easily in AMI patients (<4 h post-AMI).[59] The myocardial infarction miR-208 is identified within 1 h of AMI. Its amount reduces after 3 h but does not exist significantly distinct from controls after 24 h.[59] These findings demonstrate that miR-208 serves as a biomarker for early AMI detection, but it is not reliable as a long-termed biomarker. Other miRNAs miR-499, miR-1, or miR-133 are used in this detection.

Arrhythmia

Arrhythmia is a collection of symptoms that alter the normal heartbeats dramatically including heartbeat irregularities (dysrhythmia), too slow (bradycardia), and too fast (tachycardia) beats. Modification of many circulating miRNAs around in arrhythmia. (**Figure 2, Table 3**) Atrial fibrillation (AF) is a severe and acute arrhythmia found in most CVDS patients and is an increased risk for HF. Atrial fibrillation leads to ventricular arrhythmia promotes a high risk of stroke or ultimately leads to death.[60] The occurrence of AF rate increases 5%–15% in people over 80 years old age. [61] Alterations of many miRNA expressions are possibly related to AF. The miR-29 silencing leads to mRNAs guided fibrosis-promoting proteins for contributing apoptosis.[62-64] The miR-208b overexpression is bound to human cardiac tissue from AF samples.[65] The post-operative AF risk is expected to increase by the miR483 presence in the serum.[66] Two circulatory mRNAs miR-23a and miR-26a are responsible for the post-operative AF development possibly.[67] Importantly, miRNAs are subject to control atrial modification using regulating Ca2+ channel protein expression. A regulator of cardiac hypertrophy related to marked miR-328 overexpression [68] is found in the atrial tissues of AF patients. The presence of miR-328 sequence isolates cardiac L type Ca+2 Channel subunits, CACNA1C and CACNB2, as its important targets. The miR-328 upregulation also establishes its role as an effective AF.[69, 70] The miR-499 leads to play an important role in the electrical modification by influencing small conductance expression of Ca2+-enhanced K+ channel isoform

3.[71] The miR-208b increased expression in HL-1 atrial myocytes or primary cardiomyocytes identified from chronic AF patients decreases the CACNA1C and CACNB2 function and the Sarco/endoplasmic reticulum Ca2+-ATPase (SERCA2).[72] Cardiomyocytes from AF patients overexpress miR-30d considerably with decreased levels of CACNA1C and K+-inwardly-rectifying channel, subfamily J3 (KCNJ3) mRNA, and protein. The miR-30 upregulation *in vitro* reduces KCNJ3 expression along with a synergistic decrease of acetylcholine-sensitive inward-rectifier K+ current, whereas miR-30d silencing increased KCNJ3 expression.[73]

Table-3. Effective miRNAs in arrhythmia

miRNA	Purpose	Pathology	Modulated regulation
miRNA 1	Diagnosis: SVT	TACH	Down
mIR-133	Diagnosis: VT	TACH	Up
miR-328	Diagnosis: AF		
	Diagnosis: AF	AF	Up
miR-23a	Diagnosis: POAF	POAF	Down
miR-26a	Diagnosis: POAF	POAF	Down
miR-483	Predict: POAF	CABG	Up
miR-150	Diagnosis: AF	HF	Down

SVT: Supra ventricular tachycardia, VT : Ventricular Tachycardia, AF: Atrial Fibrillation, POAF: Postoperative Atrial Fibrillation, TACH ; Tachycardia, CABG: coronary artery bypass grafting; HF: heart failure.

The miR-21 upregulation in atria of both clinical and animal studies promotes structural modification of the atrial myocardium.[74] The preferential miR-21 upregulation is found in cardiac dysfunction.[75] Moreover, miR-21 expression influences the extracellular signal-guided/mitogen-increase protein kinase pathway for activating cardiac structure and function ultimately.[76] A 3.2-fold reduction of miR-150 in platelets and a 1.5-fold reduction in serum isolate during the analysis of AF patient serum relative to cardiac disease-free controls.[77] Perhaps silencing of miR-150 in platelets is related to several methods along with AF development like inflammation, platelet function.[77, 78] The arrhythmias, paroxysmal, or constant tachycardia is common in pediatric patients that allows cardiac modification and HF. The insertion of miR-1 into the diagnostic model improves the existence of supraventricular tachycardia symptoms. For higher accuracy, miR-133 acts as a biomarker in ventricular tachycardia improvement. Supraventricular tachycardia reduces miR-1 at a trace level. Conversely, an increased miR-133 is found in ventricular tachycardia. Therefore, improved miR-1 is well reported in AMI models and is linked generally to ischemic arrhythmia.[79]

Pulmonary hypertension

The features of pulmonary arterial hypertension improve the resistive and reactive ingredients of pulmonary vascular impedance that initiates right ventricular failure ultimately.[80, 81] More than 30 circulating miRNAs are related to the development and progression of pulmonary arterial hypertension.[82] The amount of many miRNAs interrelated with pulmonary vascular resistance index, pulmonary arterial dispensability, and reduce oxygen and/or inhaled nitric oxide sensitivity.

Challenges

miRNAs control numerous biological processes. Their levels of expression are the instrument for several human disease detections. There is no doubt that miRNAs are crucial fine tuner for cardiovascular disease and lead to instigate various aspects of cardiovascular biology. Despite our experimental experience, miRNAs should be greatly improved in biology, many intensive investigations are required for miRNA isolation and validation as diagnostic biomarkers and prognosis of cardiovascular diseases.

Many problems and obstacles are required to be addressed to characterize miRNAs for reliability in diagnostic markers or therapeutic candidates for CVD. A conflicting finding of miRNA modulation related to cardiac pathologies has occurred. These findings lead to implement the distinct characters in sample/ patient number (i.e., low *n* values), sampling time, miRNA quantification, miRNA normalization co-morbidities.[83, 84] Further clinical experiments related to sample numbers can be attributed by a mixture of many studies via meta-analysis. However, routine steps, material accumulation, storage, and miRNA operating protocols can be attributed to a narrow range for miRNA experimental variability. To improve miRNA identity related to CVD it is essential to gather data for many time points after symptom onset and repeat detailed medical histories for providing essential confounders.[84] A wide variety of protocols have been emerged for estimating miRNAs amount: qRT-PCR, droplet digital PCR, quantitative stem-loop) and chip-based digital PCR, as well as RNA seq and microarrays.[85-88] However, a common set of miRNA normalization, a variety of statistical analyses, internal spike miRNA controls, selection of primer set, and efficient amplification are established for reduction of repeat inter-study variability.

An alteration of dozens of circulating miRNAs expression is linked to many cardiac pathologies. However, a limited miRNA number can be used for diagnostic markers. This restriction of miRNA expression is also found by non-cardiac conditions (*eg*, cancer, infection, drug use, *etc.*) and eliminates mechanistic link to the heart. Moreover, diversified miRNA numbers lead to implementation as early diagnostic markers due to their link to CVD or valuable therapeutic candidates. However, four miRNAs miR-1, miR-133a, miR-208a/b, and miR-499 exist in cardiac tissue predominantly and/ or necessary for heart development, function, or repair of heart-specific damage. Despite many valuable restrictions of miRNA-applicable therapeutics have been isolated. Elimination of these restrictions will be an essential challenge in miRNAs for diagnostic and therapeutic applications. Therefore, additional miRNA recognition and development is the urgent need for human clinical trials for the treatment of CVD.

Coronary Artery Diseases

Cardiovascular diseases (CVD) are a larger umbrella accommodating almost all sorts of heart diseases that provide any kind of defect big or small in the heart. It mainly includes heart failure, angina, arrhythmia, valvular heart disease, and coronary artery disease (CAD) or other conditional defects. On the other hand, major heart failure is occasionally referred to as congestive heart failure. It occurs when your heart muscle doesn't pump blood and it should. Certain preexisting conditions like tight and narrowed arteries in your heart (CAD) or high blood pressure gradually make your heart too weak or stiff to fill and pump efficiently. The above-defined heart-related diseases are reported here in a cumulative manner. However, coronary artery disease is unique, devastated, and predominant relative to other diseases. Here, we cannot deprive our serious miRNA readers by describing CAD in its brief contest. Therefore, a separate section is allotted for the detailed description and miRNA contribution below.

Pathogenesis

Important pathologic methods for CAD are of large and medium-sized coronary artery atherosclerosis. The enriched cholesterol amount conjugates with low-density lipoprotein. The strong low-density lipoprotein enrichment allows a unique opportunity of infiltration into the artery wall and improves oxidation.[89] These proteins transport to smooth muscles initially from the tunica media to the artery intima.[90] These instigations of smooth muscle generate fibrotic extracellular matrix (ECM) for modulating the lipid-rich fatty streak into the advanced lesion.[91] The ECM creates the fibrous cap for maintaining the strong mechanical stability of the plaque. Further, calcification, neovascularization involves plaque structure. As the plaque size enlarges the bloodstream produced oxygen, it does not export all lesion areas. As a result, the inner section transforms hypoxic, and the neo-vessels are also transformed into tiny hemorrhages inside the plaque for enlarging its size rapidly. [92]

In the majority, ischemia and infarction are responsible for physical disruption of the lesion fibrous cap. They lead thrombogenic material to activate with blood cells.[93] This continuous interaction generates thrombosis for preventing blood flow in the artery. Inflammatory cells also fluctuate the plaque concentration by releasing pro-inflammatory cytokines, proteases, coagulating factors, and vasoactive molecules. These molecules silence the formation of a stable fibrous cap, degrade the collagen in the cap and instigate the clot formation.[94]

Nowadays, the participation of miRNAs in CAD pathogenesis instigates different lessons for diagnosis and prognostic markers for cardiovascular disease (CVD). Now, miRNAs lead to either a beneficial or detrimental path for almost every step of atherogenesis, for instance, endothelial damage and dysfunction, monocyte-wall invasion and activation, lipoprotein formation, plaque stability, modification of CV system, platelet and vascular smooth muscle cell function.[95]

Currently, circulating miRNAs are of general interest as a source of early biomarkers in many human diseases.[96-98] They are extracted from whole blood, peripheral blood mononuclear cells, platelets, serum, plasma, and other body fluids. [99] The miRNAs function as an early biomarker in CAD. The miRNA expression from body fluids has an intensive role for early detection, diagnosis, severity estimated markers, and prognostic indicators.

Early diagnosis modalities

Common diagnosis of CAD depends on the visualization of the anatomy of coronary artery and heart function. Coronary angiography is a valuable standard method for CAD diagnosis.[100] However, coronary angiography is properly estimated for early CAD diagnosis. Its fluctuations in two opposite directions also consider the technical factors and multiplicity of coronary anatomy and plaque configuration.[101, 102]

Further, this technique is also complicated by the influence of local anesthesia. It also involves the use of contrast material, increased nephropathy, infection, local vascular injury, myocardial infarction, stroke, and finally death.[103, 104]. Moreover, it is also untraceable via compromising its value in the non-flow limiting CAD in women.[105] Of late, the emergence of noninvasive techniques either imaging or non-imaging holds the key to great prospects.[106] In this regard microRNA might be important to overlook these bottlenecks.

miRNAs as early biomarkers

MicroRNAs are early important biomarkers due to their cell-specific sensitivity, distribution, abundance, and stability in clinical specimens.[107] The improved aspects of miRNA-mediated early diagnostics improve the powerful molecular issues for generating clinical significance from patient samples. MicroRNAs are also useful for stratifying the CAD patients and finally the ACS type in different groups. Three microRNAs, miRNA-25-3p, miRNA-221-3p, and miRNA-374b-5p were linked to STEMI strongly and miRNAs 221-3p and 483-5p are related to NSTEMI considerably.[108]

Early detection of coronary artery disease

A sensitive, delicate, and specific screening at a low cost and invasiveness are necessary for early and improve CAD monitoring. The microRNAs express and liberate from platelet, monocyte, endothelial cells at the early stage of CAD. Two miRNAs miRNA-31 and miRNA-720 in circulating blood are the most useful components for early CAD detection. These miRNAs control early endothelial progenitor cell (EPC) function using inhibition of FAT4 and thromboxane A2 receptors that exist early in EPC's of CAD patients.[109] Interestingly, their abundance is reduced to a low level dramatically in CAD patients relative to non-CAD patients.

On the separate hand, three isoforms of miRNA-181in coronary artery disease are derived from the inhibition of the monocyte. Mainly, miRNA-181-a is linked to CAD ultimately after the occurrence of different risk factors: obesity and metabolic syndrome, etc.[110] The miRNA-181 linked to the inflammatory toll-like receptor and nuclear factor κB signaling perhaps uses a crucial biomarker for early obesity detection bound to coronary artery disease. However, miRNA-181 is also controlled by other decreased toll-like receptor signaling factors.

The plasma miRNA- 208a is an instrumental and promising agent for an early biomarker after the onset of myocardial infarction. The highest peak of miRNA- 208a is found earlier than the existence of conventional biomarkers (cTnI and CK-MB mass). It implies that miRNAs can be as early biomarkers in an emergent role than routine marker evolving.[111]

Differentiate CAD from non-CAD

The microRNAs are broadly classified into two different categories based on the diagnostic values (AUC). The "bad" miRNAs ranging the classified power (AUC, 0.5-0.6) and some miRNAs marks as "excellent" from others with power (AUC, 0.9-1.0).

Different miRNAs are essential for CAD patients from non-CAD. Three plasma miRNAs: miRNA-765, miRNA-149, and miRNA-424 are analyzed in CAD patients with non-CAD controls. All three miRNAs evaluate critical results to differentiate stable and unstable CAD from respective controls. AUC value of decreased plasma miRNA-149 categorizes stable and unstable CAD patients from non-CAD (0.938 and 0.951) controls. The miRNA-765 overexpression also separates CAD from non-CAD patients.[112]

Discriminatory powers of the plasma miRNAs miRNA-149 and miRNA-765 categorizes fragile CAD from the controls with AUC values of 0.972 and 0.977, respectively. While with 0.959 and 0.938 AUC values for miRNA-765 and miRNA-149 generates steady stable CAD from controls respectively. Based on this significant division of power, however, plasma levels of miRNA-765 are related to age in all groups distinctly. This finally influences the ideal biomarker features. Conversely, plasma levels of miRNA-149 are not statistically significant.

Further miRNA-133a separates CAD from non-CAD individuals and increases the expected demographical data including age, sex, smoke, hypertension, diabetes, hyperlipidemia, and cardiac troponin I (cTnI).[113] The miRNA-133a exhibits an AUC value of 0.741, 0.785, and 0.918, independently in the clinical data of cTnI. Curiously, adding miRNA-133a clinical data and cTnI induces the AUC values 0.942 and 0.925 dramatically. On the other hand, plasma miRNA-133a expression is essential for the diagnosis of unstable CAD (AUC = 0.906).[114] A mixture of two miRNAs, miRNA-1 and miRNA-126 induce identified efficiency of unstable CAD from controls. Moreover, miRNA-1 and miRNA-126 perhaps separate both stable and unstable CAD from the controls individually with a potent of $\geq$ 0.85 value of AUC. Both miRNAs are overexpressed in CAD patients.[115]

Sufficient power for CAD distinction from non-CAD is also found via miRNA-206 and miRNA-574-5p function in plasma with an AUC value of 0.607

and 0.699.[116] Their proper target locus possibly participates during the onset and development of CAD for proliferation our understanding to detect them for early CAD diagnosis.

Multiple miRNAs are different statistically between CAD and non- CAD without ROC curve analysis. The miRNAs, miRNA- 34a, miRNA-21, miRNA-23a, miRNA-30a, miRNA-106, miRNA-34a, and miRNA-21 are accumulated at a higher level in the CAD patient plasma relative to controls[117], while miRNA-23a are downregulated among CAD patients (all $P<0.01$). The concentration comparison of miRNA-135a to miRNA-147 in PBMC finds a 19 folds higher accumulation in CAD patients relative to controls. The modulation in PBMC miRNA signature in CAD patients is perhaps bound to an alteration in intracellular cadherin/Wnt signaling as shown by the miRNA/target gene/biological function linkage analysis.[118] Recognition of a panel of PBMC miRNA like miRNA-24, miRNA-33, miRNA-103a, and miRNA-122 promotes diagnostic accuracy of CAD (AUC=0.911, 95% CI 0.880-0.942).[119] A mixture of three miRNAs including miRNA-155,-145, and let-7c generate a better categorization than the single miRNA alone.[120]

Severity assessment of mRNA

Two anatomical tools produced by a Synergy between the involvement of percutaneous coronary with Taxus and cardiac surgery (SYNTAX). The gensini is estimated for CAD severity.[121] In this regard, several circulating miRNAs are related to the severity of CAD. They are linked to stenosis amount, the complexity of stenosis, and steady-state firmness of CAD. The Circulating miRNA-133a is linked to the complex severity of coronary artery stenosis and the level of stenosis.

The amount of circulating miRNA-133a is upregulated strongly in CAD patients along with stenosis of a coronary artery rather than non-coronary heart disease (CHD) patients. It is also interrelated with gensini scores considerably and appears to be a better indicator of crude estimation relative to cTnI.[113] However, miRNA-133a couldn't distinguish at a trace level stenosis from non-CHD individuals. Further, miRNA-208a[122], miRNA-155 [123] and miRNA-223[124] are interrelated to gensini scores.

The miRNA-145 plasma levels are linked to the number of diseased vessels, SYNTAX score, and firmness of the plaques. A reduced level of miRNA-145 in patients along with three-vessel disease and high SYNTAX score is found rather than with one or two-vessel disease and low or intermediate SYNTAX score. However, the amount of miRNAs-145 in patients with one-vessel and two-vessel disease and/or between low score and intermediate score groups are not different distinctly.[125] Further, miRNAs-214 allows it to relate with the SYNTAX score. [126] Often a lion shares of CAD patients have a sudden cardiac arrest based on unstable plaques rupturing.[127] This procedure leads to subtotal or total occlusion and ACS. Similarly, non-invasive biomarkers perhaps detect a severe form of clinically demanded CAD.

In this consequence, the expression of miRNAs-122, -140- 3p, -720, -2861, and -3149 have been overexpressed dramatically in the ACS group relative to non-ACS groups and have strong potency for patient identification. The discriminatory powers of these miRNAs are larger than the AUC of 0.8 except for miRNAs-3149 that is 0.670. An improved category of power is executed using the panel of miRNAs-122, -2861, and -3149 relative to using it alone.[128] Other set miRNAs miRNA-106b, miRNA-25, miRNA-92a, miRNA-21, miRNA-590-5p, miRNA-126 and miRNA-451 also categorize ACS from non-ACS.[129] Similarly, a bonding of miRNA-483-5p and miRNA-451a can differentiate plaque rapture with an excellent power class, AUC (0.982; CI: 0.907-0.999). Moreover, a panel of miRNA-483-5p and miRNA-155-5p is also executed with the highest AUC (0.898; CI: 0.790-0.962).[130] The miRNA-638 is predicted of plaque instability for carotid artery individually.[131]

In contrast, 6 miRNAs in serum miRNA-92a and miRNA-122 could not discriminate ACS from non-ACS. However, a minor fraction of miRNA-486 and -92a levels demonstrate the clean discriminatory power of ACS from non-ACS by analysis of lipoprotein.[132] The amount of high-density lipoprotein-2 (HDL-2) miRNA-92a and HDL-3 miRNA-486 perhaps subdivide the ACS and non-ACS close to an accuracy of 84% with adjusting parameters for age, gender, and serum lipids. One of the drastic forms of CAD is coronary bifurcation lesion. It is a challenging lesion in percutaneous coronary intervention (PCI) medicine based on the rate of re-stenosis and dramatic adverse cardiac events.[133] As a consequence, whether the lesion is bifurcated or not is critical for effective management of CAD is determined. The miRNA-30-d is overexpressed and miRNA-1246 under-expressed in bifurcated relative to patients with non-bifurcated lesions.[134]

Further, miRNAs are interrelated with different plaques. For example, strongly calcified plaque and weakly calcified plaque have different arrays of clinical outcome and miRNAs. These miRNAs are linked to the calcification amount that possibly has importance for estimating the CAD severity. The miRNA-21 distribution in macrophages of non-calcified coronary artery lesions is elevated markedly with an AUC value of 0.655.[135] The biomarkers are also found in classify calcified from non-calcified lesions. Out of 8 miRNAs, additional validation of miRNA-2861, miRNA-3135b, and miRNA-191-3p exhibit an improved classification power.[136]

Prognostic markers

Despite scanty information that has been delivered so far about miRNAs relationship with CAD prognosis, nevertheless several miRNAs perhaps act as potential detecting agents. Their capability related to all aspects of CAD progression including vascular performance and cardiac remodeling either in beneficial or detrimental direction, function them potent of the future consequence of the diseases.[95] To support this,10 miRNAs from plasma and microvesicles (MV): miRNA-126, miRNA-222, miRNA-let7d, miRNA-21, miRNA-20a, miRNA-27a, miRNA-92a have established their early diagnosis role. Three miRNAs miRNA-17, miRNA-130, and miRNA-199a are required in vascular activities. However, a possible link between cardiovascular events and plasma miRNA levels are established. Conversely, miRNA-126, and miRNA-199a overexpressed in circulating MVs are related to a reduced level of adverse CV event rate distinctly. [137]

Similarly, miRNA-126 in serum is not a useful prognostic marker of CAD, even with relationship ACS and stable CAD groups.[138] The miRNA-197 and miRNA-223 overexpression cannot firmly detect future cardiovascular death. The prognostic value of cardiomyocyte-enriched miRNAs Including miRNA-1, miRNA-133a, miRNA-133b, miRNA-208a, miRNA-208b, and miRNA-499 exhibit among ACS patients. Among them, only miRNA-133a and miRNA-208b amount are linked to the risk of death distinctly.[139]

Association of restenosis with miRNA expression

Restenosis is a common harmful event of endovascular procedure. It causes the recurrence of tightening and narrowing of a blood vessel. This is known as in-stent restenosis (ISR) if restenosis appears after stenting.[140] Usually, the most appropriate threshold value for restenosis is a ≥50% narrowing of blood vessels.[141] The microRNAs are also interrelated to the emergence of restenosis in CAD patients. The downregulation of circulating miRNA-143 and miRNA-145 levels are related to ISR emergence and possibly function as ideal non-invasive biomarkers for ISR.[142] Further, the miRNA-93-5p expression is an independent indicator of ISR after accounting for conventional CAD risk factors.[143] Three miRNAs miRNA-181b, miRNA-185, and miRNA-155 expression separate ISR patients from non-ISR.

The miRNA-181b is reduced strongly, while two miRNA-185 and miRNA-155 are overexpressed in ISR patients relative to non-ISR.[144] The functions and implementations of different selected miRNAs that establish their diagnostic and prognostic purpose for CAD are reported. (**Table 4**)

==

Table 4. Selective miRNAs from different studies of coronary artery diseases

miRNA	Study Population	Sample	Extraction Methods	Alteration	Role
miR-31	CAD (n=20) vs	Plasma	qRT-PCR	Down	prognosis,
miR-720.	HC (n=15)	EPCs		Down	diagnosis
miR-181a	Obese (n= 21) vs. non-obese (n=1)	monocytes	Microarray qRT-PCR	Down	diagnosis
miR-149	CAD (n= 95) vs.	plasma	qRT-PCR	Down	diagnosis
miR-424	HC (n=32)			Down	
miR-765				Up	

miR-765	SCAD (n= 37)	Plasma	Microarray,	Up	diagnosis
miR-149	UCAD (n=32) vs. HC (n=20)		qRT-PCR	Down	
miR-133	AMI (n=13), AP (n=176), vs. HC (n=127)	Plasma	qRT-PCR	Up	diagnosis
miR-135a	SAP (n= 25)	PBMC	qRT-PCR	Up	diagnosis
miR-147	UAP (n= 25) vs controls (n= 20)			Down	
miR-1	SAP (n= 34)	Plasma	qRT-PCR	Up	Severity evaluation
miR-126	UAP (n=19) vs.			Up	
miR-483	non-CAD (n= 20)			Up	
miR-133a				Up	
miR-126	CAD (n=36) vs.	Serum/ plasma	qRT-PCR	Down	diagnosis
miR-17	non-CAD (n=17)			Down	
miR-92a				Down	
miR-155				Down	
miR-145				Down	
miR-133a				Up	
miR-208a				Up	
miR-206	CAD (n=67)	Plasma	Microarray	Up	diagnosis
miR-574	vs. non-CAD (n=67)		qRT-PCR	Up	
miR-34a	CAD (n=32)	Plasma	Microarray	Up	diagnosis
miR-21	non-CAD (n=20)		qRT-PCR	Up	
miR-23a				Up	
miR-2861	CAD (n=90),	Plasma	Microarray	Up	Severity
miR-3135b	vs. non-CAD (n=70)		qRT-PCR	Up	assessment

miR-191				Up	
miR-126	CAD (n=176)	Plasma, MVs	qRT-PCR	Up	prognostic
miR-199a		Exosomes		Up	
miR-197	(ACS, SAP) (n=873)	Serum	qRT-PCR	Up	prognostic
miR-223				Up	
miR-133a	ACS (n=444)	Plasma	qRT-PCR	Up	Prognostic,
miR-208b				Up	diagnostic
miR-208a	CHD (n=290)	Plasma	qRT-PCR	Up	Severity
	vs. HC (n=110)				
miR-155	CHD (n=300)	Serum	qRT-PCR	Up	Severity
	vs. HC (n=100)				
miR-483	SCAD (n=59)	Plasma	qRT-PCR	Up	Severity
miR-451a				Down	
miR-155				Up	
miR-486a	CAD (n=95) SAP	Lipo-protein	qRT PCR	Up	Severity
miR-92a	(n=30) UAP (n=39) MI (n=26) vs. HC (n=16)	fractions		Up	
miR-100	ISR (n=51) non-ISR	Plasma	qRT PCR	Down	Severity
miR-143	(n=130) vs HC (n=52)			Down	
miR-145				Down	
miR-21				Up	
miR-425	ISR (n=39) vs.	Plasma	miRNA	Up	Severity
miR-93	non-ISR (n=39)		PCR array	Up	

miR-181b	ISR (n=6) vs	Plasma/	qRT-PCR	Down	Restenosis
miR-155	non-ISR (N=43)	cell culture		Up	
miR-185				Up	

*Abbreviations - ACS: Acute Coronary Syndromes, AP: Angina Pectoris, CAD: Coronary Artery Disease, CHD: Coronary Heart Disease, ISR: In-Stent Restenosis, HC: Health Control, MI: Myocardial Infarction, miR: microRNA, MV: Micro vesicles, n: number of participants, qRT-PCR: quantitative Real-Time Polymerase Chain Reaction, SAP: Stable Angina Pectoris, UAP: Unstable Angina Pectoris, RNU: small non-coding RNAs.

miRNAs as possible biomarkers

Parameters influencing miRNAs levels

Heparin injection before blood sample collection is highly sensitive for modifying the miRNAs expression.[145, 146] Collection of blood samples within heparinized test tube mislead miRNA estimation and isolation.[147] In fact, endogenous heparin has a great influence on miRNA measurement.[148]

Further, antiplatelet therapy has also influenced miRNAs expression. The miRNA-92a and miRNA-19b extracted from platelets are sensitive and fragile to aspirin therapy.[149] The platelet-derived miRNAs, miRNA-223, miRNA-191, and other miRNAs, miRNA-126, and miRNA-150 from plasma levels are decreased markedly under antiplatelet treatment.[150] Therefore, extreme precaution is required before collected blood *in vivo* for miRNA extraction along with heparin and anti-platelet injection Truly, heparinase enzyme administration in blood sample collection reduces the heparin influence.[151, 152]

Estimation of miRNA amount changes in CSD and CAD due to different medication intakes including statins and angiotensin-converting enzyme (ACE) inhibitors.[153] These observations lead to emphasize the importance of the drug- and metabolite-mediated miRNA quantification in various clinical settings. An inconsistency is detected essentially for further data collection that influences how miRNA levels in circulating blood occur. Considerably, the amounts of biomarkers are involved and influenced their elimination rapidly. Cardiac miRNA expresses bonds strongly with renal function revealing that the renal function perhaps affects the plasma miRNAs.[154]

The high-altitude hypoxic environments in plasma miRNA collections are affected strongly. Recently, it is reported that 175 differently expressed miRNAs collected at high altitudes are also related to red blood cell counts and hemoglobin values.[155] Co-variability of estimated miRNA along with demographic factors is also found. For instance, miRNA-126-5p and miRNA-92a-3p co-variables with two parameters age and serum creatinine.[156]

Limitations for miRNA biomarkers

The time-consuming and laborious isolation, identification, and quantitation procedures are the main drawbacks of using miRNAs as biomarkers for clinical analysis. Moreover, proper optimization and validation are required for their extraction and estimation.[157]

Excess of trace amounts of tissues and their disease specificity is the most critical hurdle for apparent miRNA expression in different diseased states. For instance, miRNA-141 in pregnant women is increased during different cancers derived from epithelial, breast, colon, and lung [158] in CAD.

miRNA Therapeutics

Selective miRNAs in the Cardiovascular System

A wide range of miRNAs have been contributed to cardiovascular disease development and their progression including acute myocardial infarction (AMI), fibrosis, heart failure, and atherosclerosis (**Figure 3**) etc. The following parts highlight cardiovascular diseases related to miRNAs using for promising therapeutic targets.

Acute Myocardial Infarction

The predominant six selective miRNAs, miR-15, miR-24, miR-34, miR-92a, miR-199a, miR-320, and miR-590 are required for miRNAs controlling myocardial infarction. The miR-15 family is needed to control in the heart infarcted region related to ischemia/reperfusion injury and control hypoxia-induced cardiomyocyte cell death.[159] Silencing of miR-15 family members enhances postnatal cardiac regeneration.[160] Instrumentally, a pivotal mitochondrial isoenzyme the pyruvate dehydrogenase lipoamide kinase isozyme 4 is downregulated in hypertrophic remodeling. This hypertrophic and checkpoint kinase 1 are the potential target for miR-15 family members.[159, 160] MiR-34 family members are required for aging and ischemic induction. They also induce multiple events including cardiomyocyte cell death, destroy DNA damage, and influence telomere erosion. Therefore, they function as a potential controller for cardiac repair and regeneration.[161-163] Moreover, the aging-related cardiomyocyte cell death induction is related directly to miR-34a that suppresses target gene *PNUTS* (protein phosphatase 1 nuclear-targeting subunit). PNUTS. On the other hand, they are also required for controlling telomere shortening, DNA damage response, and cardiomyocyte apoptosis for enhancing recovery of Acute Myocardial Infarction (AMI).[161] An overexpression of miR-92a in ischemic tissue exposes antiangiogenic properties via inhibiting its target genes *integrin α5* and the *deacetylase sirtuin 1* and thereby downregulating ischemia-induced angiogenesis.[164-166] Two miRNAs miR-92a, miR-24 interact with sirtuin 1 target and found as a crucial regulator of endothelial cell apoptosis and angiogenesis after AMI.[167]

Fibrosis

A few selective miRNAs are involved in cardiac fibrosis either by interacting directly with the fibrotic response itself or downregulating the two miRNAs, miR-29 and miR-2 are most critical. The upregulation of MiR-29 inhibits fibrosis associated with many matrix proteins. [169] MiR-21 is overexpressed in cardiac fibroblasts during AMI or transverse aortic constriction (TAC). It also enhances the profibrotic ERK-MAP kinase signaling pathway by silencing Spry1(sprouty homolog 1).[170] This method controls fibroblast survival and growth factor secretion with the impression of apparent regulation of interstitial fibrosis and cardiac hypertrophy.[170] Despite these effects that were first not noticed in miR-21−/− mice, [171] nevertheless further results establishes that genetic and pharmacological depletion of miR-21 downregulates cardiac fibrosis.[172-175]

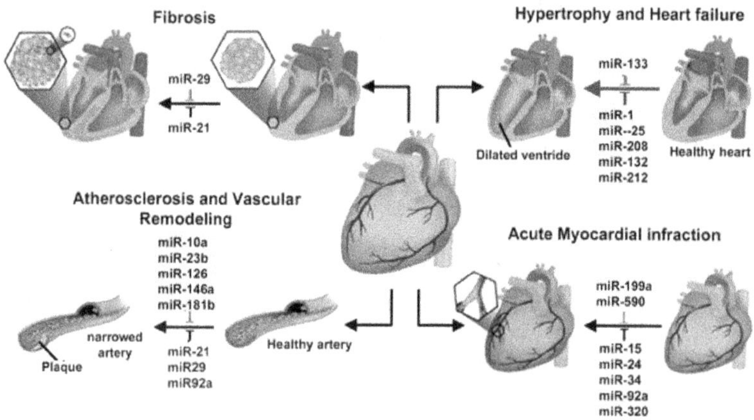

Figure-3. The function of microRNA (miRNAs) in the cardiovascular system. A sketch of miRNAs involved as therapeutic targets for cardiovascular disease treatments. miRNAs from the red lines acts upregulation of a beneficial effect, miRNAs following black lines show inhibition as a beneficial effect. Four sections denote (**A**) fibrosis, (**B**) hypertrophy and heart failure, (**C**) atherosclerosis and vascular remodeling, and (**D**) acute myocardial infarction. Enlarged pictures in **A** display an improved infiltration of fibroblasts (cells) and their compaction of extracellular matrix (blue fibers). Atherosclerosis and vascular remodeling are exhibited by an occluded cardiac artery (in belched red; **C**). (Loosely Adapted, Redraw and Modify using skeleton from Lucas *et al.*, 2018)

Hypertrophy and Heart Failure

Notable examples of miRNAs in heart failure contains many myomiRs. They include miR-1, miR-133, miR-208, miR-499, miR-25, and miR-212/132. The contracting capacity of cardiomyocytes improves using changing intracellular calcium level that appears to be an effective therapy against human heart failure. The miR-25 silencing reorganizes cardiac function using calcium handling. It is also downgraded the calcium uptake pump SERCA2a (Sarco/endoplasmic reticulum Ca^{2+}-ATPase 2a). [176] Conversely, miR-25 silencing using antagomiRs generates murine myocardium to heart failure. Moreover, it conducts embryonic transcription factor Hand2 re-expression (heart- and neural crest derivatives-expressed protein 2), thus influencing cardiac remodeling in a TAC model.[177] Moreover, miR-133 and miR-1 silence the same genomic cluster and appears to be found in human cardiac hypertrophy. Similarly, *in vitro* upregulation of miR-133 or miR-1 silence cardiac hypertrophy. In connection, miR-133 silencing using antagomiRs requires for distinct and continued stable cardiac hypertrophy.[178] Three miR-133 target genes *RhoA* (a GDP-GTP exchange protein), *Cdc42* (a signal transduction kinase), and *Nelf-A/WHSC2* (a nuclear factor) are involved in cardiac hypertrophy.[178] Also, miR-1/133 controls cardiac muscle repolarization. The distinct depletion of miR-1/133a levels increases a long-QT phenotype in mice at a low heart rate.[179] Conversely, four distinct miRNAs miR-1, miR-133, miR-212, and miR-132 are overexpressed using hypertrophic stimuli in cardiomyocytes. Therefore, they are well enough to govern the hypertrophic growth of cardiomyocytes using direct control of antihypertrophic and pro-autophagic transcription factor FoxO3 (Forkhead box protein O3).[180] Finally, miR-208 is derived from *αMHC* intron and is conducted in cardiomyocyte hypertrophy, fibrosis, and expression of βMHC (myosin heavy chain-β) in response to stress and hypothyroidism mechanisms.[181]

Atherosclerosis and Vascular Remodeling

The selective flow-regulated miRNAs miR-92a, miR-126, miR-146, and miR-181 plays a core central role in atherosclerotic disease conditions, [182] Further two additional miRNAs miR-10a and miR-23b are connected to atheroprotective.[183, 184] The selective miR-21 that expresses aberrantly in the vascular walls on balloon injury, is required for critical control during neointima lesion formation, cell proliferation, and apoptosis. [185] Moreover, miR-21–driven cellular effects isolates two target gene *PTEN* and *Bcl-2*. The age-related miRNA miR-29 represses extracellular matrix proteins, collagens. It is required for generating thereby aorta during the formation of aneurysms in advanced age.[186, 187] In this disease, miR-29 is overexpressed that silences its collagen target genes. The miR-29 silencing abrogates aortic dilation and prevents vessel rupture. Particularly, the intronic miRNAs miR-33a and miR-33b are expressed in a combinatorial manner in the sterol regulatory element-binding proteins SREBF (sterol regulatory element-binding transcription factor) 1 and SREBF 2. They also control cholesterol, fatty acid, and triglyceride homeostasis formation. Therefore, their silencing due to genetic deletion is responsible for enhancing circulatory HDL (high-density lipoprotein) cholesterol levels.[188] The expression of endothelial miR-126 exists a proliferative reserve in endothelial cells via suppression of the Notch1 inhibitor Dlk1 (δ-like 1 homolog) and thereby protecting against atherosclerotic lesion formation.[188] At last, two miRNAs miR-146a[189] and miR-181b[190] generate anti-inflammatory conditions by associating with the 3′-untranslated region of TRAF6 (tumor necrosis factor receptor-associated factor 6) target gene directly and importin α3, subsequently silencing NF-κB (nuclear factor-κB) activity.[189, 190] This trend of control is valid strongly, in which NF-κB is involved in controlling proinflammatory chemokine that conducts the adhesion of monocytes to endothelial cells. Moreover, the distinct variety of miR-10a expression controls proinflammatory endothelial phenotypes for influencing the development of atherosclerosis.[191] Here, all miRNAs act as cardiovascular systems highlighting their important targets for developing versatile therapeutic strateg1ies for treating many cardiovascular diseases.

miRNA Inhibitors in Cardiovascular Diseases

The miRNAs are postinfarction angiogenesis and remodeling, cardiac fibrosis and hypertrophy, arrhythmias, atherosclerosis, and metabolic diseases. (**Figure 4**)[168, 181, 192-194] The miRNA functions are examined either in genetically modified mice, RNA therapeutics, or viral vectors to silence or upregulate particular miRNAs. For miRNA silencing and depletion, ASO and siRNA are used extensively. Both RNAs and their chemical modifications using phosphorothioate backbones provide better stability and protection against RNases. Antisense RNAs provide total complementarity with the miRNA targets and protect against mRNA base pairing that blocks miRNA silencing. miRNA inhibitors are referred to as anti-miRs. They are subclassified into various groups based on their chemical modifications.

Antisense Oligonucleotides

The most preferred class of antimiRs are AntagomiRs and lock nucleic acid (LNA) antimiRs. Different AntagomiRs are 3′cholesterol-conjugated 2′-O-Me, 2′-fluoro or 2′O-methoxyethyl oligonucleotides. They are involved in total complementation to the mature miRNA sequence using excess phosphorothioate backbone linkages. Here, sulfur backbones exchange to the one nonbridging oxygen atoms in the phosphate group. [195] The phosphorothioate backbone enhances its stability by associating with nuclease resistance. AntagomiRs help in plasma proteins (especially albumin) binding. They also degrade renal clearance but protect pharmacokinetic properties. [196] The inclusion of cholesterol increases cellular uptake of antagomiRs.[195] Further, the inclusion 2′-O-Me, 2′-fluoro, or 2′-methoxyethyl promotes miRNA targets for their binding affinity and reduces off-target effects dramatically.

The emergence of LNA-modified antimiRs is complementary to oligonucleotide chemistry. LNAs are chemically locked nucleotides. They form a bridge connecting the 2′-oxygen and the 4′-carbon in a ribonucleotide that mimics C3′- endo conformation. A classical blending between deoxy-ribonucleotide and locked ribonucleotide is utilized and provides the most promising results in different non- human primates during clinical trials.[197, 198] LNA-based antimiRs is ≈15 to16 nucleotides long with strong target miRNA affinity. Sometimes, a shorter LNA-based antimiR is generated. They are known as 8-mer LNA-based antimiR or tiny LNAs.[199] Tiny LNAs base-pairs only to the miRNA seed sequences. It improves the simultaneous target selection of miRNA families with overlapping activities. Thus, clubbing of targeting of an entire miRNA family promotes beneficial effects during many disease conditions. Importantly, a tiny LNA clubbing to seed sequences of miR-15 family members miR-15a, 15b, 16 to 1, 16 to 2, 195, and 497 are silenced downstream targets more efficient relative to the conventional LNA-based antimiR class for targeting only one selective miRNA family member.[159] The huge LNA uptake in cardiac tissue is controlled by both short and long LNAs. These combinations define antimiRs length does not promote cellular uptake.[159] Tiny LNAs however, are always as effective as longer oligonucleotides. Invariably, miR-21 silencing using antagomiR combination always depletes cardiac fibrosis and hypertrophy, whereas the use of tiny LNAs against miR-21 cannot exert such beneficial effects.

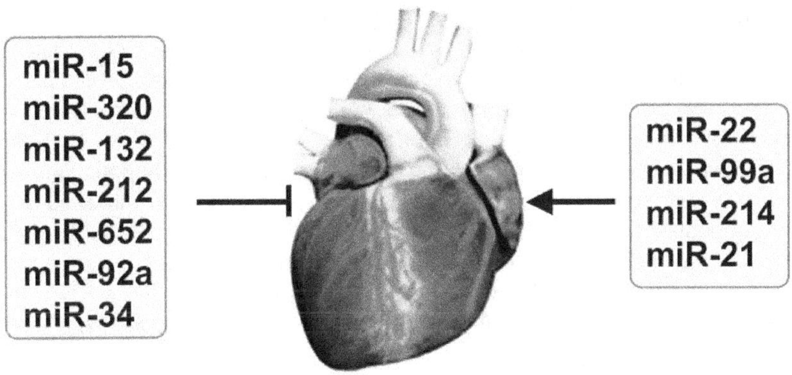

Figure-4. Probable miRNAs for therapeutic agents for cardiovascular diseases. Decreased expression (bar-head) or increased expression (arrowhead) shows beneficial or protective effects in the treatment of CVD. (Loosely Adapted, Redraw and Modify using skeleton from Shan-shan Zhou *et al.*, 2018)

Two additional strategies are executed for promoting LNA antimiRs function. One notable type of bicyclic RNA analogs is selenomethylene LNAs for high affinity, efficient metabolic stability. It also enhanced potency for miR-21 silencing in cancer cell lines *in vitro*.[200] The other strategy is known as the small RNA zipper. LNAs associated with small RNA zippers are used for base pairing to the second half of each miRNA sequence, while the first half along with a nucleotide gap to emerge space for the formation of a stable structure.[201] Small RNA zippers are utilized miR-221 and miR-17 silencing efficiently in breast cancer cell lines *in vitro*. [201] However, the use of these two new inhibitors do not exhibit any potential application for the treatment of cardiovascular diseases.

At last, peptide nucleic acids are ASOs. In the ASOs, the phosphate–sugar polynucleotide backbone is exchanged using a flexible pseudo peptide polymer looking to nucleobases. [202] The phosphodiester backbone is replaced using the multiple repetitive units of N-(2-aminoethyl) glycine. Here, the purine and pyrimidine bases are combined with a methyl carbonyl linker. Peptide nucleic acids are useful for their target sequence affinity and protect against DNases and proteinases degradation. [202] One notable example of successful peptide nucleic acids used against miRNAs *in vivo* is found by miR-155 silencing in B cells in mice. [203]

siRNAs and Other Inhibitory Strategies

The siRNAs act as miRNA blocker. These small RNA duplexes are in many cases chemically modified similar to anti-miRs for protecting the nuclease stability and increasing cellular uptake. [204] siRNA-guided silencing of a miRNA target has bound to the miRNA loop region. The miR-181a silencing using siRNAs reduces the arrhythmogenic effects of skeletal myoblast transplantation with myocardial infarction. Also, other targeting strategy involves miRNA target site blocker. They conduct complementary miRNA target site to bind with an mRNA, thereby inhibiting miRNAs efficiency from access to the target site. [205] This method leads to the protection of the most efficient miRNA targets instead of introducing all potential targets in parallel. In this instance, the target site blockers lead to preferred impairment of the miR-155 ability to target the Cebpb (CCAAT/enhancer-binding protein-β) transcript in the infantile hypothalamus.[206]

At last, miRNA silencing efficiency can be modified or inhibited using miRNA sponges derived from transgenes within cells. Sponge RNAs have 4 to 10 bp long complementary binding miRNA sequences in the 3′-untranslated regions of the RNA. The binding site is selected to link to the miRNA seed sequences or maybe depend on a particular mRNA target sequence, thereby competing with the conjugation of given miRNA targets. [206] Despite miRNA sponges require an expressed transgene construct, transportation of sponge constructs to living tissue is feasible with the viral vectors uses. An adenoviral eGFP (enhanced green fluorescence protein) sponge silences miR-133 in cardiac myocytes *in vivo* in cardiac hypertrophy. However, miRNA sponges expose a strong disadvantage in which a higher concentration of endogenous miRNA within a cell link to huge sponge concentrations. Several studies reveal that antimiRs silence miRNA efficiently in the vascular wall and heart tissue in mice.

Therapeutic Application of AntagomiRs

AntagomiRs are efficient miR-21 targets. It decreases interstitial fibrosis and cardiac hypertrophy.[170] Furthermore, antagomiRs mediated miR-92a silencing instigates neovascularization after hindlimb ischemia and leads to improve the condition of heart function after myocardial infarction in mice.[164] A transportation of microspheres encapsulated antagomiR-92a protects against adverse infarct remodification of reperfused AMI. [166] AntagomiRs governed miR-92a silencing further blocks endothelial dysfunction and atherosclerosis. [207, 208] Sharing the same seed sequences with miR-92a, MiR-25 overexpressed in the failing heart and it silent via antagomirs (300 µg per mouse; diluted in *vivo*-jetPEI solution) for rectifying cardiac function improving survival and blocks heart failure in a mouse model.[176] Despite studies with tough decoys provides the protective acts of miR-25 silencing,[209] antagomiR-25 intraperitoneal injection (80 mg/kg concentration) increases normal cardiac dysfunction. [177] However, it is not distinct whether versatile effects might be introduced to the wide variety of formulations and antagomiR-25 concentration. The antagomiRs mediated miR-320 silencing leads to heart function improvement after ischemia/reperfusion injury.[210] At last, antagomiRs guided miR-212/132 silencing rescues cardiac hypertrophy and heart failure.[180]

Therapeutic Application of LNA AntimiRs

LNA-based antimiRs are effective miRNA inhibitors in several disease models. LNAs associated with miR-29 decreases dramatically the aneurysm formation via instigating matrix synthesis and structural integrity maintenance of the vascular wall.[186, 187] Further via LNA-mediated antimiRs silencing of miR-208a allows cardiac function improvement during heart failure survivability.[211] The whole miR-34 family silencing [162] [15] or selective silencing of miR-34a alone [161, 163] decreases cell death and fibrosis after myocardial infarction and ultimately induce heart function recovery. A critical comparison between antagomiR- and LNA-guided miR-34a silencing decreases chronic myocardial infarction. A beneficial sign is found in each case.[161] However, the effectiveness of the results is restricted because different antimiRs levels are administered and following a dose-response curve is absent.[161] LNA-associated antimiRs administration of miR-34a silencing are further allowed the regeneration window extension in postnatal mice via increasing cardiomyocyte proliferation.[163] Ultimately, the above profitable impact of antagomiR-guided miR-92a silencing is also established using LNA-mediated antimiRs in the ischemia/reperfusion model.[165]

Major challenges of AntimiR Application

Despite the emergence of therapeutic improvement by miRNA silencing are well documented, nevertheless it is stated that penetrating heart tissue is still difficult relative to soft tissue organs including kidney or liver.

In general, antimiRs are injected intravenously, intraperitoneally, or subcutaneously for many weeks long silencing effectively. Many experiments inject a single or multiple doses of ≈0.5 to 25 mg/kg body weight LNA-mediated antimiR (**Table 5**). In a few cases, improved concentrations are required for antagomiRs (8–80 mg/kg body weight) injection for efficient miRNAs silencing in the heart or the vasculature (**Table 5**). The multiple injections with appropriate doses are linked not only to the chemistry but also their preferential target sequence. The miR-92a effective silencing in the heart at low concentrations, such as 0.5 mg/kg LNA-92a or 8 mg/kg antagomiR-92a is established.[165] Conversely, miR-34a or miR-29 silencing in the same experimental condition needs a much higher dose even though the same tissue is used (5 mg/kg LNA-based antimiRs, 20 mg/kg antagomiRs).[161, 187]

In initial clinical trials, antimiRs guided against the liver-enriched miR-122 are injected at 3 and 7 mg/kg body weight (5 weekly injections) concentrations,[212,] and enriched with chemical mediated miRNA administration. It leads to effective <0.01 mg/kg body weight concentrations involving in target liver-specific gene expression.[213] Overall, numerous strategies are executed appropriate doses of different miRNA silencing. The most critical issues are bioavailability and biodistribution methods for cellular uptake, the targeted miRNA levels, and its target genes. It is also noted the sequence of the miRNA and the sequence of the inhibitor itself. Bioavailability is increased via a strong suppression of circulatory clearance due to the development of potential chemical changes. As a result, antimiR efficiency is no longer restricted to excessively accumulated soft tissues like liver and kidney but also applicable to other target organs. Therefore, measuring the increased ASO analysis is not alone essential for their biological activity. [214] ASOs administer in endocytotic pathways that are classified broadly into two categories: productive and nonproductive pathways. Productive pathways allow ASO binding to the target, while administration via nonproductive pathways accumulates in late endosomes or lysosomes.[215] Despite the strategies towards pathfinding in production versus nonproduction methods are not demonstrated clearly, 1 key ESCRT-I (endosomal sorting complexes required for transport I) its agent *TSG101* particularly is a critical regulator for anti–miR-21 nonproductive uptake. [216] *TSG101* silencing influence *in vitro* and *in vivo* productive uptake. The strong uptake machinery is cell-specific based on surface protein expression and binding abilities. [215] Moreover, the most appropriate doses are responsible for miRNA expression level and its targets in a selective cell line. Nowadays, efficient silencing is estimated by isolating RNA from the entire tissue no doubt proving misleading information. The miRNAs estimation using polymerase chain reaction (PCR) method allows an excess estimation of silencing effects as PCR alters its value due to the presence of antimiR in a given tissue extract. Further Northern blot hybridization measures target silencing directly indicating and confirming the extent of silencing. Moreover, cellular heterogeneity may block the proper miR silencing assessment during biological implications. The heart carries different cell types like cardiomyocytes, endothelial cells, fibroblast, and immune cells, etc. The miRNAs and their targets are expressed differently in given cell types that do not include the entire heart tissue estimation. It gives a risky proposition for estimating a under different disease conditions.

Table 5. Different Delivery Strategies and Selective RNA Therapeutics to activate Tissue Enrichment

Strategy	miRNA Chemistry	Application	Dosing
	AntagomiR-92a encapsulated in microspheres	Myocardial infarction	0.1 mg/kg bw, single, pig intracoronary injection
	Nanoparticle packed with miR-126-5p mimics	Atherosclerosis	100 μg per injection every 3 days for 4 wks., mouse

Packaging strategies	Lipofectamine mixed with miR-181b mimics	Atherosclerosis	1 nmol mimic, 2, mouse injections per wk for 4 wk or 1 injection per wk for 12 wk, IV
	Lipid nanoparticle (cationic lipopolyamine) loaded with LNA-antimiR-145	Pulmonary hypertension	2 mg/kg bw, 3 injections (IV) for 5 wk, rat
	Transferrin receptor aptamer linked to premiR-126	Human endothelial cells, breast cancer cells	50–500 nmol/L
Receptor-mediated uptake/aptamers	Gapmers against apolipoprotein) C or transthyretin linked to galactosamine (hepatocyte-specific receptor)	Liver	3–10 mg/kg bw), single (apolipoprotein C), 6–20 mg/kg bw (transthyretin), single injection, mouse
	E-selectin targeting multistage vector microparticle combined with nanoparticle packed with miR-146a and miR-181b	Atherosclerosis	15 µg per mouse
Local activation	Light-inducible caged antagomiR-92a	Diabetic wound healing	2 µg in 25 µL, 3 treatments (1× topical, 2× ID injection), mouse
	AAV encoding miR-590 and miR-199a	Myocardial infarction	1×10^{11} viral genome particles per mouse
	Lentivirus Si-mlincRNA-p21	Atherosclerosis	1×10^{8} UT/mL in 20 µL
Vector-based enrichment	Adenovirus-driven expression of siRNA-MDRL or MDRL-lncRNA	Myocardial infarction	2×10^{11} moi virus, catheter- based injection, mouse

	Adenovirus-driven expression of CARL	Myocardial infarction	2×10^{11} moi virus, catheter- based injection, mouse
	AAV9-driven expression of miR-1	Cardiac hypertrophy	5×10^{11} viral genomes, IV injection, rat
	Catheter-based intracoronary delivery of LNA-antimiR-92a	Myocardial infarction	0.03/0.15 mg/kg bw single intracoronary infusion, pig
Delivery by Devices	LNA-antimiR-21–eluting stent	Restenosis	5 mg/kg bw, rat

AAV indicates adeno-associated virus; bw, body weight; CARL, cardiac apoptosis-related long noncoding RNA; IV, intravenous; LNA, locked nucleic acid; lncRNA, long noncoding RNA; MDRL, mitochondrial dynamic related long noncoding RNA; miRNA, microRNA; and siRNA, small interfering RNA.

===

AntimiR Toxicity

An increased amount of ASOs or antimiRs influence toxicity risk. Overall toxicity presented by oligonucleotide treatment is broadly classified into 2 important subdivisions: the hybridization-dependent and hybridization-independent toxicity. Hybridization-dependent toxicity occurs within the oligonucleotide sequence. It is influenced either by target modification (on-target effects) or by unspecified linking with similar nucleotide sequences (off-target effects). The major risk of direct target-linked toxicity is exposed and decreases considering known target information. The target information includes the mode of action, expression profile, and amount estimated during preclinical and clinical development. Often, off-target effects occur using the same family miRNAs targets due to homology of the seed sequence or by unselected association with DNA, RNA, or proteins. Although antimiRs are linked to the 3′-region of the miRNA, its ultimate results appear to be controversial. Moreover, the seed sequence-specific antimiRs appears to be the most definitive and potential choice.[199] The targeting of an entire miRNA family using tiny LNAs may be profitable but are not suitable for all miRNAs. The most notable example is found using controversial miR-21 observations.[172] Off-target effects further increase using hybridization to double-stranded DNA, particularly via site-specific mutagenesis by triplex formation. Although this event occurs, the examination of multiple oligonucleotides increases the genotoxicity by conducting routine tests like AMES and micronucleus providing negative results, and ultimately there is no example for genotoxicity.[217] However, genotoxicity experiment using modified oligonucleotides is still observed.

Hybridization-independent toxicity proceeds to a particular backbone and modulation of the oligonucleotide. Overall RNA oligonucleotides instigate the immune system using TLR (Toll-like receptor) signaling.[218] The phosphorothioate altered oligonucleotides-based activation is found at high plasma concentrations in nonhuman primates.[219-221] These results lack slow infusion relative to bolus injection for decreasing the high peak plasma amounts. An identical complement activation using oligonucleotide activation is not found in human serum.[222] Phosphorothioate-altered oligonucleotides are important for blood coagulation effect via platelet binding, therefore showing platelet activation, aggregation, and *in vitro* and *in vivo* thrombus formation.[223] Nowadays, it reveals that the modified LNA injection decreases platelet activation significantly by blocking ASOs binding to platelet proteins.[224] Nevertheless many immunologic experiments include in many preclinical and clinical programs during therapeutic development. Many LNA-modified oligonucleotides in a particular tissue induce liver transaminases ALT (alanine aminotransferase) and AST (aspartate transaminase), however even found at a low dose in mice that lead to efficient hepatotoxicity.[225] Conversely, many LNA-based oligonucleotides do not observe any hepatic toxicity sign in different *in vivo* models, hepatotoxicity appears to be a routine phenomenon but rather a sequence-specific issue. Of note, recently particular trinucleotide motifs have been isolated that appear to be an expected indicator for important hepatotoxic sequences by reducing p53 and NRF2 (nuclear factor [erythroid-derived 2]-like 2) pathway induction *in vivo*.[226]

The sign of hybridization-independent toxicities decreases reducing the effective dose additionally and by recovering the target selection in many used tissues. The present strategies decrease effective toxicological side effects by enhancing delivery protocols of antimiRs, increasing tissue-guided antimiRs uptake, or by local antimiRs induction. (**Figure 5**, **Table 5**)[227]

miRNA Overexpression in Cardiovascular diseases

Relative to normal miRNA silencing the suitable approaches for conducting miRNA overexpression is even more challenging. For this strategy, the *in vivo* synthesis of RNA duplexes known as miRNA mimics are utilized predominantly. miRNA mimics are double-stranded RNA duplexes mimicking endogenous miRNAs sequence for exchanging or enhancing miRNA concentrations in the selective tissues. The miRNA mimics require chemical modification in terms of stability and cellular uptake identical to antimiRs. The guide strand of the miRNA mimic leads to a proper functional strand complementary to seed sequences of miRNA. The presence of a 2′-F modification is responsible for guiding strand protection against degradation of exonucleases without modulating target base-pair sites. The opposite strand is prone to being fragile and is related to cholesterol for better cellular uptake. In the absence of its association with cholesterol, this normal strand is prone to rapid degradation. Numerous experimental results lead to proper therapeutic advantages of miRNA mimics for cardiovascular disease treatment. (**Table 5**)[227] The advantageous upregulation of miR-126-5p mimics leads to anti-atherosclerotic effects using endothelial cell proliferation,[188] and miR-181b mimic treatment improves glucose levels and insulin-sensitive in high-fat diet-fed animals.[190] Moreover, classical 2 miRNAs upregulation of miR-146a and miR-181b prevent activation of NF-κB and atherosclerosis in the vascular endothelium via cell-specific mechanisms.[228] The muscle-expressing miR-1 is a crucial factor of cardiac hypertrophy. The therapeutic cardiac-guided miR-1 delivery using adeno-associated virus (AAV) serotype 9 vector inhibits cardiac hypertrophy and blocks pathological remodeling.[229] At last, enhancing of miR-590-3p and miR-199a-3p levels using miRNA mimics are involved for cardiomyocyte proliferation induction and thereby enhancing cardiac regeneration after AMI. [230] Truly, high concentrations of miRNA mimics with different series of insertion in all early studies lead to achieving significant effects. However, the high dosage of mimics is reasonably degraded faster than unformulated RNA mimics using endogenous, extracellular RNases.

Despite the beneficial effect due to miRNA upregulation, a supra-physiological amount of a specific miRNA rather than systemic miRNA mimic activity allows an adverse effect. This effect is particularly found in an organ, where miRNA is not expressed endogenously.[231] Moreover, miRNA mimics leads for potential immune response activation. They lead to their proinflammatory activity using the induction of the immune system nonspecifically through toll-like receptors in a sequence-independent manner. [218] However, many approaches can overcome these problems for improving cellular uptake and tissue specificity.

The use of lentiviruses or adenoviruses vectors allows the most efficient miRNAs expression. However, their use is not well documented during targeting of the cardiovascular system especially the endothelium. Moreover, the use of lentiviruses or adenoviruses has the disadvantage too. Their precursor is upregulated for synthesizing of 2 miRNAs. However, their star strand conducts important specific functions.[188] The upregulation of two strands in a random fashion allows an unexpected response. miRNA mimics are also related to aptamers. Here RNA molecules conjugated with specific receptors and finally leads to a cell type-specific uptake. (**Figure 5**, **Table 5**)[227]

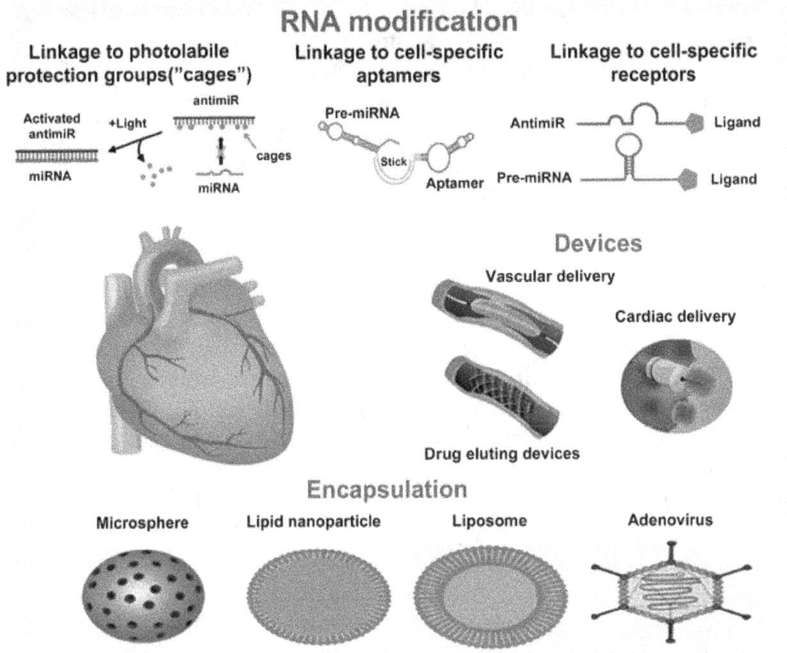

Figure-5. Improved delivery protocols for therapeutic uptake of miRNA therapeutics in cardiac tissue. miRNA therapeutics agents are modified, chemically, used for devices or encapsulated for increasing target specificity or improved cellular uptake to eliminating undesirable side effects and higher toxicity. (Loosely Adapted, Redraw and Modify using skeleton from Lucas *et al.*, 2018)

Conclusion

Numerous biological processes interrelate to cardiovascular phenotype detection and their sensitivity to injury or stress, with multiple modulations of miRNAs. The miRNAs in cardiovascular diseases produce a renewable benefit and improvement for scientific and medical communities. In the initial setup, they explain the pathologic methods using the detection of modified target transcripts. The detection of perturbed mechanisms in cardiac phenotypes perhaps leads to developing unique medicines opposite to pathogenic processes. Also, miRNAs can determine the inner factors required for vast expressivity in the inherited cardiomyopathies and link the genotype-phenotype gap, thus enhancing the patient's stratification risk.

Second, multiple circulating miRNAs from the plasma or serum modify many subtypes of patients. The potent molecules are undisputedly established the early detecting methods in borderline patients but also utilize as non-invasive early biomarkers for identifying the risk factors. However, many unavoidable restrictions still hinder their use in clinical practice. A wider and larger core principle is executed to estimate their reproducibility. Moreover, these normal and routine methods lead to detect and measure miRNA quantity.

Although a tremendous elevation of miRNA interest for early detection of cardiovascular disease and prognosis has been found, pin-pointed experiments regarding their complexity are also noticed. Nevertheless, complete elucidation of important miRNAs sample collection, improved isolation technology, and quantification methods are needed.

The emergence of novel miRNAs in different fields like molecular biology, analytical chemistry, bioinformatics, clinical trial design, epidemiology, statistics, and health-care economics is found. Estimation of large circulating miRNA family has never been participated before however, the diagnostic findings of the multiple single miRNAs have been performed so far.

The application of circulating miRNAs as promising biomarkers provides an excess module in early disease diagnosis and prognosis relative to established protein-based biomarkers. The miRNA-based methods in depicted pathophysiological pathways interfere *in vitro* with disease progression and complexities The proper utilization of miR-mimics and antagomirs in the cardiovascular field has never yet been found in clinical trials. The miRNA regulatory mechanisms and their guided gene modulation are still not understood completely. Simultaneously versatile technical methods for detecting and quantifying miRNAs in circulatory blood including RNA normalization and quantification is required for reproducible protocols.

Circulating miRNAs act as a blood-mediated biomarker in CAD. It is promising greatly for early detection, estimating its severe effects, and prognostic indicators. They have worked as "excellent" to "satisfactory" power of categorizing patients with or without CAD and patients with stable CAD or unstable CAD. Further, the group of miRNAs is not specific. An individual single miRNA perhaps over-expressed or under-expressed in various disease conditions. consequently, the development of a particular algorithm or a panel of experiments possibly improves the specificity of miRNAs. Thus, detection of CAD patients and their commercial grading of disease level are established by circulatory miRNAs. [232-236] At this ro an extensive validation of miRNAs family in large cohorts with their respective pathological and diagnostic contribution firmly displays an extraordinary finding.

In Despite different promising clinical data, particularly miRNAs used for lipid-lowering, targeting the cardiovascular system may be more problematic and challenging as an excess dose is required for therapeutic efficiency relative to their proper concentrations in the heart. Currently, different delivery methods are employed showing translation importance in the proofs-of-concept. Also, a better understanding of miRNA biology and functions is required during exploiting their therapeutic applications. Particularly in the miRNAs field, a well-established mechanism of action, a deeper understanding of cell type-specific functions, and target regulation are required. Therefore, the endpoint efficiency of initial clinical trials is of prime seeking importance. This relative cutting-edge field of research is very challenging, However, the following "Challenges are yet to be met and overcome" using multiple experiments. Hopefully, the experimental expected success leads to combat in cardiovascular disease therapy generation.

Finally, miRNAs possibly utilize as a cutting-edge component for novel tailored therapies. Numerous experiments focus on the efficiency of the miRNA treatment plan *in vivo*; however, additional experiments are needed to translate them into clinical applications. Numerous efforts must be used to examine the safety of the delivery systems, the insertions, the appropriate dosage and period related to treatment, and their existence and the protection against side effects.

References

References: chapter wise are cited before the Index at the end of the Book. (1-236)

Chapter IX

Early clinical miRNA Biomarkers and therapies in kidney

Abstract

In modern days, 1 out of 12 people in the world population has been suffering from diabetes related diseases. Diabetes Kidney disease (DKD) is a prolong life limiting disease condition related to progressive and lifelong irreversible loss of renal function. For appropriate detection, the glomerular filtration rate (eGFR) and albuminuria are measured for defining the critical DKD markers. Truly, they are not supposedly assayed the renal dysfunction, damage, and injury at an accurate rate. Recently, multiple DKD diseases have been developed due to presence of induced blood pressure that is related to blood sugar and lipid amount balances. DKD can raise and employ different agents to prevent the renin-angiotensin aldosterone methods (RAAS) for delaying its development. As a measure, for protecting normal kidney function early diagnosis and unique and many effective protocols are required for protecting kidney and treat DKD. As noted earlier, miRNAs are tiny noncoding RNAs that control gene expression via post transcriptional degradation and post translation inhibition and have been distributed nearly all human samples. Numerous urine enriched miRNAs are governed with clinical and histological DKD criteria intimately. Kidney fibrosis is the most critical situation related to the end state of different chronic kidney disease (CKD) including DKD. Changing the growth factor β (TGF β1) is required for master control of kidney fibrosis via partially regulated miRNA regulation. The participation of miRNAs in fibrosis, hypertrophy, autophagy, inflammation endo-plasmid reticulum (ER) stress, different oxidative stress, insulin resistance and podocyte injury are discussed exclusively. The miRNA involvement in these processes is germinated to offering new therapeutic agents that are further investigated for developing all important kidney damage treatment plans for DKD.

Further Diabetes Nephropathy is the most devastating disease for sudden disability and premature death. Nearly 40 % of DKD patients undergoes to generate diabetes-mediated chronic kidney disease or diabetic nephropathy (DN). The effective prognostic related tests are mostly invasive and leads to delayed development are exclusively challenging. Recently, stable microRNAs (miRNAs) are distributed ubiquitously in multiple human tissues and body fluids. Alteration expression of many urinary miRNAs require massively as a source for potent non-invasive biomarkers for kidney diseases, particularly in DN. In addition, miRNAs function as therapeutic agents and particularly potential components. The miRNAs can be utilized for the early detection of biomarkers and DN development. Recently, massive miRNA experimentations and their emerging data implement in DN pathology and systematically narrate and analyze multiple miRNAs for their effective contribution in the development of ideal therapeutic agents and its application.

Introduction

The proper features of Diabetic kidney disease (DKD) involve glomerular hypertrophy, proteinuria, decreased glomerular filtration. They also include kidney fibrosis, a main microvascular diabetes complication.[1, 2] DKD is evolved as a major public health issue. It specifically leads as more than a half of patients with type 2 diabetes (T2DM) and one third of those with type 1 diabetes (T1DM). DKD is emerged based on profound prevalence of diabetes.[3] In fact, DKD is a dominant cause of end-stage kidney disease (ESKD) that results death, however it requires renal replacement therapy, absolutely either kidney transplantation or dialysis. [4] Thus, DKD causes a considerable burden for the normal health system and problematic cost to the community.

In recent days, eGFR and albuminuria are used routinely as crucial markers for DKD at a perfect point in time.[5, 6] eGFR is measured from the estimated serum creatinine amount with proper equations that inviably lead to age, body size, and measured values based upon sex and race. The two major equations are Chronic Kidney Disease Epidemiology Collaboration (CKD-EPI) creatinine equation (2009) and alteration of Diet in Renal Disease Study (MDRD) equation. Nowadays, amount of eGFR using other biomarkers like cystatin C is evolved with rare exact accuracy. In this case, a different series of CKD-EPI estimators is investigated and finally established via the result of a cystatin C test. However, the overexpression of cystatin C calculates GFR directly. The amount of serum creatinine down estimate GFR.[7] Also the multiple criteria affecting creatinine in elderly persons lead to be more escalating because of comorbid conditions.[8] Albuminuria interprets DKD development more precisely, but devoid of any specificity and susceptibility for ESKD and gradual in eGFR declination. In T1DM, nearly one third DKD patients along with microalbuminuria causes gradual renal function declination.[9] In T2DM, a great proportion of those carrying renal disease progression are normoalbuminuric.[10-12] In summary, both eGFR and albuminuria have narrow restrictions as interpreters for kidney injury.[13] Further, a delicate and early detectable biomarker is required for monitoring the gradual loss in kidney function and to differentiate "progressors" from "non-progressors" in those with DKD.

Different consistent DKD risk factors like ethnicity and inherited genetic difference are also defined.[14] Apart from hyperglycemia, inflammation via cytokines/chemokines, advanced glycation end products (AGEs), and aberrant hemodynamics involve in pathological modifications.[15] Moreover, blood pressure control with RAAS blockers on glycemic and lipid control and lifestyle changes are found in current DKD therapies.[16-18] Despite narrow control of blood glucose along with low level of glucose, hyper-lipidaemia together with statins and of blood pressure with angiotensin-converting enzyme inhibitors (ACEIs) and angiotensin receptor blockers (ARBs), are noticed.[19-21] Moreover, a large proportion of patients also develop ESKD.[22] For the consequences, few ideal therapeutic strategies are needed. Thus, development of novel therapeutic agents and their options needs establish new molecular devices underlying the DKD development and ultimately focus on therapies development.

Accommodating all findings, ideal molecular processes of DKD can be explored for sketching features of all potential biomarkers for disease development and appropriate therapies. With the exploration of high-throughput technologies, the core and extreme role of different epigenetic mechanisms including miRNAs processes and identification have been investigated extensively.[23, 24] Circulating miRNAs are present in body fluids and are known to influence gene expression and regulation.[24] Many characters of circulatory miRNAS including abundant availability, simple complexity in appropriate tissue, stability, and evolutionary acceptance lead circulating and urinary miRNAs as potential noninvasive biomarkers in many pathophysiological conditions and disease states.[25] Importantly, a variety of miRNAs in DKD pathology claim these tiny molecules lead possibly the novel targets for therapeutic intervention.[23] Recently, many more attempts for miRNA expressions modulations are associated with one of many potential delivery approaches in DKD animal models *in vivo* have been investigated that may develop in authentic diabetes in near future .[2]

Furthermore, Diabetic nephropathy (DN) is chromic prolong and deadly complication of type 1 (T1DM) and type 2 diabetes mellitus (T2DM). This event is gradually increasing nearly 9 % of the global population. [26] DN enhances close to 40 % of T1DM sufferers. DN is also routine in T2DM patients. They are also involved at risk of renal dysfunction from another pathological symptom including ischemic nephropathy. DN is the natural cause of renal failure adopting renal replacement therapy globally [27], However, various effective procedures are isolated and block the development of DN pathophysiological changes remain elusive.

In this scenario, various risk factors are executed, including ethnicity, and inherited genetic differences in the DN development.[28, 29] Downstream contributors in the DN hierarchical branch like hyper-glycaemia and aberrant haemo-dynamics and insulin resistance, lead to intraglomerular hypertension and hyperfiltration. Hyper-glycaemia is related to advanced generation of glycation end products, renin-angiotensin system induction, enhanced cytokine production including profibrotic transforming growth factor (TGF)-β), reactive oxygen species and protein kinase C activity etc.[30, 31] Insulin resistance is linked to endothelial loss and vascular changes using nitric oxide and NF-κB beta pathways. Recently, damage in podocyte-specific insulin signaling is implicated in early DN.[32] The related histological symptoms result from different mesangial hyper-expansion, nodular glomerulosclerosis and tubulointerstitial fibrosis. [33] Their ranges relate to loss of glomerular filtration rate and ultimately, renal failure.[34]

microRNAs in noble Biomarkers in Diabetic Kidney Disease

In Early detection

Diabetic kidney disease causes utmost reduction in expected life span of diabetic patients.[35] Within a decade, the yearly symptoms of diabetic kidney disease far more than doubled, whereas today the incidence is nearly 50% reduction of all end-stage renal disease (ESRD) cases.[36] In ESRD, individual renal replacement therapy is utilized for uremia controls and various metabolic incidents of kidney failure. Ultimately <50% of patients can still alive 5 additional years post diagnosis.[37] The dramatic clinical outcomes related to diabetic kidney diseases are prolonged for raising delayed symptoms, which occasionally requires apparent 5–10 years long after the actual initiation of renal damage. Thus, an utmost symptom required for clinical management of diabetic kidney disease is needed an early and symptomized detection to attenuate loss of renal function and improved patient exclusive outcomes. Diabetic kidney disease is a unique progressive microvascular complication of diabetes that is required for the damage to the capillaries of the glomeruli. It is symptomized by stable and consistent albuminuria (>300 mg/day or >200 μg/ min), gradual reduction in glomerular filtration rate (GFR) and enhanced arterial blood pressure. The stepwise improvement in urinary albumin excretion (UAE) rates are related with diabetic kidney disease development and sequential progress of dramatic clinical forms required for renal damages. Currently, the disease is classified based on microalbuminuria states (UAE >20 μg/min and ≤199 μg/min) and macroalbuminuria (UAE ≥200 μg/min). Despite microalbuminuria state is appearing to be the most widely optimized measure for early renal damages in diabetes. Its utility as a potent clinical marker for the diagnosis and prognosis of diabetic kidney disease is exclusively restricted. Invariably, microalbuminuria is not a special marker for diabetic kidney disease, in lieu it reflects many changes in the glomerulus required for any kind of chronic kidney disease. Additionally, UAE rates in normal range in diabetic patients are oftentimes required for calculating GFR initiative of chronic kidney disease.[38, 39] Nevertheless, microalbuminuria has been required for a growing risk factor of macroalbuminuria, but not in all patients progress to this stage. A considerable proportion of patients even reduce to normo-albuminuria.[40, 41] Approximately 80% of diabetic patients with microalbuminuria increase to proteinuria over 6–14 years in early stage.[42] However, only 30–45% of micro-albumin uric patients increase to overt nephropathy over a similar period nowadays.[43] It occurs most likely due to inclusive interventions to conduct and manage hyperglycemia and hypertension. Lastly, substantial renal damage in many patients causes numerous changes in the glomerular basement membrane and alteration in inflammation. It has already occurred in many patients before the appearance of clinical symptoms of microalbuminuria.[40, 44] In accommodation, these results demonstrate that gradual increase of diabetic kidney disease presenting a clear reduction of GFR does not essentially parallel enhancement of UAE rates.

Therefore, microalbuminuria concentration in kidney is neither an accurate and proper marker of renal dysfunction, nor a stable predictor of disease enhancement for some diabetic patients. In this process, defined protocols are not yet available to isolate patients at risk for generating microvascular complications before detrimental and lethal changes occur. However, early markers for early diagnosis of disease for renal damages must produce numerous clinical advantages and benefits in protecting normal kidney function. Several urinary proteins have evolved as potent early markers for renal injury in diabetic kidney disease. Nevertheless, the broad clinical implications of these molecules are yet to wait for further validation.

Existing biomarkers for Diabetic Kidney Diseases (DKD)

The extraction, isolation, estimation, validation, and application of early biomarkers are extremely challenging due to their numerous variable aspects, like identification of proper biomarkers and their importance and relevance to the diseases, and the technological knowhow of the estimating different methods utilize for biomarker quantification.[45] The biomarkers for DKD are increasingly interesting for their utility, investigation in early predictions to most at risk for damaging in kidney function. They are also responsible for utilization, rationalization, and target care.[46] Clinically, eGFR and albuminuria are measured to define the versatile natures of DKD and to apparently isolate risk in increased progression to ESKD. However, they are not possibly measure proper the kidney function and injury in DKD.[47, 48] Routinely, different risk factors are involved in predicting kidney injury.[49] These different risk factors raise and dominate many factors including involving of age, span to diabetes, HbA1c, systolic BP (SBP), albuminuria, prior eGFR and retinopathy status.[49] However, a few approaches raise and validate many presumptive equations using many clinical data. These biomarkers can predict the marginal improvement in kidney disease detection.[50]

Nowadays, a growing number of ideal biomarkers have emerged that are involved for risk of DKD and early DKD detection for blocking the ESKD occurrence. Different existing novel biomarkers are broadly classified as: (a) Tubular biomarkers: neutrophil gelatinase-conjugated lipocalin (NGAL), α-1-microglobulin, kidney injury molecule 1(KIM-1), N-acetyl-β-D-glycosaminidase (NAG), cystatin C, and liver-type fatty acid-conjugated protein (L-FABP); (b) Glomerular biomarkers: Transferrin, immunoglobulin G, ceruloplasmin, type IV collagen, laminin, glycosaminoglycans(GAGs), lipocalin-type prostaglandin D synthase (L-PGDS), fibronectin, podocytes-podocalyxin, and vascular endothelial growth factor (VEGF); (c) Inflammatory biomarkers: TNF-α, IL-1β, IL-18, interferon gamma activated protein (IP-10), monocyte chemoattractant protein-1 (MCP-1), granulocyte colony-activating factor (G-CSF), eotaxins, RANTES (controlled on stimulation, normal T cell isolated and secreted) or CCL-5; (d) Biomarkers of oxidative: 8-oxo-7,8-dihydro-2-deoxyguanosine (8oHdG); (e) Others: podocalyxin, nephrin, AGEs.[47] Despite these entire biomarkers are responsible for DKD evaluation, however none of the markers have been improved early clinical decision-making [51], neutrophil gelatinase-conjugated lipocalin (NGAL) is a lipocalin iron-carrying protein of 25 kDa having a super family of lipocalins. Urinary NGAL concentration is the only indicator in diabetic damages compared to healthy controls.[52] It also correlates inversely with eGFR, and in a same direction with CysC, serum creatinine, and urea in T2DM patients.[53] Considerable induction in urinary NGAL concentration have been explained from normo- to micro- to macroalbuminuric groups of T1DM and T2DM patients. [54] Urinary NGAL are parallel to glomerular hyperfiltration in early clinical states of diabetes. However, there is no considerable correlation between the urinary NGAL concentration and enhanced decrease of eGFR in T2D with proteinuria after proper optimization for factors like systolic blood pressure, HbA1c, and diabetes duration etc.[54]

Several Bioinformatic studies have been predicted in considerable cohorts to isolate protein metabolite DKD biomarkers. A mass spectrometry-dependent process CKD273 combining with data on 273 urinary peptides produces a score. It is highly accurate in eGFR state cross-sectional classification.[55] A 737 samples conducting results in the Diabetic Retinopathy Candesartan Trials (DIRECT)- Protect 2 demonstrate that the CKD273 score is strongly related to incident microalbuminuria in the AER, eGFR, and other variables independent of baseline.[56] An Increased CKD273 score at baseline is related to a lower in ACR largely in the spironolactone group vs placebo.[57] However, the interaction between treatment and CKD273 is not significant statistically. In the present concept, CKD273 is determined the risk of disease progression and may also instigate treatment that is properly examined in the ongoing PRIORITY trial of 3280 participants with T2DM.[58] The remaining main Bioinformatics studies including the SUMMIT study using mass spectrometry to estimate low-molecular-weight metabolites, peptide and 63 proteins by ELISA and remaining 144 proteins as well as d Luminex in a prospective design. However, in these entire universal discovery studies, prediction is never estimated properly on top of existing clinical results.[59]

Therefore, despite few new biomarkers appears to be promising further new exploration is required before exploring in clinical practice and overcoming the problems of specificity and technical variability dominantly.[51] Therefore, it is essential to employ new cutting-edge approach for identifying novel and effective biomarkers for indicating of kidney injury in DKD accurately and "at risk" individuals.

Advantages of urinary miRNAs for kidney diseases

Recently, biopsy appeared to be a gold standard detection and prognostic test for different kidney disease, but it is an invasive and highly costly method without ignoring up to a 3% risk factor.[60] Medically, nearly all patients try to avoid biopsy for DKD detection. As an alternative, they invariably employ introduction of reverse metabolic cytopathology treatment before the structural pathology germinates. Therefore, an exclusive important early detection via markers of pathology is essential.

The renal pathology can be reflected by the urine, a clear biological body fluid.[61] It can be collected noninvasively, easily and at low cost. Different urinary biomarkers possibly increase in diabetic patients even before the symptoms of microalbuminuria and can be utilized as potential marker for identifying several kidney diseases early in normo-albuminuria conditions.[50-52, 62]

The stable Circulatory miRNAs are collected under various storage conditions like in highest PH and long-term room temperature preservation, protection against RNase degradation and multiple repeated freeze-thaw cycles.[63, 64] In recent days different clinical kits are available for rapid and accurate extraction of total miRNAs from urine, depending on the small RNAs binding to specialized material packed in columns.[63] Many signal-amplification procedure for miRNA detection have been developed in different procedures like as hybridization chain reaction, rolling circle amplification, nuclease amplification, catalyzed hairpin assembly amplification and nanomaterials-depending amplification.[65] With the invention of these new technologies, they are extremely possible to estimate different miRNA from urine. Thus, urinary miRNA detections and concentrations are noticed in relation with various important clinical characteristics including histopathological detection. It strongly reveals that urinary miRNAs could be the most promising early tool for noninvasive DKD biomarkers.[66-69]

miRNAs as potential biomarkers in diabetes

A noble class of potent biomarkers, miRNAs have exhibited tremendous promise for expecting and disease states over a decade. These unique tiny molecular RNAs that are responsive agents for gene expression by conjugating target mRNA complementary sequences of a gene.[70] miRNAs are involved in a wide spectrum of biological processes.[71] With their extensive involvement, it is not at all surprising that mis-control of numerous miRNAs is required for devastating impact the development and maintenance of many different diseases, like cancer, diabetes, complications of diabetes, neurological disorders and others.[72] A sporadic progress in miRNA research has not only exhibited a delicate and core understanding of the pathophysiology of kidney disease complexity but has also laid the foundation stone for early identification of unique and new diagnostic biomarkers. miRNA expression is differentially regulated in numerous renal diseases. Most likely miRNAs act as 'molecular signature profiles' for generating different information about disease diagnosis and risk of their gradual enhancement.[73, 74]

Urinary profiling in diabetic patients isolated fold changes of 27 miRNAs that alter according to untreated nephropathy stages.[75] The signature profile of miRNA markers could be recognized for microalbuminuria development nearly 2 years before it was diagnosed.[76] Estimation of five TGFB1-regulated miRNAs in plasma covers two groups of diabetes patients as expected. The progressor groups include people with proteinuria and normal renal function who rapidly increase in a fast rate for estimating GFR reduction. In non-progressor group it includes normal renal function people with proteinuria who maintained steadily normal renal function. They found that two miRNAs, let-7c-5p and miR-29a-3p are related to renal failure protection against quick disease progression, while two other miRNAs let-7b-5p and miR-21-5p are responsible for increased ESRD risk.[77] These studies produce excellent example that urinary and circulatory miRNAs are mis-regulated at an early stage at risk for ESRD progression in diabetic patients.

The miRNAs in different biological fluids like blood, serum, and urine [78] and havoc changes in circulatory miRNAs level have been regulated by different renal diseases.[74, 79, 80] In circulatory blood, miRNAs are protected from degradation by adopting encapsulation in micro-vesicles for example exosomes or conjugating with circulating proteins. Truly, exosomes are enriched in the urine. The exosomes are secreted from all nephrons in cells. They carry different proteins, mRNA and miRNA markers required for renal dysfunction and structural injury.[81] The serum derived circulatory micro-vesicles cannot cross the nephron due to mechanical and charge barriers in the glomerulus. It suggests that urinary exosomes emanate basically from renal cells.[82] The existence of urinary exosomes has considerable potential to demonstrate a reliable early snapshot of overall normal and diseased renal health.[83] Most importantly, urinary exosome-derived miRNAs alteration has been related with early renal damage and renal dysfunction in patients with T2D [84] and enhancement to fibrosis in lupus nephritis.[85]

MicroRNAs as Diabetic Nephropathy Biomarkers

Modern DN detection and continuous monitoring of disease development depends heavily on the isolation of urinary microalbuminuria. However, not all patients with microalbuminuria develop to convert proteinuria and nephropathy, ultimately complicating prognosis. Moreover, kidney tissue damage and activation of abnormal inflammation are found already by the onset of detectable microalbuminuria. Further, microalbuminuria is not a proper feature to the DNA, but it is absolutely a clinical hallmark of glomerular and strong specifical podocyte dysfunction. The conducting of biopsy during the recent diagnostic and prognostic protocols related to intrinsic renal disease, but these more invasive and costly experimental protocols showing only 3 % risk factors in main complications.

A new set of early miRNAs biomarkers have been coming to the limelight for developing potential and improved prognostic and diagnostic experiments for DN patients. They produce earlier warning, more appropriate information on disease progression and ideally treatment plan. Any biomarker selection plans with downstream applicability are observed. Taking consideration of experimental cost, speed, applicability, and feasibility of entry into ongoing treatment plans are necessary for optimized use of versatile resources. Nonetheless, the estimation of different levels in miRNA still produces crucial information on disease methods, DN pathology and the isolation of essential targets for therapy.

miRNA biomarker translation to clinical settings

miRNAs are delicate, highly sensitive, specific, concentrated agents for disease pathology and translatability in between preclinical and clinical environments.[86] However, miRNAs contribution as effective biomarkers for renal damages remains in the budding stage. There is a superficial overlap in between the differences in sketch design, experimental knowhow, and method of ultimate analysis. Moreover, patient variability using race, gender, environmental exposures, and dietary components influence the difference in miRNA levels.[87] Various studies using large, carefully phenotype cohorts will be essential to isolate and validate different miRNA profiles before miRNA translation profiling during clinical practice. These profiling methods consistently and delicately measured disease status. Different basic criteria including estimation of intra- and inter-individual variability, process standardization for different steps of sample collection and different protocols, RNA extraction, miRNA measurement, and analysis are also essential for establishing the minimum interlaboratory biases. Once appropriate protocols for miRNA estimation profiling are defined for interpretation. In fact, relative estimation of multiple study results will be more direct and straightforward. At that point, miRNA biomarkers demonstrate improved performance in disease detection and subsequent improvements in targeted clinical outcomes relative to modern diagnostic steps. In fact, different renal markers related to kidney damage and their improvement are restricted to urinary proteins analysis and GFR alteration.

Numerous circulatory and urinary miRNAs exhibit tremendous promise as real biomarkers improving early diagnosis, prognosis, clinical settings, and monitoring of diabetic kidney disease. The miRNA profiling produces a significant alternate method for the early detection of renal damage in diabetes. The miRNA profiling implementation during clinical management of diabetic kidney disease has a consistent potential for drastic change in patient outcomes. The invasive current gold standard of tissue biopsy, prone to sampling collection error, costly expenses, and cannot be used at early disease detections. However, the aggressive treatments to block gradual progression are initiated. The miRNA biomarkers allow a noninvasive and in expensive methods for detecting an altered approach in renal metabolism long before prominent damage occurs.

Urinary miRNAs signature in DKD

The urinary miRNAs are liberated by the nephron cells and accommodate in the downstream in the urinary tract. They are tightly arranged with membrane-bound extracellular vesicles including micro-vesicles. They are arranged by outward plasma membrane budding, exosomes, discharged from multivesicular endosomes produced in the endocytic tract and apoptotic bodies. Urinary miRNAs in few cases associate with Argonaute proteins and high-density lipoproteins (HDL).[88] The urinary miRNA-192 is intimately related to the amount of albuminuria and TGF-β1 expression in normo-albuminuria and microalbuminuria patients. It is found in a specific study of 80 T2DM patients including 30 normo-albuminuria, 30 micro-albuminuria and/or 20 macroalbuminuria individuals. It is observed that miRNA-192 possesses an area under 0.802 of AUC curve, which is increased than miRNA-194 with an 0.703 AUC and miRNA-215 with an 0.757 AUC. These assessments differentiate the normoalbuminuric group from the micro-albuminuria group. It demonstrates the novel use of urinary miRNA-192 as an early biomarker of DKD.[89]

Exosomes in urine are generated from most kidney cells. **(Figure1)** Thus, they seem to the best-studied vehicles.[90] The exosomes are responsible to transport miRNAs to distant target cells and contribute a critical mechanism for cell-to-cell distant communication.[91, 92] The dynamic expression of miRNA-29 in urine exosomes is used for early biomarker of kidney fibrosis.[93] A recent report [94] explains that the amount of urinary exosome miRNA-451-5p is upregulated in diabetic primate animals 3 weeks prior to marked albuminuria and 3 weeks before histologic al difference of kidney fibrosis. Conversely, the amount of renal miRNA-451-5 produces strongly and is inversely related to the indices of renal pathology at the time of DKD progression. As a result, urinary exosome miRNA-451-5p are released into the urine using the injured nephron in the early DKD state. Thus, its expression and progression exhibit a useful prognostic mark in the early and sophisticated noninvasive indicator of renal diseases and failures.

The urine samples are collected and estimated from a historical cohort regarding 906 eligible participants along with a diagnosis of T1DM from 1950 to 1980. Using Bayesian protocols to normalize and change raw signals, the urinary miRNA profile is estimated from 30 patients to their expression ratios. From normoalbuminuric T1DM 723 patient unique urinary miRNAs who did not generate DKD relative to patients. They generate micro-albuminuria subsequently are estimated by qPCR. Among different urinary miRNAs, eighteen miRNAs are strongly related to the sequential microalbuminuria development. In addition, 15 miRNAs expression observe gender-wise differences. The signature of different miRNAs including signature miRNA-1972, miRNA-28-3p, miRNA-30b-3p, miRNA-105-3p, miRNA-363-3p, miRNA-424-5p, miRNA-486-5p, miRNA-495, miRNA-5480-3p and for women miRNA-192-5p, miRNA-720 notice high internal evidence for the future microalbuminuria development.[95] Despite there are also few prominent restrictions in this estimation. Initially, the small number of patients producing precise estimation for expression changing is rather challenging. It is reflected in greater confidence intervals of many microRNAs and their apparent reduction of prognostic significance. Lastly, they never execute renal biopsies therefore it is hardly possible to relate the expression of urinary miRNA with tissue pathology.

Currently, Numerous studies are conducted for meta-analysis to select out critical urinary miRNA biomarker candidates in DKD. The results of meta-analysis suggest that two urinary miRNAs, miRNA-126 and miRNA-770 family are showed a well coordination dysregulation significantly DKD patient blood and urine. Therefore, they utilize as crucial diagnostic and pathogenetic implementations for DKD. Another bioinformatic study explain six stably dysregulated miRNAs miRNA-21-5p, miRNA-29a-3p, miRNA-126-3p, miRNA-192-5p, miRNA-214-3p, and miRNA-342-3p in DKD patients relative to controls. These six miRNAs are critical for DKD pathogenesis and possibly utilize for generating consistent potential biomarkers. (**Table 1**)[96]

In conclusion, different urinary miRNAs act as potential early biomarkers not only for enhancing of albuminuria like a surrogate marker of glomerular filtration barrier (GFB) integrity but may also expect the gradual development of DKD prior to the onset of GFB breakdown. Conversely, these studies still suffer with tight restrictions, using low numbers of recruited patients and not consistently in preclinical and clinical studies. Further multiple investigations including combination of discovery/validation parameters or executing larger cohorts are essential for identifying more appropriate urinary miRNA expression profiles for early diagnosis and damage stratification in patients with diabetes mellitus.

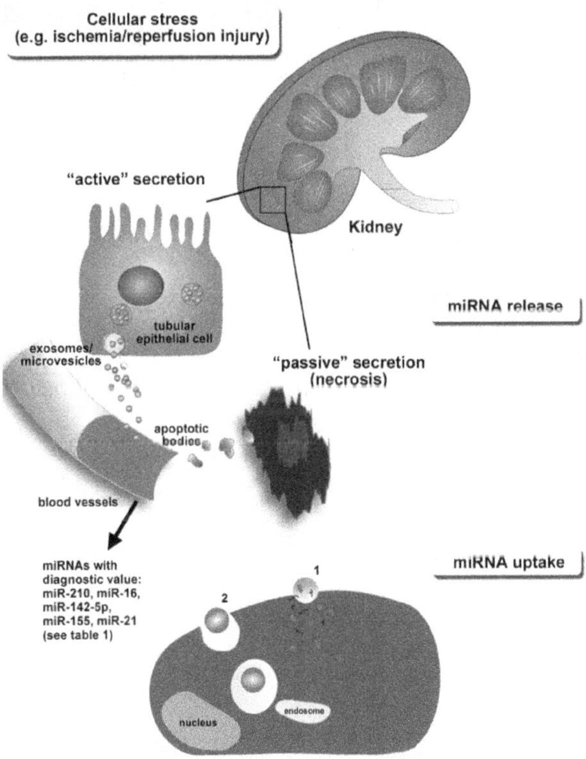

Figure-1. Cellular microRNAs are liberated from kidney cells (*e.g.*, tubular epithelial cells) to the circulation using exosomes/micro vesicles or in apoptotic bodies after prominent cellular injury. Micro vesicles (1) then fuse with the cell membranes, whereas exosomes are processed through the endosomal compartment of Kidney cells. (Loosely Adapted, Redraw and Modify using skeleton from Lorenzen and Thum, 2012)

Table 1. miRNAs for potential DKD biomarkers

miRNAs	Sample	Source	Expt. Methods	Study Population	Outcome/DKD
miR-192	80	Urinary extracellular vesicles	RT PCR	T2DM patients	miR-192 (AUC = 0.802)/Early
miR-29c	32	Urinary exosome	RT PCR	CKD patients	miR-29c ($r = -0.359$; $P < 0.05$, AUC = 0.738)/ Tubulointerstitial fibrosis
miR-451-5p	43	Urinary exosome	RNA sequencing	Diabetic rats	Increased miRNA-451-5p (>1000-fold)/ 3-6 weeks
miR-133b,	156	Urinary exosome	Bioinfo. analysis	T2DM patients	Elevated miR-133b, miR-
miR-342, MiR-30			RT PCR		342, and miR-30a ($P < 0.001$)
miR-2861, miR- 145	145	Urine	miRNA profiling, DM patients		Reduced miR-2861, miR-
1915-3p, miR-4532			urine miRNA expression ana., and in situ hydridization		1915-3p, and miR-4532 (>10-fold, $P < 0.0001$)
miR-126.	2747	Urine	Meta-analysis	DM patients	Upregulated miR-126
miR-770					(95% CI: 9.96-862623.21) and miR-770 (95% CI: 2.37-44.25)

miRNAs in DKD pathogenesis

The involvement of miRNAs in kidney diseases has been noticed strongly with a devoid of podocyte in Dicer. An important nuclease conducts in miRNA biogenesis. It demonstrates that miRNAs are involved in kidney diseases.[97-99] At least five miRNAs including miRNA-192, miRNA-194, miRNA-204, miRNA-215, and miRNA-216a are required to provide a crucial impact on kidney function in an early study. Their amounts are highly dominant in the kidney relative to other organs.[100] Thus, the role of miRNA in kidney damage is exclusively important.

Our focus on the molecular mechanisms during diabetic states results in kidney damage. It is enhanced over the last decade. Diabetic state promotes inflammation, fibrosis, and hypertrophy in renal cells via numerous cytokines and TGF-β1 growth factors.[5] The cytokines involvement and receptor growth factors instigate signal transduction cascades that cause transcription factors activation for inducing expression of inflammatory and fibrotic genes. It also responsible for changing the epigenetic states like DNA methylation, chromatin histone modifications, and noncoding RNAs including lncRNAs and miRNAs.[23] miRNAs promote diabetic states that induce pathological genes expression by mediating many post-transcriptional and post-translational mechanisms.[23] Kato et al [101] were the first to describe involvement of a specific miRNA in DKD. The first miRNA related to DKD miRNA-192 is induced *in vitro* in mesangial cells (MCs) and *in vivo* in glomeruli from type 1 streptozotocin (STZ)-promoted and type 2 db/db animal models of DKD. Strong suppression of miRNA-192 induces collagen deposition I related to TGF-β.[101] In last few decades, several miRNAs are involved in the pathogenesis as a critical member and germinate DKD via participation in fibrosis, inflammation, hypertrophy, autophagy, ER stress, oxidative stress, insulin resistance, and podocyte injury. (**Figure-2**)

miRNAs of kidney fibrosis in DKD progression

A feature for progressive tissue scarring kidney fibrosis leads to glomerular and tubulointerstitial fibrosis. It is one of the dominant pathological features of ESKD.[102] The specific steps of many molecular issues in kidney fibrosis, however the event is not sufficient to explain completely. Recently, it suggested that TGF-β works as a crucial driver of matrix synthesis, reduction of matrix degradation and myofibroblast augmentation. So, it is addressed as a master controller of the above process.[102]

The important isoform of TGF-β, TGF-β1 works via a canonical signaling procedure that is required for phosphorylation and upregulation of Smad2 and Smad3 using the TGF-β receptor 1, TGFR1 that also referred as ALK5. Smad4 then conjugates enhancement of Smad2/3 that able to generate a complex for translocating to the nucleus and specific genes transcription. However, TGF-β1 induction of TGFR1 is responsible for induction of wide range of SMAD-independent processes termed as noncanonical signaling for altering cell function. These non-SMAD methods are required for TGF-β induced kinase 1, PI3K-AKT, and Rho-like GTPase signaling processes.[102]

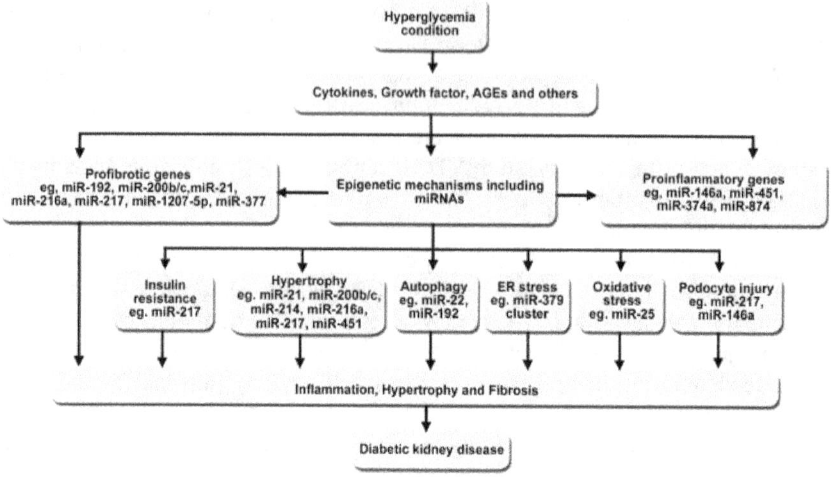

Figure-2. Processes in which miRNAs induce diabetic kidney disease (DKD) pathogenesis. Hyperglycemia influences cytokines, growth factors, and dysregulation of miRNAs. miRNAs are responsible in the DKD progression targeting multiple genes related to fibrosis, inflammation, hypertrophy, autophagy, ER stress, oxidative stress, insulin resistance, and podocyte injury. (Loosely Adapted, Redraw and Modify using skeleton from Cao *et al.*, 2018)

Recently many miRNAs control TGF-β1 during DKD development. Several miRNAs miRNA-2, miRNA-192, miRNA-200b/c, miRNA-216a, miRNA-217, and miRNA-1207-5p are activated in TGF-β1-treated murine kidney cells or in glomeruli of diabetic animal models.[103] The miRNA-192 activates Collagen 1α (Col 1α) in MCs, a critical genes related to DKD pathogenesis.[101] miRNA-192 also requires for induction of other miRNAs miRNA-216a and miRNA-217 related to DKD cellular hypertrophy.[104] miRNA-200b/c, a subset for a greater components of miRNA 200 family including miRNA-200a, miRNA-200b, miRNA-200c and miRNA-141 places in miRNA-192 downstream. In addition, miR200b/c activates Col 1α strongly, Col 4α in MCs and diabetic mice.[105] The miRNA-377 inhibits p21-activated kinase, superoxide dismutase and enhances fibronectin (FN) accommodated in DKD mouse model.[106] Moreover, miRNA-1207-5p is induced by TGF-β1 that activate TGF-β1 expression, PAI-1, and FN in MCs.[107]

In the opposite way, miRNA-200a and miRNA-141 are downregulated by TGF-β1 and resist kidneys from fibrosis by inhibiting ECM deposition. They also protect against epithelial to mesenchymal transition (EMT).[103] Moreover, Let-7 family members are also reduced in cultured human proximal tubular epithelial (HK2) cells. It also treats TGF-β1 for activation fibrosis through the TGF-β1R1 and enhanced collagen expression.[108,109] Indeed, TGF-β1 expression are controlled by miRNAs.2). TGF-β1 amount is activated by miRNA-192 and miRNA-200b/c in diabetic mice (STZ and db/db).[105] Moreover, miRNA-22 reduces bone morphogenetic protein-6 (BMP-6) and BMP-7 and further upregulates TGF-β1 signaling.[110] miRNA-433 induces TGF-β1 signaling and fibrosis by associating with antizyme inhibitor 1, an important regulator of polyamine synthesis.[111] These miRNA-controlled circuits activate TGF-β1 signaling requiring for DKD progression. On the other hand, *in vitro* functional studies isolate type II TGF-β receptor (TGFRII), Smad3, and TGFβ-1 itself as possible miRNA-23b targets. It implicates a reverse feedback loop controlling TGF-β-1 signaling.[112] Further miRNA-29b can also repress TGF-β1 expression and thus blocks the TGF-β1-dependent kidney fibrosis.

MicroRNAs in Fibrosis and chronic Diabetic Nephropathy: Detailed miR-21 contribution

TGF-β is a crucial cytokine with a tremendous basic role in Diabetes Nephropathy (DN) via its contribution in fibrosis and scarring. TGF-β signaling is controlled using miR-21 that possibly in turn control proccssing of mature miR-21 expression,[113], Moreover, the TGF-β1 3′-UTR is required for post-transcriptional protocols via miR-744.[114]

Initially, DN fibrogenesis requires E-cadherin inhibition mediating through TGF-β1. As noted above, TGF-β1 is involved in miR-192 overexpression and collagen alpha-2(I)[115] at the onset of renal injury, mouse mesangial cells. In contrast, miR-192 down expression during the mature-stage human DN renal biopsy samples packed with a low level of glomerular filtration rate and tubulointerstitial fibrosis.[116] An Identical TGF-β1-dependent reduced miR-192 level represses ZEB1 and ZEB2 expression by the help of induced transcription. Thus, it downregulates E-cadherin and exert an anti-fibrotic effect.[116] Further in other study a similar result is observed. Here TGF-β1-dependency of miR-192 reduction allows E-cadherin blockage.[117] Thus, pleiotropic contribution of kidney miR-192 gene along with their anti- and profibrotic effects that are superficially cell-type dependent.[118]

MiR-21, miR-200 and miR-29 Families in Diabetic Nephropathy

The miR-200 family of miR-200a, miR-200b, miR- 200c, miR-429 and miR-141 has been ascribed roles in maintaining epithelial differentiation, suggesting anti-fibrotic functions in DN.[119] Downregulated expression of these transcripts by TGF-β has been shown in cancer cell epithelial-to-mesenchymal transition (EMT).[120] Conversely, upregulated miR-200b, miR-200c and miR-192 expressions are observed in TGF-β-treated mouse mesangial cells.[121]

Many miR-29 family members miR-29a, miR-29b and miR-29c have shown anti-fibrotic effects in DN. A miR-29a transgene transcription induces renal function and improved podocyte viability relative to wild-type diabetic mice.[122] Functional inhibition of miR-29a activates histone deacetylase expression for leading podocyte apoptosis, sequential renal dysfunction and proteinuria.[122]

The expression of most widely studied miRNAs related to DN also alters other renal damages. Isolation of potential biomarker miRNAs can be further utilized as therapeutic agents for protecting against renal failure.

miRNAs of Inflammation in DKD progression

Induced inflammation is a dominant sign for diabetes,[123] and this proinflammatory hallmark plays a crucial important role in DKD induction and development.[124, 125] Multiple inflammatory factors like macrophage infiltration, inflammatory cytokines and different chemokines including MCP-1, IL-6, TNF-α, PAI-1, CXCR4 induce myofibroblasts at injury sites in the kidney. They also improve the MCs differentiation, glomeruli and renal tubular epithelial cells into fibroblasts that causes in improved ECM production and deposition for promoting fibrosis.[126-129]

miRNA-146a is recognized as an anti-inflammatory miRNA.[130] The miRNA-146a level upregulates both peritoneal and intrarenal macrophages in diabetic mice. The deletion of miRNA-146a allows to the improved M1 enriched marker expression and reduction of M2 markers in macrophages. It also upregulates of proinflammatory cytokines expression addressing that miRNA-146a plays a critical insulator and anti-inflammatory role during DKD pathogenesis.[131] Induction of nuclear factor-kappa B (NF-κB) is combined with inflammation during DKD development.[132-134] Moreover miRNA-451 is reduced in diabetic kidneys of mice and MCs cultured in enriched glucose conditions. It also inhibits NF-κB activity and reduces proinflammatory molecules in MCs transcription.[135] miRNA-374a is reduced considerably in DKD patient kidney. [136] Moreover, miRNA-374a inhibits inflammatory signals in DKD through reducing of IL-6, TNF-α, and MCP-1.[136] Recently, it is reported [137] that miRNA-874 is inhibited in STZ-dependent DKD rats. Further miRNA-874 upregulation with mimics attenuated the inflammatory signals via reducing IL-6, L-1β, and TNF-α.

miRNAs on hypertrophy for DKD progression

Hypertrophy is one important and crucial feature of DKD. Recently, many hypertrophy-associated miRNAs have been isolated. miRNA-216a and miRNA-217 enhance Akt, an important hypertrophy-associated kinase by focusing PTEN in the STZ and db/db diabetic model of DKD and MCs treated with TGF-β1.[104] During diabetes, miRNA-21 also focuses PTEN and upregulates Akt for expressing renal hypertrophy.[138, 139] PI3K is the upstream enhancer of Akt. miRNA-200b/c induces Akt by targeting the FOG2, a unique PI3K inhibitor.[140] The role of miRNA-451 is observed in mesangial hypertrophy.[141] It demonstrates that miRNA-451 control inversely the Ywhaz expression, a novel protein requires for induction of p38 MAPK signaling via the Ywhaz 3′UTR binding. The miRNA-451 is reduced strongly for induction of hypertrophy via Ywhaz. miRNA-34a is upregulated in MCs under high glucose conditions in db/db diabetic mice.[142] Reduction of miRNA-34a can alleviate glomerular hypertrophy via affecting of growth arrest specific 1(GAS1).[142] Currently, induced miRNA-214 level is related to reduced levels of PTEN. It also induces Akt phosphorylation to renal hypertrophy contribution in renal glomerular MCs and proximal tubular epithelial cells.[143] miRNA-181a is increased strongly in MCs exposed to high glucose and it reduces the mammalian target of rapamycin (mTOR) inhibitor Deptor and enhances mammalian target of rapamycin complex 2 (mTORC2) for negotiating TGF-β1-induced glomerular MC hypertrophy.[144]

miRNAs in Autophagy for DKD progression

A lysosome destruction pathway, autophagy plays an essential critical role in liberating protein aggregates and harmed for mitochondria, an excess organelles for maintaining intracellular homeostasis.[145] Thus, autophagy improves cellular health against different stress conditions invariably in hypoxia, ER stress, or oxidative stress.[146] The environmental condition including cellular stresses and excess nutrition intake are enhanced by multiple metabolic dysfunction including diabetes, autophagy damage via induction of mammalian target of rapamycin complex 1 (mTORC1) and the decrease of AMP-inductive kinase (AMPK) and Sirt1 activity.[147, 148] Thus, autophagy opens up an instructive and critical role in homeostasis maintenance in numerous organs, particularly metabolic organs. The blockage and impairment of autophagy is required in the pathogenesis of metabolic diseases Like DKD.[148, 149]

Nowadays, many miRNAs play an inductive role in DKD by controlling autophagy. In different miRNAs, miRNA-22 is responsible for autophagy repression. The endogenous miRNA-22 repression enhances autophagy, elevates high glucose-dependent Col 4 and the amount of α-SMA in NRK-52E cells.[150] The autophagy gene expression is reduced in kidneys of STZ-dependent and db/db diabetic mice. [151] miRNA-192 inhibitors can suppress autophagy genes in these model organisms. The miRNA-192-KO in diabetic mice reduces autophagy genes and is also attenuated. The miRNA-192 mimic oligonucleotides down regulate autophagy gene expression in mouse glomerular MCs with TGF-β1 stimulation. It reveals that miRNA-192 acts as most effective therapeutic target for DKD.[151]

The miRNA-217 level is upregulated in highly glucose-dependent podocytes. The downregulation of miRNA-217 are inverse protectively to high glucose-sensitive podocyte damage. The insulin insulation are often restored the defective autophagy processes through targeting PTEN.[152] Multiple miRNAs are important for oxidative stress and ER stress during DKD pathogenesis. Nearly 40 miRNAs including the miRNA-379 cluster are controlled by ER stress in DKD.[153] NOX4 is an important catalytic subunit of NADPH oxidase under hyperglycemia. The miRNA-25 inhibitor upregulates NOX4 mRNA and protein level in STZ-dependent diabetic models. It indicates that reduced miRNA-25 level increases NOX4 for inducing oxidative stress and renal dysfunction.[154]

Therapy development

miRNA-dependent therapies for DKD treatment

As indicated above many potential drugs agents lead to the pathogenic signaling including inflammation and TGF-β1 for DKD treatment mostly via protein-coding genes under complex development.[155] However, due to few limited number of protein-coding genes, noncoding RNAs, miRNAs are attractive proposition for most effective new drug candidates.[156] Distinct features of miRNAs including short sequence and their high homology across multiple vertebrate species make them potentially suitable as therapeutic agents.[157] Numerous studies report that miRNAs show abnormal deregulation in DKD. **(Table-2)** The critical therapeutic miRNA components are explored in fundamental and mechanistic studies using unique processes blocking DKD-related miRNAs or upregulate kidney-protecting miRNAs.[2]

Fundamentally, manipulated induction of miRNA levels are investigated by *in vivo* mimic delivery for restoring miRNA amounts or silencer to block miRNA function.[158] The miRNA mimics are double-stranded artificial oligonucleotides (oligos) that leads to organize the endogenous miRNA function. However, their novel chemical modifications improve stability and are effectively engulfed by the cells. The proper adopting strategies till now prevent the miRNA function using chemically modified anti-miRNA oligos (AMOs). The AMOs are designed against the mature miRNA sequence that shows a steady state level in circulation and cause the cell permeability. In fact, a chemically modified oligo can reduce a cluster of close to 40 miRNAs related to a decrease in glomerular ECM and hypertrophy in diabetic animals.

Moreover, miRNA inhibitor AMOs, miRNA blocks their expressions by achieving the miRNA-target sequences for capturing pathogenic miRNAs (miRNA spongites e). Short hairpin RNA plasmids to abnormal miRNA expressions occur using RNA interference, miRNA deletion, or via complementary oligonucleotides. They are done either to the 3' untranslated sequence of the mediating mRNA binding site sequence (masking approach) or to the miRNA sequence (erasers).

Table-2. Different miRNAs for DKD therapeutic targets

miRNAs	Model examined	Targets	Pathological output
miR-192	MCs, STZ-mice, db/db mice	SIP1, Zeb1	↑Col1α1 and Col1α2
	MCs, STZ-mice, db/db mice	miR-216a, miR-217	↑MC survival, Hypertrophy
	MCs, STZ-mice, db/db mice	miR-200b/c	↑TGF-β1, Col1α, Col4α
miR-216a, miR-217	MCs, STZ-mice, db/db mice	PTEN	↑MC survival, Hypertrophy
miR-200b/c	MCs, STZ-mice, db/db mice	Zeb1	↑TGF-β1, Col1α, Col4α

miR-377	MCs, STZ-mice	PAK1, SOD	↑FN
miR-1207-5p	MCs, RPTEC	G6PD, PMEPA1, PDPK1 SMAD7	↑TGF-β1, PAI-1, FN
miR-146a	STZ-mice	Traf6, Irak-1	↓TNF-α, MCP-1, IL-1β, IL-18 ↓Col1a2, Col4a2, PAI-1, TGF-β1
miR-451	db/db mice	Ywhaz, p38 MAPK	↓MC proliferation, mesangial hypertrophy
miR-21	MCs, db/db mice	PTEN	↑ Hypertrophy, COL1α2, FN ↑TGF-β1, NF-κ
miR-22	NRK-52E cells, STZ-rats	PTEN	↓Autophage, Col4, α-SMA

SIP1: SMAD-interacting protein 1; ZEB1: zinc finger E-box-binding homeobox 1; PTEN: phosphatase and tensin homolog; PAK1: p21-activated kinase; SOD: superoxide dismutase; G6PD: glucose 6 phosphate dehydrogenase; PMEPA1: prostate transmembrane protein, androgen induced 1; PDPK1: 3-phosphoinositide-dependent protein kinase-1; SMAD7: SMAD family member 7; Traf6: TNF receptor-associated factor 6; Irak-1: Interleukin-1 receptor-associated kinase 1; Ywhaz: tyrosine 3-monooxygenase/tryptophan 5-monooxygenase activation protein zeta; p38 MAPK: p38 mitogen-activated protein kinases.

==

Locked nucleic acid (LNA)-attached oligos are one of the specific and dramatic modifications for reducing miRNA levels.[159] LNA-conjugated anti-miRNA-192 strongly knockdown miRNA-192 activity and its downstream cluster including miRNA-216a, miRNA-217 and miRNA-200 family. It also knockdowns p53 and inhibits collagen, TGF-β, gene expression and FN in kidneys of diabetic mice.[160] Transportation of miRNA-21 inhibits plasmids, carrying LNA- anti-miRNA-21 into the diabetic kidneys of db/db mice at 10 weeks markedly attenuated microalbuminuria, kidney fibrosis, and inflammation at 20 weeks old mice.[161]

Nowadays, many miRNA-mediated therapeutic procedures are being explored during clinical trials. Among them, the first one is an LNA- anti-miRNA-122 (Miravirsen) for targeting hepatitis C virus RNA.[162, 163] in a phase 2 clinical trial. Miravirsen is responsible for dose-dependent antiviral therapy active over 4 weeks. The silencing agent of miRNA-21 (RG-012) has an effective effect in kidney fibrosis using kidney biopsies of model mice. However, the effect is suddenly blocked during clinical trials based on chronic toxicity. The miRNA-29 upregulation appears to be required for an instrumental anti-fibrotic effect. The miRNA-29 mimic (MRG-201) is examined during Phase II clinical trial. It involves treatment of patients carrying a predisposition for keloid formation.[164] Indeed, the anti-fibrotic effect of miRNA-29 mimic is not related to skin fibrosis but might be involved in ESKD along with DKD kidney fibrosis.

The major problem and prevention in the application of therapeutic miRNAs relies on an effective and easy delivery system. The use of naked miRNAs is degraded by cytoplasmic nucleases rapidly and discharged through renal excretion. [165, 166] In addition, miRNA insertion in live cells instigates innate immunity, resulting an unwelcomed the toxicities.[167] The miRNAs are designed and transported their respective pathways in multiple manners. Therefore, they cause various off- target gene silencing for producing toxic effects. To prevent any such off-target effects and nonspecific immunity miRNA should be modified and transported delicately and effectively.

One of the defined multiple approaches is to use chemically modified miRNAs for preventing its fragile character and immune responses. Multiple chemical modifications can be utilized including (a) ribose 2′-OH group modification, (b) backbone modification (c) peptide nucleic acids (PNA) modification and (d) LNA modification that are responsible for elimination of different off-target effects.[168] Other procedure is use of different conjugates along with small transport domains like aptamers and cell-penetrating peptides etc. Aptamers characterize single-stranded oligonucleotides carrying three-dimensional structure. They are responsible for binding receptor surfaces along with more affinity and potentiality for efficient drug delivery. Aptamers are unique for easy targeting because they are noncytotoxic, non-immunogenic, with efficient tissue transportation, easily modify and low cost.[169] The miRNAs amalgamation with specific aptamers are referred for specific target ligands of the nucleic acid to cells that allure aptamer.[170-172] The cell-inserting peptides enables to conjugates cell and endosomal membranes. With the generating of different nanotechnology, nanocarriers become attractive for miRNA delivery for inducing cellular uptake and improved delivery efficiency and elimination of toxicity.[173] Nanocarriers are stable, not risky and safe. They require easy manufacturing. Moreover, they also contain low immunogenicity, cheap, and multiple forms.[174] In near future, novel miRNA therapies along with efficient and productive delivery vehicles combine for an efficient and potent treatment plan for DKD.

MicroRNAs as DN Therapeutic Targets

In most therapeutic studies disease protecting miRNAs control oncogene expression in cancer. They are often termed as oncomiRs. The reduction of oncogenes through small molecule inhibitors has a narrow therapeutic sensitivity due to complex carcinogenesis. Such complication is possibly reflected and raised a mirror image in other multifactorial diseases including DN.

The transcription of several oncomiR mRNAs not only targets cancer alone but possibly affects multiple procedures. miR-34a is multifactorial mRNAs. It affects diversified pathways including proto-oncogene c-Met, cell cycle regulator cyclin-dependent kinase 4 and B cell lymphoma 2, an anti-apoptotic oncogenic protein etc.[175] A miR-34 sensitive mimic molecule is utilized as therapy of lung cancer causing amelioration of disease development that is related to suppression of all three potential targets.[176] In their use, the generating harmful off-target effects must also be conceived.

The current requirement for many fruitful early-stage DN treatment remains obscure. As reported above, Numerous miRNAs lead to most effective early DN biomarkers. Recognition of these miRNAs are selected depending on the function and/or GWAS study. Their outputs lead to exploration of their usefulness as potential therapeutic agents for protecting against disease progression.

Multiple miRNAs levels are effective *in vivo* for diabetic models. The miR-192 is appeared to be reduced using a locked nucleic acid-modified anti-miRNA in the renal cortex for inducing renal fibrosis symptoms.[177,178] Following the similar trend, miR-29c reduction causes protection against DN development in db/db mice.[179] A low-dose effective anti-cancer molecule paclitaxel induces renal function, prevents Smad2/3 induction, and inhibits miR-192 transcription.[180]

No renal-specific miRNAs have been identified so far, but quick uptake circulatory miRNAs using renal proximal tubular epithelial cells promotes pinpointed focused delivery to the kidney. Recently, ultrasound-microbubble-dependent gene transport has been utilized for shifting plasmids to mammalian kidneys.[181] Microbubbles, a newly developed ultrasound contrast agents known as 'theragnostic' components are used in diagnostics and nowadays in therapy based on the ultra-sound parameters used.[182] Ultrasound with great intensity requires microbubble oscillation. It causes in 'inertial cavitation' and liberates different therapeutic agent(s) conjugated to or attached within the microbubbles.[182] Microbubble technology is often used for downregulating miR-21 transcription in db/db mice ameliorating microalbuminuria, renal fibrosis, and inflammation etc.[183]

microRNAs of kidney related disorders

In renal Fibrosis

Fibrosis is one of the most dominating organ dysfunctions in diseases. It occurs either for uncontrolled reaction for chronic tissue injury or as the primary disease itself in predisposed individuals.[184] Renal fibrosis allows a tremendous accumulation of extracellular matrix that ultimately leads to the last stage of renal failure. TGF-β is a critical regulator of numerous miRNAs in renal fibrosis as in diabetic nephropathy and is a key pathological inducer of fibrotic diseases. TGF-β enhances renal fibrosis by regulating renal miR-433 expression.[185] The TGF-β/Smad pathway is the ultimate mechanism, by which miR-21 expression is upregulated in fibrotic tissues. A differential transcription of miR-205 and miR-192 occurs in IgA nephropathy and these modulations are related to disease severity and progression.[186]

In Lupus Nephritis (LN)

Lupus Nephritis (LN) has a well-known and devastating outcome-defining many problems in Systemic Lupus Erythematosus (SLE) that influences up to 60% of patients at some point of the disease. The miRNA expression profiles of renal tissue have gained enormous focus since a wide detection of differentially expressed miRNAs in lupus nephritis kidney biopsy samples from 11 patients with three controls *in vivo.*[187] Furthermore, miR-423 and miR-663 are silenced in lupus nephritis.[188] Thus, miRNA expression patterns are cell and organ specific. The levels of multiple miRNAs including miR-200a, miR-200c, miR-141, miR-429 and miR-192 are reduced in patients with active lupus nephritis than those in healthy controls.[189] In another study, the separate expression of miR-638, miR-663, miR198, miR-155, and miR-146a are found from dissected glomeruli, tubules and interstitial tissues from kidney biopsy.[189] Glomerular tissue from lupus nephritis patients exhibits an overexpression of miR-146a in lupus nephritis tubule-interstitial tissue.[190] Finally, the multiple sets of miRNAs are differentially expressed in lupus nephritis patients from different racial groups. This focus leads the potentiality of miRNAs as biomarkers for the diagnosis or prognosis of lupus nephritis.[190] Further a combination of miR-192 and miR-27b from urinary exosomes differentiate lupus patients with or without nephritis [191] while the levels of miR-221 and miR-222 in urinary sediment are negatively related to serum anti-dsDNA level in patients with active lupus nephritis.[192]

In Polycystic Kidney Diseases (PKD)

Polycystic kidney disease is a predominant reason for last stage of renal failure and is the effect of ultimate progressive growth of renal cysts. The mechanisms instigate cyst formation in the kidney by the deregulation or mutation of the effected genes polycystic kidney disease 1 (PKD1), polycystic kidney disease 2 (PKD2) and polycystic kidney and hepatic disease 1 (PKHD1).[193] The mutations are possibly either autosomal dominant or autosomal recessive. PKD leads the uncontrolled protein translation during cell division. The autosomal dominant polycystic kidney disease (ADKP) is detected at 80% of kidney patients with hypertension before the loss of kidney function.[194] The relevant miRNAs in PKD are first discovered in rat. These miRNAs play an effective contribution for therapy and early and authentic diagnosis of the disease.[195] Multiple miRNA expression patterns are found in the kidneys of a PKD rat model, where the abrupt miRNAs deregulation lead many signaling pathways, (mTOR signaling, mitogen-activated protein kinase signaling, Wnt signaling and the TGF-β pathway). They participate in PKD with the alteration of these gene expressions that promote disrupted cell division and proliferation.[196]

There are 30 miRs in the kidney that have been expressed differently in a rat model. Only miR-21 and miR-217 are recognized earlier in the kidney tissues.[197] In another report, miR-17 interacts directly with PKD2 and has a potential role in cytogenesis.[198] Moreover, miR-15 downregulates its expression in cell division cycle 25A (Cdc25A), leading to inhibit the cell cycle and proliferation. Silencing in miR-15 transcription facilitates the overexpression of Cdc25A, thus inducing cyst growth in disease conditions.[199] The expression of the PKHD 1 gene is downregulated by miR-365-1 leading in repressed cell adhesion in PKD.

In Renal Cell Carcinoma (RCC)

Renal cell carcinomas (RCC) are members of a carcinoma family that produce from the epithelium of the renal tubules and encompass for nearly 90% of all renal malignancies.[200] Renal cell carcinoma is a major invasive, popular, and well-known neoplasm of the adult kidney, considering 2%–3% of adult malignancies. Clear cell Renal Cell Carcinoma (ccRCC) is the initial and budding stage of renal cancers that is asymptomatic medically and problematic for identifying the proper diagnosis. The abnormal and abrupt miRNAs expression is related to various kinds of cancers and opens up a new avenue in renal cancer development.

Several miRNA expressions are examined in esophageal squamous cell carcinoma (ESCC) for renal cell carcinoma detection. They compare miRNA expression profiles between RCC and normal kidney tissues.[201-203] In the studies, miR-424 is linked to RCC. [201] Further, miR-203 silencing is found in different cancer types.[204-206] The miR-203 expression directs GSK-3β for inducing RCC pathway.[207] GSK-3β is established to enhance p53 mRNA translation using phosphorylating RNPC1. Thus, miR-203 is one of the potential factors either directed or established by an epigenetic mechanism or by other biological developmental processes. The miR-708 interacts BMI1 and ZEB2 and increases apoptosis in cellular pathways.[204] This miRNA controls apoptosis and cell proliferation potentially. (**Figure 3**) The miR203 drives an anti-apoptotic role like inversely controlling GSK-3β expression in a p53 reduced level. Apoptosis is instigated by miR-708 expression that acts as BMI1 and ZEB2 interactions. The effective silencing of two miRNAs miR 141 and miR-200c promotes Cancer development that enhance the amount of ZFHX 1B and a reduction in cadherin levels.

In Nephrotic Syndrome

Nephrotic syndrome is related to an enhanced permeability across the glomerular filtration barrier due to methods that change the dynamics and perm selectivity of glomerular filtration. It is the most prominent definitive clinical abnormalities for example proteinuria in the nephrotic range, hypoalbumenia, edema, and hyperlipidemia. The urinary miR-638 level is highly decreased in adult patients along with nephrotic syndrome without considering their inherent pathology.[208] However, miR-30a-p concentrations in serum are upregulated in nephrotic syndrome.[209] The concentrations of multiple miRNAs including miR-30-5p, miR-152-3p, miR-150, miR-191 and miR-19b are modulated in the serum and the urine of patients recognized nephrotic syndrome. Almost all miRNAs upregulate their expression in diseased patients. This enhanced expression is also considered to reduce for successful treatment.[209]

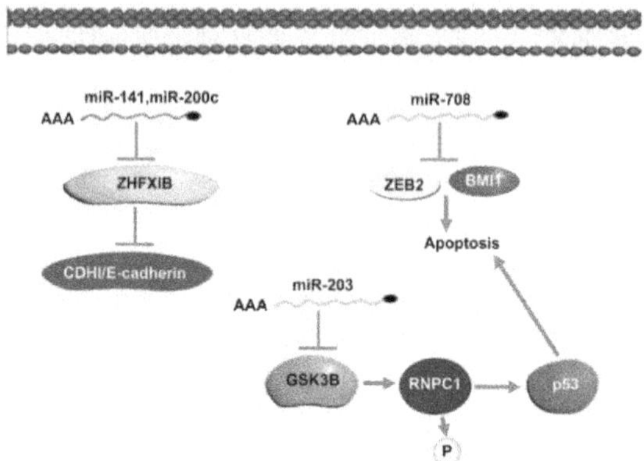

Figure-3. MicroRNA in Renal Cell Carcinoma (RCC). miR203 opens up an anti-apoptotic role via inhibiting the GSK-3β expression, causing p53 silencing. The miR708 activates Apoptosis that directs BMI1 and ZEB2 genes. The miR-141 and miR-200c silencing causes Cancer progression that increases ZFHX 1B and reduces the cadherin amounts. (Loosely Adapted, Redraw and Modify using skeleton from Mukhadi *et al.*, 2015)

In Wilms's Tumor

Wilm's tumor (WT) is also recognized as nephroblastoma. It is a renal cancer that develops predominantly in children and rarely in adults. Most prominent location of their development is embryonic kidney.[205] The miRNA expression is estimated in Wilms' nephroblastoma many occasions [206, 207, 210] Many mutations have been detected in Dicer1 (instrumental miRNA-processing enzyme) in patients with Wilm's tumor. The oncomir-1 gene is markedly induced along with Wilm's tumor patients using the transcription factor E2F3.[206]

The onco-genetic transcription factor STAT3 increases the miR-370 expression that suppresses FAM123B level for controlling the Wnt signaling pathway. The miR-562 interacts with eyes absent homolog1 (EYA1) within the 2q37.1 region.[211] EYA1 are detected to silence in Wilm's tumor regulation. Furthermore, studies of miR483-3p expression are upregulated in Wilm's tumor samples relative to the normal controls.[212] The miRNA transcription factor regulator (Six1), miR-185, is silenced in pediatric malignancies including Wilm's tumor.[213] Thus reported the role of the miR-17-92 miRNA cluster in Wilms' tumors. The miR-17-92 miRNA cluster in Wilms' tumors cluster performs a prominent role in promoting tumor cell proliferation.[207, 214] The miR-17-92 cluster is known as cell cycle-controlled locus. It governs ectopic expression of a miR-17-5p that is well enough to conduct a proliferative signal.[215] The miR-193a that silences in Wilms' tumor (WT1) gene synthesizes a transcriptional factor and functions as a master regulator for podocyte homeostasis.[216]

In IgA Nephropathy

IgA nephropathy (IgAN) is a frequently distributed inflammatory disease of renal corpuscles. It is instigated by the immunoglobulin IgA deposition in the kidney. The miRNAs contribute a potential treatment plan for inflammatory pathogenesis and immune disease including IgAN, rheumatoid arthritis, multiple sclerosis and systemic lupus erythematosus.[217-220] The expression of sixty-five miRNAs is modified differently in IgA nephropathy and roughly half are silenced strongly.[187] The expression of two urinary miRNAs miR-146a and miR-155 are involved in the amount of versatile inflammatory markers in IgA nephropathy.[189, 221] The intra-renal miRNAs expression in IgAN patients demonstrated that miR-200c is silenced, whereas multiple miRNAs miR-141, miR-192 and miR-205 are induced. Mature miR-196a expression has been reduced in IgAN patients.[186]

Moreover, the altered miR-148b expression in peripheral blood mononuclear cells possibly account for abnormal glycosylation of IgA1 IgA nephropathy. A comprehensive microarray screening demonstrates miRNA-418b changes mRNA amount encoding the C1β3GALT1 single nucleotide polymorphism (SNP) in IgAN patients during conducting pathophysiological mechanism.[221] However, its functional relationship is still controversial and dark. The reduced glycosylation of IgA1 is a consequence for its reduced expression.[218, 219] Patients with IgA nephropathy showed trace level of C1GALT1 expression correlates inversely with miR-148b expression. The miR-148b conjugates to the 3′-untranslated region of C1B3GALT mRNA and degrade. The miRNA-148b expression is excess in IgAN patients. C1β3GALT is reduced by the Th2 cytokine, IL-4 but miR-418b enhances possibly by the Th2 cytokine. Currently, it is shown that abnormal transcription of miR-205 and miR-192 in IgA nephropathy causes disease severity and gradual development.[186]

In Renal Ischemia/Reperfusion

Ischemia-reperfusion injury (IRI) is a major natural cause of acute kidney injury. It is related to delay graft function, chronic allograft injury and ultimately in death. Acute kidney injury (formerly referred as acute renal failure) is a chronic syndrome instigated by quick loss of the kidney's excretory function.[220] The IRI method requires strong changes in miRNA expression in the proximal tubule. So far, one-hundred and seventy-three miRNAs are identified in the renal cortex. Among them miRNA-132, -362, -379, -668, and -687 are responsible for consistent modification of reperfusion during 12–48 h.[222] Alternatively, similar modifications in miRNA expression are required during renal IRI in laboratory mice.[223] p53, an instrumental protein for conducting in an apoptotic mechanism. It functions as a mediator of transcriptional changes of IR injury. The plasma of acute kidney injury patients inhibits miR-16 and miR-320 expression relative to healthy controls. Conversely miR-210 is considerably overexpressed.[224] Further, the stable and consistent miR-494 expression and Neutrophil Gelatinase-Associated Lipocalin (NGAL) acts as an ideal indicator of acute kidney injury (AKI).[225] The involvement of miRNAs expression in AKI initially came from proximal tubule-specific Dicer knockout mice. The miRNAs are reduced strongly[222] and modifies miR-21 control in different group during both acute and chronic kidney injury.[226, 227]

In Allograft Acute Rejection

Acute rejection is a deadly and life risk complication related to renal transplantation. At least the expression of 17 miRNA is modified during chronic acute rejection biopsies.[228] Among them twelve miRNAs inhibit their expression in acute rejection after renal transplantation relative to normal controls. However, eight miRNAs are overexpressed.[229] Moreover, another study reveals that 20 miRNAs are differently expressed along with acute kidney allograft rejection in three patients.[230] Conversely, four distinct miRNAs: miR-182, miR-155, miR-125a, miR-146b are induced strongly together with antibody mediated rejection.[231] Lastly, urinary miRNA miR-210 is reduced in a trace level during acute kidney rejections identified during a kidney biopsy.[232]

In addition, twenty miRNAs are differentially expressed strongly during Acute Rejection (AR) allografts. Among them 12 are silenced and eight are overexpressed in AR comparable to normal allograft biopsies.[233] All differentially expressed miRNAs interact to protect acute rejections from renal transplants. Among these miRNAs, three overexpressed miRNAs miR-142-5p, miR-155 and miR-223 and three reduced miRNAs, miR-10b, miR-30a-3p and let-7c can be utilized intimately as diagnostic hallmarks of AR.[228]

Furthermore, the amount of miR-210 is linked to acute renal allograft rejection. It functions as a novel biomarker of AR.[234] The ideal and appropriate early diagnoses and effective treatments of AR using microRNA expression patterns will control enormously the mortality rates of renal transplant patients.

In Human Immunodefiency Virus Associated Nephropathy (HIVAN)

The renal diseases have been claimed for progressive recognition as utmost complication associated with Human Immunodeficiency Virus (HIV) infection worldwide. Human immunodeficiency virus-associated nephropathy (HIVAN) is the most common kidney disease in HIV/AIDS patients that is known as End-Stage Renal Disease (ESRD).[235] HIVAN is detected early with acquired immunodeficiency syndrome (AIDS) [236] and is popularly recognized as AIDS-associated nephropathy. However, it is recognized as HIVAN when the patients with full-grown AIDS are detected in asymptomatic individuals. HIVAN is the most aggressive form of focal segmental glomerulosclerosis (FSGS). It is associated with the collapse of the glomerular tuft and linked to tubule interstitial lesions. It ultimately leads to HIV-1 infection following AIDS development.[237, 238]

Many miRNAs have been shown differential expression patterns during HIV infection. It indicates that miRNAs open a new avenue for early detection in HIV related diseases.[239] In fact, different considerable deregulation of miRNAs during HIV-1 infection requires for ultimate critical role in renal disease development and progression.[239, 240] The miRNAs define early signs of host cellular dysfunction induced by HIV-1. HIV encephalitis (HIVE) is one of such recognized pathology. The modulation in miRNA expression causes apoptosis and contributes to the disease pathology effectively.[241]

In Hypertensive Nephrosclerosis

Prolong and chronic hypertension ultimately develops kidney disease that is known as hypertensive kidney disease or hypertensive nephrosclerosis. Hypertensive nephrosclerosis is a life-threatening disorder involved by vascular, glomerular, and tubulointerstitial histological tissues.[242] However, the exact molecular protocol of hypertensive nephrosclerosis is yet to be known. Interestingly, many miRNAs including miR-200a, miR-200b, miR-14, miR-429, miR-205 and miR-192 are overexpressed of the examination of hypertensive nephrosclerosis patients relative to healthy controls.[186]

Despite these differential miRNA expressions are interesting, nevertheless additional experimental results are required for the identification of miRNAs and target genes related to hypertensive nephropathy. The expression of 60 genes and 24 miRNAs are modified in human hypertensive kidneys.[243] These included nuclear receptor subfamily 4 group A member-1,-2 and -3 (NR4A1, NR4A2, NR4A3), period circulation protein homolog 1 (PER1) and salt-inducible kinase 1(SIK1) as well as Renin. The strong bonding of abnormal expression between miRNAs and genes suggest that miRNAs are the ultimate regulator of these genes in hypertensive nephrosclerosis.[243]

In Acute Pyelonephritis (APN)

Acute pyelonephritis is well known bacterial and deadliest form of urinary tract infection (UTI) that cause irreparable damage of kidney following renal failure. An earlier study demonstrated that the severity of acute nephritis cannot be detected with clinical parameters. Common allograft biopsy and urine culture are the ideal tools for detecting acute nephritis in the renal allograft. These biopsies overlap both acute nephritis and acute rejection. Differentially expressed 25 miRNAs between rejection and acute pyelonephritis are identified recently.[244] Five miRNAs like miR-145, miR-99b, let-7b-5p, miR-23b, and miR-30a are silenced strongly in acute rejection relative to normal kidney function and acute pyelonephritis. MiR-23b is downregulated many pro- inflammatory signaling pathways including IL-17, tumor necrosis factor α, IL-1-induced NF-κB, TGF- β-induced kinases and several others in human lupus and rheumatoid arthritis.[245]

In addition, miR-145 and miR-99b are involved in neutrophil differentiation and are required in gene temporal expression of multiple stages of myeloid maturation. The miR-145 and miR-99b are overexpressed slightly in acute pyelonephritis biopsies relative to baselines that dictate neutrophil-predominant inflammation in acute pyelonephritis.

In Drug Associated Nephrotoxicity

A wide section of drugs for treating ailments and diagnostic purposes produce profound toxic side effects to the kidney. The drug's natural effect exposes the kidney to a great risk and corrosively damages by few causal steps.

1) The kidneys are exposed corrosively by a high level of the drug and its metabolites at a maximum level of blood flow passes through the kidney. 2) The correct size and abnormalities of several drug or metabolite particles that produce toxic and corrosive effects to the kidney enter the renal epithelial tubular cells. These cells are in great risk for epithelium cells due to their high metabolic and toxic activity. 3) Sometimes water from GI track is reabsorbed for preparing ideal concentration of the drugs and induce the toxicity. 4) the oxidation of drugs by cytochrome p450 generates reactive oxygen species and 5) lastly ideal normal pH concentration in the kidney allows certain compounds to form insoluble crystals that can cause physical damage to the kidney.[246]

The toxicity and corrosiveness are the inherent nature of the compound of certain drugs. Chemotherapy medicine like cisplatin is a good example whose uses is restricted for developing frequent nephrotoxicity. The DNA crosslinking effect of Cisplatin produces apoptosis. The existence of miR-155 transcription induces toxic injury in the kidneys and identifies as an early biomarker for toxic injury. Mutant miR-155(−/−) mice having cisplatin drug generates more kidney injury than normal kidney in control mice.[247] It induces signaling pathways associated with apoptosis and oxidative stress. These protocols are controlled via c-Fos that are interacted directly by miR-155.[247] The miR-122 is influenced by cisplatin. Its reduction and conversely miR-34a overexpression following cisplatin treatment are identified. The miR-122 silences Foxo3 translation that is required for normal p53 activation. The miR-122 repression by cisplatin causes induction of Foxo3 activity, and ultimately instigates p53 levels and apoptosis.[248] Another drug Cyclosporine leads nephrotoxicity by inducing tubular epithelial cell epithelial-mesenchymal transition (EMT). The cyclosporine treatment causes induction of miR-494 transcription and a PTEN amount reduction *in vitro*.

Transforming Growth Factor (TGF-β)

The cytokine Transforming growth factor (TGF-β) opens a critical role in chronic kidney disease, and TGF-β is responsible for exposing an important role in glomerular cell proliferation and glomerular extracellular matrix expansion that are responsible for renal failure.[249] Two agent's miRNA and TGF-β coordinate each other for mitochondrial dysfunction, oxidative stress, and energy metabolism in oxidative stress-mediated renal injury.[250] Two miRNAs miR-21 and miR-192 strongly control TGF-β1 action via Smad3 stimulation but acing the miR-29 or miR-200 families for influencing renal fibrosis.[251] Apoptotic signaling using TGF-β allows using the transcription SMAD protein family factors with various SMADs performing different functions. SMADS works as a trimer for strong regulation of two receptor SMADS (SMAD1, SMAD2, SMAD3, SMAD5 and SMAD8/9) and one co-SMAD (SMAD4). The process is controlled by the inhibitory SMADs (SMAD6 and SMAD7) activity. TGF-β is basically exposed as a crucial mediator in the renal fibrosis pathogenesis during human kidney diseases.[252] Currently miR-192 controls E-box repressors; (ZEB1 and ZEB2) that are regulating TGF-B activated extracellular matrix proteins at the time of diabetic nephropathy. [253] TGF-β also induces ECM protein expressions, for example collagens reduction using the E-box repressor and SMAD-interacting protein 1 (SIP 1).

Further TGF-β controls ECM genes using SMADs in mesangial cells.[254] *MiR-192* and TGF-β activate miR-200 family member expression by targeting Zeb1/2. in diabetic nephropathy miR-216a and miR-217 expression in mesangial cells are contrlled by TGF-β1 via phosphatase and tensin (PTEN) silencing via AKT kinase, a crucial agent in diabetic nephropathy. [255] Mesangial cells activated with TGFB or by a strong glucose concentration using miR-377 activation.[256] This enhanced miR-377 expression activate fibronectins (ECM proteins) using p21-induced kinase (PAK1) silencing and superoxide dismutase (MnSOD) that increases fibronectin production.[256] **(Figure 4)**

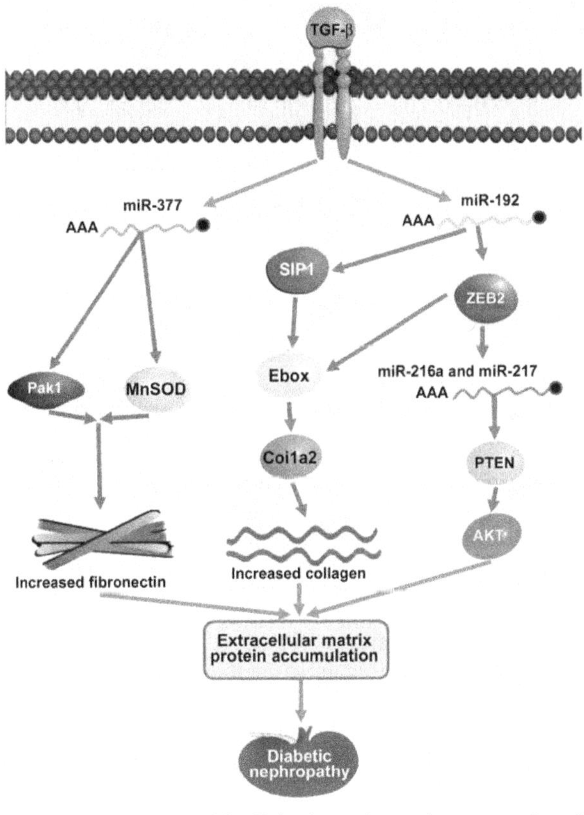

Figure-4. The miRNAs and TGF-β in diabetic nephropathy are produced by signalling processes. The TGF-β signaling pathway leads to ECM matrix protein induction. This process is strongly activated miR-192 regulation that silences E-box expression (ZEB1 and ZEB2) and SMAD-interacting protein 1 (SIP 1). It is involved in repression of increased extracellular matrix proteins at the time of diabetic nephropathy. TGF-β regulates miR-216a and miR-217 via inhibiting t of phosphatase and tensin (PTEN) using AKT kinase. Fibronectin expression is activated via p21-

activated kinase (PAK1) silencing and inhibition of superoxide dismutase (MnSOD) using miR377a. (Loosely Adapted, Redraw and Modify using skeleton from Mukhadi *et al.*, 2015)

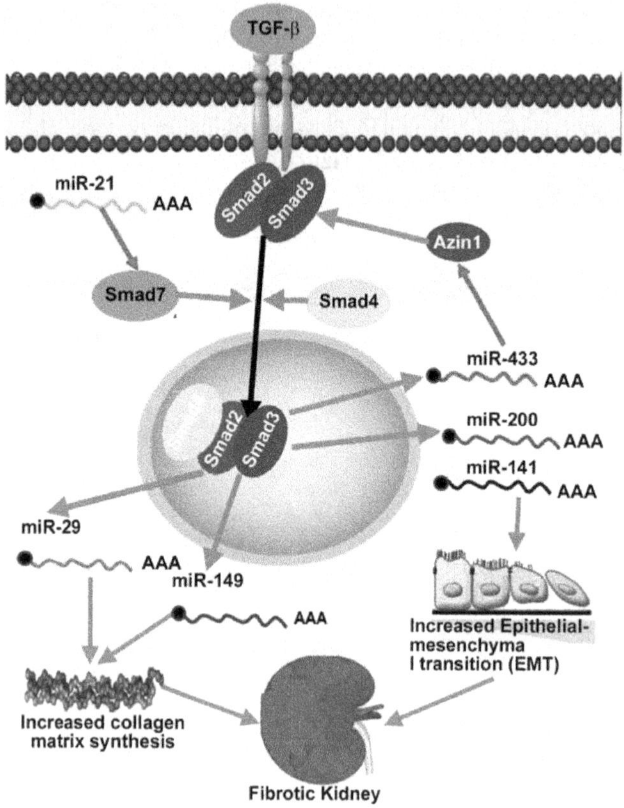

Figure-5. The existing miRNAs in renal fibrosis. The renal fibrosis activates TGF-β pathway during the silencing of SMAD inhibitor Azin transcription using miR-433. Simultaneously, miR21 reduces inhibitory Smad7 to activate TGF-β signaling. However, miR29 silent collagen synthesis; however, this miRNA function is prevented by miR-149. (Loosely Adapted, Redraw and Modify using skeleton from Mukhadi *et al.*, 2015)

TGF-β is considered a crucial mediator in renal fibrosis pathogenesis in human kidney diseases.[252] The expression of miR-433 upregulates TGF-β signaling during renal fibrosis that suppresses Azin-1, an exclusive antizyme inhibitor *in vitro* and *in vivo*. (**Figure 5**) This suggests that miR-433 may be a crucial and initiative factor for fibrosis specially miRNA in TGF-β/Smad3 driven renal fibrosis.[257] Moreover, the miRNAs miR-200a and miR-141 regulation are also essential for the activation and progression of TGF-β1 mediated EMT and fibrosis *in vitro* and *in vivo* [252], and miR-192 induces TGF-β/SMAD signaling in the fibrotic kidney both *in vivo* and *in vitro* [258] Further miR-21 expression is induced in fibrotic tissues is an additional and critical mechanisms for TGF-β/SMAD pathway For example, miR-21 is increased in response for treating with TGF-β1 or TNF-α in human renal tubular epithelial cells (TEC) *in vitro* [259] that changes promotes of miR-192 for TGF- β extracellular matrix production, inflammation and epithelial metabolic pathways and TGF-β also activates the expression of miR-200 family members for targeting Zeb1/2.[250] Similarly, miR-192 silencing reduces TGF-β1-activated collagen accumulation in TEC, while miR-192 upregulation increase TGF-β1-dependent collagen accumulation in TEC. [260] This suggests that miRNA controls multiple genes related to TGF-β signaling and specially leads in regulating TGF-β expression.

Conclusion

Diabetic kidney disease is one of the leading, devastating, and prolonged life-threatening problems for normal kidney function. Early DKD detection and identification of early symptoms for developing to ESKD leads exclusively as it considers early patient treatment plans prior to overt pathology. It likely represents proteinuria. As DKD considers complex, multifactorial disease. Truly a single miRNA profile is not well enough for detecting its accurate damage. In contrast, multiple miRNAs along with their effect in various distinct biologic processes may provide a better option for accurate diagnosis due to their versatile predictive value than a single miRNA. Invariably, many miRNAs profiling leads for demonstrating different complicated processes during DKD development. However, it is still intensely controversial if miRNA level allows the cause or DKD progressing consequence. If miRNAs refer to the progressing consequence, they act as biomarkers. If they present the cause, they may function as therapeutic components for treatment. A combination of eGFR and albuminuria and routine ideal biomarkers, multiple miRNAs demonstrate inflammation, autophagy, ER stress, fibrosis, hypertrophy, oxidative stress, insulin resistance and podocyte injury in promising accurate manners for speculating the ESKD risk. If miRNAs act as DKD cause, unique multi-miRNA-mediated therapeutic agents either restore or block their expression and level.

However, considerable consolidated biomarkers for their ideal size and proper power will be needed to induce the studies carried out to date. The mechanism rooted in mice model act as the first stage on the long road to therapeutic use. Nevertheless, the isolation of miRNAs as ideal targets for the DN improvement reflects positive outcomes for practical hope beyond unimaginable futures and their clinical developments.

References

References: chapter wise are cited before the Index at the end of the Book. (1-260)

Chapter X

Delivery of microRNAs and Anti-miR therapies

Abstract

The sporadic expansion of RNA silencing has a great impact on the biomedical diagnostic and medicine worlds. Particularly, its positive implication cannot be ignored in the microRNA-based enhancement and repression of many diseases related to gene expression. The microRNA-based drugs, antisense oligonucleotides are promising examples of new generation therapy. On one hand, nanoparticles give unprecedented opportunities for cellular delivery for miRNAs therapeutics. The tissue regeneration and other common diseases have been discussing in detail for nanoparticle delivery for miRNA-based therapeutic uses in the last few decades. Three major drawbacks of the application and delivery of therapeutics of microRNAs and anti-miR drugs are their stability in different physiological cell environments, their efficient delivery and immune responses. In the last two decades, scientists have been achieving to prevent the major challenges like intracellular delivery stability and immune response. As a success of this event, the first RNA drug was approved by the FDA. Recently, multiple microRNA drugs are in different phases of clinical trials. The new strides of delivery of microRNAs and anti-miR have triggered a wide range of expansion in the RNA-therapeutic field. Specifically, a new perspective in designing and characterization of multiple nanocarriers are described to accelerate basic research to clinical application of miRNA therapies. The success of the challenges of the numerous delivery vehicles and their transportation strategies are described both in research laboratories and clinics.

Introduction

Many genetic and RNA derived components of different diseases bring us closer to personalized medicine.[1] These experiences have been well influenced by new generation early diagnostic use. Currently, microRNA like small-molecule drugs carrying hydrophobic organic compounds inhibit the target mRNA through competitive binding. The microRNAs regulated proteins may form 2-5% 'protein pockets' predominantly of many humans' potential disease targets.[2] The miRNAs involve in several physiological responses like development, hematopoietic processes, cell proliferation/apoptosis, organogenesis, viral defense and fat metabolism.[3-6] Thus, the honest realization personalized medicines require some unique therapeutic potential like microRNA and anti-miR molecules beyond those small chemical molecules and conventional protein therapies.

There is some limitation of protein and small-molecule drugs because they cannot target every disease. Over the past two decades, miRNA and anti-miR drugs have been emerging to counteract most of the diseases. The therapeutic miRNA deliveries have several limited vehicles due to its sensitive predominant factors including naked, single-stranded RNA, nuclease degradation and immune responses etc. However, some sophisticated techniques provide few additional means of cellular entry including the escape of endosomes that transfer extracellular organic and inorganic nanoparticles in the cytosol.[7] As such, the miRNA delivery passes through a tightened channel earlier but nowadays it counts versatile materials for transportation for overcoming stringed cellular physiology. The different delivery materials including numerous carrier particles that enter through plasma pores and modulation of gene editing are described.

Delivery materials and modified microRNAs

The nanoparticles and the viral vectors claim the lion share among different miRNA's delivery vehicles. It will be a great injustice to the readers if these delivery devices are not elaborately discussed. To take these opportunities the inorganic and organic nanoparticle and viral delivery carriers are described in this separate subsection of the book.

The delivery of miRNA molecules in live cells is classified broadly into viral and non-viral vectors. Among viral vectors, the engineered adeno-associated virus is used as predominant cargo.[8] Also, the delivery of non-viral nanoparticles is widely discussed in recent days (**Table 1**). The insulted shields of nanoparticles encapsulate the small miRNA medicines to protect the RNase activity that are abundant in the cytoplasm. A special chemistry of those cases is required for cellular uptake and endosomal escape etc. It also gives high chemical stability, polymer flexibility of all nano devices.[9] Conceptually, different electrostatic condensed cationic polymers are involved that are compatible to negatively charged miRNA into nanoparticles. (**Figure 1a**)[10] These positively charged nanoparticles carry amines with physiological pH (p*K*a~7.4) for helping the rupture of the endosome by the ionic imbalance.[11, 12] The ready-made hypothesis 'proton sponge' has been employed rigorously in multiple nanodevices also.[13] Most cases, synthetic polymers including poly-L-lysine,[14] polyamidations,[15] and polyethyleneimine[16] together with natural polymer chitosan[17] have been used extensively in miRNA delivery. Their variable successes allow them to explore the use of Polyβ-amino esters that can show ease penetration and biodegradability for mRNA and miRNA delivery.[18]

A separate widely used nanoparticle for miRNA delivery is made up of Lipids and lipid-like materials. Multiple use of positive charged ionizable lipids have been involved for facilitating endosomal escape [19] and reducing toxicity.[20] On few occasions, self-assembled lipid complexes, (**Figure 1b**) are used for combining electrostatic interactions and hydrophobic interactions.[21, 22] The lipid nanoparticles carry additional hydrophobic moieties, cholesterol, PEG-lipid, and ionizable /cationic lipid that facilitate stability of nanoparticles and enhance the efficacy of miRNA-based molecules delivery.[23] However, ionizable lipid homologs are the principal criteria for promoting the efficacy of the nanoparticles. Some isolated efforts are made to synthesize semi-automated high-throughput lipid libraries. [24, 25] These highly potent lipid nanoparticles have efficiently transported a variety of RNA molecules in the liver [26, 27] and lung [28] for *in vivo* delivery.

The third alternative for more straightforward delivery is possible using conjugated bioactive ligand for targeting entry in specific cell types. Perhaps, the conjugated of *N*-acetylgalactosamine (GalNAc; **Figure 1c**) are found that target siRNA in glycoprotein receptor on hepatocytes.[29] The intravenous nanoparticle, GalNAc conjugates with penetrate subcutaneous barrier efficiently and enter inside the circulatory system and liver.[30] The cholesterol,[31] vitamin E,[32] antibodies,[33] and cell-penetrating peptides[34] are the homologous to above conjugate that can be applied. Among them triantennary GalNAc–siRNA can evolve reasonable advance momentum in clinical traction for deliver.

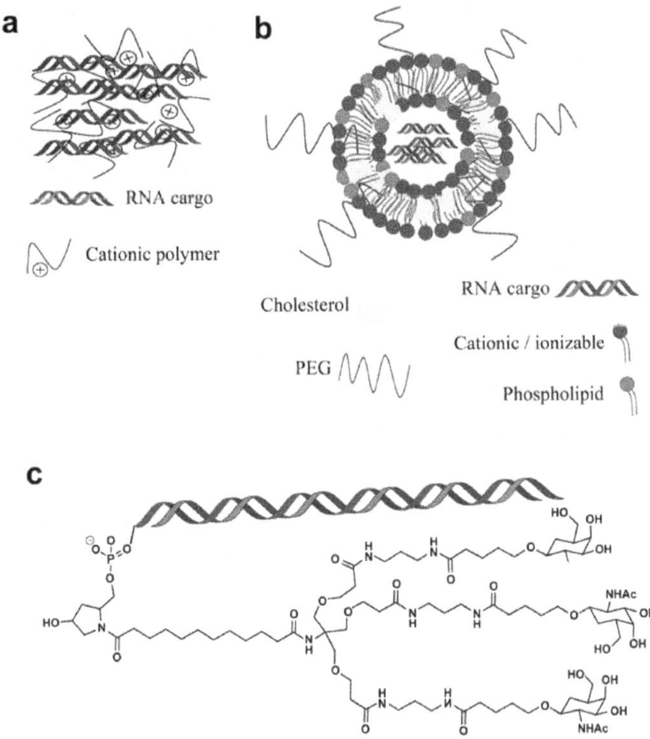

Figure-1 (a-c). Usual RNA delivery strategies. **a** Schematic sketches of RNA polymer nanoparticles and cationic polymer. **b** Graphical for RNA lipid nanoparticles, a cationic/ionizable lipid, and other hydrophobic moieties (such as cholesterol) commonly used in nanoparticle cargoes. **c** Structure of a tertiary conjugate between

N-acetylgalactosamine (GalNAc) and RNA, the cutting-edge candidate clinical trials. (Loosely Adapted, Redraw and Modify using skeleton from Kaczmarek *et al.*, 2017)

miRNA therapy and chemical modifications

The naked RNA therapeutics are vulnerable to degradation by the presence of enriched ribonucleases in the blood and ultimately phagocytosis by the reticuloendothelial system (RES). Numerous changes can promote their stability of oligonucleotides for *in vivo* delivery.

Table 1. Relative Comparison of clinically considered different RNA delivery platforms

RNA for clinical trial	delivery vehicle	Advantages	Disadvantages.
siRNA, ASO, mRNA	Naked RNA	No additional materials or synthesis required	Prone to degradation Immunogenic Difficulty entering cell Poor circulation half-life
siRNA, ASO, mRNA	Nanoparticle	Increased half-life Protection from nucleases Aids in endocytosis and endosomal escape	Elevated risk of toxicity with introducing excipient materials
siRNA, ASO	Conjugate	Defined chemical structure Ability to target specific receptors Limited toxicity due to lack of excipient materials	High doses required Dependent on chemical modifications for RNA stability

ASO antisense oligonucleotide, siRNA short interfering RNA

Antisense oligonucleotide (ASO) technology, anti-miRNA oligonucleotides (AMOs) are introduced to silence many miRNAs.[35] Numerous chemical modifications in nucleotides are used in naturally occurring miRNAs for extensive protection and anti-degradation. They include-2′-fluoro oligonucleotides (2′-F), locked nucleic acid (LNA) oligonucleotides, peptide nucleic acids (PNAs), phosphonodiamidite morpholino oligomers (PMOs) phosphonothioate containing oligonucleotides, methyl phosphonate-containing oligonucleotides, organophosphate-containing oligonucleotides, 2′-*O*-methyl-(2′-

O-Me) or 2′-*O*-methoxyethyl oligonucleotides (2′-*O*-MOE), and other chemical Cy3-, cholesterol-, biotin- and amino-modified oligonucleotides etc. (**Figure 2**)

Modifications improve the tracer function of oligonucleotides. For example (0) Unmodified RNA, (1) phosphonothioate-, (2) methyl phosphonate-, or (3) organophosphate-bearing oligonucleotides with a sulfur, methyl, or borane group, (4) 2′-*O*-methyl, (5) 2′-*O*-methoxyethyl, (6) or 2′-fluoro introduced into the 2′ oxygen of the ribose, (7) LNAs; (8) PNAs, (9) PMOs and terminal chemical modifications, including (10) Cy3-, (11) cholesterol-, (12) biotin- and (13) amino-modified oligonucleotides could enhance possible stability for *in vivo* delivery. The various Phosphonothioate-, methyl phosphonate-, or organophosphate-containing oligonucleotides substitute a sulfur, methyl, or borane group for the α-oxygen of the phosphate have been introduced to overcome the stability.[36]

Therapeutic changes in miRNA: For increasing the binding stability and better protection against nuclease degradation 2′-fluoro-oligoribonucleotides containing a fluorine molecule bound to the 2′-oxygen of the ribose [37] and 2′-*O*-methyl or 2′-*O*-methoxyethyl group of the ribose moiety with a phosphorothioate oligoribonucleotide are introduced.

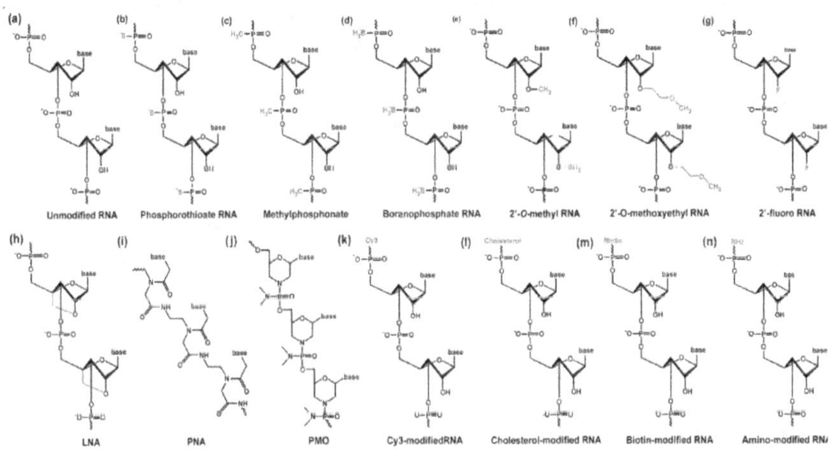

Figure-2. Structural modifications increase firmness stability, biodistribution, cellular uptake and delivery effectiveness and increase the tracer function of oligonucleotides. (**A**) Unmodified RNA; (**B**) Phosphorothioate-, (**C**) methylphosphonate-, or (**D**) boranophosphate-containing oligonucleotides containing a sulfur, methyl, or borano group etc.; (**E**) 2′-O-methyl, (**F**) 2′-O-methoxyethyl, (**G**) or 2′-fluoro introduced into the 2′ oxygen of the ribose; (**H**) LNAs; (**I**) PNAs; (**J**) PMOs; and terminal chemical modifications, including (**K**) Cy3-, (**L**) cholesterol-, (**M**) biotin- and (**N**) amino-modified oligonucleotides could increase the stability and tracer function of oligonucleotides for *in vivo* delivery systems. (Loosely Adapted, Redraw and Modify using skeleton from Yong Fu *et al.*, 2019)

A bicyclic RNA nucleotide analog, LNAs are inserted at 2′,4′ methylene bridge in the ribose.[38] An artificially synthesized polymer PNA is conjugated also the repeat N-(2-aminoethyl)-glycine units linked by peptide bonds.[39] PMOs also carry morpholine rings linked to phosphorodiamidate groups [40] at the miRNAs for protection against RNAse degradation. Apart from different terminal modifications, Cy3-, cholesterol-, biotin- and amino-modified oligonucleotides, can be added. [41, 42] Moreover multiple modifications are combined to use for better protection of miRNAs. (**Figure 3**)

The miRNA-based therapies are broadly categorized into two types for their downregulation of respective target genes: (a) miRNA suppression therapy and (b) miRNA replacement therapy. (**Figure 3**)

Suppression therapy

The downregulation of target mRNA is executed by the presence of active miRNA. In this pathway, mRNA expression is increased to a normal level. AMOs execute complementary base pairing with sense strand of miRNA that prevents interactions between miRISC and its potential target mRNA resisting mRNA degradation. For increasing inhibition efficiency, multiple chemical modifications are employed for enhancing AMOs affinity and stability through miRNA inhibitors and miRNA antagomiRs.

The anti-miRNAs are the active miRNA inhibitors that are single-stranded RNA molecules antagonistic to active miRNA sequences. These anti-miRNAs can bind with a complementary base pair of endogenous miRNAs and form a classical duplex. The duplex is attracted by the active DICER enzyme to destroy them. The active anti-miRNA is mainly used together with Lipofectamine transfection reagent for examining the loss of function miRNA experiment.

The single-stranded AntagomiRs carry out two specific oligonucleotide modifications in the 5’ and 3’ termini. A 2-Phosphorothioates is inserted at the 5′ end and the cholesterol group and 4-phosphorothioates. It is replaced by normal nucleotides at the 3′ end. Moreover, 2′-methoxy groups are exchanged with the normal nucleotides at their full-length oligos.[43] These chemical modifications prevent degradation, increase its stability and rapid cellular uptake of antagomiRs. [44] These antagomiRs are optimal for local or systemic administration to reduce the complementary endogenous miRNA actions.

The same-sized oligoribonucleotides bonded with 2′-*O*-methyl-modifications for a unique ‘miRNA mask’.[45] The miRNA mask does not target miRNA directly. In replacement the miRNA mask base pair with 3′-UTR regions of miRNA binding

sites of target gene mRNA via a full complementary sequence. The miRNA mask functions as an important supplement to AMOs, Invariably, miRNA masks are ideal for studying miRNA specific pathways that target host genes.

A plasmid-encoded miRNA copy is referred to as ‘miRNA sponges’ that can bind with complementary miRNAs seed sequences.[46] During transportation into cells, highly transcribed sponge RNAs of the seed region block a family of miRNAs carrying the same seed sequence. Thus, the introduction of the competitive inhibitors, miRNA sponges reduce identical short nucleotide fragments markedly.

Replacement therapy

The double-stranded synthesized miRNA like molecules is termed as 'miRNA mimics', which invariably activate endogenous human miRNAs and complementary base pair with target mRNAs for leading posttranscriptional silencing. The miRNA mimics with large number of chemical modifications are referred to as 'mRNA agomiR'. The antisense strand of an agomiR carried the same modifications as 'antagomiR'. These marked chemical modifications of miRNA agomiR protect the stability and activity greatly. Therefore, agomiRs are required basically for the upregulate of corresponding miRNAs in the marked tissues for the examination of miRNA function in the live cells.

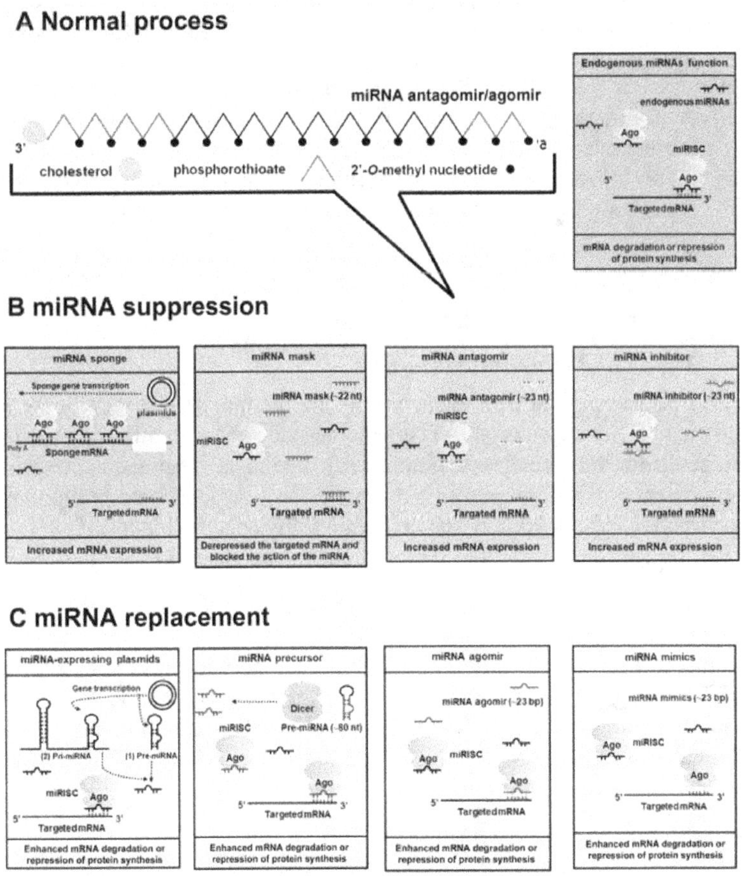

Figure-3. (a-c). Normal processing of miRNA. **(A)** Endogenous miRNA with improved structural modifications for normal processing and function; **(B)** miRNA inhibitory therapy using miRNA sponges, miRNA masks, miRNA antagomirs and miRNA inhibitors **(C)** miRNA replacement therapy using miRNA-expressing plasmids, miRNA precursors, miRNA agomirs and miRNA mimics. Each dotted lines denotes the modified structure of miRNA antagomirs and miRNA agomirs. (Loosely Adapted, Redraw and Modify using skeleton from Yong Fu *et al.*, 2019)

The pre-miRNA, the precursor of mature miRNA carries modification in single-stranded RNA fragments for activating the mature miRNAs processing chemically. Once the pre-miRNA is transported by the compatible cargo to live cells, pre-miRNA is further activated by the Dicer enzyme that cleaves pre-miRNA to generate mature miRNAs. Therefore, the pre-miRNAs can be tested with a "gain-of-function" component. These pre-miRNAs express and function in all biologically active cells.

In miRNA plasmid sequences a fluorescent reporter is initiating for expressing carrier miRNA upregulation. They can also dictate localization, transporting pathway and expression levels. For instance, *Takara Bio* constructed two-hybrid plasmids '*pmR-ZsGreen1*' and '*pmR-mCherry*' vectors that carry a selective cassette with a bright green or red fluorescent transient reporter gene respectively for marking the pathway of their transport in live tissues.

Further naked modifications although increase the stability however, they are not optimized for *in vivo* transportations. Therefore, multiple carriers are used for enhancement of miRNA delivery systems that are broadly divided into two categories and multiple sub sections which can be elaborately discussed below. **(Figure 4)**

Different therapeutic miRNAs

Tissue and temporal specificity

The tissue and temporal specificity are the major concerning criteria for therapeutic miRNA expression.[47] The genome-wide aberrant miRNA profiles are advantageous for delicate classification, early diagnosis, staging and prognosis in different diseases.[48] Although miRNAs expression is widespread in humans, but a major fraction expressed in tissue and temporally specific manner.[49]

Circulating miRNA

The miRNA-21 was the first miRNA detected in the human serum.[50] Numerous stable miRNAs exist in different organs/tissues are also present commonly in the human body fluids. The genome-wide miRNAs expression profiles are potential signals for biomarkers for the early diagnosis. For cross-remote organ communication, circulating miRNAs [51] are found in micro-vesicles or conjugated with high-density lipoprotein (HDL) particles and RNA-binding proteins [52, 53] for easy entering in the recipient cells.[54]

Cross-kingdom regulation:

During different physiological events including hematopoiesis, tumorigenesis, tumor metastasis, fat metabolism and intestinal mucosal homeostasis miRNAs are involved with the extensive workload. Their huge abundance and shorter half-life are the major advantages of developing therapy candidates. [55] However, they also face difficulties for penetrating through the phospholipid bilayer cell membranes and the vascular endothelium cells. For these consequences, they remain inside the blood storage organs, spleen or liver and finally excreted through kidneys. To overcome this flow and penetrating different physiological barriers, numerous vectors have been developed for targeted miRNAs delivery in recent days.

Adenoviral vectors:

The Adenoviridae family viruses appear to be a major skeleton for Adenoviruses (Ad) vectors. This nonenveloped virus carries 36 kb long linear double-stranded DNA with two inverted terminal repeats (ITRs) at its termini.[56]

All Adenovirus are hybrid for helper-dependent Adeno vectors (HD AdVs) that eliminate all viral proteins for silencing immunogenicity and better transgene upregulation.[57] The presence of natural hepatotropic of Adenovirus makes it advantageous specific for liver-target.[58] Thus, HD AdVs are introduced as efficient cassettes for encoding primary miRNAs in transporting to liver tissues. Recently, multiple anti-HBV pri-miRNA mimics, pri-miRNA-122/5, pri-miRNA-31/5, or pri-miRNA-31/5–8-9 are used for anti-HBV expression for blocking Hepatitis B Virus (HBV) replication.[59]

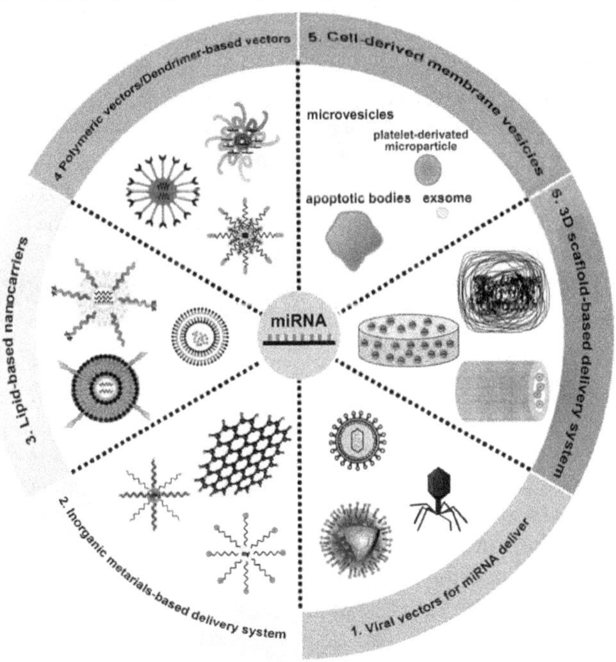

Figure-4. Multiple types of miRNA delivery vectors. Vectors are classified into two broad but distinct categories: viral vectors (1) and nonviral vectors. Nonviral vectors are divided into six broad types: (2) inorganic material-based delivery systems, (3) lipid-based nanocarriers, (4) polymeric vectors/dendrimer-based vectors, (5) cell-derived membrane vesicles and (6) 3D scaffold-based delivery systems. (Loosely Adapted, Redraw and Modify using skeleton from Yong Fu *et al.*, 2019)

Cancer-specific oncolytic adenoviruses are optimized vectors for tumor immunotherapy due to their commendable tumor-restricted replication abilities.[60] The mode of co-expression of interleukin-24 (IL-24) and miRNA-34a, formidably generates a AdCN205 oncolytic adenoviral vector executing better antitumor effects in hepatocellular carcinoma (HCC) models.[61] Despite a strong stimulation of host innate and adaptive immunity denotes a major disadvantage of HD AdVs, which may restrict its widespread applications.[61]

Adeno-associated viral vectors:

Adeno-associated viruses (AAV) are constructed from Parvoviridae family viruses. They are single-stranded DNA containing non enveloped viruses.[62] The entry after the AAV vector causes a commendable gene expression in different organs of the model organism.[63] Recently, transportation for conjugated AAV vectors and miRNA-196a downregulate Elav-like family member 2 (CELF2) in ameliorated spinal and bulbar muscular atrophy (SBMA).[64]

Retroviral vectors

Interestingly, Retroviruses (RVs) are enveloped viruses carrying two copies of a single-stranded RNA.[65] The Moloney murine leukemia virus (MMLV) is the ancestor of most Retroviruses (RVs) in which three important genes including replication machinery pol, env, and gag from the ancestral virus are flanked by long terminal repeats (LTRs).[66] Once a virus enters into host cells, the viral genome forms double-stranded DNA by the action of reverse transcriptase enzyme and then inserted into the host genome. Such integration leads to a consistent expression of the inserted gene fragment with the help of host transcriptional machinery.[67] Further, a hybrid conjugated MMLV along with miRNA-21 (MMLV-miR-21) increases miRNA-21 expression markedly in adult cardiac fibroblasts relative to MMLV entry alone.[68]

Lentiviral vectors

Similar to RVs, another virus Lentiviruses (LVs) can stably insert themselves into the genomes of recipient cells and induce consistent propagation with the help of normal host transcriptional machinery according to their needs.[69] Recently, hybrid constructs made up of the administration of a lentivirus vector with miRNA-133b increase the rapid functional recovery in spinal cord injury.[70] Another hybrid construct between lentiviral vector and miRNA-101 sponge once entered inside the intrahippocampal injection an overexpression of soluble β-amyloid precursor protein (sAPPβ) in hippocampal neurons is noticed.[71]

Despite highly efficient delivery viral vectors suffer from many physiological drawbacks and integration-related problems for instance, low loading capacity, high toxicity and strong immunogenicity etc.[72] Therefore, a large series of nonviral vectors have been explored to design and construct non-viral carriers based on their actual demands. Their less toxicity and increased biocompatibility produce these nonviral vectors advantageous over multiple efficiencies of the viral vectors.

Nano particles

Criteria of Nanoparticles (NPs) for miRNAs delivery

The abilities for the use of miRNAs as potential therapy candidates *in vitro* and *in vivo,* raise few prominent drawbacks and hurdles that can overcome to promote their clinical applications.[73, 74] The entry for naked miRNAs in the cytoplasm is extremely poor in quantity because of three predominant factors, unusual off target [75] or on-target effects,[76] quick half-lives in circulatory blood and weak stability against their degradation by many cytoplasmic RNases. [77, 78]

In vitro delivery of miR mimics and anti-miRs have been performing successfully with the help of different transfection reagents including *DharmaFECT*™ and *Lipofectamine*™ and electroporation.[79] Alternatively chemical modifications of anti-miRs and miR mimics popularly known as 'antagomiR' [80] and 'agomiR' [81] can be involved *in vivo* delivery by promoting their stability in the cell-free entry. Two locked nucleotides LNA and peptide nucleotides PNA are used at the termini of different miRNAs to protect against degradation in their naked entry.[82] Alternatively anti-miRs and miR mimics are encapsulated into insulted Nanoparticles that improve *in vivo* entry in the target cells. The encapsulated NPs also overcome most of the adverse effects of their transport in the targeted host cells.[83](**Figure 5(3)**)

The delivery of naked miRNAs hinders four major drawbacks, cellular uptake for its negatively charged groups, off-target or on-target, shorter half-life and unconditional immune response. The NPs encapsulated miRNA increase the cellular uptake while few ligands conjugate with NPs decreases the off-target harmful effects. The controlled miRNA release by many NPs avoids excess targeting of multiple genes within the cells and preventing the payload from degradation. (**Table 2**)[84, 85] The lower encapsulation causes easy water-related diffusion of miRNAs during emulsion-based NPs construction. These NPs facilitate tissue distribution and site-specific targeting of miRNAs. Furthermore, ligands of NPs induce specific target cell receptors for better uptake, receptor-mediated endocytosis and dramatically reduce the side effect.[86, 87]

Methods for Nanoparticles preparation

Many diversified techniques have been tested including single or double emulsions, nanoprecipitation, and interfacial polymerization in the generation of miRNA-loaded NPs. The optimized materials and their surface characteristics are the main criteria for their generation.[88]

Two types of Emulsion-based methods, single immersion and double emulsion are employed for the preparation of miRNA-loaded NPs powering with high-speed homogenization or ultrasonication.[89] In single-emulsion, a polymer solution using external surfactant-containing water phase forms an oil-in-water (o/w) emulsion. On the other hand, in double-emulsion technique encapsulate hydrophilic payloads with the help of two emulsification steps for instance water-in-oil-in-water (w/o/w) or oil-in-water-in-oil (o/w/o) emulsions.[88, 90]

The monomethoxy (polyethylene glycol)-poly (d,l-lactide-*co*-glycolide)-poly (l-lysine)-lactobionic acid (mPEG-PLGA-PLL-LA) [77],[91] PLGA and poly(glycerol adipate-*co*-ω-pentadecalactone) (PGA-*co*-PDL) are used in emulsion methods for miRNA delivery.[92] The preparation of miR-99a-loaded mPEG-PLGA-PLL-LA NPs is developed *via* the double emulsion method. In which, miRNA is submerged in water and immediate after into a PLL-LA solution by sonication. It reduces the surface charge from 25 mV in control NPs to 3 mV for miRNA-loaded NPs for efficient miRNA loading.

A dropwise inclusion of water molecules to a polymer solution forms a water-miscible solvent, in which polymer NPs can be formed *via* nanoprecipitation.[93, 94] For instance, miRNA-loaded PLGA/chitosan (PLGA/CS) NPs is prepared by dropwise inclusion of PLGA solution into a water solution of CS and miR-34s *via* the nanoprecipitation. This method increases the miR-34 entrapment efficiency (EE) depending on the miR-34 amount.

Table-2. The siRNA/MicroRNA-dependent Drugs Targeting multiple Diseases during Clinical Trials

Disease	Vehicles	Target	Drug's name	Sponsors
Cancer				
HC, ST, ACC, GNT	lipid nanoparticle	siRNA target PLK1	TKM 080301	Tekmira Pharmaceuticals
ST, MM, NHL	lipid nanoparticle	siRNA target MYC	DCR-MYC	Dicerna Pharmaceuticals
ST	Polymer nanoparticle	siRNA target RRM2	CALAA-01	Calando Pharmaceuticals
ST	liposome	siRNA target EphA2	siRNA-EphA2-DOPC	M.D.Anderson Cancer Center
Leukemia	neutral liposomes	antisense target GRB-2	BP1001	Bio-Path Holdings
ASC, PC	lipid nanoparticle	siRNA target PKN3	Atu027	Silence Therapeutics
PDA, PC	biodegradable polymer matrix	siRNA target K-RAS	siG12D LODER	Silenced

Glioblastoma	nanoparticle (NPs)	siRNA target p53	Temozolomide/SGT-53	SynerGene Therapeutics
Lung Cancer	DOTAP-Chol	siRNA target Fus1	Fus1/Erlotinib	Genprex
HC	liposomal Nanoparticle	siRNA target CEBPA	MTL-CEBPA	Mina Alpha
Glioblastoma	spherical gold Nanoparticle	siRNA target Bcl2L12	NU-0129	Northwestern University
IMG	nanoliposomal	siRNA target UGT1A1*28	CPT-11	University of California, San Francisco
Advanced Metastatic Cancer. ST	BIV-lipoplex	shRNA STMN1	pbi-shRNA STMN1 LP	Strike Bio
Ewing's Sarcoma	BIV-lipoplex	shRNA EWS/FLI1 type 1	pbi-shRNA EWS/FLI1 Type 1 LPX	Strike Bio
NR	polyethylenimine	eIF5AK^{50R} plasmid eIF5A siRNA	SNS01-T	Senesco Technologies
MPM, NSCLC	EDV	microRNA-16 mimic target EGFR	TargomiRs	Asbestos Diseases Research Foundation
Virus Infection				
EVD	lipid nanoparticle	siRNA target VP24, and VP35 regions, EBOV	Favipiravir polymerase inhibitor	INSERM, France
Other diseases				
Hepatic fibrosis	lipid nanoparticle	siRNA target HSP47	ND-L02 s0201 Injection	Bristol-Myers Squibb
Hypercholesterolemia	lipid Nanoparticle	siRNA target APOB	PRO-040201	Tekmira Pharmaceuticals

Source: https://clinicaltrials.gov. ACC, adrenocortical carcinoma; ASC, advanced solid cancer; GNT, gastrointestinal neuroendocrine tumors; HC, hepatocellular carcinoma; MM, mul- tiple myeloma; MPM, malignant pleural mesothelioma; NHL, non-Hodgkins lymphoma; NSCLC, non-small-cell lung cancer; PC, pancreatic cancer; PDA, pancreatic ductal adeno- carcinoma; ST, solid tumor; ALC, advanced liver cancer; SCLC, squamous cell lung cancer; IMG, intracranial malignant glioma; NR, not recorded; EGFR, epidermal growth factor receptor; GRB-2, Growth Factor Receptor Bound Protein-2; RRM2, Ribonucleotide Reductase Regulatory Subunit M2; PLK1,Polo-Like Kinase 1; HSP47, Heat Shock Protein 47; EphA2, Ephrin type-A receptor 2; eIf5A, Eukaryotic translation initiation factor 5A-1; EDV, EnGeneIC Delivery Vehicle; CEBPA,CCAAT/enhancer-binding protein alpha; BIV-lipoplex, bilamellar invaginated vesicle lipoplex; EVD, Ebola virus disease; PKN3, protein kinase N3; K-Ras oncogene, Kirsten rat sarcoma viral oncogene; APOB, apolipoprotein B; VP24, virus protein 24; VP35, virus protein 3.

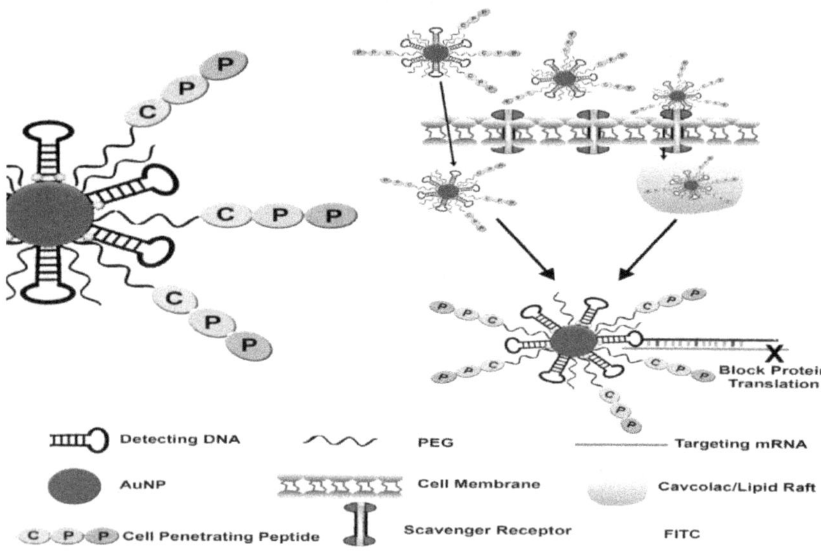

Figure-5. Detailed diagrammatic Illustration of the AuNP Beacon modification of and Cell Uptake Mechanism. (Loosely Adapted, Redraw and Modify using skeleton from Yao Xiao *et al.*, 2019)

A rare technique combining click chemistry (CC) is used in radical polymerization (CRP), electrospinning, and particle replication in the non-wetting template (PRINT) for easy and efficient preparation for miRNA-loaded NPs. This method is not used commercially to construct miRNA-loaded NPs due to the presence of harsh temperatures, extreme pH and toxic solvents. A combination of CC and CRP can allow for the making of NPs.[95, 96] On the other hand dibenzylcyclo-octyne (DBCO)-modified liposomes also bind the azide groups for expressing on cancer cells.[97] Moreover, PRINT can generate predefined shape and size of NPs by utilizing elastomeric molds with a polymer solution. They are used for examining specific tumor cell uptake [98] and brain metastasis during loading with docetaxel.[99]

Finally, the libraries of NPs syntheses are taken place by a recently developed technique of microfluidics by the rapid mixing of droplet streams into silicon or glass microchannels.[100] The microfluidic technology is further employed to modify small interfering RNA-loaded NPs (siRNA-loaded NPs),[101] but, no evidence for miRNA-loaded NPs has been found so far.

Polymeric vectors

The selection of the polymer that constitutes the polymeric NPs is of paramount interest in modern days. The polymer-based NPs employ a wide range of preparation techniques using synthetic artificial polymer and the use of natural polymer for their backbone structure. With the advent of stimuli-responsive polymers that change their physical properties like temperature or pH, polymeric NPs have been invented for an enhancing range of therapeutic applications.

Natural polymer

The animal and vegetable are the sole sources of natural polymers. Among natural polymer cargos CS and HA are the most promising carriers.[102, 103] The non-cytotoxic, non-immunogenic and highly biocompatible chitin are the main ingredient for CS.[104, 105] Two major parameters of chitin, its acidic pH and protonated amino group are antagonistic to negatively charged small miRNA and siRNA molecules carrying a high loading affinity.[106] Such advantages of CS are not sufficient to delimit its use in live tissues, because they have solid interaction with the loaded agents.[107, 108] These solid bonds prevent the efficient uptake in the cytoplasm.[107] To prevent such drawbacks the hydrophobic moieties of CS are additionally conjugated to a weak polymer/nucleic acid interaction for increasing its cytoplasmic delivery. It is found that the negatively charged polymer PLGA, derived from Lipid chains, bile acids have also been bonded with CS for promoting ease miRNA release.[109, 110]

A strong hydrophilic anionic natural polysaccharide can be identified by receptors of HA natural polymer on host cells.[111, 112] HA has been used to prepare NPs for their enormous chemical modifications. Five different carrying drug candidates, cytostatic drugs, proteins, polynucleotides, immunomodulators and imaging agents are attached to HA. HA bonds with cationic polymers or lipid molecules to deliver polynucleotides. For instance, HA associates with PEI and PEG chemically for easy and efficient release of miRNA and DOX.[113, 114] In the nano polymer, HA modifies the overall surface of the cationic polynucleotide.[115] In HA, the cationic polymeric complexes are linked with miRNAs by a charge interaction. The positive charges of HA are required for quick elimination of toxic effects.[116] Therefore, the surface modification of HA can easily mask different NPs including lipid complexes or liposomes.[115] Basically, HA is associated with positively charged liposomes[117, 118] calcium phosphate [119] and lipids on the NP surface electrostatically. A hybrid conjugate of HA-NPs with miR-145 protects against colon cancer by their overexpressed CD44 receptors. This gives the most aggressive therapeutic success observed in HA accumulation that reduces its tumorigenic growth.[86]

Artificial and synthetic polymers

Two major physiological compatible behaviors of PEI, easy and sophisticated entry in live-cell cytoplasm together with easy manufacturing and functionalization of target cells promote miRNA delivery cleanly.[73] Four separate synthetic polymers, polyethylene imines (PEIs),[120, 121] PLGA,[122, 123] poly(ε-caprolactone) (PCL),[124] and polyurethanes (PUs) [125, 126] are used for artificial polymers generation extensively. PEI is a unique artificial polymer. It has both linear and branched polycationic polymers.

Two essential components of different molecular weight in PEI, a two mJOR PHYSIcondary amine group and ethylene spacer are required for release efficiency and biocompatibility.[127, 128] In a positive change of PEI chains, pH is required for secondary amines inside the PEI that allows its electrostatic complexation.[129] The proton acceptor endogenous characters of PEI is also showing weak-base buffering properties and protection against degradation of endosomal vesicles inside the cytoplasm.[130, 131]

Two prominent tumor suppressors, miR-145 and miR-33a inside the PEI allow to estimate the *in vivo* anti-tumor activity in colon carcinoma.[121] These PEI-miRNA complexes facilitate multiple functions including reduction of tumor growth during cancer treatments [121] by targeting delivered miR-145 to metastatic breast cancer cells.[120] A branched PEI-PEG NPs can deliver miR-145 to prostate cancer actively. [132] The PEG chains that surface-altered with poly-arginine, enhance easy target in prostate cancer *in vivo,* for suppressing tumor growth and increased life span. In this context, PEI-PEG NPs also reduce toxicity and enrich their biodistribution.[133]

Two widely used synthetic biocompatible polyesters PLGA and PCL are advantageous over PEI for their lower toxicity and biodegradability.[134, 135] PLGA produces varying ratios of lactic and glycolic acid monomers that can be degraded in a variable period ranging from several months to years via implication of the Krebs cycle.[136] The polymers with higher glycoside are more hydrophilic in a shorter time. Therefore, PLGA NPs along with higher lactic acid content are hydrophobic and degrade in a longer run.

It is hard for miRNA encapsulation in the negative surface charge PLGA NPs,[137, 138] which downgrade a strong association with the phosphate groups of miRNAs. This weakness can be overcome by adding cationic compounds with PLGA. For instance, a cationic polymer in CS that keeps inside the miRNAs of PLGA NPs.[139] Similarly, protamine sulfate (PS) coating of PLGA NPs promotes the conjugation of miRNAs on the NP surface.[140]

In a separate study, fours enhancing activities of the active live cells for example colloidal activity, rapidly targeted cellular uptake, yielding low toxicity and controlled release of miRNAs are found for biocompatible PCL. The degradation product of PCL 6-hydroxyhexanoic acid is a natural metabolite in humans.[141] Moreover, using the double emulsion method, PEG-peptide-PCL releases two potential medicines to deliver miR-200c and docetaxel.[124] For delivery in tumor cells a gelatinases-cleavable peptide is used in NPs. The complex between miR-200c and spermidine increase transfection percentages.

Among Cationic PU the -short-chain of PEI nano complexes have been constructed for miR-145 that represses the stem cell niche in the brain tumor.[125] An absolute miR-145 transfection inhibits the miRNA targets Sox2 and Oct4. Intracranial insertions by injecting PU-PEI-miR-145 nano complexes also able to suppress tumor growth with simultaneous chemoradiotherapy. These nano complexes increase the potential therapeutic effect on lung adenocarcinoma,[126] with inhibition of epithelial-to-mesenchymal transition (EMT) experimentally.[126]

Inorganic nanoparticles

Four majors inorganic nanocarriers, gold, calcium phosphate, silica and iron oxides [142, 143] are applied widely due to their biocompatibility, controllable size and morphology. The iron oxide is also applied for imaging technology as an alternative for magnetic resonance imaging (MRI) in thermal therapy, in medicine and gene delivery. The use of magnetic NPs and cationic compounds together is required for high miRNA encapsulation.[144] Streptavidin-coated magnetite (Fe_3O_4) that are compatible with biotin-bound miR-335/PEI complexes in hybrid NPs are used for miRNA delivery to the human mesenchymal stem cells (hMSCs) for better targeting than PEI-miR strategy.[145] A zinc-doped Fe_3O_4 is associated with let-7a miRNA enter into glioblastoma cells via magnetoception.[146] In this loading let-7a miRNA encapsulate into NPs by an approach leading to layer-by-layer manner. This complex is required for nearly absolute cellular uptake (at nearly 98%) and downregulating multiple of the let-7a targets, PI3K and RAS.

The mesoporous silica NPs (MSPs) are unique nanostructures with high biocompatibility and stability.[147] Two encapsulated components anti-miR-221and Temozolomide (TMZ) in MSPs are introduced for drug-resistant glioma cell treatment.[148] A well-known tumor suppressor miRNA, miR-34a in MNPs downregulates neuroblastoma growth markedly.[149] Further four pinpoint advantages, delaying tumor growth, enhancing apoptosis and reducing vascularization are added in GD2-antibody conjugated NPs of live tissues.

An interesting inexpensive nontoxic, bioresorbable and easily synthesized nanocarrier is calcium phosphate (CaP) NPs. They have been widely used as DNA carriers for 40 long years. However, miRNAs may have weak interaction with CaPs for lowering spatial charge density. The long-chain miRNAs (lc-miRNAs) adding with protective thiol groups on both the 3' and the 5' miRNA termini are conjugated with PEI-CaP-NPs for efficient encapsulation.[150]

For better biocompatibility, ease of functionalization, tunable size and shape Gold NPs (Au-NPs) are used.[151, 152] Encapsulating many miRNAs to the surface of Au-NPs with the help of thiol or amino groups are also constructed. For excellent cellular uptake via endocytosis and intracellular delivery non-toxic gold-based Au-NPs is developed also.[153] The expression and protecting degradation of known tumor suppressor miR-145 inside Au-NPs are encapsulated for targeting prostate and breast cancer cells in occasional cases where miR-145 is usually repressed.[154]

Lipid-based nanoparticles

The lipid-based nanoparticles can interact with the cell membrane because lipids carry the lion share of the cell membrane components. Structurally LNPs are made up of lipid bilayer having miRNAs at their aqueous core or multi-lamellar liposomes in between lipid bilayers [155] for promoting their cellular uptake. Typically, a mixture of cationic lipids between neutral lipids and PEG is associated with miRNA-inside the LNPs.[156] For prevention of enriched RNase degradation inside cytoplasm a spontaneous electrostatic interaction between cationic lipids and negatively charged miRNA has been taking place. The cationic charge promotes its interaction with opposite-charged cellular membranes.[156] But the presence of such cationic charges in the cationic lipids increase cellular uptake but degrade cell membrane integrity for influencing of cytoplasmic vacuolization and decrease cell activity.[157] Another problem in harboring cationic charge inside the lipids is that it is spontaneously bound to negatively charge serum proteins and ultimately such usual bonding aggregates are eliminated by the liver and the spleen. For such difficulties a reduction of cationic charges has been attempted on multiple occasions.[158] Therefore, different categories of helper lipids for example cholesterol (Chol), dioleoyl phosphatidylethanolamine (DOPE) and phosphatidylcholine (PC) have been tried extensively for replacing cationic lipids. Also, PEGylated cationic/neutral LNPs are used for the treatment of liver-specific tumor formations. The PEG and miRNA-122 combination is introduced for reduction of tumor growth due to their enhanced half-life.[159]

Dendrimer-based particles

Dendrimers are 3-D hyperbranched globular nano polymeric components. Dendrimers have a greater advantage in widely used fields than other polymers due to their tight polydispersity index and change in multiple functional groups.[160]

First artificial polymers with dendritic structures were identified as PAMAM dendrimers that are developed from ethylenediamine or ammonia initiator core reagents.[161] PAMAM dendrimers have a positive charge on their surface for condensing nucleic acid molecules. PAMAM dendrimers are biodegradable polymers exhibiting low genotoxicity and cytotoxicity.

A novel carrier NGO-PEG-dendrimers are commonly used miRNA delivery. NGO-PEG-dendrimers/anti-miRNA-21 are conjugated into PAMAM dendrimers in which PEG-functionalized nanographene oxide (NGO) transformed into 2'-*O*-methyl-modified anti-miRNA-21.[162] The intravenous injection of conjugated NGO-PEG-dendrimers/anti-miRNA-21 increases in bioluminescence signals dramatically inside the tumor areas by the use of a luciferase reporter.[162]

Cell membrane vesicles devices

Intercellular communication is unique and important for cellular environmental homeostasis maintaining eukaryotic organisms. The cell-to-cell distances also find via extracellular vesicle (EVs).[163] These EVs are essential factors in intercellular communication that facilitate their application as biomarkers and medicine carriers.[164]

The DNA aptamer-modifies EVs-loaded with Cy5-labeled let-7 in AS1411 that accommodates in tumor tissues and reduces tumor growth factors during the intravenous injection.[165] On the other study, a conjugated of brain metastatic cancer cell-derived EVs and miRNA-181c destroys the blood-brain barrier (BBB)[166] for facilitating brain metastasis.

Three intracellular EVs are present in exosomes, micro-vesicles, and apoptotic bodies. Three major workloads, intercellular communication, RNA shuttling (mainly miRNA) and antigen presentation are implicated by the exosomes. Exosomes conducted intercellular communication via the transfer of bioactive miRNA molecules protect small RNAs from ribonucleases (RNases) by encapsulated exosomes in the body fluids.[167] Recently, a clean *in situ* detection method is referred as "molecular beacon" [168] in the exosome miRNAs is used for utilizing antagonized fluorescent oligonucleotides.[168]

Exosomes have two important characters, low cytotoxicity, and traceable antigenicity. They can circumvent endocytosis and escape from phagocytosis by the RES. The exosomes have high delivery efficacy. Therefore, they are ideal vehicles for nucleic acid medicinal delivery. Recently, exosomes derived from endothelial progenitor cells and containing miRNA-126-3p and 5p require for attenuation of organ injury and vascular permeability in cecal ligation and puncture (CLP)-induced sepsis.[169] Moreover, human bone marrow mesenchymal stem cells together with miRNA-375 engulf exosomes. They are restrained in islet apoptosis and primary nonfunction (PNF) at the time of islet transplantation in humanized NOD *scid* gamma models (NSG).[170] An important study reported that occasionally exosomes are used to make from bovine milk for the transportation of miRNA delivery.[171]

The shedding micro-vesicles (MV) are covered from multiple cell types in different pathological and physiological states.[172] The accumulated MVs are used for TNBS-induced rat intestinal fibrosis treatment.[173] Further the effect of conjugated MVs with miRNA-29a/c reduces tumor growth in gastric cancer (GC) cells. [174] Similarly, MVs originated from leukemia cells that can induce T cell exhaustion by delivering man functional miRNAs.[175]

The bone marrow megakaryocytes forming tiny Platelets (1 to 4 μm) are required for the maintenance of physiology hemostasis and vascular integrity.[176] The induced platelets that are released from small MV called 'microparticles' (MPs) are derived from the cytoplasmic membrane.[177] The Ago2/miRNA-223 conjugated platelet is added into the cargos for their entry of human umbilical vein endothelial cells (HUVECs).[177] Further the platelet-derived MVs together with miRNA-223 are required for lung cancer cell invasion by downregulating tumor suppressor EPB41L3.[178]

The membrane blebs the 'Apoptotic bodies' is released from apoptotic cells. The variable apoptotic bodies express phosphatidylserine as "eat-me" molecules. Sometimes apoptotic bodies derived from endothelial cells carry miRNA-126 that can trigger chemokine (C-X-C motif) ligand-12 (CXCL12) secretions and protect from atherosclerosis.[179]

3d scaffold vesicles

To circumvent mechanical barriers and promote spatiotemporal control, 3D biomaterial scaffolds are ideal for miRNAs therapeutic delivery. Multiple 3D-scaffold based delivery systems including hydrogels, electrospun fibers and porous or spongy 3D-scaffolds are noticed. The hydrophilic polymers like PEGylation hydrogels components enhance encapsulated human bone marrow generated from mesenchymal stem cells (hMSCs) for dividing into osteoblasts.[180] They release small RNAs and miRNA-20a mimics. On the other hand, the self-assembled dual-color RNA-triple-helix hydrogel carry miRNA-221. They also carry antago-miRs and miRNA-225 mimics for reducing the nearly 90% cancer shape and size in a triple-negative breast cancer model.[181]

Also, multiple electro-spun fibers are used for many different experiments. The miRNA-126 mimics construct a bilayer vascular scaffold fabricated by target carriers. They also have electro-spun fibrous membranes in the vascular endothelial cells (VECs).[182] In these compounds, the outer layer of poly(ε-caprolactone) (PCL) and gelatin introduce for mechanical stability. The complexes of miRNA-126 mimics in REDV peptide-modified trimethyl chitosan-g-poly(ethylene glycol) of the inner layer of poly (ethylene glycol)-*b*-poly(L-lactide-*co*-ε-caprolactone) (PELCL) regulate VECs responses.[182]

A binding of hyperbranched polymer (HP) with miRNA-26-moiety forms a nanoscale complex.[183] This 3D-scaffold induces defective calvaria bone regeneration in an osteoporotic model.[183] The collagen-nanohydroxyapatite miRNA-activated scaffold conjugation delivers both in human miRNA antagomiRs and miRNA mimics to mesenchymal stem cells efficiently.[184]

Oral pills and oral delivery

In 2009, the first oral delivery of a small-encapsulated particle-containing siRNA was investigated for silencing the target gene. This experiment opened up a new direction toward the most common but daunting challenge for miRNA therapeutics via the oral route. How can miRNA easily transport effective gene silencing to specific cells and tissues by oral administration? An engineered yeast is introduced selectively for oral delivery for their attraction to macrophages. The macrophages engulf and digest the yeast molecules. After engulfing the macrophages rupture the yeast digestive shells, release miRNA therapeutics and transport following through oral and digestive routes. The yeast particles sometimes carry "beta1,3-D glucan" that is promoted drug delivery and release system. One of the RNA pharma giants investigated to develop a wide spectrum of orally delivered RNAi therapeutics using the glucan particle model. The beauty of oral delivery is that it introduces as little as 20 micrograms per kilogram of body weight of model mice to silent a signal protein MAP4K4, a key player for the elimination of inflammatory response in the arthritis disease.

In vast miRNA delivery networks certain combat carriers can shut off-target gene expression efficiently. They are not equally effective for gastrointestinal diseases, including inflammatory bowel disease (IBD) and colon cancer, where delivery of medicines is inadequate for clinical benefits. The gastrointestinal tract shows the most hostile environment for delivery cargos, which appears to be a vulnerable site for considering oral administration. Despite these hurdles nowadays oral delivery is not a 'mission impossible' route for efficient and easy delivery. Various studies reported that RNA therapeutics are degraded easily by strong GI-nucleases. The miRNAs molecules are also fragile for enzymatic degradation and plasma half-lives of <1 min.[185] Sometimes the miRNA is integrated with hydrophilic and negatively charged phosphate backbone. However, these physicochemical properties for membrane permeability are extremely less. To overcome the major three hurdles enzymatic barriers, mucus barriers, and cell membrane barriers for the miRNA therapeutic agents inside oral pills can be modified dramatically. **(Figure 6)**

Enzymatic barriers

Almost nothing is known about GI-nucleases unlike well explored gastrointestinal enzymes. To protect against nuclease degradation, miRNA containing oral pills must employ multiple chemically modified carriers resisting enzymes that form a protective shield against GI-nuclease attack. Conversely strong GI-nucleases can kill the phosphodiester oligonucleotide completely within 30 mins period. Therefore, dramatic chemical modifications are indispensable for their survival. Broadly, chemical modifications are occurred based on internucleotide linkages (I) nucleobase, or (II)sugar modifications (III). As internucleotide phosphodiester linkages RNA therapeutics are required for targeting *endo*-and *exo*-nuclease cleavage. They are modified including phosphorothioate, N3' phosphoramidates, boranophosphate 2′,5′–phosphodiesters, amide linked phosphonoacetates, morpholinos and peptide nucleic acids for effective oral delivery.[186] This strategy happens to be the first choice for RNA therapeutics oral delivery. Besides, to overlook free access of nucleases to oral pills, RNA therapeutics co-precipitate with polyethyleneimine [187], chitosan [188], thiolated chitosans [189] or dendrimers[190] etc. Recently, at least EDTA, sodium dodecyl sulphate and aurintricarboxylic acid are used for a pronounced inhibitory effect against intestinal enzymes.[191] However, their rapid dilution though protects therapies against intestinal enzymes but unprotected them in the intestinal lumen. The therapeutic agents select "pilocarpine" due to its cholinergic properties that interconnect preferentially to muscarinic receptors on the crypt goblet cells.

Mucus barriers

Another huge obstacle for oral pills containing RNA drugs is the mucus gel layer. Mucus glycoproteins commonly known as 'mucins' carry cysteine-rich subdomains crosslinked with disulfide bonds. Constant water flow is a character of the intestinal epithelium for accelerating the peristaltic movement of food particles, which diffuse medicines through the mucus gel layer.[192] It is certain that without proper aid of mucolytic agents the amalgamation of therapy carriers' is losing the potential for their entry in the mucus gel layer. Alternatively, mucolytic agents for example sulfhydryl compounds are often used the cleavage of disulfide bonds inside the mucus gel layer for free therapeutic entry.

Two selective mucolytic enzymes, *papain* or *bromelain* are the other option that cleaves few protein substructures in the mucus gel layer.[193] The conjugation of two enzymes together with NPs evolves high-intensity anionic and cationic charges appears to be a direct approach.

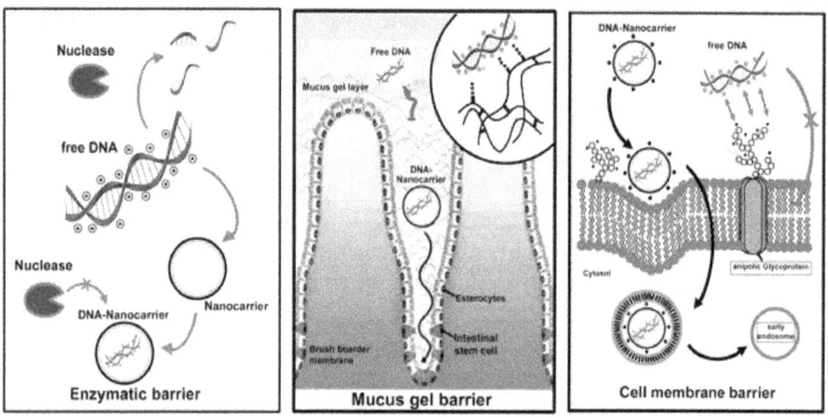

Figure-6. Different Physiological barriers for oral nucleic acid-based therapeutics in the gastrointestinal tract. (Loosely Adapted, Redraw and Modify using skeleton from O'Driscolla *et al.*, 2019)

Membrane barrier

The GI-tract epithelium lining portrays the most challenging barrier to orally administered RNA medicines. Two charged molecules inside brush border microvilli polar carbohydrates and charged amino acid side chains function as a high-density negative charge on the microvilli at the enterocytes' apical surface.[194] Another study reported that improvement of NPs membrane permeability is observed on lipophilic (I), cationic (II), and cell penetrating peptides for nanoparticle decoration. In certain cases, a combination of_intrinsic cell-penetrating properties and the cationic charges produces by cell-penetrating peptides (CPPs) arginine- and/or lysine-rich peptides. This study flashes the potentiality of oral hydrophilic macromolecular drug delivery in general [195] for more efficient oral pill formation.

In summary, it is next to impossible at this stage to draw a clear timeline for oral delivery and oral pill formulation to overcome many physiological regulatory barriers that remain outstanding. However, speed the many pre-clinical data look convincing and promising for oral pill formulation.

Running clinical trials for selective miRNA medicines

Numerous miRNA-based therapeutics have been running in different stages of clinical trials. They are the basic products of newly developed RNA pharma laboratories. Their entire list of information can be reported on the website (https://clinicaltrials.gov/ct2/home). Here, the major miRNA-based medicine candidates are in different phases of clinical trials. Among them, Miravirsen (SPC3649) (Clinical Trial.gov Identifier: NCT02452814), leads the world's prominent miRNA medicine candidate via clinical testing will be introduced for treat hepatitis C infection treatment that was in Phase II clinical trials in 2017. Locked nucleotides protect Miravirsen to their both open termini. Miravirsen contains LNA ribonucleotides whose sequences are complementary to miRNA-122.[196] Another potential therapeutic candidate an RG-101 has been in a Phase 1b clinical trial that targets miRNA-122 by a modified phosphorothioate oligonucleotide inhibitor. This molecule binds with a multivalent N-acetylgalactosamine carbohydrate structure for up-taking asialoglycoprotein receptors on the hepatocytes via complementary binding. It is effectively used for the treatment of the serious adverse conditions (SAE) of jaundice. The conjugated MRX34 molecule (Clinical Trial.gov Identifier: NCT01829971) is composed of in between a liposome and encapsulated miRNA-34a mimic that is examined for treatment of solid tumors in a multicenter Phase I trial. Also, MRG-106 (Clinical Trial.gov Identifier: NCT02580552), an anti-miRNA-155 LNA-modified antisense inhibitor is used for treatment for cutaneous T cell lymphoma and mycosis fungoides in a Phase II clinical trial. RG-125/AZD4076 (Clinical Trial.gov Identifier: NCT02612662) is an anti-miRNA-103/107 conjugated with *N*-acetylgalactosamine (GalNAc), that is useful for treating the patients with non-alcoholic fatty liver and type 2 diabetes in Phase I/II trial. MesomiR-1 (Clinical Trial.gov Identifier: NCT02369198), a miRNA-16-based miRNA mimic in nonliving bacterial minicells is associated with an anti-EGFR bispecific antibody that is the remedy for mesothelioma and non-small-cell lung cancer (NSCLC) in a Phase I study. The hybrid conjugate MRG-201 (Clinical Trials.gov Identifier: NCT0260322) is in between a miRNA-29 mimic and a cholesterol-conjugated miRNA duplex. It is the potential drug against scleroderma disease that began a Phase II clinical trial in the recent past.

Apart from therapies generation and their potential trials easy and quick isolation of extracellular miRNA from the circulatory blood, urine and sputum samples can be separated rapidly, which are described in protocols for miRNA chapters earlier. The profile of genome-wide miRNA expression appears to be an ideal biomarker for early diagnosis, staging and prognosis of the disease. For instance, miRNA7™ is the first approved commercially available kit for early diagnosis for liver cancer detection.

Conclusions

For optimized and advantageous techniques generation, vast networks for numerous miRNA-based delivery vehicles have been developed as numerous specific favorable applications. The miRNA-based delivery cargos are involved in protecting miRNA oligonucleotides from serum RNase degradation to promote targeting ability without kicking immune response related to adverse effects. The intravenous injection of local treatment is the main device for miRNA delivery. However, very few efforts have been employed for use of miRNA conjugated oral pill formation or oral delivery.

Chemical modifications and their complexities have been encouraging as main criteria for RNAs stabilization over cellular uptake and escape. The viral and nonviral vectors have disadvantages for immunogenicity and low oligonucleotide-loading capacity. Mainly nano cargos follow the quick exit route by easily accumulating in the liver, spleen and quickly liberate through the kidney. Therefore, new nanomaterials have been synthesized and new methods have been developed focusing on the targeting ability and long-term blood circulation for delivery systems. The exosomes, micro-vesicles and apoptotic bodies are the three important cell-derived membrane carriers that have been used ideal for their trace level cytotoxicity and negligible antigenicity. More importantly, a deeper and clearer understanding of physiological functions and environmental status will be required for performing the use of these vehicle generation machinery for further improvements and the most potential target generation of miRNAs carrying cargos as nearby future challenges.

In the last two decades of cargo development, conjugated RNA therapeutics have come to clinical reality. The design, selection, and chemistry of miRNAs and anti-miRs are standing at their peak point that enhances the chemical stability, easy entry in live cells, lesser immune response, longer half-life and allowing the efficient maintenance and specificity. The dramatic progress of delivery technologies deserved a great salute for their potent and biocompatibility. Undoubtedly the field of miRNA therapeutics is currently undergoing a huge expansion and running towards personalized medicine development and immunotherapy as a remedy for genetic disorders, infectious, and chronic diseases. They will encourage the continuous development of miRNA therapeutics at a same sporadic pace for several years to come.

Reference.

References: chapter wise are cited before the Index at the end of the Book. (1-196)

References

Chapter I

1. Alberti, C., and Cochella, L. (2017). A framework for understanding the roles of miRNAs in animal development. *Development* 144, 2548–2559. doi: 10.1242/dev.146613
2. Guerra-Assunção, J. A., and Enright, A. J. (2012). Large-scale analysis of microRNA evolution. *BMC Genomics* 13:218. doi: 10.1186/1471-2164-13-218
3. Hertel, J., and Stadler, P. F. (2015). The expansion of animal microRNA families revisited. *Life* 5, 905–920. doi: 10.3390/life5010905
4. Reinhart BJ, Weinstein EG, Rhoades MW, et al., MicroRNAs in plants. Genes Dev. 2002;16:1616–26.
5. 3. Rhoades MW, Reinhart BJ, Lim LP, et al. Prediction of plant microRNA targets. Cell 2002;110:513–20.
6. 4. Llave C, Kasschau KD, Rector MA, et al. Endogenous and silencing-associated small RNAs in plants. Plant Cell 2002;14: 1605–19.
7. Griffiths-Jones S, The microRNA Registry. Nucleic Acids Res 2004;32:D109–11.
8. 7. Kozomara A, Griffiths-Jones S. MiRBase: Annotating high confidence microRNAs using deep sequencing data. Nucleic Acids Res 2014;42:D68–73.
9. Hobert, O.(2008). Gene regulation by transcription factors and microRNAs. Science **319**, 1785-1786.doi:10.1126/science.1151651
10. M. A. Taylor, K. Sossey-Alaoui, C. L. Thompson, D. Danielpour, and W. P. Schiemann, "TGF-*β* upregulates miR-181a expression to promote breast cancer metastasis," *The Journal of Clinical Investigation*, vol. 123, no. 1, pp. 150–163, 2013.
11. J.-F. Chen, E. P. Murchison, R. Tang et al., "Targeted deletion of Dicer in the heart leads to dilated cardiomyopathy and heart failure," *Proceedings of the National Academy of Sciences of the United States of America*, vol. 105, no. 6, pp. 2111–2116, 2008.
12. Y. Suárez and W. C. Sessa, "MicroRNAs as novel regulators of angiogenesis," *Circulation Research*, vol. 104, no. 4, pp. 442–454, 2009.
13. M. Knöfler and J. Pollheimer, "Human placental trophoblast invasion and differentiation: a particular focus on Wnt signaling," *Frontiers in Genetics*, vol. 4, article 190, 14 pages, 2013.
14. R. Demir, Y. Seval, and B. Huppertz, "Vasculogenesis and angiogenesis in the early human placenta," *Acta Histochemica*, vol. 109, no. 4, pp. 257–265, 2007.
15. C. Qiu, G. Chen, and Q. Cui, "Towards the understanding of microRNA and environmental factor interactions and their relationships to human diseases," *Scientific Reports*, vol. 2, article 318, 2012.
16. O. Khorram, G. Han, R. Bagherpour et al., "Effect of maternal undernutrition on vascular expression of micro and messenger RNA in newborn and aging

offspring," *American Journal of Physiology - Regulatory Integrative and Comparative Physiology*, vol. 298, no. 5, pp. R1366–R1374, 2010.

17. Kouzarides T. Chromatin modifications and their function. Cell. 2007; 128:693-705.
18. Margueron R, Reinberg D. Chromatin structure and the inheritance of epigenetic information. Nat Rev Genet. 2010; 11:285-296.
19. Ruthenburg AJ, Li H, Patel DJ, Allis CD. Multivalent engagement of chromatin modifications by linked binding modules. Nat Rev Mol Cell Biol. 2007; 8:983-994.
20. Ha M, Kim VN. Regulation of microRNA biogenesis. *Nat Rev Mol Cell Biol.* (2014) 15:509–24. doi: 10.1038/nrm3838
21. Broughton JP, Lovci MT, Huang JL, Yeo GW, Pasquinelli AE. Pairing beyond the Seed Supports MicroRNA Targeting Specificity. *Mol Cell.*(2016) 64:320–33. doi: 10.1016/j.molcel.2016.09.004
22. Vasudevan S. Posttranscriptional upregulation by microRNAs. *Wiley Interdiscip Rev RNA* (2012) 3:311–30. doi: 10.1002/wrna.121
23. Makarova JA, Shkurnikov MU, Wicklein D, Lange T, Samatov TR, Turchinovich AA, et al. Intracellular and extracellular microRNA: an update on localization and biological role. *Prog Histochem Cytochem.* (2016) 51:33–49. doi: 10.1016/j.proghi.2016.06.001
24. Fu G, Brkic J, Hayder H, Peng C. MicroRNAs in human placental development and pregnancy complications. *Int J Mol Sci.* (2013) 14:5519–44. doi: 10.3390/ijms14035519
25. Tufekci KU, Oner MG, Meuwissen RL, Genc S. The role of microRNAs in human diseases. *Methods Mol Biol.* (2014) 1107:33–50. doi: 10.1007/978-1-62703-748-8_3
26. Paul P, Chakraborty A, Sarkar D, Langthasa M, Rahman M, Bari M, et al. Interplay between miRNAs and human diseases. *J Cell Physiol.*(2018) 233:2007–18. doi: 10.1002/jcp.25854
27. Hayes J, Peruzzi PP, Lawler S. MicroRNAs in cancer: biomarkers, functions and therapy. *Trends Mol Med.* (2014) 20:460–9. doi: 10.1016/j.molmed.2014.06.005
28. Wang J, Chen J, Sen S. MicroRNA as Biomarkers and Diagnostics. *J Cell Physiol.* (2016) 231:25–30. doi: 10.1002/jcp.25056
29. Lai X, Wolkenhauer O, Vera J. Understanding microRNA-mediated gene regulatory networks through mathematical modelling. *Nucleic Acids Res.* (2016) 44:6019–35. doi: 10.1093/nar/gkw550
30. Zhang JF, Fu WM, He ML, Xie WD, Lv Q, Wan G, et al. MiRNA-20a promotes osteogenic differentiation of human mesenchymal stem cells by co-regulating BMP signaling. *RNA Biol.* (2011) 8:829–38. doi: 10.4161/rna.8.5.16043
31. Ebert MS, Sharp PA. Roles for microRNAs in conferring robustness to biological processes. *Cell* (2012) 149:515–24. doi: 10.1016/j.cell.2012.04.005
32. Li Y, Liang C, Wong KC, Luo J, Zhang Z. Mirsynergy: detecting synergistic miRNA regulatory modules by overlapping neighbourhood expansion. *Bioinformatics* (2014) 30:2627–35. doi: 10.1093/bioinformatics/btu373

33. Lemus-Diaz N, Boker KO, Rodriguez-Polo I, Mitter M, Preis J, Arlt M, et al. Dissecting miRNA gene repression on single cell level with an advanced fluorescent reporter system. *Sci Rep.* (2017) 7:45197. doi: 10.1038/srep45197
34. Bose M, Bhattacharyya SN. Target-dependent biogenesis of cognate microRNAs in human cells. *Nat Commun.* (2016) 7:12200. doi: 10.1038/ncomms12200
35. Kucherenko MM, Shcherbata HR. miRNA targeting and alternative splicing in the stress response - events hosted by membrane-less compartments. *J Cell Sci.* (2018) 131:jcs202002. doi: 10.1242/jcs.2020
36. Wang K, Zhang S, Weber J, Baxter D, Galas DJ. Export of microRNAs and microRNA-protective protein by mammalian cells. *Nucleic Acids Res.* (2010) 38:7248–59. doi: 10.1093/nar/gkq601
37. Lim LP, Lau NC, Weinstein EG, Abdelhakim A, Yekta S, Rhoades MW, Burge CB, Bartel DP (2003). "The microRNAs of Caenorhabditis elegans". Genes & Development. **17** (8): 991–1008. doi:10.1101/gad.1074403.
38. Lagos-Quintana M, Rauhut R, Yalcin A, Meyer J, Lendeckel W, Tuschl T (2002). "Identification of tissue-specific microRNAs from mouse". Current Biology. **12** (9): 735–9. doi:10.1016/S0960-9822(02)00809-6.
39. *Homo sapiens* miRNAs in the miRBase at Manchester University
40. Fromm B, Billipp T, Peck LE, Johansen M, Tarver JE, King BL, et al. (2015). "A Uniform System for the Annotation of Vertebrate microRNA Genes and the Evolution of the Human microRNAome". Annual Review of Genetics. **49**: 213–42. doi:10.1146/annurev-genet-120213-092023.
41. Lewis BP, Burge CB, Bartel DP (2005). "Conserved seed pairing, often flanked by adenosines, indicates that thousands of human genes are microRNA targets". Cell. **120** (1): 15–20. doi:10.1016/j.cell.2004.12.035.
42. Friedman RC, Farh KK, Burge CB, Bartel DP (2009). "Most mammalian mRNAs are conserved targets of microRNAs". Genome Research. **19** (1): 92-105. doi:10.1101/gr.082701.108.
43. Ambros V, Bartel B, Bartel DP, Burge CB, Carrington JC, Chen X, Dreyfuss G, Eddy SR, Griffiths-Jones S, Marshall M, Matzke M, Ruvkun G, Tuschl T (2003). "A uniform system for microRNA annotation". RNA. **9** (3): 277-9. doi:10.1261/rna.2183803.
44. Griffiths-Jones S, Grocock RJ, van Dongen S, Bateman A, Enright AJ (2006). "miRBase: microRNA sequences, targets and gene nomenclature". Nucleic Acids Research. **34** (Database issue): D140-4. doi:10.1093/nar/gkj112.
45. Wright MW, Bruford EA (2011). "Naming 'junk': human non-protein coding RNA (ncRNA) gene nomenclature". Human Genomics. **5** (2): 90–8. doi:10.1186/1479-7364-5-2-90.
46. Bartel DP (2004). "MicroRNAs: genomics, biogenesis, mechanism, and function". Cell. **116** (2): 281–97. doi:10.1016/S0092-8674(04)00045-5.
47. Ambros V, Bartel B, Bartel DP, et al. A uniform system for microRNA annotation. RNA 2003;9:277–9.
48. Meyers BC, Axtell MJ, Bartel B, et al. Criteria for annotation of plant MicroRNAs. Plant Cell 2008;20:3186–90.
49. Carthew RW, Sontheimer EJ. Origins and mechanisms of miRNAs and siRNAs. Cell 2009;136:642–55.

50. Meng Y, Shao C, Wang H, et al. Are all the miRBase-registered microRNAs true? A structure- and expression-based re- examination in plants. RNA Biol 2012;9:249–53.
51. Li A, Mao L. Evolution of plant microRNA gene families. Cell Res 2007;17:212–8.
52. Roberts JT, Cooper EA, Favreau CJ, et al. Continuing analysis of microRNA origins: Formation from transposable element insertions and noncoding RNA mutations. Mob Genet Elements 2013;3:e27755.
53. Son DJ, Kumar S, Takabe W, et al. The atypical mechano-sensitive microRNA-712 derived from pre-ribosomal RNA induces endothelial inflammation and atherosclerosis. Nat Commun 2013;4:3000.
54. Meyers BC, Axtell MJ, Bartel B, et al. Criteria for annotation of plant MicroRNAs. Plant Cell 2008;20:3186–90.
55. Van Peer G, Lefever S, Anckaert J, et al., miRBase Tracker: keeping track of microRNA annotation changes. Database (Oxford) 2014;2014:1–8.
56. Zhang W, Gao S, Zhou X, et al. Multiple distinct small RNAs origin- ate from the same microRNA precursors. Genome Biol 2010;11:R81.
57. Guo L, Chen F. A challenge for miRNA: Multiple isomiRs in miRNAomics. Gene 2014;544:1–7.
58. Sripada L, Tomar D, Singh R. Mitochondria: one of the destin- ations of miRNAs. Mitochondrion 2012;12:593–9.
59. Roberts JT, Cooper EA, Favreau CJ, et al.Continuing analysis of microRNA origins: Formation from transposable element in- sertions and noncoding RNA mutations. Mob Genet Elements 2013;3:e27755.
60. Kamarajan BP, Sridhar J, Subramanian S. In silico prediction of microRNAs in plant mitochondria. Int J Bıoautomatıon 2013;16: 251–62.
61. R.C. Lee, R.L. Feinbaum, V. Ambros, The C. elegans heterochronic gene lin-4 encodes small RNAs with antisense complementarity to lin-14, Cell 75 (5) (1993) 843–854.
62. B. Wightman, I. Ha, G. Ruvkun, Posttranscriptional regulation of the heterochronic gene lin-14 by lin-4 mediates temporal pattern formation in C. elegans, Cell 75 (5) (1993) 855–862.
63. D.P. Bartel, MicroRNAs: genomics, biogenesis, mechanism, and function, Cell 116 (2) (2004) 281–297.
64. M. Chalfie, H.R. Horvitz, J.E. Sulston, Mutations that lead to reiterations in the cell lineages of C. elegans, Cell 24 (1) (1981) 59–69.
65. R. Lee, R. Feinbaum, V. Ambros, A short history of a short RNA, Cell 116 (2 Suppl.) (2004) S89–S92.
66. E.L. Ferguson, P.W. Sternberg, H.R. Horvitz, A genetic pathway for the specification of the vulval cell lineages of Caenorhabditis elegans, Nature 326 (6110) (1987) 259–267.
67. B.J. Reinhart, F.J. Slack, M. Basson, A.E. Pasquinelli, J.C. Bettinger, A.E. Rougvie, H.R. Horvitz, G. Ruvkun, The 21-nucleotide let-7 RNA regulates developmen- tal timing in Caenorhabditis elegans, Nature 403 (6772) (2000) 901–906.
68. M.C. Vella, F.J. Slack, C. elegans microRNAs, WormBook 21 (2005) 1–9.

69. M.C. Vella, E.Y. Choi, S.Y. Lin, K. Reinert, F.J. Slack, The C. elegans microRNA let- 7 binds to imperfect let-7 complementary sites from the lin-41 3′ UTR, Genes Dev.18 (2) (2004) 132–137.
70. A.E. Pasquinelli, B.J. Reinhart, F. Slack, M.Q. Martindale, M.I. Kuroda, B. Maller, D.C. Hayward, E.E. Ball, B. Degnan, P. Müller, J. Spring, A. Srinivasan, M. Fishman, J. Finnerty, J. Corbo, M. Levine, P. Leahy, E. Davidson, G. Ruvkun, Conservation of the sequence and temporal expression of let-7 heterochronic regulatory RNA, Nature 408 (6808) (2000) 86–89.
71. E.J. Kaufman, E.A. Miska, The microRNAs of Caenorhabditis elegans, Semin. Cell Dev. Biol. 21 (7) (2010) 728–737.
72. A.L. Abbott, E. Alvarez-Saavedra, E.A. Miska, N.C. Lau, D.P. Bartel, H.R. Horvitz, V. Ambros, The let-7 MicroRNA family members mir-48, mir-84, and mir-241 function together to regulate developmental timing in Caenorhabditis elegans, Dev. Cell 9 (3) (2005) 403–414.
73. S. Griffiths-Jones, H.K. Saini, S. van Dongen, A.J. Enright, miRBase: tools for microRNA genomics, Nucleic Acids Res. 36 (Database issue) (2008) D154–D158.
74. S. Griffiths-Jones, R.J. Grocock, S. van Dongen, A. Bateman, A.J. Enright, miR-Base: microRNA sequences, targets and gene nomenclature, Nucleic Acids Res. 34 (Database issue) (2006) D140–D144.
75. S. Griffiths-Jones, The microRNA Registry, Nucleic Acids Res. 32 (Database issue) (2004) D109–D111.
76. W. Xia, G. Cao, N. Shao, Progress in miRNA target prediction and identification, Sci. China C Life Sci. 52 (12) (2009) 1123–1130.
77. W. Huang da, B.T. Sherman, R.A. Lempicki, Systematic and integrative analysis of large gene lists using DAVID bioinformatics resources, Nat. Protoc. 4 (1) (2009) 44–57.
78. *Lee CT, Risom T, Strauss WM (2007). "Evolutionary conservation of microRNA regulatory circuits: an examination of microRNA gene complexity and conserved microRNA-target interactions through metazoan phylogeny". DNA and Cell Biology. 26 (4): 209–18.* doi:10.1089/dna.2006.0545.
79. *Peterson KJ, Dietrich MR, McPeek MA (2009). "MicroRNAs and metazoan macroevolution: insights into canalization, complexity, and the Cambrianexplosion". BioEssays. 31(7): 736–47.* doi:10.1002/bies.200900033.
80. *Shabalina SA, Koonin EV (2008).* "Origins and evolution of eukaryotic RNA interference". *Trends in Ecology & Evolution. 23 (10): 578–87.* doi:10.1016/j.tree.2008.06.005.
81. *Axtell MJ, Westholm JO, Lai EC (2011).* "Vive la différence: biogenesis and evolution of microRNAs in plants and animals". *Genome Biology. 12 (4): 221.* doi:10.1186/gb-2011-12-4-221.
82. *Wheeler BM, Heimberg AM, Moy VN, Sperling EA, Holstein TW, Heber S, Peterson KJ (2009). "The deep evolution of metazoan microRNAs". Evolution & Development. 11 (1): 50–68.* doi:10.1111/j.1525-142X.2008.00302.x.
83. *Pashkovskiy PP, Ryazansky SS (2013). "Biogenesis, evolution, and functions of plant microRNAs". Biochemistry. Biokhimiia. 78 (6): 627–37.* doi:10.1134/S0006297913060084.

84. *Heimberg AM, Sempere LF, Moy VN, Donoghue PC, Peterson KJ (2008).* "MicroRNAs and the advent of vertebrate morphological complexity". *Proceedings of the National Academy of Sciences of the United States of America. 105 (8): 2946-50.* Bibcode:2008PNAS..105.2946H. doi: 10.1073/pnas.0712259105.
85. *Nozawa M, Miura S, Nei M (2010).* "Origins and evolution of microRNA genes in Drosophila species". *Genome Biology and Evolution. 2: 180–9.* doi:10.1093/gbe/evq009.
86. *Allen E, Xie Z, Gustafson AM, Sung GH, Spatafora JW, Carrington JC (2004). "Evolution of microRNA genes by inverted duplication of target gene sequences in Arabidopsis thaliana". Nature Genetics. 36 (12): 1282–90.* doi: 10.1038/ng1478.
87. *Warthmann N, Das S, Lanz C, Weigel D (2008).* "Comparative analysis of the MIR319a microRNA locus in Arabidopsis and related Brassicaceae". *Molecular Biology and Evolution. 25 (5): 892 902.* doi:10.1093/molbev/msn029.
88. Caravas J, Friedrich M (2010). "Of mites and millipedes: recent progress in resolving thebase of the arthropod tree". *BioEssays*. **32** (6): 488–95. doi:10.1002/bies.201000005.
89. *Cuperus JT, Fahlgren N, Carrington JC (2011).* "Evolution and functional diversification of MIRNA genes". *The Plant Cell. 23 (2): 431-42.* doi:10.1105/tpc.110.082784.
90. *Dimond PF (2010).* "miRNAs' Therapeutic Potential". *Genetic Engineering & Biotechnology News. 30 (6): 1.*
91. Ruby, J.G., Jan, C., Player, C., Axtell, M.J., Lee, W., Nusbaum, C., Ge, H., and Bartel, D.P. (2006). Large-scale sequencing reveals 21U-RNAs and additional microRNAs and endogenous siRNAs in *C. elegans*. Cell *127*, 1193–1207.
92. Chiang, H.R., Schoenfeld, L.W., Ruby, J.G., Auyeung, V.C., Spies, N., Baek, D., Johnston, W.K., Russ, C., Luo, S., Babiarz, J.E., et al. (2010). Mammalian microRNAs: experimental evaluation of novel and previously annotated genes. Genes Dev. *24*, 992–1009.
93. Fromm, B., Billipp, T., Peck, L.E., Johansen, M., Tarver, J.E., King, B.L., Newcomb, J.M., Sempere, L.F., Flatmark, K., Hovig, E., and Peterson, K.J. (2015). A uniform system for the annotation of vertebrate microRNA genes and the evolution of the human microRNAome. Annu. Rev. Genet. *49*, 213–242.
94. Kozomara, A., and Griffiths-Jones, S. (2014). miRBase: annotating high confidence microRNAs using deep sequencing data. Nucleic Acids Res. *42*, D68–D73.
95. Friedla̎nder, M.R., Chen, W., Adamidi, C., Maaskola, J., Einspanier, R., Knespel, S., and Rajewsky, N. (2008). Discovering microRNAs from deep sequencing data using miRDeep. Nat. Biotechnol. *26*, 407–415.
96. Morin, R.D., O'Connor, M.D., Griffith, M., Kuchenbauer, F., Delaney, A., Prabhu, A.L., Zhao, Y., McDonald, H., Zeng, T., Hirst, M., et al. (2008). Application of massively parallel sequencing to microRNA profiling and discovery in human embryonic stem cells. Genome Res. *18*, 610–621.
97. Han, J., Pedersen, J.S., Kwon, S.C., Belair, C.D., Kim, Y.K., Yeom, K.H., Yang, W.Y., Haussler, D., Blelloch, R., and Kim, V.N. (2009). Posttranscriptional crossregulation between Drosha and DGCR8. Cell *136*, 75–84.

98. Bosson, A.D., Zamudio, J.R., and Sharp, P.A. (2014). Endogenous miRNA and target concentrations determine susceptibility to potential ceRNA competi- tion. Mol. Cell *56*, 347–359.
99. Denzler, R., McGeary, S.E., Title, A.C., Agarwal, V., Bartel, D.P., and Stoffel, M. (2016). Impact of microRNA levels, target-site complementarity, and coopera- tivity on competing endogenous RNA-regulated gene expression. Mol. Cell *64*, 565–579.
100. Ruby, J.G., Stark, A., Johnston, W.K., Kellis, M., Bartel, D.P., and Lai, E.C. (2007b). Evolution, biogenesis, expression, and target predictions of a substantially expanded set of *Drosophila* microRNAs. Genome Res. *17*, 1850–1864
101. Lee, R. C., Feinbaum, R. L. and Ambros, V.(1993). The C. elegans heterochronic gene lin-4 encodes small RNAs with antisense complementarity to lin-14. Cell **75**, 843-854. doi:10.1016/0092-8674(93)90529-Y
102. Reinhart, B. J., Slack, F. J., Basson, M., Pasquinelli, A. E., Bettinger, J. C., Rougvie, A. E., Horvitz, H. R. and Ruvkun, G.(2000). The 21-nucleotide let-7 RNA regulates developmental timing in Caenorhabditis elegans. Nature **403**, 901-906. doi:10.1038/35002
103. Brenner, J. L., Jasiewicz, K. L., Fahley, A. F., Kemp, B. J. andAbbott, A. L. (2010). Loss of individual microRNAs causes mutant phenotypes in sensitized genetic backgrounds in C. elegans. Curr. Biol. **20**, 1321-1325.
104. Hipfner, D. R., Weigmann, K. and Cohen, S. M (2002). The bantam gene regulates Drosophila growth. Genetics**161**, 1527-1537
105. Xu, P., Vernooy, S. Y., Guo, M. and Hay, B. A. *(2003).* The drosophila microRNA Mir-14 suppresses cell death and is required for normal fat metabolism. *Curr. Biol.* **13**, 790-795. doi:10.1016/S0960-9822(03)00250-1
106. Cochella, L. and Hobert, O. (2012a). Embryonic priming of a miRNA locus predetermines postmitotic neuronal left/right asymmetry in C. elegans. Cell 151, 1229-1242. doi:10.1016/j.cell.2012.10.049
107. Johnston, R. J., Jr. and Hobert, O. (2003). A microRNA controlling left/right neuronal asymmetry in Caenorhabditis elegans. Nature 426, 845-849. doi:10.1038/nature0225
108. Cayirlioglu, P., Kadow, I. G., Zhan, X., Okamura, K., Suh, G. S. B., Gunning, D., Lai, E. C. and Zipursky, S. L. (2008). Hybrid neurons in a microRNA mutant are putative evolutionary intermediates in insect CO2 sensory systems. Science 319, 1256-1260. doi:10.1126/science.1149483
109. Lewis, M. A., Quint, E., Glazier, A. M., Fuchs, H., De Angelis, M. H., Langford, C., van Dongen, S., Abreu-Goodger, C., Piipari, M., Redshaw, N. et al. (2009). An ENU-induced mutation of miR-96 associated with progressive hearing loss in mice. Nat. Genet. 41, 614-618. doi:10.1038/ng.369
110. Davidson, E. H. and Erwin, D. H. (2006). Gene regulatory networks and the evolution of animal body plans. Science 311, 796-800. doi:10.1126/science.1113832
111. Davidson, E. H., McClay, D. R. and Hood, L. (2003). Regulatory gene networks and the properties of the developmental process. Proc. Natl. Acad. Sci. USA 100, 1475-1480. doi:10.1073/pnas.0437746100

112. Cheloufi, S., Dos Santos, C. O., Chong, M. M. W. and Hannon, G. J. (2010). A dicer-independent miRNA biogenesis pathway that requires Ago catalysis. Nature 465, 584-589. doi:10.1038/nature09092
113. Patrick, D. M., Zhang, C. C., Tao, Y., Yao, H., Qi, X., Schwartz, R. J., Jun-Shen Huang, L. and Olson, E. N. (2010). Defective erythroid differentiation in miR-451 mutant mice mediated by 14-3-3zeta. Genes Dev. 24, 1614- 1619. doi:10.1101/gad.1942810
114. Chen, Y.-W., Song, S., Weng, R., Verma, P., Kugler, J.-M., Buescher, M., Rouam, S. and Cohen, S. M. (2014). Systematic study of drosophila microRNA functions using a collection of targeted knockout mutations. Dev. Cell 31, 784-800. doi:10.1016/j.devcel.2014.11.029
115. Miska, E. A., Alvarez-Saavedra, E., Abbott, A. L., Lau, N. C., Hellman, A. B., McGonagle, S. M., Bartel, D. P., Ambros, V. R. and Horvitz, H. R. (2007). Most Caenorhabditis elegans microRNAs are individually not essential for development or viability. PLoS Genet. 3, e215. doi:10.1371/journal.pgen.0030215
116. Park, C. Y., Jeker, L. T., Carver-Moore, K., Oh, A., Liu, H. J., Cameron, R., Richards, H., Li, Z., Adler, D., Yoshinaga, Y. et al. (2012). A resource for the conditional ablation of microRNAs in the mouse. Cell Rep. 1, 385-391. doi:10.1016/j.celrep.2012.02.008
117. Chang, S., Johnston, R. J., Jr., Frøkjaer-Jensen , C., Lockery, S. and Hobert, O. (2004). MicroRNAs act sequentially and asymmetrically to control chemosensory laterality in the nematode. Nat. Cell Biol. 430, 785- 789. doi:10.1038/nature02752
118. Landgraf, P., Rusu, M., Sheridan, R., Sewer, A., Iovino, N., Aravin, A., Pfeffer, S., Rice, A., Kamphorst, A. O., Landthaler, M. et al. (2007). A mammalian microRNA expression atlas based on small RNA library sequencing. Cell 129, 1401-1414. doi:10.1016/j.cell.2007.04.040
119. Martinez, N. J., Ow, M. C., Reece-Hoyes, J. S., Barrasa, M. I., Ambros, V. R. and Walhout, A. J. M. (2008). Genome- scale spatiotemporal analysis of Caenorhabditis elegans microRNA promoter activity. Genome Res. 18, 2005-2015. doi:10.1101/gr.083055.108
120. Ameres, S. L. and Zamore, P. D. (2013). Diversifying microRNA sequence and function. Nat. Rev. Mol. Cell Biol. 14, 475-488. doi:10.1038/nrm3611
121. Drexel, T., Mahofsky, K., Latham, R., Zimmer, M. and Cochella, L. (2016). Neuron type-specific miRNA represses two broadly expressed genes to modulate an avoidance behavior in C. elegans. Genes Dev. 30, 2042-2047. doi:10.1101/gad.287904.116
122. Alvarez-Saavedra, E. and Horvitz, H. R. (2010). Many families of C. elegans microRNAs are not essential for development or viability. Curr. Biol. 20, 367-373. doi:10.1016/j.cub.2009.12.051
123. Sokol, N. S. (2012). The role of microRNAs in muscle development. Curr. Top. Dev. Biol. 99, 59-78. doi:10.1016/B978-0-12-387038-4.00003-3
124. Cochella, L. and Hobert, O. (2012b). Diverse functions of microRNAs in nervous system development. Curr. Top. Dev. Biol. 99, 115-143. doi:10.1016/B978-0-12-387038-4.00005-7

125. Giraldez, A. J., Cinalli, R. M., Glasner, M. E., Enright, A. J., Thomson, J. M., Baskerville, S., Hammond, S. M., Bartel, D. P. and Schier, A. F. (2005). MicroRNAs regulate brain morphogenesis in zebrafish. Science 308, 833-838. doi:10.1126/science.1109020
126. Giraldez, A. J., Mishima, Y., Rihel, J., Grocock, R. J., van Dongen, S., Inoue, K., Enright, A. J. and Schier, A. F. (2006). Zebrafish MiR-430 promotes deadenylation and clearance of maternal mRNAs. Science 312, 75-79. doi:10.1126/science.1122689
127. Lund, E., Liu, M., Hartley, R. S., Sheets, M. D. and Dahlberg, J. E. (2009). Deadenylation of maternal mRNAs mediated by miR-427 in Xenopus laevis embryos. RNA 15, 2351-2363. doi:10.1261/rna.1882009
128. Papaioannou, G., Mirzamohammadi, F. and Kobayashi, T. (2014). MicroRNAs involved in bone formation. Cell. Mol. Life Sci. 71, 4747-4761. doi:10.1007/s00018-014-1700-6
129. Poy, M. N., Eliasson, L., Krutzfeldt, J., Kuwajima, S., Ma, X., MacDonald, P. E., Pfeffer, S., Tuschl, T., Rajewsky, N., Rorsman, P. et al. (2004). A pancreatic islet-specific microRNA regulates insulin secretion. Nature 432, 226- 230. doi:10.1038/nature03076
130. Wienholds, E., Kloosterman, W. P., Miska, E., Alvarez-Saavedra, E., Berezikov, E., de Bruijn, E., Horvitz, H. R., Kauppinen, S. and Plasterk, R. H. A. (2005). MicroRNA expression in zebrafish embryonic development. Science 309, 310-311. doi:10.1126/science.1114519
131. Kloosterman, W. P., Lagendijk, A. K., Ketting, R. F., Moulton, J. D. and Plasterk, R. H. A. (2007). Targeted inhibition of miRNA maturation with morpholinos reveals a role for miR-375 in pancreatic islet development. PLoS Biol. 5, e203. doi:10.1371/journal.pbio.0050203
132. Xiao, C., Calado, D. P., Galler, G., Thai, T.-H., Patterson, H. C., Wang, J., Rajewsky, N., Bender, T. P. and Rajewsky, K. (2007). MiR-150 controls B cell differentiation by targeting the transcription factor c-Myb. Cell 131, 146-159. doi:10.1016/j.cell.2007.07.021
133. Zhou, B., Wang, S., Mayr, C., Bartel, D. P. and Lodish, H. F. (2007). miR-150, a microRNA expressed in mature B and T cells, blocks early B cell development when expressed prematurely. Proc. Natl. Acad. Sci. USA 104, 7080- 7085. doi:10.1073/pnas.0702409104
134. Johnnidis, J. B., Harris, M. H., Wheeler, R. T., Stehling-Sun, S., Lam, M. H., Kirak, O., Brummelkamp, T. R., Fleming, M. D. and Camargo, F. D. (2008). Regulation of progenitor cell proliferation and granulocyte function by microRNA-223. Nature 451, 1125-1129. doi:10.1038/nature06607
135. X. Hong, L. J. Luense, L. K. McGinnis, W. B. Nothnick, and L. K. Christenson, "Dicer1 is essential for female fertility and normal development of the female reproductive system," *Endocrinology*, vol. 149, no. 12, pp. 6207–6212, 2008.
136. K. Forbes, F. Farrokhnia, J. D. Aplin, and M. Westwood, "Dicer-dependent miRNAs provide an endogenous restraint on cytotrophoblast proliferation," *Placenta*, vol. 33, no. 7, pp. 581–585, 2012.
137. S.-S. Luo, O. Ishibashi, G. Ishikawa et al., "Human villous trophoblasts express and secrete placenta-specific microRNAs into maternal circulation via exosomes," *Biology of Reproduction*, vol. 81, no. 4, pp. 717–729, 2009.

138. P. Landgraf, M. Rusu, R. Sheridan et al., "A mammalian microRNA expression atlas based on small RNA library sequencing," *Cell*, vol. 129, no. 7, pp. 1401–1414, 2007.
139. Y. Liang, D. Ridzon, L. Wong, and C. Chen, "Characterization of microRNA expression profiles in normal human tissues," *BMC Genomics*, vol. 8, article 166, 2007.
140. S. Lin, W. K. C. Cheung, S. Chen et al., "Computational identification and characterization of primate-specific microRNAs in human genome," *Computational Biology and Chemistry*, vol. 34, no. 4, pp. 232–241, 2010.
141. D. M. Morales-Prieto, S. Ospina-Prieto, W. Chaiwangyen, M. Schoenleben, and U. R. Markert, "Pregnancy-associated miRNA-clusters," *Journal of Reproductive Immunology*, vol. 97, no. 1, pp. 51–61, 2013.
142. I. Bentwich, A. Avniel, Y. Karov et al., "Identification of hundreds of conserved and nonconserved human microRNAs," *Nature Genetics*, vol. 37, no. 7, pp. 766–770, 2005.
143. G. X. Y. Zheng, A. Ravi, G. M. Gould, C. B. Burge, and P. A. Sharp, "Genome-wide impact of a recently expanded microRNA cluster in mouse," *Proceedings of the National Academy of Sciences of the United States of America*, vol. 108, no. 38, pp. 15804–15809, 2011.
144. Z. Yuan, X. Sun, D. Jiang et al., "Origin and evolution of a placental-specific microRNA family in the human genome," *BMC Evolutionary Biology*, vol. 10, no. 1, article no. 346, 2010.
145. R. B. Donker, J. F. Mouillet, T. Chu et al., "The expression profile of C19MC microRNAs in primary human trophoblast cells and exosomes," *Molecular Human Reproduction*, vol. 18, no. 8, Article ID gas013, pp. 417–424, 2012.
146. I. Flor, A. Neumann, C. Freter et al., "Abundant expression and hemimethylation of C19MC in cell cultures from placenta-derived stromal cells," *Biochemical and Biophysical Research Communications*, vol. 422, no. 3, pp. 411–416, 2012.
147. D. M. Morales-Prieto, W. Chaiwangyen, S. Ospina-Prieto et al., "MicroRNA expression profiles of trophoblastic cells," *Placenta*, vol. 33, no. 9, pp. 725–734, 2012.
148. D. Wang, W. Song, and Q. Na, "The emerging roles of placenta-specific microRNAs in regulating trophoblast proliferation during the first trimester," *The Australian & New Zealand Journal of Obstetrics & Gynaecology*, vol. 52, no. 6, pp. 565–570, 2012.
149. D. M. Morales-Prieto, E. Schleussner, and U. R. Markert, "Reduction in miR-141 is induced by leukemia inhibitory factor and inhibits proliferation in choriocarcinoma cell line JEG-3," *The American Journal of Reproductive Immunology*, vol. 66, no. 1, pp. 57–62, 2011.
150. J. Ding, F. Huang, G. Wu et al., "MiR-519d-3p suppresses invasion and migration of trophoblast cells via targeting MMP-2," *PLoS ONE*, vol. 10, no. 3, Article ID e0120321, 2015.
151. S. Ospina-Prieto, W. Chaiwangyen, J. Herrmann et al., "MicroRNA-141 is upregulated in preeclamptic placentae and regulates trophoblast invasion and intercellular communication," *Translational Research*, vol. 172, pp. 61–72, 2016

152. A. Keniry, D. Oxley, P. Monnier et al., "The H19 lincRNA is a developmental reservoir of miR-675 that suppresses growth and Igf1r," *Nature Cell Biology*, vol. 14, no. 7, pp. 659–665, 2012
153. Q. Na, D. Wang, and W. Song, "Under expression of 4 placenta-associated MicroRNAs in complete hydatidiform moles," *International Journal of Gynecological Cancer*, vol. 22, no. 6, pp. 1075–1080, 2012.
154. A. Higashijima, K. Miura, H. Mishima et al., "Characterization of placenta-specific microRNAs in fetal growth restriction pregnancy," *Prenatal Diagnosis*, vol. 33, no. 3, pp. 214–222, 2013.
155. Q. Tang, W. Wu, X. Xu et al., "miR-141 contributes to fetal growth restriction by regulating PLAG1 expression," *PLoS ONE*, vol. 8, no. 3, Article ID e58737, 2013.
156. D. M. Morales-Prieto, W. Chaiwangyen, S. Ospina-Prieto et al., "MicroRNA expression profiles of trophoblastic cells," *Placenta*, vol. 33, no. 9, pp. 725–734, 2012.
157. Y. Gu, J. Sun, L. J. Groome, and Y. Wang, "Differential miRNA expression profiles between the first and third trimester human placentas," *American Journal of Physiology - Endocrinology and Metabolism*, vol. 304, no. 8, pp. E836–E843, 2013.
158. H. Li, L. Guo, Q. Wu, J. Lu, Q. Ge, and Z. Lu, "A comprehensive survey of maternal plasma miRNAs expression profiles using high-throughput sequencing," *Clinica ChimicaActa*, vol. 413, no. 5-6, pp. 568–576, 2012.
159. J. R. Challis, C. J. Lockwood, L. Myatt, J. E. Norman, J. F. Strauss III, and F. Petraglia, "Inflammation and pregnancy," *Reproductive Sciences*, vol. 16, no. 2, pp. 206–215, 2009.
160. W. S. Webster and D. Abela, "The effect of hypoxia in development," *Birth Defects Research Part C - Embryo Today: Reviews*, vol. 81, no. 3, pp. 215–228, 2007.
161. H.-W. Chan, M. Lappas, S. W. Y. Yee, K. Vaswani, M. D. Mitchell, and G. E. Rice, "The expression of the let-7 miRNAs and Lin28 signalling pathway in human term gestational tissues," *Placenta*, vol. 34, no. 5, pp. 443–448, 2013.
162. L. Liu, Y. Wang, H. Fan et al., "MicroRNA-181a regulates local immune balance by inhibiting proliferation and immunosuppressive properties of mesenchymal stem cells," *Stem Cells*, vol. 30, no. 8, pp. 1756–1770, 2012.
163. X. Liu, Z. Zhan, L. Xu et al., "MicroRNA-148/152 impair innate response and antigen presentation of TLR-triggered dendritic cells by targeting CaMKIIα," *Journal of Immunology*, vol. 185, no. 12, pp. 7244–7251, 2010.
164. X.-M. Zhu, T. Han, X.-H. Wang et al., "Overexpression of miR-152 leads to reduced expression of human leukocyte antigen-G and increased natural killer cell mediated cytolysis in JEG-3 cells," *American Journal of Obstetrics and Gynecology*, vol. 202, no. 6, pp. 592–e7, 2010.
165. X. Liu, Z. Zhan, L. Xu et al., "MicroRNA-148/152 impair innate response and antigen presentation of TLR-triggered dendritic cells by targeting CaMKIIα," *Journal of Immunology*, vol. 185, no. 12, pp. 7244–7251, 2010. View at: Publisher Site | Google Scholar
166. X.-M. Zhu, T. Han, X.-H. Wang et al., "Overexpression of miR-152 leads to reduced expression of human leukocyte antigen-G and increased natural killer

cell mediated cytolysis in JEG-3 cells," *American Journal of Obstetrics and Gynecology*, vol. 202, no. 6, pp. 592–e7, 2010.

167. R. B. Donker, J.-F. Mouillet, D. M. Nelson, and Y. Sadovsky, "The expression of Argonaute2 and related microRNA biogenesis proteins in normal and hypoxic trophoblasts," *Molecular Human Reproduction*, vol. 13, no. 4, pp. 273–279, 2007.
168. J. F. Mouillet, T. Chu, C. A. Hubel, D. M. Nelson, W. T. Parks, and Y. Sadovsky, "The levels of hypoxia-regulated microRNAs in plasma of pregnant women with fetal growth restriction," *Placenta*, vol. 31, no. 9, pp. 781–784, 2010.
169. M. A. Maccani, M. Avissar-Whiting, C. E. Banister, B. McGonnigal, J. F. Padbury, and C. J. Marsit, "Maternal cigarette smoking during pregnancy is associated with downregulation of miR-16, miR-21 and miR-146a in the placenta," *Epigenetics*, vol. 5, no. 7, pp. 583–589, 2010.
170. F. Colleoni, N. Padmanabhan, H.-W. Yung et al., "Suppression of mitochondrial electron transport chain function in the hypoxic human placenta: a role for miRNA-210 and protein synthesis inhibition," *PLoS ONE*, vol. 8, no. 1, Article ID e55194, 2013.
171. R. Goyal, R. Lister, A. Leitzke, D. Goyal, C. P. Gheorghe, and L. D. Longo, "Antenatal maternal hypoxic stress: Adaptations of the placental renin-angiotensin system in the mouse," *Placenta*, vol. 32, no. 2, pp. 134–139, 2011.
172. P. Kumar, Y. Luo, C. Tudela, J. M. Alexander, and C. R. Mendelson, "The c-myc-regulated microRNA-17~92 (mir-17~92) and miR-106a~363 clusters target hCYP19A1 and hGCM1 to inhibit human trophoblast differentiation," *Molecular and Cellular Biology*, vol. 33, no. 9, pp. 1782–1796, 2013.
173. K. Forbes, F. Farrokhnia, J. D. Aplin, and M. Westwood, "Dicer-dependent miRNAs provide an endogenous restraint on cytotrophoblast proliferation," *Placenta*, vol. 33, no. 7, pp. 581–585, 2012.
174. W.-L. Gao, M. Liu, Y. Yang et al., "The imprinted H19 gene regulates human placental trophoblast cell proliferation via encoding miR-675 that targets Nodal Modulator 1(NOMO1)," *RNA Biology*, vol. 9, no. 7, pp. 1002–1010, 2012.
175. G. Fu, G. Ye, L. Nadeem et al., "MicroRNA-376c impairs transforming growth factor-β and nodal signaling to promote trophoblast cell proliferation and invasion," *Hypertension*, vol. 61, no. 4, pp. 864–872, 2013.
176. L. Luo, G. Ye, L. Nadeem et al., "Microrna-378a-5p promotes trophoblast cell survival, migration and invasion by targeting Nodal," *Journal of Cell Science*, vol. 125, no. 13, pp. 3124–3132, 2012.
177. J. F. Mouillet, R. B. Donker, T. Mishima, T. Cronqvist, T. Chu, and Y. Sadovsky, "The unique expression and function of miR-424 in human placental trophoblasts," *Biology of Reproduction*, vol. 89, no. 2, p. 25, 2013.
178. Y. Zou, Z. Jiang, X. Yu et al., "MiR-101 regulates apoptosis of trophoblast HTR-8/SVneo cells by targeting endoplasmic reticulum (ER) protein 44 during preeclampsia," *Journal of Human Hypertension*, vol. 28, no. 10, pp. 610–616, 2014.
179. X. Zhu, Y. Yang, T. Han et al., "Suppression of microRNA-18a expression inhibits invasion and promotes apoptosis of human trophoblast cells by targeting the estrogen receptor α gene," *Molecular medicine reports*, vol. 12, no. 2, pp. 2701–2706, 2015.

180. F. Jiang, J. Li, G. Wu et al., "Upregulation of microRNA-335 and microRNA-584 contributes to the pathogenesis of severe preeclampsia through downregulation of endothelial nitric oxide synthase," *Molecular Medicine Reports*, vol. 12, no. 4, pp. 5383–5390, 2015.
181. T. Lu, W. Lu, and L. Zhao, "MicroRNA-137 affects proliferation and migration of placenta trophoblast cells in preeclampsia by targeting ERRα," *Reproductive Sciences*, vol. 24, no. 1, pp. 85–96, 2017.
182. X. Li, C. Li, X. Dong, and W. Gou, "MicroRNA-155 inhibits migration of trophoblast cells and contributes to the pathogenesis of severe preeclampsia by regulating endothelial nitric oxide synthase," *Molecular Medicine Reports*, vol. 10, no. 1, pp. 550–554, 2014.
183. R. Demir, Y. Seval, and B. Huppertz, "Vasculogenesis and angiogenesis in the early human placenta," *Acta Histochemica*, vol. 109, no. 4, pp. 257–265, 2007.
184. L. Su, S. Zhao, M. Zhu, and M. Yu, "Differential expression of microRNAs in porcine placentas on Days 30 and 90 of gestation," *Reproduction, Fertility and Development*, vol. 22, no. 8, pp. 1175–1182, 2010.
185. F. Hong, Y. Li, and Y. Xu, "Decreased placental miR-126 expression and vascular endothelial growth factor levels in patients with pre-eclampsia," *Journal of International Medical Research*, vol. 42, no. 6, pp. 1243–1251, 2014.
186. M. Yang, Y. Chen, L. Chen et al., "MiR-15b-AGO2 play a critical role in HTR8/SVneo invasion and in a model of angiogenesis defects related to inflammation," *Placenta*, vol. 41, pp. 62–73, 2016.
187. Y. Zhang, Z. Diao, L. Su et al., "MicroRNA-155 contributes to preeclampsia by down-regulating CYR61," *American Journal of Obstetrics and Gynecology*, vol. 202, no. 5, pp. 466–e7, 2010.
188. W. Cheng, T. Liu, F. Jiang et al., "microRNA-155 regulates angiotensin II type 1 receptor expression in umbilical vein endothelial cells from severely pre-eclamptic pregnant women," *International Journal of Molecular Medicine*, vol. 27, no. 3, pp. 393–399, 2011.
189. Y. Dai, Z. Diao, H. Sun, R. Li, Z. Qiu, and Y. Hu, "MicroRNA-155 is involved in the remodelling of human-trophoblast-derived HTR-8/SVneo cells induced by lipopolysaccharides," *Human Reproduction*, vol. 26, no. 7, pp. 1882–1891, 2011.
190. Y. Dai, Z. Qiu, Z. Diao et al., "MicroRNA-155 inhibits proliferation and migration of human extravillous trophoblast derived HTR-8/SVneo cells via down-regulating cyclin D1," *Placenta*, vol. 33, no. 10, pp. 824–829, 2012.
191. X. Li, C. Li, X. Dong, and W. Gou, "MicroRNA-155 inhibits migration of trophoblast cells and contributes to the pathogenesis of severe preeclampsia by regulating endothelial nitric oxide synthase," *Molecular Medicine Reports*, vol. 10, no. 1, pp. 550–554, 2014.
192. D.-C. Lee, R. Romero, J.-S. Kim et al., "miR-210 targets iron-sulfur cluster scaffold homologue in human trophoblast cell lines: siderosis of interstitial trophoblasts as a novel pathology of preterm preeclampsia and small-for-gestational-age pregnancies," *The American Journal of Pathology*, vol. 179, no. 2, pp. 590–602, 2011.

193. L. Anton, A. O. Olarerin-George, N. Schwartz et al., "MiR-210 inhibits trophoblast invasion and is a serum biomarker for preeclampsia," *The American Journal of Pathology*, vol. 183, no. 5, pp. 1437–1445, 2013.
194. Y. Zhang, M. Fei, and G. Xue, "Elevated levels of hypoxia-inducible microRNA-210 in pre-eclampsia: new insights into molecular mechanisms for the disease," *Journal of Cellular and Molecular Medicine*, vol. 16, no. 2, pp. 249–259, 2012.
195. S. Muralimanoharan, A. Maloyan, J. Mele, C. Guo, L. G. Myatt, and L. Myatt, "MIR-210 modulates mitochondrial respiration in placenta with preeclampsia," *Placenta*, vol. 33, no. 10, pp. 816–823, 2012.
196. S. E. Kopriva, V. L. Chiasson, B. M. Mitchell, and P. Chatterjee, "TLR3-Induced Placental miR-210 Down-Regulates the STAT6/Interleukin-4 Pathway," *PLoS ONE*, vol. 8, no. 7, Article ID e67760, 2013.
197. R. Luo, X. Shao, P. Xu et al., "MicroRNA-210 contributes to preeclampsia by downregulating potassium channel modulatory factor 1," *Hypertension*, vol. 64, no. 4, pp. 839–845, 2014.
198. R. Luo, Y. Wang, P. Xu et al., "Hypoxia-inducible miR-210 contributes to preeclampsia via targeting thrombospondin type I domain containing 7A," *Scientific Reports*, vol. 6, Article ID 19588, 2016.
199. L. Lázár, B. Nagy, A. Molvarec, A. Szarka, and J. Rigó Jr., "Role of hsa-miR-325 in the etiopathology of preeclampsia," *Molecular Medicine Reports*, vol. 6, no. 3, pp. 597–600, 2012.
200. P. Li, W. Guo, J. Zhao et al., "MicroRNA-29b contributes to pre-eclampsia through its effects on apoptosis, invasion and angiogenesis of trophoblast cells," *Clinical Science*, vol. 124, no. 1, pp. 27–40, 2013.
201. Y. Bai, W. Yang, H.-X. Yang et al., "Downregulated miR-195 detected in preeclamptic placenta affects trophoblast cell invasion via modulating ActRIIA expression," *PLoS ONE*, vol. 7, no. 6, Article ID e38875, 2012.
202. L. Luo, G. Ye, L. Nadeem et al., "Microrna-378a-5p promotes trophoblast cell survival, migration and invasion by targeting Nodal," *Journal of Cell Science*, vol. 125, no. 13, pp. 3124–3132, 2012.
203. S. Chen, G. Zhao, H. Miao et al., "MicroRNA-494 inhibits the growth and angiogenesis-regulating potential of mesenchymal stem cells," *FEBS Letters*, vol. 589, no. 6, pp. 710–717, 2015.
204. E. Hu, L. Ding, H. Miao et al., "MiR-30a attenuates immunosuppressive functions of IL-1β-elicited mesenchymal stem cells via targeting TAB3," *FEBS Letters*, vol. 589, no. 24, pp. 3899–3907, 2015
205. W. Wang, L. Feng, H. Zhang et al., "Preeclampsia up-regulates angiogenesis-associated microRNA (i.e., miR-17, -20a, and -20b) that target ephrin-B2 and EPHB4 in human placenta," *The Journal of Clinical Endocrinology & Metabolism*, vol. 97, no. 6, pp. E1051–E1059, 2012.
206. T. Cindrova-Davies, E. A. Herrera, Y. Niu, J. Kingdom, D. A. Giussani, and G. J. Burton, "Reduced cystathionine γ-lyase and increased miR-21 expression are associated with increased vascular resistance in growth-restricted pregnancies: hydrogen sulfide as a placental vasodilator," *American Journal of Pathology*, vol. 182, no. 4, pp. 1448–1458, 2013.

207. J. E. Cartwright, R. J. Keogh, and M. C. van Patot, "Hypoxia and placental remodelling," in *Hypoxia and the Circulation*, vol. 618 of *Advances in Experimental Medicine and Biology*, pp. 113–126, Springer, Boston, Mass, USA, 2007.
208. J. F. Mouillet, T. Chu, C. A. Hubel, D. M. Nelson, W. T. Parks, and Y. Sadovsky, "The levels of hypoxia-regulated microRNAs in plasma of pregnant women with fetal growth restriction," *Placenta*, vol. 31, no. 9, pp. 781–784, 2010.
209. M. A. Maccani, J. F. Padbury, and C. J. Marsit, "miR-16 and miR-21 expression in the placenta is associated with fetal growth," *PLoS ONE*, vol. 6, no. 6, Article ID e21210, 2011.
210. K. Mayor-Lynn, T. Toloubeydokhti, A. C. Cruz, and N. Chegini, "Expression profile of microRNAs and mRNAs in human placentas from pregnancies complicated by preeclampsia and preterm labor," *Reproductive Sciences*, vol. 18, no. 1, pp. 46–56, 2011.
211. W. Ventura, K. Koide, K. Hori et al., "Placental expression of microRNA-17 and -19b is down-regulated in early pregnancy loss," *European Journal of Obstetrics Gynecology and Reproductive Biology*, vol. 169, no. 1, pp. 28–32, 2013.
212. X. Wang, B. Li, J. Wang et al., "Evidence that miR-133a causes recurrent spontaneous abortion by reducing HLA-G expression," *Reproductive BioMedicine Online*, vol. 25, no. 4, pp. 415–424, 2012.
213. K. Umemura, S. I. Ishioka, T. Endo, Y. Ezaka, M. Takahashi, and T. Saito, "Roles of microRNA-34a in the pathogenesis of placenta accreta," *Journal of Obstetrics and Gynaecology Research*, vol. 39, no. 1, pp. 67–74, 2013.
214. V. C. Sandrim, N. Eleuterio, E. Pilan, J. E. Tanus-Santos, K. Fernandes, and R. Cavalli, "Plasma levels of increased miR-195-5p correlates with the sFLT-1 levels in preeclampsia," *Hypertension in Pregnancy*, vol. 35, no. 2, pp. 150–158, 2016
215. Q. Yang, J. Lu, S. Wang, H. Li, Q. Ge, and Z. Lu, "Application of next-generation sequencing technology to profile the circulating microRNAs in the serum of preeclampsia versus normal pregnant women," *Clinica Chimica Acta*, vol. 412, no. 23-24, pp. 2167–2173, 2
216. R. B. Donker, J. F. Mouillet, T. Chu et al., "The expression profile of C19MC microRNAs in primary human trophoblast cells and exosomes," *Molecular Human Reproduction*, vol. 18, no. 8, Article ID gas013, pp. 417–424, 2012.
217. K. Miura, A. Higashijima, Y. Murakami et al., "Circulating chromosome 19 miRNA cluster microRNAs in pregnant women with severe pre-eclampsia," *Journal of Obstetricsand Gynaecology Research*, vol. 41, no. 10, pp. 1526–1532, 2015
218. I. Hromadnikova, K. Kotlabova, J. Doucha, K. Dlouha, and L. Krofta, "Absolute and relative quantification of placenta-specific microRNAs in maternal circulation with placental insufficiency—related complications," *Journal of Molecular Diagnostics*, vol. 14, no. 2, pp. 160–167, 2012.
219. I. Hromadnikova, K. Kotlabova, M. Ondrackova et al., "Expression profile of C19MC microRNAs in placental tissue in pregnancy-related complications," *DNA and Cell Biology*, vol. 34, no. 6, pp. 437–457, 2015.
220. S. Gilad, E. Meiri, Y. Yogev et al., "Serum microRNAs are promising novel biomarkers," *PLoS ONE*, vol. 3, no. 9, Article ID e3148, 2008.

221. Z. Yu, S. Han, P. Hu et al., "Potential role of maternal serum microRNAs as a biomarker for fetal congenital heart defects," *Medical Hypotheses*, vol. 76, no. 3, pp. 424–426, 2011.
222. A. Chakrabarty, S. Tranguch, T. Daikoku, K. Jensen, H. Furneaux, and S. K. Dey, "MicroRNA regulation of cyclooxygenase-2 during embryo implantation," *Proceedings of the National Academy of Sciences of the United States of America*, vol. 104, no. 38, pp. 15144–15149, 2007
223. H.-F. Xia, X.-H. Jin, P.-P. Song, Y. Cui, C.-M. Liu, and X. Ma, "Temporal and spatial regulation of Let-7a in the uterus during embryo implantation in the rat," *Journal of Reproduction and Development*, vol. 56, no. 1, pp. 73–78, 2010.
224. T.-Y. Fu, C.-T. Lin, and P.-C. Tang, "Steroid hormone-regulated let-7b mediates cell proliferation and basigin expression in the mouse endometrium," *Journal of Reproduction and Development*, vol. 57, no. 5, pp. 627–635, 2011.
225. K. Qian, L. Hu, H. Chen et al., "Hsa-miR-222 is involved in differentiation of endometrial stromal cells in vitro," *Endocrinology*, vol. 150, no. 10, pp. 4734–4743, 2009.
226. N. E. Renthal, C.-C. Chen, K. C. Williams, R. D. Gerard, J. Prange-Kiel, and C. R. Mendelson, "MiR-200 family and targets, ZEB1 and ZEB2, modulate uterine quiescence and contractility during pregnancy and labor," *Proceedings of the National Academy of Sciences of the United States of America*, vol. 107, no. 48, pp. 20828–20833, 2010.
227. K. C. Williams, N. E. Renthal, J. C. Condon, R. D. Gerard, and C. R. Mendelson, "MicroRNA-200a serves a key role in the decline of progesterone receptor function leading to term and preterm labor," *Proceedings of the National Academy of Sciences of the United States of America*, vol. 109, no. 19, pp. 7529–7534, 2012.
228. K. C. Williams, N. E. Renthal, R. D. Gerard, and C. R. Mendelson, "The microRNA (miR)-199a/214 cluster mediates opposing effects of progesterone and estrogen on uterine contractility during pregnancy and labor," *Molecular Endocrinology*, vol. 26, no. 11, pp. 1857–1867, 2012.
229. J.-L. Liu, X.-H. Liang, R.-W. Su et al., "Combined analysis of microRNome and 3′-UTRome reveals a species-specific regulation of progesterone receptor expression in the endometrium of rhesus monkey," *Journal of Biological Chemistry*, vol. 287, no. 17, pp. 13899–13910, 2012.
230. Rasko JE, Wong JJ-L. Nuclear microRNAs in normal hemopoiesis and cancer. J Hematol Oncol. 2017;10:8.
231. Huang, V. Endogenous miRNAa: miRNA-Mediated Gene Upregulation. RNA Activation. Singapore: Springer; 2017. p. 65–79.
232. Ramchandran, Ramani, and Pradeep Chaluvally-Raghavan. miRNA-mediated RNA activation in mammalian cells. RNA Activation. Singapore: Springer; 2017. p. 81–89.
233. Subramanian S Ajay, Brian D Athey, Inhan Lee. Unified translation repression mechanism for microRNAs and upstream AUGs. BMC Genomics. 2010;11(1): 155.
234. Li, L-C. Small RNA-Guided Transcriptional Gene Activation (RNAa) in Mammalian Cells. RNA Activation. Singapore: Springer; 2017. p. 1-20.

235. Li L-C. Chromatin remodeling by the small RNA machinery in mammalian cells. Epigenetics. 2014;9:45–52.
236. Li L-C, Okino ST, Zhao H, Pookot D, Place RF, Urakami S, Enokida H, Dahiya R. Small dsRNAs induce transcriptional activation in human cells. Proc Natl Acad Sci. 2006;103:17337–42.
237. Meng X, Jiang Q, Chang N, Wang X, Liu C, Xiong J, Cao H, Liang Z. Small activating RNA binds to the genomic target site in a seed-region-dependent manner. Nucleic Acids Res. 2016;44:2274–82.
238. Robb GB, Brown KM, Khurana J, Rana TM. Specific and potent RNAi in the nucleus of human cells. Nat Struct Mol Biol. 2005;12:133–7.
239. Morris KV, Chan SWL, Jacobsen SE, Looney DJ. Small interfering RNA-induced transcriptional gene silencing in human cells. Science. 2004;305:1289–92.
240. Ting AH, Schuebel KE, Herman JG, Baylin SB. Short dsRNA induces transcriptional gene silencing in human cancer cells in the absence of DNA methylation. Nat Genet. 2005;37:906.
241. Schwartz JC, Younger ST, Nguyen N-B, Hardy DB, Monia BP, Corey DR, Janowski BA. Antisense transcripts are targets for activating small RNAs. Nat Struct Mol Biol. 2008;15:842–8.
242. Matsui M, Chu Y, Zhang H, Gagnon KT, Shaikh S, Kuchimanchi S, Manoharan M, Corey DR, Janowski BA. Promoter RNA links transcriptional regulation of inflammatory pathway genes. Nucleic Acids Res. 2013;41:10086–109.
243. Pisignano G, Napoli S, Magistri M, Mapelli SN, Pastori C, Di Marco S, Civenni G, Albino D, Enriquez C, Allegrini S. A promoter-proximal transcript targeted by genetic polymorphism controls E-cadherin silencing in human cancers. Nat Commun. 2017;8:15622
244. Huang V, Place RF, Portnoy V, Wang J, Qi Z, Jia Z, Yu A, Shuman M, Yu J, Li L-C. Upregulation of Cyclin B1 by miRNA and its implications in cancer. Nucleic Acids Res.2011;40:1695–707.
245. Morris KV, Santoso S, Turner A-M, Pastori C, Hawkins PG. Bidirectional transcription directs both transcriptional gene activation and suppression in human cells. PLoS Genet. 2008;4:e1000258.
246. Yue X, Schwartz JC, Chu Y, Younger ST, Gagnon KT, Elbashir S, Janowski BA, Corey DR. Transcriptional regulation by small RNAs at sequences downstream from 3′ gene termini. Nat Chem Biol. 2010;6:621–9.
247. Zhang Y, Fan M, Zhang X, Huang F, Wu K, Zhang J, Liu J, Huang Z, Luo H, Tao L. Cellular microRNAs up-regulate transcription via interaction with promoter TATA-box motifs. RNA. 2014;20:1878–89.
248. Li Z, Lan X, Han R, Wang J, Huang Y, Sun J, Guo W, Chen H. miR-2478 inhibits TGFβ1 expression by targeting the transcriptional activation region downstream of the TGFβ1 promoter in dairy goats. Sci Rep. 2017;7:42627.
249. Zhang Y, Zhang H. RNAa Induced by TATA Box-Targeting MicroRNAs. RNA Activation. Singapore: Springer; 2017. p. 91-111.
250. Paugh SW, Coss DR, Bao J, Laudermilk LT, Grace CR, Ferreira AM, Waddell MB, Ridout G, Naeve D, Leuze M. MicroRNAs form triplexes with double stranded DNA at sequence-specific binding sites; a eukaryotic mechanism via

which microRNAs could directly alter gene expression. PLoS Comput Biol. 2016;12:e1004744.

251. Toscano-Garibay JD, Aquino-Jarquin G. Transcriptional regulation mechanism mediated by miRNA–DNA• DNA triplex structure stabilized by Argonaute. Bioch Biophys Acta, Gene Regul Mech. 2014;1839:1079–83.
252. Kanak M, Alseiari M, Balasubramanian P, Addanki K, Aggarwal M, Noorali S, Kalsum A, Mahalingam K, Pace G, Panasik N. Triplex-forming MicroRNAs form stable complexes with HIV-1 provirus and inhibit its replication. Appl Immunohistochem Mol Morphol. 2010;18:532–45.
253. XiaoM,LiJ,LiW,WangY,WuF,XiY,ZhangL,DingC,LuoH,LiY.MicroRNAs activate gene transcription epigenetically as an enhancer trigger. RNA Biol. 2016:00–0.
254. Sojka DK, Bruniquel D, Schwartz RH, Singh NJ. IL-2 secretion by CD4+ T cells in vivo is rapid, transient, and influenced by TCR-specific competition. J Immunol. 2004;172:6136–43.
255. Majid S, Dar AA, Saini S, Yamamura S, Hirata H, Tanaka Y, Deng G, Dahiya R. MicroRNA-205–directed transcriptional activation of tumor suppressor genes in prostate cancer. Cancer. 2010;116:5637–49.
256. Kreis S, Philippidou D, Margue C, Behrmann I. IL-24: a classic cytokine and/or a potential cure for cancer? J Cell Mol Med. 2008;12:2505–10.
257. Ribeiro-Dias F, Gomes RS, de Lima Silva LL, dos Santos JC, Joosten LA. Interleukin 32: a novel player in the control of infectious diseases. J Leukoc Biol. 2017;101:39–52.
258. Kumar S, Lanckacker E, Dentener M, Bracke K, Provoost S, De Grove K, Brusselle G, Wouters E, Maes T, Joos G. Aggravation of allergic airway inflammation by cigarette smoke in mice is CD44-dependent. PLoS One. 2016;11:e0151113.
259. Johansson MW. Eosinophil activation status in separate compartments and association with asthma. Front Med. 2017;4:75.
260. Larsen MT, Hother C, Häger M, Pedersen CC, Theilgaard-Mönch K, Borregaard N, et al. MicroRNA Profiling in Human Neutrophils during Bone Marrow Granulopoiesis and InVivo Exudation. PLoS ONE. 2013;8:e58454.
261. Sun SM, Dijkstra MK, Bijkerk AC, Brooimans RA, Valk PJ, Erkeland SJ, Löwenberg B, Jongen-Lavrencic M. Transition of highly specific microRNA expression patterns in association with discrete maturation stages of human granulopoiesis. Br J Haematol. 2011;155:395–8.
262. Wong JJ, Ritchie W, Gao D, Lau KA, Gonzalez M, Choudhary A, Taft RJ, Rasko JE, Holst J. Identification of nuclear-enriched miRNAs during mouse granulopoiesis. J Hematol Oncol. 2014;7:42.
263. Zardo G, Ciolfi A, Vian L, Starnes LM, Billi M, Racanicchi S, Maresca C, Fazi F, Travaglini L, Noguera N. Polycombs and microRNA-223 regulate human granulopoiesis by transcriptional control of target gene expression. Blood. 2012;119:4034–46.
264. Cheng C, Gerstein M. Modeling the relative relationship of transcription factor binding and histone modifications to gene expression levels in mouse embryonic stem cells. Nucleic Acids Res. 2012; 40:553-568.

265. Ernst J, Kellis M. Discovery and characterization of chromatin states for systematic annotation of the human genome. Nat Biotechnol. 2010; 28:817-825.
266. Chen, Y.-W., Song, S., Weng, R., Verma, P., Kugler, J.-M., Buescher, M., Rouam, S. and Cohen, S. M. (2014). Systematic study of drosophila microRNA functions using a collection of targeted knockout mutations. Dev. Cell 31, 784-800. doi:10.1016/j.devcel.2014.11.029
267. Chen, K. and Rajewsky, N. (2007). The evolution of gene regulation by transcriptionfactors and microRNAs. Nat. Rev. Genet. 8, 93-103. doi:10.1038/nrg1990
268. N. Soleymanlou, I. Jurisica, O. Nevo et al., "Molecular evidence of placental hypoxia in preeclampsia," *Journal of Clinical Endocrinology and Metabolism*, vol. 90, no. 7, pp. 4299–4308, 2005.
269. M. Ream, A. M. Ray, R. Chandra, and D. M. Chikaraishi, "Early fetal hypoxia leads to growth restriction and myocardial thinning," *American Journal of Physiology – Regulatory Integrative and Comparative Physiology*, vol. 295, no. 2, pp. R583–R595, 2008.
270. Li Z, Lan X, Han R, Wang J, Huang Y, Sun J, Guo W, Chen H. miR-2478 inhibits TGFβ1 expression by targeting the transcriptional activation region downstream of the TGFβ1 promoter in dairy goats. Sci Rep. 2017;7:42627.
271. Matsui M, Chu Y, Zhang H, Gagnon KT, Shaikh S, Kuchimanchi S, Manoharan M, Corey DR, Janowski BA. Promoter RNA links transcriptional regulation of inflammatory pathway genes. Nucleic Acids Res. 2013;41:10086–109.
272. Ribeiro-Dias F, Gomes RS, de Lima Silva LL, dos Santos JC, Joosten LA. Interleukin 32: a novel player in the control of infectious diseases. J Leukoc Biol. 2017;101:39–52.
273. Zhang Y, Hu JF, Wang H, Cui J, Gao S, Hoffman AR, Li W. CRISPR Cas9-guided chromatin immunoprecipitation identifies miR483 as an epigenetic modulator of IGF2 imprinting in tumors. Oncotarget. 2017;8:34177–90.
274. E. Van Rooij, S. Kauppinen. Development of microRNA therapeutics is coming of age. EMBO Mol Med. 2014;6:851-864.
275. Ross JP, Kassir Z. The varied roles of nuclear argonaute-small RNA complexes and avenues for therapy. Mol Ther-Nucleic Acids. 2014;3:e203.

Chapter II

1. A.E. Pasquinelli, B.J. Reinhart, F. Slack, M.Q. Martindale, M.I. Kuroda, B. Mal ler, D.C. Hayward, E.E. Ball, B. Degnan, P. Müller, *et al.* Conservation of the sequence and temporal expression of let-7 heterochronic regulatory RNA. Nature, 408 (2000), pp. 86-89
2. Reinhart, B.J., Slack, F.J., Basson, M., Pasquinelli, A.E., Bettinger, J.C.,Rougvie, A.E., Horvitz, H.R., and Ruvkun, G. (2000). The 21-nucleotide let-7RNA regulates developmental timing in Caenorhabditis elegans. Nature 403, 901–906.
3. Baek, D., Villén, J., Shin, C., Camargo, F.D., Gygi, S.P., and Bartel, D.P. (2008).The impact of microRNAs on protein output. Nature 455, 64–71.
4. Lim, L.P., Lau, N.C., Garrett-Engele, P., Grimson, A., Schelter, J.M., Castle, J.,Bartel, D.P., Linsley, P.S., and Johnson, J.M. (2005). Microarray analysisshows that some microRNAs downregulate large numbers of target mRNAs.Nature 433, 769–773

5. Selbach, M., Schwanha¨usser, B., Thierfelder, N., Fang, Z., Khanin, R., andRajewsky, N. (2008). Widespread changes in protein synthesis induced bymicroRNAs. Nature 455, 58–63.
6. Filipowicz, W., Bhattacharyya, S.N., and Sonenberg, N. (2008). Mechanisms ofpost-transcriptional regulation by microRNAs: are the answers in sight? Nat.Rev. Genet. 9, 102–114.
7. Hausser, J., Landthaler, M., Jaskiewicz, L., Gaidatzis, D., and Zavolan, M.(2009). Relative contribution of sequence and structure features to themRNA binding of Argonaute/EIF2C–miRNA complexes and the degradationof miRNA targets. Genome Res.19, 2009–2020.
8. Hendrickson, D.G., Hogan, D.J., McCullough, H.L., Myers, J.W., Herschlag,D., Ferrell, J.E., and Brown, P.O. (2009). Concordant regulation of translationand mRNA abundance for hundreds of targets of a human microRNA. PLoS Biol. 7, e1000238
9. Rigoutsos, I. (2009). New tricks for animal microRNAS: targeting of aminoacid coding regions at conserved and nonconserved sites. Cancer Res. 69, 3245–3248.
10. Friedman, R.C., Farh, K.K., Burge, C.B., and Bartel, D.P. (2009). Most mammalian mRNAs are conserved targets of microRNAs. Genome Res.19, 92–105.
11. Bushati, N., and Cohen, S.M. (2007). microRNA functions. Annu. Rev. Cell Dev.Biol. 23, 175–205
12. Chang, T.C., and Mendell, J.T. (2007). microRNAs in vertebrate physiology andhuman disease. Annu. Rev. Genomics Hum. Genet. 8, 215–239.
13. Garzon, R., Calin, G.A., and Croce, C.M. (2009). MicroRNAs in cancers. Annu.Rev. Med. 60, 167–179.
14. Carthew, R.W., and Sontheimer, E.J. (2009). Origins and mechanisms of miRNAs and siRNAs. Cell 136, 642–655.
15. Davis, B.N., and Hata, A. (2009). Regulation of MicroRNA biogenesis: A miRiad of mechanisms. Cell Commun. Signal. 7, 18.
16. O'Brien, J.; Hayder, H.; Zayed, Y.; Peng, C. Overview of MicroRNA Biogenesis,Mechanisms of Actions, and Circulation. *Front. Endocrinol.* **2018**, *9*, 402.
17. Kai, K.; Dittmar, R.L.; Sen, S. Secretory microRNAs as biomarkers of cancer. *Semin. Cell Dev. Biol.* **2018**, *78*, 22–36.
18. Chevillet, J.R.; Lee, I.; Briggs, H.A.; He, Y.; Wang, K. Issues and prospects of microRNA-based biomarkers in blood and other body fluids. *Molecules* **2014**, *19*, 6080–6105.
19. Bartel, D.P. MicroRNAs: Target recognition and regulatory functions. *Cell* **2009**, *136*, 215–233.
20. Huntzinger, E.; Izaurralde, E. Gene silencing by microRNAs: Contributions of translational repression and mRNA decay. *Nat. Rev. Genet.* **2011**, *12*, 99–110.
21. Eichhorn, S.W.; Guo, H.; McGeary, S.E.; Rodriguez-Mias, R.A.; Shin, C.; Baek, D.; Hsu, S.H.; Ghoshal, K.; Villen, J.; Bartel, D.P. mRNA destabilization is the dominant effect of mammalian microRNAs by the time substantial repression ensues. *Mol. Cell* **2014**, *56*, 104–115.

22. Helwak, A.; Kudla, G.; Dudnakova, T.; Tollervey, D. Mapping the human miRNA interactome by CLASH reveals frequent noncanonical binding. *Cell* **2013**, *153*, 654–665.
23. Jonas, S.; Izaurralde, E. Towards a molecular understanding of microRNA-mediated gene silencing. *Nat. Rev. Genet.* **2015**, *16*, 421–433.
24. Zen, K.; Zhang, C.Y. Circulating microRNAs: A novel class of biomarkers to diagnose and monitor human cancers. *Med. Res. Rev.* **2012**, *32*, 326–348.
25. Weber, J.A.; Baxter, D.H.; Zhang, S.; Huang, D.Y.; Huang, K.H.; Lee, M.J.; Galas, D.J.; Wang, K. The microRNA spectrum in 12 body fluids. *Clin. Chem.* **2010**, *56*, 1733–1741.
26. Wang, K.; Zhang, S.; Weber, J.; Baxter, D.; Galas, D.J. Export of microRNAs and microRNA-protective protein by mammalian cells. *Nucleic Acids Res.* **2010**, *38*, 7248–7259.
27. Zubakov, D.; Boersma, A.W.; Choi, Y.; van Kuijk, P.F.; Wiemer, E.A.; Kayser, M. MicroRNA markers for forensic body fluid identification obtained from microarray screening and quantitative RT-PCR confirmation. *Int. J. Legal. Med.* **2010**, *124*, 217–226.
28. Hanson, E.K.; Lubenow, H.; Ballantyne, J. Identification of forensically relevant body fluids using a panel of differentially expressed microRNAs. *Anal. Biochem.***2009**, *387*, 303–314.
29. O'Driscoll, L. Expanding on exosomes and ectosomes in cancer. *N. Engl. J. Med.* **2015**, *372*, 2359–2362.
30. Vickers, K.C.; Palmisano, B.T.; Shoucri, B.M.; Shamburek, R.D.; Remaley, A.T. MicroRNAs are transported in plasma and delivered to recipient cells by high-density lipoproteins. *Nat. Cell. Biol.* **2011**, *13*, 423–433.
31. Wang, K.; Li, H.; Yuan, Y.; Etheridge, A.; Zhou, Y.; Huang, D.; Wilmes, P.; Galas, D. The complex exogenous RNA spectra in human plasma: An interface with human gut biota? *PLoS ONE* **2012**, *7*, e51009.
32. Semenov, D.V.; Baryakin, D.N.; Brenner, E.V.; Kurilshikov, A.M.; Vasiliev, G.V.; Bryzgalov, L.A.; Chikova, E.D.; Filippova, J.A.; Kuligina, E.V.; Richter, V.A. Unbiased approach to profile the variety of small non-coding RNA of human blood plasma with massively parallel sequencing technology. *Expert Opin. Biol. Ther.* **2012**, *12* (Suppl. S1), S43–S51.
33. Arroyo, J.D.; Chevillet, J.R.; Kroh, E.M.; Ruf, I.K.; Pritchard, C.C.; Gibson, D.F.; Mitchell, P.S.; Bennett, C.F.; Pogosova-Agadjanyan, E.L.; Stirewalt, D.L.; et al. Argonaute2 complexes carry a population of circulating microRNAs independent of vesicles in human plasma. *Proc. Natl. Acad. Sci. USA* **2011**, *108*, 5003–5008.
34. Cortez, M.A.; Bueso-Ramos, C.; Ferdin, J.; Lopez-Berestein, G.; Sood, A.K.; Calin, G.A. MicroRNAs in body fluids–the mix of hormones and biomarkers. *Nat. Rev. Clin. Oncol.* **2011**, *8*, 467–477.
35. Weickmann, J.L.; Glitz, D.G. Human ribonucleases. Quantitation of pancreatic-like enzymes in serum, urine, and organ preparations. *J. Biol. Chem.* **1982**, *257*, 8705–8710.
36. Baggish, A.L.; Hale, A.; Weiner, R.B.; Lewis, G.D.; Systrom, D.; Wang, F.; Wang, T.J.; Chan, S.Y. Dynamic regulation of circulating microRNA during

acute exhaustive exercise and sustained aerobic exercise training. *J. Physiol.* **2011**, *589*, 3983–3994.

37. Clauss, S.; Wakili, R.; Hildebrand, B.; Kaab, S.; Hoster, E.; Klier, I.; Martens, E.; Hanley, A.; Hanssen, H.; Halle, M.; et al. MicroRNAs as Biomarkers for Acute Atrial Remodeling in Marathon Runners (The miRathon Study—A Sub-Study of the Munich Marathon Study). *PLoS ONE* **2016**, *11*, e0148599.
38. Cui, S.; Sun, B.; Yin, X.; Guo, X.; Chao, D.; Zhang, C.; Zhang, C.Y.; Chen, X.; Ma, J. Time-course responses of circulating microRNAs to three resistance training protocols in healthy young men. *Sci. Rep.* **2017**, *7*, 2203.
39. Rodriguez A, Griffiths-Jones S, Ashurst JL, et al. Identification of mammalianmicroRNA host genes and transcription units. Genome Res. 2004;14:1902–1910.
40. Lagos-Quintana M, Rauhut R, Meyer J, et al. New microRNAs from mouse and human. RNA. 2003; 9: 175–179.
41. Lau NC, Lim LP, Weinstein EG, et al. An abundant class of tiny RNAs with probableregulatory roles in Caenorhabditis elegans. Science. 2001; 294: 858–862.
42. Saini HK, Griffiths-Jones S, Enright AJ. Genomic analysis of human microRNA transcripts. Proc Natl Acad Sci U S A. 2007; 104: 17719–17724.
43. Treiber, T.; Treiber, N.; Meister, G. Regulation of microRNA biogenesis and its crosstalk with other cellular pathways. *Nat. Rev. Mol. Cell Biol.* **2019**, *20*, 5–20.
44. Ha, M.; Kim, V.N. Regulation of microRNA biogenesis. *Nat. Rev. Mol. Cell Biol.* **2014**, *15*, 509–524
45. Gebert, L.F.R.; MacRae, I.J. Regulation of microRNA function in animals. *Nat. Rev. Mol. Cell Biol.* **2019**, *20*, 21–37.
46. Suzuki, H.I.; Miyazono, K. Emerging complexity of microRNA generation cascades. *J. Biochem.* **2011**, *149*, 15–25.
47. Kim, Y.K.; Kim, V.N. Processing of intronic microRNAs. *EMBO J.* **2007**, *26*, 775-783.
48. Becker, W.R.; Ober-Reynolds, B.; Jouravleva, K.; Jolly, S.M.; Zamore, P.D.; Greenleaf, W.J. High-Throughput Analysis Reveals Rules for Target RNA Binding and Cleavage by AGO2. *Mol. Cell* **2019**, *75*, 741–755.
49. Agarwal, V.; Bell, G.W.; Nam, J.W.; Bartel, D.P. Predicting effective microRNA target sites in mammalian mRNAs. *Elife* **2015**, *4*.
50. McGeary, S.E.; Lin, K.S.; Shi, C.Y.; Pham, T.; Bisaria, N.; Kelley, G.M.; Bartel, D.P. The biochemical basis of microRNA targeting efficacy. *Science* **2019**, *10*.
51. Van Kouwenhove, M.; Kedde, M.; Agami, R. MicroRNA regulation by RNA-binding proteins and its implications for cancer. *Nat. Rev. Cancer* **2011**, *11*, 644–656.
52. Hansen, T.B.; Jensen, T.I.; Clausen, B.H.; Bramsen, J.B.; Finsen, B.; Damgaard, C.K.; Kjems, J. Natural RNA circles function as efficient microRNA sponges. *Nature* **2013**, *495*, 384–388.
53. Memczak, S.; Jens, M.; Elefsinioti, A.; Torti, F.; Krueger, J.; Rybak, A.; Maier, L.; Mackowiak, S.D.; Gregersen, L.H.; Munschauer, M.; et al. Circular RNAs are a large class of animal RNAs with regulatory potency. *Nature* **2013**, *495*, 333–338.

54. Szabo, L.; Salzman, J. Detecting circular RNAs: Bioinformatic and experimental challenges. *Nat. Rev. Genet.* **2016**, *17*, 679–692.
55. Bosson, A.D.; Zamudio, J.R.; Sharp, P.A. Endogenous miRNA and target concentrations determine susceptibility to potential ceRNA competition. *Mol. Cell* **2014**, *56*, 347–359.
56. Schwarz DS, Hutvagner G, Du T, et al. Asymmetry in the assembly of the RNAi enzyme complex. Cell. 2003;115: 199–208.
57. Czech B, Zhou R, Erlich Y, et al. Hierarchical rules for Argonaute loading in Drosophila. Mol Cell. 2009;36:445–456.
58. Okamura K, Liu N, Lai EC. Distinct mechanisms for microRNA strand selection by Drosophila Argonautes. Mol Cell. 2009;36: 431–444.
59. Castilla-Llorente V, Spraggon L, Okamura M, et al. Mammalian GW220/TNGW1 is essential for the formation of GW/P bodies containing miRISC. J Cell Biol. 2012;198:529–544.
60. Ruby, J.G.; Jan, C.H.; Bartel, D.P. Intronic microRNA precursors that bypass Drosha processing. *Nature* **2007**, *448*, 83–86.
61. Okamura, K.; Hagen, J.W.; Duan, H.; Tyler, D.M.; Lai, E.C. The mirtron pathway generates microRNA-class regulatory RNAs in Drosophila. *Cell* **2007**, *130*, 89–100.
62. Berezikov, E.; Chung, W.J.; Willis, J.; Cuppen, E.; Lai, E.C. Mammalian mirtron genes. *Mol. Cell* **2007**, *28*, 328–336.
63. Flynt, A.S.; Greimann, J.C.; Chung, W.J.; Lima, C.D.; Lai, E.C. MicroRNA biogenesis via splicing and exosome-mediated trimming in Drosophila. *Mol. Cell* **2010**, *38*, 900–907.
64. Babiarz, J.E.; Hsu, R.; Melton, C.; Thomas, M.; Ullian, E.M.; Blelloch, R. A role for noncanonical microRNAs in the mammalian brain revealed by phenotypic differences in Dgcr8 versus Dicer1 knockouts and small RNA sequencing. *RNA* **2011**, *17*, 1489–1501.
65. Hasler, D.; Lehmann, G.; Murakawa, Y.; Klironomos, F.; Jakob, L.; Grasser, F.A.; Rajewsky, N.; Landthaler, M.; Meister, G. The Lupus Autoantigen La Prevents Mis-channeling of tRNA Fragments into the Human MicroRNA Pathway. *Mol. Cell* **2016**, *63*, 110–124.
66. Saraiya, A.A.; Wang, C.C. snoRNA, a novel precursor of microRNA in Giardia lamblia. *PLoS Pathog.* **2008**, *4*, e1000224.
67. Ender, C.; Krek, A.; Friedlander, M.R.; Beitzinger, M.; Weinmann, L.; Chen, W.; Pfeffer, S.; Rajewsky, N.; Meister, G. A human snoRNA with microRNA-like functions. *Mol. Cell* **2008**, *32*, 519–528.
68. Scott, M.S.; Avolio, F.; Ono, M.; Lamond, A.I.; Barton, G.J. Human miRNA precursors with box H/ACA snoRNA features. *PLoS Comput. Biol.* **2009**, *5*, e1000507.
69. Taft, R.J.; Glazov, E.A.; Lassmann, T.; Hayashizaki, Y.; Carninci, P.; Mattick, J.S. Small RNAs derived from snoRNAs. *RNA* **2009**, *15*, 1233–1240.
70. Xie, M.; Li, M.; Vilborg, A.; Lee, N.; Shu, M.D.; Yartseva, V.; Sestan, N.; Steitz, J.A. Mammalian $5'$-capped microRNA precursors that generate a single microRNA. *Cell* **2013**, *155*, 1568–1580.
71. Cifuentes, D.; Xue, H.; Taylor, D.W.; Patnode, H.; Mishima, Y.; Cheloufi, S.; Ma, E.; Mane, S.; Hannon, G.J.; Lawson, N.D.; et al. A novel miRNA processing

pathway independent of Dicer requires Argonaute2 catalytic activity. *Science* **2010**, *328*, 1694–1698.

72. Cheloufi, S.; Dos Santos, C.O.; Chong, M.M.; Hannon, G.J. A dicer-independent miRNA biogenesis pathway that requires Ago catalysis. *Nature* **2010**, *465*, 584–589.
73. Zamudio, J.R.; Kelly, T.J.; Sharp, P.A. Argonaute-bound small RNAs from promoter-proximal RNA polymerase II. *Cell* **2014**, *156*, 920–934.
74. Denli AM, Tops BB, Plasterk RH, Ketting RF, Hannon GJ. Processing of primary microRNAs by the Microprocessor complex. *Nature* (2004) 432:231–5. doi: 10.1038/nature03049
75. Alarcon CR, Lee H, Goodarzi H, Halberg N, Tavazoie SF. N6-methyladenosine marks primary microRNAs for processing. *Nature* (2015) 519:482–5. doi: 10.1038/nature14281
76. Han J, Lee Y, Yeom KH, Kim YK, Jin H, Kim VN. The Drosha-DGCR8 complex in primary microRNA processing. *Genes Dev.* (2004) 18:3016–27. doi: 10.1101/gad.1262504
77. Okada C, Yamashita E, Lee SJ, Shibata S, Katahira J, Nakagawa A, et al. A high-resolution structure of the pre-microRNA nuclear export machinery. *Science* (2009) 326:1275–9. doi: 10.1126/science.1178705
78. Zhang H, Kolb FA, Jaskiewicz L, Westhof E, Filipowicz W. Single processing center models for human Dicer and bacterial RNase III. *Cell* (2004) 118:57–68. doi: 10.1016/j.cell.2004.06.017
79. Yoda M, Kawamata T, Paroo Z, Ye X, Iwasaki S, Liu Q, et al. ATP-dependent human RISC assembly pathways. *Nat Struct Mol Biol.* (2010) 17:17–23. doi: 10.1038/nsmb.1733
80. Meijer HA, Smith EM, Bushell M. Regulation of miRNA strand selection: follow the leader? *Biochem Soc Trans.* (2014) 42:1135–40. doi: 10.1042/BST20140142
81. Khvorova A, Reynolds A, Jayasena SD. Functional siRNAs and miRNAs exhibit strand bias. *Cell* (2003) 115:209–16. doi: 10.1016/S0092-8674(03)00801-8
82. Ha M, Kim VN. Regulation of microRNA biogenesis. *Nat Rev Mol Cell Biol.* (2014) 15:509–24. doi: 10.1038/nrm3838
83. Ruby JG, Jan CH, Bartel DP. Intronic microRNA precursors that bypass Drosha processing. *Nature* (2007) 448:83–6. doi: 10.1038/nature 05983
84. Babiarz JE, Ruby JG, Wang Y, Bartel DP, Blelloch R. Mouse ES cells express endogenous shRNAs, siRNAs, and other Microprocessor-independent, Dicer-dependent small RNAs. *Genes Dev.* (2008) 22:2773–85. doi: 10.1101/gad.1705308
85. Xie M, Li M, Vilborg A, Lee N, Shu MD, Yartseva V, et al. Mammalian 5′-capped microRNA precursors that generate a single microRNA. *Cell* (2013) 155:1568–80. doi: 10.1016/j.cell. 2013. 11.027
86. Yang JS, Maurin T, Robine N, Rasmussen KD, Jeffrey KL, Chandwani R, et al. Conserved vertebrate mir-451 provides a platform for Dicer-independent, Ago2-mediated microRNA biogenesis. *Proc Natl Acad Sci USA.* (2010) 107:15163–8. doi: 10.1073/pnas. 1006432107
87. Cheloufi S, Dos Santos CO, Chong MMW, Hannon GJ. A dicer-independent miRNA biogenesis pathway that requires Ago catalysis. *Nature* (2010) 465:584–9. doi: 10.1038/nature 09092

88. Li, X., Cassidy, J.J., Reinke, C.A., Fischboeck, S., and Carthew, R.W. (2009).A microRNA imparts robustness against environmental fluctuation duringdevelopment. Cell 137, 273–282.
89. Xu, N., Papagiannakopoulos, T., Pan, G., Thomson, J.A., and Kosik, K.S.(2009). MicroRNA-145 regulates OCT4, SOX2, and KLF4 and represses pluri-potency in human embryonic stem cells. Cell 137, 647–658.
90. Thomson, J.M., Newman, M., Parker, J.S., Morin-Kensicki, E.M., Wright, T.,and Hammond, S.M. (2006). Extensive post-transcriptional regulation ofmicroRNAs and its implications for cancer. Genes Dev. 20, 2202–2207
91. Lu, J., Getz, G., Miska, E.A., Alvarez-Saavedra, E., Lamb, J., Peck, D., Sweet-Cordero, A., Ebert, B.L., Mak, R.H., Ferrando, A.A., et al. (2005). MicroRNAexpression profiles classify human cancers. Nature435, 834–838
92. Guil, S., and Ca´ceres, J.F. (2007). The multifunctional RNA-binding proteinhnRNP A1 is required for processing of miR-18a. Nat. Struct. Mol. Biol.14,591–596
93. Yang, W., Chendrimada, T.P., Wang, Q., Higuchi, M., Seeburg, P.H., Shiekhattar, R., and Nishikura, K. (2006). Modulation of microRNA processing andexpression through RNA editing by ADAR deaminases. Nat. Struct. Mol.Biol.13, 13–21.
94. Kawahara, Y., Zinshteyn, B., Sethupathy, P., Iizasa, H., Hatzigeorgiou, A.G.,and Nishikura, K. (2007). Redirection of silencing targets by adenosine-to-inosine editing of miRNAs. Science315, 1137–1140
95. Fukuda, T., Yamagata, K., Fujiyama, S., Matsumoto, T., Koshida, I.,Yoshimura, K., Mihara, M., Naitou, M., Endoh, H., Nakamura, T., et al.(2007). DEAD-box RNA helicase subunits of the Drosha complex are requiredfor processing of rRNA and a subset of microRNAs. Nat. Cell Biol.9, 604–611
96. Suzuki, H.I., Yamagata, K., Sugimoto, K., Iwamoto, T., Kato, S., and Miyazono,K. (2009). Modulation of microRNA processing by p53. Nature 460, 529–533.
97. Davis, B.N., Hilyard, A.C., Lagna, G., and Hata, A. (2008). SMAD proteins control DROSHA-mediated microRNA maturation. Nature 454, 56–61.
98. Yamagata, K., Fujiyama, S., Ito, S., Ueda, T., Murata, T., Naitou, M.,Takeyama, K., Minami, Y., O'Malley, B.W., and Kato, S. (2009). Maturation of microRNA is hormonally regulated by a nuclear receptor. Mol. Cell 36, 340–347.
99. Michlewski, G., Guil, S., Semple, C.A., and Ca´ceres, J.F. (2008). Posttranscriptional regulation of miRNAs harboring conserved terminal loops. Mol. Cell 32, 383–393
100. Trabucchi, M., Briata, P., Garcia-Mayoral, M., Haase, A.D., Filipowicz, W.,Ramos, A., Gherzi, R., and Rosenfeld, M.G. (2009). The RNA-binding protein KSRP promotes the biogenesis of a subset of microRNAs. Nature 459,1010–1014.
101. Hwang, H.W., Wentzel, E.A., and Mendell, J.T. (2009). Cell-cell contact globally activates microRNA biogenesis. Proc. Natl. Acad. Sci. USA 106, 7016–7021.
102. Gruber, J.J., Zatechka, D.S., Sabin, L.R., Yong, J., Lum, J.J., Kong, M., Zong,W.X., Zhang, Z., Lau, C.K., Rawlings, J., et al. (2009). Ars2 links the nuclear cap-binding complex to RNA interference and cell proliferation. Cell 138, 328–339.

103. Sabin, L.R., Zhou, R., Gruber, J.J., Lukinova, N., Bambina, S., Berman, A., Lau, C.K., Thompson, C.B., and Cherry, S. (2009). Ars2 regulates both miRNA- and siRNA- dependent silencing and suppresses RNA virus infection in Drosophila. Cell 138, 340–351
104. Kim, V.N., Han, J., and Siomi, M.C. (2009). Biogenesis of small RNAs in animals. Nat. Rev. Mol. Cell Biol. 10, 126–139
105. Pawlicki, J.M., and Steitz, J.A. (2010). Nuclear networking fashions pre-messenger RNA and primary microRNA transcripts for function. Trends Cell Biol. 20, 52–61
106. Grimm, D., Streetz, K.L., Jopling, C.L., Storm, T.A., Pandey, K., Davis, C.R., Marion, P., Salazar, F., and Kay, M.A. (2006). Fatality in mice due to over saturation of cellular microRNA/short hairpin RNA pathways. Nature 441, 537–541
107. Diederichs, S., Jung, S., Rothenberg, S.M., Smolen, G.A., Mlody, B.G., and Haber, D.A. (2008). Coexpression of Argonaute-2 enhances RNA interference toward perfect match binding sites. Proc. Natl. Acad. Sci. USA 105, 9284–9289.
108. Han, J., Pedersen, J.S., Kwon, S.C., Belair, C.D., Kim, Y.K., Yeom, K.H., Yang, W.Y., Haussler, D., Blelloch, R., and Kim, V.N. (2009). Posttranscriptional cross regulation between Drosha and DGCR8. Cell 136, 75–84.
109. Shenoy, A., and Blelloch, R. (2009). Genomic analysis suggests that mRNA destabilization by the Microprocessor is specialized for the autoregulation of Dgcr8. PLoS ONE 4, e6971
110. Lee, E.J., Baek, M., Gusev, Y., Brackett, D.J., Nuovo, G.J., and Schmittgen, T.D. (2008). Systematic evaluation of microRNA processing patterns in tissues, cell lines, and tumors. RNA 14, 35–42
111. Obernosterer, G., Leuschner, P.J., Alenius, M., and Martinez, J. (2006). Post-transcriptional regulation of microRNA expression. RNA 12, 1161–1167.
112. Forman, J.J., Legesse-Miller, A., and Coller, H.A. (2008). A search for conserved sequences in coding regions reveals that the let-7 microRNA targets Dicer within its coding sequence. Proc. Natl. Acad. Sci. USA 105, 14879–14884
113. Roush, S., and Slack, F.J. (2008). The let-7 family of microRNAs. Trends Cell Biol.18, 505–516
114. Wu, L., and Belasco, J.G. (2005). Micro-RNA regulation of the mammalian lin-28 gene during neuronal differentiation of embryonal carcinoma cells. Mol. Cell. Biol. 25, 9198–9208.
115. Heo, I., Joo, C., Cho, J., Ha, M., Han, J., and Kim, V.N. (2008). Lin28 mediates the terminal uridylation of let-7 precursor MicroRNA. Mol. Cell 32, 276–28
116. Newman, M.A., Thomson, J.M., and Hammond, S.M. (2008). Lin-28 interaction with the Let-7 precursor loop mediates regulated microRNA processing. RNA 14, 1539–1549.
117. Rybak, A., Fuchs, H., Smirnova, L., Brandt, C., Pohl, E.E., Nitsch, R., and Wulczyn, F.G. (2008). A feedback loop comprising lin-28 and let-7 controlspre-let-7 maturation during neural stem-cell commitment. Nat. Cell Biol.10,987–993
118. Viswanathan, S.R., Daley, G.Q., and Gregory, R.I. (2008). Selective blockade of microRNA processing by Lin28. Science 320, 97–100.

119. Hagan, J.P., Piskounova, E., and Gregory, R.I. (2009). Lin28 recruits the TUTase Zcchc11 to inhibit let-7 maturation in mouse embryonic stem cells. Nat. Struct. Mol. Biol.16, 1021–1025
120. Heo, I., Joo, C., Kim, Y.K., Ha, M., Yoon, M.J., Cho, J., Yeom, K.H., Han, J., and Kim, V.N. (2009). TUT4 in concert with Lin28 suppresses microRNA biogenesis through pre-microRNA uridylation. Cell138, 696–708
121. Lehrbach, N.J., Armisen, J., Lightfoot, H.L., Murfitt, K.J., Bugaut, A., Balasubramanian, S., and Miska, E.A. (2009). LIN-28 and the poly(U) polymerasePUP-2 regulate let-7 microRNA processing in Caenorhabditis elegans. Nat. Struct. Mol. Biol.16, 1016–1020.
122. Melo, S.A., Ropero, S., Moutinho, C., Aaltonen, L.A., Yamamoto, H., Calin, G.A., Rossi, S., Fernandez, A.F., Carneiro, F., Oliveira, C., et al. (2009). ATARBP2 mutation in human cancer impairs microRNA processing and DICER1 function. Nat. Genet. 41, 365–370
123. Paroo, Z., Ye, X., Chen, S., and Liu, Q. (2009). Phosphorylation of the human microRNA-generating complex mediates MAPK/Erk signaling. Cell 139,112–122
124. Wang, Y., Sheng, G., Juranek, S., Tuschl, T., and Patel, D.J. (2008). Structure of the guide-strand-containing argonaute silencing complex. Nature 456,209–213.
125. Katoh, T., Sakaguchi, Y., Miyauchi, K., Suzuki, T., Kashiwabara, S., Baba, T., and Suzuki, T. (2009). Selective stabilization of mammalian microRNAs by 30adenylation mediated by the cytoplasmic poly(A) polymerase GLD-2. Genes Dev. 23, 433–438
126. Landgraf, P., Rusu, M., Sheridan, R., Sewer, A., Iovino, N., Aravin, A., Pfeffer,S., Rice, A., Kamphorst, A.O., Landthaler, M., et al. (2007). A mammalian microRNA expression atlas based on small RNA library sequencing. Cell 129, 1401–1414
127. Azuma-Mukai, A., Oguri, H., Mituyama, T., Qian, Z.R., Asai, K., Siomi, H., andSiomi, M.C. (2008). Characterization of endogenous human Argonautes and their miRNA partners in RNA silencing. Proc. Natl. Acad. Sci. USA 105, 7964–7969.
128. Ambros, V. (2008). The evolution of our thinking about microRNAs. Nat. Med. 14, 1036–1040
129. Hwang, H.W., Wentzel, E.A., and Mendell, J.T. (2007). A hexa-nucleotide element directs microRNA nuclear import. Science 315, 97–10
130. Diederichs, S., Jung, S., Rothenberg, S.M., Smolen, G.A., Mlody, B.G., andHaber, D.A. (2008). Coexpression of Argonaute-2 enhances RNA interference toward perfect match binding sites. Proc. Natl. Acad. Sci. USA 105, 9284–9289.
131. Qi, H.H., Ongusaha, P.P., Myllyharju, J., Cheng, D., Pakkanen, O., Shi, Y., Lee, S.W., Peng, J., and Shi, Y. (2008). Prolyl 4-hydroxylation regulates Argonaute 2 stability. Nature 455, 421–424
132. Gibbings, D.J., Ciaudo, C., Erhardt, M., and Voinnet, O. (2009). Multivesicular bodies associate with components of miRNA effector complexes and modulate miRNA activity. Nat. Cell Biol. 11, 1143–1149.

133. Lee, Y.S., Pressman, S., Andress, A.P., Kim, K., White, J.L., Cassidy, J.J., Li,X., Lubell, K., Lim do, H., Cho, I.S., et al. (2009). Silencing by small RNAs is linked to endosomal trafficking. Nat. Cell Biol. 11, 1150–1156
134. Chi, S.W., Zang, J.B., Mele, A., and Darnell, R.B. (2009). Argonaute HITS-CLIP decodes microRNA-mRNA interaction maps. Nature 460, 479–486
135. Mencia A, Modamio-Hoybjor S, Redshaw N, et al. Mutations in the seed region of human miR-96 are responsible for non- syndromic progressive hearing loss. Nat Genet. 2009; 41: 609–613.
136. Hanson EK, Lubenow H, Ballantyne J. Identification of forensically relevant body fluids using a panel of differentially expressed microRNAs. Anal Biochem. 2009; 87: 303–314.
137. Weber JA, Baxter DH, Zhang SL, et al. The microRNA spec- trum in 12 body fluids. Clin Chem. 2010; 56: 1733–1741.
138. Zen K, Zhang CY. Circulating microRNAs: a novel class of biomarkers to diagnose and monitor human cancers. Med Res Rev. 2012;32:326–348.
139. Etheridge A, Lee I, Hood L, et al. Extracellular microRNA: a new source of biomarkers. Mutat Res. 2011; 717: 85–90.
140. Kosaka N, Iguchi H, Yoshioka Y, et al. Secretory mechanisms and intercellular transfer of microRNAs in living cells. J Biol Chem. 2010;285:17442–17452.
141. Vickers KC, Palmisano BT, Shoucri BM, et al. MicroRNAs are transported in plasma and delivered to recipient cells by high- density lipoproteins. Nat Cell Biol. 2011;13:423–433.
142. Wang K, Zhang SL, Weber J, et al. Export of microRNAs and microRNA-protective protein by mammalian cells. Nucleic Acids Res. 2010;38:7248–7259.
143. Chen X, Ba Y, Ma L, et al. Characterization of microRNAs in serum: a novel class of biomarkers for diagnosis of cancer and other diseases. Cell Res. 2008; 18: 997–1006.
144. Mitchell PS, Parkin RK, Kroh EM, et al. Circulating microRNAs as stable blood-based markers for cancer detection. Proc Natl Acad Sci U S A. 2008;105:10513–10518.
145. Gibbings DJ, Ciaudo C, Erhardt M, et al. Multivesicular bodies associate with components of miRNA effector complexes and modulate miRNA activity. Nat Cell Biol. 2009;11:1143–1149.
146. Valadi H, Ekstrom K, Bossios A, et al. Exosome-mediated trans- fer of mRNAs and microRNAs is a novel mechanism of genetic exchange between cells. Nat Cell Biol. 2007;9:654–659.
147. Liu CJ, Kao SY, Tu HF, et al. Increase of microRNA miR-31 level in plasma could be a potential marker of oral cancer. Oral Dis. 2010;16:360–364.
148. Murata K, Yoshitomi H, Tanida S, et al. Plasma and synovial fluid microRNAs as potential biomarkers of rheumatoid arthritis and osteoarthritis. Arthritis Res Ther. 2010;12:R86.
149. Zen K, Zhang CY. Circulating microRNAs: a novel class of biomarkers todiagnose and monitor human cancers. Med Res Rev. 2012; 32:326–348.
150. Mladinov D, Liu Y, Mattson DL, et al. MicroRNAs contribute to the maintenance of cell-type-specific physiological characteristics: miR-192 targets Naþ/Kþ-ATPase beta1. Nucleic Acids Res. 2013;41:1273–1283.

151. Morin RD, O'Connor MD, Griffith M, et al. Application of massively parallel sequencing to microRNA profiling and discovery in human embryonic stem cells. Genome Res. 2008;18: 610–621.
152. Mortarino M, Gioia G, Gelain ME, et al. Identification of suit- able endogenous controls and differentially expressed micro- RNAs in canine fresh-frozen and FFPE lymphoma samples. Leukemia Res. 2010;34:1070–1077.
153. Mott JL, Kobayashi S, Bronk SF, et al. mir-29 regulates Mcl-1 protein expression and apoptosis. Oncogene. 2007;26:6133–6140.
154. Murata K, Yoshitomi H, Tanida S, et al. Plasma and synovial fluid microRNAs as potential biomarkers of rheumatoid arthritis and osteoarthritis. Arthritis Res Ther. 2010;12:R86.
155. Navarro A, Gaya A, Martinez A, et al. MicroRNA expression profiling in classic Hodgkin lymphoma. Blood. 2008;111: 2825–2832.
156. Nishiyama T, Kaneda R, Ono T, et al. miR-142-3p is essential for hematopoiesis and affects cardiac cell fate in zebrafish. Bio- chem Biophys Res Commun. 2012;425:755–761.
157. Mencia A, Modamio-Hoybjor S, Redshaw N, et al. Mutations in the seed region of human miR-96 are responsible for non- syndromic progressive hearing loss. Nat Genet. 2009;41: 609–613
158. Noguchi S, Mori T, Hoshino Y, et al. MicroRNAs as tumour suppressors in canine and human melanoma cells and as a prog- nostic factor in canine melanomas [published online December 8, 2011]. Vet Comp Oncol.
159. O'Connell RM, Chaudhuri AA, Rao DS, et al. MicroRNAs enriched in hematopoietic stem cells differentially regulate long-term hematopoietic output. Proc Natl Acad Sci U S A. 2010;107:14235–14240.
160. 127. O'Connell RM, Kahn D, Gibson WS, et al. MicroRNA-155 pro- motes autoimmune inflammation by enhancing inflammatory T cell development. Immunity. 2010;33:607–619.
161. Lund E, Guttinger S, Calado A, et al. Nuclear export of micro- RNA precursors. Science. 2004;303:95–98.
162. Okamura K, Hagen JW, Duan H, et al. The mirtron pathway generates microRNA-class regulatory RNAs in Drosophila. Cell. 2007;130:89–100.
163. Okamura K, Liu N, Lai EC. Distinct mechanisms for microRNA strand selection by Drosophila Argonautes. Mol Cell. 2009;36: 431–444.
164. Osokine I, Hsu R, Loeb GB, et al. Unintentional miRNA abla- tion is a risk factor in gene knockout studies: a short report. PLoS Genet. 2008;4:e34.
165. Otsuka M, Jing Q, Georgel P, et al. Hypersusceptibility to vesi- cular stomatitis virus infection in Dicer1-deficient mice is due to impaired miR24 and miR93 expression. Immunity. 2007;27: 123–134.
166. Papagiannakopoulos T, Shapiro A, Kosik KS. MicroRNA-21 targets a network of key tumor-suppressive pathways in glio- blastoma cells. Cancer Res. 2008;68:8164–8172.
167. Pasquinelli AE, Reinhart BJ, Slack F, et al. Conservation of the sequence and temporal expression of let-7 heterochronic regula- tory RNA. Nature. 2000;408:86–89.

168. Pauley KM, Satoh M, Chan AL, et al. Upregulated miR-146a expression in peripheral blood mononuclear cells from rheuma- toid arthritis patients. Arthritis Res Ther. 2008;10:R101.
169. Pedersen IM, Cheng G, Wieland S, et al. Interferon modulation of cellular microRNAs as an antiviral mechanism. Nature. 2007; 449:919–922.
170. Pfeffer S, Zavolan M, Grasser FA, et al. Identification of virus- encoded microRNAs. Science. 2004;304:734–736.
171. Podolska A, Anthon C, Bak M, et al. Profiling microRNAs in lung tissue from pigs infected with Actinobacillus pleuropneu- moniae. BMC Genomics. 2012;13:459.
172. Potenza N, Papa U, Mosca N, et al. Human microRNA hsa-miR- 125a-5p interferes with expression of hepatitis B virus surface antigen. Nucleic Acids Res. 2011;39:5157–5163.
173. Poy MN, Eliasson L, Krutzfeldt J, et al. A pancreatic islet- specific microRNA regulates insulin secretion. Nature. 2004; 432:226–230.
174. Recchiuti A, Krishnamoorthy S, Fredman G, et al. MicroRNAs in resolution of acute inflammation: identification of novel resolve in D1-miRNA circuits. FASEB J. 2011; 25:544–560.
175. 141. Reinhart BJ, Slack FJ, Basson M, et al. The 21-nucleotide let-7 RNA regulates developmental timing in Caenorhabditis elegans. Nature. 2000;403:901906.
176. Robinson VL. Rethinking the central dogma: noncoding RNAs are biologically relevant. Urol Oncol. 2009; 27: 304–306.
177. Rodriguez A, Griffiths-Jones S, Ashurst JL, et al. Identification of mammalian microRNA host genes and transcription units. Genome Res. 2004; 14:1902–1910.
178. Rodriguez A, Vigorito E, Clare S, et al. Requirement of bic/ microRNA-155 for normal immune function. Science. 2007; 316:608–611
179. Bernstein E, Kim SY, Carmell MA, et al. Dicer is essential for mouse development. Nat Genet. 2003; 35: 215–217.
180. Ketting RF, Fischer SEJ, Bernstein E, et al. Dicer functions in RNA interference and in synthesis of small RNA involved in developmental timing in C-elegans. Genes Dev. 2001;15:2654–2659.
181. Wienholds E, Koudijs MJ, van Eeden FJM, et al. The microRNA-producing enzyme Dicer1 is essential for zebrafish development. Nat Genet. 2003;35:217–218.
182. Andl T, Murchison EP, Liu F, et al. The miRNA-processing enzyme Dicer is essential for the morphogenesis and maintenance of hair follicles. Curr Biol. 2006;16:1041–1049.
183. Harfe BD, McManus MT, Mansfield JH, et al. The RNaseIII enzyme Dicer is required for morphogenesis but not patterning of the vertebrate limb. Proc Natl Acad Sci U S A. 2005;102: 10898–10903.
184. Harris KS, Zhang Z, McManus MT, et al. Dicer function is essential for lung epithelium morphogenesis. Proc Natl Acad Sci U S A. 2006;103:2208–2213.
185. Kloosterman WP, Plasterk RH. The diverse functions of micro- RNAs in animal development and disease. Dev Cell. 2006;11: 441–450.

186. Yi R, O'Carroll D, Pasolli HA, et al. Morphogenesis in skin is governed by discrete sets of differentially expressed microRNAs. Nat Genet. 2006;38:356–362.
187. Melton C, Blelloch R. MicroRNA regulation of embryonic stem cell self-renewal and differentiation. Adv Exp Med Biol. 2010; 695:105–117.
188. Kanellopoulou C, Muljo SA, Kung AL, et al. Dicer-deficient mouse embryonic stem cells are defective in differentiation and centromeric silencing. Genes Dev. 2005;19:489–501
189. Wang Y, Baskerville S, Shenoy A, et al. Embryonic stem cell– specific microRNAs regulate the G1-S transition and promote rapid proliferation. Nat Genet. 2008;40:1478–1483.
190. Li MA, He L. MicroRNAs as novel regulators of stem cell pluripotency and somatic cell reprogramming. Bioessays. 2012;34: 670–680.
191. Marson A, Levine SS, Cole MF, et al. Connecting microRNA genes to the core transcriptional regulatory circuitry of embryonic stem cells. Cell. 2008;134:521–533.
192. Kanellopoulou C, Muljo SA, Kung AL, et al. Dicer-deficient mouse embryonic stem cells are defective in differentiation and centromeric silencing. Genes Dev. 2005;19:489–501.
193. Wang YM, Medvid R, Melton C, et al. DGCR8 is essential for microRNA biogenesis and silencing of embryonic stem cell self-renewal. Nat Genet. 2007;39:380–385.
194. Melton C, Judson RL, Blelloch R. Opposing microRNA families regulate self-renewal in mouse embryonic stem cells. Nature. 2010;463:621–626.
195. Chaudhuri AA, So AY, Sinha N, et al. MicroRNA-125b potenti- ates macrophage activation. J Immunol. 2011;187:5062–5068.
196. Shaham L, Binder V, Gefen N, et al. MiR-125 in normal and malignant hematopoiesis. Leukemia. 2012;26:2011–2018.
197. Enomoto Y, Kitaura J, Hatakeyama K, et al. Emu/miR-125b transgenic mice develop lethal B-cell malignancies. Leukemia. 2011;25:1849–1856.
198. O'Connell RM, Chaudhuri AA, Rao DS, et al. MicroRNAs enriched in hematopoietic stem cells differentially regulate long-term hematopoietic output. Proc Natl Acad Sci U S A. 2010;107:14235–14240.
199. Hobert O. Architecture of a microRNA-controlled gene regulatory network that diversifies neuronal cell fates. Cold Spring Harb Symp Quant Biol. 2006;71:181–1 88.
200. Zhao Y, Ransom JF, Li A, et al. Dysregulation of cardiogenesis, cardiac conduction, and cell cycle in mice lacking miRNA-1-2. Cell. 2007;129:303–317.
201. Zhao Y, Samal E, Srivastava D. Serum, response factor regulates a muscle-specific microRNA that targets Hand2 during cardiogenesis. Nature. 2005;436:214–220.
202. Abdellatif M. Differential expression of microRNAs in different disease states. Circ Res. 2012;110:638–650.
203. Ikeda S, He AB, Kong SW, et al. MicroRNA-1 negatively regu- lates expression of the hypertrophy-associated calmodulin and Mef2a genes. Mol Cell Biol. 2009;29:2193–2204.

204. van Rooij E, Sutherland LB, Qi X, et al. Control of stress- dependent cardiac growth and gene expression by a microRNA. Science. 2007;316:575–579.
205. Liu N, Bezprozvannaya S, Williams AH, et al. MicroRNA-133a regulates cardiomyocyte proliferation and suppresses smooth muscle gene expression in the heart. Genes Dev. 2008;22: 3242–3254.
206. Lena AM, Shalom-Feuerstein R, Cervo PRD, et al. miR-203 represses 'stemness'
207. Yi R, Poy MN, Stoffel M, et al. A skin microRNA promotes differentiation by repressing 'stemness'. Nature. 2008;452:225–229.
208. Antonini D, Russo MT, De Rosa L, et al. Transcriptional repres- sion of miR-34 family contributes to p63-mediated cell cycle progression in epidermal cells. J Invest Dermatol. 2010;130: 1249–1257.
209. Schneider MR. MicroRNAs as novel players in skin develop- ment, homeostasis and disease. Br J Dermatol. 2012;166: 22–28.
210. Ebert MS, Sharp PA. Roles for microRNAs in conferring robust- ness to biological processes. Cell. 2012;149:515–524.
211. Stark A, Brennecke J, Bushati N, et al. Animal microRNAs confer robustness to gene expression and have a significant impact on 3' UTR evolution. Cell. 2005;123:1133–1146.
212. Bhaskaran M, Wang Y, Zhang HH, et al. MicroRNA-127 modu- lates fetal lung development. Physiol Genomics. 2009;37: 268–278
213. Lu Y, Thomson JM, Wong HY, et al. Transgenic over- expression of the microRNA miR-17-92 cluster promotes prolif- eration and inhibits differentiation of lung epithelial progenitor cells. Dev Biol. 2007;310:442–453.
214. Poy MN, Eliasson L, Krutzfeldt J, et al. A pancreatic islet- specific microRNA regulates insulin secretion. Nature. 2004; 432:226–230.
215. Esau C, Davis S, Murray SF, et al. miR-122 regulation of lipid metabolism revealed by in vivo antisense targeting. Cell Metab. 2006;3:87–98.
216. Esau C, Kang X, Peralta E, et al. MicroRNA-143 regulates adipocyte differentiation. J Biol Chem. 2004;279:52361–52365.
217. Mladinov D, Liu Y, Mattson DL, et al. MicroRNAs contribute to the maintenance of cell-type-specific physiological characteristics: miR-192 targets Naþ/Kþ-ATPase beta1. Nucleic Acids Res. 2013;41:1273–1283
218. Calin GA, Sevignani C, Dan Dumitru C, et al. Human microRNA genes are frequently located at fragile sites and genomic regions involved in cancers. Proc Natl Acad Sci U S A. 2004;101: 2999–3004.
219. Calin GA, Dumitru CD, Shimizu M, et al. Frequent deletions and down-regulation of micro-RNA genes miR15 and miR16 at 13q14 in chronic lymphocytic leukemia. Proc Natl Acad Sci U S A. 2002;99:15524–15529.
220. Bommer GT, Gerin I, Feng Y, et al. p53-mediated activation of miRNA34 candidate tumor-suppressor genes. Curr Biol. 2007; 17:1298–1307.
221. Schickel R, Boyerinas B, Park SM, et al. MicroRNAs: key play- ers in the immune system, differentiation, tumorigenesis and cell death. Oncogene. 2008;27:5959–5974.
222. Yamakuchi M, Ferlito M, Lowenstein CJ. miR-34a repression of SIRT1 regulates apoptosis. Proc Natl Acad Sci U S A. 2008;105: 13421–13426.

223. Welch C, Chen Y, Stallings RL. MicroRNA-34a functions as a potential tumor suppressor by inducing apoptosis in neuroblas- toma cells. Oncogene. 2007;26:5017–5022
224. Ji Q, Hao XB, Zhang M, et al. MicroRNA miR-34 inhibits human pancreatic cancer tumor-initiating cells. PloS One. 2009;4:e6816.
225. Johnson SM, Grosshans H, Shingara J, et al. RAS is regulated by the let-7 microRNA family. Cell. 2005;120:635–647.
226. Lee YS, Dutta A. The tumor suppressor microRNA let-7 represses the HMGA2 oncogene. Genes Dev. 2007;21:1025–1030.
227. Sayed D, Abdellatif M. MicroRNAs in development and disease. Physiol Rev. 2011;91:827–887.
228. Farazi TA, Spitzer JI, Morozov P, et al. miRNAs in human cancer. J Pathol. 2011;223:102–115.
229. Kumar MS, Lu J, Mercer KL, et al. Impaired microRNA processing enhances cellular transformation and tumorigenesis. Nat Genet. 2007;39:673–677.
230. Kumar MS, Pester RE, Chen CY, et al. Dicerl functions as a haplo insufficient tumor suppressor. Genes Dev. 2009;23:2700–2704.
231. Eis PS, Tam W, Sun L, et al. Accumulation of miR-155 and BIC RNA in human B cell lymphomas. Proc Natl Acad Sci U S A. 2005;102:3627–3632.
232. He L, Thomson JM, Hemann MT, et al. A microRNA polycistron as a potential human oncogene. Nature. 2005;435:828–833.
233. Costinean S, Zanesi N, Pekarsky Y, et al. Pre–B cell proliferation and lymphoblastic leukemia/high-grade lymphoma in E mu-miR155 transgenic mice. Proc Natl Acad Sci U S A. 2006;103:7024–7029.
234. Sayed D, Abdellatif M. MicroRNAs in development and disease. Physiol Rev. 2011;91:827–887.
235. Farazi TA, Spitzer JI, Morozov P, et al. miRNAs in human cancer. J Pathol. 2011;223:102–115
236. Chan JA, Krichevsky AM, Kosik KS. MicroRNA-21 is an antia- poptotic factor in human glioblastoma cells. Cancer Res. 2005;65: 6029–6033
237. Chen Y, Liu W, Chao TF, et al. MicroRNA-21 down-regulates the expression of tumor suppressor PDCD4 in human glioblastoma cell T98G. Cancer Lett. 2008;272:197–205.
238. Papagiannakopoulos T, Shapiro A, Kosik KS. MicroRNA-21 targets a network of key tumor-suppressive pathways in glio- blastoma cells. Cancer Res. 2008;68:8164–8172.
239. Huse J, Brennan C, Hambardzumyan D, et al. The PTEN- regulating microRNA miR-26a is amplified in high-grade glioma and facilitates gliomagenesis in vivo. Genes Dev. 2009;23: 1327–1337.
240. Kim H, Huang W, Jiang X, et al. Integrative genome analysis eveals an oncomir/oncogene cluster regulating glioblastoma survivorship. Proc Natl Acad Sci U S A. 2010;107:2183–2188.
241. Gao J, Liu QG. The role of miR-26 in tumors and normal tissues [review]. Oncol Lett. 2011;2:1019–1023.
242. Khodyrev DS, Pronina IV, Rykov SV, et al. Involvement of methylation of group of miRNA genes in regulation of expression of RAR-beta2 and NKIRAS1 target genes in lung cancer. Mol Biol. 2012;46:693–704.

243. Lehmann U, Hasemeier B, Christgen M, et al. Epigenetic inac- tivation of microRNA gene hsa-mir-9-1 in human breast cancer. J Pathol. 2008;214:17–24.
244. Saito Y, Liang G, Egger G, et al. Specific activation of microRNA-127 with downregulation of the proto-oncogene BCL6 by chromatin-modifying drugs in human cancer cells. Cancer Cell. 2006;9:435–443.
245. Lu J, Getz G, Miska EA, et al. MicroRNA expression profiles classify human cancers. Nature. 2005;435:834–838.
246. Pfeffer S, Zavolan M, Grasser FA, et al. Identification of virus- encoded microRNAs. Science. 2004;304:734–736
247. Kluiver J, Haralambieva E, de Jong D, et al. Lack of BIC and microRNA miR-155 expression in primary cases of Burkitt lymphoma. Genes Chromosomes Cancer. 2006;45:147–153.
248. Xia T, O'Hara A, Araujo I, et al. EBV microRNAs in primary lymphomas and targeting of CXCL-11 by ebv-mir-BHRF1-3. Cancer Res. 2008;68:1436–1442.
249. Cameron JE, Fewell C, Yin QY, et al. Epstein-Barr virus growth/ latency III program alters cellular microRNA expression. Virology. 2008;382:257–266.
250. Navarro A, Gaya A, Martinez A, et al. MicroRNA expression profiling in classic Hodgkin lymphoma. Blood. 2008;111: 2825–2832.
251. Triboulet R, Mari B, Lin YL, et al. Suppression of microRNA- silencing pathway by HIV-1 during virus replication. Science. 2007;315:1579–1582.
252. Huang JL, Wang FX, Argyris E, et al. Cellular microRNAs contribute to HIV-1 latency in resting primary CD4(þ) T lymphocytes. Nat Med. 2007;13:1241–1247.
253. Umbach JL, Cullen BR. The role of RNAi and microRNAs in animal virus replication and antiviral immunity. Genes Dev. 2009;23:1151–1164.
254. Wang X, Ye L, Hou W, et al. Cellular microRNA expression correlates with susceptibility of monocytes/macrophages to HIV-1 infection. Blood. 2009;113:671–674.
255. Jopling CL. Liver-specific microRNA-122 biogenesis and function. RNA Biol. 2012;9:137–142.
256. Klase Z, Houzet L, Jeang KT. MicroRNAs and HIV-1: complex interactions. J Biol Chem. 2012;287:40884–40890.
257. Lanford RE, Hildebrandt-Eriksen ES, Petri A, et al. Therapeutic silencing of microRNA-122 in primates with chronic hepatitis C virus infection. Science. 2010;327:198–201.
258. Androulidaki A, Iliopoulos D, Arranz A, et al. The kinase Akt1 controls macrophage response to lipopolysaccharide by regulat- ing microRNAs. Immunity. 2009;31:220–231.
259. BoldinMP,TaganovKD,RaoDS,etal.miR-146aisasignificant brake on autoimmunity, myeloproliferation, and cancer in mice. J Exp Med. 2011;208:1189–1201.
260. Kinjyo I, Hanada T, Inagaki-Ohara K, et al. SOCS1/JAB is a neg- ative regulator of LPS-induced macrophage activation. Immunity. 2002;17:583–591.
261. Rodriguez A, Vigorito E, Clare S, et al. Requirement of bic/ microRNA-155 for normal immune function. Science. 2007; 316:608–611.
262. Singh Y, Kaul V, Mehra A, et al. Mycobacterium tuberculosis controls microRNA-99b (miR-99b) expression in infected murine dendritic cells to modulate host immunity. J Biol Chem. 2013;288:5056–5061.

263. Wu ZW, Lu HF, Sheng JF, et al. Inductive microRNA-21 impairs anti-mycobacterial responses by targeting IL-12 and Bcl-2. FEBS Lett. 2012;586:2459–2467.
264. Liu PT, Wheelwright M, Teles R, et al. MicroRNA-21 targets the vitamin D–dependent antimicrobial pathway in leprosy. Nat Med. 2012;18:267–273.
265. Junker A. Pathophysiology of translational regulation by microRNAs in multiple sclerosis. FEBS Lett. 2011;585:3738–3746.
266. Junker A, Krumbholz M, Eisele S, et al. MicroRNA profiling of multiple sclerosis lesions identifies modulators of the regulatory protein CD47. Brain. 2009;132:3342–3352.
267. O'Connell RM, Kahn D, Gibson WS, et al. MicroRNA-155 pro- motes autoimmune inflammation by enhancing inflammatory T cell development. Immunity. 2010;33:607–619.
268. Pauley KM, Satoh M, Chan AL, et al. Upregulated miR-146a expression in peripheral blood mononuclear cells from rheuma- toid arthritis patients. Arthritis Res Ther. 2008;10:R101.
269. Shen N, Liang D, Tang Y, et al. MicroRNAs—novel regulators of systemic lupus erythematosus pathogenesis. Nat Rev Rheumatol. 2012;8:701–709.
270. Tang YJ, Luo XB, Cui HJ, et al. MicroRNA-146a contributes to abnormal activation of the type I interferon pathway in human lupus by targeting the key signaling proteins. Arthritis Rheum. 2009;60:1065–1075.
271. Lewis AP, Jopling CL. Regulation and biological function of the liver-specific miR-122. Biochem Soc Trans. 2010;38: 1553–1557.
272. Esau C, Davis S, Murray SF, et al. miR-122 regulation of lipid metabolism revealed by in vivo antisense targeting. Cell Metab. 2006;3:87–98.
273. Elmen J, Lindow M, Schutz S, et al. LNA-mediated microRNA silencing in nonhuman primates. Nature. 2008;452:896–899.
274. Rottiers V, Naar AM. MicroRNAs in metabolism and metabolic disorders. Nat Rev Mol Cell Biol. 2012;13:239–250.
275. Vickers KC, Palmisano BT, Shoucri BM, et al. MicroRNAs are transported in plasma and delivered to recipient cells by high- density lipoproteins. Nat Cell Biol. 2011;13:423–433.
276. Mencia A, Modamio-Hoybjor S, Redshaw N, et al. Mutations in the seed region of human miR-96 are responsible for non- syndromic progressive hearing loss. Nat Genet. 2009;41: 609–613.
277. Lewis MA, Quint E, Glazier AM, et al. An ENU-induced mutation of miR-96 associated with progressive hearing loss in mice. Nat Genet. 2009;41:614–618.
278. Beetz C, Schule R, Deconinck T, et al. REEP1 mutation spectrum and genotype/phenotype correlation in hereditary spastic paraple- gia type 31. Brain. 2008;131:1078–1086.
279. Zuchner S, Wang GF, Tran-Viet KN, et al. Mutations in the novel mitochondrial protein REEP1 cause hereditary spastic paraplegia type 31. Am J Hum Genet. 2006;79:365–369.
280. Li YJ, Lin L, Jin P. The microRNA pathway and fragile X mental retardation protein. Biochim Biophys Acta. 2008;1779: 702–705.

281. Shiohama A, Sasaki T, Noda S, et al. Molecular cloning and expression analysis of a novel gene DGCR8 located in the DiGeorge syndrome chromosomal region. Biochem Bioph Res Commun. 2003;304:184–190.
282. Bottini S, Hamouda-Tekaya N, Mategot R, Zaragosi LE, Audebert S, Pisano S, et al. Post-transcriptional gene silencing mediated by microRNAs is controlled by nucleoplasmic Sfpq. *Nat Commun.* (2017) 8:1189. doi: 10.1038/s41467-017-01126-x
283. Nam JW, Rissland OS, Koppstein D, Abreu-Goodger C, Jan CH, Agarwal V, et al. Global analyses of the effect of different cellular contexts on microRNA targeting. *Mol Cell.* (2014) 53:1031–43. doi: 10.1016/j.molcel.2014.02.013
284. Blazie SM, Geissel HC, Wilky H, Joshi R, Newbern J, Mangone M. Alternative polyadenylation directs tissue-specific miRNA targeting in *Caenorhabditis elegans* Somatic Tissues. *Genetics* (2017) 206:757–74. doi: 10.1534/genetics.116.196774
285. Bose M, Barman B, Goswami A, Bhattacharyya SN. Spatiotemporal uncoupling of MicroRNA-mediated translational repression and target RNA degradation controls microRNP recycling in mammalian cells. *Mol Cell Biol.* (2017) 37:e00464–16. doi: 10.1128/MCB.00464-16
286. Gibbings DJ, Ciaudo C, Erhardt M, Voinnet O. Multivesicular bodies associate with components of miRNA effector complexes and modulate miRNA activity. *Nat Cell Biol.* (2009) 11:1143–9. doi: 10.1038/ncb1929
287. Barman B, Bhattacharyya SN. mRNA targeting to endoplasmic reticulum precedes ago protein interaction and microRNA (miRNA)-mediated translation repression in mammalian cells. *J Biol Chem.* (2015) 290:24650–6. doi: 10.1074/jbc.C115.661868
288. Nishi K, Takahashi T, Suzawa M, Miyakawa T, Nagasawa T, Ming Y, et al. Control of the localization and function of a miRNA silencing component TNRC6A by Argonaute protein. *Nucleic Acids Res* (2015) 43:9856–73. doi: 10.1093/nar/gkv1026
289. Detzer A, Engel C, Wunsche W, Sczakiel G. Cell stress is related to re-localization of Argonaute 2 and to decreased RNA interference in human cells. *Nucleic Acids Res.* (2011) 39:2727–41. doi: 10.1093/nar/gkq1216
290. Barrey E, Saint-Auret G, Bonnamy B, Damas D, Boyer O, Gidrol X. Pre-microRNA and mature microRNA in human mitochondria. *PLoS ONE* (2011) 6:e20220. doi: 10.1371/journal.pone.0020220
291. Zhang X, Zuo X, Yang B, Li Z, Xue Y, Zhou Y, et al. MicroRNA directly enhances mitochondrial translation during muscle differentiation. *Cell* (2014) 158:607–19. doi: 10.1016/j.cell.2014.05.047
292. Miao L, Yao H, Li C, Pu M, Yao X, Yang H, et al. A dual inhibition: microRNA-552 suppresses both transcription and translation of cytochrome P450 2E1. *Biochim Biophys Acta* (2016) 1859:650–62. doi: 10.1016/j.bbagrm.2016.02.016
293. Gagnon KT, Li L, Chu Y, Janowski BA, Corey DR. RNAi factors are present and active in human cell nuclei. *Cell Rep.* (2014) 6:211–21. doi:10.1016/ j.celrep.2013. 12.013

294. Liu J, Valencia-Sanchez MA, Hannon GJ, Parker R. MicroRNA-dependent localization of targeted mRNAs to mammalian P-bodies. *Nat Cell Biol.* (2005) 7:719–23. doi: 10.1038/ncb1274
295. Parker R, Sheth U. P bodies and the control of mRNA translation and degradation. *Mol Cell.* (2007) 25:635–46. doi: 10.1016/j.molcel.2007.02.011
296. Eulalio A, Behm-Ansmant I, Schweizer D, Izaurralde E. P-body formation is a consequence, not the cause, of RNA-mediated gene silencing. *Mol Cell Biol.* (2007) 27:3970–81. doi: 10.1128/MCB.00128-07
297. Rajgor D, Mellad JA, Soong D, Rattner JB, Fritzler MJ, Shanahan CM. Mammalian microtubule P-body dynamics are mediated by nesprin-1. *J Cell Biol.* (2014) 205:457–75. doi: 10.1083/jcb.201306076
298. Carbonaro M, O'Brate A, Giannakakou P. Microtubule disruption targets HIF-1alpha mRNA to cytoplasmic P-bodies for translational repression. *J Cell Biol.* (2011) 192:83–99. doi: 10.1083/jcb.201004145
299. Pillai RS, Bhattacharyya SN, Artus CG, Zoller T, Cougot N, Basyuk E, et al. Inhibition of translational initiation by Let-7 MicroRNA in human cells. *Science* (2005) 309:1573–6. doi: 10.1126/science.1115079
300. Molotski N, Soen Y. Differential association of micro*RNA*s with polysomes reflects distinct strengths of interactions with their mRNA targets. *RNA* (2012) 18:1612–23. doi: 10.1261/rna.033142.112
301. Wee LM, Flores-Jasso CF, Salomon WE, Zamore PD. Argonaute divides its RNA guide into domains with distinct functions and RNA-binding properties. *Cell* (2012) 151:1055–67. doi: 10.1016/j.cell.2012.10.036
302. de Rie D, Abugessaisa I, Alam T, Arner E, Arner P, Ashoor H, et al. An integrated expression atlas of miRNAs and their promoters in human and mouse. *Nat Biotechnol.* (2017) 35:872–8. doi: 10.1038/nbt.3947
303. Lai X, Wolkenhauer O, Vera J. Understanding microRNA-mediated gene regulatory networks through mathematical modelling. *Nucleic Acids Res.* (2016) 44:6019–35. doi: 10.1093/nar/gkw550
304. Zhang JF, Fu WM, He ML, Xie WD, Lv Q, Wan G, et al. MiRNA-20a promotes osteogenic differentiation of human mesenchymal stem cells by co-regulating BMP signaling. *RNA Biol.* (2011) 8:829–38. doi: 10.4161/rna.8.5.16043
305. Shu J, Silva B, Gao T, Xu Z, Cui J. Dynamic and modularized microRNA regulation and its implication in human cancers. *Sci Rep.* (2017) 7:13356. doi: 10.1038/s41598-017-13470-5
306. Siciliano V, Garzilli I, Fracassi C, Criscuolo S, Ventre S, di Bernardo D. MiRNAs confer phenotypic robustness to gene networks by suppressing biological noise. *Nat Commun.* (2013) 4:2364. doi: 10.1038/ncomms3364
307. Schmiedel JM, Klemm SL, Zheng Y, Sahay A, Blüthgen N, Marks DS, et al. MicroRNA control of protein expression noise. *Science* (2015) 348:128–32. doi: 10.1126/science.aaa1738
308. Ebert MS, Sharp PA. Roles for microRNAs in conferring robustness to biological processes. *Cell* (2012) 149:515–24. doi: 10.1016/j.cell.2012.04.005
309. Lalanne JB, Taggart JC, Guo MS, Herzel L, Schieler A, Li GW. Evolutionary convergence of pathway-specific enzyme expression stoichiometry. *Cell* (2018) 173:749–761. doi: 10.1016/j.cell.2018.03.007

310. Schwanhäusser B, Busse D, Li N, Dittmar G, Schuchhardt J, Wolf J, et al. Global quantification of mammalian gene expression control. *Nature* (2011) 473:337–42. doi: 10.1038/nature10098
311. Denzler R, McGeary SE, Title AC, Agarwal V, Bartel DP, Stoffel M. Impact of MicroRNA levels, target-site complementarity, and cooperativity on competing endogenous RNA-regulated gene expression. *Mol Cell* (2016) 64:565–79. doi: 10.1016/j.molcel.2016.09.027
312. Kartha RV, Subramanian S. Competing endogenous RNAs (ceRNAs): new entrants to the intricacies of gene regulation. *Front Genet.* (2014) 5:8. doi: 10.3389/fgene.2014.00008
313. Barrett SP, Salzman J. Circular RNAs: analysis, expression and potential functions. *Development* (2016) 143:1838–47. doi: 10.1242/dev.128074
314. Yu CY, Li TC, Wu YY, Yeh CH, Chiang W, Chuang CY, et al. The circular RNA circBIRC6 participates in the molecular circuitry controlling human pluripotency. *Nat Commun.* (2017) 8:1149. doi: 10.1038/s41467-017-01216-w
315. Sambandan S, Akbalik G, Kochen L, Rinne J, Kahlstatt J, Glock C, et al. Activity-dependent spatially localized miRNA maturation in neuronal dendrites. *Science* (2017) 355:634–7. doi: 10.1126/science.aaf8995
316. Chou CH, Shrestha S, Yang CD, Chang NW, Lin YL, Liao KW, et al. miRTarBase update 2018: a resource for experimentally validated microRNA-target interactions. *Nucleic Acids Res.* (2018) 46:D296–D302. doi: 10.1093/nar/gkx1067
317. Elkayam E, Faehnle CR, Morales M, Sun J, Li H, Joshua-Tor L. Multivalent recruitment of human argonaute by GW182. *Mol Cell.* (2017) 67:646–658.e3. doi: 10.1016/j.molcel.2017.07.007
318. Li Y, Liang C, Wong KC, Luo J, Zhang Z. Mirsynergy: detecting synergistic miRNA regulatory modules by overlapping neighbourhood expansion. *Bioinformatics* (2014) 30:2627–35. doi: 10.1093/bioinformatics/btu373
319. Lemus-Diaz N, Boker KO, Rodriguez-Polo I, Mitter M, Preis J, Arlt M, et al. Dissecting miRNA gene repression on single cell level with an advanced fluorescent reporter system. *Sci Rep.* (2017) 7:45197. doi: 10.1038/srep45197
320. Bose M, Bhattacharyya SN. Target-dependent biogenesis of cognate microRNAs in human cells. *Nat Commun.* (2016) 7:12200. doi: 10.1038/ncomms12200
321. Pons-Espinal M, de Luca E, Marzi MJ, Beckervordersandforth R, Armirotti A, Nicassio F, et al. Synergic functions of miRNAs determine neuronal fate of adult neural stem cells. *Stem Cell Rep.* (2017) 8:1046–61. doi: 10.1016/j.stemcr.2017.02.012
322. Sahu M, Mallick B. Deciphering synergistic regulatory networks of microRNAs in hESCs and fibroblasts. *Int J Biol Macromol.* (2018) 113:1279–86. doi: 10.1016/j.ijbiomac.2018.03.061
323. Abbas T, Dutta A. p21 in cancer: intricate networks and multiple activities. *Nat Rev Cancer* (2009) 9:400–14. doi: 10.1038/nrc2657
324. Wu S, Huang S, Ding J, Zhao Y, Liang L, Liu T, et al. Multiple microRNAs modulate p21Cip1/Waf1 expression by directly targeting its 3′ untranslated region. *Oncogene* (2010) 29:2302–8. doi: 10.1038/onc.2010.34

325. Yao W, Guo G, Zhang Q, Fan L, Wu N, Bo Y. The application of multiple miRNA response elements enables oncolytic adenoviruses to possess specificity to glioma cells. *Virology* (2014). 458–459, 69–82. doi: 10.1016/j.virol.2014.04.007
326. Kucherenko MM, Shcherbata HR. miRNA targeting and alternative splicing in the stress response - events hosted by membrane-less compartments. *J Cell Sci.* (2018) 131:jcs202002. doi: 10.1242/jcs.202002
327. Wang K, Zhang S, Weber J, Baxter D, Galas DJ. Export of microRNAs and microRNA-protective protein by mammalian cells. *Nucleic Acids Res.* (2010) 38:7248–59. doi: 10.1093/nar/gkq601
328. Leung AK, Calabrese JM, Sharp PA. Quantitative analysis of Argonaute protein reveals microRNA-dependent localization to stress granules. *Proc Natl Acad Sci USA.* (2006) 103:18125–30. doi: 10.1073/pnas.0608845103
329. Protter DS, Parker R. Principles and Properties of Stress Granules. *Trends Cell Biol.* (2016) 26:668–79. doi: 10.1016/j.tcb.2016.05.004
330. Robinson VL. Rethinking the central dogma: noncoding RNAs are biologically relevant. Urol Oncol. 2009;27:304–306.
331. Osokine I, Hsu R, Loeb GB, et al. Unintentional miRNA abla- tion is a risk factor in gene knockout studies: a short report. PLoS Genet. 2008;4:e34.
332. Rossbach M. Small non-coding RNAs as novel therapeutics. Curr Mol Med. 2010;10:361–368.
333. van Rooij E, Marshall WS, Olson EN. Toward microRNA-based therapeutics for heart disease: the sense in antisense. Circ Res. 2008;103:919–928.

Chapter III

1. Bhaumik, S.R.; Malik, S. Diverse regulatory mechanisms of eukaryotic transcriptional activation by the proteasome complex. *Crit. Rev. Biochem. Mol. Biol.* **2008**, *43*, 419–433.
2. Durairaj, G.; Malik, S.; Bhaumik, S.R. Eukaryotic gene expression by RNA polymerase II. In *Gene Regulation, Epigenetics, and Hormone Signaling*; Mandal, S.S., Ed.; Wiley-Blackwell: Oxford, UK, 2017; Volume 1, pp. 1–28. ISBN 978-3-527-32281-7.
3. Shilatifard, A.; Conaway, R.C.; Conaway, J.W. The RNA polymerase II elongation complex. *Annu. Rev. Biochem* **2003**, *72*, 693–715.
4. Katayama, S.; Tomaru, Y.; Kasukawa, T.; Waki, K.; Nakanishi, M.; Nakamura, M.; Nishida, H.; Yap, C.C.; Suzuki, M.; Kawai, J. Antisense transcription in the mammalian transcriptome. *Science* **2005**, *309*, 1564–1566.
5. He, Y.; Vogelstein, B.; Velculescu, V.E.; Papadopoulos, N.; Kinzler, K.W. The antisense transcriptomes of human cells. *Science* **2008**, *322*, 1855–1857.
6. Ferrari, N.; Seguin, R.; Renzi, P. Oligonucleotides: A multi-targeted approach for the treatment of respiratory diseases. *Future Med. Chem.* **2011**, *3*, 1647–1662.
7. Lu, X.; Yu, Q.; Binder, G.K.; Chen, Z.; Slepushkina, T.; Rossi, J.; Dropulic, B. Antisense-mediated inhibition of human immunodeficiency virus (HIV) replication by use of an HIV type 1-based vector results in severely attenuated mutant incapable of developing resistance. *J. Virol.* **2004**, *78*, 7079–7088.
8. DeVos, S.L.; Miller, T.M. Antisense oligonucleotides: Treating neurodegeneration at the level of RNA. *Neurotherapeutics* **2013**, *10*, 486–497.

9. Gao, Z.; Cooper, T.A. Antisense oligonucleotides: Rising stars in eliminating RNA toxicity in myotonic dystrophy. *Hum. Gene Ther.* **2013**, *24*, 499–507.
10. Peters L, Meister G. Argonaute proteins: mediators of RNA silencing. Mol Cell. 2007;26:611–23.
11. Azlan A, Dzaki N, Azzam G. Argonaute: the executor of small RNA function. J Genet Genomics. 2016;43:481–94.
12. Hock J, Meister G. The Argonaute protein family. Genome Biol. 2008;9:210.
13. Hutvagner G, Simard MJ. Argonaute proteins: key players in RNA silencing. Nat Rev Mol Cell Biol. 2008;9:22–32.
14. Grivna ST, Beyret E, Wang Z, Lin H. A novel class of small RNAs in mouse spermatogenic cells. Genes Dev. 2006;20:1709–14.
15. Aravin A, Gaidatzis D, Pfeffer S, Lagos-Quintana M, Landgraf P, Iovino N, et al. A novel class of small RNAs bind to MILI protein in mouse testes. Nature. 2006;442:203–7.
16. Girard A, Sachidanandam R, Hannon GJ, Carmell MA. A germline- specific class of small RNAs binds mammalian Piwi proteins. Nature. 2006;442:199–202.
17. Lau NC, Seto AG, Kim J, Kuramochi-Miyagawa S, Nakano T, Bartel DP, et al. Characterization of the piRNA complex from rat testes. Science. 2006;313:363–7.
18. Grillari J, Grillari-Voglauer R. Novel modulators of senescence, aging, and longevity: small non-coding RNAs enter the stage. Exp Gerontol. 2010;45:302–11.
19. Klattenhoff C, Theurkauf W. Biogenesis and germline functions of piRNAs. Development. 2008;135:3–9.
20. Hartig JV, Tomari Y, Forstemann K. piRNAs--the ancient hunters of genome invaders. Genes Dev. 2007;21:1707–13.
21. Houwing S, Kamminga LM, Berezikov E, Cronembold D, Girard A, van den Elst H, et al. A role for Piwi and piRNAs in germ cell maintenance and transposon silencing in zebrafish. Cell. 2007;129:69–82.
22. Lin H, Yin H. A novel epigenetic mechanism in Drosophila somatic cells mediated by Piwi and piRNAs. Cold Spring Harb Symp Quant Biol. 2008;73:273–81.
23. Lee EJ, Banerjee S, Zhou H, Jammalamadaka A, Arcila M, Manjunath BS, et al. Identification of piRNAs in the central nervous system. RNA. 2011;17:1090–9.
24. Kowalczykiewicz, D., Pawlak, P., Lechniak, D., and Wrzesinski, J. (2012). Altered expression of porcine Piwi genes and piRNA during development. PLoS ONE 7, e43816.
25. Brennecke, J., Aravin, A.A., Stark, A., Dus, M., Kellis, M., Sachidanandam, R., and Hannon, G.J. (2007). Discrete small RNA-generating loci as master regulators of transposon activity in Drosophila. Cell 128, 1089–1103.
26. Li S, Xu Z, Sheng J. tRNA-derived small RNA: a novel regulatory small non-coding RNA. Genes. 2018;9(5):246.
27. FuH, FengJ, LiuQ, SunF, TieY, ZhuJ, XingR, SunZ, ZhengX. Stress induces tRNA cleavage by angiogenin in mammalian cells. FEBS Lett. 2009; 583(2):437–42.

28. Thompson DM, Parker R. The RNase Rny1p cleaves tRNAs and promotes cell death during oxidative stress in Saccharomyces cerevisiae. J Cell Biol. 2009; 185(1):43–50.
29. Joechl C, Rederstorff M, Hertel J, Stadler PF, Hofacker IL, Schrettl M, Haas H, Huettenhofer A. Small ncRNA transcriptome analysis from Aspergillus fumigatus suggests a novel mechanism for regulation of protein synthesis. Nucleic Acids Res. 2008; 36(8): 2677–89.
30. Li Y, Luo J, Zhou H, Liao J-Y, Ma L-M, Chen Y-Q, Qu L-H. Stress-induced tRNA-derived RNAs: a novel class of small RNAs in the primitive eukaryote Giardia lamblia. Nucleic Acids Res. 2008; 36(19): 6048–55.
31. Thompson DM, Lu C, Green PJ, Parker R. tRNA cleavage is a conserved response to oxidative stress in eukaryotes. RNA-a Publication of the RNA Society. 2008;14(10):2095–103.
32. Hsieh L-C, Lin S-I, Shih AC-C, Chen J-W, Lin W-Y, Tseng C-Y, Li W-H, Chiou T- J. Uncovering small RNA-mediated responses to phosphate deficiency in Arabidopsis by deep sequencing. Plant Physiol. 2009;151(4):2120–32.
33. Honda S, Loher P, Shigematsu M, Palazzo JP, Suzuki R, Imoto I, Rigoutsos I, Kirino Y. Sex hormone-dependent tRNA halves enhance cell proliferation in breast and prostate cancers. Proc Natl Acad Sci U S A. 2015;112(29):E3816–25.
34. Li Z, Ender C, Meister G, Moore PS, Chang Y, John B. Extensive terminal and asymmetric processing of small RNAs from rRNAs, snoRNAs, snRNAs, and tRNAs. Nucleic Acids Res. 2012;40(14):6787–99.
35. Maraia RJ, Lamichhane TN. 3'processing of eukaryotic precursor tRNAs. Wiley Interdiscip Rev RNA. 2011;2(3):362–75.
36. Babiarz JE, Ruby JG, Wang Y, Bartel DP, Blelloch R. Mouse ES cells express endogenous shRNAs, siRNAs, and other microprocessor-independent, dicer-dependent small RNAs. Genes Dev. 2008;22(20):2773–85.
37. Liao J-Y, Ma L-M, Guo Y-H, Zhang Y-C, Zhou H, Shao P, Chen Y-Q, Qu L-H. Deep sequencing of human nuclear and cytoplasmic small RNAs reveals an unexpectedly complex subcellular distribution of miRNAs and tRNA 3'trailers. PLoS One. 2010;5(5):e10563.
38. Goodarzi H, Liu X, Nguyen HCB, Zhang S, Fish L, Tavazoie SF. Endogenous tRNA-derived fragments suppress breast Cancer progression via YBX1 displacement. Cell. 2015;161(4):790–802.
39. Katayama, S.; Tomaru, Y.; Kasukawa, T.; Waki, K.; Nakanishi, M.; Nakamura, M.; Nishida, H.; Yap, C.C.; Suzuki, M.; Kawai, J. Antisense transcription in the mammalian transcriptome. *Science* **2005**, *309*, 1564–1566.
40. He, Y.; Vogelstein, B.; Velculescu, V.E.; Papadopoulos, N.; Kinzler, K.W. The antisense transcriptomes of human cells. *Science* **2008**, *322*, 1855–1857.
41. Hongay, C.F.; Grisafi, P.L.; Galitski, T.; Fink, G.R. Antisense transcription controls cell fate in Saccharomyces cerevisiae. *Cell* **2006**, *127*, 735–745.
42. Crooke, R.M. Antisense oligonucleotides as therapeutics for hyperlipidaemias. *Expert Opin. Biol. Ther.* **2005**, *5*, 907–917.
43. Luther, H.P. Role of endogenous antisense RNA in cardiac gene regulation. *J. Mol. Med.* **2005**, *83*, 26–32.

44. Ulanova, M.; Schreiber, A.D.; Befus, A.D. The future of antisense oligonucleotides in the treatment of respiratory diseases. *BioDrugs* **2006**, *20*, 1–11.
45. Popescu, F.D. Antisense- and RNA interference-based therapeutic strategies in allergy. *J. Cell. Mol. Med.* **2005**, *9*, 840–853.
46. Mattick, J.S.; Makunin, I.V. Non-coding RNA. *Hum. Mol. Genet.* **2006**, *15*, R17–R29.
47. Ferrari, N.; Seguin, R.; Renzi, P. Oligonucleotides: A multi-targeted approach for the treatment of respiratory diseases. *Future Med. Chem.* **2011**, *3*, 1647–1662.
48. Lu, X.; Yu, Q.; Binder, G.K.; Chen, Z.; Slepushkina, T.; Rossi, J.; Dropulic, B. Antisense-mediated inhibition of human immunodeficiency virus (HIV) replication by use of an HIV type 1-based vector results in severely attenuated mutant incapable of developing resistance. *J. Virol.* **2004**, *78*, 7079–7088.
49. DeVos, S.L.; Miller, T.M. Antisense oligonucleotides: Treating neurodegeneration at the level of RNA. *Neurotherapeutics* **2013**, *10*, 486–497.
50. Gao, Z.; Cooper, T.A. Antisense oligonucleotides: Rising stars in eliminating RNA toxicity in myotonic dystrophy. *Hum. Gene Ther.* **2013**, *24*, 499–507.
51. Derrien, T.; Johnson, R.; Bussotti, G.; Tanzer, A.; Djebali, S.; Tilgner, H.; Guernec, G.; Martin, D.; Merkel, A.; Knowles, D.G.; et al. The GENCODE v7 catalog of human long noncoding RNAs: Analysis of their gene structure, evolution, and expression. *Genome Res.* **2012**, *22*, 1775–1789.
52. Djebali, S.; Davis, C.A.; Merkel, A.; Dobin, A.; Lassmann, T.; Mortazavi, A.; Tanzer, A.; Lagarde, J.; Lin, W.; Schlesinger, F.; et al. Landscape of transcription in human cells. *Nature* **2012**, *489*, 101–108.
53. Kim, T.; Xu, Z.; Clauder-Munster, S.; Steinmetz, L.M.; Buratowski, S. Set3 HDAC mediates effects of overlapping noncoding transcription on gene induction kinetics. *Cell* **2012**, *150*, 1158–1169.
54. Carrozza, M.J.; Li, B.; Florens, L.; Suganuma, T.; Swanson, S.K.; Lee, K.K.; Shia, W.J.; Anderson, S.; Yates, J.; Washburn, M.P.; et al. Histone H3 methylation by Set2 directs deacetylation of coding regions by Rpd3S to suppress spurious intragenic transcription. *Cell* **2005**, *123*, 581–592.
55. Kaplan, C.D.; Laprade, L.; Winston, F. Transcription elongation factors repress transcription initiation from cryptic sites. *Science* **2003**, *301*, 1096–1099.
56. Whitehouse, I.; Rando, O.J.; Delrow, J.; Tsukiyama, T. Chromatin remodelling at promoters suppresses antisense transcription. *Nature* **2007**, *450*, 1031–1035.
57. Su, W.Y.; Li, J.T.; Cui, Y.; Hong, J.; Du, W.; Wang, Y.C.; Lin, Y.W.; Xiong, H.; Wang, J.L.; Kong, X.; et al. Bidirectional regulation between WDR83 and its natural antisense transcript DHPS in gastric cancer. *Cell Res.* **2012**, *22*, 1374–1389.
58. Wilkening, S.; Pelechano, V.; Järvelin, A.I.; Tekkedil, M.M.; Anders, S.; Benes, V.; Steinmetz, L.M. An efficient method for genome-wide polyadenylation site mapping and RNA quantification. *Nucleic Acids Res.* **2013**, *41*, 1–8.
59. Pelechano, V.; Wei, W.; Steinmetz, L.M. Extensive transcriptional heterogeneity revealed by isoform profiling. *Nature* **2013**, *497*, 127–131.
60. Guttman, M.; Rinn, J.L. Modular regulatory principles of large non-coding RNAs. *Nature* **2012**, *482*, 339–346.

61. Yap, K.L.; Li, S.; Muñoz-Cabello, A.M.; Raguz, S.; Zeng, L.; Mujtaba, S.; Gil, J.; Walsh, M.J.; Zhou, M.M. Molecular interplay of the noncoding RNA ANRIL and methylated histone H3 lysine 27 by polycomb CBX7 in transcriptional silencing of INK4a. *Mol. Cell* **2010**, *38*, 662–674.
62. Guttman, M.; Donaghey, J.; Carey, B.W.; Garber, M.; Grenier, J.K.; Munson, G.; Young, G.; Lucas, A.B.; Ach, R.; Bruhn, L.; et al. lincRNAs act in the circuitry controlling pluripotency and differentiation. *Nature* **2011**, *477*, 295–300.
63. Malik, S.; Durairaj, G.; Bhaumik, S.R. Mechanisms of antisense transcription initiation from the 3′-end of the GAL10 coding sequence in vivo. *Mol. Cell. Biol.* **2013**, *33*, 3549–3567.
64. Uprety, B.; Kaja, A.; Ferdoush, J.; Sen, R.; Bhaumik, S.R. Regulation of antisense transcription by NuA4 histone acetyltransferase and other chromatin regulatory factors. *Mol. Cell. Biol.* **2016**, *36*, 992–1006
65. Pelechano,V.;Steinmetz,L.M.Generegulationbyantisensetranscription.*NatRev.Genet.***2013**,*14*,880–893.
66. Tufarelli, C.; Stanley, J.A.; Garrick, D.; Sharpe, J.A.; Ayyub, H.; Wood, W.G.; Higgs, D.R. Transcription of antisense RNA leading to gene silencing and methylation as a novel cause of human genetic disease. *Nat. Genet.* **2003**, *34*, 157–165.
67. Lister,R.;Pelizzola,M.;Dowen,R.H.;Hawkins,R.D.;Hon,G.;Tonti-Filippini,J.;Nery,J.R.;Lee,L.;Ye,Z.; Ngo, Q.-M.; et al. Human DNA methylomes at base resolution show widespread epigenomic differences. *Nature* **2009**, *462*, 315–322.
68. Lyle, R.; Watanabe, D.; Vruchte, D.; Lerchner, W.; Smrzka, O.W.; Wutz, A.; Schageman, J.; Hahner, L.; Davies, C.; Barlow, D.P. The imprinted antisense RNA at the Igf2r locus overlaps but does not imprint Mas1. *Nat. Genet.* **2000**, *25*, 19–21.
69. Latos, P.; Pauler, F.M.; Koerner, M.V.; Ş‚energin, H.B.; Hudson , Q.J. Stocsits, R.R.; Allhoff, W.; Stricker, S.H.; Klement, R.M.; Warczok, K.E.; et al. Airn transcriptional overlap; but not its lncRNA products; induces imprinted Igf2r silencing. *Science* **2012**, *338*, 1469–1472.
70. Guil,S.;Esteller,M.cis acting noncoding RNAs: Friends and foes. *Nat.Struct.Mol.Biol.* **2012**,*19*,1068–1075.
71. Gupta, R.A.; Shah, N.; Wang, K.C.; Kim, J.; Horlings, H.M.; Wong, D.J.; Tsai, M.C.; Hung, T.; Argani, P.; Rinn, J.L.; et al. Long non-coding RNA HOTAIR reprograms chromatin state to promote cancer metastasis. *Nature* **2010**, *464*, 1071–1076.
72. Zhao,J.;Ohsumi,T.K.;Kung,J.T.;Ogawa,Y.;Grau,D.J.;Sarma,K.;Song,J.J.;Kingston,R.E.;Borowsky,M.; Lee, J.T. Genome-wide identification of polycomb-associated RNAs by RIP–seq. *Mol. Cell* **2010**, *40*, 939–953.
73. Houseley, J.; Rubbi, L.; Grunstein, M.; Tollervey, D.; Vogelauer, M. A ncRNA modulates histone modification and mRNA induction in the yeast GAL gene cluster. *Mol. Cell* **2008**, *32*, 685–695.
74. Pinskaya, M.; Gourvennec, S.; Morillon, A. H3 lysine 4 di- and tri-methylation deposited by cryptic transcription attenuates promoter activation. *EMBO J.* **2009**, *28*, 1697–1707.

75. Martianov, I.; Ramadass, A.; Serra Barros, A.; Chow, N.; Akoulitchev, A. Repression of the human dihydrofolate reductase gene by a non-coding interfering transcript. *Nature* **2007**, *445*, 666–670.
76. Belotserkovskii,B.P.;DeSilva,E.;Tornaletti,S.;Wang,G.;Vasquez,K.M.;Hanawalt,P.C.Atriplex-forming sequence from the human c-MYC promoter interferes with DNA transcription. *J. Biol. Chem.* **2007**, *282*, 32433–32441.
77. Schmitz,K.M.;Mayer,C.;Postepska,A.;Grummt,I.InteractionofnoncodingRNAwiththerDNApromoter mediates recruitment of DNMT3b and silencing of rRNA genes. *Genes Dev.* **2010**, *24*, 2264–2269.
78. Yu,W.;Gius,D.;Onyango,P.; Muldoon-Jacobs, K.; Karp,J.; Feinberg,A.P.; Cui,H. Epigenetic silencing of tumour suppressor gene p15 by its antisense RNA. *Nature* **2008**, *451*, 202–206.
79. Matsui, M.; Chu, Y.; Zhang, H.; Gagnon, K.T.; Shaikh, S.; Kuchimanchi, S.; Manoharan, M.; Corey, D.R.; Janowski, B.A. Promoter RNA links transcriptional regulation of inflammatory pathway genes. *Nucleic Acids Res.* **2013**, *41*, 10086–10109.
80. O'Leary,V.B.;Ovsepian,S.V.;Carrascosa,L.G.;Buske,F.A.;Radulovic,V.;Niyazi, M.;Moertl,S.;Trau, M.; Atkinson, M.J.; Anastasov, N. PARTICLE, a triplex-forming long ncRNA, regulates locus-specific methylation in response to low-dose irradiation. *Cell Rep.* **2015**, *11*, 474–485.
81. Postepska-Igielska, A.; Giwojna, A.; GasriPlotnitsky, L.; Schmitt, N.; Dold, A.; Ginsberg, D.; Grummt, I. LncRNA Khps1 regulates expression of the proto-oncogene SPHK1 via triplex-mediated changes in chromatin structure. *Mol. Cell* **2015**, *60*, 626–636.
82. Zhang,X.; Li, H.; Burnett, J.C.; Rossi, J.J. The role of antisense long non-coding RNA in small RNA-triggered gene activation. *RNA* **2014**, *20*, 1916–1928.
83. Xue, Z.; Ye, Q.; Anson, S.R.; Yang, J.; Xiao, G.; Kowbel, D.; Glass, N.L.; Crosthwaite, S.K.; Liu, Y. Transcriptional interference by antisense RNA is required for circadian clock function. *Nature* **2014**, *514*, 650–653.
84. Stojic, L.; Niemczyk, M.; Orjalo, A.; Ito, Y.; Ruijter, A.E.; Uribe-Lewis, S.; Joseph, N.; Weston, S.; Menon, S.; Odom, D.T.; et al. Transcriptional silencing of long noncoding RNA GNG12-AS1 uncouples its transcriptional and product-related functions. *Nat. Commun.* **2016**, *7*, 1–14.
85. Munroe, S.H.; Lazar, M.A. Inhibition of cerb A mRNA splicing by a naturally occurring antisense RNA. *J. Biol. Chem.* **1991**, *266*, 22083–22086.
86. Hastings, M.L.; Ingle, H.A.; Lazar, M.A.; Munroe, S.H. Post-transcriptional regulation of thyroid hormone receptor expression by cis-acting sequences and a naturally occurring antisense RNA. *J. Biol. Chem.* **2000**, *75*, 11507–11513.
87. Beltran, M.; Puig, I.; Pena, C.; Garcia, J.M.; Alvarez, A.B.; Pena, R.; Bonilla, F.; deHerreros, A.G. A natural antisense transcript regulates Zeb2/Sip1 gene expression during Snail1-induced epithelial-mesenchymal transition. *Genes Dev.* **2008**, *22*, 756–769.
88. Faghihi, M.A.; Modarresi, F.; Khalil, A.M.; Wood, D.E.; Sahagan, B.G.; Morgan, T.E.; Finch, C.E.; St Laurent III, G.; Kenny, P.J.; Wahlestedt, C. Expression of a noncoding RNA is elevated in Alzheimer's disease and drives rapid feed-forward regulation of beta-secretase. *Nat. Med.* **2008**, *14*, 723–730.

89. Carrieri, C.; Cimatti, L.; Biagioli, M.; Beugnet, A.; Zucchelli, S.; Fedele, S.; Pesce, E.; Ferrer, I.; Collavin, L.; Santoro, C.; et al. Long non-coding antisense RNA controls Uchl1 translation through an embedded SINEB2 repeat. *Nature* **2012**, *491*, 454–457.
90. Piatek, M.J.; Henderson, V.; Fearn, A.; Chaudhry, B.; Werner, A. Ectopically expressed Slc34a2a sense-antisense transcripts cause a cerebellar phenotype in zebrafish embryos depending on RNA complementarity and Dicer. *PLoS ONE.* **2017**, *12*, 1–17.
91. Tam, O.H.; Aravin, A.A.; Stein, P.; Girard, A.; Murchison, E.P.; Cheloufi, S.; Hodges, E.; Anger, M.; Sachidanandam, R.; Schultz, R.M.; et al. Pseudogene-derived small interfering RNAs regulate gene expression in mouse oocytes. *Nature* **2008**, *453*, 534–538.
92. Eckstein F. Developments in RNA chemistry, a personal view. Biochimie. 2002; 84:841–848.
93. Geary RS, Yu RZ, Levin AA. Pharmacokinetics of phosphorothioate antisense oligodeoxynucleotides. Curr Opin Investig Drugs. 2001; 2:562–573.
94. Watanabe TA, Geary RS, Levin AA. Plasma protein binding of an antisense oligonucleotide targeting human ICAM-1 (ISIS 2302). Oligonucleotides. 2006; 16:169–180.
95. Stein CA, Krieg AM. Problems in interpretation of data derived from in vitro and in vivo use of antisense oligodeoxynucleotides. Antisense Res Dev. 1994; 4:67–69.
96. While direct binding to proteins is generally considered an unwanted off-target effect in the gene silencing field it has spawned a field of its own. See for example Thiel KW, Giangrande PH. Therapeutic applications of DNA and RNA aptamers. Oligonucleotides. 2009; 19:209–222.
97. Chiang MY, Chan H, Zounes MA, et al. Antisense oligonucleotides inhibit intercellular adhesion molecule 1 expression by two distinct mechanisms. J Biol Chem. 1991; 266:18162–18171.
98. Williams DM, Benseler F, Eckstein F. Properties of 2′-fluorothymidine-containing oligonucleotides: interaction with restriction endonuclease EcoRV. Biochemistry. 1991; 30:4001– 4009.
99. Pieken WA, Olsen DB, Aurup H, et al. Structure-function relationship of hammerhead ribozymes as probed by 2′-modifications. Nucleic Acids Symp Ser. 1991:51–53.]
100. Kawasaki AM, Casper MD, Freier SM, et al. Uniformly modified 2′-deoxy-2′-fluoro- phosphorothioate oligonucleotides as nuclease-resistant antisense compounds with high affinity and specificity for RNA targets. J Med Chem. 1993; 36:831–841.
101. Monia BP, Lesnik EA, Gonzalez C, et al. Evaluation of 2′-modified oligonucleotides containing 2′- deoxy gaps as antisense inhibitors of gene expression. J Biol Chem. 1993; 268:14514–14522.
102. Freier SM, Altmann KH. The ups and downs of nucleic acid duplex stability: structure-stability studies on chemically modified DNA:RNA duplexes. Nucl Acids Res. 1997; 25:4429–4443.
103. Martin P. A New Access to 2′-O-Alkylated Ribonucleosides and Properties of 2′-O-Alkylated Oligoribonucleotides. Helv Chim Acta. 1995; 78:486–504.

104. Koshkin AA, Singh SK, Nielsen P, et al. LNA (Locked Nucleic Acids): Synthesis of the adenine, cytosine, guanine, 5-methylcytosine, thymine and uracil bicyclonucleoside monomers, oligomerisation, and unprecedented nucleic acid recognition. Tetrahedron. 1998; 54:3607–3630.
105. Obika S, Nanbu D, Hari Y, et al. Stability and structural features of the duplexes containing nucleoside analogues with a fixed N-type conformation, 2′-O,4′-C-methyleneribonucleosides. Tetrahedron Lett. 1998; 39:5401–5404.
106. Braasch DA, Corey DR. Locked nucleic acid (LNA): Fine-tuning the recognition of DNA and RNA. Chem Biol. 2001; 8:1–7.
107. Leumann CJ. DNA analogues: From supramolecular principles to biological properties. Bioorg Med Chem. 2002; 10:841–854.
108. Obika S. Development of bridged nucleic acid analogues for antigene technology. Chem Pharm Bull (Tokyo). 2004; 52:1399–1404.
109. Christensen NK, Petersen M, Nielsen P, et al. A Novel Class of Oligonucleotide Analogues Containing 2′-O,3′-C-Linked [3.2. 0]Bicyclo-arabino-nucleoside Monomers: Synthesis, Thermal Affinity Studies, and Molecular Modeling. J Am Chem Soc. 1998; 120:5458–5463.
110. Rajwanshi VK, Hakansson AE, Sorensen MD, et al. The Eight Stereoisomers of LNA (Locked Nucleic Acid): A Remarkable Family of Strong RNA Binding Molecules. Angew Chem Int Ed Engl. 2000; 39:1656–1659.
111. Morita K, Hasegawa C, Kaneko M, et al. 2′-O,4′-C-ethylene-bridged nucleic acids (ENA): highly nuclease-resistant and thermodynamically stable oligonucleotides for antisense drug. Bioorg Med Chem Lett. 2002; 12:73–76.
112. Wu H, Lima WF, Zhang H, et al. Determination of the Role of the Human RNase H1 in the Pharmacology of DNA-like Antisense Drugs. J Biol Chem. 2004; 279:17181–17189.
113. Minshull J, Hunt T. The use of single-stranded DNA and RNase H to promote quantitative ‘hybrid arrest of translation’ of mRNA/DNA Hybrids in reticulocyte lysate cell-free translations. Nucl Acids Res. 1986; 14:6433–6451.
114. Nakamura H, Oda Y, Iwai S, et al. How does RNase H recognize a DNA.RNA hybrid? Proc Natl Acad Sci U S A. 1991; 88:11535–11539.
115. Wahlestedt C, Salmi P, Good L, et al. Potent and nontoxic antisense oligonucleotides containing locked nucleic acids. Proc Natl Acad Sci U S A. 2000; 97:5633–5638.
116. Damha MJ, Wilds CJ, Noronha A, et al. Hybrids of RNA and arabinonucleic acids (ANA and 2′F- ANA) are substrates of Ribonuclease H. J Am Chem Soc. 1998; 120:12976–12977.
117. Watts JK, Damha MJ. 2′F-Arabinonucleic acids (2′F-ANA) -- History, properties, and new frontiers. Can J Chem. 2008; 86:641–656.
118. Nielsen PE, Egholm M, Berg RH, et al. Sequence-Selective Recognition of DNA by Strand Displacement with a Thymine-Substituted Polyamide. Science. 1991; 254:1497–1500.
119. Bentin T, Nielsen PE. Enhanced peptide nucleic acid binding to supercoiled DNA: possible implications for DNA “breathing” dynamics. Biochemistry. 1996; 35:8863–8869.]

120. Smulevitch SV, Simmons CG, Norton JC, et al. Enhancement of strand invasion by oligonucleotides through manipulation of backbone charge. Nat Biotech. 1996; 14:1700–1704.
121. Summerton J, Weller D. Morpholino antisense oligomers: design, preparation, and properties. Antisense Nucleic Acid Drug Dev. 1997; 7:187–195.
122. Corey DR, Abrams JM. Morpholino antisense oligonucleotides: tools for investigating vertebrate development. Genome Biol. 2001; 2:REVIEWS1015.
123. Bill BR, Petzold AM, Clark KJ, et al. A primer for morpholino use in zebrafish. Zebrafish. 2009; 6:69–77.
124. Pan Q, Shai O, Lee LJ, et al. Deep surveying of alternative splicing complexity in the human transcriptome by high-throughput sequencing. Nat Genet. 2008; 40:1413–1415.
125. Aartsma-Rus A, van Ommen GJ. Antisense-mediated exon skipping: a versatile tool with therapeutic and research applications. RNA. 2007; 13:1609–1624.
126. Bauman J, Jearawiriyapaisarn N, Kole R. Therapeutic potential of splice-switching oligonucleotides. Oligonucleotides. 2009; 19:1–13.
127. Dominski Z, Kole R. Restoration of correct splicing in thalassemic pre-mRNA by antisense oligonucleotides. Proc Natl Acad Sci U S A. 1993; 90:8673–8677.
128. Hua Y, Sahashi K, Hung G, et al. Antisense correction of SMN2 splicing in the CNS rescues necrosis in a type III SMA mouse model. Genes Dev. 2010; 24:1634–1644.
129. Goemans NM, Tulinius M, van den Akker JT, et al. Systemic administration of PRO051 in Duchenne's muscular dystrophy. N Engl J Med. 2011; 364:1513–1522.
130. Aartsma-Rus A, van Ommen GJ. Less is more: therapeutic exon skipping for Duchenne muscular dystrophy. Lancet Neurol. 2009; 8:873–875.]
131. Alter J, Lou F, Rabinowitz A, et al. Systemic delivery of morpholino oligonucleotide restores dystrophin expression body-wide and improves dystrophic pathology. Nat Med. 2006; 12:175–177.
132. Suwanmanee T, Sierakowska H, Fucharoen S, et al. Repair of a splicing defect in erythroid cells from patients with beta-thalassemia/HbE disorder. Mol Ther. 2002; 6:718–726.
133. T, Sierakowska H, Lacerra G, et al. Restoration of human beta-globin gene expression in murine and human IVS2-654 thalassemic erythroid cells by free uptake of antisense oligonucleotides. Mol Pharmacol. 2002; 62:545–553.
134. Wan J, Sazani P, Kole R. Modification of HER2 pre-mRNA alternative splicing and its effects on breast cancer cells. Int J Cancer. 2009; 124:772–777.
135. Bauman JA, Li SD, Yang A, et al. Anti-tumor activity of splice-switching oligonucleotides. Nucl Acids Res. 2010; 38:8348–8356.
136. Garzon R, Marcucci G, Croce CM. Targeting microRNAs in cancer: rationale, strategies and challenges. Nat Rev Drug Discov. 2010; 9:775–789.
137. Soifer HS, Rossi JJ, Saetrom P. MicroRNAs in disease and potential therapeutic applications. Mol Ther. 2007; 15:2070–2079.
138. Lu M, Zhang Q, Deng M, et al. An analysis of human microRNA and disease associations. PLoS One. 2008; 3:e3420.]
139. Jiang Q, Wang Y, Hao Y, et al. miR2Disease: a manually curated database for microRNA deregulation in human disease. Nucl Acids Res. 2009; 37:D98–104.

140. Esau CC. Inhibition of microRNA with antisense oligonucleotides. Methods. 2008; 44:55–60.
141. Davis S, Propp S, Freier SM, et al. Potent inhibition of microRNA in vivo without degradation. Nucl Acids Res. 2009; 37:70–77.
142. Montgomery RL, van Rooij E. Therapeutic advances in MicroRNA targeting. J Cardiovasc Pharmacol. 2011; 57:1–7.]
143. Girard M, Jacquemin E, Munnich A, et al. miR-122, a paradigm for the role of microRNAs in the liver. J Hepatol. 2008; 48:648–656.
144. Krutzfeldt J, Rajewsky N, Braich R, et al. Silencing of microRNAs in vivo with 'antagomirs'. Nature. 2005:438.
145. Esau C, Davis S, Murray SF, et al. miR-122 regulation of lipid metabolism revealed by in vivo antisense targeting. Cell Metabolism. 2006; 3:87–98.
146. Elmen J, Lindow M, Schutz S, et al. LNA-mediated microRNA silencing in non-human primates. Nature. 2008; 452:896–U810.
147. Lanford RE, Hildebrandt-Eriksen ES, Petri A, et al. Therapeutic silencing of microRNA-122 in primates with chronic hepatitis C virus infection. Science. 2010; 327:198–201.
148. Fabani MM, Gait MJ. miR-122 targeting with LNA/2′-O-methyl oligonucleotide mixers, peptide nucleic acids (PNA), and PNA-peptide conjugates. RNA. 2008; 14:336–346.
149. Orom UA, Kauppinen S, Lund AH. LNA-modified oligonucleotides mediate specific inhibition of microRNA function. Gene. 2006; 372:137–141.
150. Meister G, Landthaler M, Dorsett Y, et al. Sequence-specific inhibition of microRNA- and siRNA- induced RNA silencing. RNA. 2004; 10:544–550.
151. Kloosterman WP, Lagendijk AK, Ketting RF, et al. Targeted inhibition of miRNA maturation with morpholinos reveals a role for miR-375 in pancreatic islet development. PLoS Biol. 2007; 5:e203.
152. Watts JK, Deleavey GF, Damha MJ. Chemically modified siRNA: tools and applications. Drug Discov Today. 2008; 13:842–855.
153. Morrissey DV, Lockridge JA, Shaw L, et al. Potent and persistent in vivo anti-HBV activity of chemically modified siRNAs. Nat Biotech. 2005; 23:1002–1007.
154. Morrissey DV, Blanchard K, Shaw L, et al. Activity of stabilized short interfering RNA in a mouse model of hepatitis B virus replication. Hepatology. 2005; 41:1349–1356.
155. Allerson CR, Sioufi N, Jarres R, et al. Fully 2′-Modified Oligonucleotide Duplexes with Improved in Vitro Potency and Stability Compared to Unmodified Small Interfering RNA. J Med Chem. 2005; 48:901–904.
156. Braasch DA, Paroo Z, Constantinescu A, et al. Biodistribution of phosphodiester and phosphorothioate siRNA. Bioorg Med Chem Lett. 2004; 14:1139–1143
157. Koller E, Propp S, Murray H, et al. Competition for RISC binding predicts in vitro potency of siRNA. Nucl Acids Res. 2006; 34:4467–4476.
158. Deleavey GF, Watts JK, Alain T, et al. Synergistic effects between analogs of DNA and RNA improve the potency of siRNA-mediated gene silencing. Nucl Acids Res. 2010; 38:4547–4557.
159. Marques JT, Williams BRG. Activation of the mammalian immune system by siRNAs. Nat Biotech. 2005; 23:1399–1405.

160. Robbins M, Judge A, MacLachlan I. siRNA and Innate Immunity. Oligonucleotides. 2009; 19:89– 102.
161. Olejniczak M, Galka P, Krzyzosiak WJ. Sequence-non-specific effects of RNA interference triggers and microRNA regulators. Nucl Acids Res. 2010; 38:1–16.
162. Kortylewski M, Swiderski P, Herrmann A, et al. In vivo delivery of siRNA to immune cells by conjugation to a TLR9 agonist enhances antitumor immune responses. Nat Biotechnol. 2009; 27:925–932.
163. Gantier MP, Tong S, Behlke MA, et al. Rational design of immunostimulatory siRNAs. Mol Ther. 2010; 18:785–795.]
164. Khairuddin N, Gantier MP, Blake SJ, et al. siRNA-induced immunostimulation through TLR7 promotes antitumoral activity against HPV-driven tumors in vivo. Immunol Cell Biol. 2011 in press. 10.1038/icb.2011.1019
165. Judge A, MacLachlan I. Overcoming the innate immune response to small interfering RNA. Hum Gene Ther. 2008; 19:111–124.
166. Judge AD, Bola G, Lee ACH, et al. Design of Noninflammatory Synthetic siRNA Mediating Potent Gene Silencing in Vivo. Mol Ther. 2006; 13:494–505.
167. Corey DR. Chemical modification: the key to clinical application of RNA interference? J Clin Invest. 2007; 117:3615–3622.
168. Watts JK, Corey DR. Clinical status of duplex RNA. Bioorg Med Chem Lett. 2010; 20:3203–3207.
169. Kim DH, Behlke MA, Rose SD, et al. Synthetic dsRNA Dicer substrates enhance RNAi potency and efficacy. Nat Biotech. 2005; 23:222–226.
170. Collingwood MA, Rose SD, Huang L, et al. Chemical modification patterns compatible with high potency dicer-substrate small interfering RNAs. Oligonucleotides. 2008; 18:187–200.
171. Yuan B, Latek R, Hossbach M, et al. siRNA Selection Server: an automated siRNA oligonucleotide prediction server. Nucl Acids Res. 2004; 32:W130–134.
172. Boese Q, Leake D, Reynolds A, et al. Mechanistic insights aid computational short interfering RNA design. Methods Enzymol. 2005; 392:73–96.
173. Huesken D, Lange J, Mickanin C, et al. Design of a genome-wide siRNA library using an artificial neural network. Nat Biotech. 2005; 23:995–1001.
174. Sczakiel G, Homann M, Rittner K. Computer-aided search for effective antisense RNA target sequences of the human immunodeficiency virus type 1. Antisense Res Dev. 1993; 3:45–52.
175. Stull RA, Taylor LA, Szoka FC Jr. Predicting antisense oligonucleotide inhibitory efficacy: a computational approach using histograms and thermodynamic indices. Nucl Acids Res. 1992; 20:3501–3508
176. Sczakiel G. Theoretical and experimental approaches to design effective antisense oligonucleotides. Front Biosci. 2000; 5:D194–201.]
177. Rittner K, Burmester C, Sczakiel G. In vitro selection of fast-hybridizing and effective antisense RNAs directed against the human immunodeficiency virus type 1. Nucl Acids Res. 1993; 21:1381–1387.
178. Ho SP, Bao Y, Lesher T, et al. Mapping of RNA accessible sites for antisense experiments with oligonucleotide libraries. Nat Biotechnol. 1998; 16:59–63.
179. Kronenwett R, Haas R, Sczakiel G. Kinetic selectivity of complementary nucleic acids: bcr-abl- directed antisense RNA and ribozymes. J Mol Biol. 1996; 259:632–644.

180. Milner N, Mir KU, Southern EM. Selecting effective antisense reagents on combinatorial oligonucleotide arrays. Nat Biotechnol. 1997; 15:537–541.
181. Rudnick SI, Swaminathan J, Sumaroka M, et al. Effects of local mRNA structure on posttranscriptional gene silencing. Proc Natl Acad Sci U S A. 2008; 105:13787–13792.
182. Bacon TA, Wickstrom E. Walking along human c-myc mRNA with antisense oligodeoxynucleotides: maximum efficacy at the 5′ cap region. Oncogene Res. 1991; 6:13–19.
183. Stewart AJ, Canitrot Y, Baracchini E, et al. Reduction of expression of the multidrug resistance protein (MRP) in human tumor cells by antisense phosphorothioate oligonucleotides. Biochem Pharmacol. 1996; 51:461–469.
184. Bergan R, Hakim F, Schwartz GN, et al. Electroporation of synthetic oligodeoxynucleotides: a novel technique for ex vivo bone marrow purging. Blood. 1996; 88:731–741.
185. Spiller DG, Giles RV, Grzybowski J, et al. Improving the intracellular delivery and molecular efficacy of antisense oligonucleotides in chronic myeloid leukemia cells: a comparison of streptolysin-O permeabilization, electroporation, and lipophilic conjugation. Blood. 1998; 91:4738–4746.
186. Li, S., editor. Electroporation Protocols: Preclinical and Clinical Gene Medicine (Methods in Molecular Biology. Vol. 423. Humana Press; New York: 2008.
187. Stein CA, Hansen JB, Lai J, et al. Efficient gene silencing by delivery of locked nucleic acid antisense oligonucleotides, unassisted by transfection reagents. Nucl Acids Res. 2009
188. Zhang Y, Qu Z, Kim S, et al. Down-modulation of cancer targets using locked nucleic acid (LNA)-based antisense oligonucleotides without transfection. Gene Ther. 18:326–333.
189. Wang J, Lu Z, Wientjes MG, et al. Delivery of siRNA therapeutics: barriers and carriers. Aaps J. 2010; 12:492–503.
190. Juliano R, Alam MR, Dixit V, et al. Mechanisms and strategies for effective delivery of antisense and siRNA oligonucleotides. Nucl Acids Res. 2008; 36:4158–4171.
191. Levin AA. A review of issues in the pharmacokinetics and toxicology of phosphorothioate antisense oligonucleotides. Biochim Biophys Acta. 1999; 1489:69–84.
192. Koller E, Vincent TM, Chappell A, et al. Mechanisms of single-stranded phosphorothioate modified antisense oligonucleotide accumulation in hepatocytes. Nucl Acids Res. 2011; 39:4795–4807.
193. Whitehead KA, Langer R, Anderson DG. Knocking down barriers: advances in siRNA delivery. Nat Rev Drug Discov. 2009; 8:129–138.
194. Grimm D, Kay MA. Therapeutic application of RNAi: is mRNA targeting finally ready for prime time? J Clin Invest. 2007; 117:3633–3641.
195. De Paula D, Bentley MVLB, Mahato RI. Hydro-phobization and bioconjugation for enhanced siRNA delivery and targeting. RNA. 2007; 13:431–456.
196. Akhtar S, Benter IF. Nonviral delivery of synthetic siRNAs *in vivo*. J Clin Invest. 2007; 117:3623–3632.
197. Li W, Szoka FC Jr. Lipid-based Nanoparticles for Nucleic Acid Delivery. Pharm Res. 2007; 24:438–449.

198. Tseng YC, Mozumdar S, Huang L. Lipid-based systemic delivery of siRNA. Adv Drug Deliv Rev. 2009; 61:721–731.]
199. Eguchi A, Meade BR, Chang YC, et al. Efficient siRNA delivery into primary cells by a peptide transduction domain-dsRNA binding domain fusion protein. Nat Biotech. 2009; 27:567–571.
200. Papahadjopoulos D, Allen TM, Gabizon A, et al. Sterically stabilized liposomes: improvements in pharmacokinetics and antitumor therapeutic efficacy. Proc Natl Acad Sci U S A. 1991; 88:11460–11464.]
201. Davis ME, Zuckerman JE, Choi CHJ, et al. Evidence of RNAi in humans from systemically administered siRNA via targeted nanoparticles. Nature. 2010; 464:1067–1070.
202. Soutschek J, Akinc A, Bramlage B, et al. Therapeutic silencing of an endogenous gene by systemic administration of modified siRNAs. Nature. 2004; 432:173–178.
203. Semple SC, Akinc A, Chen J, et al. Rational design of cationic lipids for siRNA delivery. Nat Biotech. 2010; 28:172–176.
204. Dassie JP, Liu XY, Thomas GS, et al. Systemic administration of optimized aptamer-siRNA chimeras promotes regression of PSMA-expressing tumors. Nat Biotech. 2009; 27:839–849.
205. Song E, Zhu P, Lee SK, et al. Antibody mediated in vivo delivery of small interfering RNAs via cell-surface receptors. Nat Biotech. 2005; 23:709–717.
206. Sato Y, Murase K, Kato J, et al. Resolution of liver cirrhosis using vitamin A–coupled liposomes to deliver siRNA against a collagen-specific chaperone. Nat Biotech. 2008; 26:431–442.
207. Kim SS, Ye C, Kumar P, et al. Targeted delivery of siRNA to macrophages for anti-inflammatory treatment. Mol Ther. 2010; 18:993–1001.
208. Bennett CF, Swayze EE. RNA targeting therapeutics: molecular mechanisms of antisense oligonucleotides as a therapeutic platform. Annu Rev Pharmacol Toxicol. 2010; 50:259–293.
209. Gagnon KT. HD Therapeutics - CHDI Fifth Annual Conference (Meeting Report). IDrugs. 2010; 13:219–223.
210. Kunze, D.; Kraemer, K.; Fuessel, S. Antisense Oligonucleotides: Insights from Preclinical Studies and Clinical Trials. In: Erdmann, VA.; Barciszewski, J., editors. RNA Technologies and Their Applications. Springer-Verlag; Berlin: 2010. p. 285-303.
211. Haussecker D. The Business of RNAi Therapeutics. Hum Gene Ther. 2008; 19:451–462.
212. Malik, S.; Bhaumik, S.R. Rad26p,a transcription-coupled repair factor,promotes the eviction and prevents the reassociation of histone H2A-H2B dimer during transcriptional elongation *in vivo. Biochemistry* **2012**, *51*, 5873–5875.
213. Malik, S.; Bhaumik, S.R. Regulation of active genome integrity and expression. *Nucleus* **2014**, *5*, 520–526.
214. Chaurasia, P.; Sen ,R.; Pandita, T.K.; Bhaumik, S.R. Preferential DNA double-strand break repair at the active gene. *J. Biol. Chem.* **2012**, *287*, 36414–36422.
215. Chaurasia, P.; Sen, R. ;Bhaumik, S.R. Functional analysis of Rad14p, a DNA damage recognition factor in nucleotide excision repair, in regulation of transcription in vivo. *J. Biol. Chem.* **2013**, *288*, 793–806.

216. Sancar, A.; Lindsey-Boltz, L.A.; Unsal-Kacmaz, K.; Linn, S. Molecular mechanisms of mammalian DNA repair and the DNA damage checkpoints. *Annu. Rev. Biochem.* **2004**, *73*, 39–85.
217. Wang, X.; Arai, S.; Song, X.; Reichart, D.; Du, K.; Pascual, G.; Tempst, P.; Rosenfeld, M.G.; Glass, C.K.; Kurokawa, R. Induced ncRNAs allosterically modify RNA-binding proteins in cis to inhibit transcription. *Nature* **2008**, *454*, 126–130.
218. Wang, X.T.; Song, X.Y.; Glass, C.K.; Rosenfeld, M.G. The Long Arm of Long Noncoding RNAs: Roles as Sensors Regulating Gene Transcriptional Programs. *Cold Spring Harb. Perspect. Biol.* **2011**, *3*, 1–14.
219. Huarte, M.; Guttman, M.; Feldser, D.; Garber, M.; Koziol, M.J.; Kenzelmann-Broz, D.; Khalil, A.M.; Zuk, O.; Amit, I.; Rabani, M.; et al. A Large Intergenic Noncoding RNA Induced by p53 Mediates Global Gene Repression in the p53 Response. *Cell* **2010**, *142*, 409–419.
220. Congrains, A.; Kamide, K.; Ohishi, M.; Rakugi, H. ANRIL: Molecular mechanisms and implications in human health. *Int. J. Mol. Sci.* **2013**, *14*, 1278–1292.
221. Wan, G.; Mathur, R.; Hu, X.; Liu, Y.; Zhang, X.; Peng, G.; Lu, X. Long non-coding RNA ANRIL(CDKN2B-AS) is induced by the ATM-E2F1 signaling pathway. *Cell Signal* **2013**, *25*, 1086–1095.
222. Qi, W.; Wang, R.; Chen, H.; Wang, X.; Xiao, T.; Boldogh, I.; Ba, X.; Han, L.; Zeng, X. BRG1 promotes the repair of DNA double-strand breaks by facilitating the placement of RPA with RAD51. *J. Cell Sci.* **2015**, *128*, 317–330.
223. Cajigas, I.; Leib, D.E.; Cochrane, J.; Luo, H.; Swyter, K.R.; Chen, S.; Clark, B.S.; Thompson, J.; Yates, J.R.; Kingston, R.E.; et al. Evf2 lncRNA/BRG1/DLX1 interactions reveal RNA-dependent inhibition of chromatin remodeling. *Development* **2015**, *142*, 2641–2652.
224. Kwon, S.J.; Park, J.H.; Park, E.J.; Lee, S.A.; Lee, H.S.; Kang, S.W.; Kwon, J.A TM-mediated phosphorylation of the chromatin remodeling enzyme BRG1 modulates DNA double-strand break repair. *Oncogene* **2015**, *34*, 303–313.
225. Hung, T.; Wang, Y.; Lin, M.F.; Koegel, A.K.; Kotake, Y.; Grant, G.D.; Horlings, H.M.; Shah, N.; Umbricht, C.; Wang, P.; et al. Extensive and coordinated transcription of noncoding RNAs within cell-cycle promoters. *Nat. Genet.* **2011**, *43*, 621–629.
226. Sharma, V.; Misteli, T. Noncoding RNAs in DNA damage and repair. *FEBS Lett.* **2013**, *587*, 1832–1839.
227. Liu, Y.; Lu, X. Non-coding RNAs in DNA damage response. *Am .J. Cancer Res.* **2012**, *2*, 658–675.
228. Dianatpour, A.; Ghafouri-Fard, S. The role of Long non-coding RNAs in the repair of DNA double strand breaks. *Int. J. Mol. Cell Med.* **2017**, *6*, 1–12.
229. Gupta, R.A.; Shah, N.; Wang, K.C.; Kim, J.; Horlings, H.M.; Wong, D.J.; Tsai, M.C.; Hung, T.; Argani, P.; Rinn, J.L.; et al. Long non-coding RNA HOTAIR reprograms chromatin state to promote cancer metastasis. *Nature* **2010**, *464*, 1071–1076.
230. Rinn, R.L.; Kertesz, M.; Wang, J.K.; Squazzo, S.L.; Xu, X.; Brugmann, S.A.; Goodnough, L.H.; Helms, J.A.; Farnham, P.J.; Segal, E.; et al. Functional

demarcation of active and silent chromatin domains in human HOX loci by noncoding RNAs. *Cell* **2007**, *129*, 1311–1323.

231. Avazpour, N.; Hajjari, M.; Birgani, M.T. HOTAIR:A Promising Long Non-coding RNA with Potential Role in Breast Invasive Carcinoma. *Front. Genet.* **2017**, *8*, 170.
232. Kogo, R.; Shimamura, T.; Mimori, K.; Kawahara, K.; Imoto, S.; Sudo, T.; Tanaka, F.; Shibata, K.; Suzuki, A.; Komune, S.; et al. Long noncoding RNA HOTAIR regulates polycomb dependent chromatin modification and is associated with poor prognosis in colorectal cancers. *Cancer Res.* **2011**, *71*, 6320–6326.
233. Geng, Y.;Xie, S.; Li, Q.; Ma, J.; Wang, G. Large intervening non-coding RNA HOTAIR is associated with hepatocellular carcinoma progression. *J. Int. Med. Res.* **2011**, *39*, 2119–2128.
234. Kim, K.; Jutooru, I.; Chadalapaka, G.; Johnson, G.; Frank, J.; Burghardt, R.; Kim, S.; Safe, S. HOTAIR is a negative prognostic factor and exhibits pro-oncogenic activity in pancreatic cancer. *Oncogene* **2013**, *32*, 1616–1625.]
235. Alahari, S.V.; Eastlack, S.C.; Alahari, S.K. Role of long non coding RNAs in Neoplasia: Special emphasis on prostate cancer. *Int. Rev. Cell Mol. Biol.* **2016**, *324*, 229–254.
236. Pan, Y.; Pan, Y.; Cheng, Y.; Yang, F.; Yao, Z.; Wang, O. Knockdown of LncRNA MAPT-AS1 inhibits proliferation and migration and sensitizes cancer cells to paclitaxel by regulating MAPT expression in ER-negative breast cancers. *Cell Biosci.* **2018**, *8*, 1–13.
237. Mahmoudi, S.; Henriksson, S.; Farnebo, L.; Roberg, K.; Farnebo, M. WRAP53 promotes cancer cell survival and is a potential target for cancer therapy. *Cell Death Dis.* **2011**, *2*, e114.
238. Fang, Y.; Wang, J.; Wu, F.; Song, Y.; Zhao, S.; Zhang, Q. Long non-coding RNA HOXA-AS2 promotes proliferation and invasion of breast cancer by acting as a miR-520c-3p sponge. *Oncotarget* **2017**, *8*, 46090–46103.
239. Liu, Z.; Chen, Z.; Fan, R.; Jiang, B.; Chen, X.; Chen, Q.; Nie, F.; Lu, K.; Sun, M. Over-expressed long noncoding RNA HOXA11-AS promotes cell cycle progression and metastasis in gastric cancer. *Mol. Cancer* **2017**, *16*, 1–9.
240. Ma, P.; Xu, T.; Huang, M.; Shu, Y. Increased expression of LncRNA PANDAR predicts a poor prognosis in gastric cancer. *Biomed. Pharmacother.* **2016**, *78*, 172–176.
241. Sang, Y.; Tang, J.; Li, S.; Tang, X.; Cheng, C.; Luo, Y.; Qian, X.; Deng, L.; Liu, L.; Lv, X. Lnc RNA PANDAR regulates the G1/S transition of breast cancer cells by suppressing p16INK4A expression. *Sci. Rep.* **2016**, *6*, 1–10.
242. Han, L.; Zhang, E.; Yin, D.; Kong, R.; Xu, T.; Chen, W.; Xia, R.; Shu, Y.; De, W. Low expression of long noncoding RNA PANDAR predicts a poor prognosis of non-small cell lung cancer and affects cell apoptosis by regulating Bcl-2. *Cell Death Dis.* **2015**, *6*, 1–12.
243. Zhang, J.J.; Wang, D.D.; Du, C.X.; Wang, Y. Long Noncoding RNA ANRIL Promotes Cervical Cancer Development by Acting as a Sponge of miR-186. *Oncol. Res.* **2018**, *26*, 345–352.

244. Naemura, M.; Murasaki, C.; Inoue, Y.; Okamoto, H.; Kotake, Y. Long Non-coding RNA ANRIL Regulates Proliferation of Non-small Cell Lung Cancer and Cervical Cancer Cells. *Anticancer Res.* **2015**, *35*, 5377–5382.
245. Bhan, A.; Soleimani, M.; Mandal, S.S. Long Noncoding RNA and Cancer: A New Paradigm. *Cancer Res.* **2017**, *77*, 3965–3981.
246. Arun, G.; Diermeier, S.D.; Spector, D.L. Therapeutic targeting of long non-coding RNAs in cancer. *Trends Mol. Med.* **2018**, *24*, 257–277.
247. Lin, C.; Yang, L. Long noncoding RNA in Cancer: Wiring signaling circuitry. *Trends Cell Biol.* **2018**, *28*, 287–301.
248. Bolha, L.; Ravnik-Glavac, M.; Glavac, D. Long noncoding RNAs as biomarkers in cancer. *Dis. Markers* **2017**, *2017*, 7243968.
249. Parsons, C.; Tayoun, A.M.; Benando, B.D.; Ragusa, G.; Dorvil, R.F.; Rourke, E.A.; Connor, K.O.; Reed, I.G.; Alexander, A.; Willetts, L.; et al. The role of long noncoding RNAs in cancer metastasis. *J. Cancer Metastasis Treat.* **2018**, *4*, 1–23.
250. Tano, K.; Akimitsu, N. Long non-coding RNAs in cancer progression. *Front. Genet.* **2012**,*3*,1–6.
251. Faghihi, M.A.; Zhang, M.; Huang, J.; Modarresi, F.; Van der Brug, M.P.; Nalls, M.A.; Cookson, M.R.; St-Laurent III, G.; Wahlestedt, C. Evidence for natural antisense transcript-mediated inhibition of microRNA function. *Genome Biol.* **2010**, *11*, 1–13.
252. Modarresi, F.; Faghihi, M.A.; Patel, N.S.; Sahagan, B.G.; Wahlestedt, C.; Lopez-Toledano, M.A. Knockdown of BACE1-AS nonprotein-coding transcript modulates betaamyloid- related hippocampal neurogenesis. *Int. J. Alzheimers Dis.* **2011**, *2011*, 1–11.
253. Zhang, M.; Cai, F.; Zhang, S.; Zhang, S.; Song, W. Overexpression of ubiquitin carboxyl-terminal hydrolase L1 (UCHL1) delays Alzheimer's progression in vivo. *Sci. Rep.* **2015**, *4*, 1–6.
254. Carriieri, C.; Forrest, A.R.R.; Santoro, C.; Persichetti, F.; Carninci, P.; Zucchelli, S.; Gustincich, S. Expression analysis of the long non-coding RNA antisense to Uchl1 (AS Uchl1) during dopaminergic cells' differentiation in vitro and in neurochemical models of Parkinson's disease. *Front. Cell. Neurosci.* **2015**, *9*, 114.
255. Scheele, C.; Petrovic, N.; Faghihi, M.A.; Lassmann, T.; Fredriksson, K.; Rooyackers, O.; Wahlestedt, C.; Good, L.; Timmons, J.A. The human PINK1 locus is regulated in vivo by a non-coding natural antisense RNA during modulation of mitochondrial function. *BMC Genomics* **2007**, *8*, 74.
256. Chung, D.W.; Rudnicki, D.D.; Yu, L.; Margolis, R.L. A natural antisense transcript at the Huntington's disease repeat locus regulates HTT expression. *Hum. Mol. Genet.* **2011**, *20*, 3467–3477.
257. Johnson, R. Long non-coding RNAs in Huntington's disease neurodegeneration. *Neurobiol Dis.* **2012**, *46*, 245–254.
258. Khalil, A.M.; Guttman, M.; Huarte, M.; Garber, M.; Raj, A.; Rivea Morales, D.; Thomas, K.; Presser, A.; Bernstein, B.E.; van Oudenaarden, A.; et al. Many human large intergenic noncoding RNAs associate with chromatin-modifying complexes and affect gene expression. *Proc. Natl. Acad. Sci. USA* **2009**, *106*, 11667–11672.

259. Zhang, E.B.; Yin, D.D.; Sun, M.; Kong, R.; Liu, X.-H.; You, L.-H.; Han, L.; Xia, R.; Wang, K.-M.; Yang, J.-S.; et al. P53-regulated long non-coding RNA TUG1 affects cell proliferation in human non-small cell lung cancer; partly through epigenetically regulating HOXB7 expression. *Cell Death Dis.* **2014**, *5*, 1–12.
260. Young, T.L.; Matsuda, T.; Cepko, C.L. The noncoding RNA taurine upregulated gene 1 is required for differentiation of the murine retina. *Curr. Biol.* **2005**, *15*, 501–512.
261. Wu, P.; Zuo, X.; Deng, H.; Liu, X.; Liu, L.; Ji, A. Roles of long noncoding RNAs in brain development; functional diversification and neurodegenerative diseases. *Brain Res. Bull.* **2013**, *97*, 69–80.
262. Kim, J.; Kim, K.M.; Noh, J.H.; Yoon, J.H.; Abdelmohsen, K.; Gorospe, M. Long noncoding RNAs in diseases of aging. *Biochim. Biophys. Acta* **2016**, *1859*, 209–221.
263. Cipolla, G.A.; deOliveira, J.C.; Salviano-Silva, A.; Lobo-Alves, S.C.; Lemos, D.S.; Oliveira, L.C.; Jucoski, T.S.; Mathias, C.; Pedroso, G.A.; Zambalde, E.P.; et al. Long non-coding RNAs in multifactorial diseases: Another Layer of complexity. *Noncoding RNA* **2018**, *4*, 13.
264. Kim, C.; Kang, D.; Lee, E.K.; Lee, J.S. Long Noncoding RNAs and RNA-Binding Proteins in Oxidative Stress, Cellular Senescence, and Age-Related Diseases. *Oxid. Med. Cell Longev.* **2017**, *2017*, 1–21.
265. Thomas, A.A.; Feng, B.; Chakrabarti, S. ANRIL: A Regulator of VEGF in Diabetic Retinopathy. *Invest Ophthalmol. Vis Sci.* **2017**, *58*, 470–480.
266. Qiu, G.Z.; Tian, W.; Fu, H.T.; Li, C.P.; Liu, B. Long noncoding RNA-MEG3 is involved in diabetes mellitus-related microvascular dysfunction. *Biochem. Biophys. Res. Commun.* **2016**, *471*, 135–141.
267. Ma, C.; Luo, H.; Liu, B.; Li, F.; Tschope, C.; Fa, X. Long noncoding RNAs:A new player in the prevention and treatment of diabetic cardiomyopathy? *Diabetes Metab. Res. Rev.* **2018**, *34*, e3056.
268. Gong, Q.; Su, G. Roles of miRNAs and long noncoding RNAs in the progression of diabetic retinopathy. *Biosci. Rep.* **2017**, *37*, 1–18.
269. Raut, S.K.; Khulla, M. The big entity of new RNA world: Long non-coding RNAs in microvascular complications of diabetes. *Front. Endocrinol.* **2018**, *9*, 1–10.
270. Zhang, Z.; Gao, W.; Long, Q.-Q.; Zhang, J.; Li, Y.-F.; Liu, D.-C.; Yan,J.-J.; Yang, Z.-J.; Wang, L.-S. Increased plasma levels of lncRNA H19 and LIPCAR are associated with increased risk of coronary artery disease in a Chinese population. *Sci. Rep.* **2017**, *7*, 1–9.
271. Liu, L.; An, X.; Li, Z.; Song, Y.; Li, L.; Zuo, S.; Liu, N.; Yang, G.; Wang, H.; Cheng, X.; et al.The H19 long noncoding RNA is a novel negative regulator of cardiomyocyte hypertrophy. *Cardiovascular Res.* **2016**, *111*, 56–65.
272. Pan, .X. LncRNA H19 promotes at herosclerosis by regulating MAPK and NF-kB signaling pathway.*Eur.Rev. Med. Pharmacol. Sci.* **2017**, *21*, 322–328.
273. Zhuo, Y.; Zeng, Q.; Zhang, P.; Li, G.; Xie, Q.; Cheng, Y. Functional polymorphism of lncRNA MALAT1 contributes to pulmonary arterial hypertension susceptibility in Chinese people. *Clin. Chem. Lab. Med.* **2017**, *55*, 38–46.

274. Zhang, M.; Gu, H.; Chen, J.; Zhou, X. Involvement of long noncoding RNA MALAT1 in the pathogenesis of diabetic cardiomyopathy. *Int. J. Cardiol.* **2016**, *202*, 753–755.
275. Zhang, X.; Tang, X.; Hamblin, M.H.; Yin, K.-J. Long non-coding RNA Malat1 regulates angiogenesis in hindlimb ischemia. *Int. J. Mol. Sci.* **2018**, *19*, 1723.
276. Vausort, M.; Wagner, D.R.; Devaux, Y. Long noncoding RNAs in patients with acute myocardial infarction. *Circulation Res.* **2014**, *115*, 668–677.
277. Murray, R.; Bryant, J.; Titcombe, P.; Barton, S.J.; Inskip, H.; Harvey, N.C.; Cooper, C.; Lillycrop, K.; Hanson, M.; Godfrey, K.M. DNA methylation at birth within the promoter of ANRIL predicts markers of cardiovascular risk at 9 years. *Clin. Epigenetics* **2016**, *8*, 1–6.
278. Cheng, J.; Cai, M.-Y.; Chen, Y.-N.; Li, Z.-C.; Tang, S.-S.; Yang, X.-L.; Chen, C.; Liu, X.; Xiong, X.-D. Variants in ANRIL gene correlated with its expression contribute to myocardial infarction risk. *Oncotarget* **2017**, *8*, 12607–12619.
279. Schulz, S.; Seitter, L.; Werdan, K.; Hofmann, B.; Schaller, H.-G.; Schlitt, A.; Reichert, S. Single nucleotide polymorphisms in long noncoding RNA, ANRIL, are not associated with severe periodontitis but with adverse cardiovascular events among patients with cardiovascular disease. *J. Periodontal Res.* **2018**, *53*, 714–720.
280. Arbiol-Roca, A.; Padró-Miquel, A.; Vidal-Alabró, A.; Hueso, M.; Fontova, P.; Bestard, O.; Rama, I.; Torras, J.; Grinyó, J.M.; Alía-Ramos, P.; et al. ANRIL as a genetic marker for cardiovascular events in renal transplant patients—An observational follow-up cohort study. *Transpl. Int.* **2018**, *31*, 1018–1027.
281. Gao, L.; Liu, Y.; Guo, S.; Yao, R.; Wu, L.; Xiao, L.; Wang, Z.; Liu, Y.; Zhang, Y. Circulating long noncoding RNA HOTAIR is an essential mediator of acute myocardial infarction. *Cell Physiol. Biochem.* **2017**, *44*, 1497–1508.
282. Jiang, Y.; Mo, H.; Luo, J.; Zhao, S.; Liang, S.; Zhang, M.; Yuan, J. HOTAIR is a potential novel biomarker in patients with congenital heart diseases. *BioMed Res. Int.* **2018**, *2018*, 2850657.
283. Haemmig, S.; Simion, V.; Yang, D.; Deng, Y.; Feinberg, M.W. Long noncoding RNAs in cardiovascular disease, diagnosis, and therapy. *Curr. Opin. Cardiol.* **2017**, *32*, 776–783.
284. Shen, S.; Jiang, H.; Bei, Y.; Xiao, J.; Li, X. Long Non-Coding RNAs in Cardiac Remodeling. *Cell Physiol. Biochem.* **2017**, *41*, 1830–1837.
285. Gomes, C.P.C.; Spencer, H.; Ford, K.L.; Michel, L.Y.M.; Baker, A.H.; Emanueli, C.; Balligand, J.L.; Devaux, Y. The Function and Therapeutic Potential of Long Non-coding RNAs in Cardiovascular Development and Disease. *Mol. Ther Nucleic Acids* **2017**, *8*, 494–507.
286. Chen, X.; He, L.; Zhao, Y.; Li, Y.; Zhang, S.; Sun, K.; So, K.; Chen, F.; Zhou, L.; Lu, L.; et al. Malat1 regulates myogenic differentiation and muscle regeneration through modulating MyoD transcriptional activity. *Cell Discov.* **2017**, *3*, 1–23.
287. Watts, R.; Johnsen, V.L.; Shearer, J.; Hittel, D.S. Myostatin-induced inhibition of the long noncoding RNA Malat1 is associated with decreased myogenesis. *Am. J. Physiol. Cell Physiol.* **2013**, *304*, C995–C1001.

288. Thomas, M.; Langley, B.; Berry, C.; Sharma, M.; Kirk, S.; Bass, J.; Kambadur, R. Myostatin, a negative regulator of muscle growth, functions by inhibiting myoblast proliferation. *J. Biol. Chem.* **2000**, *275*, 40235–40243.
289. Wang, Y.; Pang, W.J.; Wei, N.; Xiong, Y.; Wu, W.J.; Zhao, C.Z.; Shen, Q.W.; Yang, G.S. Identification, stability and expression of Sirt1 antisense long non-coding RNA. *Gene* **2014**, *539*, 117–124.
290. Wang, G.; Wang, Y.; Xiong, Y.; Chen, X.; Ma, M.; Cai, R.; Gao, Y.; Sun, Y.; Yang, G.; Pang, W. Sirt1 AS lncRNA interacts with its mRNA to inhibit muscle formation by attenuating function of miR-34a. *Sci. Rep.* **2016**, *6*, 1–13.
291. Lim, Y.-H.; Kwon, D.-H.; Kim, J.; Park, W.J.; Kook, H.; Kim, Y.-K. Identification of long noncoding RNAs involved in muscle differentiation. *PLoS ONE* **2018**, *13*, 1–15.
292. Li, Y.; Chen, X.; Sun, H.; Wang, H. Long non-coding RNAs in the regulation of skeletal myogenesis and muscle diseases. *Cancer Lett.* **2018**, *417*, 58–64.
293. Stuhlmüller, B.; Kunisch, E.; Franz, J.; Martinez-Gamboa, L.; Hernandez, M.M.; Pruss, A.; Ulbrich, N.; Erdmann, V.A.; Burmester, G.R.; Kinne, R.W. Detection of oncofetal h19 RNA in rheumatoid arthritis synovial tissue. *Am. J. Pathol.* **2003**, *163*, 901–911.
294. Song, J.; Kim, D.; Han, J.; Kim, Y.; Lee, M.; Jin, E.J. PBMC an dexosome-derived Hotair is a critical regulator and potent marker for rheumatoid arthritis. *Clin. Exp. Med.* **2015**, *15*, 121–126.
295. Pan, F.; Zhu, L.; Lv, H.; Pei, C. Quercetin promotes the apoptosis of fibroblast-likes ynoviocytes in rheumatoid arthritis by upregulating lncRNA MALAT1. *Int. J. Mol. Med.* **2016**, *38*, 1507–1514.
296. Chaichian S, Shafabakhsh R, Mirhashemi SM, Moazzami B, Asemi Z. Circular RNAs: a novel biomarker for cervical cancer. J Cell Physiol. 2020;235(2):718–24.
297. Grivna ST, Beyret E, Wang Z, Lin H. A novel class of small RNAs in mouse spermatogenic cells. Genes Dev. 2006;20(13):1709–14.
298. Lau NC, Seto AG, Kim J, Kuramochi-Miyagawa S, Nakano T, Bartel DP, Kingston RE. Characterization of the piRNA complex from rat testes. Science. 2006;313(5785):363–7.
299. Mirzaei H, Ferns GA, Avan A, Mobarhan MG. Cytokines and MicroRNA in coronary artery disease. Adv Clin Chem. 2017;82:47–70.
300. Gholamin S, Mirzaei H. GD2-targeted immunotherapy and poten- tial value of circulating microRNAs in neuroblastoma. J Cell Physiol. 2018;233(2):866–79.
301. Keshavarzi M, Sorayayi S, Jafar Rezaei M, Mohammadi M, Ghaderi A, Rostamzadeh A, Masoudifar A, Mirzaei H. MicroRNAs-based imaging techniques in cancer diagnosis and therapy. J Cell Biochem. 2017;118(12):4121–8.
302. Lin H. piRNAs in the germ line. Science. 2007;316(5823):397.
303. Wei JW, Huang K, Yang C, Kang CS. Non-coding RNAs as regulators in epigenetics (Review). Oncol Rep. 2017;37(1):3–9.
304. Hock J, Meister G. The Argonaute protein family. Genome Biol. 2008;9(2):210.
305. Ishizu H, Siomi H, Siomi MC. Biology of PIWI-interacting RNAs: new insights into biogenesis and function inside and outside of germlines. Genes Dev. 2012;26(21):2361–73.

306. Ross RJ, Weiner MM, Lin H. PIWI proteins and PIWI-interacting RNAs in the soma. Nature. 2014;505(7483):353–9.
307. Stuwe E, Toth KF, Aravin AA. Small but sturdy: small RNAs in cellular memory and epigenetics. Genes Dev. 2014;28(5):423–31.
308. Cheng J, Guo JM, Xiao BX, Miao Y, Jiang Z, Zhou H, Li QN. piRNA, the new non-coding RNA, is aberrantly expressed in human cancer cells. Clin Chim Acta. 2011;412(17–18):1621–5.
309. Yu Y, Xiao J, Hann SS. The emerging roles of PIWI-interacting RNA in human cancers. Cancer Manag Res. 2019;11:5895–909.
310. Gan H, Lin X, Zhang Z, Zhang W, Liao S, Wang L, et al. piRNA profiling during specific stages of mouse spermatogenesis. RNA. 2011;17:1191–203.
311. Brennecke J, Aravin AA, Stark A, Dus M, Kellis M, Sachidanandam R, et al. Discrete small RNA-generating loci as master regulators of transposon activity in Drosophila. Cell. 2007;128:1089–103.
312. Frank F, Sonenberg N, Nagar B. Structural basis for 5′-nucleotide base-specific recognition of guide RNA by human AGO2. Nature. 2010;465:818–22.
313. Kawaoka S, Izumi N, Katsuma S, Tomari Y. 3′ end formation of PIWI-interacting RNAs in vitro. Mol Cell. 2011;43:1015–22.
314. Iwasaki YW, Siomi MC, Siomi H. PIWI-Interacting RNA: Its Biogenesis and functions. Annu Rev Biochem. 2015;84:405–33.
315. Aravin AA, Sachidanandam R, Bourc'his D, Schaefer C, Pezic D, Toth KF, et al. A piRNA pathway primed by individual transposons is linked to de novo DNA methylation in mice. Mol Cell. 2008;31:785–99.
316. Gou LT, Dai P, Yang JH, Xue Y, Hu YP, Zhou Y, et al. Pachytene piRNAs instruct massive mRNA elimination during late spermiogenesis. Cell Res. 2014;24:680–700.
317. Zheng K, Xiol J, Reuter M, Eckardt S, Leu NA, McLaughlin KJ, et al. Mouse MOV10L1 associates with Piwi proteins and is an essential component of the Piwi-interacting RNA (piRNA) pathway. Proc Natl Acad Sci U S A. 2010;107:11841–6.
318. Watanabe T, Tomizawa S, Mitsuya K, Totoki Y, Yamamoto Y, Kuramochi-Miyagawa S, et al. Role for piRNAs and noncoding RNA in de novo DNA methylation of the imprinted mouse Rasgrf1 locus. Science. 2011;332:848–52.
319. Mohn, F., Sienski, G., Handler, D., and Brennecke, J. (2014). The rhino-deadlock- cutoff complex licenses noncanonical transcription of dual-strand piRNA clusters in Drosophila. Cell 157, 1364–1379.
320. Lim, R.S., and Kai, T. (2015). A piece of the pi(e): the diverse roles of animal piRNAs and their PIWI partners. Semin. Cell Dev. Biol. 47-48, 17–31.
321. Zhang, Z., Wang, J., Schultz, N., Zhang, F., Parhad, S.S., Tu, S., Vreven, T., Zamore, P.D., Weng, Z., and Theurkauf, W.E. (2014). The HP1 homolog rhino anchors a nuclear complex that suppresses piRNA precursor splicing. Cell 157, 1353–1363.
322. Chen, Y.A., Stuwe, E., Luo, Y., Ninova, M., Le Thomas, A., Rozhavskaya, E., Li, S., Vempati, S., Laver, J.D., Patel, D.J., et al. (2016). Cutoff suppresses RNA polymerase II termination to ensure expression of piRNA precursors. Mol. Cell 63, 97–109.

323. Strässer, K., Masuda, S., Mason, P., Pfannstiel, J., Oppizzi, M., Rodriguez-Navarro, S., Rondón, A.G., Aguilera, A., Struhl, K., Reed, R., and Hurt, E. (2002). TREX is a conserved complex coupling transcription with messenger RNA export. Nature 417, 304–308.
324. Reed, R. (2003). Coupling transcription, splicing and mRNA export. Curr. Opin. Cell Biol. 15, 326–331.
325. Masuda, S., Das, R., Cheng, H., Hurt, E., Dorman, N., and Reed, R. (2005). Recruitment of the human TREX complex to mRNA during splicing. Genes Dev. 19, 1512–1517.
326. Zhang, F., Wang, J., Xu, J., Zhang, Z., Koppetsch, B.S., Schultz, N., Vreven, T., Meignin, C., Davis, I., Zamore, P.D., et al. (2012). UAP56 couples piRNA clusters to the perinuclear transposon silencing machinery. Cell 151, 871–884.
327. Houwing, S., Kamminga, L.M., Berezikov, E., Cronembold, D., Girard, A., van den Elst, H., Filippov, D.V., Blaser, H., Raz, E., Moens, C.B., et al. (2007). A role for Piwi and piRNAs in germ cell maintenance and transposon silencing in zebrafish. Cell 129, 69–82.
328. Kawaoka, S., Izumi, N., Katsuma, S., and Tomari, Y. (2011). 3^0 End formation of PIWI-interacting RNAs in vitro. Mol. Cell 43, 1015–1022.
329. Juliano, C.E., Reich, A., Liu, N., Götzfried, J., Zhong, M., Uman, S., Reenan, R.A., Wessel, G.M., Steele, R.E., and Lin, H. (2014). PIWI proteins and PIWI-interacting RNAs function in Hydra somatic stem cells. Proc. Natl. Acad. Sci. USA 111, 337–342.
330. Horwich, M.D., Li, C., Matranga, C., Vagin, V., Farley, G., Wang, P., and Zamore, P.D. (2007). The Drosophila RNA methyltransferase, DmHen1, modifies germline piRNAs and single-stranded siRNAs in RISC. Curr. Biol. 17, 1265–1272.
331. Kirino, Y., and Mourelatos, Z. (2007). The mouse homolog of HEN1 is a potential methylase for Piwi-interacting RNAs. RNA 13, 1397–1401.
332. Kamminga, L.M., Luteijn, M.J., den Broeder, M.J., Redl, S., Kaaij, L.J.T., Roovers, E.F., Ladurner, P., Berezikov, E., and Ketting, R.F. (2010). Hen1 is required for oocyte development and piRNA stability in zebrafish. EMBO J. 29, 3688–3700.
333. Montgomery, T.A., Rim, Y.S., Zhang, C., Dowen, R.H., Phillips, C.M., Fischer, S.E.J., and Ruvkun, G. (2012). PIWI associated siRNAs and piRNAs specifically require the Caenorhabditis elegans HEN1 ortholog henn-1. PLoS Genet. 8, e1002616.
334. Billi, A.C., Alessi, A.F., Khivansara, V., Han, T., Freeberg, M., Mitani, S., and Kim, J.K. (2012). The Caenorhabditis elegans HEN1 ortholog, HENN-1, methylates and stabilizes select subclasses of germline small RNAs. PLoS Genet. 8, e1002617.
335. Xiol, J., Spinelli, P., Laussmann, M.A., Homolka, D., Yang, Z., Cora, E., Couté, Y., Conn, S., Kadlec, J., Sachidanandam, R., et al. (2014). RNA clamping by Vasa assem- bles a piRNA amplifier complex on transposon transcripts. Cell 157, 1698–1711.
336. Zhang, Z., Xu, J., Koppetsch, B.S., Wang, J., Tipping, C., Ma, S.M., Weng, Z., Theurkauf, W.E., and Zamore, P.D. (2011). Heterotypic piRNA Ping-Pong

requires Qin, a protein with both E3 ligase and tudor domains. Mol. Cell 44, 572–584.

337. Williams, Z., Morozov, P., Mihailovic, A., Lin, C., Puvvula, P.K., Juranek, S., Rosenwaks, Z., and Tuschl, T. (2015). Discovery and characterization of piRNAs in the human fetal ovary. Cell Rep. 13, 854–863.
338. Rouget, C., Papin, C., Boureux, A., Meunier, A.C., Franco, B., Robine, N., Lai, E.C., Pelisson, A., and Simonelig, M. (2010). Maternal mRNA deadenylation and decay by the piRNA pathway in the early Drosophila embryo. Nature 467, 1128–1132.
339. Ayarpadikannan, S., and Kim, H.S. (2014). The impact of transposable elements in genome evolution and genetic instability and their implications in various diseases. Genomics Inform. 12, 98–104.
340. Hedges, D.J., and Deininger, P.L. (2007). Inviting instability: gransposable elements, double-strand breaks, and the maintenance of genome integrity. Mutat. Res. 616, 46–59.
341. Sugimoto, K., Kage, H., Aki, N., Sano, A., Kitagawa, H., Nagase, T., Yatomi, Y., Ohishi, N., and Takai, D. (2007). The induction of H3K9 methylation by PIWIL4 at the $p16^{Ink4a}$ locus. Biochem. Biophys. Res. Commun. 359, 497–502.
342. Lu, Y., Zhang, K., Li, C., Yao, Y., Tao, D., Liu, Y., Zhang, S., and Ma, Y. (2012). Piwil2 suppresses p53 by inducing phosphorylation of signal transducer and activator of transcription 3 in tumor cells. PLoS ONE 7, e30999.
343. Liu, Y., Zhang, J., Li, A., Zhang, Y., Li, Y., Yuan, X., He, Z., Liu, Z., and Tuo, S. (2019). Identification of PIWI-interacting RNA modules by weighted correlation network analysis. Cluster Comput. 22, 707–717.
344. Iqbal, M.A., Arora, S., Prakasam, G., Calin, G.A., and Syed, M.A. (2019). MicroRNA in lung cancer: role, mechanisms, pathways and therapeutic relevance. Mol. Aspects Med. 70, 3–20.
345. Di, X., Jin, X., Li, R., Zhao, M., and Wang, K. (2019). circRNAs and lung cancer: bio- markers and master regulators. Life Sci. 220, 177–185.
346. Mei, Y., Wang, Y., Kumari, P., Shetty, A.C., Clark, D., Gable, T., MacKerell, A.D., Ma, M.Z., Weber, D.J., Yang, A.J., et al. (2015). A piRNA-like small RNA interacts with and modulates p-ERM proteins in human somatic cells. Nat. Commun. 6, 7316.
347. Peng, L., Song, L., Liu, C., Lv, X., Li, X., Jie, J., Zhao, D., and Li, D. (2016). piR-55490 inhibits the growth of lung carcinoma by suppressing mTOR signaling. Tumour Biol. 37, 2749–2756.
348. Zhang, S.J., Yao, J., Shen, B.Z., Li, G.B., Kong, S.S., Bi, D.D., Pan, S.H., and Cheng, B.L. (2018). Role of piwi-interacting RNA-651 in the carcinogenesis of non-small cell lung cancer. Oncol. Lett. 15, 940–946.
349. Daugaard, I., Venø, M.T., Yan, Y., Kjeldsen, T.E., Lamy, P., Hager, H., Kjems, J., and Hansen, L.L. (2017). Small RNA sequencing reveals metastasis-related microRNAs in lung adenocarcinoma. Oncotarget 8, 27047–27061.
350. Reeves, M.E., Firek, M., Jliedi, A., and Amaar, Y.G. (2017). Identification and char- acterization of RASSF1C piRNA target genes in lung cancer cells. Oncotarget 8, 34268–34282.

351. Cheng, J., Deng, H., Xiao, B., Zhou, H., Zhou, F., Shen, Z., and Guo, J. (2012). piR- 823, a novel non-coding small RNA, demonstrates in vitro and in vivo tumor sup- pressive activity in human gastric cancer cells. Cancer Lett. 315, 12–17.
352. Cui, L., Lou, Y., Zhang, X., Zhou, H., Deng, H., Song, H., Yu, X., Xiao, B., Wang, W., and Guo, J. (2011). Detection of circulating tumor cells in peripheral blood from pa- tients with gastric cancer using piRNAs as markers. Clin. Biochem. 44, 1050–1057.
353. Rizzo, F., Rinaldi, A., Marchese, G., Coviello, E., Sellitto, A., Cordella, A., Giurato, G., Nassa, G., Ravo, M., Tarallo, R., et al. (2016). Specific patterns of PIWI-interacting small noncoding RNA expression in dysplastic liver nodules and hepatocellular carcinoma. Oncotarget 7, 54650–54661.
354. Tang, X., Xie, X., Wang, X., Wang, Y., Jiang, X., and Jiang, H. (2018). The combina- tion of piR-823 and eukaryotic initiation factor 3 B (EIF3B) activates hepatic stellate cells via upregulating TGF-b1 in liver fibrogenesis. Med. Sci. Monit. 24, 9151–9165.
355. Law, P.T., Qin, H., Ching, A.K., Lai, K.P., Co, N.N., He, M., Lung, R.W., Chan, A.W., Chan, T.F., and Wong, N. (2013). Deep sequencing of small RNA transcriptome re- veals novel non-coding RNAs in hepatocellular carcinoma. J. Hepatol. 58, 1165– 1173.
356. Weng, W., Liu, N., Toiyama, Y., Kusunoki, M., Nagasaka, T., Fujiwara, T., Wei, Q., Qin, H., Lin, H., Ma, Y., and Goel, A. (2018). Novel evidence for a PIWI-interacting RNA (piRNA) as an oncogenic mediator of disease progression, and a potential prognostic biomarker in colorectal cancer. Mol. Cancer 17, 16.
357. Mai, D., Ding, P., Tan, L., Zhang, J., Pan, Z., Bai, R., Li, C., Li, M., Zhou, Y., Tan, W., et al. (2018). PIWI-interacting RNA-54265 is oncogenic and a potential therapeutic target in colorectal adenocarcinoma. Theranostics 8, 5213–5230.
358. Yin, J., Jiang, X.Y., Qi, W., Ji, C.G., Xie, X.L., Zhang, D.X., Cui, Z.J., Wang, C.K., Bai, Y., Wang, J., and Jiang, H.Q. (2017). piR-823 contributes to colorectal tumorigenesis by enhancing the transcriptional activity of HSF1. Cancer Sci. 108, 1746–1756.
359. Qu, A., Wang, W., Yang, Y., Zhang, X., Dong, Y., Zheng, G., Wu, Q., Zou, M., Du, L., Wang, Y., and Wang, C. (2019). A serum piRNA signature as promising non- invasive diagnostic and prognostic biomarkers for colorectal cancer. Cancer Manag. Res. 11, 3703–3720.
360. Vychytilova-Faltejskova, P., Stitkovcova, K., Radova, L., Sachlova, M., Kosarova, Z., Slaba, K., Kala, Z., Svoboda, M., Kiss, I., Vyzula, R., et al. (2018). Circulating PIWI- interacting RNAs piR-5937 and piR-28876 are promising diagnostic biomarkers of colon cancer. Cancer Epidemiol. Biomarkers Prev. 27, 1019–1028.
361. Ng, K.W., Anderson, C., Marshall, E.A., Minatel, B.C., Enfield, K.S., Saprunoff, H.L., Lam, W.L., and Martinez, V.D. (2016). Piwi-interacting RNAs in cancer: emerging functions and clinical utility. Mol. Cancer 15, 5.
362. Iliev, R., Fedorko, M., Machackova, T., Mlcochova, H., Svoboda, M., Pacik, D., Dolezel, J., Stanik, M., and Slaby, O. (2016). Expression levels of PIWI-interacting RNA, piR-823, are deregulated in tumor tissue, blood serum and urine of patients with renal cell carcinoma. Anticancer Res. 36, 6419–6423.

363. Zhao, C., Tolkach, Y., Schmidt, D., Toma, M., Muders, M.H., Kristiansen, G., Müller, S.C., and Ellinger, J. (2019). Mitochondrial PIWI-interacting RNAs are novel biomarkers for clear cell renal cell carcinoma. World J. Urol. 37, 1639–1647.
364. Li, Y., Wu, X., Gao, H., Jin, J.M., Li, A.X., Kim, Y.S., Pal, S.K., Nelson, R.A., Lau, C.M., Guo, C., et al. (2015). Piwi-interacting RNAs (piRNAs) are dysregulated in renal cell carcinoma and associated with tumor metastasis and cancer-specific sur- vival. Mol. Med. 21, 381–388.
365. Drake, C.G., Lipson, E.J., and Brahmer, J.R. (2014). Breathing new life into immu- notherapy: review of melanoma, lung and kidney cancer. Nat. Rev. Clin. Oncol. 11, 24–37.
366. Chu, H., Hui, G., Yuan, L., Shi, D., Wang, Y., Du, M., Zhong, D., Ma, L., Tong, N., Qin, C., et al. (2015). Identification of novel piRNAs in bladder cancer. Cancer Lett. 356 (2 Pt B), 561–567.
367. Voduc, K.D., Cheang, M.C., Tyldesley, S., Gelmon, K., Nielsen, T.O., and Kennecke, H. (2010). Breast cancer subtypes and the risk of local and regional relapse. J. Clin. Oncol. 28, 1684–1691.
368. Hawkes, W.C., Printsev, I., and Alkan, Z. (2012). Selenoprotein W depletion induces a p53- and p21-dependent delay in cell cycle progression in RWPE-1 prostate epithelial cells. J. Cell. Biochem. 113, 61–69.
369. Tan, L., Mai, D., Zhang, B., Jiang, X., Zhang, J., Bai, R., Ye, Y., Li, M., Pan, L., Su, J., et al. (2019). PIWI-interacting RNA-36712 restrains breast cancer progression and chemoresistance by interaction with SEPW1 pseudogene SEPW1P RNA. Mol. Cancer 18, 9.
370. Fu, A., Jacobs, D.I., Hoffman, A.E., Zheng, T., and Zhu, Y. (2015). PIWI-interacting RNA 021285 is involved in breast tumorigenesis possibly by remodeling the cancer epigenome. Carcinogenesis 36, 1094–1102.
371. Lee, Y.J., Moon, S.U., Park, M.G., Jung, W.Y., Park, Y.K., Song, S.K., Ryu, J.G., Lee, Y.S., Heo, H.J., Gu, H.N., et al. (2016). Multiplex bioimaging of piRNA molecular pathway-regulated theragnostic effects in a single breast cancer cell using a piRNA molecular beacon. Biomaterials 101, 143–155.
372. Zhang, H., Ren, Y., Xu, H., Pang, D., Duan, C., and Liu, C. (2013). The expression of stem cell protein Piwil2 and piR-932 in breast cancer. Surg. Oncol. 22, 217–223.
373. Huang, G., Hu, H., Xue, X., Shen, S., Gao, E., Guo, G., Shen, X., and Zhang, X. (2013). Altered expression of piRNAs and their relation with clinicopathologic features of breast cancer. Clin. Transl. Oncol. 15, 563–568.
374. Hashim, A., Rizzo, F., Marchese, G., Ravo, M., Tarallo, R., Nassa, G., Giurato, G., Santamaria, G., Cordella, A., Cantarella, C., and Weisz, A. (2014). RNA sequencing identifies specific PIWI-interacting small non-coding RNA expression patterns in breast cancer. Oncotarget 5, 9901–9910.
375. Koduru, S.V., Tiwari, A.K., Leberfinger, A., Hazard, S.W., Kawasawa, Y.I., Mahajan, M., and Ravnic, D.J. (2017). A comprehensive NGS data analysis of differentially regulated miRNAs, piRNAs, lncRNAs and sn/snoRNAs in triple negative breast cancer. J. Cancer 8, 578–596.

376. Singh, G., Roy, J., Rout, P., and Mallick, B. (2018). Genome-wide profiling of the PIWI-interacting RNA-mRNA regulatory networks in epithelial ovarian cancers. PLoS ONE 13, e0190485.
377. Alsop, K., Fereday, S., Meldrum, C., deFazio, A., Emmanuel, C., George, J., Dobrovic, A., Birrer, M.J., Webb, P.M., Stewart, C., et al. (2012). BRCA mutation fre- quency and patterns of treatment response in BRCA mutation-positive women with ovarian cancer: a report from the Australian Ovarian Cancer Study Group. J. Clin. Oncol. 30, 2654–2663.
378. Ostrom, Q.T., Bauchet, L., Davis, F.G., Deltour, I., Fisher, J.L., Langer, C.E., Pekmezci, M., Schwartzbaum, J.A., Turner, M.C., Walsh, K.M., et al. (2014). The epidemiology of glioma in adults: a "state of the science" review. Neuro-oncol. 16, 896–913.
379. Jacobs, D.I., Qin, Q., Lerro, M.C., Fu, A., Dubrow, R., Claus, E.B., DeWan, A.T., Wang, G., Lin, H., and Zhu, Y. (2016). PIWI-interacting RNAs in gliomagenesis: ev- idence from post-GWAS and functional analyses. Cancer Epidemiol. Biomarkers Prev. 25, 1073–1080.
380. Shen, S., Yu, H., Liu, X., Liu, Y., Zheng, J., Wang, P., Gong, W., Chen, J., Zhao, L., and Xue, Y. (2018). PIWIL1/piRNA-DQ593109 regulates the permeability of the blood- tumor barrier via the MEG3/miR-330-5p/RUNX3 axis. Mol. Ther. Nucleic Acids 10, 412–425.
381. Jacobs, D.I., Qin, Q., Fu, A., Chen, Z., Zhou, J., and Zhu, Y. (2018). piRNA-8041 is downregulated in human glioblastoma and suppresses tumor growth in vitro and in vivo. Oncotarget 9, 37616–37626.
382. Roy, J., and Mallick, B. (2018). Investigating piwi-interacting RNA regulome in human neuroblastoma. Genes Chromosomes Cancer 57, 339–349.
383. Li,B.,Hong,J.,Hong,M.,Wang,Y.,Yu,T.,Zang,S.,andWu,Q.(2019).piRNA-823 delivered by multiple myeloma-derived extracellular vesicles promoted tumorigenesis through re-educating endothelial cells in the tumor environment. Oncogene 38, 5227–5238.
384. Siegel, R.L., Miller, K.D., and Jemal, A. (2016). Cancer statistics, 2016. CA Cancer J. Clin. 66, 7–30.
385. Ai, L., Mu, S., Sun, C., Fan, F., Yan, H., Qin, Y., Cui, G., Wang, Y., Guo, T., Mei, H., et al. (2019). Myeloid-derived suppressor cells endow stem-like qualities to multiple myeloma cells by inducing piRNA-823 expression and DNMT3B activation. Mol. Cancer 18, 88.
386. Yan, H., Wu, Q.L., Sun, C.Y., Ai, L.S., Deng, J., Zhang, L., Chen, L., Chu, Z.B., Tang, B., Wang, K., et al. (2015). piRNA-823 contributes to tumorigenesis by regulating de novo DNA methylation and angiogenesis in multiple myeloma. Leukemia 29, 196–206.
387. Marur, S., and Forastiere, A.A. (2016). Head and neck squamous cell carcinoma: up- date on epidemiology, diagnosis, and treatment. Mayo Clin. Proc. 91, 386–396.
388. Firmino,N.,Martinez,V.D.,Rowbotham,D.A.,Enfield,K.S.S.,Bennewith,K.L.,an d Lam, W.L. (2016). HPV status is associated with altered PIWI-interacting RNA expression pattern in head and neck cancer. Oral Oncol. 55, 43–48.
389. Krishnan, A.R., Korrapati, A., Zou, A.E., Qu, Y., Wang, X.Q., Califano, J.A., Wang- Rodriguez, J., Lippman, S.M., Hovell, M.F., and Ongkeko, W.M. (2017).

Smoking status regulates a novel panel of PIWI-interacting RNAs in head and neck squamous cell carcinoma. Oral Oncol. 65, 68–75.

390. Saad, M.A., Ku, J., Kuo, S.Z., Li, P.X., Zheng, H., Yu, M.A., Wang-Rodriguez, J., and Ongkeko, W.M. (2019). Identification and characterization of dysregulated P- element induced wimpy testis-interacting RNAs in head and neck squamous cell carcinoma. Oncol. Lett. 17, 2615–2622.
391. Ta, H.T., Dass, C.R., Choong, P.F., and Dunstan, D.E. (2009). Osteosarcoma treatment: state of the art. Cancer Metastasis Rev. 28, 247–263.
392. Das, B., Roy, J., Jain, N., and Mallick, B. (2019). Tumor suppressive activity of PIWI- interacting RNA in human fibrosarcoma mediated through repression of RRM2. Mol. Carcinog. 58, 344–357.
393. Hanahan, D., and Weinberg, R.A. (2011). Hallmarks of cancer: the next generation. Cell 144, 646–674.
394. Li S, Xu Z, Sheng J. tRNA-derived small RNA: a novel regulatory small non-coding RNA. Genes. 2018;9(5):246.
395. Fu H, Feng J, Liu Q, Sun F, Tie Y, Zhu J, Xing R, Sun Z, Zheng X. Stress induces tRNA cleavage by angiogenin in mammalian cells. FEBS Lett. 2009; 583(2):437–42.
396. Thompson DM, Parker R. The RNase Rny1p cleaves tRNAs and promotes cell death during oxidative stress in Saccharomyces cerevisiae. J Cell Biol. 2009; 185(1):43–50.
397. Yamasaki S, Ivanov P, G-f H, Anderson P. Angiogenin cleaves tRNA and promotes stress-induced translational repression. J Cell Biol. 2009;185(1):35–42.
398. Lee SR, Collins K. Starvation-induced cleavage of the tRNA anticodon loop in Tetrahymena thermophila. J Biol Chem. 2005;280(52):42744–9.
399. Joechl C, Rederstorff M, Hertel J, Stadler PF, Hofacker IL, Schrettl M, Haas H, Huettenhofer A. Small ncRNA transcriptome analysis from Aspergillus fumigatus suggests a novel mechanism for regulation of protein synthesis. Nucleic Acids Res. 2008;36(8):2677–89.
400. Li Y, Luo J, Zhou H, Liao J-Y, Ma L-M, Chen Y-Q, Qu L-H. Stress-induced tRNA-derived RNAs: a novel class of small RNAs in the primitive eukaryote Giardia lamblia. Nucleic Acids Res. 2008;36(19):6048–55.
401. Honda S, Loher P, Shigematsu M, Palazzo JP, Suzuki R, Imoto I, Rigoutsos I, Kirino Y. Sex hormone-dependent tRNA halves enhance cell proliferation in breast and prostate cancers. Proc Natl Acad Sci U S A. 2015;112(29):E3816–25.
402. Kumar P, Mudunuri SB, Anaya J, Dutta A. tRFdb: a database for transfer RNA fragments. Nucleic Acids Res. 2015;43(D1):D141–5.
403. Lee YS, Shibata Y, Malhotra A, Dutta A. A novel class of small RNAs: tRNA-derived RNA fragments (tRFs). Genes Dev. 2009;23(22):2639–49.
404. Li Z, Ender C, Meister G, Moore PS, Chang Y, John B. Extensive terminal and asymmetric processing of small RNAs from rRNAs, snoRNAs, snRNAs, and tRNAs. Nucleic Acids Res. 2012;40(14):6787–99.
405. Maraia RJ, Lamichhane TN. 3'processing of eukaryotic precursor tRNAs. Wiley Interdiscip Rev RNA. 2011;2(3):362–75.
406. Bartel DP. MicroRNAs: genomics, biogenesis, mechanism, and function. Cell. 2004;116(2):281–97.

407. Chen X, Liang H, Zhang J, Zen K, Zhang C-Y. microRNAs are ligands of toll-like receptors. RNA-a Publication of the RNA Society. 2013;19(6):737–9.
408. Yanbo Wang, Hongwei Liang, Fangfang Jin, Xin Yan, Guifang Xu, Huanhuan Hu, Gaoli Liang, Shoubin Zhan, Xiuting Hu, Quan Zhao, Yuan Liu, Zhen-You Jiang, Chen-Yu Zhang, Xi Chen, Ke Zen. Injured liver-released miRNA-122 elicits acute pulmonary inflammation via activating alveolar macrophage TLR7 signaling pathway. Proceedings of the National Academy of Sciences. 2019;116(13):6162-6171.
409. Maute RL, Schneider C, Sumazin P, Holmes A, Califano A, Basso K, Dalla-Favera R. tRNA-derived microRNA modulates proliferation and the DNA damage response and is down-regulated in B cell lymphoma. Proc Natl Acad Sci U S A. 2013;110(4):1404–9.
410. Shao Y, Sun Q, Liu X, Wang P, Wu R, Ma Z. tRF-Leu-CAG promotes cell proliferation and cell cycle in non-small cell lung cancer. Chem Biol Drug Des. 2017;90(5):730–8.
411. Kumar P, Anaya J, Mudunuri SB, Dutta A. Meta-analysis of tRNA derived RNA fragments reveals that they are evolutionarily conserved and associate with AGO proteins to recognize specific RNA targets. BMC Biol. 2014;12:14.
412. Haussecker D, Huang Y, Lau A, Parameswaran P, Fire AZ, Kay MA. Human tRNA-derived small RNAs in the global regulation of RNA silencing. RNA-a Publication of the Rna Society. 2010;16(4):673–95.
413. Goodarzi H, Liu X, Nguyen HCB, Zhang S, Fish L, Tavazoie SF. Endogenous tRNA-derived fragments suppress breast Cancer progression via YBX1 displacement. Cell. 2015;161(4):790–802.
414. Ivanov P, Emara MM, Villen J, Gygi SP, Anderson P. Angiogenin-induced tRNA fragments inhibit translation initiation. Mol Cell. 2011;43(4):613–23.
415. Ivanov P, O'Day E, Emara MM, Wagner G, Lieberman J, Anderson P. G-quadruplex structures contribute to the neuroprotective effects of angiogenin-induced tRNA fragments. Proc Natl Acad Sci U S A. 2014; 111(51):18201–6.
416. Gebetsberger J, Wyss L, Mleczko AM, Reuther J, Polacek N. A tRNA-derived fragment competes with mRNA for ribosome binding and regulates translation during stress. RNA Biol. 2017;14(10):1364–73.
417. Schorn AJ, Gutbrod MJ, LeBlanc C, Martienssen R. LTR-retrotransposon control by tRNA-derived small RNAs. Cell. 2017;170(1):61–+.
418. Ruggero K, Guffanti A, Corradin A, Sharma VK, De Bellis G, Corti G, Grassi A, Zanovello P, Bronte V, Ciminale V, et al. Small noncoding RNAs in cells transformed by human T-cell leukemia virus type 1: a role for a tRNA fragment as a primer for reverse transcriptase. J Virol. 2014;88(7):3612–22.
419. Deng J, Ptashkin RN, Chen Y, Cheng Z, Liu G, Phan T, Deng X, Zhou J, Lee I, Lee YS, et al. Respiratory syncytial virus utilizes a tRNA fragment to suppress antiviral responses through a novel targeting mechanism. Mol Ther. 2015; 23(10):1622–9.
420. Wang Q, Lee I, Ren J, Ajay SS, Lee YS, Bao X. Identification and functional characterization of tRNA-derived RNA fragments (tRFs) in respiratory syncytial virus infection. Mol Ther. 2013;21(2):368–79.
421. Zhou J, Liu S, Chen Y, Fu Y, Silver AJ, Hill MS, Lee I, Lee YS, Bao X. Identification of two novel functional tRNA-derived fragments induced in

response to respiratory syncytial virus infection. J Gen Virol. 2017;98(7):1600–10.
422. Thompson DM, Parker R. Stressing out over tRNA cleavage. Cell. 2009;138(2): 215–9.
423. Emara MM, Ivanov P, Hickman T, Dawra N, Tisdale S, Kedersha N, Hu G-F, Anderson P. Angiogenin-induced tRNA-derived stress-induced RNAs promote stress-induced stress granule assembly. J Biol Chem. 2010;285(14):10959–68.
424. Pekarsky Y, Balatti V, Palamarchuk A, Rizzotto L, Veneziano D, Nigita G, Rassenti LZ, Pass HI, Kipps TJ, Liu C-G, et al. Dysregulation of a family of short noncoding RNAs, tsRNAs, in human cancer. Proc Natl Acad Sci U S A. 2016;113(18):5071–6.
425. Zhao C, Tolkach Y, Schmidt D, Kristiansen G, Mueller SC, Ellinger J. 5 '-tRNA halves are dysregulated in clear cell renal cell carcinoma. J Urol. 2018;199(2):378–83.
426. Greenway MJ, Andersen PM, Russ C, Ennis S, Cashman S, Donaghy C, Patterson V, Swingler R, Kieran D, Prehn J, et al. ANG mutations segregate with familial and 'sporadic' amyotrophic lateral sclerosis. Nat Genet. 2006; 38(4):411–3.
427. Sebastia J, Kieran D, Breen B, King MA, Netteland DF, Joyce D, Fitzpatrick SF, Taylor CT, Prehn JHM. Angiogenin protects motoneurons against hypoxic injury. Cell Death Differ. 2009;16(9):1238–47.
428. Kieran D, Sebastia J, Greenway MJ, King MA, Connaughton D, Concannon CG, Fenner B, Hardiman O, Prehn JHM. Control of Motoneuron survival by Angiogenin. J Neurosci. 2008;28(52):14056–61.
429. Wu D, Yu WH, Kishikawa H, Folkerth RD, Iafrate AJ, Shen YP, Xin WM, Sims K, Hu GF. Angiogenin loss-of-function mutations in amyotrophic lateral sclerosis. Ann Neurol. 2007;62(6):609–17.
430. Saikia M, Jobava R, Parisien M, Putnam A, Krokowski D, Gao X-H, Guan B-J, Yuan Y, Jankowsky E, Feng Z, et al. Angiogenin-cleaved tRNA halves interact with cytochrome c, protecting cells from apoptosis during osmotic stress. Mol Cell Biol. 2014;34(13):2450–63.
431. Hanada T, Weitzer S, Mair B, Bernreuther C, Wainger BJ, Ichida J, Hanada R, Orthofer M, Cronin SJ, Komnenovic V, et al. CLP1 links tRNA metabolism to progressive motor-neuron loss. Nature. 2013;495(7442):474–80.
432. Karaca E, Weitzer S, Pehlivan D, Shiraishi H, Gogakos T, Hanada T, Jhangiani SN, Wiszniewski W, Withers M, Campbell IM, et al. Human CLP1 mutations Alter tRNA biogenesis, affecting both peripheral and central nervous system function. Cell. 2014;157(3):636–50.
433. Schaffer AE, Eggens VRC, Caglayan AO, Reuter MS, Scott E, Coufal NG, Silhavy JL, Xue Y, Kayserili H, Yasuno K, et al. CLP1 founder mutation links tRNA splicing and maturation to cerebellar development and neurodegeneration. Cell. 2014;157(3):651–63.
434. Anderson P, Ivanov P. tRNA fragments in human health and disease. FEBS Lett. 2014;588(23):4297–304.
435. Guzzi N, Ciesla M, Phuong Cao Thi N, Lang S, Arora S, Dimitriou M, Pimkova K, Sommarin MNE, Munita R, Lubas M, et al. Pseudo-uridylation of tRNA-derived fragments steers translational control in stem cells. Cell. 2018;173(5): 1204–1216.e26.

436. Kim HK, Fuchs G, Wang S, Wei W, Zhang Y, Park H, Roy-Chaudhuri B, Li P, Xu J, Chu K, et al. A transfer-RNA-derived small RNA regulates ribosome biogenesis. Nature. 2017;552(7683):57–+.
437. Dhahbi JM, Spindler SR, Atamna H, Yamakawa A, Boffelli D, Mote P, Martin DIK. 5'tRNA halves are present as abundant complexes in serum, concentrated in blood cells, and modulated by aging and calorie restriction. BMC Genomics. 2013;14:14.
438. Dhahbi JM, Spindler SR, Atamna H, Boffelli D, Martin DI. Deep sequencing of serum small RNAs identifies patterns of 5′ tRNA half and YRNA fragment expression associated with breast Cancer. Biomarker Cancer. 2014;6:37–47.
439. Zhang YF, Zhang Y, Shi JC, Zhang H, Cao ZH, Gao X, Ren WH, Ning YN, Ning LN, Cao YJ, et al. Identification and characterization of an ancient class of small RNAs enriched in serum associating with active infection. J Mol Cell Biol. 2014;6(2):172–4.
440. Peng H, Shi J, Zhang Y, Zhang H, Liao S, Li W, Lei L, Han C, Ning L, Cao Y, et al. A novel class of tRNA-derived small RNAs extremely enriched in mature mouse sperm. Cell Res. 2012;22(11):1609–12.
441. Chen Q, Yan M, Cao Z, Li X, Zhang Y, Shi J, Feng G-h, Peng H, Zhang X, Zhang Y, et al. Sperm ts-RNAs contribute to intergenerational inheritance of an acquired metabolic disorder. Science. 2016;351(6271):397–400.
442. Sharma U, Conine CC, Shea JM, Boskovic A, Derr AG, Bing XY, Belleannee C, Kucukural A, Serra RW, Sun F, et al. Biogenesis and function of tRNA fragments during sperm maturation and fertilization in mammals. Science. 2016;351(6271):391–6.
443. Rinaldi, C.; Wood, M.J.A. Antisense oligonucleotides: The next frontier for treatment of neurological disorders. *Nat. Rev. Neurol.* **2018**, *14*, 9–21.
444. Sarkar A, Volff JN, Vaury C. piRNAs and their diverse roles: a transposable element-driven tactic for gene regulation? FASEB J. 2017;31:436–46.
445. Zamore PD. Somatic piRNA biogenesis. EMBO J. 2010;29:3219–21.
446. Li C, Vagin VV, Lee S, Xu J, Ma S, Xi H, et al. Collapse of germline piRNAs in the absence of Argonaute3 reveals somatic piRNAs in flies. Cell. 2009;137: 509–21.
447. Zhong F, Zhou N, Wu K, Guo Y, Tan W, Zhang H, et al. A SnoRNA-derived piRNA interacts with human interleukin-4 pre-mRNA and induces its decay in nuclear exosomes. Nucleic Acids Res. 2015;43:10474–91.
448. Ogawa Y, Taketomi Y, Murakami M, Tsujimoto M, Yanoshita R. Small RNA transcriptomes of two types of exosomes in human whole saliva determined by next generation sequencing. Biol Pharm Bull. 2013;36:66–75.
449. Sharma AK, Nelson MC, Brandt JE, Wessman M, Mahmud N, Weller KP, et al. Human CD34(+) stem cells express the hiwi gene, a human homologue of the Drosophila gene piwi. Blood. 2001;97:426–34.
450. Juliano C, Wang J, Lin H. Uniting germline and stem cells: the function of Piwi proteins and the piRNA pathway in diverse organisms. Annu Rev Genet. 2011;45:447–69.
451. Litwin M, Dubis J, Arczynska K, Piotrowska A, Frydlewicz A, Karczewski M, et al. Correlation of HIWI and HILI expression with Cancer stem cell markers in colorectal Cancer. Anticancer Res. 2015;35:3317–24.

452. Liu W, Gao Q, Chen K, Xue X, Li M, Chen Q, et al. Hiwi facilitates chemoresistance as a cancer stem cell marker in cervical cancer. Oncol Rep. 2014;32:1853–60.
453. Liang D, Yang Y, Liu Y. The role Hiwi gene in the maintenance of lung cancer stem cell populations. Neoplasma. 2013;22. https://doi.org/10.4149/neo_2014_022.
454. Liang D, Fang Z, Dong M, Liang C, Xing C, Zhao J, et al. Effect of RNA interference related HiWi gene expression on the proliferation and apoptosis of lung cancer stem cells. Oncol Lett. 2012;4:146–50.
455. Rosenkranz D. piRNA cluster database: a web resource for piRNA producing loci. Nucleic Acids Res. 2016;44:D223–30.

Chapter IV

1. O'Brien, J.; Hayder, H.; Zayed, Y.; Peng, C. Overview of MicroRNA Biogenesis,Mechanisms of Actions, and Circulation. *Front. Endocrinol.* **2018**, *9*, 402
2. Kai, K.; Dittmar, R.L.; Sen, S. Secretory microRNAs as biomarkers of cancer. *Semin. Cell Dev. Biol.* **2018**, *78*, 22–36.
3. Chevillet, J.R.; Lee, I.; Briggs, H.A.; He, Y.; Wang, K. Issues and prospects of microRNA-based biomarkers in blood and other body fluids. *Molecules* **2014**, *19*, 6080 6105.
4. Bartel, D.P. MicroRNAs: Target recognition and regulatory functions. *Cell* **2009**, *136*, 215–233.
5. Huntzinger, E.; Izaurralde, E. Gene silencing by microRNAs: Contributions of translational repression and mRNA decay. *Nat. Rev. Genet.* **2011**, *12*, 99–110.
6. Eichhorn, S.W.; Guo, H.; McGeary, S.E.; Rodriguez-Mias, R.A.; Shin, C.; Baek, D.; Hsu, S.H.; Ghoshal, K.; Villen, J.; Bartel, D.P. mRNA destabilization is the dominant effect of mammalian microRNAs by the time substantial repression ensues. *Mol. Cell* **2014**, *56*,104–115.
7. Helwak, A.; Kudla, G.; Dudnakova, T.; Tollervey, D. Mapping the human miRNA interactome by CLASH reveals frequent noncanonical binding. *Cell* **2013**, *153*, 654–665.
8. Jonas, S.; Izaurralde, E. Towards a molecular understanding of microRNA-mediated gene silencing. *Nat. Rev. Genet.* **2015**, *16*, 421–433.
9. Zen, K.; Zhang, C.Y. Circulating microRNAs: A novel class of biomarkers to diagnose and monitor human cancers. *Med. Res. Rev.* **2012**, *32*, 326–348.
10. Weber, J.A.; Baxter, D.H.; Zhang, S.; Huang, D.Y.; Huang, K.H.; Lee, M.J.; Galas, D.J.; Wang, K. The microRNA spectrum in 12 body fluids. *Clin. Chem.* **2010**, *56*, 1733–1741.
11. Wang, K.; Zhang, S.; Weber, J.; Baxter, D.; Galas, D.J. Export of microRNAs and microRNA-protective protein by mammalian cells. *Nucleic Acids Res.* **2010**, *38*, 7248–7259.
12. Zubakov, D.; Boersma, A.W.; Choi, Y.; van Kuijk, P.F.; Wiemer, E.A.; Kayser, M. MicroRNA markers for forensic body fluid identification obtained from microarray screening and quantitative RT-PCR confirmation. *Int. J. Legal. Med.* **2010**, *124*, 217–226.

13. Hanson, E.K.; Lubenow, H.; Ballantyne, J. Identification of forensically relevant body fluids using a panel of differentially expressed microRNAs. *Anal. Biochem.* **2009**, *387*, 303–314.
14. Raposo, G.; Stoorvogel, W. Extracellular vesicles: Exosomes, microvesicles, and friends. *J. Cell Biol.* **2013**, *200*, 373–383.
15. O'Driscoll, L. Expanding on exosomes and ectosomes in cancer. *N. Engl. J. Med.* **2015**, *372*, 2359–2362.
16. Vickers, K.C.; Palmisano, B.T.; Shoucri, B.M.; Shamburek, R.D.; Remaley, A.T. MicroRNAs are transported in plasma and delivered to recipient cells by high-density lipoproteins. *Nat. Cell. Biol.* **2011**, *13*, 423–433.
17. Arroyo, J.D.; Chevillet, J.R.; Kroh, E.M.; Ruf, I.K.; Pritchard, C.C.; Gibson, D.F.; Mitchell, P.S.; Bennett, C.F.; Pogosova-Agadjanyan, E.L.; Stirewalt, D.L.; et al. Argonaute2 complexes carry a population of circulating microRNAs independent of vesicles in human plasma. *Proc. Natl. Acad. Sci. USA* **2011**, *108*, 5003–5008.
18. Cortez, M.A.; Bueso-Ramos, C.; Ferdin, J.; Lopez-Berestein, G.; Sood, A.K.; Calin, G.A. MicroRNAs in body fluids–the mix of hormones and biomarkers. *Nat. Rev. Clin. Oncol.* **2011**, *8*, 467–477.
19. Weickmann, J.L.; Glitz, D.G. Human ribonucleases. Quantitation of pancreatic-like enzymes in serum, urine, and organ preparations. *J. Biol. Chem.* **1982**, *257*, 8705–8710.
20. Valadi, H.; Ekström, K.; Bossios, A.; Sjöstrand, M.; Lee, J.J.; Lötvall, J.O. Exosome-mediated transfer of mRNAs and microRNAs is a novel mechanism of genetic exchange between cells. *Nat. Cell Biol.* **2007**, *9*, 654–659.
21. Morello, M.; Minciacchi, V.R.; de Candia, P.; Yang, J.; Posadas, E.; Kim, H.; Griffiths, D.; Bhowmick, N.; Chung, L.W.; Gandellini, P.; et al. Large oncosomes mediate intercellular transfer of functional microRNA. *Cell Cycle* **2013**, *12*, 3526–3536.
22. Kharaziha, P.; Ceder, S.; Li, Q.; Panaretakis, T. Tumor cell-derived exosomes: Amessage in a bottle. *Biochim. Biophys. Acta* **2012**, *1826*, 103–111.
23. Taylor, C.R. Introduction to Predictive Biomarkers: Definitions and Characteristics. In *Predictive Biomarkers in Oncology: Applications in Precision Medicine*; Badve, S., Kumar, G.L., Eds.; Springer International Publishing: Cham, Switzerland, 2019; pp. 3 18.
24. Solé, C.; Moline, T.; Vidal, M.; Ordi-Ros, J.; Cortés-Hernández, J. An Exosomal Urinary miRNA Signature for Early Diagnosis of Renal Fibrosis in Lupus Nephritis. *Cells* **2019**, *8*, 773.
25. Gillespie, P.; Ladame, S.; O'Hare, D. Molecular methods in electrochemical microRNA detection. *Analyst* **2018**, *144*, 114–129.
26. Chen, Y.X.; Huang, K.J.; Niu, K.X. Recent advances in signal amplification strategy based on oligonucleotide and nanomaterials for microRNA detection-a review. *Biosens. Bioelectron.* **2018**, *99*, 612–624.
27. Yuan, Y.H.; Chi, B.Z.; Wen, S.H.; Liang, R.P.; Li, Z.M.; Qiu, J.D. Ratiometric electrochemical assay for sensitive detecting microRNA based on dual-amplification mechanism of duplex-specific nuclease and hybridization chain reaction. *Biosens. Bioelectron.* **2018**, *102*, 211–216.

28. Chen, C.; Tan, R.; Wong, L.; Fekete, R.; Halsey, J. Quantitation of microRNAs by real time RT-qPCR. *Methods Mol. Biol.* **2011**, *687*, 113–134.
29. Chen, C.; Ridzon, D.A.; Broomer, A.J.; Zhou, Z.; Lee, D.H.; Nguyen, J.T.; Barbisin, M.; Xu, N.L.; Mahuvakar, V.R.; Andersen, M.R.; et al. Real-time quantification of microRNAs by stem-loop RT-PCR. *Nucleic Acids Res.* **2005**, *33*, e179.
30. Pacak, A.; Barciszewska-Pacak, M.; Swida-Barteczka, A.; Kruszka, K.; Sega, P.; Milanowska, K.; Jakobsen, I.; Jarmolowski, A.; Szweykowska-Kulinska, Z. Heat Stress Affects Pi-related Genes Expression and Inorganic Phosphate Deposition/Accumulation in Barley. *Front. Plant Sci.* **2016**, *7*, 926.
31. Barciszewska-Pacak, M.; Milanowska, K.; Knop, K.; Bielewicz, D.; Nuc, P.; Plewka, P.; Pacak, A.M.; Vazquez, F.; Karlowski, W.; Jarmolowski, A.; et al. Arabidopsis microRNA expression regulation in a wide range of abiotic stress responses. *Front. Plant Sci.* **2015**, *6*, 410.
32. Kruszka, K.; Pacak, A.; Swida-Barteczka, A.; Nuc, P.; Alaba, S.; Wroblewska, Z.; Karlowski, W.; Jarmolowski, A.; Szweykowska-Kulinska, Z. Transcriptionally and post-transcriptionally regulated microRNAs in heat stress response in barley. *J. Exp. Bot.* **2014**, *65*, 6123–6135.
33. Smoczynska, A.; Sega, P.; Stepien, A.; Knop, K.; Jarmolowski, A.; Pacak, A.; Szweykowska-Kulinska, Z. miRNA Detection by Stem-Loop RT-qPCR in Studying microRNA Biogenesis and microRNA Responsiveness to Abiotic Stresses. *Methods Mol. Biol.* **2019**, *1932*, 131–150.
34. Pall,G.S.;Hamilton,A.J.Improvednorthernblotmethodforenhanceddetectionofsmall RNA.*Nat.Protoc.* **2008**, *3*, 1077–1084.
35. Wang,X.;Tong,Y.;Wang,S.RapidandaccuratedetectionofplantmiRNAsbyliquidnorthernhybridization. *Int. J. Mol. Sci.* **2010**, *11*, 3138–3148.
36. Varallyay,E.;Burgyan,J.;Havelda,Z.DetectionofmicroRNAsbyNorthernblotanalysesusin LNAprobes. *Methods* **2007**, *43*, 140–145.
37. Javelle,M.;Timmermans,M.C.InsitulocalizationofsmallRNAsinplantsbyusingLNAprobes.*Nat.Protoc.* **2012**, *7*, 533–541.
38. Brown, R. A. M. *et al.* Total RNA extraction from tissues for microRNA and target gene expression analysis: not all kits are created equal. *BMC biotechnology* **18**, 16–16, https://doi.org/10.1186/s12896-018-0421-6 (2018).
39. Ramón-Núñez, L. A. *et al.* Comparison of protocols and RNA carriers for plasma miRNA isolation. Unraveling RNA carrier influence on miRNA isolation. *Plos One* **12**, e0187005, https://doi.org/10.1371/journal.pone.0187005 (2017)
40. El-Khoury, V., Pierson, S., Kaoma, T., Bernardin, F. & Berchem, G. Assessing cellular and circulating miRNA recovery: the impact of the RNA isolation method and the quantity of input material. *Sci Rep* **6**, 19529, https://doi.org/10.1038/srep19529 (2016).
41. Moret, I. *et al.* Assessing an improved protocol for plasma microRNA extraction. *Plos one* **8**, e82753–e82753, https://doi.org/10.1371/journal.pone.0082753 (2013).
42. Andreasen, D. *et al.* Improved microRNA quantification in total RNA from clinical samples. *Methods* **50**, S6–S9, https://doi.org/10.1016/j.ymeth.2010.01.006 (2010)

43. Brunet-Vega, A. *et al.* Variability in microRNA recovery from plasma: Comparison of five commercial kits. *Analytical Biochemistry* **488**, 28–35, https://doi.org/10.1016/j.ab.2015.07.018 (2015)
44. Meerson, A. & Ploug, T. Assessment of six commercial plasma small RNA isolation kits using qRT-PCR and electrophoretic separation: higher recovery of microRNA following ultracentrifugation. *Biology Methods and Protocols* **1**, bpw003–bpw003, https://doi.org/10.1093/biomethods/bpw003 (2016).
45. Bustin, S. A. *et al.* The MIQE Guidelines: Minimum Information for Publication of Quantitative Real-Time PCR Experiments. *Clinical Chemistry* **55**, 611, https://doi.org/10.1373/clinchem.2008.112797 (2009).
46. Becker, C., Hammerle-Fickinger, A., Riedmaier, I. & Pfaffl, M. W. mRNA and microRNA quality control for RT-qPCR analysis. *Methods* **50**, 237–243, https://doi.org/10.1016/j.ymeth.2010.01.010 (2010).
47. Kim, Y.-K., Yeo, J., Kim, B., Ha, M. & Kim, V. N. Short Structured RNAs with Low GC Content Are Selectively Lost during Extraction from a Small Number of Cells. *Molecular Cell* **46**, 893–895, https://doi.org/10.1016/j.molcel.2012.05.036 (2012).
48. El-Khoury, V., Pierson, S., Kaoma, T., Bernardin, F. & Berchem, G. Assessing cellular and circulating miRNA recovery: the impact of the RNA isolation method and the quantity of input material. *Scientific Reports* **6**, 19529, https://doi.org/10.1038/srep19529 (2016).
49. Ford, K. L. *et al.* Optimisation of laboratory methods for whole transcriptomic RNA analyses in human left ventricular biopsies and blood samples of clinical relevance. *Plos One* **14**, e0213685, https://doi.org/10.1371/journal.pone.0213685 (2019).
50. McAlexander, M. A., Phillips, M. J. & Witwer, K. W. Comparison of Methods for miRNA Extraction from Plasma and Quantitative Recovery of RNA from Cerebrospinal Fluid. *Frontiers in genetics* **4**, 83–83, https://doi.org/10.3389/fgene.2013.00083 (2013).
51. Gautam, A. *et al.* Identification of extracellular miRNA in archived serum samples by next-generation sequencing from RNA extracted using multiple methods. *Molecular Biology Reports* **43**, 1165–1178, https://doi.org/10.1007/s11033-016-4043-6 (2016).
52. Ioannidis, J., Risse, J. & Donadeu, F. X. In Mic*ro*RNA Pro*to*cols (ed. Shao-Yao Ying) 65–77 (Springer New York, 2018).
53. Garcia-Elias, A. *et al.* Defining quantification methods and optimizing protocols for microarray hybridization of circulating microRNAs. *Scientific Reports* **7**, 7725, https://doi.org/10.1038/s41598-017-08134-3 (2017).
54. Schipor, S. *et al.* A comparative analysis of three methods used for RNA quantitation. *Romanian Reports in Physics* **68**, 1178–1188 (2016).
55. Aranda, R., Dineen, S. M., Craig, R. L., Guerrieri, R. A. & Robertson, J. M. Comparison and evaluation of RNA quantification methods using viral, prokaryotic, and eukaryotic RNA over a 104 concentration range. *Analytical Biochemistry* **387**, 122–127, https://doi.org/10.1016/j.ab.2009.01.003 (2009).
56. Ambros V (2004) The functions of animal microRNAs. Nature 431: 350–5.

57. Bissels U, Wild S, Tomiuk S, Holste A, Hafner M, et al. (2009) Absolute quantification of microRNAs by using a universal reference. RNA 15: 2375–2384.
58. Arefian E, Kiani J, Soleimani M, Shariati SA, Aghaee-Bakhtiari SH, et al. (2011) Analysis of microRNA signatures using size-coded ligation-mediated PCR. Nucleic Acids Res 39: e80.
59. Chen C, Ridzon DA, Broomer AJ, Zhou Z, Lee DH, Nguyen JT, Barbisin M, Xu NL, Mahuvakar VR, Andersen MR, Lao KQ, Livak KJ, Gue- gler KJ: **Real-time quantification of microRNAs by stem-loop RT-PCR.** *Nucleic Acids Res* 2005, **33(20):**e179.
60. 32. Tang F, Hajkova P, Barton SC, Lao K, Surani MA: **MicroRNA expression profiling of single whole embryonic stem cells.** *Nucleic Acids Res* 2006, **34(2):**e9.
61. Hafner M, Renwick N, Brown M, Mihailović A, Holoch D, Lin C, et al. RNA-ligase-dependent biases in miRNA representation in deep-sequenced small RNA cDNA libraries. RNA. 2011;17(9):1697–712.
62. Sorefan K, Pais H, Hall AE, Kozomara A, Griffiths-Jones S, Moulton V, et al. Reducing ligation bias of small RNAs in libraries for next generation sequencing. Silence. 2012;3(1):4.
63. Fuchs RT, Sun Z, Zhuang F, Robb GB. Bias in ligation-based small RNA sequencing library construction is determined by adaptor and RNA structure. PLoS One. 2015;10(5):e0126049 Zhang B, editor.
64. Giraldez MD, Spengler RM, Etheridge A, Godoy PM, Barczak AJ, Srinivasan S, et al. Comprehensive multi-center assessment of small RNA-seq methods for quantitativemiRNA profiling. Nat Biotechnol. 2018;36(8):746–57.
65. Li Y, Kowdley KV. Method for microRNA isolation from clinical serum samples. Anal Biochem. 2012;431(1):69–75.
66. McAlexander MA, Phillips MJ, Witwer KW. Comparison of methods for miRNA extraction from plasma and quantitative recovery of RNA from cerebrospinal fluid. Front Genet. 2013;4:83.
67. Li X, Mauro M, Williams Z. Comparison of plasma extracellular RNA isolation kits reveals kit-dependent biases. Biotechniques. 2015;59(1):13–7.
68. Meerson A, Ploug T. Assessment of six commercial plasma small RNA isolation kits using qRT-PCR and electrophoretic separation: higher recovery of microRNA followingultracentrifugation. Biol Methods Protoc. 2016;1(1): bpw003.
69. Lusk RW. Diverse and widespread contamination evident in the unmapped depths of high throughput sequencing data. PLoS One. 2014;9(10):e110808.
70. Kang W, Bang-Berthelsen CH, Holm A, Houben AJS, Müller AH, Thymann T, et al. Survey of 800+ data sets from human tissue and body fluid reveals xenomiRs are likely artifacts. RNA. 2017;23(4):433–45.
71. Witwer KW. Alternative miRNAs? Human sequences misidentified as plant miRNAs in plant studies and in human plasma. F1000Research. 2018;7:244.
72. Baran-Gale J, Kurtz CL, Erdos MR, Sison C, Young A, Fannin EE, et al. Addressing Bias in small RNA library preparation for sequencing: a new protocol recovers MicroRNAs that evade capture by current methods. Front Genet. 2015;6:352.

73. Shore S, Henderson JM, Lebedev A, Salcedo MP, Zon G, McCaffrey AP, et al. Small RNA library preparation method for next-generation sequencing using chemical modifications to prevent adapter dimer formation. Mahalingam R, editor. PLoS One. 2016;11(11):e0167009.
74. Dard-Dascot C, Naquin D, d'Aubenton-Carafa Y, Alix K, Thermes C, van Dijk E. Systematic comparison of small RNA library preparation protocols for next-generationsequencing. BMC Genomics. 2018;19(1):118.
75. Barberán-Soler S, Vo JM, Hogans RE, Dallas A, Johnston BH, Kazakov SA. Decreasing miRNA sequencing bias using a single adapter and circularization approach. Genome Biol. 2018;19(1):105.
76. Coenen-Stass AML, Magen I, Brooks T, Ben-Dov IZ, Greensmith L, Hornstein E, et al. Evaluation of methodologies for microRNA biomarker detection by next generationsequencing. RNA Biol. 2018;15(8):1133–45.
77. Buschmann D, Haberberger A, Kirchner B, Spornraft M, Riedmaier I, Schelling G, et al. Toward reliable biomarker signatures in the age of liquid biopsies - how to standardize the small RNA-Seq workflow. Nucleic Acids Res. 2016;44: 5995–6018.
78. Chen,C., Ridzon,D.A., Broomer,A.J., Zhou,Z., Lee,D.H., Nguyen,J.T., Barbisin,M., Xu,N.L., Mahuvakar,V.R., Andersen,M.R. et al. (2005) Real-time quantification of microRNAs by stem-loop RT-PCR. Nucleic Acids Res., 33, e179.
79. Lu,C., Tej,S.S., Luo,S., Haudenschild,C.D., Meyers,B.C. and Green,P.J. (2005) Elucidation of the small RNA component of the transcriptome. Science, 309, 1567–1569.
80. Landgraf,P., Rusu,M., Sheridan,R., Sewer,A., Iovino,N., Aravin,A., Pfeffer,S., Rice,A., Kamphorst,A.O., Landthaler,M. et al. (2007) A mammalian microRNA expression atlas based on small RNA library sequencing. Cell, 129, 1401–1414.
81. Neely,L.A., Patel,S., Garver,J., Gallo,M., Hackett,M., McLaughlin,S., Nadel,M., Harris,J., Gullans,S. and Rooke,J. (2006) A single-molecule method for the quantitation of microRNA gene expression. Nat. Methods, 3, 41–46.
82. Sempere,L.F., Freemantle,S., Pitha-Rowe,I., Moss,E., Dmitrovsky,E. and Ambros,V. (2004) Expression profiling of mammalian microRNAs uncovers a subset of brain-expressed microRNAs with possible roles in murine and human neuronal differentiation. Genome Biol., 5, R13.
83. Cummins,J.M., He,Y., Leary,R.J., Pagliarini,R., Diaz,L.A. Jr, Sjoblom,T., Barad,O., Bentwich,Z., Szafranska,A.E., Labourier,E. et al. (2006) The colorectal microRNAome. Proc. Natl Acad. Sci. USA, 103, 3687–3692.
84. Lu,J., Getz,G., Miska,E.A., Alvarez-Saavedra,E., Lamb,J., Peck,D., Sweet-Cordero,A., Ebert,B.L., Mak,R.H., Ferrando,A.A. et al. (2005) MicroRNA expression profiles classify human cancers. Nature, 435, 834–838.
85. Castoldi,M., Schmidt,S., Benes,V., Noerholm,M., Kulozik,A.E., Hentze,M.W. and Muckenthaler,M.U. (2006) A sensitive array for microRNA expression profiling (miChip) based on locked nucleic acids (LNA). RNA, 12, 913–920.
86. Liu,C.G., Calin,G.A., Meloon,B., Gamliel,N., Sevignani,C., Ferracin,M., Dumitru,C.D., Shimizu,M., Zupo,S., Dono,M. et al. (2004) An oligonucleotide microchip for genome-wide microRNA profiling in human and mouse tissues. Proc. Natl Acad. Sci. USA, 101, 9740–9744.

87. Thomson,J.M., Parker,J., Perou,C.M. and Hammond,S.M. (2004) A custom microarray platform for analysis of microRNA gene expression. Nat. Methods, 1, 47–53.
88. Wang,H., Ach,R.A. and Curry,B. (2007) Direct and sensitive miRNA profiling from low-input total RNA. RNA, 13, 151–159.
89. Nelson,P.T., Baldwin,D.A., Scearce,L.M., Oberholtzer,J.C., Tobias,J.W. and Mourelatos,Z. (2004) Microarray-based, high- throughput gene expression profiling ofmicroRNAs. Nat. Methods, 1, 155–161.
90. Fan,J.B., Chee,M.S. and Gunderson,K.L. (2006) Highly parallel genomic assays. Nat. Rev. Genet., 7, 632–644.
91. Fan,J.B., Oliphant,A., Shen,R., Kermani,B.G., Garcia,F., Gunderson,K.L., Hansen,M., Steemers,F., Butler,S.L., Deloukas,P. et al. (2003) Highly parallel SNP genotyping. Cold Spring Harb. Symp. Quant. Biol., 68, 69–78.
92. Fan,J.B., Yeakley,J.M., Bibikova,M., Chudin,E., Wickham,E., Chen,J., Doucet,D., Rigault,P., Zhang,B., Shen,R. et al. (2004) A versatile assay for high-throughput gene expression profiling on universal array matrices. Genome Res., 14, 878–885.
93. Fan,J.B., Gunderson,K.L., Bibikova,M., Yeakley,J.M., Chen,J., Wickham Garcia,E., Lebruska,L.L., Laurent,M., Shen,R. and Barker,D. (2006) Illumina universal bead arrays. Methods Enzymol., 410, 57–73.
94. Bibikova,M., Talantov,D., Chudin,E., Yeakley,J.M., Chen,J., Doucet,D., Wickham,E., Atkins,D., Barker,D., Chee,M. et al. (2004) Quantitative gene expression profiling in formalin-fixed, paraffin-embedded tissues using universal bead arrays. Am. J. Pathol., 165, 1799–1807.
95. de Planell-Saguer, M. & Rodicio, M. C. Analytical aspects of microRNA in diagnostics: a review. *Anal. Chim. Acta* **699**, 134–152, https://doi.org/10.1016/j.aca.2011.05.025 (2011).
96. Li, Y. & Kowdley, K. V. MicroRNAs in common human diseases. *Genomics, Proteomics Bioinf.* **10**, 246–253, https://doi.org/10.1016/j. gpb.2012.07.005 (2012).
97. Wark, A. W., Lee, H. J. & Corn, R. M. Multiplexed detection methods for profiling microRNA expression in biological samples. *Angew. Chem., Int. Ed* **47**, 644–652, https://doi.org/10.1002/anie.200702450 (2008).
98. Zhang, P. *et al.* Multiplex ligation-dependent probe amplification (MLPA) for ultrasensitive multiplexed microRNA detection using ribonucleotide-modified DNA probes. *Chem. Commun.* **49**, 10013–10015, https://doi.org/10.1039/c3cc45760e (2013).
99. de Lencastre, A. *et al.* MicroRNAs both promote and antagonize longevity in C. elegans. *Curr. Biol.* **20**, 2159–2168 (2010).
100. Kaufman, E. J. & Miska, E. A. The microRNAs of Caenorhabditis elegans. *Semin. Cell Dev. Biol.* **21**, 728–737, https://doi. org/10.1016/j.semcdb.2010.07.001 (2010).
101. Liu, Y. Q., Zhang, M., Yin, B. C. & Ye, B. C. Attomolar ultrasensitive microRNA detection by DNA-scaffolded silver-nanocluster probe based on isothermal amplification. *Anal. Chem.* **84**, 5165–5169, https://doi.org/10.1021/ac300483f (2012).

102. Zhao, Y., Zhou, L. & Tang, Z. Cleavage-based signal amplification of RNA. *Nat. Commun.* **4**, 1493, https://doi.org/10.1038/ ncomms2492 (2013).
103. Yin, B. C., Liu, Y. Q. & Ye, B. C. One-step, multiplexed fluorescence detection of microRNAs based on duplex-specific nuclease signal amplification. *J. Am. Chem. Soc.* **134**, 5064–5067, https://doi.org/10.1021/ja300721s (2012).
104. Jia, H., Li, Z., Liu, C. & Cheng, Y. Ultrasensitive detection of microRNAs by exponential isothermal amplification. *Angew. Chem., Int. Ed* **49**, 5498–5501, https://doi.org/10.1002/anie.201001375 (2010).
105. Na, J., Shin, G. W., Jung, G. Y. & Jung, G. Y. A robust and simple-to-design multiplex DNA methylation assay based on MS-MLPA- CE-SSCP. *Analyst.* **138**, 6969–6976, https://doi.org/10.1039/c3an01178j (2013).
106. Hwang, H. S., Shin, G. W., Chung, B., Na, J. & Jung, G. Y. Multiplex and quantitative pathogen detection with high-resolution capillary electrophoresis-based single-strand conformation polymorphism. *Methods Mol. Biol.* **919**, 155–163, https://doi. org/10.1007/978-1-62703-029-8_15 (2013).
107. Shin, G. W. *et al.* A novel pathogen detection system based on high-resolution CE-SSCP using a triblock copolymer matrix. *J. Sep. Sci* **33**, 1639–1643, https://doi.org/10.1002/jssc.200900871 (2010).
108. Shin, G. W., Cho, Y. S., Hwang, H. S., Park, J. H. & Jung, G. Y. A two-step quantitative pathogen detection system based on capillary electrophoresis. *Anal. Biochem.* **383**, 31–37, https://doi.org/10.1016/j.ab.2008.08.021 (2008).
109. Shin, G. W. *et al.* A new single-step quantitative pathogen detection system: Template-tagging followed by multiplex asymmetric PCR using common primers and CE-SSCP. *Electrophoresis* **30**, 2728–2736 (2009).
110. Kramer, M. F. Stem-loop RT-qPCR for miRNAs. *Curr Protoc Mol Biol*. Chapter 15, Unit 15. 10., doi:https://doi. org/10.1002/0471142727.mb1510s95 (2011).
111. Jung, U., Jiang, X., Kaufmann, S. H. & Patzel, V. A universal TaqMan-based RT-PCR protocol for cost-efficient detection of small noncoding RNA. *RNA.* **19**, 1864–1873, https://doi.org/10.1261/rna.040501.113 (2013).
112. Shin, G. W. *et al*. Precise expression profiling by stuffer-free multiplex ligation-dependent probe amplification. *Anal. Chem.* **85**, 9383–9389, https://doi.org/10.1021/ ac402314h (2013).
113. Van Ness, J., Van Ness, L. K. & Galas, D. J. Isothermal reactions for the amplification of oligonucleotides. *Proc. Natl. Acad. Sci. USA* **100**, 4504–4509,https://doi.org/10.1073/pnas.0730811100 (2003)
114. Liu, H., Li, L., Wang, Q., Duan, L. & Tang, B. Graphene fluorescence switch-based cooperative amplification: a sensitive and accurate method to detection microRNA. *Anal. Chem.* **86**, 5487–5493, https://doi.org/10.1021/ac500752t (2014).
115. Tauro, B. J. et al. Comparison of ultracentrifugation, density gradient separation, and immunoaffinity capture methods for isolating human colon cancer cell line LIM1863-derived exosomes. Methods 56, 293–304 (2012).
116. Stoorvogel, W. Functional transfer of microRNA by exosomes. Blood 119, 646648 (2012).
117. Villarroya-Beltri, C. et al. Sumoylated hnRNPA2B1 controls the sorting of miRNAs into exosomes through binding to specific motifs. Nat. Commun. 4, 2980 (2013).

118. Schrauder, M. G. et al. Circulating micro-RNAs as potential blood-based markers for early stage breast cancer detection. PLoS ONE 7, e29770 (2012).
119. Arata, H., Komatsu, H., Han, A., Hosokawa, K. & Maeda, M. Rapid microRNA detection using power-free microfluidic chip: coaxial stacking effect enhances the sandwich hybridization. Analyst 137, 3234–3237 (2012).
120. Zhang, Z. & Nagrath, S. Microfluidics and cancer: are we there yet ? Biomed. Microdevices 15, 595–609 (2013).
121. Go, D. B., Atashbar, M. Z., Ramshani, Z. & Chang, H. C. Surface acoustic wave devices for chemical sensing and microfluidics: a review and perspective. Anal. Methods 9, 4112–4134 (2017).
122. Slouka, Z., Senapati, S. & Chang, H. C. Microfluidic systems with ion-selection membranes. Annu. Rev. Anal. Chem. 7, 317–335 (2014).
123. Senapati, S. et al. An ion-exchange nanomembrane sensor for detection of nucleic acids using a surface charge inversion phenomenon. Biosens. Bioelectron. 60, 92–100 (2014).
124. Xu,W.;Zhou,Y.;Xu,G.;Geng,B.;Cui,Q.Transcriptomeanalysisrevealsnon-identicalmicroRNAprofiles between arterial and venous plasma. *Oncotarget* **2017**, *8*, 28471–28480
125. Zhou, X.; Wen, W.; Shan, X.; Zhu, W.; Xu, J.; Guo, R.; Cheng, W.; Wang, F.; Qi, L.W.; Chen, Y.; et al. A six-microRNA panel in plasma was identified as a potential biomarker for lung adenocarcinoma diagnosis. *Oncotarget* **2017**, *8*, 6513–6525.
126. Monzo,M.;Santasusagna,S.;Moreno,I.;Martinez,F.;Hernández,R.;Muñoz,C.;Castellano,J.J.;Moreno, J.; Navarro, A. Exosomal microRNAs isolated from plasma of mesenteric veins linked to liver metastases in resected patients with colon cancer. *Oncotarget* **2017**, *8*, 30859–30869.
127. Sempere, L. F.;Freemantle, S.;Pitha-Rowe, I.; Moss, E.; Dmitrovsky, E.; Ambros, V. Expression profiling of mammalian microRNAs uncovers a subset of brain-expressed microRNAs with possible roles in murine and human neuronal differentiation. *Genome Biol.* **2004**, *5*, R13.
128. Valoczi, A.; Hornyik, C.; Varga, N.; Burgyan, J.; Kauppinen, S.; Havelda, Z. Sensitive and specific detection of microRNAs by northern blot analysis using LNA-modified oligonucleotide probes. *Nucleic Acids Res.* **2004**, *32*, e175.
129. Kloosterman,W.P.; Wienholds, E.; de Bruijn, E.; Kauppinen, S. ;Plasterk, R. H. In situ detection of miRNAs in animal embryos using LNA-modified oligonucleotide probes. *Nat. Methods* **2006**, *3*, 27–29.
130. Chen, C.; Ridzon, D.A.; Broomer, A.J.; Zhou, Z.; Lee, D.H.; Nguyen, J.T.; Barbisin, M.; Xu, N.L.; Mahuvakar, V.R.; Andersen, M.R.; et al. Real-time quantification of microRNAsby stem-loop RT-PCR. *Nucleic Acids Res.* **2005**, *33*, e179.
131. Wang, J.; Chen, J.; Chang, P.; LeBlanc, A.; Li, D.; Abbruzzesse, J.L.; Frazier, M.L.; Killary, A.M.; Sen, S. MicroRNAs in plasma of pancreatic ductal adenocarcinoma patients as novel blood-based biomarkers of disease. *Cancer Prev. Res.* **2009**, *2*, 807–813.
132. Thomson, J.M.; Parker, J.; Perou, C.M.; Hammond, S.M. A custom microarray platform for analysis of microRNA gene expression. *Nat. Methods* **2004**, *1*, 47–53.

133. Wang,J.;Raimondo,M.;Guha,S.;Chen,J.;Diao,L.;Dong,X.;Wallace,M.B.;Killary, A.M.;Frazier,M.L.; Woodward, T.A.; et al. Circulating microRNAs in Pancreatic Juice as Candidate Biomarkers of Pancreatic Cancer. *J. Cancer* **2014**, *5*, 696–705.
134. Wang, J.; Paris, P.L.; Chen, J.; Ngo, V.; Yao, H.; Frazier, M.L.; Killary, A.M.; Liu, C.-G.; Liang, H.; Mathy, C.; et al. Next generation sequencing of pancreatic cyst fluid microRNAs from low grade-benign and high grade-invasive lesions. *Cancer Lett.* **2015**, *356*, 404–409.
135. Sato, F.; Tsuchiya, S.;Terasawa, K.; Tsujimoto, G. Intra-platform repeatability an dinter-platform comparability of microRNA microarray technology. *PLoS ONE* **2009**, *4*, e5540.
136. Witwer, K.W. Circulating microRNA biomarker studies: Pitfalls and potential solutions. *Clin. Chem.***2015**, *61*, 56–63.
137. Wang, H.; Peng, R.; Wang, J.; Qin, Z.; Xue, L. Circulating microRNAs as potential cancer biomarkers: The advantage and disadvantage. *Clin. Epigenetics* **2018**, *10*, 59.
138. Binderup, H.G.; Madsen, J.S.; Heegaard, N.H.H.; Houlind, K.; Andersen, R.F.; Brasen, C.L. Quantification of microRNA levels in plasma—Impact of preanalytical and analytical conditions. *PLoS ONE* **2018**, *13*, e0201069
139. Madadi, S.; Soleimani, M. Comparison of miR-16 and cel-miR-39 as reference controls for serum miRNA normalization in colorectal cancer. *J. Cell. Biochem.* **2019**, *120*, 4802–4803.
140. Schwarzenbach, H.; da Silva, A.M.; Calin, G.; Pantel, K. Data Normalization Strategies for MicroRNA Quantification. *Clin. Chem.* **2020**, *61*, 1333–1342.
141. Wang, J.; Chen, J.; Sen, S. MicroRNA as Biomarkers and Diagnostics. *J. Cell. Physiol.***2016**, *231*, 25–30.
142. Cao, D.-D.; Li, L.; Chan, W.-Y. MicroRNAs: Key Regulators in the Central Nervous System and Their Implication in Neurological Diseases. *Int. J. Mol. Sci.* **2016**, *17*, 842.
143. Sun, X.Y.; Zhang, J.; Niu, W.; Guo, W.; Song, H.T.; Li, H.Y.; Fan, H.M.; Zhao, L.; Zhong, A.F.; Dai, Y.H.; et al. A preliminary analysis of microRNA as potential clinical biomarker for schizophrenia. *Am. J. Med Genet. Part B Neuropsychiatr. Genet.* **2015**, *168*, 170–178.
144. Montagnana, M.; Benati, M.; Danese, E.; Minicozzi, A.M.; Paviati, E.; Gusella, M.; Pasini, F.; Bovo, C.; Guidi, G.C.; Lippi, G. Plasma Expression Levels of Circulating miR-21 are not Useful for Diagnosing and Monitoring Colorectal Cancer. *Clin. Lab.* **2016**, *62*, 967–970.
145. Etheridge, A.; Lee, I.; Hood, L.; Galas, D.; Wang, K. Extracellular microRNA: A new source of biomarkers. *Mutat. Res.* **2011**, *717*, 85–90.
146. Williams, G.; Uchimoto, M.L.; Coult, N.; World, D.; Beasley, E. Body fluid mixtures: Resolution using forensic microRNA analysis. *Forensic Sci. Int. Genet. Suppl. Ser.* **2013**, *4*, e292.
147. Barger, J.F.; Nana-Sinkam, S.P. MicroRNA as tools and therapeutics in lung cancer. *Respir. Med.* **2015**,*109*, 803–812.
148. Alexandrov, P.N.; Dua, P.; Hill ,J.M.; Bhattacharjee, S.; Zhao, Y.; Lukiw, W.J. microRNA (miRNA) speciation in Alzheimer's disease (AD) cerebrospinal fluid (CSF) and extracellular fluid (ECF). *Int. J. Biochem. Mol. Biol.* **2012**, *3*, 365–373.

149. Muller, M.; Kuiperij, H.B.; Claassen, J.A.; Kusters, B.; Verbeek, M.M. MicroRNAs in Alzheimer's disease: Differential expression in hippocampus and cell-free cerebrospinal fluid. *Neurobiol. Aging* **2014**, *35*, 152–158.
150. Cogswell, J.P.; Ward, J.; Taylor, I.A.; Waters, M.; Shi, Y.; Cannon, B.; Kelnar ,K.; Kemppainen, J.; Brown, D.; Chen, C.; et al. Identification of miRNA changes in Alzheimer's disease brain and CSF yields putative biomarkers and insights into disease pathways. *J. Alzheimers Dis.* **2008**, *14*, 27–41.
151. Jarry, J.; Schadendorf, D.; Greenwood, C.; Spatz, A.; vanKempen, L.C. The validity of circulating microRNAs in oncology: Five years of challenges and contradictions. *Mol. Oncol.* **2014**, *8*, 819–829.
152. Hrustincova, A.; Votavova, H.; Dostalova Merkerova, M. Circulating MicroRNAs: Methodological Aspects in Detection of These Biomarkers. *Folia Biol.* **2015**, *61*, 203–218.
153. Pritchard, C.C.; Cheng, H.H.; Tewari, M. MicroRNA profiling: Approaches and considerations. *Nat. Rev. Genet.* **2012**, *13*, 358–369.

Chapter V

1. World Health Organization. (2011). *Global Status Report on Alcohol and Health* (Geneva: WHO Press), 1–85.
2. Leshner, A. I. (1997). Addiction is a brain disease, and it matters. *Science* 278, 45–47.
3. Ron, D., and Messing, R. O. (2011). Signaling pathways mediating alcohol effects. *Curr. Top. Behav. Neurosci.* doi:10.1007/7854_2011_161.
4. Miranda, R. C., Pietrzykowski, A. Z., Tang, Y., Sathyan, P., Mayfield, D., Keshavarzian, A., Sampson, W., and Hereld, D. (2010). MicroRNAs: master regulators of ethanol abuse and toxicity. *Alcohol. Clin. Exp. Res.* 34, 575–587.
5. Ambros, V. (2001). microRNAs: tiny regulators with great potential. *Cell* 107, 823–826.
6. Filipowicz,W., Bhattacharyya, S.N., and Sonenberg, N. (2008). Mechanisms of post-transcriptional regulation by microRNAs: are the answers in sight. *Nat. Rev. Genet.* 9, 102–114.
7. Cheng, L.-C., Pastrana, E., Tava- zoie, M., and Doetsch, F. (2009). miR-124 regulates adult neurogenesis in the subventricular zone stem cell niche. *Nat. Neurosci.* 12, 399–408.
8. Schaefer, A., O'Carroll, D., Tan, C. L., Hillman, D., Sugimori, M., Llinas, R., and Greengard, P. (2007). Cerebellar neurodegeneration in the absence of microRNAs. *J. Exp. Med.* 204, 1553–1558.
9. Bushati, N., and Cohen, S. M. (2008). MicroRNAs in neuro-degeneration. *Curr. Opin. Neurobiol.* 18, 292–296.
10. Fiore, R., Siegel, G., and Schratt, G. (2008). MicroRNA function in neu- ronal development, plasticity and disease. *Biochim. Biophys. Acta* 1779, 471–478.
11. Schratt, G. (2009). microRNAs at the synapse. *Nat. Rev. Neurosci.* 10, 842–849.
12. Huang, W., and Li, M. D. (2008). Nicotine modulates expression of miR-140*, which targets the 3′ - untranslated region of dynamin 1 gene (Dnm1). *Int. J. Neuropsy- chopharmacol.* 12, 537.
13. Chandrasekar, V., and Dreyer, J.-L. (2009). microRNAs miR-124, let- 7d and miR-181a regulate cocaine- induced plasticity. *Mol. Cell. Neu- rosci.* 42, 350–362.

14. He, Y., Yang, C., Kirkmire, C. M., and Wang, Z. J. (2010). Regula- tion of opioid tolerance by let- 7 family microRNA targeting the opioid receptor. *J. Neurosci.* 30, 10251–10258.
15. Sathyan, P., Golden, H. B., and Miranda, R. C. (2007). Competing interactions between micro-RNAs determine neural progenitor survival and proliferation after ethanol exposure: evidence from an ex vivo model of the fetal cerebral cortical neuroepithelium. *J. Neurosci.* 27, 8546–8557.
16. Pietrzykowski, A. Z., Friesen, R. M., Martin, G. E., Puig, S. I., Nowak, C. L., Wynne, P. M., Siegelmann, H. T., and Treistman, S. N. (2008). Posttranscriptional regulation of BK channel splice variant stability by miR-9 underlies neuroadaptation to alcohol. *Neuron* 59, 274–287.
17. Miranda, R. C., Pietrzykowski, A. Z., Tang, Y., Sathyan, P., Mayfield, D., Keshavarzian, A., Sampson, W., and Hereld, D. (2010). MicroRNAs: master regulators of ethanol abuse and toxicity. *Alcohol. Clin. Exp. Res.* 34, 575–587.
18. Lewohl, J. M., Nunez, Y. O., Dood, P. R., Tiwari, G. R., Harris, R. A., and Mayfield, R. D. (2011). Up-regulation of MicroRNAs in brain of human alcoholics. *Alcohol. Clin. Exp. Res.* 35, 1928–1937.
19. Youssef-Elabd EM, McGee KC, Tripathi G, Aldaghri N, Abdalla MS, Sharada HM, et al. Acute and chronic saturated fatty acid treatment as a key instigator of the TLR-mediated inflammatory response in human adipose tissue, in vitro. The Journal of nutritional biochemistry. 2012;23(1):39–50.
20. Blüher M. The inflammatory process of adipose tissue. Pediatric endocrinology reviews: PER. 2008;6(1):24–31.
21. Kanasaki K, Koya D. Biology of obesity: lessons from animal models of obesity. BioMed Research International. 2011;2011.
22. Kahn SE, Hull RL, Utzschneider KM. Mechanisms linking obesity to insulin resistance and type 2 diabetes. Nature. 2006;444(7121):840.
23. Foley NH, O'Neill LA. miR-107: a Toll-like receptor-regulate\\d miRNA dysregulated in obesity and type II diabetes. Journal of leukocyte biology. 2012;92(3):521–7.
24. Trajkovski M, Hausser J, Soutschek J, Bhat B, Akin A, Zavolan M, et al. MicroRNAs 103 and 107 regulate insulin sensitivity. Nature. 2011;474 (7353):649.
25. Chen T, Huang Z, Wang L, Wang Y, Wu F, Meng S, et al. MicroRNA-125a-5p partly regulates the inflammatory response, lipid uptake, and ORP9 expression in oxLDL-stimulated monocyte/macrophages. Cardiovascular research. 2009;83(1):131–9.
26. Tang X, Muniappan L, Tang G, Özcan S. Identification of glucose-regulated miRNAs from pancreatic β cells reveals a role for miR-30d in insulin transcription. Rna. 2009;15(2):287–93.
27. Zaragosi L-E, Wdziekonski B, Le Brigand K, Villageois P, Mari B, Waldmann R, et al. Small RNA sequencing reveals miR-642a-3p as a novel adipocyte-specific microRNA and miR-30 as a key regulator of human adipogenesis. Genome biology. 2011;12(7):R64.

28. Yadav, S., Pandey, A., Shukla, A., Tal- welkar, S. S., Kumar, A., Pant, A. B., and Parmar, D. (2011). miR-497 and miR-302b regulate ethanol-induced neuronal cell death through BCL2 protein and cyclin D2. *J. Biol. Chem.* 286, 37347–37357.
29. Liu, J., Lewohl, J. M., Harris, R. A., Iyer, V. R., Dodd, P. R., Randall, P. K., and Mayfield, R. D. (2006b). Patterns of gene expression in the frontal cortex discriminate alcoholic from nonalcoholic individuals. *Neuropsychopharmacology* 31, 1574–1582.
30. Johansson, S., Ekström, T. J., Marinova, Z., Ökvist, A., Sheedy, D., Garrick, T., Harper, C., Kuzmin, A., Yakovleva, T., and Bakalkin, G. (2008). Dysregulation of cell death machinery in the prefrontal cortex of human alcoholics. *Int. J. Neuropsychopharmacol.* 12, 109.
31. Crews, F. T., Zou, J., and Qin, L. (2011). Induction of innate immune genes in brain create the neurobiology of addiction. *Brain Behav. Immun.* 25(Suppl. 1), S4–S12.
32. Kelley, K. W., and Dantzer, R. (2011). Alcoholism and inflammation: neuroimmunology of behavioral and mood disorders. *Brain Behav. Immun.* 25, S13–S20.
33. Yakovleva, T., Bazov, I., Watanabe, H., Hauser, K. F., and Bakalkin, G. (2011). Transcriptional control of maladaptive and protective responses in alcoholics: a role of the NF-κB system. *Brain Behav. Immun.* 25(Suppl. 1), S29–S38.
34. O'Neill, L. A. J. (2006). How Toll-like receptors signal: what we know and what we don't know. *Curr. Opin. Immunol.* 18, 3–9.
35. Pastor, I. J., Laso, F. J., Avila, J. J., Rodríguez, R. E., and González- Sarmiento, R. (2000). Polymor- phism in the interleukin-1 receptor antagonist gene is associated with alcoholism in Spanish men. *Alcohol. Clin. Exp. Res.* 24, 1479–1482.
36. Pastor, I. J., Laso, F. J., Romero, A., and González-Sarmiento, R. (2005). Interleukin-1 gene cluster polymorphisms and alcoholism in Spanish men. *Alcohol Alcohol.* 40, 181–186.
37. Edenberg, H. J., Xuei, X., Wetherill, L. F., Bierut, L., Bucholz, K., Dick, D. M., Hesselbrock, V., Kuperman, S., Por- jesz, B., Schuckit, M. A., Tischfield, J. A., Almasy, L. A., Nurnberger, J. I. Jr., and Foroud, T. (2008). Association of NFKB1, which encodes a subunit of the transcription factor NF-kappaB, with alcohol dependence. *Hum. Mol. Genet.* 17, 963–970.
38. Saiz, P. A., Garcia-Portilla, M. P., Flo- rez, G., Corcoran, P., Arango, C., Morales, B., Leza, J. C., Alvarez, S., Díaz, E. M., Alvarez, V., Coto, E., Nogueiras, L., and Bobes, J. (2009). Polymorphisms of the IL-1 gene complex are associated with alcohol dependence in Spanish Caucasians: data from an association study. *Alcohol. Clin. Exp. Res.* 33, 2147–2153.
39. O'Neill, L. A., Sheedy, F. J., and McCoy, C. E. (2011). MicroRNAs: the fine-tuners of Toll-like receptor signalling. *Nat. Rev. Immunol.* 11, 163–175.
40. Hou, J., Wang, P., Lin, L., Liu, X., Ma, F., An, H., Wang, Z., and Cao, X. (2009). MicroRNA-146a feedback inhibits RIG-I-dependent Type I IFN pro- duction in macrophages by targeting TRAF6, IRAK1, and IRAK2. *J. Immunol.* 183, 2150–2158.

41. Sonkoly, E., Ståhle, M., and Pivarcsi, A. (2008). MicroRNAs and immunity: novel players in the regulation of normal immune function and inflammation. *Semin. Cancer Biol.* 18, 13140.
42. Russo, S. J., Mazei-Robison, M. S., Ables, J. L., and Nestler, E. J. (2009). Neurotrophic factors and structural plasticity in addiction. *Neurophar- macology* 56(Suppl. 1), 73–82.
43. Robison, A. J., and Nestler, E. J. (2011). Transcriptional and epigenetic mechanisms of addiction. *Nat. Rev. Neurosci.* 12, 623–637.
44. Hyman, S. E., Malenka, R. C., and Nestler, E. J. (2006). Neural mechanisms of addiction: the role of reward-related learning and memory. *Annu. Rev. Neurosci.* 29, 565–598.
45. Sutton, M. A., and Schuman, E. M. (2006). Dendritic protein synthesis, synaptic plasticity, and memory. *Cell* 127, 49–58.
46. Schratt, G. (2009). microRNAs at the synapse. *Nat. Rev. Neurosci.* 10, 842–849.
47. Lugli, G., Torvik, V. I., Larson, J., and Smalheiser, N. R. (2008). Expression of microRNAs and their precursors in synaptic fractions of adult mouse forebrain. *J. Neurochem.* 106, 650–661.
48. Siegel, G., Obernosterer, G., Fiore, R., Oehmen, M., Bicker, S., Christensen, M., Khudayberdiev, S., Leuschner, P. F., Busch, C. J., Kane, C., Hübel, K., Dekker, F., Hedberg, C., Rengarajan, B., Drepper, C., Waldmann, H., Kauppinen, S., Greenberg, M. E., Draguhn, A., Rehmsmeier, M., Martinez, J., and Schratt, G. M. (2009). A functional screen implicates microRNA- 138-dependent regulation of the depalmitoylation enzyme APT1 in dendritic spine morphogenesis. *Nat. Cell Biol.* 11, 705–716.
49. Eipper-Mains, J. E., Kiraly, D. D., Palakodeti, D., Mains, R. E., Eip- per, B. A., and Graveley, B. R. (2011). microRNA-Seq reveals cocaine-regulated expression of striatal microRNAs. *RNA* 17, 1529–1543.
50. Ceman, S., and Saugstad, J. (2011). MicroRNAs: meta-controllers of gene expression in synaptic activity emerge as genetic and diagnostic markers of human disease. *Pharmacol. Ther.* 130, 26–37.
51. Pillai, R. S. (2005). Inhibition of translational initiation by Let-7 MicroRNA in human cells. *Science* 309, 1573–1576.
52. Contet, C., Kieffer, B. L., and Befort, K. (2004). Mu opioid receptor: a gateway to drug addiction. *Curr. Opin. Neurobiol.* 14, 370–378.
53. Wang, W. C. H., Juan, A. H., Pane- bra, A., and Liggett, S. B. (2011). MicroRNA let-7 establishes expression of β2-adrenergic receptors and dynamically down-regulates agonist-promoted down-regulation. *Proc. Natl. Acad. Sci. U.S.A.* 108, 6246–6251.
54. Otis, J. M., and Mueller, D. (2011). Inhibition of β-adrenergic receptors induces a persistent deficit in retrieval of a cocaine-associated memory providing protection against reinstatement. *Neuropsychopharmacology* 36, 1912–1920.
55. Chandrasekar, V., and Dreyer, J.-L. (2011). Regulation of MiR-124, Let- 7d, and MiR-181a in the accumbens affects the expression, extinction, and reinstatement of cocaine- induced conditioned place preference. *Neuropsychopharmacology* 36, 1149–1164.
56. Saba, L. M., Bennett, B., Hoffman, P. L., Barcomb, K., Ishii, T., Kechris, K., and Tabakoff, B. (2011). A systems genetic analysis of alcohol drinking by mice, rats

and men: influence of brain GABAergic transmission. *Neuropharmacology* 60, 1269–1280.
57. Huang, W., and Li, M. D. (2008). Nicotine modulates expression of miR-140∗, which targets the 3′ - untranslated region of dynamin 1 gene (Dnm1). *Int. J. Neuropsy- chopharmacol.* 12, 537.
58. Ferguson, S. M., Brasnjo, G., Hayashi, M., Wölfel, M., Collesi, C., Giovedi, S., Raimondi, A., Gong, L. W., Ariel, P., Paradise, S., O'Toole, E., Flavell, R., Cremona, O., Miesen- böck, G., Ryan, T. A., and De Camilli, P. (2007). A selective activity- dependent requirement for dynamin 1 in synaptic vesicle endocytosis. *Science* 316, 570–574.
59. Etheridge, N., Lewohl, J. M., Mayfield, R. D., Harris, R. A., and Dodd, P. R. (2009). Synaptic proteome changes in the superior frontal gyrus and occipital cortex of the alcoholic brain. *Proteomics Clin. Appl.* 3, 730–742.
60. Gorini, G., Ponomareva, O., Shores, K. S., Person, M. D., Harris, R. A., and Mayfield, R. D. (2010). Dynamin-1 co-associates with native mouse brain BKCa channels: proteomics analysis of synaptic protein complexes. *FEBS Lett.* 584, 845–851.
61. Simon, D. J., Madison, J. M., Con- ery, A. L., Thompson-Peer, K. L., Soskis, M., Ruvkun, G. B., Kaplan, J. M., and Kim, J. K. (2008). The MicroRNA miR-1 regulates a MEF- 2-dependent retrograde signal at neuromuscular junctions. *Cell* 133, 903–915.
62. Maze, I., and Nestler, E. J. (2011). The epigenetic landscape of addiction. *Ann. N. Y. Acad. Sci.* 1216, 99–113.
63. Robison, A. J., and Nestler, E. J. (2011). Transcriptional and epigenetic mechanisms of addiction. *Nat. Rev. Neurosci.* 12, 623–637.
64. Wong, C. C. Y., Mill, J., and Fernan- des, C. (2011). Drugs and addiction: an introduction to epigenetics. *Addiction* 106, 480–489.
65. Bönsch, D., Lenz, B., Reulbach, U., Kornhuber, J., and Bleich, S. (2004). Homocysteine associated genomic DNA hypermethylation in patients with chronic alcoholism. *J. Neural Transm.* 111, 1611–1616.
66. Bönsch, D., Lenz, B., Fiszer, R., Frieling, H., Kornhuber, J., and Bleich, S. (2006). Lowered DNA methyltransferase (DNMT-3b) mRNA expression is associated with genomic DNA hypermethylation in patients with chronic alcoholism. *J. Neural Transm.* 113, 1299–1304.
67. Muschler, M. A. N., Hillemacher, T., Kraus, C., Kornhuber, J., Bleich, S., and Frieling, H. (2010). DNA methylation of the POMC gene promoter is associated with craving in alcohol dependence. *J. Neural Transm.* 117, 513–519.
68. Heberlein, A., Muschler, M., Frieling, H., Behr, M., Eberlein, C., Wilhelm, J., Gröschl, M., Kornhuber, J., Ble- ich, S., and Hillemacher, T. (2011). Epigenetic down regulation of nerve growth factor during alcohol withdrawal. *Addict Biol.* doi:10.1111/ j.1369-1600.2010.00307.x
69. Taqi, M. M., Bazov, I., Watanabe, H., Sheedy, D., Harper, C., Alkass, K., Druid, H., Wentzel, P., Nyberg, F., Yakovleva, T., and Bakalkin, G. (2011). Prodynorphin CpG-SNPs associated with alcohol dependence: elevated methylation in the brain of human alcoholics. *Addict. Biol.* 16, 499–509.
70. Sato, F., Tsuchiya, S., Meltzer, S. J., and Shimizu, K. (2011). MicroRNAs and epigenetics. *FEBS J.* 278, 1598–1609.

71. Varambally, S., Cao, Q., Mani, R.-S., Shankar, S., Wang, X., Ateeq, B., Laxman, B., Cao, X., Jing, X., Ram- narayanan, K., Brenner, J. C., Yu, J., Kim, J. H., Han, B., Tan, P., Kumar-Sinha, C., Lonigro, R. J., Palanisamy, N., Maher, C. A., and Chinnaiyan, A. M. (2008). Genomic loss of microRNA-101 leads to overexpression of histone methyltransferase EZH2 in cancer. *Science* 322, 1695–1699.
72. Friedman, J. M., Liang, G., Liu, C.- C., Wolff, E. M., Tsai, Y. C., Ye, W., Zhou, X., and Jones, P. A. (2009). The putative tumor sup- pressor microRNA-101 modulates the cancer epigenome by repressing the polycomb group protein EZH2. *Cancer Res.* 69, 2623–2629.
73. Wellner, U., Schubert, J., Burk, U. C., Schmalhofer, O., Zhu, F., Son- ntag, A., Waldvogel, B., Vannier, C., Darling, D., zur Hausen, A., Brunton, V. G., Morton, J., San- som, O., Schüler, J., Stemmler, M. P., Herzberger, C., Hopt, U., Keck, T., Brabletz, S., and Brabletz, T. (2009). The EMT-activator ZEB1 promotes tumorigenicity by repressing stemness-inhibiting microRNAs. *Nat. Cell Biol.* 11, 1487–1495.
74. Tsuruta, T., Kozaki, K.-I., Uesugi, A., Furuta, M., Hirasawa, A., Imoto, I., Susumu, N., Aoki, D., and Inazawa, J. (2011). miR-152 is a tumor suppressor microRNA that is silenced by DNA hypermethylation in endometrial cancer. *Cancer Res.* 71, 6450–6462.
75. Rodríguez-Paredes, M., and Esteller, M. (2011). Cancer epigenetics reaches mainstream oncology. *Nat. Med.* 17, 330–339.
76. Tsai, H.-C., and Baylin, S. B. (2011). Cancer epigenetics: linking basic biology to clinical medicine. *Cell Res.* 21, 502–517.
77. Pandey, S. C., Ugalc, R., Zhang, H., Tang, L., and Prakash, A. (2008). Brain chromatin remodeling: a novel mechanism of alcoholism. *J. Neurosci.* 28, 3729–3737.
78. Kozaki, K.-I., Imoto, I., Mogi, S., Omura, K., and Inazawa, J. (2008). Exploration of tumor-suppressive microRNAs silenced by DNA hyper-methylation in oral cancer. *Cancer Res.* 68, 2094–2105.
79. Furuta, M., Kozaki, K.-I., Tanaka, S., Arii, S., Imoto, I., and Inazawa, J. (2010). miR-124 and miR-203 are epigenetically silenced tumor- suppressive microRNAs in hepato- cellular carcinoma. *Carcinogenesis* 31, 766–776.
80. Haslam, D.W.; James, W.P.T. Obesity. *Lancet* **2005**, *366*, 1197–1209.
81. van Vliet-Ostaptchouk JV, Snieder H, Lagou V. Gene–lifestyle interactions in obesity. Current nutrition reports. 2012;1(3):184–96.
82. Chopra M, Galbraith S, Darnton-Hill I. A global response to a global problem: the epidemic of overnutrition. Bulletin of the World Health Organization. 2002;80:952–8.
83. Youssef-Elabd EM, McGee KC, Tripathi G, Aldaghri N, Abdalla MS, Sharada HM, et al. Acute and chronic saturated fatty acid treatment as a key instigator of the TLR-mediated inflammatory response in human adipose tissue, in vitro. The Journal of nutritional biochemistry. 2012;23(1):39–50.
84. Blüher M. The inflammatory process of adipose tissue. Pediatric endocrinology reviews: PER. 2008;6(1):24–31
85. Kanasaki K, Koya D. Biology of obesity: lessons from animal models of obesity. BioMed Research International. 2011;2011.

86. Kahn SE, Hull RL, Utzschneider KM. Mechanisms linking obesity to insulin resistance and type 2 diabetes. Nature. 2006;444(7121):840.
87. Muurling M, Jong MC, Mensink RP, Hornstra G, Dahlmans VE, Pijl H, et al. A low-fat diet has a higher potential than energy restriction to improve high-fat diet-induced insulin resistance in mice. Metabolism-Clinical and Experimental. 2002;51(6):695–701.
88. Bartel DP. MicroRNAs: genomics, biogenesis, mechanism, and function. cell. 2004;116(2):281-97.
89. Deiuliis J. MicroRNAs as regulators of metabolic disease: pathophysiologic significance and emerging role as biomarkers and therapeutics. International journal of obesity. 2016;40(1):88.
90. Zhu H, Shyh-Chang N, Segrè AV, Shinoda G, Shah SP, Einhorn WS, et al. The Lin28/let-7 axis regulates glucose metabolism. Cell. 2011;147(1):81–94.
91. Sun T, Fu M, Bookout AL, Kliewer SA, Mangelsdorf DJ. MicroRNA let-7 regulates 3T3-L1 adipogenesis. Molecular endocrinology. 2009;23(6):925–31.
92. Viswanathan SR, Daley GQ, Gregory RI. Selective blockade of microRNA processing by Lin28. Science. 2008;320(5872):97–100.
93. Berndt, S.I.; Gustafsson, S.; Mägi, R.; Ganna, A.; Wheeler, E.; Feitosa, M.F.; Justice, A.E.; Monda, K.L.; Croteau-Chonka, D.C.; Day, F.R.; et al. Genome-wide meta-analysis identifies 11 new loci for anthropometric traits and provides insights into genetic architecture. *Nat. Genet.* **2013**, *45*, 501–512.
94. Kunej, T.; Jevsinek Skok, D.; Zorc, M.; Ogrinc, A.; Michal, J.J.; Kovac, M.; Jiang, Z. Obesity gene atlas in mammals. *J. Genom.* **2013**, *1*, 45–55.
95. Loos, R.J.F.; Yeo, G.S.H. The bigger picture of FTO: The first GWAS-identified obesity gene. *Nat. Rev. Endocrinol.* **2014**, *10*, 51–61.
96. Jiang, Q.; Wang, Y.; Hao, Y.; Juan, L.; Teng, M.; Zhang, X.; Li, M.; Wang, G.; Liu, Y. miR2Disease: A manually curated database for microRNA deregulation in human disease. *Nucleic Acids Res.* **2009**, *37*, D98–D104.
97. Dumortier, O.; Hinault, C.; Van Obberghen, E. MicroRNAs and metabolism crosstalk in energy homeostasis. *Cell Metab.* **2013**, *18*, 312–324.
98. Palmer, J.D.; Soule, B.P.; Simone, B.A.; Zaorsky, N.G.; Jin, L.; Simone, N.L. MicroRNA expression altered by diet: Can food be medicinal? *Ageing Res. Rev.* **2014**, *17*, 16–24.
99. Derghal, A.; Djelloul, M.; Trouslard, J.; Mounien, L. The Role of MicroRNA in the Modulation of the Melanocortinergic System. *Front. Neurosci.* **2017**, *11*, 181.
100. Morton, G.J.; Cummings, D.E.; Baskin, D.G.; Barsh, G.S.; Schwartz, M.W. Central nervous system control of food intake and body weight. *Nature* **2006**, *443*, 289–295.
101. Balthasar, N.; Coppari, R.; McMinn, J.; Liu, S.M.; Lee, C.E.; Tang, V.; Kenny, C.D.; McGovern, R.A.; Chua, S.C., Jr.; Elmquist, J.K.; et al. Leptin receptor signaling in POMC neurons is required for normal body weight homeostasis. *Neuron* **2004**, *42*, 983–991.
102. Coppari, R.; Ichinose, M.; Lee, C.E.; Pullen, A.E.; Kenny, C.D.; McGovern, R.A.; Tang, V.; Liu, S.M.; Ludwig, T.; Chua, S.C.; et al. The hypothalamic arcuate nucleus: A key site for mediating leptin's effects on glucose homeostasis and locomotor activity. *Cell Metab.* **2005**, *1*, 63–72.

103. Dhillon, H.; Zigman, J.M.; Ye, C.; Lee, C.E.; McGovern, R.A.; Tang, V.; Kenny, C.D.; Christiansen, L.M.; White, R.D.; Edelstein, E.A.; et al. Leptin directly activates SF1 neurons in the VMH, and this action by leptin is required for normal body-weight homeostasis. *Neuron* **2006**, *49*, 191–203.
104. Landrier, J.-F.; Karkeni, E.; Marcotorchino, J.; Bonnet, L.; Tourniaire, F. Vitamin D modulates adipose tissue biology: Possible consequences for obesity? *Proc. Nutr. Soc.* **2016**, *75*, 38–46.
105. Landrier, J.-F.; Marcotorchino, J.; Tourniaire, F. Lipophilic micronutrients and adipose tissue biology. *Nutrients* **2012**, *4*, 1622–1649.
106. Vienberg, S.; Geiger, J.; Madsen, S.; Dalgaard, L.T. MicroRNAs in metabolism. *Acta. Physiol. (Oxf.)* **2017**, *219*, 346–361.
107. Ortega, F.J.; Moreno-Navarrete, J.M.; Pardo, G.; Sabater, M.; Hummel, M.; Ferrer, A.; Rodriguez-Hermosa, J.I.; Ruiz, B.; Ricart, W.; Peral, B.; et al. MiRNA expression profile of human subcutaneous adipose and during adipocyte differentiation. *PLoS ONE* **2010**, *5*, e9022.
108. Martinelli, R.; Nardelli, C.; Pilone, V.; Buonomo, T.; Liguori, R.; Castanò, I.; Buono, P.; Masone, S.; Persico, G.; Forestieri, P.; et al. miR-519d overexpression is associated with human obesity. *Obesity (Silver Spring)* **2010**, *18*, 2170–2176.
109. Arner, P.; Kulyté, A. MicroRNA regulatory networks in human adipose tissue and obesity. *Nat. Rev. Endocrinol.* **2015**, *11*, 276–288.
110. Heneghan, H.M.; Miller, N.; Kerin, M.J. Role of microRNAs in obesity and the metabolic syndrome. *Obes. 'Rev.* **2010**, *11*, 354–361.
111. Maurizi, G.; Babini, L.; Della Guardia, L. Potential role of microRNAs in the regulation of adipocytes liposecretion and adipose tissue physiology. *J. Cell. Physiol.* **2018**.
112. Vohl, M.-C.; Sladek, R.; Robitaille, J.; Gurd, S.; Marceau, P.; Richard, D.; Hudson, T.J.; Tchernof, A. A survey of genes differentially expressed in subcutaneous and visceral adipose tissue in men. *Obes. Res.* **2004**, *12*, 1217–1222.
113. Klöting, N.; Berthold, S.; Kovacs, P.; Schön, M.R.; Fasshauer, M.; Ruschke, K.; Stumvoll, M.; Blüher, M. MicroRNA expression in human omental and subcutaneous adipose tissue. *PLoS ONE* **2009**, *4*, e4699.
114. Heneghan, H.M.; Miller, N.; McAnena, O.J.; O'Brien, T.; Kerin, M.J. Differential miRNA expression in omental adipose tissue and in the circulation of obese patients identifies novel metabolic biomarkers. *J. Clin. Endocrinol. Metab.* **2011**, *96*, E846–E850.
115. Keller, P.; Gburcik, V.; Petrovic, N.; Gallagher, I.J.; Nedergaard, J.; Cannon, B.; Timmons, J.A. Gene-chip studies of adipogenesis-regulated microRNAs in mouse primary adipocytes and human obesity. *BMC Endocr. Disord.* **2011**, *11*, 7.
116. Kristensen, M.M.; Davidsen, P.K.; Vigelsø, A.; Hansen, C.N.; Jensen, L.J.; Jessen, N.; Bruun, J.M.; Dela, F.; Helge, J.W. miRNAs in human subcutaneous adipose tissue: Effects of weight loss induced by hypocaloric diet and exercise. *Obesity (Silver Spring)* **2017**, *25*, 572–580.
117. Meerson, A.; Traurig, M.; Ossowski, V.; Fleming, J.M.; Mullins, M.; Baier, L.J. Human adipose microRNA-221 is upregulated in obesity and affects fat metabolism downstream of leptin and TNF-α. *Diabetologia* **2013**, *56*, 1971–1979.

118. Li, S.; Chen, X.; Zhang, H.; Liang, X.; Xiang, Y.; Yu, C.; Zen, K.; Li, Y.; Zhang, C.-Y. Differential expression of microRNAs in mouse liver under aberrant energy metabolic status. *J. Lipid Res.* **2009**, *50*, 1756–1765.
119. Iacomino, G.; Siani, A. Role of microRNAs in obesity and obesity-related diseases. *Genes Nutr.* **2017**, *12*, 23.
120. Chartoumpekis, D.V.; Zaravinos, A.; Ziros, P.G.; Iskrenova, R.P.; Psyrogiannis, A.I.; Kyriazopoulou, V.E.; Habeos, I.G. Differential expression of microRNAs in adipose tissue after long-term high-fat diet-induced obesity in mice. *PLoS ONE* **2012**, *7*, e34872.
121. Seeger, T.; Fischer, A.; Muhly-Reinholz, M.; Zeiher, A.M.; Dimmeler, S. Long-term inhibition of miR-21 leads to reduction of obesity in db/db mice. *Obesity (Silver Spring)* **2014**, *22*, 2352–2360.
122. Manning, P.; Munasinghe, P.E.; Bellae Papannarao, J.; Gray, A.R.; Sutherland, W.; Katare, R. Acute Weight Loss Restores Dysregulated Circulating MicroRNAs in Individuals Who Are Obese. *J. Clin. Endocrinol. Metab.* **2019**, *104*, 1239–1248.
123. Atkin, S.L.; Ramachandran, V.; Yousri, N.A.; Benurwar, M.; Simper, S.C.; McKinlay, R.; Adams, T.D.; Najafi-Shoushtari, S.H.; Hunt, S.C. Changes in Blood microRNA Expression and Early Metabolic Responsiveness 21 Days Following Bariatric Surgery. *Front. Endocrinol. (Lausanne)* **2018**, *9*, 773.
124. Turchinovich, A.; Weiz, L.; Langheinz, A.; Burwinkel, B. Characterization of extracellular circulating microRNA. *Nucleic Acids Res.* **2011**, *39*, 7223–7233.
125. Cortez, M.A.; Calin, G.A. MicroRNA identification in plasma and serum: A new tool to diagnose and monitor diseases. *Expert Opin. Biol. Ther.* **2009**, *9*, 703–711.
126. Guay, C.; Regazzi, R. Exosomes as new players in metabolic organ cross-talk. *Diabetes Obes. Metab.* **2017**, *19*, 137–146.
127. Deng, Z.; Poliakov, A.; Hardy, R.W.; Clements, R.; Liu, C.; Liu, Y.; Wang, J.; Xiang, X.; Zhang, S.; Zhuang, X.; et al. Adipose tissue exosome-like vesicles mediate activation of macrophage-induced insulin resistance. *Diabetes* **2009**, *58*, 2498–2505.
128. Thomou, T.; Mori, M.A.; Dreyfuss, J.M.; Konishi, M.; Sakaguchi, M.; Wolfrum, C.; Rao, T.N.; Winnay, J.N.; Garcia-Martin, R.; Grinspoon, S.K.; et al. Adipose-derived circulating miRNAs regulate gene expression in other tissues. *Nature* **2017**, *542*, 450–455.
129. Flehmig, G.; Scholz, M.; Klöting, N.; Fasshauer, M.; Tönjes, A.; Stumvoll, M.; Youn, B.-S.; Blüher, M. Identification of adipokine clusters related to parameters of fat mass, insulin sensitivity and inflammation. *PLoS ONE* **2014**, *9*, e99785.
130. Maurizi, G.; Della Guardia, L.; Maurizi, A.; Poloni, A. Adipocytes properties and crosstalk with immune system in obesity-related inflammation. *J. Cell. Physiol.* **2018**, *233*, 88–97.]
131. Xu, L.; Shi, C.; Xu, G.; Chen, L.; Zhu, L.; Zhu, L.; Guo, X.; Xu, M.; Ji, C. TNF-α, IL-6, and leptin increase the expression of miR-378, an adipogenesis-related microRNA in human adipocytes. *Cell Biochem. Biophys.* **2014**, *70*, 771–776.
132. Du, H.; Fu, Z.; He, G.; Wang, Y.; Xia, G.; Fang, M.; Zhang, T. MicroRNA-218 targets adiponectin receptor 2 to regulate adiponectin signaling. *Mol. Med. Rep.* **2015**, *11*, 4701–4705.

133. Derghal, A.; Djelloul, M.; Airault, C.; Pierre, C.; Dallaporta, M.; Troadec, J.-D.; Tillement, V.; Tardivel, C.; Bariohay, B.; Trouslard, J.; et al. Leptin is required for hypothalamic regulation of miRNAs targeting POMC 3′UTR. *Front. Cell. Neurosci.* **2015**, *9*, 172.
134. Benoit, C.; Ould-Hamouda, H.; Crepin, D.; Gertler, A.; Amar, L.; Taouis, M. Early leptin blockade predisposes fat-fed rats to overweight and modifies hypothalamic microRNAs. *J. Endocrinol.* **2013**, *218*, 35–47.
135. Crépin, D.; Benomar, Y.; Riffault, L.; Amine, H.; Gertler, A.; Taouis, M. The over-expression of miR-200a in the hypothalamus of ob/ob mice is linked to leptin and insulin signaling impairment. *Mol. Cell. Endocrinol.* **2014**, *384*, 1–11.
136. Vinnikov, I.A.; Hajdukiewicz, K.; Reymann, J.; Beneke, J.; Czajkowski, R.; Roth, L.C.; Novak, M.; Roller, A.; Dörner, N.; Starkuviene, V.; et al. Hypothalamic miR-103 protects from hyperphagic obesity in mice. *J. Neurosci.* **2014**, *34*, 10659–10674.
137. Croizier, S.; Park, S.; Maillard, J.; Bouret, S.G. Central Dicer-miR-103/107 controls developmental switch of POMC progenitors into NPY neurons and impacts glucose homeostasis. *eLife* **2018**, 7.
138. Derghal, A.; Djelloul, M.; Azzarelli, M.; Degonon, S.; Tourniaire, F.; Landrier, J.-F.; Mounien, L. MicroRNAs are involved in the hypothalamic leptin sensitivity. *Epigenetics* **2018**, *13*, 1127–1140.
139. Spiegelman, B.M.; Flier, J.S. Adipogenesis and Obesity: Rounding Out the Big Picture. *Cell* **1996**, *87*, 377–389.
140. Price, N.L.; Fernández-Hernando, C. miRNA regulation of white and brown adipose tissue differentiation and function. *Biochim. Biophys. Acta* **2016**, *1861*, 2104–2110.
141. Price, N.L.; Holtrup, B.; Kwei, S.L.; Wabitsch, M.; Rodeheffer, M.; Bianchini, L.; Suárez, Y.; Fernández-Hernando, C. SREBP-1c/MicroRNA 33b Genomic Loci Control Adipocyte Differentiation. *Mol. Cell. Biol.* **2016**, *36*, 1180–1193.
142. Xu, P.; Vernooy, S.Y.; Guo, M.; Hay, B.A. The Drosophila microRNA Mir-14 suppresses cell death and is required for normal fat metabolism. *Curr. Biol.* **2003**, *13*, 790–795.
143. Teleman, A.A.; Maitra, S.; Cohen, S.M. Drosophila lacking microRNA miR-278 are defective in energy homeostasis. *Genes Dev.* **2006**, *20*, 417–422.
144. Esau, C.; Kang, X.; Peralta, E.; Hanson, E.; Marcusson, E.G.; Ravichandran, L.V.; Sun, Y.; Koo, S.; Perera, R.J.; Jain, R.; et al. MicroRNA-143 regulates adipocyte differentiation. *J. Biol. Chem.* **2004**, *279*, 52361–52365.
145. Takanabe, R.; Ono, K.; Abe, Y.; Takaya, T.; Horie, T.; Wada, H.; Kita, T.; Satoh, N.; Shimatsu, A.; Hasegawa, K. Up-regulated expression of microRNA-143 in association with obesity in adipose tissue of mice fed high-fat diet. *Biochem. Biophys. Res. Commun.* **2008**, *376*, 728–732.
146. Lee, E.K.; Lee, M.J.; Abdelmohsen, K.; Kim, W.; Kim, M.M.; Srikantan, S.; Martindale, J.L.; Hutchison, E.R.; Kim, H.H.; Marasa, B.S.; et al. miR-130 suppresses adipogenesis by inhibiting peroxisome proliferator-activated receptor gamma expression. *Mol. Cell. Biol.* **2011**, *31*, 626–638.
147. Wang, Q.; Li, Y.C.; Wang, J.; Kong, J.; Qi, Y.; Quigg, R.J.; Li, X. miR-17-92 cluster accelerates adipocyte differentiation by negatively regulating tumor-suppressor Rb2/p130. *Proc. Natl. Acad. Sci. USA* **2008**, *105*, 2889–2894.

148. Ling, H.-Y.; Wen, G.-B.; Feng, S.-D.; Tuo, Q.-H.; Ou, H.-S.; Yao, C.H.; Zhu, B.-Y.; Gao, Z.-P.; Zhang, L.; Liao, D.-F. MicroRNA-375 promotes 3T3-L1 adipocyte differentiation through modulation of extracellular signal-regulated kinase signalling. *Clin. Exp. Pharmacol. Physiol.* **2011**, *38*, 239–246.
149. Tseng, Y.-H.; Cypess, A.M.; Kahn, C.R. Cellular bioenergetics as a target for obesity therapy. *Nat. Rev. Drug Discov.* **2010**, *9*, 465–482.
150. Marcotorchino, J.; Tourniaire, F.; Landrier, J.-F. Vitamin D, adipose tissue, and obesity. *Horm. Mol. Biol. Clin. Investig.* **2013**, *15*, 123–128.
151. Ge, Q.; Brichard, S.; Yi, X.; Li, Q. microRNAs as a New Mechanism Regulating Adipose Tissue Inflammation in Obesity and as a Novel Therapeutic Strategy in the Metabolic Syndrome. *J. Immunol. Res.* **2014**, *2014*.
152. Hulsmans, M.; De Keyzer, D.; Holvoet, P. MicroRNAs regulating oxidative stress and inflammation in relation to obesity and atherosclerosis. *FASEB J.* **2011**, *25*, 2515–2527.
153. Strum, J.C.; Johnson, J.H.; Ward, J.; Xie, H.; Feild, J.; Hester, A.; Alford, A.; Waters, K.M. MicroRNA 132 regulates nutritional stress-induced chemokine production through repression of SirT1. *Mol. Endocrinol.* **2009**, *23*, 1876–1884.
154. Arner, E.; Mejhert, N.; Kulyté, A.; Balwierz, P.J.; Pachkov, M.; Cormont, M.; Lorente-Cebrián, S.; Ehrlund, A.; Laurencikiene, J.; Hedén, P.; et al. Adipose tissue microRNAs as regulators of CCL2 production in human obesity. *Diabetes* **2012**, *61*, 1986–1993.
155. Chou, W.-W.; Wang, Y.-T.; Liao, Y.-C.; Chuang, S.-C.; Wang, S.-N.; Juo, S.-H.H. Decreased microRNA-221 is associated with high levels of TNF-α in human adipose tissue-derived mesenchymal stem cells from obese woman. *Cell. Physiol. Biochem.* **2013**, *32*, 127–137.
156. Lorente-Cebrián, S.; Mejhert, N.; Kulyté, A.; Laurencikiene, J.; Åström, G.; Hedén, P.; Rydén, M.; Arner, P. MicroRNAs regulate human adipocyte lipolysis: Effects of miR-145 are linked to TNF-α. *PLoS ONE* **2014**, *9*, e86800.
157. Karkeni, E.; Astier, J.; Tourniaire, F.; El Abed, M.; Romier, B.; Gouranton, E.; Wan, L.; Borel, P.; Salles, J.; Walrand, S.; et al. Obesity-associated Inflammation Induces microRNA-155 Expression in Adipocytes and Adipose Tissue: Outcome on Adipocyte Function. *J. Clin. Endocrinol. Metab.* **2016**, *101*, 1615–1626.
158. Xie, H.; Lim, B.; Lodish, H.F. MicroRNAs induced during adipogenesis that accelerate fat cell development are downregulated in obesity. *Diabetes* **2009**, *58*, 1050–1057.
159. Zhu, L.; Chen, L.; Shi, C.-M.; Xu, G.-F.; Xu, L.-L.; Zhu, L.-L.; Guo, X.-R.; Ni, Y.; Cui, Y.; Ji, C. MiR-335, an adipogenesis-related microRNA, is involved in adipose tissue inflammation. *Cell Biochem. Biophys.* **2014**, *68*, 283–290.
160. Kim, C.; Lee, H.; Cho, Y.M.; Kwon, O.-J.; Kim, W.; Lee, E.K. TNFα-induced miR-130 resulted in adipocyte dysfunction during obesity-related inflammation. *FEBS Lett.* **2013**, *23*, 3853–3858.
161. Shi, C.; Zhu, L.; Chen, X.; Gu, N.; Chen, L.; Zhu, L.; Yang, L.; Pang, L.; Guo, X.; Ji, C.; et al. IL-6 and TNF-α induced obesity-related inflammatory response through transcriptional regulation of miR-146b. *J. Interferon Cytokine Res.* **2014**, *34*, 342–348.
162. Karkeni, E.; Bonnet, L.; Marcotorchino, J.; Tourniaire, F.; Astier, J.; Ye, J.; Landrier, J.-F. Vitamin D limits inflammation-linked microRNA expression in

adipocytes in vitro and in vivo: A new mechanism for the regulation of inflammation by vitamin D. *Epigenetics* **2018**, *13*, 156–162.

163. Milagro FI, Campion J, Cordero P, Goyenechea E, Gomez-Uriz AM, Abete I, Zulet MA, Martinez JA (2011) A dual epigenomic approach for the search of obesity biomarkers: DNA methylation in relation to diet-induced weight loss. FASEB J 25(4):1378– 1389. https://doi.org/10.1096/fj.10-170365
164. Milagro FI, Miranda J, Portillo MP, Fernandez-Quintela A, Campion J, Martinez JA (2013) High-throughput sequencing of microRNAs in peripheral blood mononuclear cells: identification of potential weight loss biomarkers. PLoS ONE 8(1):e54319. https://doi.org/10.1371/journal.pone.0054319
165. Marques-Rocha JL, Milagro FI, Mansego ML, Zulet MA, Bres- san J, Martinez JA (2016) Expression of inflammation-related miRNAs in white blood cells from subjects with metabolic syn- drome after 8 wk of following a Mediterranean diet-based weight loss program. Nutrition 32(1):48–55. https://doi.org/10.1016/j.nut.2015.06.008
166. Garcia-Lacarte M, Martinez JA, Zulet MA, Milagro FI (2018) Implication of miR-612 and miR-1976 in the regulation of TP53 and CD40 and their relationship in the response to spe- cific weight-loss diets. PLoS ONE 13(8):e0201217. https://doi. org/10.1371/journal.pone.0201217
167. Garcia-Lacarte M, Mansego ML, Zulet MA, Martinez JA, Milagro FI (2019) miR-1185-1 and miR-548q are biomarkers of response to weight loss and regulate the expression of GSK3B. Cells 8:12. https://doi.org/10.3390/cells8121548
168. Beurel E (2011) Regulation by glycogen synthase kinase-3 of inflammation and T cells in CNS diseases. Front Mol Neurosci 4:18. https://doi.org/10.3389/fnmol.2011.00018
169. Hess AL, Larsen LH, Udesen PB, Sanz Y, Larsen TM, Dal- gaard LT (2020) Levels of circulating miR-122 are associated with weight loss and metabolic syndrome. Obes (Silver Spring) 28(3):493–501. https://doi.org/10.1002/oby.22704
170. Cai C, Lin Y, Yu C (2019) Circulating miRNAs as novel diag- nostic biomarkers in nonalcoholic fatty liver disease: a system- atic review and meta-analysis. Can J Gastroenterol Hepatol 2019:2096161. https://doi.org/10.1155/2019/2096161
171. 67. Shi Z, Zhao C, Guo X, Ding H, Cui Y, Shen R, Liu J (2014) Differential expression of microRNAs in omental adipose tis- sue from gestational diabetes mellitus subjects reveals miR-222 as a regulator of ERalpha expression in estrogen-induced insu- lin resistance. Endocrinology 155(5):1982–1990. https://doi. org/10.1210/en.2013-2046
172. Ortega FJ, Mercader JM, Moreno-Navarrete JM, Rovira O, Guerra E, Esteve E, Xifra G, Martinez C, Ricart W, Rieusset J, Rome S, Karczewska-Kupczewska M, Straczkowski M, Fernan- dez-Real JM (2014) Profiling of circulating microRNAs reveals common microRNAs linked to type 2 diabetes that change with insulin sensitization. Diabetes Care 37(5):1375–1383. https://doi. org/10.2337/dc13-1847
173. Filardi T, Catanzaro G, Mardente S, Zicari A, Santangelo C, Lenzi A, Morano S, Ferretti E (2020) Non-coding RNA: role in gestational diabetes pathophysiology and complications. Int J Mol Sci 21:11. https://doi.org/10.3390/ijms21114020

174. Ortega FJ, Mercader JM, Catalan V, Moreno-Navarrete JM, Pueyo N, Sabater M, Gomez-Ambrosi J, Anglada R, Fernan- dez-Formoso JA, Ricart W, Fruhbeck G, Fernandez-Real JM (2013) Targeting the circulating microRNA signature of obesity. Clin Chem 59(5):781–792. https://doi.org/10.1373/clinc hem.2012.195776
175. Giardina S, Hernandez-Alonso P, Diaz-Lopez A, Salas-Huetos A, Salas-Salvado J, Bullo M (2019) Changes in circulating miRNAs in healthy overweight and obese subjects: effect of diet com- position and weight loss. Clin Nutr 38(1):438–443. https://doi. org/10.1016/j.clnu.2017.11.014
176. Assmann TS, Riezu-Boj JI, Milagro FI, Martinez JA (2020) Circulating adiposity-related microRNAs as predictors of the response to a low-fat diet in subjects with obesity. J Cell Mol Med 24(5):2956–2967. https://doi .org/10.1111/jcmm.14920
177. Gurha P, Wang T, Larimore AH, Sassi Y, Abreu-Goodger C, Ramirez MO, Reddy AK, Engelhardt S, Taffet GE, Wehrens XH, Entman ML, Rodriguez A (2013) microRNA-22 pro- motes heart failure through coordinate suppression of PPAR/ ERR-nuclear hormone receptor transcription. PLoS ONE 8(9):e75882. https://doi.org/10.1371/journal.pone.0075882
178. Kurylowicz A, Owczarz M, Polosak J, Jonas MI, Lisik W, Jonas M, Chmura A, Puzianowska-Kuznicka M (2016) SIRT1 and SIRT7 expression in adipose tissues of obese and normal- weight individuals is regulated by microRNAs but not by meth- ylation status. Int J Obes (Lond) 40(11):1635–1642. https://doi. org/10.1038/ijo.2016.131
179. Manning P, Munasinghe PE, Bellae Papannarao J, Gray AR, Sutherland W, Katare R (2019) Acute weight loss restores dysregulated circulating microRNAs in individuals who are obese. J Clin Endocrinol Metab 104(4):1239–1248. https://doi. org/10.1210/jc.2018-00684
180. Cannataro R, Perri M, Gallelli L, Caroleo MC, De Sarro G, Cione E (2019) Ketogenic diet acts on body remodeling and microRNAs expression profile. Microrna 8(2):116–126. https ://doi.org/10.2174/2211536608666181126093903
181. Margolis LM, Rivas DA, Pasiakos SM, McClung JP, Ceglia L, Fielding RA (2017) Upregulation of circulating myomiR following short-term energy restriction is inversely associated with whole body protein synthesis. Am J Physiol Regul Integr Comp Physiol 313(3):R298–R304. https://doi.org/10.1152/ajpregu.00054.2017
182. Donghui T, Shuang B, Xulong L, Meng Y, Yujing G, Yujie H, Juan L, Dongsheng Y (2019) Improvement of microvascular endothelial dysfunction induced by exercise and diet is associ- ated with microRNA-126 in obese adolescents. Microvasc Res 123:86–91. https://doi.org/10.1016/j.mvr.2018.10.009
183. Arner P, Kulyte A (2015) MicroRNA regulatory networks in human adipose tissue and obesity. Nat Rev Endocrinol 11(5):276–288. https://doi.org/10.1038/nrendo.2015.25
184. Hulsmans M, Sinnaeve P, Van der Schueren B, Mathieu C, Jans- sens S, Holvoet P (2012) Decreased miR-181a expression in monocytes of obese patients is associated with the occurrence of metabolic syndrome and coronary artery disease. J Clin Endocrinol Metab 97(7):E1213-1218. https://doi.org/10.1210/jc.2012-1008

185. Nunez Lopez YO, Coen PM, Goodpaster BH, Seyhan AA (2017) Gastric bypass surgery with exercise alters plasma microRNAs that predict improvements in cardiometabolic risk. Int J Obes (Lond) 41(7):1121–1130. https://doi.org/10.1038/ijo.2017.84
186. Atkin SL, Ramachandran V, Yousri NA, Benurwar M, Simper SC, McKinlay R, Adams TD, Najafi-Shoushtari SH, Hunt SC (2018) Changes in blood microRNA expression and early meta- bolic responsiveness 21 days following bariatric surgery. Front Endocrinol (Lausanne) 9:773. https://doi.org/10.3389/fendo.2018.00773
187. Wensveen FM, Jelencic V, Valentic S, Sestan M, Wensveen TT, Theurich S, Glasner A, Mendrila D, Stimac D, Wunder- lich FT, Bruning JC, Mandelboim O, Polic B (2015) NK cells link obesity-induced adipose stress to inflammation and insulin resistance. Nat Immunol 16(4):376–385. https://doi.org/10.1038/ ni.3120
188. Giroud M, Karbiener M, Pisani DF, Ghandour RA, Beranger GE, Niemi T, Taittonen M, Nuutila P, Virtanen KA, Langin D, Scheideler M, Amri EZ (2016) Let-7i-5p represses brite adipo- cyte function in mice and humans. Sci Rep 6:28613. https://doi. org/10.1038/srep28613
189. Latreille M, Hausser J, Stutzer I, Zhang Q, Hastoy B, Gargani S, Kerr-Conte J, Pattou F, Zavolan M, Esguerra JL, Eliasson L, Rulicke T, Rorsman P, Stoffel M (2014) MicroRNA-7a regulates pancreatic beta cell function. J Clin Invest 124(6):2722–2735. https://doi.org/10.1172/JCI73066
190. Hubal MJ, Nadler EP, Ferrante SC, Barberio MD, Suh JH, Wang J, Dohm GL, Pories WJ, Mietus-Snyder M, Freishtat RJ (2017) Circulating adipocyte-derived exosomal MicroRNAs associated with decreased insulin resistance after gastric bypass. Obcs (Sil- vcr Spring) 25(1):102 110. https://doi.org/10.1002/oby.21709
191. Bae Y U ,Kim Y, Lee H, Kim H, Jeon J S, Noh H, Han D C, Ryu S, Kwon SH (2019) Bariatric surgery alters microRNA content of circulating exosomes in patients with obesity. Obes (Silver Spring) 27(2):264–271. https://doi.org/10.1002/oby.22379.
192. Jensen MD, Ryan DH, Apovian CM, Ard JD, Comuzzie AG, Donato KA, Hu FB, Hubbard VS, Jakicic JM, Kushner RF, Loria CM, Millen BE, Nonas CA, Pi-Sunyer FX, Stevens J, Stevens VJ, Wadden TA, Wolfe BM, Yanovski SZ, Jordan HS, Kendall KA, Lux LJ, Mentor-Marcel R, Morgan LC, Trisolini MG, Wnek J, Anderson JL, Halperin JL, Albert NM, Bozkurt B, Brindis RG, Curtis LH, DeMets D, Hochman JS, Kovacs RJ, Ohman EM, Pressler SJ, Sellke FW, Shen WK, Smith SC Jr, Tomaselli GF, American College of Cardiology, American Heart Association Task Force on Practice G, Obesity S (2014) 2013 AHA/ACC/TOS guideline for the management of overweight and obesity in adults: a report of the American College of Cardiology/American Heart Association Task Force on Practice Guidelines and The Obesity Society. Circulation 129(25 Suppl 2):S102-138. https:// doi.org/10.1161/01.cir.0000437739.71477.ee
193. McCafferty BJ, Hill JO, Gunn AJ (2020) Obesity: scope, lifestyle interventions, and medical management. Tech Vasc Interv Radiol 23(1):100653. https://doi.org/10.1016/j.tvir.2020.100653
194. Chin SH, Kahathuduwa CN, Binks M (2016) Physical activity and obesity: what we know and what we need to know. Obes Rev 17(12):1226–1244. https://doi.org/10.1111/obr.12460

195. Foster D, Sanchez-Collins S, Cheskin LJ (2017) Multidiscipli- nary team-based obesity treatment in patients with diabetes: current practices and the state of the science. Diabetes Spectr 30(4):244–249. https://doi.org/10.2337/ds17-0045
196. Trumbo P, Schlicker S, Yates AA, Poos M, Food, Nutrition Board of the Institute of Medicine TNA (2002) Dietary reference intakes for energy, carbohydrate, fiber, fat, fatty acids, choles- terol, protein and amino acids. J Am Diet Assoc 102(11):1621– 1630. https://doi.org/10.1016/s0002-8223(02)90346-9
197. Yannakoulia M, Poulimeneas D, Mamalaki E, Anastasiou CA (2019) Dietary modifications for weight loss and weight loss maintenance. Metabolism 92:153–162. https://doi.org/10.1016/j. metabol.2019.01.001
198. Arterburn DE, Courcoulas AP (2014) Bariatric surgery for obe- sity and metabolic conditions in adults. BMJ 349:g3961. https:// doi.org/10.1136/bmj.g3961
199. Abell TL, Minocha A (2006) Gastrointestinal complications of bariatric surgery: diagnosis and therapy. Am J Med Sci 331(4):214–218. https://doi.org/10.1097/00000441-20060 4000-00008
200. Svensson PA, Anveden A, Romeo S, Peltonen M, Ahlin S, Burza MA, Carlsson B, Jacobson P, Lindroos AK, Lonroth H, Maglio C, Naslund I, Sjoholm K, Wedel H, Soderpalm B, Sjostrom L, Carlsson LM (2013) Alcohol consumption and alcohol prob- lems after bariatric surgery in the Swedish obese subjects study. Obes (Silver Spring) 21(12):2444–2451. https://doi.org/10.1002/ oby.20397
201. Bouchard L, Rabasa-Lhoret R, Faraj M, Lavoie ME, Mill J, Perusse L, Vohl MC (2010) Differential epigenomic and tran- scriptomic responses in subcutaneous adipose tissue between low and high responders to caloric restriction. Am J Clin Nutr 91(2):309–320. https://doi.org/10.3945/ajcn.2009.28085
202. Campion J, Milagro FI, Martinez JA (2009) Individuality and epigenetics in obesity. Obes Rev 10(4):383–392. https://doi. org/10.1111/j.1467-789X.2009.00595.x
203. Milagro FI, Campion J, Cordero P, Goyenechea E, Gomez-Uriz AM, Abete I, Zulet MA, Martinez JA (2011) A dual epigenomic approach for the search of obesity biomarkers: DNA methylation in relation to diet-induced weight loss. FASEB J 25(4):1378– 1389. https://doi.org/10.1096/fj.10-170365
204. Goni L, Cuervo M, Milagro FI, Martinez JA (2015) Future perspectives of personalized weight loss interventions based on nutrigenetic, epigenetic, and metagenomic data. J Nutr 146(4):905S-912S. https://doi.org/10.3945/jn.115.218354

Chapter VI

1. Nana-Sinkam SP, Croce CM. MicroRNA regulation of tumorigenesis, cancer progression and interpatient heterogeneity: towards clinical use. Genome Biol. 2014;15:445.
2. Boon RA, Dimmeler S. MicroRNAs in myocardial infarction. Nat Rev Cardiol. 2015;12:135–142.
3. Issler O, Chen A. Determining the role of microRNAs in psychiatric disorders. Nat Rev Neurosci. 2015;16:201–212.
4. Liao Q, Wang B, Li X, Jiang G. miRNAs in acute myeloid leukemia. Oncotarget. 2017;8:3666–3682.

5. Garo LP, Murugaiyan G. Contribution of microRNAs to autoimmune diseases. Cell Mol Life Sci. 2016;73:2041–2051.
6. Valadi H, Ekström K, Bossios A, Sjöstrand M, Lee JJ, Lötvall JO. Exosome-mediated transfer of mRNAs and microRNAs is a novel mechanism of genetic exchange between cells. Nat Cell Biol. 2007;9:654–659.
7. Zhang Y, Liu D, Chen X, et al. Secreted monocytic miR-150 enhances targeted endothelial cell migration. Mol Cell. 2010;39:133–144.
8. Hergenreider E, Heydt S, Tréguer K, et al. Atheroprotective communication between endothelial cells and smooth muscle cells through miRNAs. Nat Cell Biol. 2012;14:249–256.
9. Mitchell PS, Parkin RK, Kroh EM, et al. Circulating microRNAs as stable blood-based markers for cancer detection. Proc Natl Acad Sci U S A. 2008;105:10513–10518.
10. Adams BD, Parsons C, Walker L, Zhang WC, Slack FJ. Targeting noncoding RNAs in disease. J Clin Invest. 2017;127:761–771.
11. Mehta A, Baltimore D. MicroRNAs as regulatory elements in immune system logic. Nat Rev Immunol. 2016;16:279–294.
12. Liu J, Shi K, Chen M, et al. Elevated miR-155 expression induces immunosuppression via CD39(+) regulatory T-cells in sepsis patient. Int J Infect Dis. 2015;40:135–141.
13. Sheng B, Zhao L, Zang X, Zhen J, Chen W. miR-375 ameliorates sepsis by downregulating miR-21 level via inhibiting JAK2-STAT3 signaling. Biomed Pharmacother. 2017;86:254–261.
14. Wang H, Bei Y, Shen S, et al. miR-21-3p controls sepsis-associated cardiac dysfunction via regulating SORBS2. J Mol Cell Cardiol. 2016;94:43 53.
15. Goodwin AJ, Guo C, Cook JA, Wolf B, Halushka PV, Fan H. Plasma levels of microRNA are altered with the development of shock in human sepsis: an observational study. Crit Care. 2015;19:440.
16. Wang HJ, Zhang PJ, Chen WJ, Feng D, Jia YH, Xie LX. Four serum microRNAs identified as diagnostic biomarkers of sepsis. J Trauma Acute Care Surg. 2012;73:850–854.
17. Wang JF, Yu ML, Yu G, et al. Serum miR-146a and miR-223 as potential new biomarkers for sepsis. Biochem Biophys Res Commun. 2010;394:184–188.
18. van der Heide V, Möhnle P, Rink J, Briegel J, Kreth S. Down-regulation of microRNA-31 in CD4+ T cells contributes to immunosuppression in human sepsis by promoting TH2 skewing. Anesthesiology. 2016;124:908–922.
19. Möhnle P, Schütz SV, van der Heide V, et al. MicroRNA-146a controls Th1-cell differentiation of human CD4+ T lymphocytes by targeting PRKCε. Eur J Immunol. 2015;45:260–272.
20. Zhou J, Chaudhry H, Zhong Y, et al. Dysregulation in microRNA expression in peripheral blood mononuclear cells of sepsis patients is associated with immunopathology. Cytokine. 2015;71:89–100.
21. Roderburg C, Luedde M, Vargas Cardenas D, et al. Circulating microRNA-150 serum levels predict survival in patients with critical illness and sepsis. PLoS One. 2013;8:e54612.

22. Tacke F, Roderburg C, Benz F, et al. Levels of circulating miR-133a are elevated in sepsis and predict mortality in critically ill patients. Crit Care Med. 2014;42:1096–1104.
23. Wang L, Wang HC, Chen C, et al. Differential expression of plasma miR-146a in sepsis patients compared with non-sepsis-SIRS patients. Exp Ther Med. 2013;5:1101–1104.
24. Ma Y, Vilanova D, Atalar K, et al. Genome-wide sequencing of cellular microRNAs identifies a combinatorial expression signature diagnostic of sepsis. PLoS One. 2013;8:e75918.
25. Han Y, Dai QC, Shen HL, Zhang XW. Diagnostic value of elevated serum miRNA-143 levels in sepsis. J Int Med Res. 2016;44:875–881.
26. Neudecker V, Brodsky KS, Kreth S, Ginde AA, Eltzschig HK. Emerging roles for microRNAs in perioperative medicine. Anesthesiology. 2016;124:489–506.
27. Ceppi M, Pereira PM, Dunand-Sauthier I, et al. MicroRNA-155 modulates the interleukin-1 signaling pathway in activated human monocyte-derived dendritic cells. Proc Natl Acad Sci U S A. 2009;106:2735–2740.
28. Yamagishi M, Nakano K, Miyake A, et al. Polycomb-mediated loss of miR-31 activates NIK-dependent NF-κB pathway in adult T cell leukemia and other cancers. Cancer Cell. 2012;21:121–135.
29. Wang X, Wang X, Liu X, et al. miR-15a/16 are upregulated in the serum of neonatal sepsis patients and inhibit the LPS-induced inflammatory pathway. Int J Clin Exp Med. 2015;8:5683–5690.
30. Yang L, Boldin MP, Yu Y, et al. miR-146a controls the resolution of T cell responses in mice. J Exp Med. 2012;209:1655–1670.
31. Li T, Morgan MJ, Choksi S, Zhang Y, Kim YS, Liu ZG. MicroRNAs modulate the noncanonical transcription factor NF-kappaB pathway by regulating expression of the kinase IKKalpha during macrophage differentiation. Nat Immunol. 2010;11:799–805.
32. Lawrence T, Fong C. The resolution of inflammation: anti-inflammatory roles for NF-kappaB. Int J Biochem Cell Biol. 2010;42:519–523.
33. Hayden MS, West AP, Ghosh S. NF-kappaB and the immune response. Oncogene. 2006;25:6758–6780.
34. Agweyu A, Kibore M, Digolo L, Kosgei C, Maina V, Mugane S, Muma S, Wachira J, Waiyego M and Maleche-Obimbo E: Prevalence and correlates of treatment failure among Kenyan children hospitalised with severe community-acquired pneu- monia: A prospective study of the clinical effectiveness of WHO pneumonia case management guidelines. Trop Med Int Health 19: 1310-1320, 2014.
35. Bhutta ZA, Das JK, Walker N, Rizvi A, Campbell H, Rudan I and Black RE; Lancet Diarrhoea and Pneumonia Interventions Study Group: Interventions to address deaths from childhood pneumonia and diarrhoea equitably: What works and at what cost? Lancet 381: 1417-1429, 2013.
36. Kim SJ, Kim K, Park SB, Hong DJ and Jhun BW: Outcomes of early administration of cidofovir in non-immunocompromised patients with severe adenovirus pneumonia. PLoS One 10: e0122642, 2015.
37. Nascimento-Carvalho CM: Etiology of childhood community acquired pneumonia and its implications for vaccination. Braz J Infect Dis 5: 87-97, 2001.

38. Pfuntner A, Wier LM, Stocks C (2013) Most frequent conditions in U.S. hospitals, 2010. Healthcare Cost and Utilization Project Statistical Brief #148 (Agency for Healthcare Research and Quality, Rockville, MD).
39. Marriott HM, Dockrell DH (2006) Streptococcus pneumoniae: The role of apoptosis in host defense and pathogenesis. Int J Biochem Cell Biol **38**:1848–1854.
40. Matuschak GM, Lechner AJ (2010) Acute lung injury and the acute respiratory distress syndrome: Pathophysiology and treatment. Mo Med **107**:252–258.
41. Jamieson AM, et al.(2013) Role of tissue protection in lethal respiratory viral-bacterial coinfection. Science **340**:1230–1234
42. Noble PW, Barkauskas CE, Jiang D (2012) Pulmonary fibrosis: Patterns and perpetrators. J Clin Invest **122**:2756–2762.
43. Steele MP, Schwartz DA (2013) Molecular mechanisms in progressive idiopathic pulmonary fibrosis. Annu Rev Med **64**:265–276.
44. Baddour JA, Sousounis K, Tsonis PA (2012) Organ repair and regeneration: An overview. Birth Defects Res C Embryo Today **96**:1–29.
45. Johnson R, Halder G (2014) The two faces of Hippo: Targeting the Hippo pathway for regenerative medicine and cancer treatment. Nat Rev Drug Discov **13**:63–79.
46. Bartel DP (2004) MicroRNAs: Genomics, biogenesis, mechanism, and function. Cell**116**:281–297.
47. Porrello ER (2013) microRNAs in cardiac development and regeneration. Clin Sci (Lond)**125**:151–166.
48. Tian Y, et al. (2011) Regulation of lung endoderm progenitor cell behavior by miR302/367. Development **138**:1235–1245.
49. Barkauskas CE, et al. (2013) Type 2 alveolar cells are stem cells in adult lung. J Clin Invest**123**:3025–3036.
50. Jain R, et al. (2015) Plasticity of Hopx(+) type I alveolar cells to regenerate type II cells in the lung. Nat Commun **6**:6727.
51. Rock JR, et al. (2011) Multiple stromal populations contribute to pulmonary fibrosis without evidence for epithelial to mesenchymal transition. Proc Natl Acad Sci USA **108**:E1475–E1483.
52. Kim SJ, Kim K, Park SB, Hong DJ and Jhun BW: Outcomes of early administration of cidofovir in non-immunocompromised patients with severe adenovirus pneumonia. PLoS One 10: e0122642, 2015.
53. Nascimento-Carvalho CM: Etiology of childhood community acquired pneumonia and its implications for vaccination. Braz J Infect Dis 5: 87-97, 2001.
54. Elenius V, Peltola V, Ruuskanen O, Ylihärsilä M and Waris M: Plasma procalcitonin levels in children with adenovirus infec- tion. Arch Dis Child 97: 582-583, 2012.
55. Girouard G, Garceau R, Thibault L, Bourque C, Bastien N and Li Y: Province-wide adenovirus type 3 outbreak with severe cases in New Brunswick. Can J Infect Dis Med Microbiol 22: e4-e6, 2011.
56. Lynch JP III, Fishbein M and Echavarria M: Adenovirus. Semin Respir Crit Care Med 32: 494-511, 2011

57. Lynch JP III and Kajon AE: Adenovirus: Epidemiology, global spread of novel serotypes, and advances in treatment and preven- tion. Semin Respir Crit Care Med 37: 586-602, 2016.
58. Rudan I, Chan KY, Zhang JSF, Theodoratou E, Feng XL, Salomon JA, Lawn JE, Cousens S, Black RE, Guo Y, *et al*: Causes of deaths in children younger than 5 years in China in 2008. Lancet 375: 1083-1089, 2010.
59. Lee J, Choi EH and Lee HJ: Clinical severity of respiratory adenoviral infection by serotypes in Korean children over 17 consecutive years (1991-2007). J Clin Virol 49: 115-120, 2010.
60. Liao J, Liu R, Yin LH and Pu YP: Expression profiling of exosomal miRNAs derived from human esophageal cancer cells by Solexa high-throughput sequencing. Int J Mol Sci 15: 15530-15551, 2014.
61. Pacholewska A, Kraft M, Gerber V and Jagannathan V: Differential expression of serum MicroRNAs supports $CD4^+$ T cell differentiation into Th2/Th17 cells in severe equine asthma. Genes (Basel) 8: pii: E383, 2017.
62. Tan KS, Choi H, Jiang X, Yin L, Seet JE, Patzel V, Engelward BP and Chow VT: Micro-RNAs in regenerating lungs: An integrative systems biology analysis of murine influenza pneumonia. BMC Genomics 15: 587, 2014.
63. Tan KS, Choi H, Jiang X, Yin L, Ju ES, Patzel V, Engelward BP and Chow VT: Micro-RNAs in regenerating lungs: An integrative systems biology analysis of murine influenza pneumonia. BMC Genomics 15: 1-13, 2014.
64. Li YJ, Ping C, Tang J and Zhang W: MicroRNA-455 suppresses non-small cell lung cancer through targeting ZEB1. Cell Biol Int 40: 621-628, 2016.
65. Rajasekaran S, Pattarayan D, Rajaguru P, Sudhakar Gandhi PS, Thimmulappa RK. MicroRNA regulation of acute lung injury and acute respiratory distress syndrome. J Cell Physiol. 2016;231:2097–2106.
66. Guo Z, Wen Z, Qin A, et al. Antisense oligonucleotide treatment enhances the recovery of acute lung injury through IL-10-secreting M2-like macrophage-induced expansion of CD4+ regulatory T cells. J Immunol. 2013;190:4337–4348.
67. Rao R, Nagarkatti P, Nagarkatti M. Role of miRNA in the regulation of inflammatory genes in staphylococcal enterotoxin B-induced acute inflammatory lung injury and mortality. Toxicol Sci. 2015;144:284–297.
68. Zeng Z, Gong H, Li Y, et al. Upregulation of miR-146a contributes to the suppression of inflammatory responses in LPS-induced acute lung injury. Exp Lung Res. 2013;39:275–282.
69. Vergadi E, Vaporidi K, Theodorakis EE, et al. Akt2 deficiency protects from acute lung injury via alternative macrophage activation and miR-146a induction in mice. J Immunol. 2014;192:394–406.
70. Guo Z, Gu Y, Wang C, et al. Enforced expression of miR-125b attenuates LPS-induced acute lung injury. Immunol Lett. 2014;162:18–26.
71. Starkey Lewis PJ, Dear J, Platt V, et al. Circulating microRNAs as potential markers of human drug-induced liver injury. Hepatology. 2011;54:1767–1776.
72. Starkey Lewis PJ, Merz M, Couttet P, et al. Serum microRNA biomarkers for drug-induced liver injury. Clin Pharmacol Ther. 2012;92:291–293.
73. Yamada H, Suzuki K, Ichino N, et al. Associations between circulating microRNAs (miR-21, miR-34a, miR-122 and miR-451) and non-alcoholic fatty liver. Clin Chim Acta. 2013;424:99–103.

74. Bala S, Tilahun Y, Taha O, et al. Increased microRNA-155 expression in the serum and peripheral monocytes in chronic HCV infection. J Transl Med. 2012;10:151.
75. Neudecker V, Haneklaus M, Jensen O, et al. Myeloid-derived miR-223 regulates intestinal inflammation via repression of the NLRP3 inflammasome. J Exp Med. 2017;214:1737–1752.
76. D'Alessandra Y, Devanna P, Limana F, et al. Circulating microRNAs are new and sensitive biomarkers of myocardial infarction. Eur Heart J. 2010;31:2765–2773.
77. Zile MR, Mehurg SM, Arroyo JE, Stroud RE, DeSantis SM, Spinale FG. Relationship between the temporal profile of plasma microRNA and left ventricular remodeling in patients after myocardial infarction. Circ Cardiovasc Genet. 2011;4:614–619.
78. Wang GK, Zhu JQ, Zhang JT, et al. Circulating microRNA: a novel potential biomarker for early diagnosis of acute myocardial infarction in humans. Eur Heart J. 2010;31:659–666.
79. Li C, Fang Z, Jiang T, et al. Serum microRNAs profile from genome-wide serves as a fingerprint for diagnosis of acute myocardial infarction and angina pectoris. BMC Med Genomics. 2013;6:16.
80. Zampetaki A, Willeit P, Tilling L, et al. Prospective study on circulating microRNAs and risk of myocardial infarction. J Am Coll Cardiol. 2012;60:290–299.
81. Schulte C, Molz S, Appelbaum S, et al. miRNA-197 and miRNA-223 predict cardiovascular death in a cohort of patients with symptomatic coronary artery disease. PLoS One. 2015;10:e0145930.
82. Vogel B, Keller A, Frese KS, et al. Multivariate miRNA signatures as biomarkers for non-ischaemic systolic heart failure. Eur Heart J. 2013;34:2812–2822.
83. Cakmak HA, Coskunpinar E, Ikitimur B, et al. The prognostic value of circulating microRNAs in heart failure: preliminary results from a genome-wide expression study. J Cardiovasc Med (Hagerstown). 2015;16:431–437.
84. Phillips CJ. Health economic and quality of life considerations in the management of pain. Drugs. 2003;63:43–46.
85. Andersen HH, Duroux M, Gazerani P. MicroRNAs as modulators and biomarkers of inflammatory and neuropathic pain conditions. Neurobiol Dis. 2014;71:159–168.
86. Bai G, Ambalavanar R, Wei D, Dessem D. Downregulation of selective microRNAs in trigeminal ganglion neurons following inflammatory muscle pain. Mol Pain. 2007;3:15.
87. von Schack D, Agostino MJ, Stuart Murray B, et al. Dynamic changes in the microRNA expression profile reveal multiple regulatory mechanisms in the spinal nerve ligation model of neuropathic pain. PLoS One. 2011;6:e17670.
88. Ho Kim S, Kim SH, Chung JM. An experimental model for peripheral neuropathy produced by segmental spinal nerve ligation in the rat. Pain. 1992;50:355–363.
89. Komori N, Takemori N, Kim HK, et al. Proteomics study of neuropathic and nonneuropathic dorsal root ganglia: altered protein regulation following segmental spinal nerve ligation injury. Physiol Genomics. 2007;29:215–23

90. Costigan M, Befort K, Karchewski L, et al. Replicate high-density rat genome oligonucleotide microarrays reveal hundreds of regulated genes in the dorsal root ganglion after peripheral nerve injury. BMC Neurosci. 2002;3:16.
91. Gold MS, Gebhart GF. Nociceptor sensitization in pain pathogenesis. Nat Med. 2010;16:1248–1257.
92. Basbaum AI, Bautista DM, Scherrer G, Julius D. Cellular and molecular mechanisms of pain. Cell. 2009;139:267–284.
93. Zhao J, Lee MC, Momin A, et al. Small RNAs control sodium channel expression, nociceptor excitability, and pain thresholds. J Neurosci. 2010;30:10860–10871.
94. Ni J, Gao Y, Gong S, Guo S, Hisamitsu T, Jiang X. Regulation of μ-opioid type 1 receptors by microRNA134 in dorsal root ganglion neurons following peripheral inflammation. Eur J Pain. 2013;17:313–323.
95. Bali KK, Hackenberg M, Lubin A, Kuner R, Devor M. Sources of individual variability: miRNAs that predispose to neuropathic pain identified using genome-wide sequencing. Mol Pain. 2014;10:22.
96. Lehmann SM, Krüger C, Park B, et al. An unconventional role for miRNA: let-7 activates Toll-like receptor 7 and causes neurodegeneration. Nat Neurosci. 2012;15:827–835.
97. Park CK, Xu ZZ, Berta T, et al. Extracellular microRNAs activate nociceptor neurons to elicit pain via TLR7 and TRPA1. Neuron. 2014;82:47–54.
98. Orlova IA, Alexander GM, Qureshi RA, et al. MicroRNA modulation in complex regional pain syndrome. J Transl Med. 2011;9:195.
99. Pohl K-W, Yeol J-F, Ongl W-Y. MicroRNA changes in the mouse prefrontal cortex after inflammatory pain. Eur J Pain. 2011;15:801.e1–801.e12.
100. Wei Q, et al. Targeted deletion of Dicer from proximal tubules protects against renal ischemia-reperfusion injury. J Am Soc Nephrol. 2010;21(5):756–61. doi:10.1681/asn.2009070718.
101. Wang JF, et al. Screening plasma miRNAs as biomarkers for renal ischemia-reperfusion injury in rats. Med Sci Monit. 2014;20:283–9. doi:10.12659/msm.889937.
102. Wang N, et al. Urinary microRNA-10a and microRNA-30d serve as novel, sensitive and specific biomarkers for kidney injury. PLoS One. 2012;7(12):e51140. doi:10.1371/journal.pone.0051140.
103. Ho J, et al. The pro-apoptotic protein Bim is a microRNA target in kidney progenitors. J Am Soc Nephrol. 2011;22(6):1053–63. doi:10.1681/asn.2010080841.
104. Kaucsar T, et al. Activation of the miR-17 family and miR-21 during murine kidney ischemia-reperfusion injury. Nucleic Acid Ther. 2013;23(5):344–54. doi:10.1089/nat.2013.0438.
105. Ma L, et al. Changes of miRNA-17-5p, miRNA-21 and miRNA-106a level during rat kidney ischemia-reperfusion injury. Zhonghua yi xue za zhi. 2015;95(19):1488–92.
106. Liang S, Wang W, Gou X. MicroRNA 26a modulates regulatory T cells expansion and attenuates renal ischemia-reperfusion injury. Mol Immunol. 2015;65(2):321–7. doi:10.1016/j.molimm.2015.02.003.

107. Aguado-Fraile E, et al. A pilot study identifying a set of microRNAs as precise diagnostic biomarkers of acute kidney injury. PLoS One. 2015;10(6): e0127175. doi:10.1371/journal.pone.0127175.
108. Liu F, et al. Upregulation of microRNA-210 regulates renal angiogenesis mediated by activation of VEGF signaling pathway under ischemia/perfusion injury in vivo and in vitro. Kidney Blood Press Res. 2012;35(3):182–91. doi:10.1159/000331054.
109. Lee CG, et al. Discovery of an integrative network of microRNAs and transcriptomics changes for acute kidney injury. Kidney Int. 2014;86(5): 943–53. doi:10.1038/ki.2014.117.
110. Aguado-Fraile E, et al. miR-127 protects proximal tubule cells against ischemia/reperfusion: identification of kinesin family member 3B as miR-127 target. PLoS One. 2012;7(9):e44305. doi:10.1371/journal.pone.0044305.
111. Godwin JG, et al. Identification of a microRNA signature of renal ischemia reperfusion injury. Proc Natl Acad Sci U S A. 2010;107(32):14339–44. doi:10.1073/pnas.0912701107.
112. Jenkins RH, et al. miR-192 induces G2/M growth arrest in aristolochic acid nephropathy. Am J Pathol. 2014;184(4):996–1009. doi:10.1016/j.ajpath.2013.12.028.
113. Kito N, Endo K, Ikesue M, Weng H, Iwai N. miRNA profiles of tubular cells: diagnosis of kidney injury. Biomed Res Int. 2015;2015:465479. doi:10.1155/2015/465479.
114. Chen R, et al. Regulation of IKKbeta by miR-199a affects NF-kappaB activity in ovarian cancer cells. Oncogene. 2008;27(34):4712–23. doi:10.1038/onc.2008.112.
115. Kim S, et al. MicroRNA miR-199a* regulates the MET proto-oncogene and the downstream extracellular signal-regulated kinase 2 (ERK2). J Biol Chem. 2008;283(26):18158–66. doi:10.1074/jbc.M800186200.
116. Rane S, et al. Downregulation of miR-199a derepresses hypoxia-inducible factor-1 alpha and Sirtuin 1 and recapitulates hypoxia preconditioning in cardiac myocytes. Circ Res. 2009;104(7):879–86. doi:10.1161/circresaha.108.193102.
117. Cantaluppi V, et al. Micro-vesicles derived from endothelial progenitor cells protect the kidney from ischemia-reperfusion injury by microRNA-dependent reprogramming of resident renal cells. Kidney Int. 2012;82(4):412–27. doi:10.1038/ki.2012.105
118. Denby L, et al. miR-21 and miR-214 are consistently modulated during renal injury in rodent models. Am J Pathol. 2011;179(2):661–72. doi:10.1016/j.ajpath.2011.04.021.
119. Lorenzen JM, et al. Circulating miR-210 predicts survival in critically ill patients with acute kidney injury. Clin J Am Soc Nephrol. 2011;6(7):1540–6. doi:10.2215/cjn.00430111.
120. Denby L, et al. MicroRNA-214 antagonism protects against renal fibrosis. J Am Soc Nephrol. 2014;25(1):65–80. doi:10.1681/asn.2013010072.
121. Denby L, et al. miR-21 and miR-214 are consistently modulated during renal injury in rodent models. Am J Pathol. 2011;179(2):661–72. doi:10.1016/j.ajpath.2011.04.021.

122. Lorenzen JM, et al. Circulating miR-210 predicts survival in critically ill patients with acute kidney injury. Clin J Am Soc Nephrol. 2011;6(7):1540–6. doi:10.2215/cjn.00430111.
123. Lan YF, et al. MicroRNA-494 reduces ATF3 expression and promotes AKI. J Am Soc Nephrol. 2012;23(12):2012–23. doi:10.1681/asn.2012050438.
124. Bhatt K, et al. MicroRNA-687 induced by hypoxia-inducible factor-1targets phosphatase and tension homolog in renal ischemia-reperfusion injury. J Am Soc Nephrol. 2015;26(7):1588–96. doi:10.1681/asn. 2014050463.
125. Janssen HL, et al. Treatment of HCV infection by targeting microRNA. N Engl J Med. 2013;368(18):1685–94. doi:10.1056/NEJMoa1209026.
126. Bosnjak ZJ, Logan S, Liu Y, Bai X. Recent insights into molecular mechanisms of propofol- induced developmental neurotoxicity: implications for the protective strategies. Anesth Analg. 2016;123:1286–1296.
127. Song Q, Ma YL, Song JQ, et al. Sevoflurane induces neurotoxicity in young mice through FAS/FASL signaling. Genet Mol Res. 2015;14:18059–18068.
128. Aydin M, Deveci U. Preedy VR. Pathophysiology of ketamine neurotoxicity. In: Neuropathology of Drug Addictions and Substance Misuse. 2016:London, UK: Elsevier, 563–72.
129. Twaroski D, Bosnjak ZJ, Bai X. MicroRNAs: new players in anesthetic-induced developmental neurotoxicity. Pharm Anal Acta. 2015;6:357.
130. Twaroski DM, Yan Y, Olson JM, Bosnjak ZJ, Bai X. Down-regulation of microRNA-21 is involved in the propofol-induced neurotoxicity observed in human stem cell-derived neurons. Anesthesiology. 2014;121:786–800.
131. Sun W, Pei L. microRNA expression profiling of propofol-treated developing rat hippocampal astrocytes. DNA Cell Biol. 2015;34:511–523.
132. Yan H, Xu T, Zhao H, Lee KC, Wang HY, Zhang Y. Isoflurane increases neuronal cell death vulnerability by downregulating miR-214. PLoS One. 2013;8:e55276.
133. Luo T, Yin S, Shi R, et al. miRNA expression profile and involvement of Let-7d-APP in aged rats with isoflurane-induced learning and memory impairment. PLoS One. 2015;10:e0119336.
134. Jiang XL, Du BX, Chen J, Liu L, Shao WB, Song J. MicroRNA-34a negatively regulates anesthesia-induced hippocampal apoptosis and memory impairment through FGFR1. Int J Clin Exp Pathol. 2014;7:6760–6767.
135. Huang C, Zhang X, Zheng J, Chen C, Chen Y, Yi J. Upregulation of miR-137 protects anesthesia-induced hippocampal neurodegeneration. Int J Clin Exp Pathol. 2014;7:5000–5007.
136. Bolkier Y, Nevo-Caspi Y, Salem Y, Vardi A, Mishali D, Paret G. Micro-RNA-208a, -208b, and -499 as biomarkers for myocardial damage after cardiac surgery in children. Pediatr Crit Care Med. 2016;17:e193–e197.
137. Yao Y, Du J, Cao X, et al. Plasma levels of microRNA-499 provide an early indication of perioperative myocardial infarction in coronary artery bypass graft patients. PLoS One. 2014;9:e104618.
138. Zhu W, Yang L, Shan H, et al. MicroRNA expression analysis: clinical advantage of propranolol reveals key microRNAs in myocardial infarction. PLoS One. 2011;6:e14736.

139. Izzotti A, Pulliero A. The effects of environmental chemical carcinogens on the microRNA machinery. Int J Hyg Environ Health. 2014;217:601–627.
140. Rupaimoole R, Slack FJ. MicroRNA therapeutics: towards a new era for the management of cancer and other diseases. Nat Rev Drug Discov. 2017;16:203–222.
141. Tutar L, Tutar E, Özgür A, Tutar Y. Therapeutic targeting of microRNAs in cancer: future perspectives. Drug Dev Res. 2015;76:382–388.
142. Janssen HL, Reesink HW, Lawitz EJ, et al. Treatment of HCV infection by targeting microRNA. N Engl J Med. 2013;368:1685–1694.

Chapter VII

1. WHO (2018a). *Global Health Estimates 2016: Disease Burden by Cause, Age, Sex, by Country and by Religion, 2000-2016.* Geneva: World Health Organization.
2. Kullolli M., Knouf E., Arampatzidou M., Tewari M., Pitteri S. J. (2014). Intact microRNA analysis using high resolution mass spectrometry. *J. Am. Soc. Mass Spectr.* 25 80–87. 10.1007/s13361-013-0759-x
3. Lanciotti R. S., Kosoy O. L., Laven J. J., Velez J. O., Lambert A. J., Johnson A. J., et al. (2008). Genetic and serologic properties of Zika virus associated with an epidemic, Yap State, Micronesia, 2007. *Emerg. Infect. Dis. J.* 14 1232–1239. 10.3201/eid1408.080287
4. Ito N., Moseley G. W., Sugiyama M. (2016). The importance of immune evasion in the pathogenesis of rabies virus. *J. Vet. Med. Sci.* 78 1089–1098. 10.1292/jvms.16-0092
5. Patro A. R. K. (2019). Subversion of immune response by human cytomegalovirus. *Front. Immunol.* 10:1155. 10.3389/fimmu.2019.01155
6. Stewart C. R., Marsh G. A., Jenkins K. A., Gantier M. P., Tizard M. L., Middleton D., et al. (2013). Promotion of Hendra virus replication by microRNA 146a. *J. Virol.* 87 3782–3791. 10.1128/JVI.01342-12
7. Biswas S., Haleyurgirisetty M., Lee S., Hewlett I., Devadas K. (2019). Development and validation of plasma miRNA biomarker signature panel for the detection of early HIV-1 infection. *EBioMedicine*42 307–316. 10.1016/j.ebiom.2019.04.023
8. Scheller N., Herold S., Kellner R., Bertrams W., Jung A. L., Janga H., et al. (2019). Proviral microRNAs detected in extracellular vesicles from bronchoalveolar lavage fluid of patients with influenza virus-induced acute respiratory distress syndrome. *J. Infect. Dis.* 219 540–543. 10.1093/infdis/jiy554
9. Hasegawa K., Perez-Losada M., Hoptay C. E., Epstein S., Mansbach J. M., Teach S. J., et al. (2018). RSV vs. rhinovirus bronchiolitis: difference in nasal airway microRNA profiles and NFkappaB signaling. *Pediatr. Res.* 83 606–614. 10.1038/pr.2017.309
10. Hou X., Liang Y., Chen J., Wei Y., Zeng P., Wang L., et al. (2017). Expression profiling of cellular microRNA in asymptomatic HBsAg carriers and chronic hepatitis b patients. *Biomed Res. Int.*2017:6484835. 10.1155/2017/6484835

11. Chen X., Ba Y., Ma L., Cai X., Yin Y., Wang K., et al. (2008). Characterization of microRNAs in serum: a novel class of biomarkers for diagnosis of cancer and other diseases. *Cell Res.* 18 997–1006. 10.1038/cr.2008.282
12. Farr R. J., Januszewski A. S., Joglekar M. V., Liang H., McAulley A. K., Hewitt A. W., et al. (2015a). A comparative analysis of high-throughput platforms for validation of a circulating microRNA signature in diabetic retinopathy. *Sci. Rep.* 5:10375. 10.1038/srep10375
13. Broekema NM, Imperiale MJ. miRNA regulation of BK polyomavirus replication during early infection. *Proc Natl Acad Sci U S A*(2013) 110:8200–5. doi:10.1073/pnas.1301907110
14. Bellutti F, Kauer M, Kneidinger D, Lion T, Klein R. Identification of RISC-associated adenoviral microRNAs, a subset of their direct targets, and global changes in the targetome upon lytic adenovirus 5 infection. *J Virol* (2015) 89:1608–27. doi:10.1128/JVI.02336-14
15. Hu M, Wang C, Li W, Lu W, Bai Z, Qin D, et al. A KSHV microRNA directly targets G protein-coupled receptor kinase 2 to promote the migration and invasion of endothelial cells by inducing CXCR2 and activating AKT sig- naling. *PLoS Pathog* (2015) 11:e1005171. doi:10.1371/journal.ppat.1005171
16. Kawano Y, Iwata S, Kawada J, Gotoh K, Suzuki M, Torii Y, et al. Plasma viral microRNA profiles reveal potential biomarkers for chronic active Epstein- Barr virus infection. *J Infect Dis* (2013) 208:771–9. doi:10.1093/infdis/jit222
17. Kim S, Lee S, Shin J, Kim Y, Evnouchidou I, Kim D, et al. Human cytomegalovirus microRNA miR-US4-1 inhibits CD8+ T cell responses by targeting the aminopeptidase ERAP1. *Nat Immunol* (2011) 12:984–91. doi:10.1038/ni.2097
18. Liu Y, Sun J, Zhang H, Wang M, Gao GF, Li X. Ebola virus encodes a miR- 155 analog to regulate importin-α5 expression. *Cell Mol Life Sci* (2016) 73:3733–44. doi:10.1007/s00018-016-2215-0
19. Liu X, Happel C, Ziegelbauer JM. Kaposi's sarcoma-associated herpesvirus microRNAs target GADD45B To protect infected cells from cell cycle arrest and apoptosis. *J Virol* (2017) 91:e02045–16. doi:10.1128/JVI.02045-16
20. Ho B-CC, Yu I-S, Lu L-F, Rudensky A, Chen H-Y, Tsai C-W, et al. Inhibition of miR-146a prevents enterovirus-induced death by restoring the production of type I interferon. *Nat Commun* (2014) 5:3344. doi:10.1038/ncomms4344
21. Chen Y, Gao D-Y, Huang L. In vivo delivery of miRNAs for cancer therapy: challenges and strategies. *Adv Drug Deliv Rev* (2015) 81:128–41. doi:10.1016/j.addr.2014.05.009
22. Beg MS, Brenner AJ, Sachdev J, Borad M, Kang Y-K, Stoudemire J, et al. Phase I study of MRX34, a liposomal miR-34a mimic, administered twice weekly in patients with advanced solid tumors. *Invest New Drugs* (2017) 35:180–8. doi:10.1007/s10637-016-0407-y
23. Reid G, Pel ME, Kirschner MB, Cheng YY, Mugridge N, Weiss J, et al. Restoring expression of miR-16: a novel approach to therapy for malignant pleural mesothelioma. *Ann Oncol* (2013) 24:3128–35. doi:10.1093/annonc/ mdt412
24. Ghorpade, D. S., Leyland, R., Kurowska-Stolarska, M., Patil, S. A., and Balajia, K. N. (2012). MicroRNA-155 is required for *Mycobacterium bovis* BCG-

mediated apoptosis of macrophages. *Mol. Cell. Biol.* 12, 2239–2253. doi: 10.1128/MCB. 06597- 11

25. Vegh, P., Magee, D. A., Nalpas, N. C., Bryan, K., McCabe, M. S., John, A., et al. (2015). MicroRNA profiling of the bovine alveolar macrophage response to *Mycobacterium bovis* infection suggests pathogen survival is enhanced by microRNA regulation of endocytosis and lysosome trafficking. *Tuberculosis* 95, 60–67. doi: 10.1016/j.tube.2014.10.011
26. Zheng, D., Yu, Y., Li, M., Wang, G., Chen, R., Fan, G., et al. (2015). Inhibition of microRNA 195 prevents apoptosis and multiple-organ injury in mouse models of sepsis. *J. Infect. Dis.* 213, 1661–1670. doi: 10.1093/infdis/ jiv760
27. Kumar, M., Sahu, S. K., Kumar, R., Subuddhi, A., Maji, R. K., Jana, K., et al. (2015). MicroRNA let-7 modulates the immune response to *Mycobacterium tuberculosis* infection via control of A20, an inhibitor of the NF-kB pathway. *Cell Host Microbe* 11, 345–356. doi: 10.1016/j.chom.2015.01.007
28. Rajaram, M. V., Ni, B., Morris, J. D., Brooks, M. N., Carlson, T. K., Bakthavachalu, B., et al. (2011). *Mycobacterium tuberculosis* lipomannan blocks TNF biosynthesis by regulating macrophage MAPK-activated protein kinase 2(MK2) and microRNA miR-125b. *Proc. Natl. Acad. Sci. U.S.A.* 108, 17408–17413. doi: 10.1073/pnas.1112660108
29. McGregor, R. A., and Choi, M. S. (2011). MicroRNAs in the regulation of adipogenesis and obesity. *Curr. Mol. Med.* 11, 304–316. doi: 10.2174/ 156652411795677990
30. Graff, J. W., Dickson, A. M., Clay, G., McCaffrey, A. P., and Wilson, M. E. (2012). Identifying functional microRNAs in macrophages with polarized phenotypes. *J. Biol. Chem.* 287, 21816–21825. doi: 10.1074/jbc.M111. 327031
31. Spizzo, R., Nicoloso, M. S., Lupini, L., Lu, Y., Fogarty, J., Rossi, S., et al. (2010). MiR- 145 participates with TP53 in a death-promoting regulatory loop and targets estrogen receptor-alpha in human breast cancer cells. *Cell Death Differ.* 17, 246–254. doi: 10.1038/cdd.2009.117
32. Keane, J., Remold, H. G., and Kornfeld, H. (2000). Virulent *Mycobacterium tuberculosis* strains evade apoptosis of infected alveolar macrophages. *J. Immunol.* 164, 2016–2020. doi: 10.4049/jimmunol.164.4.2016
33. Arcila, M. L., Sanchez, M. D., Ortiz, B., Barrera, L. F., Garcia, L. F., and Rojas, M. (2007). Activation of apoptosis, but not necrosis, during *Mycobacterium tuberculosis* infection correlated with decreased bacterial growth: role of TNF-alpha, IL-10, caspases and phospholipase A2. *Cell. Immunol.* 249, 80–93. doi: 10.1016/j.cellimm.2007.11.006
34. Starczynowski, D. T., Kuchenbauer, F., Argiropoulos, B., Sung, S., Morin, R., Muranyi, A., et al. (2010). Identification of miR-145 and miR-146a as mediators of the 5q- syndrome phenotype. *Nat. Med.* 16, 49–58. doi: 10.1038/nm. 2054
35. O'Neill, L. A., Sheedy, F. J., and McCoy, C. E. (2011). MicroRNAs: the fine tuners of Toll-like receptor signaling. *Nat. Rev. Immunol.* 11, 163–175. doi: 10.1038/ nri2957
36. Schepeler, T., Holm, A., Halvey, P., Nordentoft, I., Lamy, P., Riising, E. M., et al. (2012). Attenuation of the beta-catenin/TCF4 complex in colorectal cancer cells induces several growth-suppressive microRNAs that target cancer promoting genes. *Oncogene* 31, 50–60. doi: 10.1038/onc.2011.453

37. Kumarswamy, R., Mudduluru, G., Ceppi, P., Muppala, S., Kozlowski, M., Niklinski, J., et al. (2012). MicroRNA-30a inhibits epithelial-to-mesenchymal transition by targeting Snail and is downregulated in non-small cell lung cancer. *Int. J. Cancer* 130, 2044–2053. doi: 10.1002/ijc.26218
38. Kahlert, C., Klupp, F., Brand K., Lasitschka, F., Diederichs, S., Kirchberg, J., et al. (2011). Invasion front specific expression and prognostic significance of microRNA in colorectal liver metastases. *Cancer Sci.* 102, 1799–1807. doi: 10.1111/j.1349-7006.2011.02023.x
39. Rajaram, M. V., Ni, B., Morris, J. D., Brooks, M. N., Carlson, T. K., Bakthavachalu, B., et al. (2011). *Mycobacterium tuberculosis* lipomannan blocks TNF biosynthesis by regulating macrophage MAPK-activated protein kinase 2(MK2) and microRNA miR-125b. *Proc. Natl. Acad. Sci. U.S.A.* 108, 17408–17413. doi: 10.1073/pnas.1112660108
40. Ni, B., Rajaram, M. V. S., Lafuse, W. P., Landes, M. B., and Schlesinger, L. S. (2014). *Mycobacterium tuberculosis* decreases human macrophage IFN- g responsiveness through miR-132 and miR-26a. *J. Immunol.* 193, 4537–4547. doi: 10.4049/jimmunol.1400124
41. Zhang, Z. M., Zhang, A. R., Xu, M., Lou, J., and Qiu, W. Q. (2017). TLR-4/miRNA- 32-5p/FSTL1 signaling regulates mycobacterial survival and inflammatory responses in *Mycobacterium tuberculosis* infected macrophages. *Exp. Cell Res.* 352, 313–321. doi: 10.1016/j.yexcr.2017.02.025
42. Saltini C. Chemotherapy and diagnosis of tuberculosis. Respir Med. 2006; 100:2085–97.
43. Bhavsar AP, Guttman JA, Finlay BB. Manipulation of host-cell pathways by bacterial pathogens. Nature. 2007;449:827–34.
44. Mehta MD, Liu PT. microRNAs in mycobacterial disease: friend or foe? Front Genet. 2014;5:231.
45. Bezman NA, Cedars E, Steiner DF, Blelloch R, Hesslein DG, Lanier LL. Distinct requirements of microRNAs in NK cell activation, survival, and function. J Immunol (Baltimore, Md: 1950). 2010;185:3835–46.
46. O'Connell RM, Rao DS, Chaudhuri AA, Baltimore D. Physiological and pathological roles for microRNAs in the immune system. Nat Rev Immunol. 2010;10:111–22.
47. Maertzdorf J, Repsilber D, Parida SK, Stanley K, Roberts T, Black G, et al. Human gene expression profiles of susceptibility and resistance in tuberculosis. Genes Immun. 2011;12:15–22.
48. Marin ND, Paris SC, Rojas M, Garcia LF. Functional profile of CD4+ and CD8+ T cells in latently infected individuals and patients with active TB. Tuberculosis (Edinburgh, Scotland). 2013;93:155–66.
49. Sharbati J, Lewin A, Kutz-Lohroff B, Kamal E, Einspanier R, Sharbati S. Integrated microRNA-mRNA-analysis of human monocyte derived macrophages upon Mycobacterium avium subsp hominissuis infection. PloS one. 2011;6:e20258.
50. Lui YL, Tan TL, Woo WH, Timms P, Hafner LM, Tan KH, et al. Enterovirus71 (EV71) utilise host microRNAs to mediate host immune system enhancing survival during infection. PLoS One. 2014;9:e102997.

51. Du C, Liu C, Kang J, Zhao G, Ye Z, Huang S, et al. MicroRNA miR-326 regulates TH- 17 differentiation and is associated with the pathogenesis of multiple sclerosis. Nat Immunol. 2009;10:1252–9.
52. Ni B, Rajaram MV, Lafuse WP, Landes MB, Schlesinger LS. Mycobacterium tuberculosis decreases human macrophage IFN-gamma responsiveness through miR-132 and miR-26a. J Immunol (Baltimore, Md: 1950). 2014; 193:4537–47.
53. Xie T, Liang J, Guo R, Liu N, Noble PW, Jiang D. Comprehensive microRNA analysis in bleomycin-induced pulmonary fibrosis identifies multiple sites of molecular regulation. Physiol Genomics. 2011;43:479–87.
54. Fu Y, Yi Z, Wu X, Li J, Xu F. Circulating microRNAs in patients with active pulmonary tuberculosis. J Clin Microbiol. 2011;49:4246–51.
55. Rome S. Are extracellular microRNAs involved in type 2 diabetes and related pathologies? Clin Biochem. 2013;46:937–45.
56. Yi Z, Fu Y, Ji R, Li R, Guan Z. Altered microRNA signatures in sputum of patients with active pulmonary tuberculosis. PLoS One. 2012;7:e43184.
57. Ma F, Xu S, Liu X, Zhang Q, Xu X, Liu M, et al. The microRNA miR-29 controls innate and adaptive immune responses to intracellular bacterial infection by targeting interferongamma. Nat Immunol. 2011; 12:861–9.
58. Maudet C, Mano M, Eulalio A. MicroRNAs in the interaction between host and bacterial pathogens. FEBS Lett. 2014;588:4140–7.
59. Fu Y, Yi Z, Li J, Li R. Deregulated microRNAs in CD4+ T cells from individuals with latent tuberculosis versus active tuberculosis. J Cell Mol Med. 2014;18:503–13.
60. Kleinsteuber K, Heesch K, Schattling S, Kohns M, Sander-Julch C, Walzl G, et al. Decreased expression of miR-21, miR-26a, miR-29a, and miR-142-3p in CD4(+) T cells and peripheral blood from tuberculosis patients. PLoS One. 2013;8:e61609.
61. Wu Z, Lu H, Sheng J, Li L. Inductive microRNA-21 impairs anti-mycobacterial responses by targeting IL-12 and Bcl-2. FEBS Lett. 2012;586:2459–67.
62. Harapan H, Fitra F, Ichsan I, Mulyadi M, Miotto P, Hasan NA, et al. The roles of microRNAs on tuberculosis infection: meaning or myth? Tuberculosis (Edinburgh, Scotland). 2013;93:596–605.
63. Rajaram MV, Ni B, Morris JD, Brooks MN, Carlson TK, Bakthavachalu B, et al. Mycobacterium tuberculosis lipomannan blocks TNF biosynthesis by regulating macrophage MAPK-activated protein kinase 2 (MK2) and microRNA miR-125b. Proc Natl Acad Sci U S A. 2011;108:17408–13.
64. Ottenhoff TH. New pathways of protective and pathological host defense to mycobacteria. Trends Microbiol. 2012;20:419–28.
65. Pennini ME, Pai RK, Schultz DC, Boom WH, Harding CV. Mycobacterium tuberculosis 19-kDa lipoprotein inhibits IFN-gamma-induced chromatin remodeling of MHC2TA by TLR2 and MAPK signaling. Journal of immunology (Baltimore, Md : 1950). 2006;176:4323–30.
66. Tili E, Michaille JJ, Cimino A, Costinean S, Dumitru CD, Adair B, et al. Modulation of miR-155 and miR-125b levels following lipopolysaccharide/TNF-alpha stimulation and their possible roles in regulating the response to endotoxin shock. J Immunol (Baltimore, Md: 1950). 2007;179:5082–9.

67. Kumar R, Halder P, Sahu SK, Kumar M, Kumari M, Jana K, et al. Identification of a novel role of ESAT-6-dependent miR-155 induction during infection of macrophages with Mycobacterium tuberculosis. Cell Microbiol. 2012;14:1620–31.
68. Ghorpade DS, Leyland R, Kurowska-Stolarska M, Patil SA, Balaji KN. MicroRNA-155 is required for Mycobacterium bovis BCG-mediated apoptosis of macrophages. Mol Cell Biol. 2012;32:2239–53.
69. Wang J, Yang K, Zhou L, Minhaowu W. Y, Zhu M, et al. MicroRNA-155 promotes autophagy to eliminate intracellular mycobacteria by targeting Rheb. PLoS Pathog. 2013;9:e1003697.
70. Wan G, Xie W, Liu Z, Xu W, Lao Y, Huang N, et al. Hypoxia-induced MIR155 is a potent autophagy inducer by targeting multiple players in the MTOR pathway. Autophagy. 2014;10:70–9.
71. WHO global tuberculosis control report 2010. Summary. Cent Eur J Public Health. 2010;18:237. https://www.ncbi.nlm.nih.gov/pubmed/21361110.
72. Liu Y, Wang X, Jiang J, Cao Z, Yang B, Cheng X. Modulation of T cell cytokine production by miR-144* with elevated expression in patients with pulmonary tuberculosis. Mol Immunol. 2011;48:1084–90.
73. Wang C, Yang S, Sun G, Tang X, Lu S, Neyrolles O, et al. Comparative miRNA expression profiles in individuals with latent and active tuberculosis. PLoS One. 2011;6:e25832.
74. Spinelli SV, Diaz A, D'Attilio L, Marchesini MM, Bogue C, Bay ML, et al. Altered microRNA expression levels in mononuclear cells of patients with pulmonary and pleural tuberculosis and their relation with components of the immune response. Mol Immunol. 2013;53:265–9.
75. Jacobsen M, Repsilber D, Gutschmidt A, Neher A, Feldmann K, Mollenkopf HJ, et al. Candidate biomarkers for discrimination between infection and disease caused by Mycobacterium tuberculosis. Journal of molecular medicine (Berlin, Germany). 2007;85:613–21.
76. Qin Y, Wang Q, Zhou Y, Duan Y, Gao Q. Inhibition of IFN-gamma-induced nitric oxide dependent Antimycobacterial activity by miR-155 and C/ EBPbeta. Int J Mol Sci. 2016;17:535.
77. Dorhoi A, Iannaccone M, Farinacci M, Fae KC, Schreiber J, Moura-Alves P, et al. MicroRNA-223 controls susceptibility to tuberculosis by regulating lung neutrophil recruitment. J Clin Invest. 2013;123:4836–48.
78. Li S, Yue Y, Xu W, Xiong S. MicroRNA-146a represses mycobacteria-induced inflammatory response and facilitates bacterial replication via targeting IRAK-1 and TRAF- 6. PLoS One. 2013;8:e81438.
79. Astarie-Dequeker C, N'Diaye EN, Le Cabec V, Rittig MG, Prandi J, Maridonneau-Parini I. The mannose receptor mediates uptake of pathogenic and nonpathogenic mycobacteria and bypasses bactericidal responses in human macrophages. Infect Immun. 1999;67:469–77.
80. Spinelli SV, Diaz A, D'Attilio L, Marchesini MM, Bogue C, Bay ML, et al. Altered microRNA expression levels in mononuclear cells of patients with pulmonary and pleural tuberculosis and their relation with components of the immune response. Mol Immunol. 2013;53:265–9.

81. Espert L, Beaumelle B, Vergne I. Autophagy in Mycobacterium tuberculosis and HIV infections. Front Cell Infect Microbiol. 2015;5:49.
82. Russell DG. Mycobacterium tuberculosis and the intimate discourse of a chronic infection. Immunol Rev. 2011;240:252–68.
83. Vergne I, Gilleron M, Nigou J. Manipulation of the endocytic pathway and phagocyte functions by Mycobacterium tuberculosis lipoarabinomannan. Front Cell Infect Microbiol. 2014;4:187.
84. Lamb CA, Yoshimori T, Tooze SA. The autophagosome: origins unknown, biogenesis complex. Nature reviews. Mol Cell Biol. 2013;14:759–74.
85. Sabir N, Hussain T, Shah SZA, Peramo A, Zhao D. Zhou X miRNAs in Tuberculosis: New Avenues for Diagnosis and Host-Directed Therapy Frontiers in microbiology. 2018;9:602.
86. Velayati AA, Masjedi MR, Farnia P, Tabarsi P, Ghanavi J, ZiaZarifi AH, et al. Emergence of new forms of totally drug-resistant tuberculosis bacilli: super extensively drugresistant tuberculosis or totally drug-resistant strains in Iran. Chest. 2009;136:420–5
87. Chen Z, Wang T, Liu Z, Zhang G, Wang J, Feng S, et al. Inhibition of autophagy by MiR-30A induced by mycobacteria tuberculosis as a possible mechanism of immune escape in human macrophages. Jpn J Infect Dis. 2015;68:420–4.
88. Grassmann R, Jeang KT. The roles of microRNAs in mammalian virus infection. Biochim Biophys Acta. 2008;1779:706–11.
89. Xiao B, Liu Z, Li BS, Tang B, Li W, Guo G, et al. Induction of microRNA-155 during helicobacter pylori infection and its negative regulatory role in the inflammatory response. J Infect Dis. 2009;200:916–25.
90. Miotto P, Mwangoka G, Valcntc IC, Norbis L, Sotgiu G, Bosu R, et al. miRNA signatures in sera of patients with active pulmonary tuberculosis. PLoS One. 2013;8:e80149.
91. Maertzdorf J, Weiner J 3rd, Mollenkopf HJ, Network TB, Bauer T, Prasse A, et al. Common patterns and disease-related signatures in tuberculosis and sarcoidosis. Proc Natl Acad Sci U S A. 2012;109:7853–8.
92. Wu J, Lu C, Diao N, Zhang S, Wang S, Wang F, et al. Analysis of microRNA expression profiling identifies miR-155 and miR-155* as potential diagnostic markers for active tuberculosis: a preliminary study. Hum Immunol. 2012;73:31–7.
93. Guo, L., Zhao, J., Qu, Y., Yin, R., Gao, Q., Ding, S., et al. (2016). MicroRNA-20a inhibits autophagic process by targeting ATG7 and ATG16L1 and favors mycobacterial survival in macrophage cells. *Front. Cell. Infect. Microbiol.* 6:134. doi: 10.3389/fcimb.2016.00134
94. Chatterjee, S., Dwivedi, V. P., Singh, Y., Siddiqui, I., Sharma, P., Van Kaer, L., et al. (2011). Early secreted antigen ESAT-6 of *Mycobacterium tuberculosis* promotes protective T helper 17 cell responses in a Toll-Like receptor- 2-dependent manner. *PLoS Pathog.* 7:e1002378. doi: 10.1371/journal.ppat.1002378
95. Fu, Y., Yi, Z., Wu, X., Wu, X., Li, J., Xu, F., et al. (2011). Circulating microRNAs in patients with active pulmonary tuberculosis. *J. Clin. Microbiol.* 49, 4246–4251. doi: 10.1128/JCM.05459-11

96. Huang, Y., Shen, X. J., Zou, Q., Shen, X. J., Zou, Q., Wang, S. P., et al. (2011). Biological functions of microRNAs: a review. *J. Physiol. Biochem.* 67, 129–139. doi: 10.1007/s13105-010-0050-6
97. Li, D., Wang, T., Song, X., Qucuo, M., Yang, B., Zhang, J., et al. (2011). Genetic study of two single nucleotide polymorphisms within corresponding microRNAs and susceptibility to tuberculosis in a Chinese Tibetan and Han population. *Hum. Immunol.* 72, 598–602. doi: 10.1016/j.humimm.2011.03.004
98. Liu, Y. H., Wang, X. J., Jiang, J., Cao, Z. H., Yang, B. F., and Cheng, X. (2011). Modulation of T cell cytokine production by miR-144* with elevated expression in patients with pulmonary tuberculosis. *Mol. Immunol.* 48, 1084–1090.
99. Belver, L., Papavasiliou, F. N., and Ramiro, A. R. (2011). MicroRNA control of lymphocyte differentiation and function. *Curr. Opin. Immunol.* 23, 368–373. doi: 10.1016/j.coi.2011.02.001
100. Gantier, M. P. (2010). New perspectives in MicroRNA regulation of innate immunity. *J. Interferon Cytokine Res.* 30, 283–289. doi: 10.1089/jir.2010.0037
101. Liston, A., Linterman, M., and Lu, L. F. (2010). MicroRNA in the adaptive immune system, in sickness and in health. *J. Clin. Immunol.* 30, 339–346. doi: 10.1007/ s10875- 010- 9378- 5
102. Ma, F., Xu, S., Liu, X., Zhang, Q., Xu, X., Liu, M., et al. (2011). The microRNA miR- 29 controls innate and adaptive immune responses to intracellular bacterial infection by targeting interferon-γ. *Nat. Immunol.* 12, 861–869. doi: 10.1038/ ni.2073
103. Taganov, D., Boldin, M. P., Chang, K., and Baltimore, D. (2006). NF-κB-dependent induction of microRNA miR-146, an inhibitor targeted to signaling proteins of innate immune responses Konstantin. *Proc. Natl. Acad. Sci. U.S.A.* 33, 12481–12486. doi: 10.1073/pnas.0605298103
104. Li, S., Yue, Y., Xu, W., and Xiong, S. (2013). MicroRNA-146a represses mycobacteria-induced inflammatory response and facilitates bacterial replication via targeting IRAK-1 and TRAF-6. *PLoS One* 12:e81438. doi: 10.1371/journal.pone.0081438
105. Hacker, H., and Karin, M. (2006). Regulation and function of IKK and IKK-related kinases. *Sci. STKE* 2006:re13. doi: 10.1126/stke.3572006re13
106. Faraoni, I., Antonetti, F. R., Cardone, J., and Bonmassar, E. (2009). MiR-155 gene: a typical multifunctional microRNA. *Biochim. Biophys. Acta* 1792, 497–505. doi: 10.1016/j.bbadis.2009.02.013
107. Kutty, R. K., Nagineni, C. N., Samuel, W., Vijayasarathy, C., Hooks, J. J., and Redmond, T. M. (2010). Inflammatory cytokines regulate microRNA-155 expression in human retinal pigment epithelial cells by activating JAK/STAT pathway. *Biochem. Biophys. Res. Commun.* 402, 390–395. doi: 10.1016/j.bbrc. 2010.10.042
108. Wang, P., Hou, J., Lin, L., Wang, C., Liu, X., Li, D., et al. (2010). Inducible microRNA-155 feedback promotes type I IFN signaling in antiviral innate immunity by targeting suppressor of cytokine signaling 1. *J. Immunol.* 185, 6226–6233. doi: 10.4049/jimmunol.1000491
109. Savan, R. (2014). Post-transcriptional regulation of interferons and their signaling pathways. *J. Interferon Cytokine Res.* 5, 318–329. doi: 10.1089/jir.2013.0117

110. Forster, S. C., Michelle, D. T., and Hertzog, P. J. (2015). MicroRNA as type i interferon-regulated transcripts and modulators of the innate immune response. *Front. Immunol.* 6:334. doi: 10.3389/fimmu.2015.00334
111. Wang, J., Wu, M., Wen, J., Yang, K., Li, M., Zhan, X., et al. (2014). MicroRNA-155 induction by *Mycobacterium bovis* BCG enhances ROS production through targeting SHIP. *Mol. Immunol.* 62, 29–36. doi: 10.1016/j.molimm.2014. 05.012
112. O'Connell, R. M., Taganov, K. D., Boldin, M. P., Cheng, G., and Baltimore, D. (2007). MicroRNA-155 is induced during the macrophage inflammatory response. *Proc. Natl. Acad. Sci. U.S.A.* 5, 1604–1609. doi: 10.1073/pnas. 0610731104
113. Iwai, H., Funatogawa, K., Matsumura, K., Kato-Miyazawa, M., Kirikae, F., Kiga, K., et al. (2015). MicroRNA-155 knockout mice are susceptible to *Mycobacterium tuberculosis* infection. *Tuberculosis* 95, 246–250. doi: 10.1016/j.tube.2015.03.006
114. Rothchild, A. C., Sissonsa, J. R., Shafiani, S., Plaisier, C., Mina, D., Maia, D., et al. (2016). MiR-155–regulated molecular network orchestrates cell fate in the innate and adaptive immune response to *Mycobacterium tuberculosis*. *Proc. Natl. Acad. Sci. U.S.A.* 113, E6172–E6181.
115. Huang, J., Junhua, J., Xu, W., Zhao, H., Zhang, C., Shi, Y., et al. (2015). MiR-155 is up-regulated in patients with active tuberculosis and inhibits apoptosis of monocytes by targeting FOXO3. *Mol. Med. Rep.* 12, 7102–7108. doi: 10.3892/ mmr.2015.4250
116. Zhang, C., Xi, X., Wang, Q., Jiao, J., Zhang, L., and Zhao, H. (2015). The association between serum miR-155 and natural killer cells from tuberculosis patients. *Int. J. Clin. Exp. Med.* 8, 9168–9172.
117. Mayr, C., Hemann, M. T., and Bartel, D. P. (2007). Disrupting the pairing between let-7 and Hmga2 enhances oncogenic transformation. *Science* 315, 1576–1579. doi: 10.1126/science.1137999
118. Pillai, R. S., Bhattacharyya, S. N., and Filipowicz, W. (2007). Repression of protein synthesis by miRNAs: how many mechanisms? *Trends Cell Biol.* 17, 118–126. doi: 10.1016/j.tcb.2006.12.007
119. Dorhoi, A., Iannaccone, M., Farinacci, M., Faé, K. C., Schreiber, J., Moura-Alves, P., et al. (2013). MicroRNA-223 controls susceptibility to tuberculosis by regulating lung neutrophil recruitment. *J. Clin. Invest.* 123, 4836–4848. doi: 10.1172/ JCI67604
120. Wang, J. Y., Jia, Z., Wei, B., Zhou, Y., Niu, C., Bai, S., et al. (2017). MicroRNA-27a restrains the immune response to *Mycobacterium tuberculosis* infection by targeting IRAK4, a promoter of the NF-κB pathway. *Int. J. Clin. Exp. Pathol.* 10, 9894
121. Zhang, G., Liu, X., Wang, W., Cai, Y., Li, S., and Chen, Q. (2016). Down-regulation of miR-20a-5p triggers cell apoptosis to facilitate mycobacterial clearance through targeting JNK2 in human macrophages. *Cell Cycle* 15, 2527–2538. doi: 10.1080/15384101.2016.1215386 –9901.
122. Wang, Q., Liu, S., Tang, Y., Liu, Q., and Yao, Y. (2014). MPT64 protein from *Mycobacterium tuberculosis* inhibits apoptosis of macrophages through NF-kB-miRNA21-Bcl-2 pathway. *PLoS One* 9:e100949. doi: 10.1371/journal. pone.0100949

123. Lamb, C. A., Yoshimori, T., and Tooze, S. A. (2013). The autophagosome: origins unknown, biogenesis complex. *Nat. Rev. Mol. Cell Biol.* 14, 759–774. doi: 10.1038/nrm3696
124. Huang, J., and Brumell, J. H. (2014). Bacteria-autophagy interplay: a battle for survival. *Nat. Rev. Microbiol.* 12, 101–114. doi: 10.1038/nrmicro3160
125. Weiss, G., and Schaible, U. E. (2015). Macrophage defense mechanisms against intracellular bacteria. *Immunol. Rev.* 264, 182–203. doi: 10.1111/imr. 12266
126. Ahmad, S. (2011). Pathogenesis, immunology, and diagnosis of latent *Mycobacterium tuberculosis* infection. *Clin. Dev. Immunol.* 2011:814943. doi: 10.1155/2011/814943
127. Vergne, I., Chua, J., Singh, S. B., and Deretic, V. (2004). Cell biology of *Mycobacterium tuberculosis* phagosome. *Annu. Rev. Cell Dev. Biol.* 20, 367–394. doi: 10.1146/annurev.cellbio.20.010403.114015
128. Russell, D. G. (2011). *Mycobacterium tuberculosis* and the intimated is course of a chronic infection. *Immunol. Rev.* 240, 252–268. doi: 10.1111/j.1600-065X.2010. 00984.x
129. Espert, L., Beaumelle, B., and Vergne, I. (2015). Autophagy in *Mycobacterium tuberculosis* and HIV infections. *Front. Cell. Infect. Microbiol.* 5:49. doi: 10.3389/ fcimb.2015.00049
130. Deretic, V., Saitoh, T., and Akira, S. (2013). Autophagy in infection, inflammation and immunity. *Nat. Rev. Immunol.* 13, 722–737. doi: 10.1038/nri3532
131. Gu, X., Gao, Y., Mu, D. G., and Fu, E. Q. (2017). MiR-23a-5p modulates mycobacterial survival and autophagy during *Mycobacterium tuberculosis* infection through TLR2/MyD88/NF-κB pathway by targeting TLR2. *Exp. Cel Res.* 354, 71–77. doi: 10.1016/j.yexcr.2017.03.039
132. Guo, L., Zhao, J., Qu, Y., Yin, R., Gao, Q., Ding, S., et al. (2016). MicroRNA-20a inhibits autophagic process by targeting ATG7 and ATG16L1 and favors mycobacterial survival in macrophage cells. *Front. Cell. Infect. Microbiol.* 6:134. doi: 10.3389/fcimb.2016.00134
133. Bettencourt, P., Marion, S., Pires, D., Santos, L. F., Lastrucci, C., Carmo, N., et al. (2013). Actin-binding protein regulation by microRNAs as a novel microbial strategy to modulate phagocytosis by host cells: the case of N-Wasp and miR-142-3p. *Front. Cell. Infect. Microbiol.* 3:19. doi: 10.3389/fcimb.2013.00019
134. Lee, Y., Han, J., Yeom, K. H., Jin, H., and Kim, V. N. (2006). Drosha in primary microRNA processing. *Cold Spring Harb. Symp. Quant. Biol.* 71, 51–57. doi: 10.1101/sqb.2006.71.041
135. Meister, G., Landthaler, M., Dorsett, Y., and Tuschl, T. (2004). Sequence-specific inhibition of microRNA- and siRNA-induced RNA silencing. *RNA* 10, 544–550. doi: 10.1261/rna.5235104
136. Grimm, D., Streetz, K. L., Jopling, C. L., Storm, T. A., Pandey, K., Davis, C. R., et al. (2006). Fatality in mice due to oversaturation of cellular microRNA/short hairpin RNA pathways. *Nature* 7092, 537–541. doi: 10.1038/nature04791
137. Baumann, V., and Winkler, J. (2014). MiRNA-based therapies: strategies and delivery platforms for oligonucleotide and non-oligonucleotide agents. *Future Med. Chem.* 17, 1967–1984. doi: 10.4155/fmc.14.116

138. Uprichard, S. L. (2005). The therapeutic potential of RNA interference. *FEBS Lett.* 579, 5996–6007. doi: 10.1016/j.febslet.2005.08.004

139. Singh, Y., Kau, V., Mehra, A., Chatterjee, S., Tousif, S., Dwivedi, V. P., et al. (2013). *Mycobacterium tuberculosis* controls MicroRNA-99b (miR-99b) expression in infected murine dendritic cells to modulate host immunity. *J. Biol. Chem.* 288, 5056–5061. doi: 10.1074/jbc.C112.439778

140. Fu, X. D., Zeng, L., Liu, Z., Ke, X., Lei, L., and Li, G. (2016). MicroRNA-206 regulates the secretion of inflammatory cytokines and MMP9 expression by targeting TIMP3 in *Mycobacterium tuberculosis*–infected THP-1 human macrophages. *Biochem. Biophys. Res. Commun.* 477, 167–173. doi: 10.1016/j.bbrc.2016.06.038

141. Rosas-Taraco, A. G., Higgins, D. M., Sanchez-Campillo, J., Lee, E. J., Orme, I. M., and González-Juarrero, M. (2011). Local pulmonary immunotherapy with siRNA targeting TGFbeta1 enhances antimicrobial capacity in *Mycobacterium tuberculosis* infected mice. *Tuberculosis* 91, 98–106. doi: 10.1016/j.tube.2010.11.004

142. Bryniarski, K., Ptak, W., Jayakumar, A., Püllmann, K., Caplan, M. J., Chairoungdua, A., et al. (2013). Antigen-specific, antibody-coated, exosome-like nanovesicles deliver suppressor T-cell microRNA-150 to effector T cells to inhibit contact sensitivity. *J. Allergy Clin. Immunol.* 132, 170–181. doi: 10.1016/j.jaci.2013.04.048

143. Almanza, G., Anufreichik, V., Rodvold, J. J., Chiu, K. T., DeLaney, A., Akerset, J. C., et al. (2013). Synthesis and delivery of short, noncoding RNA by B lymphocytes. *Proc. Natl. Acad. Sci. U.S.A.* 110, 20182–20187. doi: 10.1073/pnas.1311145110

144. Tay, H. L., Kaiko, G. E., Plank, M., Li, J., Maltby, S., Essilfie, A. T., et al. (2015). Antagonism of miR-328 increases the antimicrobial function of macrophages and neutrophils and rapid clearance of nontypeable *Haemophilus influenzae* (NTHi) from infected lung. *PLoS Pathog.* 11:e1004549. doi: 10.1371/journal.ppat.1004549

145. Dacoba, T. G., Olivera, A., Torres, D., Crecente-Campo, J., and Alonso, M. J. (2017). Modulating the immune system through nanotechnology. *Semin. Immunol.* 34, 78–102. doi: 10.1016/j.smim.2017.09.007

146. Vázquez-Hernández, F., Granada-Ramírez, D. A., Arias-Cerón, J. S., Rodriguez-Fragoso, P., Mendoza-Álvarez, J. G. E., Ramón-Gallegos, A., et al. (2017). "Use of nanostructured materials in drug delivery," *Nanobiomaterials,* ed. R. Narayan (Sawston: Woodhead Publishing), 503–549.

147. Ayer, M., and Harm-Anton, K. (2017). Cell-mediated delivery of synthetic nano- and microparticles. *J. Control. Release* 259, 92–104. doi: 10.1016/j.jconrel.2017.01.048

148. Parodi, A., Molinaro, R., Sushnitha, M., Evangelopoulos, M., Martinez, J. O., Arrighetti, N., et al. (2017). Bio-inspired engineering of cell- and virus-like nanoparticles for drug delivery. *Biomaterials* 147, 155–168. doi: 10.1016/j.biomaterials.2017.09.020

149. Jiang, B., Yang, J., Rahoui, N., Taloub, N., and Huang, Y. D. (2017). Functional polymer materials affecting cell attachment. *Adv. Colloid Interface Sci.* 250, 185–194. doi: 10.1016/j.cis.2017.09.002

150. Tashima, T. (2016). Intelligent substance delivery into cells using cell-penetrating peptides. *Bioorg. Med. Chem. Lett.* 27, 121–130. doi: 10.1016/j.bmcl.2016.11.083
151. Prosperi, D., Colombo, M., Zanonia, I., and Granucci, F. (2017). Drug nanocarriers to treat autoimmunity and chronic inflammatory diseases. *Semin. Immunol.* 34, 61–67. doi: 10.1016/j.smim.2017.08.010
152. Zhou, Y., Zhang, L., Zhao, W., Wu, Y., Zhu, C., and Yang, Y. (2013). Nanoparticle- mediated delivery of TGF-β1 miRNA plasmid for preventing flexor tendon adhesion formation. *Biomaterials* 34, 8269–8278. doi: 10.1016/j.biomaterials. 2013.07.072
153. Pegtel, D. M., Cosmopoulos, K., Thorley-Lawson, D. A., van Eijndhoven, M. A., Hopmans, E. S., Lindenberg, J. L., et al. (2010). Functional delivery of viral miRNAs via exosomes. *Proc. Natl. Acad. Sci. U.S.A.* 107, 6328–6333. doi: 10.1073/pnas.0914843107
154. Duan, W., Yang, T., Zhou, S. F., Wang, Z. L., Zhou, Z. W., and He, Z. X. (2015). Novel targeting of PEGylated liposomes for codelivery of TGF-beta1 siRNA and four antitubercular drugs to human macrophages for the treatment of mycobacterial infection: a quantitative proteomic study. *Drug Des. Dev. Ther.* 9, 4441–4470. doi: 10.2147/DDDT.S79369
155. Moore, L. B., Sawyer, A. J., Saucier-Sawyer, J., Saltzman, W. M., and Kyriakides, T. R. (2016). Nanoparticle delivery of miR-223 to attenuate macrophage fusion. *Biomaterials* 89, 127–135. doi: 10.1016/j.biomaterials.2016.02.036
156. Tran, T., Krishnan, S., and Amiji, M. M. (2016). MicroRNA-223 induced repolarization of peritoneal macrophages using CD44 targeting hyaluronic acid nanoparticles for anti-inflammatory effects. *PLoS One* 11:e152024. doi: 10.1371/ journal.pone.0152024
157. Wang, J., Wu, M., Wen, J., Yang, K., Li, M., Zhan, X., et al. (2014). MicroRNA-155 induction by *Mycobacterium bovis* BCG enhances ROS production through targeting SHIP. *Mol. Immunol.* 62, 29–36. doi: 10.1016/j.molimm.2014. 05.012
158. Spizzo, R., Nicoloso, M. S., Lupini, L., Lu, Y., Fogarty, J., Rossi, S., et al. (2010). MiR- 145 participates with TP53 in a death-promoting regulatory loop and targets estrogen receptor-alpha in human breast cancer cells. *Cell Death Differ.* 17, 246–254. doi: 10.1038/cdd.2009.117
159. Belver, L., Papavasiliou, F. N., and Ramiro, A. R. (2011). MicroRNA control of lymphocyte differentiation and function. *Curr. Opin. Immunol.* 23, 368–373. doi: 10.1016/j.coi.2011.02.001
160. McGregor, R. A., and Choi, M. S. (2011). MicroRNAs in the regulation of adipogenesis and obesity. *Curr. Mol. Med.* 11, 304–316. doi: 10.2174/ 156652411795677990
161. Graff, J. W., Dickson, A. M., Clay, G., McCaffrey, A. P., and Wilson, M. E. (2012). Identifying functional microRNAs in macrophages with polarized phenotypes. *J. Biol. Chem.* 287, 21816–21825. doi: 10.1074/jbc.M111. 327031
162. Schaefer JS. MicroRNAs: how many in inflammatory bowel disease? *Curr Opin Gastroenterol* 32: 258–266, 2016. doi:10.1097/MOG. 0000000000000284.
163. Koukos G, Polytarchou C, Kaplan JL, Morley-Fletcher A, Gras- Miralles B, Kokkotou E, Baril-Dore M, Pothoulakis C, Winter HS, Iliopoulos D. MicroRNA-

124 regulates STAT3 expression and is down- regulated in colon tissues of pediatric patients with ulcerative colitis. *Gastroenterology* 145: 842–852.e2, 2013. doi:10.1053/j.gastro.2013.07. 001.

164. Koukos G, Polytarchou C, Kaplan JL, Oikonomopoulos A, Ziring D, Hommes DW, Wahed R, Kokkotou E, Pothoulakis C, Winter HS, Iliopoulos D. A microRNA signature in pediatric ulcerative colitis: deregulation of the miR-4284/CXCL5 pathway in the intestinal epithelium. *Inflamm Bowel Dis* 21: 996 – 1005, 2015. doi:10.1097/MIB. 0000000000000339.

165. Pierdomenico M, Cesi V, Cucchiara S, Vitali R, Prete E, Costanzo M, Aloi M, Oliva S, Stronati L. NOD2 is regulated by Mir-320 in physio- logical conditions but this control is altered in inflamed tissues of patients with inflammatory bowel disease. *Inflamm Bowel Dis* 22: 315–326, 2016. doi:10.1097/MIB.0000000000000659.

166. Huang Z, Shi T, Zhou Q, Shi S, Zhao R, Shi H, Dong L, Zhang C, Zeng K, Chen J, Zhang J. miR-141 regulates colonic leukocytic traf- ficking by targeting CXCL12 during murine colitis and human Crohn's disease. *Gut* 63: 1247–1257, 2014. doi:10.1136/gutjnl-2012-304213.

167. Dai X, Chen X, Chen Q, Shi L, Liang H, Zhou Z, Liu Q, Pang W, Hou D, Wang C, Zen K, Yuan Y, Zhang CY, Xia L. MicroRNA-193a-3p reduces intestinal inflammation in response to microbiota via down- regulation of colonic PepT1. *J Biol Chem* 290: 16,099–16,115, 2015. doi:10.1074/jbc.M115.659318.

168. ChenY,XiaoY,GeW,ZhouK,WenJ,YanW,WangY,WangB, Qu C, Wu J, Xu L, Cai W. miR-200b inhibits TGF-□1-induced epithe- lial-mesenchymal transition and promotes growth of intestinal epithelial cells. *Cell Death Dis* 4: e541, 2013. doi:10.1038/cddis.2013.22.

169. Wu W, He C, Liu C, Cao AT, Xue X, Evans-Marin HL, Sun M, Fang L, Yao S, Pinchuk IV, Powell DW, Liu Z, Cong Y. miR-10a inhibits dendritic cell activation and Th1/Th17 cell immune responses in IBD. *Gut* 64: 1755–1764, 2015. doi:10.1136/gutjnl-2014-307980.

170. Chen B, She S, Li D, Liu Z, Yang X, Zeng Z, Liu F. Role of miR-19a targeting TNF-□ in mediating ulcerative colitis. *Scand J Gastroenterol* 48: 815–824, 2013. doi:10.3109/00365521.2013.800991.

171. Wu F, Zikusoka M, Trindade A, Dassopoulos T, Harris ML, Bayless TM, Brant SR, Chakravarti S, Kwon JH. MicroRNAs are differentially expressed in ulcerative colitis and alter expression of macrophage inflam- matory peptide-2 alpha. *Gastroenterology* 135: 1624–1635.e24, 2008. doi:10.1053/j.gastro.2008.07.068.

172. Thorlacius-Ussing G, Schnack Nielsen B, Andersen V, Holmstrøm K, Pedersen AE. Expression and Localization of miR-21 and miR-126 in Mucosal Tissue from Patients with Inflammatory Bowel Disease. *Inflamm Bowel Dis* 23: 739–752, 2017.doi:10.1097/MIB.0000000000001086.

173. He C, Shi Y, Wu R, Sun M, Fang L, Wu W, Liu C, Tang M, Li Z, Wang P, Cong Y, Liu Z. miR-301a promotes intestinal mucosal inflammation through induction of IL-17A and TNF-□ in IBD. *Gut* 65: 1938– 1950, 2016. doi:10.1136/gutjnl-2015-309389.

174. Li M, Zhang S, QiuY,HeY,ChenB,MaoR,CuiY,ZengZ,Chen M. Upregulation of miR-665 promotes apoptosis and colitis in inflamma- tory bowel disease by

repressing the endoplasmic reticulum stress com- ponents XBP1 and ORMDL3. *Cell Death Dis* 8: e2699, 2017. doi:10. 1038/cddis.2017.76.

175. Min M, Peng L, Yang Y, Guo M, Wang W, Sun G. MicroRNA-155 is involved in the pathogenesis of ulcerative colitis by targeting FOXO3a. *Inflamm Bowel Dis* 20: 652–659, 2014. doi:10.1097/MIB. 0000000000000009.
176. Law IK, Bakirtzi K, Polytarchou C, Oikonomopoulos A, Hommes D, Iliopoulos D, Pothoulakis C. Neurotensin—regulated miR-133□ is in- volved in proinflammatory signalling in human colonic epithelial cells and in experimental colitis. *Gut* 64: 1095–1104, 2015. doi:10.1136/gutjnl- 2014-307329.
177. Bakirtzi K, Law IK, Xue X, Iliopoulos D, Shah YM, Pothoulakis C. Neurotensin promotes the development of colitis andntestinal angiogen- esis via HIF-1□-miR-210 signaling. *J Immunol* 196: 4311–4321, 2016. doi:10.4049/jimmunol.1501443.
178. Shi T, Xie Y, Fu Y, Zhou Q, Ma Z, Ma J, Huang Z, Zhang J, Chen J. The signaling axis of microRNA-31/interleukin-25 regulates Th1/Th17- mediated inflammation response in colitis. *Mucosal Immunol* 10: 983– 995, 2017. doi:10.1038/mi.2016.102.
179. Zhao Y, Ma T, Chen W, Chen Y, Li M, Ren L, Chen J, Cao R, Feng Y, Zhang H, Shi R. MicroRNA-124 promotes intestinal inflammation by targeting aryl hydrocarbon receptor in Crohn's disease. *J Crohn's Colitis* 10: 703–712, 2016. doi:10.1093/ecco-jcc/jjw010.
180. Wang H, Chao K, Ng SC, Bai A H, Yu Q, Yu J, Li M, Cui Y, Chen M, Hu JF, Zhang S. Pro-inflammatory miR-223 mediates the cross-talk between the IL23 pathway and the intestinal barrier in inflammatory bowel disease. *Genome Biol* 17: 58, 2016. doi:10.1186/s13059-016-0901-8.
181. Wu F, Zhang S, Dassopoulos T, Harris ML, Bayless TM, Meltzer SJ, Brant SR, Kwon JH. Identification of microRNAs associated with ileal and colonic Crohn's disease. *Inflamm Bowel Dis* 16: 1729–1738, 2010. doi:10.1002/ibd.21267.
182. Ando Y, Mazzurana L, Forkel M, Okazaki K, Aoi M, Schmidt PT, Mjösberg J, Bresso F. Downregulation of microRNA-21 in colonic CD3□ T cells in UC remission. *Inflamm Bowel Dis* 22: 2788 –2793, 2016. doi:10.1097/MIB.00000 00000000969.
183. Yang Y, Ma Y, Shi C, Chen H, Zhang H, Chen N, Zhang P, Wang F, Yang J, Yang J, Zhu Q, Liang Y, Wu W, Gao R, Yang Z, Zou Y, Qin H. Overexpression of miR-21 in patients with ulcerative colitis impairs intestinal epithelial barrier function through targeting the Rho GTPase RhoB. *Biochem Biophys Res Commun* 434: 746–752, 2013. doi:10.1016/ j.bbrc.2013.03.122.
184. ShiC,LiangY,YangJ,XiaY,ChenH,HanH,YangY,WuW,Gao R, Qin H. MicroRNA-21 knockout improve the survival rate in DSS induced fatal colitis through protecting against inflammation and tissue injury. *PLoS One* 8: e66814, 2013. doi:10.1371/journal.pone.0066814.
185. **Schaefer JS,** Attumi T, Opekun AR, Abraham B, Hou J, Shelby H, Graham DY, Streckfus C, Klein JR. MicroRNA signatures differentiate Crohn's disease from ulcerative colitis. *BMC Immunol* 16: 5, 2015. doi:10.1186/s12865-015-0069-0.
186. Cai M, Chen S, Hu W. MicroRNA-141 is involved in ulcerative colitis pathogenesis via aiming at CXCL5. *J Interferon Cytokine Res* 37: 415– 420, 2017. doi:10.1089/jir.2017.0019.

187. Huang Z, Shi T, Zhou Q, Shi S, Zhao R, Shi H, Dong L, Zhang C, Zeng K, Chen J, Zhang J. miR-141 regulates colonic leukocytic traf- ficking by targeting CXCL12□ during murine colitis and human Crohn's disease. *Gut* 63: 1247–1257, 2014. doi:10.1136/gutjnl-2012-304213.
188. Bumcrot D, Manoharan M, Koteliansky V, Sah DW. RNAi therapeu- tics: a potential new class of pharmaceutical drugs. *Nat Chem Biol* 2: 711–719, 2006. doi:10.1038/nchembio839.
189. Marafini I, Di Fusco D, Calabrese E, Sedda S, Pallone F, Monteleone G. Antisense approach to inflammatory bowel disease: prospects and challenges. *Drugs* 75: 723–730, 2015. doi:10.1007/s40265-015-0391-0.
190. Pan S, Yang X, Jia Y, Li Y, Chen R, Wang M, Cai D, Zhao R. Intravenous injection of microvesicle-delivery miR-130b alleviates high- fat diet-induced obesity in C57BL/6 mice through translational repression of PPAR-□. *J Biomed Sci* 22: 86, 2015. doi:10.1186/s12929-015-0193-4.
191. Yang J, Zhou CZ, Zhu R, Fan H, Liu XX, Duan XY, Tang Q, Shou ZX, Zuo DM. Microvesicles shuttled miR-200b attenuate experimental colitis associated intestinal fibrosis by inhibiting the development of EMT. *J Gastroenterol Hepatol* 32: 1966–1974, 2017. doi:10.1111/jgh.13797.
192. Yu X, Odenthal M, Fries JW. Exosomes as miRNA carriers: formation- function- future. *Int J Mol Sci* 17: 2028, 2016. doi:10.3390/ijms17122028.
193. Wu F, Guo NJ, Tian H, Marohn M, Gearhart S, Bayless TM, Brant SR, Kwon JH. Peripheral blood microRNAs distinguish active ulcerative colitis and Crohn's disease. *Inflamm Bowel Dis* 17: 241–250, 2011. doi:10.1002/ibd.21450.
194. Lewis A, Mehta S, Hanna LN, Rogalski LA, Jeffery R, Nijhuis A, Kumagai T, Biancheri P, Bundy JG, Bishop CL, Feakins R, Di Sabatino A, Lee JC, Lindsay JO, Silver A. Low serum levels of microRNA-19 are associated with a stricturing Crohn's disease phe- notype. *Inflamm Bowel Dis* 21: 1926–1934,2015. doi:10.1097/MIB. 0000000000000443.
195. Geremia A, Biancheri P, Allan P, et al. Innate and adaptive immunity in inflammatory bowel disease. Autoimmun Rev 2014;13:3–10.
196. Cuthbert AP, Fisher SA, Mirza MM, et al. The contribution of NOD2 gene mutations to the risk and site of disease in inflammatory bowel disease. Gastroenterology 2002;122:867–74.
197. Chen Y, Wang C, Liu Y, et al. miR-122 targets NOD2 to decrease intestinal epithelial cell injury in Crohn's disease. Biochem Biophys Res Commun 2013;438:133–9.
198. Ghorpade DS, Sinha AY, Holla S, et al. NOD2-nitric oxide-responsive microRNA-146a activates Sonic hedgehog signaling to orchestrate inflammatory responses in murine model of inflammatory bowel disease. J Biol Chem 2013;288:33037–48.
199. Chuang AY, Chuang JC, Zhai Z, et al. NOD2 expression is regulated by microRNAs in colonic epithelial HCT116 cells. Inflamm Bowel Dis 2014;20:126–35.
200. He X, Jing Z, Cheng G. MicroRNAs: new regulators of Toll-like receptor signalling pathways. Biomed Res Int 2014;2014:945169.
201. Thai T-H, Calado DP, Casola S, et al. Regulation of the germinal center response by microRNA-155. Science 2007;316:604–8.

202. Boldin MP, Taganov KD, Rao DS, et al. miR-146a is a significant brake on autoimmunity, myeloproliferation, and cancer in mice. J Exp Med 2011;208:1189–201.
203. Taganov KD, Boldin MP, Chang K-J, et al. NF-kappaB-dependent induction of microRNA miR-146, an inhibitor targeted to signaling proteins of innate immune responses. Proc Natl Acad Sci USA 2006;103:12481–6.
204. Ceppi M, Pereira PM, Dunand-Sauthier I, et al. MicroRNA-155 modulates the interleukin-1 signaling pathway in activated human monocyte-derived dendritic cells. Proc Natl Acad Sci USA 2009;106:2735–40.
205. O'Connell RM, Taganov KD, Boldin MP, et al. MicroRNA-155 is induced during the macrophage inflammatory response. Proc Natl Acad Sci USA 2007;104:1604–9.
206. Rodriguez A, Vigorito E, Clare S, et al. Requirement of bic/microRNA-155 for normal immune function. Science 2007;316:608–11.
207. Lu L-F, Thai T-H, Calado DP, et al. Foxp3-dependent microRNA155 confers competitive fitness to regulatory T cells by targeting SOCS1 protein. Immunity 2009;30:80–91.
208. Xu C, Ren G, Cao G, et al. miR-155 regulates immune modulatory properties of mesenchymal stem cells by targeting TAK1-binding protein 2. J Biol Chem 2013;288:11074–9.
209. Schulte LN, Westermann AJ, Vogel J. Differential activation and functional specialization of miR-146 and miR-155 in innate immune sensing. Nucleic Acids Res 2013;41:542–53.
210. Mazumder A, Bose M, Chakraborty A, et al. A transient reversal of miRNA-mediated repression controls macrophage activation. EMBO Rep 2013;14:1008–16.
211. Androulidaki A, Iliopoulos D, Arranz A, et al. The kinase Akt1 controls macrophage response to lipopolysaccharide by regulating microRNAs. Immunity 2009;31:220–31.
212. Chen X-M, Splinter PL, O'Hara SP, et al. A cellular micro-RNA, let-7i, regulates Toll-like receptor 4 expression and contributes to cholangiocyte immune responses against Cryptosporidium parvum infection. J Biol Chem 2007;282:28929–38.
213. Qi J, Qiao Y, Wang P, et al. microRNA-210 negatively regulates LPS-induced production of proinflammatory cytokines by targeting NF-κB1 in murine macrophages. FEBS Lett 2012;586:1201–7.
214. Bazzoni F, Rossato M, Fabbri M, et al. Induction and regulatory function of miR-9 in human monocytes and neutrophils exposed to proinflammatory signals. Proc Natl Acad Sci USA 2009;106:5282–7.
215. Morrison, C.R.; Plante, K.S.; Heise, M.T. Chikungunya Virus: Current Perspectives on a Reemerging Virus. *Microbiol. Spectr.* **2016**, *4*, 143–161.
216. Robinson, M.C. An epidemic of virsus disease in Southern Province, Tanganyika Territory, in 1952–53, Clinical Features. *Trans. R. Soc. Trop. Med. Hyg.* **1955**, *49*, 28–32.
217. Lumsden, W.H.R. An epidemic of virus disease in Southern Province, Tanganyika Territory, in 1952–53. II. General description and epidemiology. *Trans. R. Soc. Trop. Med. Hyg.* **1955**, *49*, 33–57.

218. Weaver, S.C.; Forrester, N.L. Chikungunya: Evolutionary history and recent epidemic spread. *Antivir. Res.* **2015**, *120*, 32–39.
219. Cunha, R.V.D.; Trinta, K.S. Chikungunya virus: Clinical aspects and treatment–A Review. *Mem. Inst. Oswaldo Cruz.* **2017**, *112*, 523–531.
220. Ganesan, V.K.; Duan, B.; Reid, S.P. Chikungunya Virus: Pathophysiology, Mechanism, and Modeling. *Viruses* **2017**, *9*, 368.
221. Rocha, V.F.D.; de Oliveira, A.H.P.; Bandeira, A.C.; Sardi, S.I.; Garcia, R.F.; Magalhães, S.A.; Sampaio, C.A.; Campos Soares, G. Chikungunya Virus Infection Associated with Encephalitis and Anterior Uveitis. *Ocul. Immunol. Inflamm.* **2018**, *26*, 677–679.
222. Mehta, R.; Gerardin, P.; Brito CAAd Soares, C.N.; Ferreira, M.L.B.; Solomon, T. The neurological complications of chikungunya virus: A systematic review. *Rev. Med. Virol.* **2018**, *28*, e1978.
223. Girardi, E.; Lopez, P.; Pfeffer, S. On the Importance of Host MicroRNAs During Viral Infection. *Front. Genet.* **2018**, *9*, 439.
224. Bruscella, P.; Bottini, S.; Baudesson, C.; Pawlotsky, J.M.; Feray, C.; Trabucchi, M. Viruses and miRNAs: More Friends than Foes. *Front. Microbiol.* **2017**, *8*, 824
225. Ardekani, A.M.; Naeini, M.M. The Role of MicroRNAs in Human Diseases. *Avicenna J. Med. Biotechnol.* **2010**, *2*, 161–179.
226. Sharma, A.; Balakathiresan, N.S.; Maheshwari, R.K. Chikungunya Virus Infection Alters Expression of MicroRNAs Involved in Cellular Proliferation, Immune Response and Apoptosis. *Intervirology* **2015**, *58*, 332–341.
227. Nan, Y.; Wu, C.; Zhang, Y.J. Interplay between Janus Kinase/Signal Transducer and Activator of Transcription Signaling Activated by Type I Interferons and Viral Antagonism. *Front. Immunol.* **2017**, *8*, 1758.
228. Lee, W.S.; Webster, J.A.; Madzokere, E.T.; Stephenson, E.B.; Herrero, L.J. Mosquito antiviral defense mechanisms: A delicate balance between innate immunity and persistent viral infection. *Parasit Vectors* **2019**, *12*, 165.
229. Fros, J.J.; Liu, W.J.; Prow, N.A.; Geertsema, C.; Ligtenberg, M.; Vanlandingham, D.L.; Schnettler, E.; Vlak, J.M.; Suhrbier, A.; Khromykh, A.A.; et al. Chikungunya virus nonstructural protein 2 inhibits type I/II interferon-stimulated JAK-STAT signaling. *J. Virol.* **2010**, *84*, 10877–10887.
230. Parashar, D.; Paingankar, M.S.; More, A.; Patil, P.; Amdekar, S. Altered microRNA expression signature in Chikungunya-infected mammalian fibroblast cells. *Virus Genes.* **2018**, *54*, 502–513.
231. Varghese, F.S.; Thaa, B.; Amrun, S.N.; Simarmata, D.; Rausalu, K.; Nyman, T.A.; Merits, A.; McInerney, G.M.; Ng, L.F.P.; Ahola, T. The Antiviral Alkaloid Berberine Reduces Chikungunya Virus-Induced Mitogen-Activated Protein Kinase Signaling. *J. Virol.* **2016**, *90*, 9743–9757.
232. Cimmino, A.; Calin, G.A.; Fabbri, M.; Iorio, M.V.; Ferracin, M.; Shimizu, M.; Wojcik, S.E.; Aqeilan, R.I.; Zupo, S.; Dono, M.; et al. miR-15 and miR-16 induce apoptosis by targeting BCL2. *Proc. Natl. Acad. Sci. USA* **2005**, *102*, 13944–13949.
233. Li, G.; Qiu, Z. Deletion of miR-15 Protects Against Rheumatoid Arthritis via Deregulating its Target Gene BCL2L2 and Repressing NF-κB Pathway. *Ann. Clin. Lab. Sci.* **2019**, *49*, 581–589.

234. Moran-Moguel, M.C.; Petarra-Del Rio, S.; Mayorquin-Galvan, E.E.; Zavala-Cerna, M.G. Rheumatoid Arthritis and miRNAs: A Critical Review through a Functional View. *J. Immunol. Res.* **2018**, *2018*, 2474529.
235. Zhu, G.; Chai, J.; Ma, L.; Duan, H.; Zhang, H. Downregulated microRNA-32 expression induced by high glucose inhibits cell cycle progression via PTEN upregulation and Akt inactivation in bone marrow-derived mesenchymal stem cells. *Biochem. Biophys. Res. Commun.* **2013**, *433*, 526–531.
236. Saxena, T.; Tandon, B.; Sharma, S.; Chameettachal, S.; Ray, P.; Ray, A.R.; Kulshreshtha, R. Combined miRNA and mRNA signature identifies key molecular players and pathways involved in chikungunya virus infection in human cells. *PLoS ONE* **2013**, *8*, e79886.
237. Uhrlaub, J.L.; Pulko, V.; DeFilippis, V.R.; Broeckel, R.; Streblow, D.N.; Coleman, G.D.; Park, B.S.; Lindo, J.F.; Vickers, I.; Anzinger, J.J.; et al. Dysregulated TGF-beta Production Underlies the Age-Related Vulnerability to Chikungunya Virus. *PLoS Pathog.* **2016**, *12*, e1005891.
238. Selvamani, S.P.; Mishra, R.; Singh, S.K. Chikungunya virus exploits miR-146a to regulate NF-kappaB pathway in human synovial fibroblasts. *PLoS ONE* **2014**, *9*, e103624.
239. Nahand, J.S.; Karimzadeh, M.R.; Nezamnia, M.; Fatemipour, M.; Khatami, A.; Jamshidi, S.; Moghoofei, M.; Taghizadieh, M.; Hajighadimi, S.; Shafiee, A.; et al. The role of miR-146a in viral infection. *IUBMB Life* **2020**, *72*, 343–360.
240. Agrawal, M.; Pandey, N.; Rastogi, M.; Dogra, S.; Singh, S.K. Chikungunya virus modulates the miRNA expression patterns in human synovial fibroblasts. *J. Med. Virol.* **2020**, *92*, 139–148.
241. Bartok, B.; Boyle, D.L.; Liu, Y.; Ren, P.; Ball, S.T.; Bugbee, W.D.; Rommel, C.; Firestein, G.S. PI3 kinase delta is a key regulator of synoviocyte function in rheumatoid arthritis. *Am. J. Pathol.* **2012**, *180*, 1906–1916.
242. Malemud, C.J. The PI3K/Akt/PTEN/mTOR pathway: A fruitful target for inducing cell death in rheumatoid arthritis? *Future Med. Chem.* **2015**, *7*, 1137–1147.
243. Smith, M.V.; Lee, M.J.; Islam, A.S.; Rohrer, J.L.; Goldberg, V.M.; Beidelschies, M.A.; Greenfield, E.M. Inhibition of the PI3K-Akt signaling pathway reduces tumor necrosis factor-alpha production in response to titanium particles in vitro. *J. Bone Joint Surg. Am.* **2007**, *89*, 1019–1027.
244. Thaa, B.; Biasiotto, R.; Eng, K.; Neuvonen, M.; Götte, B.; Rheinemann, L.; Mutso, M.; Utt, A.; Varghese, F.; Balistreri, G.; et al. Differential Phosphatidylinositol-3-Kinase-Akt-mTOR Activation by Semliki Forest and Chikungunya Viruses Is Dependent on nsP3 and Connected to Replication Complex Internalization. *J. Virol.* **2015**, *89*, 11420–11437.
245. Yang, R.; Xu, X.; Li, H.; Chen, J.; Xiang, X.; Dong, Z.; Zhang, D. p53 induces miR199a-3p to suppress SOCS7 for STAT3 activation and renal fibrosis in UUO. *Sci. Rep.* **2017**, *7*, 43409.
246. Kuchipudi, S.V. The Complex Role of STAT3 in Viral Infections. *J. Immunol. Res.* **2015**, *2015*, 272359.
247. Gao, G.; Luo, H. The ubiquitin-proteasome pathway in viral infections. *Can. J. Physiol. Pharmacol.* **2006**, *84*, 5–14.

248. Calistri, A.; Munegato, D.; Carli, I.; Parolin, C.; Palu, G. The ubiquitin-conjugating system: Multiple roles in viral replication and infection. *Cells* **2014**, *3*, 386–417.
249. Liu, J.; Huang, X.; Hao, S.; Wang, Y.; Liu, M.; Xu, J.; Zhang, X.; Yu, T.; Gan, S.; Dai, D.; et al. Peli1 negatively regulates noncanonical NF-kappaB signaling to restrain systemic lupus erythematosus. *Nat. Commun.* **2018**, *9*, 1136.
250. Rocha, V.F.D.; de Oliveira, A.H.P.; Bandeira, A.C.; Sardi, S.I.; Garcia, R.F.; Magalhães, S.A.; Sampaio, C.A.; Campos Soares, G. Chikungunya Virus Infection Associated with Encephalitis and Anterior Uveitis. *Ocul. Immunol. Inflamm.* **2018**, *26*, 677–679.
251. Mehta, R.; Gerardin, P.; Brito CAAd Soares, C.N.; Ferreira, M.L.B.; Solomon, T. The neurological complications of chikungunya virus: A systematic review. *Rev. Med. Virol.* **2018**, *28*, e1978.
252. Slota, J.A.; Booth, S.A. MicroRNAs in Neuroinflammation: Implications in Disease Pathogenesis, Biomarker Discovery and Therapeutic Applications. *Noncoding RNA* **2019**, *5*, 35.
253. Li, J.; Song, Q.; Shao, L.; Zhang, L.L.; Guo, X.H.; Mao, Y.J. MiR-124a inhibits proliferation and invasion of rheumatoid arthritis synovial fibroblasts. *Eur. Rev. Med. Pharmacol. Sci.* **2018**, *22*, 4581–4588.
254. Tchankouo-Nguetcheu, S.; Bourguet, E.; Lenormand, P.; Rousselle, J.C.; Namane, A.; Choumet, V. Infection by chikungunya virus modulates the expression of several proteins in Aedes aegypti salivary glands. *Parasit Vectors* **2012**, *5*, 1–11.
255. Dubey, S.K.; Shrinet, J.; Sunil, S. Aedes aegypti microRNA, miR-2944b-5p interacts with 3'UTR of chikungunya virus and cellular target vps-13 to regulate viral replication. *PLoS Negl. Trop Dis.* **2019**, *13*, e0007429.
256. Shrinet, J.; Jain, S.; Jain, J.; Bhatnagar, R.K.; Sunil, S. Next generation sequencing reveals regulation of distinct Aedes microRNAs during chikungunya virus development. *PLoS Negl. Trop. Dis.* **2014**, *8*, e2616.
257. Maharaj, P.D.; Widen, S.G.; Huang, J.; Wood, T.G.; Thangamani, S. Discovery of mosquito saliva microRNAs during CHIKV infection. *PLoS Negl. Trop. Dis.* **2015**, *9*, e0003386.
258. Yen, P.S.; Chen, C.H.; Sreenu, V.; Kohl, A.; Failloux, A.B. Assessing the Potential Interactions between Cellular miRNA and Arboviral Genomic RNA in the Yellow Fever Mosquito, Aedes aegypti. *Viruses* **2019**, *11*, 540.
259. Goupil, B.A.; McNulty, M.A.; Martin, M.J.; McCracken, M.K.; Christofferson, R.C.; Mores, C.N. Novel Lesions of Bones and Joints Associated with Chikungunya Virus Infection in Two Mouse Models of Disease: New Insights into Disease Pathogenesis. *PLoS ONE* **2016**, *11*, e0155243.
260. Manimunda, S.P.; Vijayachari, P.; Uppoor, R.; Sugunan, A.P.; Singh, S.S.; Rai, S.K.; Sudeep, A.B.; Muruganandam, N.; Chaitanya, I.K.; Guruprasad, D.R. Clinical progression of chikungunya fever during acute and chronic arthritic stages and the changes in joint morphology as revealed by imaging. *Trans. R. Soc. Trop. Med. Hyg.* **2010**, *104*, 392–399.
261. Sims, N.A.; Gooi, J.H. Bone remodeling: Multiple cellular interactions required for coupling of bone formation and resorption. *Semin. Cell Dev. Biol.* **2008**, *19*, 444–451.

262. Friedenstein, A.J.; Petrakova, K.V.; Kurolesova, A.I.; Frolova, G.P. Heterotopic of bone marrow. Analysis of precursor cells for osteogenic and hematopoietic tissues. *Transplantation* **1968**, *6*, 230–247.
263. Parfitt, A.M. Osteonal and Hemi-Osteonal Remodeling: The Spatial and Temporal Framework for Signal Traffic in Adult Human Bone. *J. Cell Biochem.* **1994**, *55*, 273–286.
264. Leong,K.-Y.B.;Chan,Y.-H.;Abdullah,W.M.A.N.W.;Lim,E.;Lai,K.-S.MicroRNAsinBoneDiseases:Progress and Prospects. In *Transcriptional and Post-Transcriptional Regulation*; IntechOpen: London, UK, 2018.
265. Eskildsen,T.;Taipaleenmäki,H.;Stenvang,J.;Abdallah,B.M.;Ditzel,N.;Nossent,A.Y.;Bak,M.;Kauppinen,S.; Kassem, M. MicroRNA-138 regulates osteogenic differentiation of human stromal (mesenchymal) stem cells in vivo. *Proc. Natl. Acad. Sci. USA* **2011**, *108*, 6139–6144.
266. Rodriguez-Carballo,E.;Gamez,B.;Ventura,F.p38MAPKSignalinginOsteoblastDifferentiation.*Front.Cell Dev. Biol.* **2016**, *4*, 40.
267. Pathak,H.;Mohan,M.C.;Ravindran,V.Chikungunyaarthritis.*Clin.Med.***2019**,*19*,381–385.
268. Kumaresan, J.A.; Mecaskey, J.W. The global elimination of blinding trachoma: Progress and promise. *Am. J. Trop Med. Hyg.* **2003**, *69*, 24–28.
269. WHO simplified trachoma grading system. *Community Eye Health* **2004**, *17*, 68.
270. Derrick, T.; Last, A.R.; Burr, S.E.; Roberts, C.H.; Nabicassa, M.; Cassama, E.; Bailey, R.L.; Mabey, D.C.; Burton, M.J.; Holland, M.J. Inverse relationship between microRNA-155 and -184 expression with increasing conjunctival inflammation during ocular Chlamydia trachomatis infection. *BMC Infect. Dis.* **2016**, *16*, 60.
271. Derrick, T.; Roberts, C.; Rajasekhar, M.; Burr, S.E.; Joof, H.; Makalo, P.; Bailey, R.L.; Mabey, D.C.; Burton, M.J.; Holland, M.J. Conjunctival MicroRNA expression in inflammatory trachomatous scarring. *PLoS Negl. Trop Dis.* **2013**, *7*, e2117.
272. Burton, M.J.; Rajak, S.N.; Bauer, J.; Weiss, H.A.; Tolbert, S.B.; Shoo, A.; Habtamu, E.; Manjurano, A.; Emerson, P.M.; Mabey, D.C.; et al. Conjunctival transcriptome in scarring trachoma. *Infect. Immun.* **2011**, *79*, 499–511
273. Landgraf, P.; Rusu, M.; Sheridan, R.; Sewer, A.; Iovino, N.; Aravin, A.; Pfeffer, S.; Rice, A.; Kamphorst, A.O.; Landthaler, M.; et al. A mammalian microRNA expression atlas based on small RNA library sequencing. *Cell* **2007**, *129*, 1401–1414.
274. Cheng, D.; Fan, J.; Ma, Y.; Zhou, Y.; Qin, K.; Shi, M.; Yang, J. LncRNA SNHG7 promotes pancreatic cancer proliferation through ID4 by sponging miR-342–3p. *Cell Biosci.* **2019**, *9*, 28.
275. Wang, L.; Xia, J.W.; Ke, Z.P.; Zhang, B.H. Blockade of NEAT1 represses inflammation response and lipid uptake via modulating miR-342–3p in human macrophages THP-1 cells. *J. Cell Physiol.* **2019**, *234*, 5319–5326.
276. Pekow, J.; Hutchison, A.L.; Meckel, K.; Harrington, K.; Deng, Z.; Talasila, N.; Rubin, D.T.; Hanauer, S.B.; Hurst, R.; Umanskiy, K.; et al. miR-4728–3p Functions as a Tumor Suppressor in Ulcerative Colitis-associated Colorectal

Neoplasia Through Regulation of Focal Adhesion Signaling. *Inflamm. Bowel. Dis.* **2017**, *23*, 1328–1337.
277. Ryan, D.G.; Oliveira-Fernandes, M.; Lavker, R.M. MicroRNAs of the mammalian eye display distinct and overlapping tissue specificity. *Mol. Vis.* **2006**, *12*, 1175–1184.
278. Iliff, B.W.; Riazuddin, S.A.; Gottsch, J.D. A single-base substitution in the seed region of miR-184 causes EDICT syndrome. *Invest. Ophthalmol Vis. Sci* **2012**, *53*, 348–353.
279. Hughes, A.E.; Bradley, D.T.; Campbell, M.; Lechner, J.; Dash, D.P.; Simpson, D.A.; Willoughby, C.E. Mutation altering the miR-184 seed region causes familial keratoconus with cataract. *Am. J. Hum. Genet.* **2011**, *89*, 628–633.
280. Bykhovskaya, Y.; Caiado Canedo, A.L.; Wright, K.W.; Rabinowitz, Y.S.C. 57 C > T Mutation in MIR 184 is Responsible for Congenital Cataracts and Corneal Abnormalities in a Five-generation Family from Galicia, Spain. *Ophthalmic Genet.* **2013**, *36*, 244–247.
281. Takahashi, Y.; Chen, Q.; Rajala, R.V.; Ma, J.X. MicroRNA-184 modulates canonical Wnt signaling through the regulation of frizzled-7 expression in the retina with ischemia-induced neovascularization. *FEBS Lett.* **2015**, *589*, 1143–1149.
282. Enk, C.D. Onchocerciasis–river blindness. *Clin. Derm.* **2006**, *24*, 176–180.
283. Lagatie,O.;BatsaDebrah,L.;Debrah,A.;Stuyver,L.J.Plasma-derivedparasiticmicroRNAshaveinsufficient concentrations to be used as diagnostic biomarker for detection of Onchocerca volvulus infection or treatment monitoring using LNA-based RT-qPCR. *Parasitol Res.* **2017**, *116*, 1013–1022.
284. Borup,L.H.;Peters,J.S.;Sartori,C.R.Onchocerciasis(riverblindness).*Cutis***2003**,*72* ,297–302.
285. Hotez,P.J.;Alvarado,M.;Basanez,M.G.;Bolliger,I.;Bourne,R.;Boussinesq,M.;Brooker,S.J.;Brown,A.S.; Buckle, G.; Budke, C.M.; et al. The global burden of disease study 2010: Interpretation and implications for the neglected tropical diseases. *PLoS Negl. Trop Dis.* **2014**, *8*, e2865.
286. Hall,L.R.;Pearlman,E.Pathogenesisofonchocercalkeratitis(Riverblindness).*Clin. Microbiol.Rev.***1999**,*12*, 445–453.
287. Cotton,J.A.;Bennuru,S.;Grote,A.;Harsha,B.;Tracey,A.;Beech,R.;Doyle,S.R.;Dunn,M.;Hotopp,J.C.; Holroyd, N.; et al. The genome of Onchocerca volvulus, agent of river blindness. *Nat. Microbiol.* **2016**, *2*, 16216.
288. Cupp, E.W.; Sauerbrey, M.; Richards, F. Elimination of human onchocerciasis: History of progress and current feasibility using ivermectin (Mectizan((R))) monotherapy. *Acta Trop* **2011**, *12*, S100–S108.
289. Turner,H.C.;Churcher,T.S.;Walker,M.;OseiAtweneboana,M.Y.;Prichard,R.K.;Basanez,M.G.Uncertainty surrounding projections of the long-term impact of ivermectin treatment on human onchocerciasis. *PLoS Negl. Trop Dis.* **2013**, *7*, e2169.
290. WHO (1992) Lymphatic filariasis: the disease and its control. Fifth report of the WHO Expert Committee on Filariasis. 0512-3054 (Print)0512-3054 (Linking). 1–71 p.
291. Scott AL (2000) Lymphatic-dwelling Filariae. In: Nutman TB, editor. Lymphatic Filariasis. London: Imperial College Press. pp. 5–40.

292. Simonsen PE (2009) Filariases. In: Cook G, Manson P, Zumla A, editors. Manson's Tropical Diseases. 22 ed: Saunders, Elsevier. pp. 1477–1513.
293. Kumaraswami V (2000) The Clinical Manifestations of Lymphatic Filariasis. In: Nutman TB, editor. Lymphatic Filariasis. London: Imperial College Press. pp. 103–126.
294. Molyneux DH, Taylor MJ (2001) Current status and future prospects of the Global Lymphatic Filariasis Programme. Curr Opin Infect Dis 14: 155–159.
295. Addiss DG, Dreyer G (2000) Treatment of Lymphatic Filariasis. In: Nutman TB, editor. Lymphatic Filariasis. London: Imperial College Press. pp. 151–200.
296. Bourguinat C, Ardelli BF, Pion SD, Kamgno J, Gardon J, et al. (2008) P-glycoprotein-like protein, a possible genetic marker for ivermectin resistance selection in *Onchocerca volvulus*. Mol Biochem Parasitol 158: 101–111.
297. Poole CB, Davis PJ, Jin J, McReynolds LA (2010) Cloning and bioinformatic identification of small RNAs in the filarial nematode, *Brugia malayi*. Mol Biochem Parasitol 169: 87–94.
298. Ghedin E, Wang S, Spiro D, Caler E, Zhao Q, et al. (2007) Draft genome of the filarial nematode parasite *Brugia malayi*. Science 317: 1756–1760.
299. Lewis BP, Burge CB, Bartel DP (2005) Conserved seed pairing, often flanked by adenosines, indicates that thousands of human genes are microRNA targets. Cell 120: 15–20.
300. Ambros V, Bartel B, Bartel DP, Burge CB, Carrington JC, et al. (2003) A uniform system for microRNA annotation. RNA 9: 277–279.
301. Ibanez-Ventoso C, Vora M, Driscoll M (2008) Sequence relationships among *C. elegans*, *D. melanogaster* and human microRNAs highlight the extensive conservation of microRNAs in biology. PLoS One 3: e2818.
302. Winter AD, Weir W, Hunt M, Berriman M, Gilleard JS, et al. (2012) Diversity in parasitic nematode genomes: the microRNAs of Brugia pahangi and Haemonchus contortus are largely novel. BMC Genomics 13: 4.
303. Sapio MR, Hilliard MA, Cermola M, Favre R, Bazzicalupo P (2005) The Zona Pellucida domain containing proteins, CUT-1, CUT-3 and CUT-5, play essential roles in the development of the larval alae in *Caenorhabditis elegans*. Dev Biol 282: 231–245.
304. Leaman D, Chen PY, Fak J, Yalcin A, Pearce M, et al. (2005) Antisense-mediated depletion reveals essential and specific functions of microRNAs in *Drosophila* development. Cell 121: 1097–1108.
305. Conradt B, Horvitz HR (1998) The *C. elegans* protein EGL-1 is required for programmed cell death and interacts with the Bcl-2-like protein CED-9. Cell 93: 519–529.
306. Zhang L, Hammell M, Kudlow BA, Ambros V, Han M (2009) Systematic analysis of dynamic miRNA-target interactionsduring C. elegans development. Development 136: 3043–3055.
307. Hafner M, Renwick N, Brown M, Mihailovic A, Holoch D, et al. (2011) RNA-ligase-dependent biases in miRNA representation in deep-sequenced small RNA cDNA libraries. RNA 17: 1697–1712. View Article Google Scholar.
308. Zhuang F, Fuchs RT, Sun Z, Zheng Y, Robb GB (2012) Structural bias in T4 RNA ligase-mediated 3′-adapter ligation. Nucleic acids research 40: e54.

309. Simonsen PE (2009) Filariases. In: Cook G, Manson P, Zumla A, editors. Manson's Tropical Diseases. 22 ed: Saunders, Elsevier. pp. 1477–1513.
310. Ponnudurai T, Denham DA, Rogers R (1975) Studies on *Brugia pahangi* 9. The longevity of microfilariae transfused from cat to cat. J Helminthol 49: 25– 30.
311. de Lencastre A, Pincus Z, Zhou K, Kato M, Lee SS, et al. (2010) MicroRNAs both promote and antagonize longevity in *C. elegans*. Curr Biol 20: 2159– 2168.
312. Karp X, Hammell M, Ow MC, Ambros V (2011) Effect of life history on microRNA expression during *C. elegans* development. RNA 17: 639–651.
313. Le Bechec A, Portales-Casamar E, Vetter G, Moes M, Zindy PJ, et al. (2011) MIR@NT@N: a framework integrating transcription factors, microRNAs and their targets to identify sub-network motifs in a meta-regulation network model. BMC Bioinformatics 12: 67.
314. Shalgi R, Lieber D, Oren M, Pilpel Y (2007) Global and local architecture of the mammalian microRNA-transcription factor regulatory network. PLoS Comput Biol 3: e131.
315. Martinez NJ, Ow MC, Barrasa MI, Hammell M, Sequerra R, et al. (2008) A *C. elegans* genome-scale microRNA network contains composite feedback motifs with high flux capacity. Genes Dev 22: 2535–2549.
316. Krutzfeldt J, Rajewsky N, Braich R, Rajeev KG, Tuschl T, et al. (2005) Silencing of microRNAs in vivo with 'antagomirs'. Nature 438: 685–689. View Article . Google Scholar
317. Obad S, dos Santos CO, Petri A, Heidenblad M, Broom O, et al. (2011) Silencing of microRNA families by seed-targeting tiny LNAs. Nat Genet 43: 371– 378.
318. Castanotto D, Rossi JJ (2009) The promises and pitfalls of RNA-interference-based therapeutics. Nature 457: 426–433.
319. Simarro P, Cecchi G, Franco J, Paone M, Diarra A, et al. (2012) Estimating and Mapping the Population at Risk of Sleeping Sickness. PLoS Negl Trop Dis 6: e1859.
320. Simarro PP, Diarra A, Ruiz Postigo JA, Franco JR, Jannin JG (2011) The human African trypanosomiasis control and surveillance programme of the World Health Organization 2000–2009: the way forward. PLoS Negl Trop Dis 5: e1007.
321. Dumas M, Bouteille B (1996) Human African trypanosomiasis. Compte-Rendu de la Société de Biologie: 395–408.
322. Jamonneau V, Ravel S, Garcia A, Koffi M, Truc P, et al. (2004) Characterization of *Trypanosoma brucei* s.l. infecting asymptomatic sleeping-sickness patients in Côte d'Ivoire: a new genetic group? Ann Trop Med Parasitol 98: 329–337.
323. Jamonneau V, Ilboudo H, Kaboré J, Kaba D, Koffi M, et al. (2012) Untreated human infections by *Trypanosoma brucei gambiense* are not 100% fatal. PLoS Negl Trop Dis 6: e1691.
324. Liston A, Papadopoulou AS, Danso-Abeam D, Dooley J (2012) MicroRNA-29 in the adaptive immune system: setting the threshold. Cell Mol Life Sci 69: 3533–3541.
325. Leonardo TR, Schultheisz HL, Loring JF, Laurent LC (2012) The functions of microRNAs in pluripotency and reprogramming. Nat Cell Biol 14: 1114–1121. View Article .Google Scholar.
326. Amiel J, de Pontual L, Henrion-Caude A (2012) miRNA, development and disease. Adv Genet 80: 1–36.

327. Boeri M, Pastorino U, Sozzi G (2012) Role of microRNAs in lung cancer: microRNA signatures in cancer prognosis. Cancer J 18: 268–274.
328. Healy NA, Heneghan HM, Miller N, Osborne CK, Schiff R, et al. (2012) Systemic mirnas as potential biomarkers for malignancy. Int J Cancer 131: 2215–2222.
329. Creemers EE, Tijsen AJ, Pinto YM (2012) Circulating microRNAs: novel biomarkers and extracellular.
330. Keller A, Leidinger P, Bauer A, Elsharawy A, Haas J, et al. (2011) Toward the blood-borne miRNome of human diseases. Nat Methods 8: 841–843
331. Jamonneau V, Bucheton B, Kabore J, Ilboudo H, Camara O, et al. (2010) Revisiting the immune trypanolysis test to optimise epidemiological surveillance and control of sleeping sickness in West Africa. PLoS Negl Trop Dis 4: e917.
332. Wang J, He Q, Han C, Gu H, Jin L, et al. (2012) p53-facilitated miR-199a-3p regulates somatic cell reprogramming. Stem Cells 30: 1405–1413.
333. Huang Y, Chen HC, Chiang CW, Yeh CT, Chen SJ, et al. (2012) Identification of a two-layer regulatory network of proliferation-related microRNAs in hepatoma cells. Nucleic Acids Res 40: 10478–10493.
334. Hamad S, Satoh K, Fujibuchi W, Hirota M, Kanno A, et al. (2012) MiR-126 acts as a tumor suppressor in pancreatic cancer cells via the regulation of ADAM9. Mol Cancer Res 10: 3–10.
335. Guan P, Yin Z, Li X, Wu W, Zhou B (2012) Meta-analysis of human lung cancer microRNA expression profiling studies comparing cancer tissues with normal tissues. J Exp Clin Cancer Res 31: 54
336. Liu Y, Wang X, Jiang J, Cao Z, Yang B, et al. (2011) Modulation of T cell cytokine production by miR-144* with elevated expression in patients with pulmonary tuberculosis. Mol Immunol 48: 1084–1090.
337. Vickers KC, Shoucri BM, Levin MG, Wu H, Pearson DS, et al.. (2012) MicroRNA-27b is a regulatory hub in lipid metabolism and is altered in dyslipidemia. Hepatology.
338. Biyashev D, 2687. Veliceasa D, Topczewski J, Topczewska JM, Mizgirev I, et al. (2012) miR-27b controls venous specification and tip cell fate. Blood 119: 2679–2687
339. Ergonul O. Crimean–Congo hemorrhagic fever virus: new outbreaks, new discoveries. Curr Opin Virol. 2012;2:215-220.
340. Zivcec M, Scholte FE, Spiropoulou CF, Spengler JR, Gergeron É, et al. Molecular insights into Crimean-Congo hemorrhagic fever virus. Viruses. 2016;8:106.
341. Burt FJ, Swanepoel R, Shieh WJ, et al. Immunohistochemical and in situ localization of Crimean-Congo hemorrhagic fever (CCHF) virus in human tissues and implications for CCHF pathogenesis. Arch Path Lab Med. 1997;121:839-846.
342. Trobaugh DW, Klimstra WB. MicroRNA regulation of RNA virus replication and pathogenesis. Trends Mol Med. 2017;23:80-93.
343. Liu X, Zhan Z, Xu L, et al. MicroRNA-148/ 152 impair innate response and antigen presentation of TLR-triggered dendritic cells by targeting CaMKIIalpha. J Immunol. 2010;185:7244-7251.

344. Núñez-Hernández F, Pérez LJ, Muñoz M, et al. Identification of microRNAs in PCV2 subclinically infected pigs by high throughput sequencing. Vet Res. 2015;46:18.
345. Ma Z, Liu T, Huang W, et al. MicroRNA regulatory pathway analysis identifies miR-142-5p as a negative regulator of TGF-β pathway via targeting SMAD3. Oncotarget. 2016;7:71504-71513.
346. Yang D, Qiu GB. Biological functions of miR-590 and its role in carcinogenesis. Front Lab Med. 2017;1:173-176.
347. Papa A, Bino S, Velo E, et al. Cytokine levels in Crimean-Congo hemorrhagic fever. J Clin Virol: Off Publ Pan Am Soc Clin Virol. 2006;36:272-276.
348. Daubney R, Hudson JR, Garnham PC. 1931. Enzootic hepatitis or rift valley fever. An undescribed virus disease of sheep cattle and man from east africa. J Pathol 34:545–579. https://doi.org/10.1002/path.1700340418.
349. Gerdes GH. 2004. Rift Valley fever. Rev Sci Tech 23:613–623. https://doi .org/10.20506/rst.23.2.1500.
350. Smithburn KC. 1949. Rift Valley fever; the neurotropic adaptation of the virus and the experimental use of this modified virus as a vaccine. Br J Exp Pathol 30:1–16.
351. Al-Hazmi M, Ayoola EA, Abdurahman M, Banzal S, Ashraf J, El-Bushra A, Hazmi A, Abdullah M, Abbo H, Elamin A, Al-Sammani el T, Gadour M, Menon C, Hamza M, Rahim I, Hafez M, Jambavalikar M, Arishi H, Aqeel A. 2003. Epidemic Rift Valley fever in Saudi Arabia: a clinical study of severe illness in humans. Clin Infect Dis 36:245–252. https://doi.org/10.1086/ 345671.
352. LaBeaud AD, Muiruri S, Sutherland LJ, Dahir S, Gildengorin G, Morrill J, Muchiri EM, Peters CJ, King CH. 2011. Postepidemic analysis of Rift Valley fever virus transmission in northeastern Kenya: a village cohort study. PLoS Negl Trop Dis 5:e1265. https://doi.org/10.1371/journal.pntd.0001265.
353. Madani TA, Al-Mazrou YY, Al-Jeffri MH, Mishkhas AA, Al-Rabeah AM, Turkistani AM, Al-Sayed MO, Abodahish AA, Khan AS, Ksiazek TG, Shobokshi O. 2003. Rift Valley fever epidemic in Saudi Arabia: epidemiological, clinical, and laboratory characteristics. Clin Infect Dis 37: 1084 –1092. https://doi.org/10.1086/378747.
354. Pepin M, Bouloy M, Bird BH, Kemp A, Paweska J. 2010. Rift Valley fever virus(Bunyaviridae: phlebovirus): an update on pathogenesis, molecular epidemiology, vectors, diagnostics and prevention. Vet Res 41:61. https://doi.org/10.1051/vetres/2010033.
355. Kreher F, Tamietti C, Gommet C, Guillemot L, Ermonval M, Failloux AB, Panthier JJ, Bouloy M, Flamand M. 2014. The Rift Valley fever accessory proteins NSm and P78/NSm-GN are distinct determinants of virus prop- agation in vertebrate and invertebrate hosts. Emerg Microbes Infect 3:e71. https://doi.org/10.1038/emi.2014.71.
356. Bouloy M, Weber F. 2010. Molecular biology of Rift valley fever virus. Open Virol J 4:8 –14. https://doi.org/10.2174/1874357901004020008.
357. Ikegami T. 2012. Molecular biology and genetic diversity of Rift Valley fever virus. Antiviral Res 95:293–310. https://doi.org/10.1016/j.antiviral

358. Kakumani PK, Ponia SS, S RK, Sood V, Chinnappan M, Banerjea AC, et al. Role of RNA interference (RNAi) in dengue virus replication and identification of NS4B as an RNAi suppressor. J Virol. 2013;87(16):8870–83.
359. Ouyang X, Jiang X, Gu D, Zhang Y, Kong SK, Jiang C, et al. Dysregulated serum miRNA profile and promising biomarkers in dengue-infected patients. Int J Med Sci. 2016;13(3):195–205.
360. Hussain M, Asgari S. MicroRNA-like viral small RNA from dengue virus 2 autoregulates its replication in mosquito cells. Proc Natl Acad Sci U S A. 2014;111(7):2746–51.
361. Hussain M, Torres S, Schnettler E, Funk A, Grundhoff A, Pijlman GP, et al. West Nile virus encodes a microRNA-like small RNA in the 3′ untranslated region which up-regulates GATA4 mRNA and facilitates virus replication in mosquito cells. Nucleic Acids Res. 2012;40(5):2210–23.
362. Li X, Fu Z, Liang H, Wang Y, Qi X, Ding M, et al. H5N1 influenza virus-specific miRNA-like small RNA increases cytokine production and mouse mortality via targeting poly(rC)-binding protein 2. Cell Res. 2018;28(2):157–71.
363. Weitao W, He Z, Jing Q, Hu Y, Lin C, Zhou R, et al. Cellular microRNA-miR-548g-3p modulates the replication of dengue virus. J Inf Secur. 2015;70(6):631–40.
364. Castillo JA, Castrillón JC, Diosa-Toro M, Betancur JG, St Laurent G 3rd, Smit JM, et al. Complex interaction between dengue virus replication and expression of miRNA-133a. BMC Infect Dis. 2016;16:29.
365. Castrillón-Betancur JC, Urcuqui-Inchima S. Overexpression of miR-484 and miR-744 in Vero cells alters dengue virus replication. Mem Inst Oswaldo Cruz. 2017;112(4):281–91.
366. Yan H, Zhou Y, Liu Y, Deng Y, Puthiyakunnon S, Chen X. miR-252 of the Asian tiger mosquito *Aedes albopictus* regulates dengue virus replication by suppressing the expression of the dengue virus envelope protein. J Med Virol. 2014;86(8):1428–36.
367. Chen Y, Maguire T, Marks RM. Demonstration of binding of dengue virus envelope protein to target cells. J Virol. 1996;70(12):8765–72.
368. Girardi E, López P, Pfeffer S. On the importance of host microRNAs during viral infection. Front Genet. 2018;9:439.
369. Escalera-Cueto M, Medina-Martínez I, del Angel RM, Berumen-Campos J, Gutiérrez-Escolano AL, Yocupicio-Monroy M. Let-7c overexpression inhibits dengue virus replication in human hepatoma Huh-7 cells. Virus Res. 2015;196:105–12.
370. Yen YT, Chen HC, Lin YD, Shieh CC, Wu-Hsieh BA. Enhancement by tumor necrosis factor alpha of dengue virus-induced endothelial cell production of reactive nitrogen and oxygen species is key to hemorrhage development. J Virol. 2008;82(24):12312–24.
371. Marianneau P, Cardona A, Edelman L, Deubel V, Desprès P. Dengue virus replication in human hepatoma cells activates NF-kappaB which in turn induces apoptotic cell death. J Virol. 1997;71(4):3244–9.
372. Jan JT, Chen BH, Ma SH, Liu CI, Tsai HP, Wu HC, et al. Potential dengue virus-triggered apoptotic pathway in human neuroblastoma cells: arachidonic acid,

superoxide anion, and NF-kappaB are sequentially involved. J Virol. 2000;74(18):8680–91.

373. Dalrymple NA, Mackow ER. Roles for endothelial cells in dengue virus infection. Adv Virol. 2012;2012:840654.

374. Huang YH, Lei HY, Liu HS, Lin YS, Liu CC, Yeh TM. Dengue virus infect human endothelial cells and induces IL-6 and IL-8 production. Am J Trop Med Hyg. 2000;63(1–2):71–5.

375. Tseng CK, Lin CK, Wu YH, Chen YH, Chen WC, Young KC, et al. Human heme oxygenase 1 is a potential host cell factor against dengue virus replication. Sci Rep. 2016;6:32176.

376. Zhu X, He Z, Hu Y, Wen W, Lin C, Yu J, et al. MicroRNA-30e* suppresses dengue virus replication by promoting NF-κB-dependent IFN production. PLoS Negl Trop Dis. 2014;8(8):e3088.

377. Vallabhapurapu S, Karin M. Regulation and function of NF-κB transcription factors in the immune system. Annu Rev Immunol. 2009;27(1):693–733.

378. Smith JL, Jeng S, McWeeney SK, Hirsch AJ. A microRNA screen identifies the Wnt signaling pathway as a regulator of the interferon response during flavivirus infection. J Virol. 2017;91(8):e02388–16.

379. Wu N, Gao N, Fan D, Wei J, Zhang J, An J. miR-223 inhibits dengue virus replication by negatively regulating the microtubule-destabilizing protein STMN1 in EAhy926 cells. Microbes Infect. 2014;16(11):911–22.

380. McCartney EM, Helbig KJ, Narayana SK, Eyre NS, Aloia AL, Beard MR. Signal transducer and activator of transcription 3 is a proviral host factor for hepatitis C virus. Hepatology. 2013;58(5):1558–68.

381. Agis-Juárcz RA, Galván I, Medina F, Daikoku T, Padmanabhan R, Ludert J, et al. Polypyrimidine tract-binding protein is relocated to the cytoplasm and is required during dengue virus infection in Vero cells. J Gen Virol. 2009;90:2893–901.

382. De Nova-Ocampo M, Villegas-Sepúlveda N, del Angel RM. Translation elongation factor-1α, la, and PTB interact with the 3′ untranslated region of dengue 4 virus RNA. Virology. 2002;295(2):337–47.

383. Diosa-Toro M, Echavarría-Consuegra L, Flipse J, Fernández GJ, Kluiver J, van den Berg A, et al. MicroRNA profiling of human primary macrophages exposed to dengue virus identifies miRNA-3614-5p as antiviral and regulator of ADAR1 expression. PLoS Negl Trop Dis. 2017;11(10):e0005981.

384. Wu S, He L, Li Y, Wang T, Feng L, Jiang L, et al. miR-146a facilitates replication of dengue virus by dampening interferon induction by targeting TRAF6. J Inf Secur. 2013;67(4):329–41.

385. Pu J, Wu S, Xie H, Li Y, Yang Z, Wu X, et al. miR-146a inhibits dengue-virus-induced autophagy by targeting TRAF6. Arch Virol. 2017;162(12):3645–59.

386. Kanokudom S, Vilaivan T, Wikan N, Thepparit C, Smith DR, Assavalapsakul W. miR-21 promotes dengue virus serotype 2 replication in HepG2 cells. Antivir Res. 2017;142:169–77.

387. Miranda KC, Huynh T, Tay Y, Ang Y-S, Tam W-L, Thomson AM, et al. A pattern-based method for the identification of microRNA binding sites and their corresponding heteroduplexes. Cell. 2006;126(6):1203–17.

388. Vaughn DW, Green S, Kalayanarooj S, Innis BL, Nimmannitya S, Suntayakorn S, et al. Dengue viremia titer, antibody response pattern, and virus serotype correlate with disease severity. J Infect Dis. 2000;181(1):2–9.
389. Wang WK, Chao DY, Kao CL, Wu HC, Liu YC, Li CM, et al. High levels of plasma dengue viral load during defervescence in patients with dengue hemorrhagic fever: implications for pathogenesis. Virology. 2003;305(2):330–8.
390. Libraty DH, Endy TP, Houng HS, Green S, Kalayanarooj S, Suntayakorn S, et al. Differing influences of virus burden and immune activation on disease severity in secondary dengue-3 virus infections. J Infect Dis. 2002;185(9):1213–21.
391. Martina B. Dengue pathogenesis: a disease driven by the host response. Sci Prog. 2014;97:197–214.
392. Cardier JE, Mariño E, Romano E, Taylor P, Liprandi F, Bosch N, et al. Proinflammatory factors present in sera from patients with acute dengue infection induce activation and apoptosis of human microvascular endothelial cells: possible role of TNF-α in endothelial cell damage in dengue. Cytokine. 2005;30(6):359–65.
393. Chen RF, Yang KD, Lee IK, Liu JW, Huang CH, Lin CY, et al. Augmented miR-150 expression associated with depressed SOCS1 expression involved in dengue haemorrhagic fever. J Inf Secur. 2014;69(4):366–74.
394. Wang P, Hou J, Lin L, Wang C, Liu X, Li D, et al. Inducible microRNA-155 feedback promotes type I IFN signaling in antiviral innate immunity by targeting suppressor of cytokine signaling 1. J Immunol. 2010;185(10):6226.
395. Qi Y, Li Y, Zhang L, Huang J. MicroRNA expression profiling and bioinformatic analysis of dengue virus-infected peripheral blood mononuclear cells. Mol Med Rep. 2013;7(3):791–8.
396. Nguyen TH, Liu X, Su ZZ, Hsu AC, Foster PS, Yang M. Potential role of microRNAs in the regulation of antiviral responses to influenza infection. Front Immunol. 2018;9:1541.
397. Fernandez-Pineiro I, Badiola I, Sanchez A. Nanocarriers for microRNA delivery in cancer medicine. Biotechnol Adv. 2017;35(3):350–60.

Chapter VIII

1. Elia L, Contu R, Quintavalle M, Varrone F, Chimenti C, Russo MA, *et al.* Reciprocal regulation of microRNA-1 and insulin-like growth factor-1 signal transduction cascade in cardiac and skeletal muscle in physiological and pathological conditions. Circulation 2009; 120: 2377–85.
2. Gupta SK, Foinquinos A, Thum S, Remke J, Zimmer K, Bauters C, *et al.* Preclinical development of a microRNA-based therapy for elderly patients with myocardial infarction. J Am Coll Cardiol 2016; 68: 1557–71.
3. Marques FZ, Campain AE, Tomaszewski M, Zukowska-Szczechowska E, Yang YH, Charchar FJ, *et al* mRNA overexpression in human hypertensive kidneys and a role for microRNAs. Hypertension 2011; 58: 1093–8.
4. Barwari T, Joshi A, Mayr M. MicroRNAs in cardiovascular disease. J Am Coll Cardiol 2016; 68: 2577–84.
5. Thum, T. MicroRNA therapeutics in cardiovascular medicine. *EMBO Mol Med.* 2012;4:3–14.

6. Boon, RA, Dimmeler, S. MicroRNAs in myocardial infarction. *Nat Rev Cardiol.* 2015;12:135–142.
7. Zampetaki A, Willeit P, Drozdov I, Kiechl S, Mayr M. Profiling of circulating microRNAs: from single biomarkers to re-wired networks. Cardiovasc Res 2012; 93: 555–62.
8. Kosaka N, Iguchi H, Yoshioka Y, Takeshita F, Matsuki Y, Ochiya T. Secretory mechanisms and intercellular transfer of microRNAs in living cells. J Biol Chem 2010; 285: 17442–52.
9. Leonard BL, Smaill BH, LeGrice IJ. Structural remodeling and mechanical function in heart failure. Microsc Microanal 2012; 18: 50–67.
10. Schulte C, Westermann D, Blankenberg S, Zeller T. Diagnostic and prognostic value of circulating microRNAs in heart failure with preserved and reduced ejection fraction. World J Cardiol 2015; 7: 843–60.
11. Ovchinnikova ES, Schmitter D, Vegter EL, Ter Maaten JM, Valente MA, Liu LC, *et al.* Signature of circulating microRNAs in patients with acute heart failure. Eur J Heart Fail 2016; 18: 414–23.
12. Sygitowicz G, Tomaniak M, Blaszczyk O, Koltowski L, Filipiak KJ, Sitkiewicz D. Circulating microribonucleic acids mir-1, mir-21 and mir- 208a in patients with symptomatic heart failure: preliminary results. Arch Cardiovasc Dis 2015; 108: 634–42.
13. Endo K, Naito Y, Ji X, Nakanishi M, Noguchi T, Goto Y, *et al.* MicroRNA 210 as a biomarker for congestive heart failure. Biol Pharm Bull 2013; 36: 48–54.
14. Seronde MF, Vausort M, Gayat E, Goretti E, Ng LL, Squire IB, *et al.* Circulating microRNAs and outcome in patients with acute heart failure. PLoS One 2015; 10: e0142237.
15. Tijsen AJ, Creemers EE, Moerland PD, de Windt LJ, van der Wal AC, Kok WE, *et al.* Mir423-5p as a circulating biomarker for heart failure. Circ Res 2010; 106: 1035–9.
16. Goren Y, Kushnir M, Zafrir B, Tabak S, Lewis BS, Amir O. Serum levels of microRNAs in patients with heart failure. Eur J Heart Fail 2012; 14: 147–54.
17. Bayes-Genis A, Lanfear DE, de Ronde MWJ, Lupon J, Leenders JJ, Liu Z, *et al.* Prognostic value of circulating microRNAs on heart failure-related morbidity and mortality in two large diverse cohorts of general heart failure patients. Eur J Heart Fail 2018; 20: 67–75.
18. van Boven N, Kardys I, van Vark LC, Akkerhuis KM, de Ronde MWJ, Khan MAF, *et al.* Serially measured circulating microRNAs and adverse clinical outcomes in patients with acute heart failure. Eur J Heart Fail 2018; 20: 89–96.
19. Watson CJ, Gupta SK, O'Connell E, Thum S, Glezeva N, Fendrich J, *et al.* MicroRNA signatures differentiate preserved from reduced ejection fraction heart failure. Eur J Heart Fail 2015; 17: 405–15.
20. Akat KM, Moore-McGriff D, Morozov P, Brown M, Gogakos T, Correa Da Rosa J, *et al.* Comparative RNA-sequencing analysis of myocardial and circulating small RNAs in human heart failure and their utility as biomarkers. Proc Natl Acad Sci U S A 2014; 111: 11151–6.
21. Gidlof O, Smith JG, Miyazu K, Gilje P, Spencer A, Blomquist S, *et al.* Circulating cardio-enriched microRNAs are associated with long-term prognosis following myocardial infarction. BMC Cardiovasc Disorder 2013; 13: 12.

22. Marfella R, Di Filippo C, Potenza N, Sardu C, Rizzo MR, Siniscalchi M, *et al.* Circulating microRNA changes in heart failure patients treated with cardiac resynchronization therapy: responders *vs* non- responders. Eur J Heart Fail 2013; 15: 1277–88.
23. Xiao J, Gao R, Bei Y, Zhou Q, Zhou Y, Zhang H, *et al.* Circulating mir- 30d predicts survival in patients with acute heart failure. Cell Physiol Biochem 2017; 41: 865–74.
24. Melman YF, Shah R, Danielson K, Xiao J, Simonson B, Barth A, *et al.* Circulating microRNA-30d is associated with response to cardiac resynchronization therapy in heart failure and regulates cardiomyocyte apoptosis: A translational pilot study. Circulation 2015; 131: 2202–16.
25. Fox CS, Coady S, Sorlie PD, D'Agostino RB, Pencina MJ, Vasan RS, *et al.* Increasing cardiovascular disease burden due to diabetes mellitus - the framingham heart study. Circulation 2007; 115: 1544–50.
26. Sun Y. Myocardial repair/remodelling following infarction: Roles of local factors. Cardiovasc Res 2009; 81: 482–90.
27. Takemura G, Nakagawa M, Kanamori H, Minatoguchi S, Fujiwara H. Benefits of reperfusion beyond infarct size limitation Cardiovasc Res 2009; 83: 269–76.
28. Eefting F, Rensing B, Wigman J, Pannekoek WJ, Liu WM, Cramer MJ, *et al.* Role of apoptosis in reperfusion injury. Cardiovasc Res 2004; 61: 414–26.
29. Liu X, Dong Y, Chen S, Zhang G, Zhang M, Gong Y, *et al.* Circulating microRNA-146a and microRNA-21 predict left ventricular remodeling after ST-elevation myocardial infarction. Cardiology 2015; 132: 233– 41.
30. Cheng Y, Tan N, Yang J, Liu X, Cao X, He P, *et al.* A translational study of circulating cell-free microRNA-1 in acute myocardial infarction. Clin Sci (Lond) 2010; 119: 87–95.
31. Ai J, Zhang R, Li Y, Pu J, Lu Y, Jiao J, *et al.* Circulating microRNA-1 as a potential novel biomarker for acute myocardial infarction. Biochem Biophys Res Commun 2010; 391: 73–7.
32. Danowski N, Manthey I, Jakob HG, Siffert W, Peters J, Frey UH. Decreased expression of mir-133a but not of mir-1 is associated with signs of heart failure in patients undergoing coronary bypass surgery. Cardiology 2013; 125: 125–30.
33. Zile MR, Mehurg SM, Arroyo JE, Stroud RE, DeSantis SM, Spinale FG. Relationship between the temporal profile of plasma microRNA and left ventricular remodeling in patients after myocardial infarction. Circ Cardiovasc Genet 2011; 4: 614–19.
34. Pan Z, Sun X, Shan H, Wang N, Wang J, Ren J, *et al.* MicroRNA-101 inhibited postinfarct cardiac fibrosis and improved left ventricular compliance via the FBJ osteosarcoma oncogene/transforming growth factor-beta1 pathway. Circulation 2012; 126: 840–50.
35. van Rooij E, Sutherland LB, Thatcher JE, DiMaio JM, Naseem RH, Marshall WS, *et al.* Dysregulation of microRNAs after myocardial Sci U S A 2008; 105: 13027–32.
36. Widera C, Gupta SK, Lorenzen JM, Bang C, Bauersachs J, Bethmann K, *et al.* Diagnostic and prognostic impact of six circulating microRNAs in acute coronary syndrome. J Mol Cell Cardiol 2011; 51: 872–5.

37. Grabmaier U, Clauss S, Gross L, Klier I, Franz WM, Steinbeck G, *et al.* Diagnostic and prognostic value of mir-1 and mir-29b on adverse ventricular remodeling after acute myocardial infarction–the sitagrami-mir analysis. Int J Cardiol 2017; 244: 30–36.
38. Chen JF, Mandel EM, Thomson JM, Wu Q, Callis TE, Hammond SM, *et al.* The role of microRNA-1 and microRNA-133 in skeletal muscle proliferation and differentiation. Nat Genet 2006; 38: 228–33.
39. Kondkar AA, Abu-Amero KK. Utility of circulating microRNAs as clinical biomarkers for cardiovascular diseases. Biomed Res Int 2015; 2015: 821823.
40. Wang R, Li N, Zhang Y, Ran Y, Pu J. Circulating microRNAs are promising novel biomarkers of acute myocardial infarction. Intern Med 2011; 50: 1789–95.
41. Bauters C, Kumarswamy R, Holzmann A, Bretthauer J, Anker SD, Pinet F, *et al.* Circulating mir-133a and mir-423-5p fail as biomarkers for left ventricular remodeling after myocardial infarction. Int J Cardiol 2013; 168: 1837–40.
42. Zhang Y, Liu YJ, Liu T, Zhang H, Yang SJ. Plasma microRNA-21 is a potential diagnostic biomarker of acute myocardial infarction. Eur Rev Med Pharmacol Sci 2016; 20: 323–9.
43. Bialek S, Gorko D, Zajkowska A, Koltowski L, Grabowski M, Stachurska A, *et al.* Release kinetics of circulating mirRNA-208a in the early phase of myocardial infarction. Kardiol Pol 2015; 73: 613– 9.
44. Lv P, Zhou M, He J, Meng W, Ma X, Dong S, *et al.* Circulating mir-208b and mir-34a are associated with left ventricular remodeling after acute myocardial infarction. Int J Mol Sci 2014; 15: 5774–88.
45. Corsten MF, Dennert R, Jochems S, Kuznetsova T, Devaux Y, Hofstra L, *et al.* Circulating microRNA-208b and microRNA-499 reflect myocardial damage in cardiovascular disease. Circ Cardiovasc Genet 2010; 3: 499–506.
46. Liu X, Fan Z, Zhao T, Cao W, Zhang L, Li H, *et al.* Plasma mir-1, mir- 208, mir-499 as potential predictive biomarkers for acute myocardial infarction: an independent study of Han population. Exp Gerontol 2015; 72: 230–8.
47. Shalaby SM, El-Shal AS, Shoukry A, Khedr MH, Abdelraheim N. Serum miRNA-499 and miRNA-210: a potential role in early diagnosis of acute coronary syndrome. IUBMB Life 2016; 68: 673–82.
48. Yao Y, Du J, Cao X, Wang Y, Huang Y, Hu S, *et al.* Plasma levels of microRNA-499 provide an early indication of perioperative myocardial infarction in coronary artery bypass graft patients. PLoS One 2014; 9: e104618.
49. Bauters C, Kumarswamy R, Holzmann A, Bretthauer J, Anker SD, Pinet F, *et al.* Circulating mir-133a and mir-423-5p fail as biomarkers for left ventricular remodeling after myocardial infarction. Int J Cardiol 2013; 168: 1837–40.
50. Nabialek E, Wanha W, Kula D, Jadczyk T, Krajewska M, Kowalowka A, *et al.* Circulating microRNAs (mir-423-5p, mir-208a and mir-1) in acute myocardial infarction and stable coronary heart disease. Minerva Cardioangiol 2013; 61: 627–37.
51. Eryilmaz U, Akgullu C, Beser N, Yildiz O, Kurt Omurlu I, Bozdogan B. Circulating microRNAs in patients with ST-elevation myocardial infarction. Anatol J Cardiol 2016; 16: 392–6.
52. Han C, Shen JK, Hornicek FJ, Kan Q, Duan Z. Regulation of microRNA-1 (mir-1) expression in human cancer. Biochim Biophys Acta 2017; 1860: 227–32.

53. Parkes JE, Day PJ, Chinoy H, Lamb JA. The role of microRNAs in the idiopathic inflammatory myopathies. Curr Opin Rheumatol 2015; 27: 608–15.
54. Coenen-Stass AML, Wood MJA, Roberts TC. Biomarker potential of extracellular miRNAs in duchenne muscular dystrophy. Trends Mol Med 2017; 23: 989–1001.
55. Horak M, Novac J, Bienertova-Vasku J. Muscle specific MicroRNAs in skeletal muscle development. Dev Biol 2016; 410: 1–13.
56. Creemers EE, Tijsen AJ, Pinto YM. Circulating microRNAs: Novel biomarkers and extracellular communicators in cardiovascular disease? Circ Res 2012; 110: 483–95.
57. Sun T, Dong YH, Du W, Shi CY, Wang K, Tariq MA, *et al.* The role of microRNAs in myocardial infarction: from molecular mechanism to clinical application. Int J Mol Sci 2017; 18. pii: E745. doi: 10.3390/ ijms18040745.
58. Malizia AP, Wang DZ. MicroRNAs in cardiomyocyte development. Wiley Interdiscip Rev Syst Biol Med 2011; 3: 183–90.
59. Wang GK, Zhu JQ, Zhang JT, Li Q, Li Y, He J, *et al.* Circulating microRNA: A novel potential biomarker for early diagnosis of acute myocardial infarction in humans. Eur Heart J 2010; 31: 659–66.
60. Sistino JJ. Epidemiology of cardiovascular disease in the last decade: Treatment options and implications for perfusion in the 21st century. Perfusion 2003; 18: 73–7.
61. den Hoed M, Eijgelsheim M, Esko T, Brundel BJ, Peal DS, Evans DM, *et al.* Identification of heart rate-associated loci and their effects on cardiac conduction and rhythm disorders. Nat Genet 2013; 45: 621–31.
62. Dawson K, Wakili R, Ordog B, Clauss S, Chen Y, Iwasaki Y, *et al.* MicroRNA29: A mechanistic contributor and potential biomarker in atrial fibrillation. Circulation 2013; 127 : 1466-77, 75e1-28.
63. Hale CS, Levis WR. MicroRNA-29 and an integrated understanding of atrial fibrillation. J. Drug Dermatol 2013; 12: 1083
64. Cushing L, Kuang PP, Qian J, Shao F, Wu J, Little F, *et al.* Mir-29 is a major regulator of gene associated with pulmonary fibrosis Am. J. Respir Cell Mol Biol 2011; 45: 287–94.
65. Oliveira-Carvalho V, Carvalho VO, Bocchi EA. The emerging role of mir-208a in the heart. DNA Cell Biol 2013; 32: 8–12.
66. Harling L, Lambert J, Ashrafian H, Darzi A, Gooderham NJ, Athannasiou T. Elevated serum microRNA 483-5p levels may predict patients at risk of post-operative atrial fibrillation. Eur. J. Cadiothorac Surg 2017; 51: 73–78.
67. Feldman A, Moreira DAR, Gun C, Wang HL, Hirata MH, de Freitas Germano J, *et al.* Analysis of circulating mir-1, mir-23a, and mir- grafting surgery. Ann Hum Genet 2017; 81: 99–105.
68. Li C, Li X, Gao X, Zhang R, Zhang Y, Liang H, *et al.* MicroRNA-328 as a regulator of cardiac hypertrophy. Int J Cardiol 2014; 173: 268–76.
69. Lu Y, Hou S, Huang D, Luo X, Zhang J, Chen J, *et al* analysis of circulating microRNAs and their effects on ion channels in Chinese atrial fibrillation patients. Int J Clin Exp Med 2015; 8: 845–53.

70. Lu Y, Zhang Y, Wang N, Pan Z, Gao X, Zhang F, *et al.* MicroRNA-328 contributes to adverse electrical remodeling in atrial fibrillation. Circulation 2010; 122: 2378–87.
71. Ling TY, Wang XL, Chai Q, Lau TW, Koestler CM, Park SJ, *et al.* Regulation of the sk3 channel by microRNA-499--potential role in atrial fibrillation. Heart Rhythm 2013; 10: 1001-9.
72. Canon S, Caballero R, Herraiz-Martinez A, Perez-Hernandez M, Lopez B, Atienza F, *et al.* Mir-208b upregulation interferes with calcium handling in hl-1 atrial myocytes: Implications in human chronic atrial fibrillation. J. Mol. Cardiol 2016; 99: 162-73.
73. Morishima M, Iwata E, Nakada C, Tsukamoto Y, Takanari H, Miyamoto S, *et al* myocardial electrical remodeling of the G-protein-gated K^+ channel, *I*K.Ach. Circ J 2016; 80: 1346–55.
74. Adam O, Lohfelm B, Thum T, Gupta SK, Puhl SL, Schafers HJ, *et al.* 2012; 107: 278.
75. 85 Engelhardt S, Hein L, Wiesmann F, Lohse MJ. Progressive hypertrophy and heart failure in beta1-adrenergic receptor transgenic mice. Proc Natl Acad Sci U S A 1999; 96: 7059–64.
76. 86 Thum T, Gross C, Fiedler J, Fischer T, Kissler S, Bussen M, *et al.* MicroRNA-21 contributes to myocardial disease by stimulating map
77. 87 Goren Y, Meiri E, Hogan C, Mitchell H, Lebanony D, Salman N, *et al.* Relation of reduced expression of mir-150 in platelets to atrial 2014; 113: 976–81.
78. 88 Schotten U, Verheule S, Kirchhof P, Goette A. Pathophysiological mechanisms of atrial fibrillation: a translational appraisal. Physiol Rev 2011; 91: 265–325.
79. 89 Lu Y, Zhang Y, Shan H, Pan Z, Li X, Li B, *et al.* MicroRNA-1 downregulation by propranolol in a rat model of myocardial infarction: A new mechanism for ischaemic cardioprotection. Cardiovasc Res 2009; 84: 434–41.
80. O'Rourke MF. Vascular impedance in studies of arterial and cardiac function. Physiol Rev 1982; 62: 570–623.
81. Wang Z, Chesler NC. Pulmonary vascular wall stiffness: An important contributor to the increased right ventricular afterload with pulmonary hypertension. Pulm Circ 2011; 1: 212–23.
82. Sucharov CC, Sucharov J, Karimpour-Fard A, Nunley K, Stauffer BL, Miyamoto SD. Micro-RNA expression in hypoplastic left heart syndrome. J Card Fail 2015; 21: 83–8.
83. Tijsen AJ, Pinto YM, Creemers EE. Circulating microRNAs as diagnostic biomarkers for cardiovascular diseases. Am J Physiol Heart Circ Physiol 2012; 303: H1085–95.
84. Navickas R, Gal D, Laucevicius A, Taparauskaite A, Zdanyte M, Holvoet P. Identifying circulating microRNAs as biomarkers of cardiovascular disease: a systematic review. Cardiovasc Res 2016; 111: 322–37.
85. de Planell-Saguer M, Rodicio MC. Detection methods for microRNAs in clinic practice. Clin Biochem 2013; 46: 869–78.

86. Robinson S, Follo M, Haenel D, Mauler M, Stallmann D, Heger LA, *et al*. Chip-based digital pcr as a novel detection method for quantifying microRNAs in acute myocardial infarction patients. Acta Pharmacol Sin 2018; 39: 1217–27.
87. Sourvinou IS, Markou A, Lianidou ES. Quantification of circulating miRNAs in plasma: effect of preanalytical and analytical parameters on their isolation and stability. J Mol Diagn 2013; 15: 827–34.
88. Miotto E, Saccenti E, Lupini L, Callegari E, Negrini M, Ferracin M. Quantification of circulating miRNAs by droplet digital pcr: Comparison of evagreen- and taqman-based chemistries. Cancer Epidemiol Biomarkers Prev 2014; 23: 2638–42.
89. Camejo G H-CE, Wiklund O, Bondjers G. Association of apo B lipoproteins with arterial proteoglycans: patho- logical significance and molecular basis. . Atherosclerosis. 1998; 139(2):205-22.
90. Mason DP KR, Hasenstab D, Bowen-Pope DF, Seif- ert RA, Coats S, Hawkins SM & Clowes AW. Matrix me- talloproteinase-9 overexpression enhances vascular smooth muscle cell migration and alters remodeling in the injured rat carotid artery. Circulation research. 1999;85(12):1179-85.
91. Raines EW FN. Thematic review series: The immune system and atherogenesis: Cytokines affecting endothelial and smooth muscle cells in vascular disease Journal of lipid research. 2005; 46(6):1081-92.
92. Packard RR LP. Inflammation in atherosclerosis: from vascular biology to biomarker discovery and risk prediction. Clinical chemistry. 2008;54(1):24-38.
93. Libby P. Current concepts of the pathogen- esis of the acute coronary syndromes. Circulation 2001;104(3):365-72.
94. Ait-Oufella H, Taleb S, Mallat Z, Tedgui A. Recent Advances on the Role of Cytokines in Atherosclerosis. Arteriosclerosis, Thrombosis, and Vascular Biology. 2011;31:969-79
95. N. Papageorgiou DT, M. Charakida. Prognostic role of miRNAs in coronary artery disease. Curr Top Med Chem. 2013;13:1540-7.
96. Tai-You Ha. MicroRNAs in Human Diseases: From Cancer to Cardiovascular Disease. Immune Netw 2011 11(3):135-54.
97. Galimberti D VC, Fenoglio C, Serpente M, Ghezzi L, Cioffi SM, Arighi A, Fumagalli 1, Scarpini E. Circulating miRNAs as potential biomarkers in Alzheimer's disease. J Alzheimers Dis. 2014;42(4):1261-7.
98. Khoo SK PD, Kang UJ, Resau JH, Berryhill B, Linder J, Forsgren L, Neuman LA, Tan AC. Plasma-based circulating MicroRNA biomarkers for Parkinson's disease. J Parkin- sons Dis. 2012; 2(4):321-31.
99. Jessica A. Weber DHB, Shile Zhang, David Y. Huang, Kuo How Huang, Ming Jen Lee, David J. Galas, Kai Wang. The MicroRNA Spectrum in 12 Body Fluids. Clinical Chemistry 2010; 56(11):1733-41.
100. Evidence based Practice Center Systematic Review Protocol Noninvasive Testing for Coronary Artery Disease December 18, 2014.
101. Fuster V. Acute coronary syndromes: the degree and morphology of coronary stenoses. J Am Coll Cardiol. 1999;34(7):1854-6.
102. Topol EJ NS. Our preoccupation with coronary luminology. The dissociation between clinical and angio- graphic findings in ischemic heart disease. Circulation. 1995;92(8):2333-42.

103. Tavakol M AS, Brener SJ. . Risks and complications of coronary angiography: a comprehensive review. Global journal of health science. 2012;4(1):65-93.
104. Al Adas Z, Lodewyk K, Robinson D, Qureshi S, Kabbani LS, Sullivan B, et al. Contrast-induced nephropathy after peripheral vascular intervention: Long-term renal out- come and risk factors for progressive renal dysfunction. Journal of vascular surgery. 2019;69(3):913-20.
105. Verena Stangl VW, Gert Baumann, Karl Stangl. Current diagnostic concepts to detect coronary artery disease in women. European Heart Journal , . 2008;29:707-17.
106. Sajjadieh A HA, Keivani M, Asoodeh A, Pourmoghad- das M, Sanei H. . Diagnostic performance of 64-row coronary CT angiography in detecting significant stenosis as compared with conventional invasive coronary angiography. ARYA Atherosclerosis. 2013;9(2):157-63.
107. Trzybulska D, Vergadi E, Tsatsanis C. miRNA and Other Non-Coding RNAs as Promising Diagnostic Markers. EJIF- CC. 2018;29(3):221-6.
108. Ward JA, Esa N, Pidikiti R, Freedman JE, Keaney JF, Tanriverdi K, et al. Circulating Cell and Plasma microRNA Profiles Differ between Non-ST-Segment and ST-Segment- Elevation Myocardial Infarction. Family medicine & medical science research. 2013;2(2):108-.
109. Wang H-W, Huang T-S, Lo H-H, Huang P-H, Lin C-C, Chang S-J, et al. Deficiency of the MicroRNA-31–MicroRNA-720 Pathway in the Plasma and Endothelial Progenitor Cells From Patients With Coronary Artery Disease. Arterioscler Thromb Vasc Biol. 2014;34:857-69.
110. Hulsmans M, Sinnaeve P, Schueren BVd, Mathieu C, Janssens S, Holvoet P. Decreased miR-181a Expression in Monocytes of Obese Patients Is Associated with the Occurrence of Metabolic Syndrome and Coronary Artery Disease. J Clin Endocrinol Metab 2012;97:E1213-E8.
111. Bialek S, Gorko D, Zajkowska A, Koltowski L, Grabows- ki M, Stachurska A, et al. Release kinetics of circulating miRNA-208a in the early phase of myocardial infarction. Kardiologia polska. 2015;73(8):613-9.
112. Md Sayed AS, K. Xia, F. Li, X. Deng, U. Salma, T. Li, H. Deng. The diagnostic value of circulating microRNAs for middle-aged (40-60-year-old) coronary artery disease patients. Clinics 2015;70(4):257-63.
113. Wang F, Long G, Zhao C, Li H, Chaugai S, Wang Y, et al. Plasma microRNA-133a is a new marker for both acute myocardial infarction and underlying coronary artery stenosis. Journal of Translational Medicine 2013;11:222.
114. D'Alessandra Y, Carena MC, Spazzafumo L, Martinelli F, Bassetti B, Devanna P, et al. Diagnostic Potential of Plasmatic MicroRNA Signatures in Stable and Unstable Angina. PLoS ONE (): . 2013;8(11):e80345.
115. D'Alessandra Y PG, Capogrossi MC. . MicroRNAs and myocardial infarction Curr Opin Cardiol 2012;27:228-35.
116. Zhou J, Shao G, Chen X, Yang X, Huang X, Peng P, et al. MicroRNA 206 and MicroRNA 574-5p are highly expression in coronary artery disease. Biosci Rep 2015.
117. Han H, Qu G, Han C, Wang Y, Sun T, Li F, et al. MiR-34a, miR-21 and miR-23a as potential biomarkers for coronary artery disease: a pilot microarray study and

confirmation in a 32 patient cohort. Experimental & Molecular Medicine 2015;47:e138.

118. Hoekstra M, Lans CACvd, Halvorsen B, Gullestad L, Kuiper J, Aukrust P, et al. The peripheral blood mononuclear cell microRNA signature of coronary artery disease. Biochemical and Biophysical Research Communications 2010;394:792-7.

119. Dong J, Liang YZ, Zhang J, Wu LJ, Wang S, Hua Q, et al. Potential Role of Lipometabolism-Related MicroRNAs in Peripheral Blood Mononuclear Cells as Biomarkers for Coronary Artery Disease. Journal of atherosclerosis and thrombosis. 2017;24(4):430-41.

120. Faccini J, Ruidavets J-B, Cordelier P, Martins F, Maoret J-J, Bongard V, et al. Circulating miR-155, miR-145 and let- 7c as diagnostic biomarkers of the coronary artery disease. Scientific Reports. 2017;7:42916.

121. Sinning C, Lillpopp L, Appelbaum S, Ojeda F, Zeller T, Schnabel R, et al. Angiographic score assessment im- proves cardiovascular risk prediction: the clinical value of SYNTAX and Gensini application. Clinical Research in Cardiology. 2013;102(7):495-503.

122. Zhang Y, Li HH, Yang R, Yang BJ, Gao ZY. Association between circulating microRNA-208a and severity of coronary heart disease. 2017;77(5):379-84.

123. Qiu XK, Ma J. Alteration in microRNA-155 level correspond to severity of coronary heart disease. Scandinavian journal of clinical and laboratory investigation. 2018;78(3):219-23.

124. Guo JF, Zhang Y, Zheng QX, Zhang Y, Zhou HH, Cui LM. Association between elevated plasma microRNA-223 content and severity of coronary heart disease. Scandinavian journal of clinical and laboratory investigation. 2018;78(5):373-8.

125. Gao H, Guddeti RR, Matsuzawa Y, Liu L-P, Li-Xiao, Guo S, et al. Plasma Levels of microRNA-145 Are Associated with Severity of Coronary Artery Disease. PLoS ONE 2015;10(5):e0123477.

126. Lu H-Q, Liang C, He Z-Q, Fan M, Wu Z-G. Circulating miR-214 is associated with the severity of coronary artery disease. Journal of Geriatric Cardiology. 2013;10:34–8.

127. Myerburg RJ IAJ, Mitrani RM. Frequency of sud- den cardiac death and profiles of risk. Am J Cardiol 1997;80:10-9.

128. Li X, Yang Y, Wang L, Qiao S, Lu X, Wu Y, et al. Plasma miR-122 and miR-3149 Potentially Novel Biomarkers for Acute Coronary Syndrome. PLoS ONE 2015;10(5):e0125430.

129. Ren J, Zhang J, Xu N, Han G, Geng Q, Song J, et al. Signature of Circulating MicroRNAs as Potential Biomarkers in Vulnerable Coronary Artery Disease. PLoS ONE 2013;8(12):e80738.

130. Li S, Lee C, Song J, Lu C, Liu J, Cui Y, et al. Circulating microRNAs as potential biomarkers for coronary plaque rupture. Oncotarget. 2017;8(29):48145-56.

131. Luque A, Farwati A, Krupinski J, Aran JM. Association between low levels of serum miR-638 and atherosclerotic plaque vulnerability in patients with high-grade carotid stenosis. Journal of neurosurgery. 2018:1-8

132. Niculescu LS, Simionescu N, Sanda GM, Carnuta MG, Stancu CS, Popescu AC, et al. MiR-486 and miR-92a Identified in Circulating HDL Discriminate between

Stable and Vulnerable Coronary Artery Disease Patients. PLoS ONE 2015;10(10):e0140958.
133. C. Frangos SN, N. Piazza Impact of bifurcation le- sions on angiographic characteristics and procedural success in primary percutaneous coronary intervention for ST-segment elevation myocardial infarction. Archives of Cardiovascular Diseases. 2011;104(4):234-41.
134. Liu Y, Chen S, Zhang J, Shoujie Shan, Chen L, Wang R, et al. Analysis of Serum MicroRNAs as Potential Biomarker in Coronary Bifurcation Lesion. Disease Markers. 2015;2015.
135. Fan X, EnshiWang, Wang X, Cong X, Chen X. MicroR- NA-21 is a unique signature associated with coronary plaque instability in humans by regulating matrix metal- loproteinase-9 via reversion-inducing cysteine-rich protein with Kazal motifs. Experimental and Molecular Pathology 2014;96:242-9.
136. Liu W, Ling S, Sun W, Liu T, Li Y, Zhong G, et al. Circulating microRNAs correlated with the level of coronary artery calcification in symptomatic patients. Scientific Re- ports. 2015;5:16099.
137. Jansen F, Yang X, Proebsting S, Hoelscher M, Przybilla D, Baumann K, et al. MicroRNA Expression in Circulating Microvesicles Predicts Cardiovascular Events in Patients with Coronary Artery Disease. J Am Heart Assoc. 2014;3:e001249.
138. Schulte C, Molz S, Appelbaum S, Karakas M, Oje- da F, Lau DM, et al. miRNA-197 and miRNA-223 Predict Cardiovascular Death in a Cohort of Patients with Symptomatic Coronary Artery Disease. PLoS ONE. 2015;10(12):e0145930.
139. Widera C, Gupta SK, Lorenzen JM, Bang C, Bauer- sachs J, Bethmann K, et al. Diagnostic and prognostic impact of six circulating microRNAs in acute coronary syndrome. Journal of Molecular and Cellular Cardiology. 2011;51 872-5.
140. Bennett MR. In-stent stenosis: pathology and implications for the development of drug eluting stents. Heart (British Cardiac Society). 2003;89(2):218-24.
141. Hamid H, Coltart J. 'Miracle stents'--a future without restenosis. McGill journal of medicine : MJM : an international forum for the advancement of medical sciences by students. 2007;10(2):105-11.
142. He M, Gong Y, Shi J, Pan Z, Zou H, Sun D, et al. Plasma microRNAs as potential noninvasive biomarkers for in- stent restenosis. PLoS One. 2014;9(11):e112043.
143. O'Sullivan JF, Neylon A, Fahy EF, Yang P, McGorrian C, Blake GJ. MiR-93-5p is a novel predictor of coronary instent restenosis. Heart Asia. 2019;11(1):e011134.
144. Fejes Z, Czimmerer Z, Szuk T, Poliska S, Horvath A, Balogh E, et al. Endothelial cell activation is attenuated by everolimus via transcriptional and posttranscriptional regulatory mechanisms after drug-eluting coronary stent- ing. PLoS ONE 2018;13(6):e0197890.
145. Kaudewitz D, Lee R, Willeit P, McGregor R, Markus HS, Kiechl S, et al. Impact of intravenous heparin on quantification of circulating microRNAs in patients with coronary artery disease. Thromb Haemost. 2013;110(3):609-15.
146. Boeckel JN TC, Leistner D, Leistner D, Zeiher AM, Fich- tlscherer S, Dimmeler S. Heparin selectively affects the quantification of microRNAs in human blood samples. Clin Chem. 2013;59:1125-7.

147. Basso D, Padoan A, Laufer T, Aneloni V, Moz S, Schro- ers H, et al. Relevance of pre-analytical blood management on the emerging cardiovascular protein biomarkers TWEAK and HMGB1 and on miRNA serum and plasma profiling. Clinical biochemistry. 2017;50(4-5):186-93.
148. Boileau A, Lino Cardenas CL, Lindsay ME, Devaux Y. Endogenous Heparin Interferes with Quantification of MicroRNAs by RT-qPCR. Clin Chem. 2018;64(5):863-5.
149. Russo I, Penna C, Musso T, Popara J, Alloatti G, Cavalot F, et al. Platelets, diabetes and myocardial ischemia/reperfusion injury. Cardiovascular diabetology. 2017;16(1):71-.
150. Willeit P, Zampetaki A, Dudek K, Kaudewitz D, King A, Kirkby NS, et al. Circulating microRNAs as novel biomarkers for platelet activation. Circ Res. 2013;112(4):595-600.
151. Li S, Zhang F, Cui Y, Wu M, Lee C, Song J, et al. Modified high-throughput quantification of plasma microRNAs in heparinized patients with coronary artery disease using heparinase. Biochem Biophys Res Commun. 2017;493(1): 556-61.
152. Kondratov K, Kurapeev D, Popov M, Sidorova M, Minasian S, Galagudza M, et al. Heparinase treatment of heparin- contaminated plasma from coronary artery bypass grafting patients enables reliable quantification of microRNAs. Biomolecular detection and quantification. 2016;8:9-14.
153. Weber M BM, Patel RS, Quyyumi AA, Bao G, Searles CD. MicroRNA Expression Profile in CAD Patients and the Impact of ACEI/ARB. Cardiol Res Pract. 2011;2011.
154. Gidlöf O AP, van der Pals J, Götberg M, Erlinge D. Cardiospecific microRNA plasma levels correlate with troponin and cardiac function in patients with ST elevation myocardial infarction, are selectively dependent on renal elimination, and can be detected in urine samples. Cardiology. 2011;118(4):217-26.
155. Yan Y, Shi Y, Wang C, Guo P, Wang J, Zhang C-Y, et al. In- fluence of a high-altitude hypoxic environment on human plasma microRNA profiles. Scientific Reports 2015; 5.
156. Singh N, Heggermont W, Fieuws S, Vanhaecke J, Van- Cleemput J, DeGeest B. Endothelium-enriched microRNAsasdiagnostic biomarkers for cardiac allo-graft vasculopathy. J Heart Lung Transplant 2015;34:1376-84.
157. Kuwabara Y, Ono K, Horie T, Nishi H, Nagao K, Kinoshita M, et al. Increased microRNA-1 and microRNA-133a levels in serum of patients with cardiovascular disease indicate myocardial damage. Circulation: Cardiovascular Genetics. 2011; 4(4):446-54.
158. Witwer KW. Circulating microRNA biomarker studies: pitfalls and potential solutions. Clin Chem. 2015;61(1):56-63.
159. Hullinger, TG, Montgomery, RL, Seto, AG, Dickinson, BA, Semus, HM, Lynch, JM, Dalby, CM, Robinson, K, Stack, C, Latimer, PA, Hare, JM, Olson, EN, van Rooij, E. Inhibition of miR-15 protects against cardiac ischemic injury. *Circ Res*. 2012;110:71–81.
160. Porrello, ER, Mahmoud, AI, Simpson, E, Johnson, BA, Grinsfelder, D, Canseco, D, Mammen, PP, Rothermel, BA, Olson, EN, Sadek, HA. Regulation of neonatal and adult mammalian heart regeneration by the miR-15 family. *Proc Natl Acad Sci USA*. 2013;110:187–192.

161. Boon, RA, Iekushi, K, Lechner, S et al,. MicroRNA-34a regulates cardiac ageing and function. *Nature*. 2013;495:107–110.
162. Bernardo, BC, Gao, XM, Winbanks, CE, Boey, EJH, Tham, YK, Kiriazis, H, Gregorevic, P, Obad, S, Kauppinen, S, Du, XJ, Lin, RCY, McMullen, JR. Therapeutic inhibition of the miR-34 family attenuates pathological cardiac remodeling and improves heart function. *Proc Natl Acad Sci USA*. 2012;109:17615–17620.
163. Yang, Y, Cheng, HW, Qiu, Y, Dupee, D, Noonan, M, Lin, YD, Fisch, S, Unno, K, Sereti, KI, Liao, R. MicroRNA-34a plays a key role in cardiac repair and regeneration following myocardial infarction. *Circ Res*. 2015;117:450–459.
164. Bonauer, A, Carmona, G, Iwasaki, Met al.. MicroRNA-92a controls angiogenesis and functional recovery of ischemic tissues in mice. *Science*. 2009;324:1710–1713.
165. Hinkel, R, Penzkofer, D, Zühlke, S, Fischer, A, Husada, W, Xu, QF, Baloch, E, van Rooij, E, Zeiher, AM, Kupatt, C, Dimmeler, S. Inhibition of microRNA-92a protects against ischemia/reperfusion injury in a large-animal model. *Circulation*. 2013;128:1066–1075.
166. Bellera, N, Barba, I, Rodriguez-Sinovas, A, Ferret, E, Asín, MA, Gonzalez-Alujas, MT, Pérez-Rodon, J, Esteves, M, Fonseca, C, Toran, N, Garcia Del Blanco, B, Pérez, A, Garcia-Dorado, D. Single intracoronary injection of encapsulated antagomir-92a promotes angiogenesis and prevents adverse infarct remodeling. *J Am Heart Assoc*. 2014;3:e000946.
167. Fiedler, J, Jazbutyte, V, Kirchmaier, BCet al.. MicroRNA-24 regulates vascularity after myocardial infarction. *Circulation*. 2011;124:720–730.
168. Thum, T. Noncoding RNAs and myocardial fibrosis. *Nat Rev Cardiol*. 2014;11:655–663.
169. van Rooij, E, Sutherland, LB, Thatcher, JE, DiMaio, JM, Naseem, RH, Marshall, WS, Hill, JA, Olson, EN. Dysregulation of microRNAs after myocardial infarction reveals a role of miR-29 in cardiac fibrosis. *Proc Natl Acad Sci USA*. 2008;105:13027–13032.
170. Thum, T, Gross, C, Fiedler, Jet al.. MicroRNA-21 contributes to myocardial disease by stimulating MAP kinase signalling in fibroblasts. *Nature*. 2008;456:980–984.
171. Patrick, DM, Montgomery, RL, Qi, X, Obad, S, Kauppinen, S, Hill, JA, van Rooij, E, Olson, EN. Stress-dependent cardiac remodeling occurs in the absence of microRNA-21 in mice. *J Clin Invest*. 2010;120:3912–3916.
172. Thum, T, Chau, N, Bhat, B, Gupta, SK, Linsley, PS, Bauersachs, J, Engelhardt, S. Comparison of different miR-21 inhibitor chemistries in a cardiac disease model. *J Clin Invest*. 2011;121:461–462; author reply 462–463.
173. Gupta, SK, Itagaki, R, Zheng, X et al. miR-21 promotes fibrosis in an acute cardiac allograft transplantation model. *Cardiovasc Res*. 2016;110:215–226.
174. Liu, G, Friggeri, A, Yang, Y, Milosevic, J, Ding, Q, Thannickal, VJ, Kaminski, N, Abraham, E. miR-21 mediates fibrogenic activation of pulmonary fibroblasts and lung fibrosis. *J Exp Med*. 2010;207:1589–1597.
175. Zhang, J, Jiao, J, Cermelli, S, Muir, K, Jung, KH, Zou, R, Rashid, A, Gagea, M, Zabludoff, S, Kalluri, R, Beretta, L. miR-21 inhibition reduces liver fibrosis and

prevents tumor development by inducing apoptosis of CD24+ progenitor cells. *Cancer Res*. 2015;75:1859–1867.

176. Wahlquist, C, Jeong, D, Rojas-Muñoz, A, Kho, C, Lee, A, Mitsuyama, S, van Mil, A, Park, WJ, Sluijter, JPG, Doevendans, PAF, Hajjar, RJ, Mercola, M. Inhibition of miR-25 improves cardiac contractility in the failing heart. *Nature*. 2014;508:531–535.
177. Dirkx, E, Gladka, MM, Philippen, LEet al.. Nfat and miR-25 cooperate to reactivate the transcription factor Hand2 in heart failure. *Nat Cell Biol*. 2013;15:1282–1293.
178. Carè, A, Catalucci, D, Felicetti, F et al,. MicroRNA-133 controls cardiac hypertrophy. *Nat Med*. 2007;13:613–618.
179. Besser, J, Malan, D, Wystub, K, Bachmann, A, Wietelmann, A, Sasse, P, Fleischmann, BK, Braun, T, Boettger, T. MiRNA-1/133a clusters regulate adrenergic control of cardiac repolarization. *PLoS One*. 2014;9:e113449.
180. Ucar, A, Gupta, SK, Fiedler, Jet al.. The miRNA-212/132 family regulates both cardiac hypertrophy and cardiomyocyte autophagy. *Nat Commun*. 2012;3:1078.
181. van Rooij, E, Sutherland, LB, Qi, X, Richardson, JA, Hill, J, Olson, EN. Control of stress-dependent cardiac growth and gene expression by a microRNA. *Science*. 2007;316:575–579.
182. Boon, RA, Hergenreider, E, Dimmeler, S. Atheroprotective mechanisms of shear stress-regulated microRNAs. *Thromb Haemost*. 2012;108:616–620.
183. Qin, X, Wang, X, Wang, Y, Tang, Z, Cui, Q, Xi, J, Li, YS, Chien, S, Wang, N. MicroRNA-19a mediates the suppressive effect of laminar flow on cyclin D1 expression in human umbilical vein endothelial cells. *Proc Natl Acad Sci USA*. 2010;107:3240–3244.
184. Wang, KC, Garmire, LX, Young, A, Nguyen, P, Trinh, A, Subramaniam, S, Wang, N, Shyy, JY, Li, YS, Chien, S. Role of microRNA-23b in flow-regulation of Rb phosphorylation and endothelial cell growth. *Proc Natl Acad Sci USA*. 2010;107:3234–3239.
185. Ji, R, Cheng, Y, Yue, J, Yang, J, Liu, X, Chen, H, Dean, DB, Zhang, C. MicroRNA expression signature and antisense-mediated depletion reveal an essential role of MicroRNA in vascular neointimal lesion formation. *Circ Res*. 2007;100:1579–1588.
186. Maegdefessel, L, Azuma, J, Toh, R, Merk, DR, Deng, A, Chin, JT, Raaz, U, Schoelmerich, AM, Raiesdana, A, Leeper, NJ, McConnell, M V, Dalman, RL, Spin, JM, Tsao, PS. Inhibition of microRNA-29b reduces murine abdominal aortic aneurysm development. *J Clin Invest*. 2012;122:497–506.
187. Boon, RA, Seeger, T, Heydt, S, Fischer, A, Hergenreider, E, Horrevoets, AJG, Vinciguerra, M, Rosenthal, N, Sciacca, S, Pilato, M, van Heijningen, P, Essers, J, Brandes, RP, Zeiher, AM, Dimmeler, S. MicroRNA-29 in aortic dilation: implications for aneurysm formation. *Circ Res*. 2011;109:1115–1119.
188. Schober, A, Nazari-Jahantigh, M, Wei, Y, Bidzhekov, K, Gremse, F, Grommes, J, Megens, RTA, Heyll, K, Noels, H, Hristov, M, Wang, S, Kiessling, F, Olson, EN, Weber, C. MicroRNA-126-5p promotes endothelial proliferation and limits atherosclerosis by suppressing Dlk1. *Nat Med*. 2014;20:368–376.

189. Li, K, Ching, D, Luk, FS, Raffai, RL. Apolipoprotein E enhances microRNA-146a in monocytes and macrophages to suppress nuclear factor-κB-driven inflammation and atherosclerosis. *Circ Res*. 2015;117:e1–e11.
190. Sun, X, He, S, Wara, AKM, Icli, B, Shvartz, E, Tesmenitsky, Y, Belkin, N, Li, D, Blackwell, TS, Sukhova, GK, Croce, K, Feinberg, MW. Systemic delivery of microRNA-181b inhibits nuclear factor-κB activation, vascular inflammation, and atherosclerosis in apolipoprotein E-deficient mice. *Circ Res*. 2014;114:32–40.
191. Fang, Y, Shi, C, Manduchi, E, Civelek, M, Davies, PF. MicroRNA-10a regulation of proinflammatory phenotype in athero-susceptible endothelium *in vivo* and *in vitro*. *Proc Natl Acad Sci USA*. 2010;107:13450–13455.
192. Thum, T. MicroRNA therapeutics in cardiovascular medicine. *EMBO Mol Med*. 2012;4:3–14.
193. Boon, RA, Dimmeler, S. MicroRNAs in myocardial infarction. *Nat Rev Cardiol*. 2015;12:135–142.
194. van Rooij, E, Olson, EN. MicroRNA therapeutics for cardiovascular disease: opportunities and obstacles. *Nat Rev Drug Discov*. 2012;11:860–872.
195. Krützfeldt, J, Rajewsky, N, Braich, R, Rajeev, KG, Tuschl, T, Manoharan, M, Stoffel, M. Silencing of microRNAs in vivo with "antagomirs". *Nature*. 2005;438:685–689.
196. Lennox, KA, Behlke, MA. A direct comparison of anti-microRNA oligonucleotide potency. *Pharm Res*. 2010;27:1788–1799.
197. Elmén, J, Lindow, M, Schütz, S, Lawrence, M, Petri, A, Obad, S, Lindholm, M, Hedtjärn, M, Hansen, HF, Berger, U, Gullans, S, Kearney, P, Sarnow, P, Straarup, EM, Kauppinen, S. LNA-mediated microRNA silencing in non-human primates. *Nature*. 2008;452:896–899.
198. Elmén, J, Lindow, M, Silahtaroglu, A, Bak, M, Christensen, M, Lind-Thomsen, A, Hedtjärn, M, Hansen, JB, Hansen, HF, Straarup, EM, McCullagh, K, Kearney, P, Kauppinen, S. Antagonism of microRNA-122 in mice by systemically administered LNA-antimiR leads to up-regulation of a large set of predicted target mRNAs in the liver. *Nucleic Acids Res*. 2008;36:1153–1162.
199. Obad, S, dos Santos, CO, Petri, A, Heidenblad, M, Broom, O, Ruse, C, Fu, C, Lindow, M, Stenvang, J, Straarup, EM, Hansen, HF, Koch, T, Pappin, D, Hannon, GJ, Kauppinen, S. Silencing of microRNA families by seed-targeting tiny LNAs. *Nat Genet*. 2011;43:371–378.
200. Nahar, S, Singh, A, Morihiro, K, Moai, Y, Kodama, T, Obika, S, Maiti, S. Systematic evaluation of biophysical and functional characteristics of selenomethylene-locked nucleic acid-mediated inhibition of miR-21. *Biochemistry*. 2016;55:7023–7032.
201. Meng, L, Liu, C, Lü, Jet al.. Small RNA zippers lock miRNA molecules and block miRNA function in mammalian cells. *Nat Commun*. 2017;8:13964.
202. Paulasova, P, Pellestor, F. The peptide nucleic acids (PNAs): a new generation of probes for genetic and cytogenetic analyses. *Ann Génétique*. 2004;47:349–358.
203. Fabani, MM, Abreu-Goodger, C, Williams, D, Lyons, PA, Torres, AG, Smith, KG, Enright, AJ, Gait, MJ, Vigorito, E. Efficient inhibition of miR-155 function

in vivo by peptide nucleic acids. *Nucleic Acids Res*. 2010;38:4466–4475. doi: 10.1093/nar/gkq160.

204. Behlke, MA. Chemical modification of siRNAs for in vivo use. *Oligonucleotides*. 2008;18:305–319.
205. Choi, WY, Giraldez, AJ, Schier, AF. Target protectors reveal dampening and balancing of Nodal agonist and antagonist by miR-430. *Science*. 2007;318:271–274.
206. Messina, A, Langlet, F, Chachlaki, K, Roa, J, Rasika, S, Jouy, N, Gallet, S, Gaytan, F, Parkash, J, Tena-Sempere, M, Giacobini, P, Prevot, V. A microRNA switch regulates the rise in hypothalamic GnRH production before puberty. *Nat Neurosci*. 2016;19:835–844.
207. Iaconetti, C, Polimeni, A, Sorrentino, S, Sabatino, J, Pironti, G, Esposito, G, Curcio, A, Indolfi, C. Inhibition of miR-92a increases endothelial proliferation and migration in vitro as well as reduces neointimal proliferation in vivo after vascular injury. *Basic Res Cardiol*. 2012;107:296.
208. Loyer, X, Potteaux, S, Vion, AC, Guérin, CL, Boulkroun, S, Rautou, PE, Ramkhelawon, B, Esposito, B, Dalloz, M, Paul, JL, Julia, P, Maccario, J, Boulanger, CM, Mallat, Z, Tedgui, A. Inhibition of microRNA-92a prevents endothelial dysfunction and atherosclerosis in mice. *Circ Res*. 2014;114:434–443.
209. Jeong, D, Yoo, J, Lee, P, Kepreotis, SV, Lee, A, Wahlquist, C, Brown, BD, Kho, C, Mercola, M, Hajjar, RJ. miR-25 tough decoy enhances cardiac function in heart failure. *Mol Ther*. 2017.
210. Ren, XP, Wu, J, Wang, X, Sartor, MA, Jones, K, Qian, J, Nicolaou, P, Pritchard, TJ, Fan, GC. MicroRNA-320 is involved in the regulation of cardiac ischemia/reperfusion injury by targeting heat-shock protein 20. *Circulation*. 2009;119:2357–2366.
211. Montgomery, RL, Hullinger, TG, Semus, HM, Dickinson, BA, Seto, AG, Lynch, JM, Stack, C, Latimer, PA, Olson, EN, van Rooij, E. Therapeutic inhibition of miR-208a improves cardiac function and survival during heart failure. *Circulation*. 2011;124:1537–1547.
212. Janssen, HLA, Reesink, HW, Lawitz, EJ, Zeuzem, S, Rodriguez-Torres, M, Patel, K, van der Meer, AJ, Patick, AK, Chen, A, Zhou, Y, Persson, R, King, BD, Kauppinen, S, Levin, AA, Hodges, MR. Treatment of HCV infection by targeting microRNA. *N Engl J Med*. 2013;368:1685–1694.
213. Jayaraman, M, Ansell, SM, Mui, BLet al.. Maximizing the potency of siRNA lipid nanoparticles for hepatic gene silencing in vivo. *Angew Chem Int Ed Engl*. 2012;51:8529–8533.
214. Geary, RS, Wancewicz, E, Matson, J, Pearce, M, Siwkowski, A, Swayze, E, Bennett, F. Effect of dose and plasma concentration on liver uptake and pharmacologic activity of a 2'-methoxyethyl modified chimeric antisense oligonucleotide targeting PTEN. *Biochem Pharmacol*. 2009;78:284–291.
215. Crooke, ST, Wang, S, Vickers, TA, Shen, W, Liang, XH. Cellular uptake and trafficking of antisense oligonucleotides. *Nat Biotechnol*. 2017;35:230–237.
216. Wagenaar, TR, Tolstykh, T, Shi, C et al,. Identification of the endosomal sorting complex required for transport-I (ESCRT-I) as an important modulator of anti-miR uptake by cancer cells. *Nucleic Acids Res*. 2015;43:1204–1215.

217. Berman, CL, Barros, SA, Galloway, SM, Kasper, P, Oleson, FB, Priestley, CC, Sweder, KS, Schlosser, MJ, Sobol, Z. OSWG recommendations for genotoxicity testing of novel oligonucleotide-based therapeutics. *Nucleic Acid Ther.* 2016;26:73–85.
218. Peacock, H, Fucini, R V, Jayalath, P, Ibarra-Soza, JM, Haringsma, HJ, Flanagan, WM, Willingham, A, Beal, PA. Nucleobase and ribose modifications control immunostimulation by a microRNA-122-mimetic RNA. *J Am Chem Soc.* 2011;133:9200–9203.
219. Monteith, DK, Geary, RS, Leeds, JM, Johnston, J, Monia, BP, Levin, AA. Preclinical evaluation of the effects of a novel antisense compound targeting C-raf kinase in mice and monkeys. *Toxicol Sci.* 1998;46:365–375.
220. Levin, AA. A review of the issues in the pharmacokinetics and toxicology of phosphorothioate antisense oligonucleotides. *Biochim Biophys Acta.* 1999;1489:69–84.
221. Henry, SP, Beattie, G, Yeh, G, Chappel, A, Giclas, P, Mortari, A, Jagels, MA, Kornbrust, DJ, Levin, AA. Complement activation is responsible for acute toxicities in rhesus monkeys treated with a phosphorothioate oligodeoxynucleotide. *Int Immunopharmacol.* 2002;2:1657–1666.
222. Henry, SP, Jagels, MA, Hugli, TE, Manalili, S, Geary, RS, Giclas, PC, Levin, AA. Mechanism of alternative complement pathway dysregulation by a phosphorothioate oligonucleotide in monkey and human serum. *Nucleic Acid Ther.* 2014;24:326–335.
223. Flierl, U, Nero, TL, Lim, Bet al.. Phosphorothioate backbone modifications of nucleotide-based drugs are potent platelet activators. *J Exp Med.* 2015;212:129–137.
224. Sewing, S, Roth, AB, Winter, M, Dieckmann, A, Bertinetti-Lapatki, C, Tessier, Y, McGinnis, C, Huber, S, Koller, E, Ploix, C, Reed, JC, Singer, T, Rothfuss, A. Assessing single-stranded oligonucleotide drug-induced effects in vitro reveals key risk factors for thrombocytopenia. *PLoS One.* 2017;12:e0187574.
225. Swayze, EE, Siwkowski, AM, Wancewicz, EV, Migawa, MT, Wyrzykiewicz, TK, Hung, G, Monia, BP, Bennett, CF. Antisense oligonucleotides containing locked nucleic acid improve potency but cause significant hepatotoxicity in animals. *Nucleic Acids Res.* 2007;35:687–700.
226. Burdick, AD, Sciabola, S, Mantena, SR, Hollingshead, BD, Stanton, R, Warneke, JA, Zeng, M, Martsen, E, Medvedev, A, Makarov, SS, Reed, LA, Davis, JW, Whiteley, LO. Sequence motifs associated with hepatotoxicity of locked nucleic acid–modified antisense oligonucleotides. *Nucleic Acids Res.* 2014;42:4882–4891.
227. Lucas, T, Dimmeler, S. RNA therapeutics for treatment of cardiovascular diseases. *Circ Res.* 2016;119:794–797.
228. Ma, S, Tian, XY, Zhang, Y, Mu, C, Shen, H, Bismuth, J, Pownall, HJ, Huang, Y, Wong, WT. E-selectin-targeting delivery of microRNAs by microparticles ameliorates endothelial inflammation and atherosclerosis. *Sci Rep.* 2016;6:22910.
229. Karakikes, I, Chaanine, AH, Kang, S, Mukete, BN, Jeong, D, Zhang, S, Hajjar, RJ, Lebeche, D. Therapeutic cardiac-targeted delivery of miR-1 reverses pressure

overload-induced cardiac hypertrophy and attenuates pathological remodeling. *J Am Heart Assoc*. 2013;2:e000078–e000078.
230. Eulalio, A, Mano, M, Dal Ferro, M, Zentilin, L, Sinagra, G, Zacchigna, S, Giacca, M. Functional screening identifies miRNAs inducing cardiac regeneration. *Nature*. 2012;492:376–381.
231. van Rooij, E, Kauppinen, S. Development of microRNA therapeutics is coming of age. *EMBO Mol Med*. 2014;6:851–864.
232. D'Alessandra Y PG, Capogrossi MC. . MicroRNAs and myocardial infarction Curr Opin Cardiol 2012;27:228-35.
233. Dong J, Liang YZ, Zhang J, Wu LJ, Wang S, Hua Q, et al. Potential Role of Lipometabolism-Related MicroRNAs in Peripheral Blood Mononuclear Cells as Biomarkers for Coronary Artery Disease. Journal of atherosclerosis and thrombosis. 2017;24(4):430-41.
234. Faccini J, Ruidavets J-B, Cordelier P, Martins F, Maoret J-J, Bongard V, et al. Circulating miR-155, miR-145 and let- 7c as diagnostic biomarkers of the coronary artery dis- ease. Scientific Reports. 2017;7:42916.
235. Ren J, Zhang J, Xu N, Han G, Geng Q, Song J, et al. Signature of Circulating MicroRNAs as Potential Biomark- ers in Vulnerable Coronary Artery Disease. PLoS ONE 2013;8(12):e80738.
236. 68. Li S, Lee C, Song J, Lu C, Liu J, Cui Y, et al. Circulating microRNAs as potential biomarkers for coronary plaque rupture. Oncotarget. 2017;8(29):48145-56.

Chapter IX

1. Reidy K, Kang HM, Hostetter T, Susztak K. Molecular mechanisms of diabetic kidney disease. *J Clin Invest*. 2014;124:2333-2340.
2. Kato M, Natarajan R. MicroRNAs in diabetic nephropathy: functions, biomarkers, and therapeutic targets. *Ann N Y Acad Sci*. 2015;1353:72-88.
3. Gnudi L, Coward R, Long DA. Diabetic nephropathy: perspective on novel molecular mechanisms. *Trends Endocrinol Metab*.2016;27:820-830.
4. Fineberg D, Jandeleit-Dahm KA, Cooper ME. Diabetic nephropathy: diagnosis and treatment. *Nat Rev Endocrinol*. 2013;9:713-723.
5. Kanwar YS, Sun L, Xie P, Liu FY, Chen S. A glimpse of various pathogenetic mechanisms of diabetic nephropathy. *Annu Rev Pathol*. 2011;6:395-423.
6. Levey AS, Becker C, Inker LA. Glomerular filtration rate and albuminuria for detection and staging of acute and chronic kidney dise ase in adults: a systematic review. *JAMA*. 2015;313:837- 846.
7. Stevens LA, Coresh J, Schmid CH, et al. Estimating GFR using serum cystatin C alone and in combination with serum creatinine: a pooled analysis of 3,418 individuals with CKD. *Am J Kidney Dis*. 2008;51:395-406.
8. Odden MC, Shlipak MG, Tager IB. Serum creatinine and func- tional limitation in elderly persons. *J Gerontol A Biol Sci Med Sci*. 2009;64:370-376.
9. Krolewski AS. Progressive renal decline: the new paradigm of diabetic nephropathy in type 1 diabetes. *Diabetes Care*. 2015;38:954-962.
10. Macisaac RJ, Jerums G. Diabetic kidney disease with and without albuminuria. *Curr Opin Nephrol Hypertens*. 2011;20:246-257.

11. Nosadini R, Velussi M, Brocco E, et al. Course of renal function in type 2 diabetic patients with abnormalities of albumin excretion rate. *Diabetes*. 2000;49:476-484.
12. Araki S-i, Haneda M, Sugimoto T, et al. Factors associated with frequent remission of microalbuminuria in patients with type 2 diabetes. *Diabetes*. 2005;54:2983-2987.
13. Uwaezuoke SN. The role of novel biomarkers in predicting diabetic nephropathy: a review. *Int J Nephrol Renovasc Dis*. 2017;10:221-231.
14. Parving HH, Lewis JB, Ravid M, Remuzzi G, Hunsicker LG. Prevalence and risk factors for microalbuminuria in a referred co- hort of type II diabetic patients: a global perspective. *Kidney Int*. 2006;69:2057-2063.
15. Forbes JM, Cooper ME. Mechanisms of diabetic complications. *Physiol Rev*. 2013;93:137-188.
16. Nathan DM, Genuth S, Lachin J, et al; Diabetes Control and Complications Trial Research Group. The effect of intensive treatment of diabetes on the development and progression of long- term complications in insulin-dependent diabetes mellitus. *N Engl J Med*. 1993;329:977-986.
17. Patel A, MacMahon S, Chalmers J, et al; Advance Collaborative Group. Intensive blood glucose control and vascular outcomes in patients with type 2 diabetes. *N Engl J Med*. 2008;358:2560-2572.
18. Perkovic V, Heerspink HL, Chalmers J, et al. Intensive glucose control improves kidney outcomes in patients with type 2 diabe- tes. *Kidney Int*. 2013;83:517-523.
19. Lewis EJ, Hunsicker LG, Bain RP, Rohde RD. The effect of angio- tensin-converting-enzyme inhibition on diabetic nephropathy. The Collaborative Study Group. *N Engl J Med*. 1993;329:1456-1462.
20. Lewis EJ, Hunsicker LG, Clarke WR, et al. Renoprotective ef- fect of the angiotensin-receptor antagonist irbesartan in pa- tients with nephropathy due to type 2 diabetes. *N Engl J Med*. 2001;345:851-860.
21. Brenner BM, Cooper ME, de Zeeuw D, et al. Effects of losartan on renal and cardiovascular outcomes in patients with type 2 dia- betes and nephropathy. *N Engl J Med*. 2001;345:861-869.
22. Nathan DM, Bayless M, Cleary P, et al. Diabetes control and complications trial/epidemiology of diabetes interventions and complications study at 30 years: advances and contributions. *Diabetes*. 2013;62:3976-3986.
23. Kato M, Natarajan R. Diabetic nephropathy–emerging epigenetic mechanisms. *Nat Rev Nephrol*. 2014;10:517-530.
24. Thomas MC. Epigenetic Mechanisms in Diabetic Kidney Disease. *Curr Diab Rep*. 2016;16:31.
25. Saikumar J, Ramachandran K, Vaidya VS. Noninvasive micro- markers. *Clin Chem*. 2014;60:1158-1173.
26. World Health Organisation. Global status on noncommunicable diseases. Geneva: World Health Organisation; 2014.
27. Jha V, Garcia-Garcia G, Iseki K, Li Z, Naicker S, Plattner B, et al. Chronic kidney disease: global dimension and perspectives. Lancet. 2013;382(9888):260–72. doi:10.1016/S0140-6736(13)60687-X.
28. Cowie CC, Port FK, Wolfe RA, Savage PJ, Moll PP, Hawthorne VM. Disparities in incidence of diabetic end-stage renal disease according to race and type of

diabetes. N Engl J Med. 1989;321(16):1074–9. doi:10.1056/NEJM19891019 3211603.

29. Regele F, Jelencsics K, Shiffman D, Pare G, McQueen MJ, Mann JF, et al. Genome-wide studies to identify risk factors for kidney disease with a focus on patients with diabetes. Nephrol Dial Transplant. 2015;30 Suppl 4:iv26–34. doi:10.1093/ndt/gfv087.
30. Makita Z, Radoff S, Rayfield EJ, Yang Z, Skolnik E, Delaney V, et al. Advanced glycosylation end products in patients with diabetic nephropathy. N Engl J Med. 1991;325(12):836–42. doi:10.1056/ NEJM199109193251202.
31. Schrijvers BF, De Vriese AS, Flyvbjerg A. From hyperglycemia to diabetic kidney disease: the role of metabolic, hemodynamic, intra- cellular factors and growth factors/cytokines. Endocr Rev. 2004;25(6):971–1010. doi:10.1210 /er.2003-0018.
32. Coward RJ, Saleem MA. Podocytes as a target of insulin. Curr Diabetes Rev. 2011;7(1):22–7.
33. Kolset SO, Reinholt FP, Jenssen T. Diabetic nephropathy and ex- tracellular matrix. J Histochem Cytochem. 2012;60(12):976–86. doi:10.1369/00221554 12465073.
34. Rodriguez-Iturbe B, Johnson RJ, Herrera-Acosta J. Tubulointerstitial damage and progression of renal failure. Kidney Int Suppl. 2005;99:S82–6. doi:10. 1111/j.1523-1755. 2005.09915.x.
35. Afkarian M, Sachs MC, Kestenbaum B *et al.* Kidney disease and increased mortality risk in Type 2 diabetes. *J. Am. Soc. Nephrol.* 24(2), 302–308 (2013).
36. Kanwar YS, Sun L, Xie P, Liu FY, Chen S. A glimpse of various pathogenetic mechanisms of diabetic nephropathy. *Annu. Rev. Pathol.* 6, 395–423 (2011).
37. Blumenthal SS. Evolution of treatment for diabetic nephropathy: historical progression from RAAS inhibition and onward. *Postgrad. Med.* 123(6), 166–179 (2011).
38. Kramer HJ, Nguyen QD, Curhan G, Hsu CY. Renal insufficiency in the absence of albuminuria and retinopathy among adults with Type 2 diabetes mellitus. *JAMA* 289(24), 3273–3277 (2003).
39. Perkins BA, Nelson RG, Ostrander BE *et al.* Detection of renal function decline in patients with diabetes and normal or elevated GFR by serial measurements of serum cystatin C concentration: results of a 4-year follow-up study. *J. Am. Soc. Nephrol.* 16(5), 1404–1412 (2005).
40. Perkins BA, Ficociello LH, Roshan B, Warram JH, Krolewski AS. In patients with Type 1 diabetes and new- onset microalbuminuria the development of advanced chronic kidney disease may not require progression to proteinuria. *Kidney Int.* 77(1), 57–64 (2010).
41. Perkins BA, Ficociello LH, Silva KH, Finkelstein DM, Warram JH, Krolewski AS. Regression of microalbuminuria in Type 1 diabetes. *N. Engl. J. Med.* 348(23), 2285–2293 (2003).
42. Mogensen CE, Christensen CK. Predicting diabetic nephropathy in insulin-dependent patients. *N. Engl. J. Med.* 311(2), 89–93 (1984).
43. Rossing P, Hougaard P, Parving HH. Progression of microalbuminuria in Type 1 diabetes: ten-year prospective observational study. *Kidney Int.* 68(4), 1446–1450 (2005).

44. Fioretto P, Steffes MW, Mauer M. Glomerular structure in nonproteinuric IDDM patients with various levels of albuminuria. *Diabetes* 43(11), 1358–1364 (1994).
45. Asadullah K, Busch A, Gottwald M, Reinke P, Landeck L. Industry–academia collaborations for biomarkers. *Nat Rev Drug Discovery*. 2015;14:805.
46. Wong MG, Pollock CA. (2014) Biomarkers in kidney fibrosis: are they useful? *Kidney Int Suppl*. 2011;4:79-83.
47. Uwaezuoke SN. The role of novel biomarkers in predicting diabetic nephropathy: a review. *Int J Nephrol Renovasc Dis*. 2017;10:221-231.
48. Colhoun HM, Marcovecchio ML. Biomarkers of diabetic kidney disease. *Diabetologia*. 2018;61:996-1011.
49. Radcliffe NJ, Seah JM, Clarke M, MacIsaac RJ, Jerums G, Ekinci EI. Clinical predictive factors in diabetic kidney disease progres- sion. *J Diabetes Investig*. 2017;8:6-18.
50. Elley CR, Robinson T, Moyes SA, et al. Derivation and validation of a renal risk score for people with type 2 diabetes. *Diabetes Care*. 2013;36:3113-3120.
51. Papadopoulou-Marketou N, Kanaka-Gantenbein C, Marketos N, Chrousos GP, Papassotiriou I. Biomarkers of diabetic nephropa- thy: a 2017 update. *Crit Rev Clin Lab Sci*. 2017;54:326-342.
52. Woo KS, Choi JL, Kim BR, Kim JE, An WS, Han JY. Urinary neutrophil gelatinase-associated lipocalin levels in comparison with glomerular filtration rate for evaluation of renal function in patients with diabetic chronic kidney disease. *Diabetes Metab J*. 2012;36:307-313.
53. Wu J, Shao X, Lu K, et al. Urinary RBP and NGAL levels are associated with nephropathy in patients with type 2 diabetes. *Cell Physiol Biochem*. 2017;42:594-602.
54. Mahfouz MH, Assiri AM, Mukhtar MH. Assessment of neutro- phil gelatinase-associated lipocalin (ngal) and retinol-binding protein 4 (rbp4) in type 2 diabetic patients with nephropathy. *Biomark Insights*. 2016;11:31-40.
55. Argiles A, Siwy J, Duranton F, et al. CKD273, a new proteom- ics classifier assessing CKD and its prognosis. *PLoS ONE*. 2013;8:e62837.
56. Lindhardt M, Persson F, Zurbig P, et al. Urinary proteomics predict onset of microalbuminuria in normoalbuminuric type 2 diabetic patients, a sub-study of the DIRECT-Protect 2 study. *Nephrol Dial Transplant*. 2017;32:1866-1873.
57. Lindhardt M, Persson F, Oxlund C, et al. Predicting albuminuria response to spironolactone treatment with urinary proteomics in patients with type 2 diabetes and hypertension. *Nephrol Dial Transplant*. 2018;33:296-303.
58. Lindhardt M, Persson F, Currie G, et al. Proteomic prediction and Renin angiotensin aldosterone system Inhibition prevention Of early diabetic nephRopathy in TYpe 2 diabetic patients with normoalbuminuria (PRIORITY): essential study design and ra- tionale of a randomised clinical multicentre trial. *BMJ Open*. 2016;6:e010310.
59. Looker HC, Colombo M, Hess S, et al. Biomarkers of rapid chronic kidney disease progression in type 2 diabetes. *Kidney Int*. 2015;88:888-896.
60. Lopez-Giacoman S, Madero M. Biomarkers in chronic kidney disease, from kidney function to kidney damage. *World J Nephrol*. 2015;4:57-73.

61. Rysz J, Gluba-Brzozka A, Franczyk B, Jablonowski Z, Cialkowska-Rysz A. Novel biomarkers in the diagnosis of chronic kidney disease and the prediction of its outcome. *Int J Mol Sci*. 2017;18.
62. Jardine MJ, Hata J, Woodward M, et al. Prediction of kidney-related outcomes in patients with type 2 diabetes. *Am J Kidney Dis*. 2012;60:770-778.
63. Papadopoulos T, Belliere J, Bascands JL, Neau E, Klein J, Schanstra JP. miRNAs in urine: a mirror image of kidney disease? *Expert Rev Mol Diagn*. 2015;15:361-374
64. Glinge C, Clauss S, Boddum K, et al. Stability of circulating blood-based MicroRNAs - pre-analytic methodological consider- ations. *PLoS ONE*. 2017;12:e0167969.
65. Chen YX, Huang KJ, Niu KX. Recent advances in signal am- plification strategy based on oligonucleotide and nanomate- rials for microRNA detection-a review. *Biosens Bioelectron*. 2018;99:612-624.
66. Peng H, Zhong M, Zhao W, et al. Urinary miR-29 correlates with albuminuria and carotid intima-media thickness in type 2 diabetes patients. *PLoS ONE*. 2013;8:e82607.
67. Zhou J, Peng R, Li T, et al. A potentially functional polymorphism in the regulatory region of let-7a-2 is associated with an increased risk for diabetic nephropathy. *Gene*. 2013;527:456-461.
68. Eissa S, Matboli M, Aboushahba R, Bekhet MM, Soliman Y. Urinary exosomal microRNA panel unravels novel biomark- ers for diagnosis of type 2 diabetic kidney disease. *J Diabetes Complications*. 2016;30:1585-1592.
69. Jia Y, Guan M, Zheng Z, et al. miRNAs in urine extracellular ves- icles as predictors of early-stage diabetic nephropathy. *J Diabetes Res*. 2016;2016:7932765.
70. Kim DH, Saetrom P, Snove O Jr, Rossi JJ. MicroRNA- directed transcriptional gene silencing in mammalian cells. *Proc. Natl Acad. Sci. USA* 105(42), 16230–16235 (2008).
71. Chang TC, Mendell JT. microRNAs in vertebrate physiology and human disease. *Annu. Rev. Genomics. Hum. Genet.* 8, 215–239 (2007).
72. Alvarez ML, Distefano JK. The role of noncoding RNAs in diabetic nephropathy: potential applications as biomarkers for disease development and progression. *Diabetes. Res. Clin. Pract.* 99(1), 1–11 (2013).
73. Gottardo F, Liu CG, Ferracin M *et al.* Micro-RNA profiling in kidney and bladder cancers. *Urol. Oncol.* 25(5), 387–392 (2007).
74. Neal CS, Michael MZ, Pimlott LK, Yong TY, Li JY, Gleadle JM. Circulating microRNA expression is reduced in chronic kidney disease. *Nephrol. Dial. Transplant.* 26(11), 3794–3802 (2011).
75. Argyropoulos C, Wang K, McClarty S *et al.* Urinary microRNA profiling in the nephropathy of Type 1 diabetes. *PLoS ONE* 8(1), e54662 (2013).
76. Argyropoulos C, Wang K, Bernardo J *et al.* Urinary microRNA profiling predicts the development of microalbuminuria in patients with Type 1 diabetes. *J. Clin. Med.* 4(7), 1498–1517 (2015).
77. Pezzolesi MG, Satake E, McDonnell KP, Major M, Smiles AM, Krolewski AS. Circulating TGF-beta1-regulated miRNAs and the risk of rapid progression to ESRD in Type 1 diabetes. *Diabetes* 64(9), 3285–3293 (2015).

78. Cortez MA, Calin GA. MicroRNA identification in plasma and serum: a new tool to diagnose and monitor diseases. *Expert. Opin. Biol. Ther.* 9(6), 703–711 (2009).
79. Hauser S, Wulfken LM, Holdenrieder S *et al.* Analysis of serum microRNAs (miR-26a-2*, miR-191, miR-337–3p and miR-378) as potential biomarkers in renal cell carcinoma. *Cancer Epidemiol.* 36(4), 391–394 (2012).
80. Wang G, Kwan BC, Lai FM, Chow KM, Li PK, Szeto CC. Elevated levels of miR-146a and miR-155 in kidney biopsy and urine from patients with IgA nephropathy. *Dis. Markers.* 30(4), 171–179 (2011).
81. van Balkom BW, Pisitkun T, Verhaar MC, Knepper MA. Exosomes and the kidney: prospects for diagnosis and therapy of renal diseases. *Kidney Int.* 80(11), 1138–1145 (2011).
82. Gildea JJ, Seaton JE, Victor KG *et al.* Exosomal transfer from human renal proximal tubule cells to distal tubule and collecting duct cells. *Clin. Biochem.* 47(15), 89–94 (2014).
83. Miranda KC, Bond DT, McKee M *et al.* Nucleic acids within urinary exosomes/microvesicles are potential biomarkers for renal disease. *Kidney Int.* 78(2), 191–199 (2010).
84. Delic D, Eisele C, Schmid R *et al.* Urinary exosomal miRNA signature in Type II diabetic nephropathy patients. *PLoS ONE* 11(3), e0150154 (2016).
85. Sole C, Cortes-Hernandez J, Felip ML, Vidal M, Ordi-Ros J. miR-29c in urinary exosomes as predictor of early renal fibrosis in lupus nephritis. *Nephrol. Dial. Transplant.* 30(9), 1488–1496 (2015).
86. Etheridge A, Lee I, Hood L, Galas D, Wang K. Extracellular microRNA: a new source of biomarkers. *Mutat.Res.* 717(1–2), 85–90 (2011).
87. Becker N, Lockwood CM. Pre-analytical variables in miRNA analysis. *Clin. Biochem.* 46(10–11), 861–868 (2013).
88. Barreiro K, Holthofer H. Urinary extracellular vesicles. A prom- ising shortcut to novel biomarker discoveries. *Cell Tissue Res*. 2017;369:217-227.
89. Jia Y, Guan M, Zheng Z, et al. miRNAs in urine extracellular ves- icles as predictors of early-stage diabetic nephropathy. *J Diabetes Res*. 2016;2016:7932765.
90. Pomatto M, Gai C, Bussolati B, Camussi G. Extracellular vesicles in renal pathophysiology. *Front Mol Biosci*. 2017;4:37.
91. van Balkom BW, Pisitkun T, Verhaar MC, Knepper MA. Exosomes and the kidney: prospects for diagnosis and therapy of renal diseases. *Kidney Int.* 2011;80:1138-1145.
92. Tkach M, Thery C. Communication by extracellular vesi- cles: where we are and where we need to go. *Cell*. 2016;164: 1226-1232.
93. Lv LL, Cao YH, Ni HF, et al. MicroRNA-29c in urinary exosome/ microvesicle as a biomarker of renal fibrosis. *Am J Physiol Renal Physiol*. 2013;305:F1220-1227.
94. Mohan A, Singh RS, Kumari M, et al. Urinary exosomal microR- NA-451-5P is a potential early biomarker of diabetic nephropathy in rats. *PLoS ONE*. 2016;11:e0154055.
95. Argyropoulos C, Wang K, Bernardo J, et al. Urinary microrna profiling predicts the development of microalbuminuria in pa- tients with type 1 diabetes. *J Clin Med*. 2015;4:1498-1517.

96. Assmann TS, Recamonde-Mendoza M, de Souza BM, Bauer AC, Crispim D. MicroRNAs and diabetic kidney disease: Systematic review and bioinformatic analysis. *Mol Cell Endocrinol*. 2018;477:90-102.
97. Harvey SJ, Jarad G, Cunningham J, et al. Podocyte-specific dele- tion of dicer alters cytoskeletal dynamics and causes glomerular disease. *J Am Soc Nephrol*. 2008;19:2150-2158.
98. Ho J, Ng KH, Rosen S, Dostal A, Gregory RI, Kreidberg JA. Podocyte-specific loss of functional microRNAs leads to rapid glomerular and tubular injury. *J Am Soc Nephrol*. 2008;19:2069-2075.
99. Shi S, Yu L, Chiu C, et al. Podocyte-selective deletion of dicer induces proteinuria and glomerulosclerosis. *J Am Soc Nephrol*. 2008;19:2159-2169.
100. Sun Y, Koo S, White N, et al. Development of a micro-array to detect human and mouse microRNAs and characterization of ex- pression in human organs. *Nucleic Acids Res*. 2004;32:e188.
101. Kato M, Zhang J, Wang M, et al. MicroRNA-192 in diabetic kid- ney glomeruli and its function in TGF-beta-induced collagen ex- pression via inhibition of E-box repressors. *Proc Natl Acad Sci U S A*. 2007;104:3432-3437.
102. Meng XM, Nikolic-Paterson DJ, Lan HY. TGF-beta: the master regulator of fibrosis. *Nat Rev Nephrol*. 2016;12:325-338.
103. Wu H, Kong L, Zhou S, et al. The role of microRNAs in diabetic nephropathy. *J Diabetes Res*. 2014;2014:920134.
104. Kato M, Putta S, Wang M, et al. TGF-beta activates Akt kinase through a microRNA-dependent amplifying circuit targeting PTEN. *Nat Cell Biol*. 2009;11:881-889.
105. Kato M, Arce L, Wang M, Putta S, Lanting L, Natarajan R. A microRNA circuit mediates transforming growth factor-beta1 autoregulation in renal glomerular mesangial cells. *Kidney Int*. 2011;80:358-368.
106. Wang Q, Wang Y, Minto AW, et al. MicroRNA-377 is up-regu- lated and can lead to increased fibronectin production in diabetic nephropathy. *FASEB J*. 2008;22:4126-4135.
107. Alvarez ML, Khosroheidari M, Eddy E, Kiefer J. Role of mi- croRNA 1207–5P and its host gene, the long non-coding RNA Pvt1, as mediators of extracellular matrix accumulation in the kidney: implications for diabetic nephropathy. *PLoS ONE*. 2013;8:e77468.
108. Brennan EP, Nolan KA, Borgeson E, et al; GENIE Consortium. Lipoxins attenuate renal fibrosis by inducing let-7c and suppressing TGFbetaR1. *J Am Soc Nephrol*. 2013;24:627-637.
109. Wang B, Jha JC, Hagiwara S, et al. Transforming growth fac- tor-beta1-mediated renal fibrosis is dependent on the regulation of transforming growth factor receptor 1 expression by let-7b. *Kidney Int*. 2014;85:352-361.
110. Long J, Badal SS, Wang Y, Chang BH, Rodriguez A, Danesh FR. MicroRNA-22 is a master regulator of bone morpho- genetic protein-7/6 homeostasis in the kidney. *J Biol Chem*. 2013;288:36202-36214.
111. Li R, Chung AC, Dong Y, Yang W, Zhong X, Lan HY. The mi- croRNA miR-433 promotes renal fibrosis by amplifying the TGF- beta/Smad3-Azin1 pathway. *Kidney Int*. 2013;84:1129-1144.

112. Hagiwara S, McClelland A, Kantharidis P. MicroRNA in diabetic nephropathy: renin angiotensin, aGE/RAGE, and oxidative stress pathway. *J Diabetes Res.* 2013;2013:173783.
113. Davis BN, Hilyard AC, Lagna G, Hata A. SMAD proteins control DROSHA-mediated microRNA maturation. Nature. 2008;454(7200):56–61. doi:10.1038/nature07086.
114. Martin J, Jenkins RH, Bennagi R, Krupa A, Phillips AO, Bowen T, et al. Post-transcriptional regulation of transforming growth factor beta-1 by microRNA-744. PLoS One. 2011;6(10), e25044. doi:10. 1371/journal.pone.0025044.
115. Kato M, Zhang J, Wang M, Lanting L, Yuan H, Rossi JJ, et al. MicroRNA-192 in diabetic kidney glomeruli and its function in TGF-beta-induced collagen expression via inhibition of E-box re- pressors. Proc Natl Acad Sci U S A. 2007;104(9):3432–7. doi:10. 1073/pnas.0611192104.
116. Krupa A, Jenkins R, Luo DD, Lewis A, Phillips A, Fraser D. Loss of microRNA-192 promotes fibrogenesis in diabetic nephropathy. J Am Soc Nephrol. 2010;21(3):438–47. doi:10.1681/ASN. 2009050530.
117. Wang B, Herman-Edelstein M, Koh P, Burns W, Jandeleit-Dahm K, Watson A, et al. E-cadherin expression is regulated by miR-192/215 by a mechanism that is independent of the profibrotic effects of transforming growth factor-beta. Diabetes. 2010;59(7):1794–802. doi:10.2337/db09-1736.
118. Jenkins RH, Martin J, Phillips AO, Bowen T, Fraser DJ. Pleiotropy of microRNA-192 in the kidney. Biochem Soc Trans. 2012;40(4): 762–7. doi:10.1042/BST20120085.
119. Bracken CP, Khew-Goodall Y, Goodall GJ. Network-based ap- proaches to understand the roles of miR-200 and other microRNAs in cancer. Cancer Res. 2015;75(13):2594–9. doi:10. 1158/0008-5472.CAN-15-0287.
120. Park SM, Gaur AB, Lengyel E, Peter ME. The miR-200 family determines the epithelial phenotype of cancer cells by targeting the E-cadherin repressors ZEB1 and ZEB2. Genes Dev. 2008;22(7):894–907. doi:10.1101/gad.1640608.
121. Kato M, Arce L, Wang M, Putta S, Lanting L, Natarajan R. A microRNA circuit mediates transforming growth factor-beta1 auto- regulation in renal glomerular mesangial cells. Kidney Int. 2011;80(4):358–68. doi:10.1038/ki.2011.43.
122. Lin CL, Lee PH, Hsu YC, Lei CC, Ko JY, Chuang PC, et al. MicroRNA-29a promotion of nephrin acetylation ameliorates hyperglycemia-induced podocyte dysfunction. J Am Soc Nephrol. 2014;25(8):1698–709. doi:10.1681/ASN.2013050527.
123. Pickup JC. Inflammation and activated innate immunity in the pathogenesis of type 2 diabetes. *Diabetes Care*. 2004;27:813-823.
124. Tuttle KR. Linking metabolism and immunology: diabetic nephropathy is an inflammatory disease. *J Am Soc Nephrol*. 2005;16:1537-1538.
125. Mora C, Navarro JF. Inflammation and diabetic nephropathy. *Curr Diab Rep*. 2006;6:463-468.
126. Sayyed SG, Hagele H, Kulkarni OP, et al. Podocytes produce homeostatic chemokine stromal cell-derived factor-1/CXCL12, which contributes to glomerulosclerosis, podocyte loss and al- buminuria in a mouse model of type 2 diabetes. *Diabetologia*. 2009;52:2445-2454.

127. Navarro-Gonzalez JF, Mora-Fernandez C, Muros de Fuentes M, Garcia-Perez J. Inflammatory molecules and pathways in the pathogenesis of diabetic nephropathy. *Nat Rev Nephrol.* 2011;7:327-340.
128. Lee IJ, Hilliard BA, Ulas M, et al. Monocyte and plasma ex- pression of TAM ligand and receptor in renal failure: Links to unregulated immunity and chronic inflammation. *Clin Immunol.* 2015;158:231-241.
129. Vidakovic M, Grdovic N, Dinic S, Mihailovic M, Uskokovic A, Arambasic Jovanovic J. The importance of the CXCL12/CXCR87 axis in therapeutic approaches to diabetes mellitus attenuation. *Front Immunol.* 2015;6:403.
130. O'Connell RM, Rao DS, Baltimore D. microRNA regulation of inflammatory responses. *Annu Rev Immunol.* 2012;30:295-312.
131. Bhatt K, Lanting LL, Jia Y, et al. Anti-inflammatory role of MicroRNA-146a in the pathogenesis of diabetic nephropathy. *J Am Soc Nephrol.* 2016;27:2277-2288.
132. Mezzano S, Aros C, Droguett A, et al. NF-kappaB activation and overexpression of regulated genes in human diabetic nephropathy. *Nephrol Dial Transplant.* 2004;19:2505-2512.
133. Ohga S, Shikata K, Yozai K, et al. Thiazolidinedione ameliorates renal injury in experimental diabetic rats through anti-inflamma- tory effects mediated by inhibition of NF-kappaB activation. *Am J Physiol Renal Physiol.* 2007;292:F1141-1150.
134. Zhang H, Sun SC. NF-kappaB in inflammation and renal dis- eases. *Cell Biosci.* 2015;5:63.
135. Sun Y, Peng R, Peng H, et al. miR-451 suppresses the NF-kap- paB-mediated proinflammatory molecules expression through inhibiting LMP7 in diabetic nephropathy. *Mol Cell Endocrinol.* 2016;433:75-86.
136. Yang Z, Guo Z, Dong J, et al. miR-374a regulates inflammatory response in diabetic nephropathy by targeting MCP-1 expression. *Front Pharmacol.* 2018;9:900.
137. Yao T, Zha D, Gao P, Shui H, Wu X. MiR-874 alleviates renal injury and inflammatory response in diabetic nephrop- athy through targeting toll-like receptor-4. *J Cell Physiol.* 2018;234:871-879.
138. Dey N, Das F, Mariappan MM, et al. MicroRNA-21 orchestrates high glucose-induced signals to TOR complex 1, resulting in renal cell pathology in diabetes. *J Biol Chem.* 2011;286:25586-25603.
139. McClelland AD, Herman-Edelstein M, Komers R, et al. miR-21 promotes renal fibrosis in diabetic nephropathy by targeting PTEN and SMAD7. *Clin Sci (Lond).* 2015;129:1237-1249.
140. Park JT, Kato M, Yuan H, et al. FOG2 protein down-regulation by transforming growth factor-beta1-induced microRNA-200b/c leads to Akt kinase activation and glomerular mesangial hy- pertrophy related to diabetic nephropathy. *J Biol Chem.* 2013;288:22469-22480.
141. Zhang Z, Luo X, Ding S, et al. MicroRNA-451 regulates p38 MAPK signaling by targeting of Ywhaz and suppresses the me- sangial hypertrophy in early diabetic nephropathy. *FEBS Lett.* 2012;586:20-26.
142. Zhang L, He S, Guo S, et al. Down-regulation of miR-34a allevi- ates mesangial proliferation in vitro and glomerular hypertrophy in early diabetic nephropathy mice by targeting GAS1. *J Diabetes Complications.* 2014;28:259-264.

143. Bera A, Das F, Ghosh-Choudhury N, Mariappan MM, Kasinath BS, Ghosh Choudhury G. Reciprocal regulation of miR-214 and PTEN by high glucose regulates renal glomerular mesangial and proximal tubular epithelial cell hypertrophy and matrix expan- sion. *Am J Physiol Cell Physiol*. 2017;313:C430-C447.
144. Maity S, Bera A, Ghosh-Choudhury N, Das F, Kasinath BS, Choudhury GG. microRNA-181a downregulates deptor for TGFbeta-induced glomerular mesangial cell hypertrophy and ma- trix protein expression. *Exp Cell Res*. 2018;364:5-15.
145. Mizushima N, Komatsu M. Autophagy: renovation of cells and tissues. *Cell*. 2011;147:728-741.
146. Mizushima N, Levine B, Cuervo AM, Klionsky DJ. Autophagy fights disease through cellular self-digestion. *Nature*. 2008;451:1069-1075.
147. Gonzalez CD, Lee MS, Marchetti P, et al. The emerging role of au- tophagy in the pathophysiology of diabetes mellitus. *Autophagy*. 2011;7:2-11.
148. Ding Y, Choi ME. Autophagy in diabetic nephropathy. *J Endocrinol*. 2015;224:R15-30.
149. Choi A, Ryter SW, Levine B. Autophagy in human health and disease. *N Engl J Med*. 2013;368:651-662.
150. Zhang Y, Zhao S, Wu D, et al. MicroRNA-22 promotes renal tubulointerstitial fibrosis by targeting PTEN and suppressing autophagy in diabetic nephropathy. *J Diabetes Res*. 2018;2018:4728645.
151. Deshpande S, Abdollahi M, Wang M, Lanting L, Kato M, Natarajan R. Reduced autophagy by a microRNA-mediated signaling cascade in diabetes-induced renal glomerular hypertro- phy. *Sci Rep*. 2018;8:6954.
152. Sun J, Li ZP, Zhang RQ, Zhang HM. Repression of miR-217 protects against high glucose-induced podocyte injury and insu- lin resistance by restoring PTEN-mediated autophagy pathway. *Biochem Biophys Res Commun*. 2017;483:318-324.
153. Kato M, Wang M, Chen Z, et al. An endoplasmic reticulum stress- regulated lncRNA hosting a microRNA megacluster induces early features of diabetic nephropathy. *Nat Commun*. 2016;7:12864.
154. Fu Y, Zhang Y, Wang Z, et al. Regulation of NADPH oxidase activity is associated with miRNA-25-mediated NOX4 ex- pression in experimental diabetic nephropathy. *Am J Nephrol*. 2010;32:581-589.
155. Breyer MD, Susztak K. The next generation of therapeutics for chronic kidney disease. *Nat Rev Drug Discov*. 2016;15:568-588.
156. Witasp A, Van Craenenbroeck AH, Shiels PG, Ekstrom TJ, Stenvinkel P, Nordfors L. Current epigenetic aspects the clin- ical kidney researcher should embrace. *Clin Sci (Lond)*. 2017;131:1649-1667.
157. Baek D, Villen J, Shin C, Camargo FD, Gygi SP, Bartel DP. The impact of microRNAs on protein output. *Nature*. 2008;455:64-71.
158. Trionfini P, Benigni A, Remuzzi G. MicroRNAs in kidney physi- ology and disease. *Nat Rev Nephrol*. 2015;11:23-33.
159. Grünweller A, Hartmann KR. Locked nucleic acid oligonu- cleotides: the next generation of antisense agents? *BioDrugs*. 2007;21:235-243.

160. Putta S, Lanting L, Sun G, Lawson G, Kato M, Natarajan R. Inhibiting microRNA-192 ameliorates renal fibrosis in diabetic nephropathy. *J Am Soc Nephrol*. 2012;23:458-469.
161. Zhong X, Chung AC, Chen HY, et al. miR-21 is a key therapeu- tic target for renal injury in a mouse model of type 2 diabetes. *Diabetologia*. 2013;56:663-674.
162. Luna JM, Scheel TK, Danino T, et al. Hepatitis C virus RNA functionally sequesters miR-122. *Cell*. 2015;160:1099-1110.
163. Ottosen S, Parsley TB, Yang L, et al. In vitro antiviral activity and preclinical and clinical resistance profile of miravirsen, a novel anti-hepatitis C virus therapeutic targeting the human factor miR-122. *Antimicrob Agents Chemother*. 2015;59:599-608.
164. Harmanci D, Erkan EP, Kocak A, Akdogan GG. Role of the mi- croRNA-29 family in fibrotic skin diseases. *Biomedical reports*. 2017;6:599-604.
165. Raemdonck K, Vandenbroucke RE, Demeester J, Sanders NN, De Smedt SC. Maintaining the silence: reflections on long-term RNAi. *Drug Discov Today*. 2008;13:917-931.
166. Yu B, Zhao X, Lee LJ, Lee RJ. Targeted delivery systems for oli- gonucleotide therapeutics. *AAPS J*. 2009;11:195-203.
167. Meng Z, Lu M. RNA interference-induced innate immunity, off- target effect, or immune adjuvant? *Front Immunol*. 2017;8:331.
168. Lennox KA, Behlke MA. Chemical modification and design of anti-miRNA oligonucleotides. *Gene Ther*. 2011;18:1111-1120.
169. Sriramoju B, Kanwar R, Veedu RN, Kanwar JR. Aptamer-tar- geted oligonucleotide theranostics: a smarter approach for brain delivery and the treatment of neurological diseases. *Curr Top Med Chem*. 2015;15:1115-1124.
170. Rohde JH, Weigand JE, Suess B, Dimmeler S. A universal ap- tamer chimera for the delivery of functional microRNA-126. *Nucleic Acid Ther*. 2015;25:141-151.
171. Iaboni M, Russo V, Fontanella R, et al. Aptamer-miRNA-212 conjugate sensitizes NSCLC cells to TRAIL. *Mol Ther Nucleic Acids*. 2016;5:e289.
172. Esposito CL, Catuogno S, de Franciscis V. Aptamer-MiRNA conjugates for cancer cell-targeted delivery. *Methods Mol Biol*. 2016;1364:197-208.
173. Chaudhary V, Jangra S, Yadav NR. Nanotechnology based ap- proaches for detection and delivery of microRNA in healthcare and crop protection. *J Nanobiotechnology*. 2018;16:40.
174. Riley MK, Vermerris W. Recent advances in nanomaterials for gene delivery-a review. *Nanomaterials (Basel)*. 2017;7:94.
175. Ji Q, Hao X, Zhang M, Tang W, Yang M, Li L, et al. MicroRNA miR-34 inhibits human pancreatic cancer tumor-initiating cells. PLoS One. 2009;4(8), e6816. doi:10.1371/journal.pone.0006816.
176. Wiggins JF, Ruffino L, Kelnar K, Omotola M, Patrawala L, Brown D, et al. Development of a lung cancer therapeutic based on the tumor suppressor microRNA-34. Cancer Res. 2010;70(14):5923– 30. doi:10.1158/0008-5472.CAN-10-0655.
177. Putta S, Lanting L, Sun G, Lawson G, Kato M, Natarajan R. Inhibiting microRNA-192 ameliorates renal fibrosis in diabetic ne- phropathy. J Am Soc Nephrol. 2012;23(3):458–69. doi:10.1681/ ASN.2011050485.

178. Kato M, Natarajan R. MicroRNAs in diabetic nephropathy: func- tions, biomarkers, and therapeutic targets. Ann N Y Acad Sci. 2015. doi:10. 1111/nyas.12758.
179. Long J, Wang Y, Wang W, Chang BH, Danesh FR. MicroRNA-29c is a signature microRNA under high glucose conditions that targets Sprouty homolog 1, and its in vivo knockdown prevents progres- sion of diabetic nephropathy. J Biol Chem. 2011;286(13):11837– 48. doi:10.1074/jbc.M110.194969.
180. Sun L, Zhang D, Liu F, Xiang X, Ling G, Xiao L, et al. Low-dose paclitaxel ameliorates fibrosis in the remnant kidney model by down-regulating miR-192. J Pathol. 2011;225(3):364–77. doi:10. 1002/path.2961.
181. Deelman LE, Declèves AE, Rychak JJ, Sharma K. Targeted renal therapies through microbubbles and ultrasound. Adv Drug Deliv Rev. 2010;62(14):1369–77. doi:10.1016/j.addr.2010.10.002.
182. Sirsi S, Borden M. Microbubble compositions, properties and bio- medical applications. Bubble Sci Eng Technol. 2009;1(1-2):3–17. doi:10.1179/175889709X446507.
183. Zhong X, Chung AC, Chen HY, Dong Y, Meng XM, Li R, et al. miR-21 is a key therapeutic target for renal injury in a mouse model of type 2 diabetes. Diabetologia. 2013;56(3):663–74. doi:10.1007/ s00125-012-2804-x. This study describes overexpression and downregulation of miR-21 in disease models, and subsequent findings showing profibrotic and antifibrotic effects, respectively.
184. Wynn, T.A.; Ramalingam, T.R. Mechanisms of fibrosis: Therapeutic translation for fibrotic disease. Nat. Med. 2012, 18, 1028–1040.
185. Li, R.; Chung, A.C.; Dong, Y.; Zhong, X.; Lan, H.Y. miR-433 promotes renal fibrosis by targeting the TGF-β/Smad3-Azin1 pathway. Kidney Int. 2013, 84, 1129–1144.
186. Wang, B.; Herman-Edelstein, M.; Koh, P.; Burns, W.; Jandeleit-Dahm, K.; Watson, A.; Saleem, M.; Goodall, G.J.; Twigg, S.M.; Cooper, M.E.; et al. E-cadherin expression is regulated by miR- 192/215 by a mechanism that is independent of the profibrotic effects of transforming growth factor-beta. Diabetes 2010, 59, 1794–1802.
187. Dai, Y.; Sui, W.; Lan, H.; Yan, Q.; Huang, H.; Huang, Y. Comprehensive analysis of microRNA expression patterns in renal biopsies of lupus nephritis patients. Rheumatol. Int. 2009, 29, 749–754.
188. Liang, D.; Shen, N. MicroRNA Involvement in Lupus. Curr. Opin. Rheumatol. 2012, 24, 489–498.
189. Wang, G.; Tam, L.S.; Li, E.K.; Kwan, B.C.; Chow, K.M.; Luk, C.C.; Li, P.K.; Szeto, C.C. Serum and urinary free microRNA level in patients with systemic lupus erythematosus. Lupus 2011, 20, 493–500.
190. Lu, J.; Kwan, B.C.; Lai, F.M.; Tam, L.S.; Li, E.K.; Chow, K.M.; Wang, G.; Li, P.K.; Szeto, C.C. Glomerular and tubulointerstitial miR-638, miR-198 and miR-146a expression in lupus nephritis. Nephrology 2012, 17, 346–351.
191. Te, J.L.; Dozmorov, I.M.; Guthridge, J.M.; Nguyen, K.L.; Cavett, J.W.; Kelly, J.A.; Bruner, G.R.; Harley, J.B.; Ojwang, J.O. Identification of unique microRNA signature associated with lupus nephritis. PLoS ONE 2010, 5, e10344.

192. Zhou, H.; Cleary, R.C, Bogaert, Y.E. Combination of micro-RNA-192 and microRNA-27b from urinary exosomes differentiate between renal tubular damage and glomerular injury. J. Am. Soc. Nephrol. 2008, 19, 672A.
193. Chandrasekaran, K.; Karolina, D.S.; Sepramaniam, S.; Armugam, A.; Wintour, E.M.; Bertram, J.F.; Jeyaseelan, K. Role of microRNAs in kidney homeostasis and disease. Kidney Int. 2012, 81, 617–627.
194. Guan, J. Wang, G.; Tam, L.S.; Kwan, B.C.H.; Li, E.K.M.; Chow, K.M.; Li P.K.T.; Szeto, C.C. Urinary sediment ICAM-1 in lupus nephritis. Lupus 2012, 21, 1190–1195.
195. Ecder, T.; Schrier, R.W. Hypertension in autosomal-dominant polycystic kidney disease: Early occurrence and unique aspects. J. Am. Soc. Nephrol. 2001, 12, 194–200.
196. Dweep, H.; Sticht, C.; Kharkar, A.; Pandey, P.; Gretz, N. Parallel analysis of mRNA and microRNA microarray profiles to explore functional regulatory patterns in polycystic kidney disease: Using PKD/Mhm rat model. PLoS ONE 2013, 8, e53780.
197. Pandey, P.; Bros, B; Srivastava, P.K.; Bott, A.; Boehn, S.N.; Groene, H.J.; Gretz, N. Microarray-based approach identifies microRNAs and their target functional patterns in polycystic kidney disease. BMC Genomics 2008, 9, 624.
198. Tan, Y.C.; Blumenfeld, J.; Rennert, H. Autosomal dominant polycystic kidney disease: Genetics, mutations and microRNAs. Biochim. Biophys. Acta 2011, 1812, 1202–1212.
199. Sun, H.; Li, Q.W.; Lv, X.Y.; Ai, J.Z.; Yang, Q.T.; Duan, J.J.; Bian, G.H.; Xiao, Y.; Wang, Y.D.; Zhang, Z.; et al. MicroRNA-17 post-transcriptionally regulates polycystic kidney disease-2 gene and promotes cell proliferation. Mol. Biol. Rep. 2010, 37, 2951–2958.
200. Lee, S.O.; Masyuk, T.; Splinter, P.; Banales, J.M.; Masyuk, A.; Stroope, A.; Larusso, N. MicroRNA15a modulates expression of the cell-cycle regulator Cdc25A and affects hepatic cystogenesis in a rat model of polycystic kidney disease. J. Clin. Investig. 2008, 118, 3714–3724.
201. Lughezzani, G.; Jeldres, C.; Isbarn, H.; Perrotte, P.; Shariat, S.F.; Sun, M.; Widmer, H.; Arjane, P.; Peloquin, F.; Pharand, D.; et al. Tumor size is a determinant of the rate of stage T1 renal cell cancer synchronous metastasis. J. Urol. 2009, 182, 1287–1293.
202. Petillo, D.; Kort, E.J.; Anema, J.; Furge, K.A.; Yang, X.J.; Teh, B.T. MicroRNA profiling of human kidney cancer subtypes. Int. J. Oncol. 2009, 35, 109–114.
203. Fridman, E.; Dotan, Z.; Barshack, I.; David, M.B.; Dov, A.; Tabak, S.; Zion, O.; Benjamin, S.; Benjamin, H.; Kuker, H.; et al. Accurate molecular classification of renal tumors using microRNA expression. J. Mol. Diagn. 2010, 12, 687–696.
204. Youssef, Y.M.; White, N.M.; Grigull, J.; Krizova, A.; Samy, C.; Mejia-Guerrero, S.; Evans, A.; Yousef, G.M. Accurate molecular classification of kidney cancer subtypes using microRNA signature. Eur. Urol. 2011, 59, 721–730.
205. Saini, S.; Yamamura, S.; Majid, S.; Shahryari, v.; Hirata, H.; Tanaka, Y.; Dahiya, R. Regulatory role of miR-203in prostate cancer progression and metastasis. Clin. Cancer Res. 2011, 17, 5287–5298.
206. Huang, A.; Campbell, C.E.; Bonetta, L.; McAndrews-Hill, M.S.; Chilton-MacNeill, S.; Coppes, M.J.; Law, D.J.; Feinberg, A.P.; Yeger, H.; Williams,

B.R.G. Tissue, developmental, and tumor-specific expression of divergent transcripts in Wilms tumor. Science 1990, 250, 991–994.
207. Kort, E.J.; Farber, L.; Tretiakova, M.; Petillo, D.; Furge, K.A.; Yang, X.J.; Cornelius, A.; Teh, B.T. The E2F3-Oncomir-1 axis is activated in Wilms' tumor. Cancer Res. 2008, 68, 4034–4038.
208. Lorenzen, J.M.; Volkmann, I.; Fiedler, J.; Schmidt, M.; Scheffner, I.; Haller, H.; Gwinner, W.; Thum, T. Urinary miR-210 as a mediator of acute T-cell mediated rejection in renal allograft recipients. Am. J. Transplant. 2011, 11, 2221–2227.
209. Lorenzen, J.M.; Thum, T. Circulating and Urinary microRNAs in Kidney Disease. Clin. J. Am. Soc. Nephrol. 2012, 7, 1528–1533.
210. Watson, J.A.; Bryan, K.; Williams, R.; Popov, S.; Vujanic, G.; Coulomb, A.; Boccon-Gibod, L.; Graf, N.; Pritchard-Jones, K.; O'Sullivan, M. Mirna profiles as a predictor of chemoresponsiveness in Wilms' tumor blastema. PLoS ONE 2013, 8, e53417.
211. Wegert, J.; Ishaque, N.; Vardapour, R.; Geörg, C.; Gu, Z.; Bieg, M.; Ziegler, B.; Bausenwein, S.; Nourkami, N.; Ludwig, N.; et al. Mutations in the SIX1/2 pathway and the DROSHA/DGCR8 miRNA microprocessor complex underlie high-risk blastemal type Wilms tumors. Cancer Cell 2015, 27, 298–311.
212. Drake, K.M.; Ruteshouser, E.C.; Natrajan, R.; Harbor, P.; Wegert, J.; Gessler, M.; Pritchard-Jones, K.; Grundy, P.; Dome, J.; Huff, V.; et al. Loss of heterozygosity at 2q37 in sporadic Wilms' tumor: Putative role for miR-562. Clin. Cancer Res. 2009, 15, 5985–5992.
213. Veronese, A.; Lupini, L.; Consiglio, J.; Visone, R.; Ferracin, M.; Fornari, F.; Zanesi, N.; Alder, H.; D'Elia, G.; Gramantieri, L.; et al. Oncogenic role of miR-483-5p at the IGF2/483 locus. Cancer 2010, 70, 3140–3149.
214. Imam, J.S.; Buddavarapu, K.; Lee-Chang, J.S.; Ganapathy, S.; Camosy C.; Chen, Y.; Rao, M.K. MicroRNA-185 suppresses tumor growth and progression by targeting the Six1 oncogene in human cancers. Oncogene 2010, 29, 4971–4979.
215. Takakura, S.; Mitsutake, N.; Nakashima, M.; Namba, H.; Saenko, V.A.; Rogounovitch, T.I.; Nakazawa, Y.; Hayashi, T.; Ohtsuru, A.; Yamashita, S. Oncogenic role of miR-17-92 cluster in anaplastic thyroid cancer cells. Cancer Sci. 2008, 99, 1147–1154.
216. Cloonan, N.; Brown, M.K.; Steptoe, A.L.; Wani S.; Chan, W.L.; Forrest, AR.; Kolle, G.; Gabrielli, B.; Grimmond, S.M. The miR-17-5p microRNA is a key regulator of the G1/S phase cell cycle transition. Genome Biol. 2008, 9, R127.
217. Serino,G.;Sallustio,F.;Cox,S.N.;Pesce,F.;Schena,F.P.AbnormalmiR-418bexpressionpromotes aberrant glycosylation of IgA1 in IgA nephropathy. J. Am. Soc. Nephrol. 2012, 23, 814–824
218. Allen, A.C.; Topham, P.S.; Harper, S.J.; Feehally, J. Leucocyte beta 1,3-galactosyltransferase activity in IgA nephropathy. Nephrol. Dial. Transplant. 1997, 12, 701–706.
219. Qin, W.; Zhou, Q.; Yang, L.C.; Li, Z.; Su, B.H.; Luo, H.; Fan, J.M. Peripheral B lymphocyte beta1,3-galactosyltransferase and chaperone expression in immunoglobulin A nephropathy. J. Intern. Med. 2005, 258, 467–477.
220. Ohl, K.; Tenbrock. K. Inflammatory cytokines in systemic lupus erythematosus. J. Biomed. Biotechnol. 2011, doi:10.1155/2011/432595.

221. Bellomo, R.; Kellum, J.A.; Ronco, C. Acute kidney injury. Lancet 2012, 380, 756–766.
222. Wei, Q.; Bhatt, K.; He, H.Z.; Mi, Q.S.; Hasse, V.H.; Dong, Z.Targeted deletion of Dicer from proximal tubules protects against renal ischemia-reperfusion injury. J. Am. Soc. Nephrol. 2010, 21, 756–761.
223. Godwin, J.G.; Ge, X.; Stephan, K.; Jurisch, A.; Tullius, S.G.; Iacomini, J. Identification of a microRNA signature of renal ischemia reperfusion injury. Proc. Natl. Acad. Sci. USA 2010, 107, 14339–14344.
224. Lorenzen, J.M.; Volkmann, I.; Fiedler, J.; Schmidt, M.; Scheffner, I.; Haller, H.; Gwinner, W.; hum, T. Urinary miR-210 as a mediator of acute T-cell mediated rejection in renal allograft recipients. Am. J. Transplant. 2011, 11, 2221–2227.
225. Munshi, R.; Johnson, A.; Siew, E.D.; Ikizler, T.A.; Ware, L.B.; Wurfel, M.M.; Himmelfarb, J.; Zager, R.A. MCP-1 gene activation marks acute kidney injury. J. Am. Soc. Nephrol. 2011, 22, 165–175.
226. Chau, B.; Xin, C.; Hartner, J.; Ren, S.; Castano, A.; Linn, G.; Li, J.; Tran, P.; Kaimal, V.; Huang, X.; et al. MicroRNA-21 promotes fibrosis of the kidney by silencing metabolic pathways. Sci. Transl. Med. 2012, 4, doi:10.1126/scitranslmed.3003205.
227. Shapiro, M.D.; Bagley, J.; Latz, J.; Godwin ,J.G.; Ge, X.; Tullius, S.G.; Iacomini, J. MicroRNA expression data reveals a signature of kidney damage following ischemia reperfusion injury. PLoS ONE 2011, 6, e23011.
228. Anglicheaua, D.; Sharmaa, V.K.; Dinga, R.; Hummela, A.; Snopkowskia, C.; Dadhaniaa, D.; Seshane, S.V.; Suthanthirana, M. MicroRNA expression profiles predictive of human renal allograft status. Proc. Natl. Acad. Sci. USA 2009, 106, 13.
229. Sui, W.; Dai, Y.; Huang, Y.; Lan, H.; Yan, Q.; Huang, H. Microarray analysis of MicroRNA expression in acute rejection after renal transplantation. Transpl. Immunol. 2008, 19, 81.
230. Glowacki, F.; Savary, G.; Gnemmi, V.; Buob, D.; van der Hauwaert, C.; Lo-Guidice, J.M.; Bouyé, S.; Hazzan, M.; Pottier, N.; Perrais, M.; et al. Increased circulating miR-21 levels are associated with kidney fibrosis. PLoS ONE 2013, 8, e58014.
231. Zununi, S.; Ardalan, M. MicroRNA and Renal Allograft Monitoring. Nephrol. Urol. 2013, 5, 783–786.
232. Lorenzen, J.M.; Thum, T. Circulating and Urinary microRNAs in Kidney Disease. Clin. J. Am. Soc. Nephrol. 2012, 7, 1528–1533.
233. Luo, Y.; Wang, C.; Chen, X.; Zhong, T.; Cai, X.; Chen, S.; Shi, Y.; Hu, J.; Guan, X.; Xia, Z.; et al. Increased serum and urinary microRNAs in Children with idiopathic nephrotic syndrome. Clin. Chem. 2013, doi: 10.1373/clinchem.2012.195297.
234. Wang, G.; Kwan, B.C.; Lai, F.M.; Chow, K.M.; Li, P.K.; Szeto, C.C. Urinary sediment miRNA levels in adult nephrotic syndrome. Clin. Chim. Acta 2013, 15, 5–11.
235. Snyder, A.; Alsauskas, Z.C.; Leventhal, J.S.; Rosenstiel, P.E.; Gong, P.; Chan, J.J.; Barley, K; He, J.C.; Klotman, M.E.; Ross, M.J.; et al. HIV-1 viral protein r induces ERK and caspase-8-dependent apoptosis in renal tubular epithelial cells. AIDS 2010, 24, 1107–1119.

236. Rao, T.K.; Filippone, E.J.; Nicastri, A.D.; Anthony, D.; Nicastri, M.D.; Landesman, S.H.; Frank, E.; Chen, C.K.; Friedman, E.A. Associated focal and segmental glomerulosclerosis in the acquired immunodeficiency syndrome. N. Eng. J. Med., 1984, 310, 669–673.
237. Lucas, G.M.; Lau, B.; Atta, M.G.; Fine, D.M.; Keruly, J.; Moore, R.D. Chronic kidney disease incidence and progression to end-stage renal disease, in HIV-infected individuals: A tale of two races. J. Infect. Dis. 2008, 197, 1548–1557.
238. Wyatt, C.M.; Meliambro, K.; Klotman, P.E. Recent progress in HIV-Associated Nephropathy. Annu. Rev. Med. 2012, 633, 147–159.
239. Chang, J.R.; Mukerjee R.; Bagashev, A.; del Valle, L.; Chabrashvili, T.; Hawkins, B.J.; He, J.J.; Sawaya, B.E. HIV-1 Tat protein promotes neuronal dysfunction through disruption of microRNAs. J. Biol. Chem. 2011, 286, 41125–41134.
240. Gupta,A.;Nagilla,P.;Le,H.S.;Bunney,C.;Zych,C.;Thalamuthu,A.;Bar-Joseph,Z.;Mathavan,S.; Ayyavoo, V. Comparative expression profile of miRNA and mRNA in primary peripheral blood monoclear cells infected with human immunodeficiency virus (HIV-1). PLoS ONE 2011, 6, e22730.
241. Noorbakhsh, F.; Ramachandran, R.; Barsby, N.; Ellestad, K.K.; LeBlanc, A.; Dickie, P.; Baker, G.; Hollenberg, M.D.; Cohen, E.A.; Power, C. MicroRNA profiling reveals new aspects of HIV neurodegeneration: Caspase-6 regulates astrocyte survival. FASEB J. 2010, 4, 1799–1812.
242. Freedman, B.I.; Iskanda, R.S.S.; Appel, R.G. The link between hypertension and nephrosclerosis. Am. J. Kidney Dis. 1995, 25, 207–221.
243. Rollino, C.; Boero, R.; Ferro, M.; Anglesio, A.; Vaudano, G.P.; Cametti A. Acute pyelonephritis: Analysis of 52 cases. Ren. Fail. 2002, 24, 601–608.
244. Oghumu, S.; Bracewell, A.; Nori, U.; Maclean, K.H.; Balada-Lasat, J.M.; Brodsky, S.; Pelletier, R.; Henry, M.; Satoskar, A.R.; Nadasdy, T.; et al. Acute Pyelonephritis in Renal Allografts—A New Role for MicroRNAs? Transplantation 2014, 97, 559–568.
245. Zhu, S.; Pan, W.; Song, X.; Liu, Y.; Shao, X.; Tang, Y.; Liang, D.; He, D.; Wang, H.; Liu, W.; et al. The microRNA miR-23b suppresses IL-17 associated autoimmune inflammation by targeting TAB2, TAB3 and IKK-α. Nat. Med. 2013, 18, 1077–1086.
246. Saikumar, J.; Hoffmann, D.; Kim, T.M.; Gonzalez, V.R.; Zhang, Q.; Goering, P.L.; Brown, R.P.; Bijol, V.; Park, P.J.; Waikar, S.S.; et al. Expression, circulation and excretion profile microRNA-21-155 and 18a following acute kidney injury. Toxicol. Sci. 2012, 129, 256–267.
247. Pazhayattil, P.G.; Shirali, A.C. Drug-induced impairment of renal function. Int. J. Nephrol. Renovasc. Dis. 2014, 7, 457–468.
248. Pellegrini, K.L.; Han, T.; Bijol, V.; Saikumar, J.; Cracium, F.L.; Chen, W.W.; Fuscoe, J.C; Vaida, V.S. MicroRNA-155 deficient mice experience heightened kidney toxicity when dosed with cisplatin. Toxicol. Sci. 2014, 2, 484–492.
249. Lee, C.G.; Kim, J.G.; Kim, H.J.; Kwon, H.K.; Cho, I.J.; dal Choi, W.; Lee, H.W.; Kim, W.D.; Hwang, S.J.; Choi, S.; et al. Discovery of an integrative network of microRNAs and transcriptomics changes for acute kidney injury. Kidney Int. 2014, 86, 943–953.

250. Kato, M.; Zhang, J.; Wang, M.; Lanting, L.; Yuan, H.; Rossi, J.J.; Natarajan, R. MicroRNA-192 in diabetic kidney glomeruli and its function in TGF-beta-induced collagen expression via inhibition of E-box repressors. Proc. Natl. Acad. Sci. USA 2007, 104, 3432–3437.
251. Ma, L.; Qu, L. The function of microRNAs in renal development and pathophysiology. J. Genet. Genomics 2013, 40, 143–152.
252. Chandrasekaran, K.; Karolina, D.S.; Sepramaniam, S.; Armugam, A.; Wintour, E.M.; Bertram, J.F.; Jeyaseelan, K. Role of microRNAs in kidney homeostasis and disease. Kidney Int. 2012, 81, 617–627
253. Cybulsky, A.V. Growth factor pathways in proliferative glomerulonephritis. Curr. Opin. Nephrol. Hypertens. 2000, 9, 217–223.
254. Schmitt, M.J, Margue, C, Behrmann, I. and Kreis, S, MiRNA-29: A microRNA Family with Tumor-Suppressing and Immune-Modulating Properties. Curr. Mol. Med. 2013, 13, 572–585.
255. Tsuchida, K.; Nakatani, M.; Hitachi, K.; Uezumi, A.; Sunada, Y.; Ageta, H.; Inokuchi, K. Activin signaling as an emerging target therapeutic intervention. Cell Commun. Signal. 2009, doi:10.1186/1478-811X-7-15.
256. Wang, Q.; Wang, Y.; Minto, A.W.; Wang, J.; Shi, Q.; Li, X.; Quigg, R.J. MicroRNA-377 is up-regulated and can lead to increased fibronectin production in diabetic nephropathy. FASEB J. 2008, 22, 4126–4135.
257. Li, R.; Chung, A.C.; Dong, Y.; Zhong, X.; Lan, H.Y. miR-433 promotes renal fibrosis by targeting the TGF-β/Smad3-Azin1 pathway. Kidney Int. 2013, 84, 1129–1144.
258. Kato, M.; Putta, S.; Wang, M.; Yuan, H.; Lanting, L.; Nair, I.; Gunn, A.; Nakagawa, Y.; Shimano, H.; Todorov, I.; et al. TGF-beta activates Akt kinase through a microRNA-dependent amplifying circuit targeting PTEN. Nat. Cell Biol. 2009, 11, 881–889.
259. Chung, A.C.K.; Huang, X.R.; Meng, X.; Lan, H.Y. miR-192medi-ates TGF-β/Smad3-driven renal fibrosis. J. Am. Soc. Nephrol. 2010, 21, 1317–1325.
260. Fire, A.; Xu, S.; Montgomery, M.K.; Kostas, S.A.; Driver, S.E.; Mello, C.C. Potent and specific genetic interference by double-stranded RNA in Caenorhabditis elegans. Nature 1998, 391, 806–811.

Chapter X

1. Auffray C, Chen Z, Hood L. Systems medicine: the future of medical genomics and healthcare. Genome Med. 2009;1:2.
2. Verdine GL, Walensky LD. The challenge of drugging undruggable targets in cancer: lessons learned from targeting BCL-2 family members. Clin Cancer Res. 2007;13:7264–70.
3. O'Connell RM, Rao DS, Chaudhuri AA, Baltimore D. Physiological and pathological roles for microRNAs in the immune system. Nat Rev Immunol. 2010;10(2):111–22.
4. Lu J, Getz G, Miska EA, Alvarez-Saavedra E, Lamb J, Peck D, et al. MicroRNA expression profiles classify human cancers. Nature. 2005;435(7043):834–8.
5. Jeang KT. RNAi in the regulation of mammalian viral infections. BMC Biol. 2012;10:58.

6. Rottiers V, Naar AM. MicroRNAs in metabolism and metabolic disorders. Nat Rev Mol Cell Biol. 2012;13(4):239–50.
7. Sahay G, Alakhova DY, Kabanov AV. Endocytosis of nanomedicines. Control Release. 2010;145:182–95.
8. Kotterman MA, Schaffer DV. Engineering adeno-associated viruses for clinical genetherapy. Nat Rev Genet. 2014;15:445–51.
9. Anderson DG, Lynn DM, Langer R. Semi-automated synthesis and screening of a large library of degradable cationic polymers for gene delivery. Angew Chemie Int Ed. 2003;42:3153–8.
10. Pack DW, Hoffman AS, Pun S, Stayton PS. Design and development of polymers for gene delivery. Nat Rev Drug Discov. 2005;4:581–93.
11. Behr J. The proton sponge: a trick to enter cells the viruses did not exploit. Int J Chem. 1997;2:34–6.
12. Rehman ZU, Hoekstra D, Zuhorn IS. Mechanism of polyplex- and lipoplex-mediated delivery of nucleic acids: real-time visualization of transient membrane destabilization without endosomal lysis. ACS Nano. 2013;7: 3767–77.
13. Won Y-Y, Sharma R, Konieczny SF. Missing pieces in understanding the intracellular trafficking of polycation/DNA complexes. J Control Release. 2009;139:88–93.
14. Guo J, Cheng WP, Gu J, Ding C, Qu X, Yang Z, et al. Systemic delivery of therapeutic small interfering RNA using a pH-triggered amphiphilic poly-L-lysine nanocarrier to suppress prostate cancer growth in mice. Eur J Pharm Sci. 2012;45:521–32.
15. Tsutsumi T, Hirayama F, Uekama K, Arima H. Evaluation of polyamidoamine dendrimer/alpha-cyclodextrin conjugate (generation 3, G3) as a novel carrier for small interfering RNA (siRNA). J Control Release. 2007;119:349–59.
16. Boussif O, Lezoualc'h F, Zanta MA, Mergny MD, Scherman D, Demeneix B, et al. A versatile vector for gene and oligonucleotide transfer into cells in culture and in vivo: polyethylenimine. Proc Natl Acad Sci U S A. 1995;92: 7297–301.
17. Howard KA, Rahbek UL, Liu X, Damgaard CK, Glud SZ, Andersen M, et al. RNA interference in vitro and in vivo using a novel chitosan/siRNA nanoparticle system. Mol Ther. 2006;14:476–84.
18. Kaczmarek JC, Patel AK, Kauffman KJ, Fenton OS, Webber MJ, Heartlein MW, et al. Polymer-lipid nanoparticles for systemic delivery of mRNA to the lungs. Angew Chemie Int Ed. 2016;55:13808–12.
19. Schroeder A, Levins CG, Cortez C, Langer R, Anderson DG. Lipid-based nanotherapeutics for siRNA delivery. J Intern Med. 2010;267:9–21.
20. Kanasty R, Dorkin JR, Vegas A, Anderson D. Delivery materials for siRNA therapeutics. Nat Mater. 2013;12:967–77.
21. Desigaux L, Sainlos M, Lambert O, Chevre R, Letrou-Bonneval E, Vigneron J-P, et al. Self-assembled lamellar complexes of siRNA with lipidic aminoglycoside derivatives promote efficient siRNA delivery and interference. Proc Natl Acad Sci U S A. 2007;104:16534–9.
22. Leus NGJ, Morselt HWM, Zwiers PJ, Kowalski PS, Ruiters MHJ, Molema G,et al. VCAM-1 specific PEGylated SAINT-based lipoplexes deliver siRNA to activated endothelium in vivo but do not attenuate target gene expression. Int J Pharm. 2014;469:121–31.

23. Chen D, Love KT, Chen Y, Eltoukhy AA, Kastrup C, Sahay G, et al. Rapid discovery of potent siRNA-containing lipid nanoparticles enabled by controlled microfluidic formulation. J Am Chem Soc. 2012;134:6948–51.
24. Love KT, Mahon KP, Levins CG, Whitehead KA, Querbes W, Dorkin JR, et al. Lipid-like materials for low-dose, in vivo gene silencing. Proc Natl Acad Sci U S A. 2010;107:1864–9.
25. Akinc A, Zumbuehl A, Goldberg M, Leshchiner ES, Busini V, Hossain N, et al. A combinatorial library of lipid-like materials for delivery of RNAi therapeutics. Nat Biotechnol. 2008;26:561–9.
26. Dong Y, Love KT, Dorkin JR, Sirirungruang S, Zhang Y, Chen D, et al. Lipopeptide nanoparticles for potent and selective siRNA delivery in rodents and nonhuman primates. Proc Natl Acad Sci U S A. 2014;111: 3955–60.
27. Fenton OS, Kauffman KJ, McClellan RL, Appel EA, Dorkin JR, Tibbitt MW, et al. Bioinspired alkenyl amino alcohol ionizable lipid materials for highly potent in vivo mRNA delivery. Adv Mater. 2016;28:2939–43.
28. Dahlman JE, Barnes C, Khan OF, Thiriot A, Jhunjunwala S, Shaw TE, et al. In vivo endothelial siRNA delivery using polymeric nanoparticles with low molecular weight. Nat Nanotechnol. 2014;9:648–55.
29. Nair JK, Willoughby JLS, Chan A, Charisse K, Alam MR, Wang Q, et al. Multivalent N-acetylgalactosamine-conjugated siRNA localizes in hepatocytes and elicits robust RNAi-mediated gene silencing. J Am Chem Soc. 2014;136:16958–61.
30. Yu RZ, Graham MJ, Post N, Riney S, Zanardi T, Hall S, et al. Disposition and pharmacology of a GalNAc3-conjugated ASO targeting human lipoprotein(a) in mice. Mol Ther Nucleic Acids. 2016;5, e317.
31. Lorenz C, Hadwiger P, John M, Vornlocher HP, Unverzagt C. Steroid and lipid conjugates of siRNAs to enhance cellular uptake and gene silencing in liver cells. Bioorganic Med Chem Lett. 2004;14:4975–7.
32. Nishina K, Unno T, Uno Y, Kubodera T, Kanouchi T, Mizusawa H, et al. Efficient in vivo delivery of siRNA to the liver by conjugation of alpha-tocopherol. Mol Ther. 2008;16:734–40.
33. Xia C-F, Boado RJ, Pardridge WM. Antibody-mediated targeting of siRNA via the human insulin receptor using avidin − biotin technology. Mol Pharm. 2009;6:747–51.
34. Moschos SA, Jones SW, Perry MM, Williams AE, Erjefalt JS, Turner JJ, et al. Lung delivery studies using siRNA conjugated to TAT(48-60) and penetratin reveal peptide induced reduction in gene expression and induction of innate immunity. Bioconjug Chem. 2007;18:1450–9.
35. Zhang B, Farwell MA. microRNAs: a new emerging class of players for disease diagnostics and gene therapy. J Cell Mol Med. 2008;12(1):3–21.
36. Sharma VK, Watts JK. Oligonucleotide therapeutics: chemistry, delivery and clinical progress. Future Med Chem. 2015;7(16):2221–42.
37. Bernardo BC, Ooi JY, Lin RC, McMullen JR. miRNA therapeutics: a new class of drugs with potential therapeutic applications in the heart. Future Med Chem. 2015;7(13):1771–92.

38. Chabot S, Orio J, Castanier R, Bellard E, Nielsen SJ, Golzio M, et al. LNA- based oligonucleotide electrotransfer for miRNA inhibition. Mol Ther. 2012;20(8):1590–8.
39. Bertucci A, Prasetyanto EA, Septiadi D, Manicardi A, Brognara E, Gambari R, et al. Combined delivery of Temozolomide and anti-miR221 PNA using mesoporous silica nanoparticles induces apoptosis in resistant glioma cells. Small. 2015;11(42):5687–95.
40. Lennox KA, Behlke MA. Chemical modification and design of anti-miRNA oligonucleotides. Gene Ther. 2011;18(12):1111–20.
41. Guo YE, Steitz JA. 3′-biotin-tagged microRNA-27 does not associate with Argonaute proteins in cells. RNA. 2014;20(7):985–8.
42. Shingara J, Keiger K, Shelton J, Laosinchai-Wolf W, Powers P, Conrad R, et al. An optimized isolation and labeling platform for accurate microRNA expression profiling. RNA. 2005;11(9):1461–70.
43. Krutzfeldt J, Kuwajima S, Braich R, Rajeev KG, Pena J, Tuschl T, et al. Specificity, duplex degradation and subcellular localization of antagomirs. Nucleic Acids Res. 2007;35(9):2885–92.
44. Krutzfeldt J, Rajewsky N, Braich R, Rajeev KG, Tuschl T, Manoharan M, et al. Silencing of microRNAs in vivo with 'antagomirs'. Nature. 2005;438(7068): 685–9.
45. Wang Z. The principles of MiRNA-masking antisense oligonucleotides technology. Methods Mol Biol. 2011;676:43–9.
46. Kluiver J, Slezak-Prochazka I, Smigielska-Czepiel K, Halsema N, Kroesen BJ, van den Berg A. Generation of miRNA sponge constructs. Methods. 2012;58(2):113–7.
47. Wienholds E, Kloosterman WP, Miska E, Alvarez-Saavedra E, Berezikov E, de Bruijn E, et al. MicroRNA expression in zebrafish embryonic development. Science.2005;309(5732):310–1.
48. Cheng WT, Rosario R, Muthukaruppan A, Wilson MK, Payne K, Fong PC, et al. MicroRNA profiling of ovarian granulosa cell tumours reveals novel diagnostic and prognostic markers. Clin Epigenetics. 2017;9:72.
49. XuL, HuY, CaoY, LiJ, MaL, LiY, et al. An expression at last of miRNAs in Arabidopsis thaliana. Sci China Life Sci. 2018;61(2):178–89.
50. Lawrie CH, Gal S, Dunlop HM, Pushkaran B, Liggins AP, Pulford K, et al. Detection of elevated levels of tumour-associated microRNAs in serum of patients with diffuse large B-cell lymphoma. Br J Haematol. 2008;141(5):672–5.
51. Chen X, Ba Y, Ma L, Cai X, Yin Y, Wang K, et al. Characterization of microRNAs in serum: a novel class of biomarkers for diagnosis of cancer and other diseases. Cell Res. 2008;18(10):997–1006.
52. Vickers KC, Palmisano BT, Shoucri BM, Shamburek RD, Remaley AT. MicroRNAs are transported in plasma and delivered to recipient cells by high-density lipoproteins. Nat Cell Biol. 2011;13(4):423–33.
53. Skog J, Wurdinger T, van Rijn S, Meijer DH, Gainche L, Sena-Esteves M, et al. Glioblastoma microvesicles transport RNA and proteins that promote tumour growth and provide diagnostic biomarkers. Nat Cell Biol. 2008; 10(12):1470–U209.

54. Jung HJ, Suh Y. Circulating miRNAs in ageing and ageing-related diseases. J Genet Genomics. 2014;41(9):465–72.
55. Czauderna F, Fechtner M, Dames S, Aygun H, Klippel A, Pronk GJ, et al. Structural variations and stabilising modifications of synthetic siRNAs in mammalian cells. Nucleic Acids Res. 2003;31(11):2705–16.
56. Douglas JT. Adenoviral vectors for gene therapy. Mol Biotechnol. 2007;36(1):71–80.
57. Maione D, Della Rocca C, Giannetti P, D'Arrigo R, Liberatoscioli L, Franlin LL, et al. An improved helper-dependent adenoviral vector allows persistent gene expression after intramuscular delivery and overcomes preexisting immunity to adenovirus. Proc Natl Acad Sci U S A. 2001;98(11):5986–91.
58. Crowther C, Mowa B, Arbuthnot P. Hepatic delivery of artificial micro RNAs using helper-dependent adenoviral vectors. In: Shum K, Rossi J, editors. SiRNA delivery methods: methods and protocols. New York: Springer New York; 2016. p. 249–60.
59. Mowa MB, Crowther C, Ely A, Arbuthnot P. Inhibition of hepatitis B virus replication by helper dependent adenoviral vectors expressing artificial anti-HBV pri-miRs from a liver-specific promoter. Biomed Res Int. 2014;2014:718743.
60. Bressy C, Benihoud K. Association of oncolytic adenoviruses with chemotherapies: an overview and future directions. Biochem Pharmacol. 2014;90(2):97–106.
61. Lou W, Chen Q, Ma L, Liu J, Yang Z, Shen J, et al. Oncolytic adenovirus co-expressing miRNA-34a and IL-24 induces superior antitumor activity in experimental tumor model. J Mol Med (Berl). 2013;91(6):715–25.
62. Mezzina M, Merten OW. Adeno-associated viruses. Methods Mol Biol. 2011; 737:211–34.
63. Jooss K, Chirmule N. Immunity to adenovirus and adeno-associated viral vectors: implications for gene therapy. Gene Ther. 2003;10(11):955–63.
64. Miyazaki Y, Adachi H, Katsuno M, Minamiyama M, Jiang YM, Huang Z, et al. Viral delivery of miR-196a ameliorates the SBMA phenotype via the silencing of CELF2. Nat Med. 2012;18(7):1136–41.
65. Pages JC, Bru T. Toolbox for retrovectorologists. J Gene Med. 2004;6(Suppl 1):S67–82.
66. Yu YE, Choe W, Zhang WG, Stoica G, Wong PKY. Development of pathological lesions in the central nervous system of transgenic mice expressing the env gene of ts1 Moloney murine leukemia virus in the absence of the viral gag and pol genes and viral replication. J Neuro-Oncol. 1997;3(4):274–82.
67. Liu YP, Berkhout B. miRNA cassettes in viral vectors: problems and solutions. Biochim Biophys Acta. 2011;1809(11–12):732–45.
68. Ramanujam D, Sassi Y, Laggerbauer B, Engelhardt S. Viral vector-based targeting of miR-21 in cardiac nonmyocyte cells reduces pathologic remodeling of the heart. Mol Ther. 2016;24(11):1939–48.
69. Sakuma T, Barry MA, Ikeda Y. Lentiviral vectors: basic to translational. Biochem J. 2012;443(3):603–18.

70. Theis T, Yoo M, Park CS, Chen J, Kugler S, Gibbs KM, et al. Lentiviral delivery of miR-133b improves functional recovery after spinal cord injury in mice. Mol Neurobiol. 2017;54(6):4659–71.
71. Barbato C, Pezzola S, Caggiano C, Antonelli M, Frisone P, Ciotti MT, et al. A lentiviral sponge for miR-101 regulates RanBP9 expression and amyloid precursor protein metabolism in hippocampal neurons. Front Cell Neurosci. 2014;8:37.
72. Vannucci L, Lai M, Chiuppesi F, Ceccherini-Nelli L, Pistello M. Viral vectors: a look back and ahead on gene transfer technology. New Microbiol. 2013;36(1):1–22.
73. Y. Chen, D.Y. Gao, L. Huang, In vivo delivery of miRNAs for cancer therapy: challenges and strategies, Adv. Drug Deliv. Rev. 81 (2015) 128–141, https://doi.org/10.1016/j.addr.2014.05.009.
74. D.M. Pereira, P.M. Rodrigues, P.M. Borralho, C.M.P. Rodrigues, Delivering the promise of miRNA cancer therapeutics, Drug Discov. Today 18 (2013) 282–289, https://doi.org/10.1016/j.drudis.2012.10.002.
75. S. van Dongen, C. Abreu-goodger, A.J. Enright, Detecting microRNA binding and siRNA off-target effects from expression data, Nat. Methods 5 (2008) 1023–1025, https://doi.org/10.1038/nmeth.1267.
76. A. Loinger, D.Y. Shemla, D.I. Simon, H. Margalit, O. Biham, Competition between small RNAs: a quantitative view, Biophys. J. 102 (2012) 1712–1721, https://doi.org/10.1016/j.bpj.2012.01.058.
77. F. Czauderna, M. Fechtner, S. Dames, H. Aygün, A. Klippel, G.J. Pronk, K. Giese, J. Kaufmann, Structural variations and stabilising modifications of synthetic siRNAs in mammalian cells, Nucleic Acids Res. 31 (2003) 2705–2716, https://doi. org/10.1093/nar/gkg393.
78. K. Raemdonck, R.E. Vandenbroucke, J. Demeester, N.N. Sanders, S.C. De Smedt, Maintaining the silence: reflections on long-term RNAi, Drug Discov. Today 13 (2008) 917–931, https://doi.org/10.1016/j.drudis.2008.06.008.
79. J. Visvanathan, S. Lee, B. Lee, J.W. Lee, S.K. Lee, The microRNA miR-124 antagonizes the anti-neural REST/SCP1 pathway during embryonic CNS development, Genes Dev. 21 (2007) 744–749, https://doi.org/10.1101/gad.1519107.
80. C. van Solingen, L. Seghers, R. Bijkerk, J.M. Duijs, M.K. Roeten, A.M. van Oeveren- Rietdijk, H.J. Baelde, M. Monge, J.B. Vos, H.C. de Boer, P.H. Quax, Antagomir- mediated silencing of endothelial cell specific microRNA-126 impairs ischemia- induced angiogenesis, J. Cell. Mol. Med. 13 (2009) 1577–1585, https://doi.org/ 10.1111/j.1582-4934.2008.00613.x.
81. X. Chen, S. Gu, B.F. Chen, W.L. Shen, Z. Yin, G.W. Xu, J.J. Hu, T. Zhu, G. Li, C. Wan, H.W. Ouyang, T.L. Lee, W.Y. Chan, Nanoparticle delivery of stable miR- 199a-5p agomir improves the osteogenesis of human mesenchymal stem cells via the HIF1a pathway, Biomaterials. 53 (2015) 239–250, https://doi.org/10.1016/j. biomaterials.2015.02.071.
82. K.A. Lennox, M.A. Behlke, Chemical modification and design of anti-miRNA oli- gonucleotides, Gene Ther. 18 (2011) 1111–1120, https://doi.org/10.1038/gt. 2011.100.

83. P. Velpurisiva, A. Gad, B. Piel, R. Jadia, P. Rai, Nanoparticle design strategies for effective cancer immunotherapy, J. Biomed. 2 (2017) 64–77, https://doi.org/10.7150/jbm.18877.
84. P. Decuzzi, B. Godin, T. Tanaka, S.Y. Lee, C. Chiappini, X. Liu, M. Ferrari, Size and shape effects in the biodistribution of intravascularly injected particles, J. Control. Release 141 (2009) 320–327, https://doi.org/10.1016/j.jconrel.2009.10.014.
85. S.H. Kim, H. Mok, J.H. Jeong, S.W. Kim, T.G. Park, Comparative evaluation of target-specific GFP gene silencing efficiencies for antisense ODN, synthetic siRNA, and siRNA plasmid complexed with PEI-PEG-FOL conjugate, Bioconjug. Chem. 17 (2006) 241–244, https://doi.org/10.1021/bc050289f.
86. G. Liang, Y. Zhu, A. Jing, J. Wang, F. Hu, W. Feng, Z. Xiao, B. Chen, Cationic microRNA-delivering nanocarriers for efficient treatment of colon carcinoma in xenograft model, Gene Ther. 23 (2016) 829–838, https://doi.org/10.1038/gt.2016.60.
87. T.J. Lee, J.Y. Yoo, D. Shu, H. Li, J. Zhang, J.G. Yu, A.C. Jaime-Ramirez, M. Acunzo, G. Romano, R. Cui, H.L. Sun, Z. Luo, M. Old, B. Kaur, P. Guo, C.M. Croce, RNA nanoparticle-based targeted therapy for glioblastoma through inhibition of onco- genicmiR-21, Mol. Ther. 25 (2017) 1544–1555, https://doi.org/10.1016/j.ymthe. 2016.11.016.
88. J.P. Rao, K.E. Geckeler, Polymer nanoparticles: preparation techniques and size-control parameters, Prog. Polym. Sci. 36 (2011) 887–913, https://doi.org/10.1016/j.progpolymsci.2011.01.001.
89. A. Lamprecht, N. Ubrich, M. Hombreiro Pérez, C.M. Lehr, M. Hoffman, P. Maincent, Influences of process parameters on nanoparticle preparation performed by a double emulsion pressure homogenization technique, Int. J. Pharm. 196 (2000) 177–182, https://doi.org/10.1016/S0378-5173(99)00422-6.
90. Y.S. Lee, P.J. Johnson, P.T. Robbins, R.H. Bridson, Production of nanoparticles-in- microparticles by a double emulsion method: a comprehensive study, Eur. J. Pharm. Biopharm. 83 (2013) 168–173, https://doi.org/10.1016/j.ejpb.2012.10.016.
91. C. Cai, Y. Xie, L. Wu, X. Chen, H. Liu, Y. Zhou, H. Zou, D. Liu, Y. Zhao, X. Kong, P. Liu, PLGA-based dual targeted nanoparticles enhance miRNA transfection ef- ficiency in hepatic carcinoma, Sci. Rep. 7 (2017), https://doi.org/10.1038/ srep46250.
92. A. Mohamed, N.K. Kunda, K. Ross, G.A. Hutcheon, I.Y. Saleem, Polymeric nano- particles for the delivery of miRNA to treat Chronic Obstructive Pulmonary Disease (COPD), Eur. J. Pharm. Biopharm. 136 (2019) 1–8, https://doi.org/10.1016/j.ejpb.2019.01.002.
93. J. Yang, H. Liu, X. Zhang, Design, preparation and application of nucleic acid delivery carriers, Biotechnol. Adv. 32 (2014) 804–817, https://doi.org/10.1016/j.biotechadv.2013.11.004.
94. A. Bhargava-Shah, K. Foygel, R. Devulapally, R. Paulmurugan, Orlistat and anti-sense-miRNA-loaded PLGA-PEG nanoparticles for enhanced triple negative breast cancertherapy, Nanomedicine 11 (2016) 235–247, https://doi.org/10.2217/nnm. 15.193.

95. F. Degliangeli, P. Kshirsagar, V. Brunetti, P.P. Pompa, R. Fiammengo, Absolute and direct microRNA quantification using DNA–gold nanoparticle probes, J. Am. Chem. Soc. 136 (2014) 2264–2267, https://doi.org/10.1021/ja412152x.
96. S.W. Yang, T. Vosch, Rapid detection of microRNA by a silver nanocluster DNA probe, Anal. Chem. 83 (2011) 6935–6939, https://doi.org/10.1021/ac201903n.
97. H. Koo, S. Lee, J. Hee Na, S. Hwa Kim, S. Kwang Hahn, K. Choi, I. Chan Kwon, S. Young Jeong, K. Kim, Bioorthogonal copper-free click chemistry in vivo for tumor-targeted delivery of nanoparticles, Angew. Chem. Int. Ed. 124 (2012) 12006–12010, https://doi.org/10.1002/ange.201206703.
98. L.E. Roode, H. Brighton, T. Bo, J.L. Perry, M.C. Parrott, F. Kersey, J.C. Luft, J.E. Bear, J.M. DeSimone, I.J. Davis, Subtumoral analysis of PRINT nanoparticle distribution reveals targeting variation based on cellular and particle properties, Nanomed. Nanotechnol. Biol. Med. 12 (2016) 1053–1062, https://doi.org/10.1016/j.nano.2015.12.382.
99. M. Sambade, A. Deal, A. Schorzman, J.C. Luft, C. Bowerman, K. Chu, O. Karginova, A. Van Swearingen, W. Zamboni, J. DeSimone, C.K. Anders, Efficacy and pharmacokinetics of a modified acid-labile docetaxel-PRINT® nanoparticle formulation against non-small-cell lung cancer brain metastases, Nanomedicine 11 (2016) 1947–1955, https://doi.org/10.2217/nnm-2016-0147.
100. B. Herranz-Blanco, E. Ginestar, H. Zhang, J. Hirvonen, H.A. Santos, Microfluidics platform for glass capillaries and its application in droplet and nanoparticle fab- rication,Int. J. Pharm. 516 (2017) 100–105, https://doi.org/10.1016/j.ijpharm. 2016.11.024.
101. D. Chen, K.T. Love, Y. Chen, A.A. Eltoukhy, C. Kastrup, G. Sahay, A. Jeon, Y. Dong, K.A. Whitehead, D.G. Anderson, Rapid discovery of potent siRNA-containing lipid nanoparticles enabled by controlled microfluidic formulation, J. Am. Chem. Soc. 134 (2012) 6948–6951, https://doi.org/10.1021/ja301621z.
102. L. Hu, Y. Sun, Y. Wu, Advances in chitosan-based drug delivery vehicles, Nanoscale 5 (2013) 3103, https://doi.org/10.1039/c3nr00338h.
103. A. Bernkop-Schnürch, S. Dünnhaupt, Chitosan-based drug delivery systems, Eur. J. Pharm. Biopharm. 81 (2012) 463–469, https://doi.org/10.1016/J.EJPB.2012.04. 007.
104. D.F. Costa, V.P. Torchilin, Micelle-like nanoparticles as siRNA and miRNA carriers for cancer therapy, Biomed. Microdev. 20 (2018), https://doi.org/10.1007/ s10544-018-0298-0.
105. L.M. Bravo-Anaya, K.G. Fernández-Solís, J. Rosselgong, J.L.E. Nano-Rodríguez, F. Carvajal, M. Rinaudo, Chitosan-DNA polyelectrolyte complex: influence of chitosan characteristics and mechanism of complex formation, Int. J. Biol. Macromol. 126 (2019) 1037–1049, https://doi.org/10.1016/j.ijbiomac. 2019.01.008.
106. B. Layek, L. Lipp, J. Singh, Cell penetrating peptide conjugated chitosan for enhanced delivery of nucleic acid, Int. J. Mol. Sci. 16 (2015) 28912–28930, https://doi.org/10.3390/ijms161226142.
107. D. Cosco, C. Federico, J. Maiuolo, S. Bulotta, R. Molinaro, D. Paolino, P. Tassone, M. Fresta, Physicochemical features and transfection properties of chitosan/po- loxamer 188/poly(D,L-lactide-co-glycolide) nanoplexes, Int. J. Nanomed. 9 (2014) 2359–2372, https://doi.org/10.2147/IJN.S58362.

108. T.A. Sonia, C.P. Sharma, Chitosan and its derivatives for drug delivery perspective, in: R. Jayakumar, R.A.A. Muzzarelli, M. Prabaharan (Eds.), Chitosan Biomater. I. Springer, Berlin, Heidelberg, 2011, pp. 23–53, , https://doi.org/10.1007/12_ 2011_117.
109. S.Y. Chae, S. Son, M. Lee, M.K. Jang, J.W. Nah, Deoxycholic acid-conjugated chitosan oligosaccharide nanoparticles for efficient gene carrier, J. Control. Release 109 (2005) 330–344, https://doi.org/10.1016/J.JCONREL.2005.09.040.
110. R. Mandke, J. Singh, Effect of acyl chain length and unsaturation on physicochemical properties and transfection efficiency of *N*-acyl-substituted low-molecular-weight chitosan, J. Pharm. Sci. 101 (2012) 268–282, https://doi.org/10.1002/JPS.22767.
111. K.T. Dicker, L.A. Gurski, S. Pradhan-Bhatt, R.L. Witt, M.C. Farach-Carson, X. Jia, Hyaluronan: a simple polysaccharide with diverse biological functions, Acta Biomater. 10 (2014) 1558–1570, https://doi.org/10.1016/J.ACTBIO.2013.12.019.
112. A. Cadete, M.J. Alonso, Targeting cancer with hyaluronic acid-based nanocarriers: recent advances and translational perspectives, Nanomedicine 11 (2016) 2341–2357, https://doi.org/10.2217/nnm-2016-0117.
113. S. Wang, J. Zhang, Y. Wang, M. Chen, Hyaluronic acid-coated PEI-PLGA nanoparticles mediated co-delivery of doxorubicin and miR-542-3p for triple negative breast cancer therapy, Nanomed. Nanotechnol. Biol. Med. 12 (2016) 411–420, https://doi.org/10.1016/j.nano.2015.09.014.
114. S. Ganesh, A.K. Iyer, D.V. Morrissey, M.M. Amiji, Hyaluronic acid based self-as- sembling nanosystems for CD44 target mediated siRNA delivery to solid tumors, Biomaterials 34 (2013) 3489–3502, https://doi.org/10.1016/j.biomaterials.2013. 01.077.
115. D. Landesman-Milo, M. Goldsmith, S.L. Ben-Arye, B. Witenberg, E. Brown, S. Leibovitch, S. Azriel, S. Tabak, V. Morad, D. Peer, Hyaluronan grafted lipid-based nanoparticles as RNAi carriers for cancer cells, Cancer Lett. 334 (2013) 221–227, https://doi.org/10.1016/j.canlet.2012.08.024.
116. X. Sun, N. Zhang, Cationic polymer optimization for efficient gene delivery, Mini Rev. Med. Chem. 10 (2010) 108–125.
117. Z.R. Cohen, S. Ramishetti, N. Peshes-Yaloz, M. Goldsmith, A. Wohl, Z. Zibly, D. Peer, Localized RNAi therapeutics of chemoresistant grade IV glioma using hyaluronan-grafted lipid-based nanoparticles, ACS Nano 9 (2015) 1581–1591, https://doi.org/10.1021/nn506248s.
118. D. Haussecker, Current issues of RNAi therapeutics delivery and development, J. Control. Release 195 (2014) 49–54, https://doi.org/10.1016/j.jconrel.2014.07.056.
119. M.S. Lee, J.E. Lee, E. Byun, N.W. Kim, K. Lee, H. Lee, S.J. Sim, D.S. Lee, J.H. Jeong, Target-specific delivery of siRNA by stabilized calcium phosphate nanoparticles using dopa–hyaluronic acid conjugate, J. Control. Release 192 (2014) 122–130, https://doi.org/10.1016/J.JCONREL.2014.06.049.
120. [H.J. Lee, R. Namgung, W.J. Kim, J.I. Kim, I.K. Park, Targeted delivery of microRNA-145 to metastatic breast cancer by peptide conjugated branched PEI gene carrier, Macromol. Res. 21 (2013) 1201–1209, https://doi.org/10.1007/s13233-013-1161-z.

121. A.F. Ibrahim, U. Weirauch, M. Thomas, A. Gruñweller, R.K. Hartmann, A. Aigner, MicroRNA replacement therapy for miR-145 and miR-33a is efficacious in a model of colon carcinom, Cancer Res. 71 (2011) 5214–5224, https://doi.org/10.1158/ 0008-5472.CAN-10-4645.
122. D. Cosco, F. Cilurzo, J. Maiuolo, C. Federico, M.T. Di Martino, M.C. Cristiano, P. Tassone, M. Fresta, D. Paolino, Delivery of miR-34a by chitosan/PLGA nanoplexes for the anticancer treatment of multiple myeloma, Sci. Rep. 5 (2015) 1–11, https://doi.org/10.1038/srep17579.
123. C. Saraiva, D. Talhada, A. Rai, R. Ferreira, L. Ferreira, L. Bernardino, K. Ruscher, MicroRNA-124-loaded nanoparticles increase survival and neuronal differentia- tion of neural stem cells in vitro but do not contribute to stroke outcome in vivo, PLoS One 13 (2018) e0193609, https://doi.org/10.1371/ journal.pone.0193609.
124. Q. Liu, R.T. Li, H.Q. Qian, J. Wei, L. Xie, J. Shen, M. Yang, X.P. Qian, L.X. Yu, X.Q. Jiang, B.R. Liu, Targeted delivery of miR-200c/DOC to inhibit cancer stem cells and cancer cells by the gelatinases-stimuli nanoparticles, Biomaterials 34 (2013) 7191–7203, https://doi.org/10.1016/J.BIOMATERIALS.2013.06.004.
125. G.Y. Chiou, J.Y. Cherng, H.S. Hsu, M.L. Wang, C.M. Tsai, K.H. Lu, Y. Chien, S.C. Hung, Y.W. Chen, C.I. Wong, L.M. Tseng, P.I. Huang, C.C. Yu, W.H. Hsu, S.H. Chiou, Cationic polyurethanes-short branch PEI-mediated delivery of Mir145 inhibited epithelial–mesenchymal transdifferentiation and cancer stem-like prop- erties and in lung adenocarcinoma, J. Control. Release 159 (2012) 240–250, https://doi.org/10.1016/j.jconrel.2012.01.014.
126. Y.P. Yang, Y. Chien, G.Y. Chiou, J.Y. Cherng, M.L. Wang, W.-L. Lo, Y.L. Chang, P.I. Huang, Y.W. Chen, Y.H. Shih, M.T. Chen, S.H. Chiou, Inhibition of cancer stem cell-like properties and reduced chemoradioresistance of glioblastoma using microRNA145 with cationic polyurethane-short branch PEI, Biomaterials 33 (2012) 1462–1476, https://doi.org/10.1016/J.BIOMATERIALS .2011.10.071.
127. S. Venkiteswaran, T. Thomas, T.J. Thomas, Selectivity of polyethyleneimines on DNA nanoparticle preparation and gene transport, ChemistrySelect 1 (2016) 1144–1150, https://doi.org/10.1002/slct.201600026.
128. T. Bieber, H.-P. Elsässer, Preparation of a low molecular weight polyethylenimine for efficient cell transfection, Biotechniques 30 (2001) 74–81, https://doi.org/10. 2144/01301st03.
129. C. Sun, T. Tang, H. Uludağ, J.E. Cuervo, Molecular dynamics simulations of DNA/ PEI complexes: effect of PEI branching and protonation state, Biophys. J. 100 (2011) 2754–2763, https://doi.org/10.1016/j.bpj.2011.04.045.
130. L. Huang, S. Guo, Nanoparticles escaping RES and endosome: challenges for siRNA delivery for cancer therapy, J. Nanomater. 2011 (2011) 12, https://doi.org/10. 1155/2011/742895.
131. A.K. Varkouhi, M. Scholte, G. Storm, H.J. Haisma, Endosomal escape pathways for delivery of biologicals, J. Control. Release 151 (2011) 220–228, https://doi.org/ 10.1016/J.JCONREL.2010.11.004.
132. T. Zhang, X. Xue, D. He, J.T. Hsieh, A prostate cancer-targeted polyarginine-dis- ulfide linked PEI nanocarrier for delivery of microRNA, Cancer Lett. 365 (2015) 156–165, https://doi.org/10.1016/J.CANLET.2015.05.003.

133. J.V. Jokerst, T. Lobovkina, R.N. Zare, S.S. Gambhir, Nanoparticle PEGylation for imaging and therapy, Nanomedicine (Lond.) 6 (2011) 715–728, https://doi.org/ 10.2217/nnm.11.19.
134. S.Y. Wong, J.M. Pelet, D. Putnam, Polymer systems for gene delivery—past, pre- sent, and future, Prog. Polym. Sci. 32 (2007) 799–837, https://doi.org/10.1016/j. progpolymsci.2007.05.007.
135. M.C. Serrano, R. Pagani, M. Vallet-Reg, J. Penã, I. Izquierdo, M.T. Portoles, In vitro biocompatibility assessment of poly (ε-caprolactone) films using L929 mouse fi- broblasts, Biomaterials 25 (2004) 5603–5611, https://doi.org/10.1016/j. biomaterials.2004.01.037.
136. F. Danhier, E. Ansorena, J.M. Silva, R. Coco, A. Le Breton, V. Préat, PLGA-based nanoparticles: an overview of biomedical applications, J. Control. Release 161 (2012) 505–522, https://doi.org/10.1016/J.JCONREL.2012.01.043.
137. Y. Patil, J. Panyam, Polymeric nanoparticles for siRNA delivery and gene silencing, Int. J. Pharm. 367 (2009) 195–203, https://doi.org/10.1016/j.ijpharm.2008.09.039.
138. X. Yuan, B.A. Shah, N.K. Kotadia, J. Li, H. Gu, Z. Wu, The development and me- chanism studies of cationic chitosan-modified biodegradable PLGA nanoparticles for efficient siRNA drug delivery, Pharm. Res. 27 (2010) 1285–1295, https://doi. org/10.1007/s11095-010-0103-0.
139. D. Cosco, F. Cilurzo, J. Maiuolo, C. Federico, M.T. Di Martino, M.C. Cristiano, P. Tassone, M. Fresta, D. Paolino, Delivery of miR-34a by chitosan/PLGA nanoplexes for the anticancer treatment of multiple myeloma, Sci. Rep. 5 (2015) 1–11, https://doi.org/10.1038/srep17579.
140. C. Saraiva, D. Talhada, A. Rai, R. Ferreira, L. Ferreira, L. Bernardino, K. Ruscher, MicroRNA-124-loaded nanoparticles increase survival and neuronal differentia- tion of neural stem cells in vitro but do not contribute to stroke outcome in vivo, PLoS One 13 (2018) e0193609, https://doi.org/10.1371/journal.pone.0193609.
141. P. Grossen, D. Witzigmann, S. Sieber, J. Huwyler, PEG-PCL-based nanomedicines: a biodegradable drug delivery system and its application, J. Control. Release 260 (2017) 46–60, https://doi.org/10.1016/j.jconrel.2017.05.028.
142. B.S. Sekhon, S.R. Kamboj, Inorganic nanomedicine-part 1, Nanomed. Nanotechnol. Biol. Med. 6 (2010) 516–522, https://doi.org/10.1016/j.nano.2010.04.004.
143. B.S. Sekhon, S.R. Kamboj, Inorganic nanomedicine-part 2, Nanomed. Nanotechnol. Biol. Med. 6 (2010) 612–618, https://doi.org/10.1016/j.nano.2010.04.003.
144. V. Mulens, M. del P. Morales, D.F. Barber, Development of magnetic nanoparticles for cancer gene therapy: a comprehensive review, ISRN Nanomater. 2013 (2013) 114, https://doi.org/10.1155/2013/646284.
145. A. Schade, E. Delyagina, D. Scharfenberg, A. Skorska, C. Lux, R. David, G. Steinhoff, Innovative strategy for microRNA delivery in human mesenchymal stem cells via magnetic nanoparticles, Int. J. Mol. Sci. 14 (2013) 10710–10726, https://doi.org/10.3390/ijms140610710.

146. P.T. Yin, B.P. Shah, K.B. Lee, Combined magnetic nanoparticle-based microRNA and hyperthermia therapy to enhance apoptosis in brain cancer cells, Small 10 (2014) 4106–4112, https://doi.org/10.1002/smll.201400963.
147. W. Cha, R. Fan, Y. Miao, Y. Zhou, C. Qin, X. Shan, X. Wan, J. Li, Mesoporous silica nanoparticles as carriers for intracellular delivery of nucleic acids and subsequent therapeutic applications, Molecules 22 (2017), https://doi.org/10.3390/ molecules22050782.
148. A. Bertucci, E.A. Prasetyanto, D. Septiadi, A. Manicardi, E. Brognara, R. Gambari, R. Corradini, L. De Cola, Combined delivery of temozolomide and anti-miR221 PNA using mesoporous silica nanoparticles induces apoptosis in resistant glioma cells, Small 11 (2015) 5687–5695, https://doi.org/10.1002/smll.201500540.
149. A. Tivnan, W.S. Orr, V. Gubala, R. Nooney, D.E. Williams, C. McDonagh, S. Prenter, H. Harvey, R. Domingo-Fernández, I.M. Bray, O. Piskareva, C.Y. Ng, H.N. Lode, A.M. Davidoff, R.L. Stallings, Inhibition of neuroblastoma tumor growth by targeted delivery of MicroRNA-34a using anti-disialoganglioside GD2 coated nanoparticles, PLoS One 7 (2012) e38129 https://doi.org/10.1371/journal.pone. 0038129.
150. H. Jung, S.A. Kim, Y.G. Yang, H. Yoo, S.J. Lim, H. Mok, Long chain microRNA conjugates in calcium phosphate nanoparticles for efficient formulation and de- livery, Arch. Pharm. Res. 38 (2015) 705–715, https://doi.org/10.1007/s12272- 014-0451-0.
151. E.E. Connor, J. Mwamuka, A. Gole, C.J. Murphy, M.D. Wyatt, Gold nanoparticles are taken up by human cells but do not cause acute cytotoxicity, Small 1 (2005) 325–327, https://doi.org/10.1002/smll.200400093.
152. P.K. Jain, K.S. Lee, I.H. El-Sayed, M.A. El-Sayed, Calculated absorption and scat- tering properties of gold nanoparticles of different size, shape, and composition: applications in biological imaging and biomedicine, J. Phys. Chem. B 110 (2006) 7238–7248, https://doi.org/10.1021/jp057170o.
153. R. Ghosh, L.C. Singh, J.M. Shohet, P.H. Gunaratne, A gold nanoparticle platform for the delivery of functional microRNAs into cancer cells, Biomaterials 34 (2013) 807–816, https://doi.org/10.1016/j.biomaterials.2012.10.023.
154. A. Ekin, O.F. Karatas, M. Culha, M. Ozen, Designing a gold nanoparticle-based nanocarrier for microRNA transfection into the prostate and breast cancer cells, J.Gene Med. 16 (2014) 331–335, https://doi.org/10.1002/jgm.2810.
155. P. Couvreur, G. Barret, Critical review in therapeutic drug carrier System, J. Pediatr. Surg. 5 (2005) 1–20, https://doi.org/10.1615/ CritRevTherDrug CarrierSyst.v26.i6.10.
156. Y. Zhao, L. Huang, Lipid nanoparticles for gene delivery, Adv. Genet. Academic Press, 2014, pp. 13–36, https://doi.org/10.1016/B978-0-12-800148-6.00002-X.
157. H. Lv, S. Zhang, B. Wang, S. Cui, J. Yan, Toxicity of cationic lipids and cationic polymers in gene delivery, J. Control. Release 114 (2006) 100–109, https://doi.org/10.1016/j.jconrel.2006.04.014.
158. X. Cheng, R.J. Lee, The role of helper lipids in lipid nanoparticles (LNPs) designed for oligonucleotide delivery, Adv. Drug Deliv. Rev. 99 (2016) 129–137, https:// doi.org/10.1016/j.addr.2016.01.022.

159. S.H. Hsu, B. Yu, X. Wang, Y. Lu, C.R. Schmidt, R.J. Lee, L.J. Lee, S.T. Jacob, K. Ghoshal, Cationic lipid nanoparticles for therapeutic delivery of siRNA and miRNA to murine liver tumor, Nanomed. Nanotechnol. Biol. Med. 9 (2013) 1169–1180, https://doi.org/10.1016/j.nano.2013.05.007.
160. Wu LP, Ficker M, Christensen JB, Trohopoulos PN, Moghimi SM. Dendrimers in medicine: therapeutic concepts and pharmaceutical challenges. Bioconjug Chem. 2015;26(7):1198–211.
161. Abbasi E, Aval SF, Akbarzadeh A, Milani M, Nasrabadi HT, Joo SW, et al. Dendrimers: synthesis, applications, and properties. Nanoscale Res Lett. 2014;9(1):247.
162. Wang F, Zhang B, Zhou L, Shi Y, Li Z, Xia Y, et al. Imaging dendrimer- grafted graphene oxide mediated anti-miR-21 delivery with an Activatable luciferase reporter. ACS Appl Mater Interfaces. 2016;8(14):9014–21.
163. Robbins PD, Morelli AE. Regulation of immune responses by extracellular vesicles. Nat Rev Immunol. 2014;14(3):195–208.
164. EL Andaloussi S, Maeger I, Breakefield XO, Wood MJA. Extracellular vesicles: biology and emerging therapeutic opportunities. Nat Rev Drug Discov. 2013;12(5):348–58.
165. Wang YY, Chen X, Tian BQ, Liu JF, Yang L, Zeng LL, et al. Nucleolin-targeted extracellular vesicles as a versatile platform for biologics delivery to breast Cancer. Theranostics. 2017;7(5):1360–72.
166. Tominaga N, Kosaka N, Ono M, Katsuda T, Yoshioka Y, Tamura K, et al. Brain metastatic cancer cells release microRNA-181c-containing extracellular vesicles capable of destructing blood-brain barrier. Nat Commun. 2015;6:6716.
167. Ha D, Yang NN, Nadithe V. Exosomes as therapeutic drug carriers and delivery vehicles across biological membranes: current perspectives and future challenges. Acta Pharm Sin B. 2016;6(4):287–96.
168. Lee JH, Kim JA, Kwon MH, Kang JY, Rhee WJ. In situ single step detection of exosome microRNA using molecular beacon. Biomaterials. 2015;54:116–25.
169. Zhou Y, Li P, Goodwin AJ, Cook JA, Halushka PV, Chang E, et al. Exosomes from endothelial progenitor cells improve the outcome of a murine model of Sepsis. Mol Ther. 2018;26(5):1375–84.
170. Wen D, Peng Y, Liu D, Weizmann Y, Mahato RI. Mesenchymal stem cell and derived exosome as small RNA carrier and Immunomodulator to improve islet transplantation. J Control Release. 2016;238:166–75.
171. Munagala R, Aqil F, Jeyabalan J, Gupta RC. Bovine milk-derived exosomes for drug delivery. Cancer Lett. 2016;371(1):48–61.
172. Raposo G, Stoorvogel W. Extracellular vesicles: exosomes, microvesicles, and friends. J Cell Biol. 2013;200(4):373–83.
173. Yang J, Zhou CZ, Zhu R, Fan H, Liu XX, Duan XY, et al. miR-200b-containing microvesicles attenuate experimental colitis associated intestinal fibrosis by inhibiting epithelial-mesenchymal transition. J Gastroenterol Hepatol. 2017;32(12):1966–74.
174. Zhang H, Bai M, Deng T, Liu R, Wang X, Qu Y, et al. Cell-derived microvesicles mediate the delivery of miR-29a/c to suppress angiogenesis in gastric carcinoma. Cancer Lett. 2016;375(2):331–9.

175. Cui J, Li Q, Luo M, Zhong Z, Zhou S, Jiang L, et al. Leukemia cell-derived microvesicles induce T cell exhaustion via miRNA delivery. Oncoimmunology.2018;7(7):e1448330.
176. Leslie M. Cell biology. Beyond clotting: the powers of platelets. Science. 2010;328(5978):562–4.
177. Laffont B, Corduan A, Ple H, Duchez AC, Cloutier N, Boilard E, et al. Activated platelets can deliver mRNA regulatory Ago2.microRNA complexes to endothelial cells via microparticles. Blood. 2013;122(2):253–61.
178. Liang H, Yan X, Pan Y, Wang Y, Wang N, Li L, et al. MicroRNA-223 delivered by platelet-derived microvesicles promotes lung cancer cell invasion via targeting tumor suppressor EPB41L3. Mol Cancer. 2015;14:58.
179. Zernecke A, Bidzhekov K, Noels H, Shagdarsuren E, Gan L, Denecke B, et al. Delivery of microRNA-126 by apoptotic bodies induces CXCL12-dependent vascular protection. Sci Signal. 2009;2(100):ra81.
180. Nguyen MK, Jeon O, Dang PN, Huynh CT, Varghai D, Riazi H, et al. RNA interfering molecule delivery from in situ forming biodegradable hydrogels for enhancement of bone formation in rat calvarial bone defects. Acta Biomater. 2018;75:105–114.
181. Conde J, Oliva N, Atilano M, Song HS, Artzi N. Self-assembled RNA-triple-helix hydrogel scaffold for microRNA modulation in the tumour microenvironment. Nat Mater. 2016;15(3):353–63.
182. Zhou F, Jia X, Yang Y, Yang Q, Gao C, Hu S, et al. Nanofiber-mediated microRNA-126 delivery to vascular endothelial cells for blood vessel regeneration. Acta Biomater. 2016;43:303–13.
183. Zhang X, Li Y, Chen YE, Chen J, Ma PX. Cell-free 3D scaffold with two-stage delivery of miRNA-26a to regenerate critical-sized bone defects. Nat Commun. 2016;7:10376.
184. Mencia Castano I, Curtin CM, Shaw G, Murphy JM, Duffy GP, O'Brien FJ. A novel collagen-nanohydroxyapatite microRNA-activated scaffold for tissue engineering applications capable of efficient delivery of both miR-mimics and antagomiRs to human mesenchymal stem cells. J Control Release. 2015;200:42–51.
185. Godinho, B.M.D.C., Ogier, J.R., Quinlan, A., Darcy, R., Griffin, B.T., Cryan, J.F., O'Driscoll, C.M., 2014. PEGylated cyclodextrins as novel siRNA nanosystems: corre- lations between polyethylene glycol length and nanoparticle stability. Int. J. Pharm. 473 (1–2), 105–112.
186. Deleavey, G.F., Damha, M.J., 2012. Designing chemically modified oligonucleotides for targeted gene silencing. Chem. Biol. 19 (8), 937–954.
187. Banerjee, P., Weissleder, R., Bogdanov, A., 2006. Linear polyethyleneimine grafted to a hyperbranched poly(ethylene glycol)-like core: a copolymer for gene delivery. Bioconjug. Chem. 17 (1), 125–131.
188. Ragelle, H., Riva, R., Vandermeulen, G., Naeye, B., Pourcelle, V., Le Duff, C.S., D'Haese, C., Nysten, B., Braeckmans, K., Smedt, S.C. de, Jérôme, C., Préat, V., 2014. Chitosan nanoparticles for siRNA delivery: optimizing formulation to increase stability and efficiency. J. Control. Release 176, 54–63.

189. Zhang, J., Tang, C., Yin, C., 2013. Galactosylated trimethyl chitosan-cysteine nano- particles loaded with Map4k4 siRNA for targeting activated macrophages. Biomaterials 34 (14), 3667–3677.
190. Bielinska, A.U., Kukowska-Latallo, J.F., Baker, J.R., 1997. The interaction of plasmid DNA with polyamidoamine dendrimers: mechanism of complex formation and analysis of alterations induced in nuclease sensitivity and transcriptional activity of the com- plexed DNA. Biochim. Biophys. Acta 1353 (2), 180–190.
191. Loretz, B., Föger, F., Werle, M., Bernkop-Schnürch, A., 2006. Oral gene delivery: strategies to improve stability of pDNA towards intestinal digestion. J. Drug Target. 14 (5), 311-319.
192. Fabiano, A., Zambito, Y., Bernkop-Schnürch, A., 2017. About the impact of water movement on the permeation behaviour of nanoparticles in mucus. Int. J. Pharm. 517 (1–2), 279–285.
193. Müller, C., Leithner, K., Hauptstein, S., Hintzen, F., Salvenmoser, W., Bernkop-Schnürch, A., 2013. Preparation and characterization of mucus-penetrating papain/poly(acrylic acid) nanoparticles for oral drug delivery applications. J. Nanopart. Res. 15 (1), 1.
194. Bennett, K.M., Walker, S.L., Lo, D.D., 2014. Epithelial microvilli establish an electrostatic barrier to microbial adhesion. Infect. Immun. 82 (7), 2860–2871.
195. Wu, J., Zheng, Y., Liu, M., Shan, W., Zhang, Z., Huang, Y., 2018. Biomimetic viruslike and charge reversible nanoparticles to sequentially overcome mucus and epithelial bar- riers for oral insulin delivery. ACS Applied Material & Interfaces 10, 9916–992
196. Gebert LFR, Rebhan MAE, Crivelli SEM, Denzler R, Stoffel M, Hall J. Miravirsen (SPC3649) can inhibit the biogenesis of miR-122. Nucleic Acids Res. 2014;42(1):609–21.

Index

G

H

I

N

O

S

T

www.ingramcontent.com/pod-product-compliance
Lightning Source LLC
LaVergne TN
LVHW010717210926
848485LV00008B/1434/J